Means Residential Detailed Costs

Contractor's Pricing Guide 2009

 New for 2009!

Line items found in this cost data that have the icon shown above are green.

The identified items fall into a broad definition if what is considered green.

Please see page xiii for more information.

Senior Editor
Bob Mewis, CCC

Contributing Editors
Christopher Babbit
Ted Baker
Barbara Balboni
Robert A. Bastoni
John H. Chiang, PE
Gary W. Christensen
David G. Drain, PE
Cheryl Elsmore
Robert J. Kuchta
Robert C. McNichols
Melville J. Mossman, PE
Jeannene D. Murphy
Stephen C. Plotner
Eugene R. Spencer
Marshall J. Stetson
Phillip R. Waier, PE

Vice President of Operations
Dave Walsh

Senior Vice President & General Manager
John Ware

Vice President of Sales & Marketing
Sev Ritchie

Marketing Director
John M. Shea

Director of Product Development
Thomas J. Dion

Engineering Manager
Bob Mewis, CCC

Production Manager
Michael Kokernak

Production Coordinator
Jill Goodman

Technical Support
Jonathan Forgit
Mary Lou Geary
Roger Hancock
Gary L. Hoitt
Genevieve Medeiros
Debbie Panarelli
Paula Reale-Camelio
Kathryn S. Rodriguez
Sheryl A. Rose

Book & Cover Design
Norman R. Forgit

RSMeans

Means
Residential
Detailed
Costs

Contractor's
Pricing
Guide 2009

- Unit Costs for Thousands of Residential Building Components
- Cost Adjustment Factors for Your Location
- Daily Productivities & Standard Crews
- Overhead & Profit Guidance

$39.95 per copy (in United States)
Price subject to change without prior notice.

RS**Means**

Copyright © 2008
R.S. Means Company, Inc.
Construction Publishers & Consultants
63 Smiths Lane
Kingston, MA 02364-3008
781-422-5000
www.rsmeans.com

Printed in the United States of America

ISSN 1074-0481

ISBN 978-0-87629-148-1

This book is printed on recycled paper (10% PCW cover, 10% PCW text),
using soy-based printing ink. This book is recyclable.

Foreword

Our Mission

Since 1942, RSMeans has been actively engaged in construction cost publishing and consulting throughout North America.

Today, over 60 years after RSMeans began, our primary objective remains the same: to provide you, the construction and facilities professional, with the most current and comprehensive construction cost data possible.

Whether you are a contractor, an owner, an architect, an engineer, a facilities manager, or anyone else who needs a reliable construction cost estimate, you'll find this publication to be a highly useful and necessary tool.

With the constant flow of new construction methods and materials today, it's difficult to find the time to look at and evaluate all the different construction cost possibilities. In addition, because labor and material costs keep changing, last year's cost information is not a reliable basis for today's estimate or budget.

That's why so many construction professionals turn to RSMeans. We keep track of the costs for you, along with a wide range of other key information, from city cost indexes . . . to productivity rates . . . to crew composition . . . to contractor's overhead and profit rates.

RSMeans performs these functions by collecting data from all facets of the industry and organizing it in a format that is instantly accessible to you. From the preliminary budget to the detailed unit price estimate, you'll find the data in this book useful for all phases of construction cost determination.

The Staff, the Organization, and Our Services

When you purchase one of RSMeans' publications, you are, in effect, hiring the services of a full-time staff of construction and engineering professionals.

Our thoroughly experienced and highly qualified staff works daily at collecting, analyzing, and disseminating comprehensive cost information for your needs. These staff members have years of practical construction experience and engineering training prior to joining the firm. As a result, you can count on them not only for the cost figures, but also for additional background reference information that will help you create a realistic estimate.

The RSMeans organization is always prepared to help you solve construction problems through its four major divisions: Construction and Cost Data Publishing, Electronic Products and Services, Consulting and Business Solutions, and Professional Development Services.

Besides a full array of construction cost estimating books, RSMeans also publishes a number of other reference works for the construction industry. Subjects include construction estimating and project and business management; special topics such as HVAC, roofing, plumbing, and hazardous waste remediation; and a library of facility management references.

In addition, you can access all of our construction cost data electronically using *Means CostWorks®* CD or on the Web using Means CostWorks.com.

What's more, you can increase your knowledge and improve your construction estimating and management performance with an RSMeans Construction Seminar or In-House Training Program. These two-day seminar programs offer unparalleled opportunities for everyone in your organization to get updated on a wide variety of construction-related issues.

RSMeans is also a worldwide provider of construction cost management and analysis services for commercial and government owners.

In short, RSMeans can provide you with the tools and expertise for constructing accurate and dependable construction estimates and budgets in a variety of ways.

Robert Snow Means Established a Tradition of Quality That Continues Today

Robert Snow Means spent years building RSMeans, making certain he always delivered a quality product.

Today, at RSMeans, we do more than talk about the quality of our data and the usefulness of our books. We stand behind all of our data, from historical cost indexes to construction materials and techniques to current costs.

If you have any questions about our products or services, please call us toll-free at 1-800-334-3509. Our customer service representatives will be happy to assist you. You can also visit our Web site at www.rsmeans.com.

Table of Contents

How the Book Is Built: An Overview

The Construction Specifications Institute (CSI) and Construction Specifications Canada (CSC) have produced the 2004 edition of MasterFormat, an updated system of titles and numbers used extensively to organize construction information. All unit price data in the RSMeans cost data books is now arranged in the 50-division MasterFormat 2004 system.

A Powerful Construction Tool

You have in your hands one of the most powerful construction tools available today. A successful project is built on the foundation of an accurate and dependable estimate. This book will enable you to construct just such an estimate.

For the casual user the book is designed to be:

• quickly and easily understood so you can get right to your estimate.

• filled with valuable information so you can understand the necessary factors that go into the cost estimate.

For the regular user, the book is designed to be:

• a handy desk reference that can be quickly referred to for key costs.

• a comprehensive, fully reliable source of current construction costs and productivity rates, so you'll be prepared to estimate any project.

• a source book for preliminary project cost, product selections, and alternate materials and methods.

To meet all of these requirements we have organized the book into the following clearly defined sections.

New: Quick Start

See our new "Quick Start" instructions on the following page.

Estimating with RSMeans Unit Price Cost Data

Please see these steps to complete an estimate using RSMeans unit price cost data.

Unit Price Section

All cost data has been divided into the 50 divisions according to the MasterFormat system of classification and numbering. For a listing of these divisions and an outline of their subdivisions, see the Unit Price Section Table of Contents.

Reference Section

This section includes information on Equipment Rental Costs, Crew Listings, Location Factors, Reference Tables, and a listing of Abbreviations.

Equipment Rental Costs:
This section contains the average costs to rent and operate hundreds of pieces of construction equipment.

Crew Listings:
This section lists all the crews referenced in the book. For the purposes of this book, a crew is composed of more than one trade classification and/or the addition of power equipment to any trade classification. Power equipment is included in the cost of the crew. Costs are shown both with bare labor rates and with the installing contractor's overhead and profit added. For each, the total crew cost per eight-hour day and the composite cost per labor-hour are listed.

Location Factors:
You can adjust total project costs to over 900 locations throughout the U.S. and Canada by using the data in this section.

Reference Tables:
At the beginning of selected major classifications in the Unit Price Section are reference numbers shown in a shaded box. These numbers refer you to related information in the Reference Section. In this section, you'll find reference tables, explanations, and estimating information that support how we develop the unit price data, technical data, and estimating procedures.

Abbreviations:
A listing of the abbreviations used throughout this book, along with the terms they represent, is included in this section.

Index

A comprehensive listing of all terms and subjects in this book will help you quickly find what you need when you are not sure where it falls in MasterFormat.

The Scope of This Book

This book is designed to be as comprehensive and as easy to use as possible. To that end we have made certain assumptions and limited its scope in two key ways:

1. We have established material prices based on a national average.

2. We have computed labor costs based on a 7 major-region average of residential wage rates.

Project Size/Type

The material prices in Means data cost books are "contractor's prices." They are the prices that contractors can expect to pay at the lumberyards, suppliers/distributers warehouses, etc. Small orders of speciality items would be higher than the costs shown, while very large orders, such as truckload lots, would be less. The variation would depend on the size, timing, and negotiating power of the contractor. The labor costs are primarily for new construction or major renovations rather than repairs or minor alterations. With reasonable exercise of judgment, the figures can be used for any building work.

Absolute Essentials for a Quick Start

If you feel you are ready to use this book and don't think you will need the detailed instructions that begin on the following page, this Absolute Essentials for a Quick Start page is for you. These steps will allow you to get started estimating in a matter of minutes.

1 Scope

Think through the project that you will be estimating and identify the many individual work tasks that will need to be covered in your estimate.

2 Quantify

Determine the number of units that will be required for each work task that you identified.

3 Pricing

Locate individual Unit Price line items that match the work tasks you identified. The Unit Price Section Table of Contents that begins on page 1 and the Index in the back of the book will help you find these line items.

4 Multiply

Multiply the Total Incl O&P cost for a Unit Price line item in the book by your quantity for that item. The price you calculate will be an estimate for a completed item of work performed by a subcontractor. Keep adding line items in this manner to build your estimate.

5 Project Overhead

Include project overhead items in your estimate. These items are needed to make the job run and are typically, but not always, provided by the General Contractor. They can be found in Division 1. An alternate method of estimating project overhead costs is to apply a percentage of the total project cost.

Include rented tools not included in crews, waste, rubbish handling and cleanup.

6 Estimate Summary

Include General Contractor's markup on subcontractors, General Contractor's office overhead and profit, and sales tax on materials and equipment.

Adjust your estimate to the project's location by using the City Cost Indexes or Location Factors found in the Reference Section.

Editors' Note: We urge you to spend time reading and understanding the supporting material in the front of this book. An accurate estimate requires experience, knowledge and careful calculation. The more you know about how we at RSMeans developed the data, the more accurate your estimate will be. In addition, it is important to take into consideration the reference material in the back of the book such as Equipment Listings, Crew Listings, Location Factors, and Reference Numbers.

Estimating with RSMeans Unit Price Cost Data

Following these steps will allow you to complete an accurate estimate using RSMeans Unit Price cost data.

1 Scope out the project

- Identify the individual work tasks that will need to be covered in your estimate.
- The Unit Price data inside this book has been divided into 34 Divisions according to the CSI MasterFormat2004 – their titles are listed on the back cover of your book.
- Think through the project that you will be estimating and identify those CSI Divisions that you will need to use in your estimate.
- The Unit Price Section Table of Contents that begins on page 1 may also be helpful when scoping out your project.
- Experienced estimators find it helpful to begin an estimate with Division 2, estimating Division 1 after the full scope of the project is known.

2 Quantify

- Determine the number of units that will be required for each work task that you previously identified.
- Experienced estimators will include an allowance for waste in their quantities (waste is not included in RSMeans Unit Price line items unless so stated).

3 Price the quantities

- Use the Unit Price Table of Contents, and the Index, to locate the first individual Unit Price line item for your estimate.
- Reference Numbers indicated within a Unit Price section refer to additional information that you may find useful.
- The crew will tell you who is performing the work for that task. Crew codes are expanded in the Crew Listings in the Reference Section to include all trades and equipment that comprise the crew.
- The Daily Output is the amount of work the crew is expected to do in one day.
- The Labor-Hours value is the amount of time it will take for the average crew member to install one unit of measure.
- The abbreviated Unit designation indicates the unit of measure upon which the crew, productivity and prices are based.
- Bare Costs are shown for materials, labor, and equipment needed to complete the Unit Price line item. Bare costs do not include waste, project overhead, payroll insurance, payroll taxes, main office overhead, or profit.
- The Total Incl O&P cost is the billing rate or invoice amount of the installing contractor or subcontractor who performs the work for the Unit Price line item.

4 Multiply

- Multiply the total number of units needed for your project by the Total Incl O&P cost for the Unit Price line item.
- Be careful that the final unit of measure for your quantity of units matches the unit of measure in the Unit column in the book.
- The price you calculate will be an estimate for a completed item of work.
- Keep scoping individual tasks, determining the number of units required for those tasks, matching them up with individual Unit Price line items in the book, and multiplying quantities by Total Incl O&P costs. In this manner keep building your estimate.
- An estimate completed to this point in this manner will be priced as if a subcontractor, or set of subcontractors, will perform the work. The estimate does not yet include Project Overhead or Estimate Summary components such as General Contractor markups on subcontracted and self-performed work, General Contractor office overhead and profit, contingency, and location factor.

5 Project Overhead

- Include project overhead items from Division 1 – General Requirements.
- These are items that will be needed to make the job run. They are typically, but not always, provided by the General Contractor. They include, but are not limited to, such items as field personnel, insurance, performance bond, permits, testing, temporary utilities, field office and storage facilities, temporary scaffolding and platforms, equipment mobilization and demobilization, temporary roads and sidewalks, winter protection, temporary barricades and fencing, temporary security, temporary signs, field engineering and layout, final cleaning and commissioning.
- These items should be scoped, quantified, matched to individual Unit Price line items in Division 1, priced and added to your estimate.
- An alternate method of estimating project overhead costs is to apply a percentage of the total project cost, usually 5% to 15% with an average of 10%.
- Include other project related expenses in your estimate such as:
  - Rented equipment not itemized in the Crew Listings.
  - Rubbish handling throughout the project (see 02 41 19.23).

6 Estimate Summary

- Includes sales tax on materials and equipment.
  Note: Sales tax must be added for materials in subcontracted work.
- Include the General Contractor's markup on self-performed work, usually 5% to 15% with an average of 10%.
- Include the General Contractor's markup on subcontracted work, usually 5% to 15% with an average of 10%.
- Include General Contractor's main office overhead and profit.
- RSMeans gives general guidelines on the General Contractor's main office overhead.
- RSMeans gives no guidance on the General Contractor's profit.
- Markups will depend on the size of the General Contractor's operations, his projected annual revenue, the level of risk he is taking on, and on the level of competitiveness in the local area and for this project in particular.
- Include a contingency, usually 3% to 5%.
- Adjust your estimate to the project's location by using the City Cost Indexes or the Location Factors in the Reference Section.
- Look at the rules on the pages for How to Use the City Cost Indexes to see how to apply the Indexes for your location.
- When the proper Index or Factor has been identified for the project's location, convert it to a multiplier by dividing it by 100, then multiply that multiplier by your estimate total cost. Your original estimate total cost will now be adjusted up or down from the national average to a total that is appropriate for your location.

Editors' Notes:

1) *We urge you to spend time reading and understanding the supporting material in the front of this book. An accurate estimate requires experience, knowledge, and careful calculation. The more you know about how we at RSMeans developed the data, the more accurate your estimate will be. In addition, it is important to take into consideration the reference material in the back of the book such as Equipment Listings, Crew Listings, City Cost Indexes, Location Factors, and Reference Numbers.*

2) *Contractors who are bidding or are involved in JOC, DOC, SABER, or IDIQ type contracts are cautioned that workers' compensation Insurance, federal and state payroll taxes, waste, project supervision, project overhead, main office overhead, and profit are not included in bare costs. Your coefficient or multiplier must cover these costs.*

How to Use the Book: The Details

What's Behind the Numbers? The Development of Cost Data

The staff at RSMeans continuously monitors developments in the construction industry in order to ensure reliable, thorough and up-to-date cost information.

While *overall* construction costs may vary relative to general economic conditions, price fluctuations within the industry are dependent upon many factors. Individual price variations may, in fact, be opposite to overall economic trends. Therefore, costs are continually monitored and complete updates are published yearly. Also, new items are frequently added in response to changes in materials and methods.

Costs – $ (U.S.)

All costs represent U.S. national averages and are given in U.S. dollars. The RSMeans Location Factors can be used to adjust costs to a particular location. The Location Factors for Canada can be used to adjust U.S. national averages to local costs in Canadian dollars. No exchange rate conversion is necessary.

G The processes or products identified by the green symbol in this publication have been determined to be environmentally responsible and/or resource-efficient solely by the RSMeans engineering staff. The inclusion of the green symbol does not represent compliance with any specific industry association or standard.

Material Costs

The RSMeans staff contacts manufacturers, dealers, distributors, and contractors all across the U.S. and Canada to determine national average material costs. If you have access to current material costs for your specific location, you may wish to make adjustments to reflect differences from the national average.

Included within material costs are fasteners for a normal installation. RSMeans engineers use manufacturers' recommendations, written specifications, and/or standard construction practice for size and spacing of fasteners. Adjustments to material costs may be required for your specific application or location. Material costs do not include sales tax.

Labor Costs

Labor costs are based on the average of residential wages from across the U.S. for the current year. Rates, along with overhead and profit markups, are listed on the inside back cover of this book.

- If wage rates in your area vary from those used in this book, or if rate increases are expected within a given year, labor costs should be adjusted accordingly.

Labor costs reflect productivity based on actual working conditions. In addition to actual installation, these figures include time spent during a normal workday on tasks such as material receiving and handling, mobilization at site, site movement, breaks, and cleanup.

Productivity data is developed over an extended period so as not to be influenced by abnormal variations and reflects a typical average.

Equipment Costs

Equipment costs include not only rental costs, but also operating costs for equipment under normal use. Equipment and rental rates are obtained from industry sources throughout North America—contractors, suppliers, dealers, manufacturers, and distributors.

Factors Affecting Costs

Costs can vary depending upon a number of variables. Here's how we have handled the main factors affecting costs.

Quality—The prices for materials and the workmanship upon which productivity is based represent sound construction work. They are also in line with U.S. government specifications.

Overtime—We have made no allowance for overtime. If you anticipate premium time or work beyond normal working hours, be sure to make an appropriate adjustment to your labor costs.

Productivity—The productivity, daily output, and labor-hour figures for each line item are based on working an eight-hour day in daylight hours in moderate temperatures. For work that extends beyond normal work hours or is performed under adverse conditions, productivity may decrease. (See the section in "How to Use the Unit Price Pages" for more on productivity.)

Size of Project—The size, scope of work, and type of construction project will have a significant impact on cost. Economies of scale can reduce costs for large projects. Unit costs can often run higher for small projects. Costs in this book are intended for the size and type of project as previously described in "How the Book Is Built: An Overview." Costs for projects of a significantly different size or type should be adjusted accordingly.

Location—Material prices in this book are for metropolitan areas. However, in dense urban areas, traffic and site storage limitations may increase costs. Beyond a 20-mile radius of large cities, extra trucking or transportation charges may also increase the material costs slightly. On the other hand, lower wage rates may be in effect. Be sure to consider both of these factors when preparing an estimate, particularly if the job site is located in a central city or remote rural location.

In addition, highly specialized subcontract items may require travel and per-diem expenses for mechanics.

Other Factors —

- season of year
- contractor management
- weather conditions
- local union restrictions
- building code requirements
- availability of:
  - adequate energy
  - skilled labor
  - building materials
- owner's special requirements/restrictions
- safety requirements
- environmental considerations

Overhead & Profit—The extreme right-hand column of each chart gives the "Total Including O&P." These figures contain the original installing subcontractor's O&P. Therefore, it is necessary for a general contractor to add a percentage of all subcontracted items. For a detailed breakdown of O&P see the inside back cover of this book.

Tables in the reference section give details on overhead. These tables can be used by the general contractor as a guide to determine the appropriate overhead and profit markups.

General Conditions—General Conditions, or General Requirements, of the contract should also be added to the Total Cost including O&P when applicable. Costs for general Conditions are listed in Division 1 and the Reference Section of this book. General Conditions for the *Installing Contractor* may range from 0% to 10% of the Total Cost including O&P. For the *General* or *Prime Contractor,* costs for General Conditions may range from 5% to 15% of the Total Cost including O&P, with a figure of 10% as the most typical allowance.

Unpredictable Factors—General business conditions influence "in-place" costs of all items. Substitute materials and construction methods may have to be employed. These may affect the installed cost and/or life cycle costs. Such factors may be difficult to evaluate and cannot necessarily be predicted on the basis of the job's location in a particular section of the country. Thus, where these factors apply, you may find significant but unavoidable cost variations for which you will have to apply a measure of judgment to your estimate.

Rounding of Costs

In general, all unit prices in excess of $5.00 have been rounded to make them easier to use and still maintain adequate precision of the results. The rounding rules we have chosen are in the following table.

Prices from . . .	Rounded to the nearest . . .
$.01 to $5.00	$.01
$5.01 to $20.00	$.05
$20.01 to $100.00	$.50
$100.01 to $300.00	$1.00
$300.01 to $1,000.00	$5.00
$1,000.01 to $10,000.00	$25.00
$10,000.01 to $50,000.00	$100.00
$50,000.01 and above	$500.00

Final Checklist

Estimating can be a straightforward process provided you remember the basics. Here's a checklist of some of the steps you should remember to complete before finalizing your estimate.

Did you remember to . . .

- factor in the Location Factor for your locale?
- take into consideration which items have been marked up and by how much?
- mark up the entire estimate sufficiently for your purposes?
- read the background information on techniques and technical matters that could impact your project time span and cost?
- include all components of your project in the final estimate?
- double check your figures for accuracy?
- call RSMeans if you have any questions about your estimate or the data you've found in our publications?

Remember, RSMeans stands behind its publications. If you have any questions about your estimate . . . about the costs you've used from our books . . . or even about the technical aspects of the job that may affect your estimate, feel free to call the RSMeans editors at 1-800-334-3509.

Unit Price Section

Table of Contents

1

Table of Contents (cont.)

How to Use the Unit Price Pages

Important
Prices in this section are listed in two ways: as bare costs and as costs including overhead and profit of the installing contractor. In most cases, if the work is to be subcontracted, it is best for a general contractor to add an additional 10% to the figures found in the column titled "TOTAL INCL. O&P."

Unit
The unit of measure listed here reflects the material being used in the line item. For example: roof trusses are priced per each (Ea.).

Productivity
The daily output represents typical total daily amount of work that the designated crew will produce.
Labor-hours are a unit of measure for the labor involved in performing a task. To derive the total labor-hours for a task, multiply the quantity of the item involved times the labor-hour figure shown.

Line Number Determination
Each line item is identified by a unique twelve-digit number.

MasterFormat
Division (06)
06 17 53.10 5050

MasterFormat Level 2 (06 17 00)
06 17 53.10 5050

MasterFormat Level 3
06 17 53.10 5050

06 17 53.10 5050
RSMeans 12-Digit
Line Number

Description
This line item describes a common wood truss that will span 20'. It will be installed by an F-6 Crew at the rate of 62 per day, or .645 labor hours each.

RO61636 -20   **Reference Number**
These reference numbers refer to charts, tables, estimating data, cost derivations and other information which may be useful to the user of this book. This information is located in the Reference Section of this book.

Crew F-6

Crew No.	Bare Costs		Incl. Subs O & P		Cost Per Labor-Hour	
Crew F-6	Hr.	Daily	Hr.	Daily	Bare Costs	Incl. O&P
2 Carpenters	$27.95	$447.20	$47.25	$756.00	$25.27	$42.44
2 Building Laborers	20.55	328.80	34.75	556.00		
1 Equip. Oper. (crane)	29.35	234.80	48.20	385.60		
1 Hyd. Crane, 12 Ton		768.80		845.68	19.22	21.14
40 L.H., Daily Totals		$1779.60		$2543.28	$44.49	$63.58

Bare Costs are developed as follows for line no. 06 17 53.10 5050
Mat. is **Bare Material Cost ($72.00)**
Labor for Crew F-6 = Labor-hour Cost ($25.27) x Labor-hour Units (.645) = $16.30 (Rounded)
Equipment = Equipment Cost per Labor-hour ($19.27) × Labor-hour Units (.645) = $12.40
Total = Mat. Cost ($72.00) + Labor Cost ($16.30) + Equip. Cost ($12.40) = $100.70 per EA.

(**Note**: Where the crew is indicated, Equipment and Labor cost are derived from the Crew Tables. See example above.)

Total Costs Including O&P are developed as follows:
Mat. is **Bare Material Cost + 10%** = 72.00 + $7.20 = $79.20
Labor for Crew F-6 = Labor-hour Cost ($42.44) × Labor-Hour Units (.645) = $27.37 (Rounded)
Equipment = Equipment Cost per Labor-hour ($21.14) × Labor-hour Units (.645) = $13.63 (Rounded)
Total = Mat. Cost ($79.20) + Labor Cost ($27.37) + Equip. Cost ($13.63) = $121 Rounded

(**Note**: Where a crew is indicated, Equipment and Labor costs are derived from the Crew Tables. See example above. "Total incl. O&P" costs may be rounded.)

06 17 Shop-Fabricated Structural Wood

06 17 53 – Shop-Fabricated Wood Trusses

06 17 53.10 Roof Trusses		Crew	Daily Output	Labor-Hours	Unit	Material	2009 Bare Costs Labor	Equipment	Total	Total Incl O&P
0010	**ROOF TRUSSES**									
5000	Common wood, 2" x 4" metal plate connected, 24" O.C., 4/12 slope									
5010	1' overhang, 12' span	F-5	55	.582	Ea.	47	14.20		61.20	75.50
5050	20' span	F-6	62	.645		72	16.30	12.40	100.70	121
5100	24' span		60	.667		84	16.85	12.80	113.65	135
5150	26' span		57	.702		87	17.75	13.50	118.25	141
5200	28' span		53	.755		77.50	19.05	14.50	111.05	133
5240	30' span		51	.784		108	19.80	15.05	142.85	169
5250	32' span		50	.800		112	20	15.40	147.40	174
5280	34' span		48	.833		136	21	16	173	203
5350	8/12 pitch, 1' overhang, 20' span		57	.702		76.50	17.75	13.50	107.75	129
5400	24' span		55	.727		97	18.40	14	129.40	153
5450	26' span		52	.769		101	19.45	14.80	135.25	160
5500	28' span		49	.816		111	20.50	15.70	147.20	175
5550	32' span		45	.889		134	22.50	17.10	173.60	203
5600	36' span		41	.976		167	24.50	18.75	210.25	245

Division 1
General Requirements

01 11 Summary of Work

01 11 31 – Professional Consultants

01 11 31.10 Architectural Fees		Crew	Daily Output	Labor-Hours	Unit	Material	2009 Bare Costs Labor	Equipment	Total	Total Incl O&P
0010	**ARCHITECTURAL FEES** R011110-10									
0020	For new construction									
0060	Minimum				Project					4.90%
0090	Maximum									16%
0100	For alteration work, to $500,000, add to new construction fee									50%
0150	Over $500,000, add to new construction fee				↓					25%

01 11 31.20 Construction Management Fees										
0010	**CONSTRUCTION MANAGEMENT FEES**									
0060	For work to $100,000				Project					10%
0070	To $250,000									9%
0090	To $1,000,000				↓					6%

01 11 31.75 Renderings										
0010	**RENDERINGS** Color, matted, 20" x 30", eye level,									
0050	Average				Ea.	2,775			2,775	3,075

01 21 Allowances

01 21 16 – Contingency Allowances

01 21 16.50 Contingencies

		Crew	Daily Output	Labor-Hours	Unit	Material	Labor	Equipment	Total	Total Incl O&P
0010	**CONTINGENCIES**, Add to estimate									
0020	Conceptual stage				Project					20%
0150	Final working drawing stage				"					3%

01 21 63 – Taxes

01 21 63.10 Taxes

		Crew	Daily Output	Labor-Hours	Unit	Material	Labor	Equipment	Total	Total Incl O&P
0010	**TAXES** R012909-80									
0020	Sales tax, State, average				%	4.65%				
0050	Maximum R012909-85					7%				
0200	Social Security, on first $102,000 of wages						7.65%			
0300	Unemployment, combined Federal and State, minimum						2.10%			
0350	Average						6.20%			
0400	Maximum				↓		8%			

01 31 Project Management and Coordination

01 31 13 – Project Coordination

01 31 13.30 Insurance

		Crew	Daily Output	Labor-Hours	Unit	Material	Labor	Equipment	Total	Total Incl O&P
0010	**INSURANCE** R013113-40									
0020	Builders risk, standard, minimum				Job					.22%
0050	Maximum R013113-60									.59%
0200	All-risk type, minimum									.25%
0250	Maximum				↓					.62%
0400	Contractor's equipment floater, minimum				Value					.50%
0450	Maximum				"					1.50%
0600	Public liability, average				Job					1.55%
0800	Workers' compensation & employer's liability, average									
0850	by trade, carpentry, general				Payroll		18.51%			
0900	Clerical						.60%			
0950	Concrete						15.79%			
1000	Electrical						6.40%			
1050	Excavation				↓		10.34%			

01 31 Project Management and Coordination

01 31 13 – Project Coordination

01 31 13.30 Insurance

		Crew	Daily Output	Labor-Hours	Unit	Material	2009 Bare Costs Labor	Equipment	Total	Total Incl O&P
1100	Glazing				Payroll		13.82%			
1150	Insulation						15.24%			
1200	Lathing						10.71%			
1250	Masonry						14.97%			
1300	Painting & decorating						12.89%			
1350	Pile driving						22.94%			
1400	Plastering						14.62%			
1450	Plumbing						7.78%			
1500	Roofing						31.75%			
1550	Sheet metal work (HVAC)						11.09%			
1600	Steel erection, structural						38.86%			
1650	Tile work, interior ceramic						9.63%			
1700	Waterproofing, brush or hand caulking						7.27%			
1800	Wrecking						40.51%			
2000	Range of 35 trades in 50 states, excl. wrecking, min.						2.50%			
2100	Average						16.20%			
2200	Maximum						110.10%			

01 41 Regulatory Requirements

01 41 26 – Permits

01 41 26.50 Permits

		Crew	Daily Output	Labor-Hours	Unit	Material	Labor	Equipment	Total	Total Incl O&P
0010	**PERMITS**									
0020	Rule of thumb, most cities, minimum				Job					.50%
0100	Maximum				"					2%

01 54 Construction Aids

01 54 16 – Temporary Hoists

01 54 16.50 Weekly Forklift Crew

		Crew	Daily Output	Labor-Hours	Unit	Material	Labor	Equipment	Total	Total Incl O&P
0010	**WEEKLY FORKLIFT CREW**									
0100	All-terrain forklift, 45' lift, 35' reach, 9000 lb. capacity	A-3P	.20	40	Month		1,075	2,225	3,300	4,225

01 54 19 – Temporary Cranes

01 54 19.50 Daily Crane Crews

		Crew	Daily Output	Labor-Hours	Unit	Material	Labor	Equipment	Total	Total Incl O&P
0010	**DAILY CRANE CREWS** for small jobs, portal to portal									
0100	12-ton truck-mounted hydraulic crane	A-3H	1	8	Day		235	1,025	1,260	1,500

01 54 23 – Temporary Scaffolding and Platforms

01 54 23.60 Pump Staging

		Crew	Daily Output	Labor-Hours	Unit	Material	Labor	Equipment	Total	Total Incl O&P
0010	**PUMP STAGING**, Aluminum									
1300	System in place, 50' working height, per use based on 50 uses	2 Carp	84.80	.189	C.S.F.	6.40	5.25		11.65	15.95
1400	100 uses	R015423-20	84.80	.189		3.21	5.25		8.46	12.45
1500	150 uses		84.80	.189		2.15	5.25		7.40	11.25

01 54 23.70 Scaffolding

		Crew	Daily Output	Labor-Hours	Unit	Material	Labor	Equipment	Total	Total Incl O&P
0010	**SCAFFOLDING**	R015423-10								
0015	Steel tube, regular, no plank, labor only to erect & dismantle									
0091	Building exterior, wall face, 1 to 5 stories, 6'-4" x 5' frames	3 Clab	8	3	C.S.F.		61.50		61.50	104
0201	6 to 12 stories	4 Clab	8	4			82		82	139
0310	13 to 20 stories	5 Carp	8	5			140		140	236
0461	Building interior, walls face area, up to 16' high	3 Clab	12	2			41		41	69.50
0561	16' to 40' high		10	2.400			49.50		49.50	83.50

01 54 Construction Aids

01 54 23 – Temporary Scaffolding and Platforms

01 54 23.70 **Scaffolding**	Crew	Daily Output	Labor-Hours	Unit	Material	2009 Bare Costs Labor	Equipment	Total	Total Incl O&P
0801	Building interior floor area, up to 30' high	3 Clab	150	.160	C.C.F.		3.29		3.29	5.55
0901	Over 30' high	4 Clab	160	.200	"		4.11		4.11	6.95
0906	Complete system for face of walls, no plank, material only rent/mo				C.S.F.	35.50			35.50	39
0908	Interior spaces, no plank, material only rent/mo				C.C.F.	3.40			3.40	3.74
0910	Steel tubular, heavy duty shoring, buy									
0920	Frames 5' high 2' wide				Ea.	93			93	102
0925	5' high 4' wide					105			105	116
0930	6' high 2' wide					106			106	117
0935	6' high 4' wide					124			124	136
0940	Accessories									
0945	Cross braces				Ea.	18			18	19.80
0950	U-head, 8" x 8"					21.50			21.50	24
0955	J-head, 4" x 8"					15.80			15.80	17.40
0960	Base plate, 8" x 8"					17.60			17.60	19.35
0965	Leveling jack					38			38	41.50
1000	Steel tubular, regular, buy									
1100	Frames 3' high 5' wide				Ea.	71			71	78
1150	5' high 5' wide					83			83	91.50
1200	6'-4" high 5' wide					104			104	114
1350	7'-6" high 6' wide					179			179	197
1500	Accessories cross braces					20.50			20.50	22.50
1550	Guardrail post					18.60			18.60	20.50
1600	Guardrail 7' section					9.90			9.90	10.90
1650	Screw jacks & plates					27			27	29.50
1700	Sidearm brackets					43			43	47.50
1750	8" casters					35.50			35.50	39
1800	Plank 2" x 10" x 16'-0"					51			51	56
1900	Stairway section					325			325	360
1910	Stairway starter bar					36.50			36.50	40
1920	Stairway inside handrail					66			66	72.50
1930	Stairway outside handrail					98			98	108
1940	Walk-thru frame guardrail					47			47	51.50
2000	Steel tubular, regular, rent/mo.									
2100	Frames 3' high 5' wide				Ea.	5			5	5.50
2150	5' high 5' wide					5			5	5.50
2200	6'-4" high 5' wide					5.05			5.05	5.55
2250	7'-6" high 6' wide					7			7	7.70
2500	Accessories, cross braces					1			1	1.10
2550	Guardrail post					1			1	1.10
2600	Guardrail 7' section					1			1	1.10
2650	Screw jacks & plates					2			2	2.20
2700	Sidearm brackets					2			2	2.20
2750	8" casters					8			8	8.80
2800	Outrigger for rolling tower					3			3	3.30
2850	Plank 2" x 10" x 16'-0"					6			6	6.60
2900	Stairway section					40			40	44
2940	Walk-thru frame guardrail					2.50			2.50	2.75
3000	Steel tubular, heavy duty shoring, rent/mo.									
3250	5' high 2' & 4' wide				Ea.	5			5	5.50
3300	6' high 2' & 4' wide					5			5	5.50
3500	Accessories, cross braces					1			1	1.10
3600	U - head, 8" x 8"					1			1	1.10
3650	J - head, 4" x 8"					1			1	1.10

01 54 Construction Aids

01 54 23 – Temporary Scaffolding and Platforms

01 54 23.70 Scaffolding

		Crew	Daily Output	Labor-Hours	Unit	Material	2009 Bare Costs Labor	2009 Bare Costs Equipment	Total	Total Incl O&P
3700	Base plate, 8" x 8"				Ea.	1			1	1.10
3750	Leveling jack					2			2	2.20
5700	Planks, 2x10x16'-0", labor only to erect & remove to 50' H	3 Carp	72	.333			9.30		9.30	15.75
5800	Over 50' high	4 Carp	80	.400			11.20		11.20	18.90

01 54 23.80 Staging Aids

		Crew	Daily Output	Labor-Hours	Unit	Material	2009 Bare Costs Labor	2009 Bare Costs Equipment	Total	Total Incl O&P
0010	**STAGING AIDS** and fall protection equipment									
0100	Sidewall staging bracket, tubular, buy				Ea.	43			43	47.50
0110	Cost each per day, based on 250 days use				Day	.17			.17	.19
0200	Guard post, buy				Ea.	22.50			22.50	25
0210	Cost each per day, based on 250 days use				Day	.09			.09	.10
0300	End guard chains, buy per pair				Pair	31			31	34
0310	Cost per set per day, based on 250 days use				Day	.17			.17	.19
1010	Cost each per day, based on 250 days use				"	.04			.04	.04
1100	Wood bracket, buy				Ea.	16.60			16.60	18.25
1110	Cost each per day, based on 250 days use				Day	.07			.07	.07
2010	Cost per pair per day, based on 250 days use				"	.43			.43	.48
2100	Steel siderail jack, buy per pair				Pair	122			122	134
2110	Cost per pair per day, based on 250 days use				Day	.49			.49	.54
3010	Cost each per day, based on 250 days use				"	.20			.20	.22
3100	Aluminum scaffolding plank, 20" wide x 24' long, buy				Ea.	890			890	975
3110	Cost each per day, based on 250 days use				Day	3.55			3.55	3.91
4010	Cost each per day, based on 250 days use				"	.86			.86	.95
4100	Rope for safety line, 5/8" x 100' nylon, buy				Ea.	70			70	77
4110	Cost each per day, based on 250 days use				Day	.28			.28	.31
4200	Permanent U-Bolt roof anchor, buy				Ea.	35			35	38.50
4300	Temporary (one use) roof ridge anchor, buy				"	28.50			28.50	31
5000	Installation (setup and removal) of staging aids									
5010	Sidewall staging bracket	2 Carp	64	.250	Ea.		7		7	11.80
5020	Guard post with 2 wood rails	"	64	.250			7		7	11.80
5030	End guard chains, set	1 Carp	64	.125			3.49		3.49	5.90
5100	Roof shingling bracket		96	.083			2.33		2.33	3.94
5200	Ladder jack		64	.125			3.49		3.49	5.90
5300	Wood plank, 2x10x16'	2 Carp	80	.200			5.60		5.60	9.45
5310	Aluminum scaffold plank, 20" x 24'	"	40	.400			11.20		11.20	18.90
5410	Safety rope	1 Carp	40	.200			5.60		5.60	9.45
5420	Permanent U-Bolt roof anchor (install only)	2 Carp	40	.400			11.20		11.20	18.90
5430	Temporary roof ridge anchor (install only)	1 Carp	64	.125			3.49		3.49	5.90

01 54 36 – Equipment Mobilization

01 54 36.50 Mobilization or Demob.

		Crew	Daily Output	Labor-Hours	Unit	Material	2009 Bare Costs Labor	2009 Bare Costs Equipment	Total	Total Incl O&P
0010	**MOBILIZATION OR DEMOB.** (One or the other, unless noted)									
0015	Up to 25 mi haul dist (50 mi RT for mob/demob crew)									
0020	Dozer, loader, backhoe, excav., grader, paver, roller, 70 to 150 H.P.	B-34N	4	2	Ea.		45.50	117	162.50	204
0900	Shovel or dragline, 3/4 C.Y.	B-34K	3.60	2.222			50.50	218	268.50	325
1100	Small equipment, placed in rear of, or towed by pickup truck	A-3A	8	1			22	16.05	38.05	54.50
1150	Equip up to 70 HP, on flatbed trailer behind pickup truck	A-3D	4	2			44	57.50	101.50	137
2000	Mob & demob truck-mounted crane up to 75 ton, driver only	1 Eqhv	3.60	2.222			65		65	107
2200	Crawler-mounted, up to 75 ton	A-3F	2	8			208	405	613	790
2500	For each additional 5 miles haul distance, add						10%	10%		
3000	For large pieces of equipment, allow for assembly/knockdown									
3100	For mob/demob of micro-tunneling equip, see Div. 33 05 23.19									

01 54 Construction Aids

01 54 39 – Construction Equipment

01 54 39.70 Small Tools	Crew	Daily Output	Labor-Hours	Unit	Material	2009 Bare Costs Labor	Equipment	Total	Total Incl O&P
0010 **SMALL TOOLS**									
0020 As % of contractor's bare labor cost for project, minimum				Total					.50%
0100 Maximum				"					2%

01 56 Temporary Barriers and Enclosures

01 56 13 – Temporary Air Barriers

01 56 13.60 Tarpaulins	Crew	Daily Output	Labor-Hours	Unit	Material	2009 Bare Costs Labor	Equipment	Total	Total Incl O&P
0010 **TARPAULINS**									
0020 Cotton duck, 10 oz. to 13.13 oz. per S.Y., minimum				S.F.	.54			.54	.59
0050 Maximum					.53			.53	.58
0200 Reinforced polyethylene 3 mils thick, white					.15			.15	.17
0300 4 mils thick, white, clear or black					.20			.20	.22
0730 Polyester reinforced w/ integral fastening system 11 mils thick					1.07			1.07	1.18

01 71 Examination and Preparation

01 71 23 – Field Engineering

01 71 23.13 Construction Layout	Crew	Daily Output	Labor-Hours	Unit	Material	2009 Bare Costs Labor	Equipment	Total	Total Incl O&P
0010 **CONSTRUCTION LAYOUT**									
1100 Crew for layout of building, trenching or pipe laying, 2 person crew	A-6	1	16	Day		445	70	515	815
1200 3 person crew	A-7	1	24	"		710	70	780	1,250

01 74 Cleaning and Waste Management

01 74 13 – Progress Cleaning

01 74 13.20 Cleaning Up	Crew	Daily Output	Labor-Hours	Unit	Material	2009 Bare Costs Labor	Equipment	Total	Total Incl O&P
0010 **CLEANING UP**									
0020 After job completion, allow, minimum				Job					.30%
0040 Maximum				"					1%

Division 2
Existing Conditions

02 21 Surveys

02 21 13 – Site Surveys

02 21 13.09 Topographical Surveys	Crew	Daily Output	Labor-Hours	Unit	Material	2009 Bare Costs Labor	Equipment	Total	Total Incl O&P
0010 **TOPOGRAPHICAL SURVEYS**									
0020 Topographical surveying, conventional, minimum	A-7	3.30	7.273	Acre	18	215	21	254	405
0100 Maximum	A-8	.60	53.333	"	55	1,550	117	1,722	2,775

02 21 13.13 Boundary and Survey Markers

	Crew	Daily Output	Labor-Hours	Unit	Material	Labor	Equipment	Total	Total Incl O&P
0010 **BOUNDARY AND SURVEY MARKERS**									
0300 Lot location and lines, large quantities, minimum	A-7	2	12	Acre	32	355	35	422	670
0320 Average	"	1.25	19.200		51	565	56	672	1,075
0400 Small quantities, maximum	A-8	1	32	↓	68	930	70	1,068	1,700
0600 Monuments, 3' long	A-7	10	2.400	Ea.	34	71	7	112	164
0800 Property lines, perimeter, cleared land	"	1000	.024	L.F.	.03	.71	.07	.81	1.30
0900 Wooded land	A-8	875	.037	"	.05	1.06	.08	1.19	1.92

02 32 Geotechnical Investigations

02 32 13 – Subsurface Drilling and Sampling

02 32 13.10 Boring and Exploratory Drilling

	Crew	Daily Output	Labor-Hours	Unit	Material	Labor	Equipment	Total	Total Incl O&P
0010 **BORING AND EXPLORATORY DRILLING**									
0020 Borings, initial field stake out & determination of elevations	A-6	1	16	Day		445	70	515	815
0100 Drawings showing boring details				Total		185		185	270
0200 Report and recommendations from P.E.						415		415	595
0300 Mobilization and demobilization, minimum	B-55	4	4	↓		85	237	322	405
0350 For over 100 miles, per added mile		450	.036	Mile		.76	2.11	2.87	3.59
0600 Auger holes in earth, no samples, 2-1/2" diameter		78.60	.204	L.F.		4.33	12.10	16.43	20.50
0800 Cased borings in earth, with samples, 2-1/2" diameter		55.50	.288	"	20	6.15	17.10	43.25	51
1400 Borings, earth, drill rig and crew with truck mounted auger	↓	1	16	Day		340	950	1,290	1,625
1500 For inner city borings add, minimum									10%
1510 Maximum									20%

02 41 Demolition

02 41 13 – Selective Site Demolition

02 41 13.17 Demolish, Remove Pavement and Curb

	Crew	Daily Output	Labor-Hours	Unit	Material	Labor	Equipment	Total	Total Incl O&P
0010 **DEMOLISH, REMOVE PAVEMENT AND CURB** R024119-10									
5010 Pavement removal, bituminous roads, 3" thick	B-38	690	.035	S.Y.		.79	.65	1.44	2.03
5050 4" to 6" thick		420	.057			1.30	1.06	2.36	3.34
5100 Bituminous driveways		640	.038			.85	.70	1.55	2.19
5200 Concrete to 6" thick, hydraulic hammer, mesh reinforced		255	.094			2.13	1.75	3.88	5.50
5300 Rod reinforced	↓	200	.120	↓		2.72	2.23	4.95	7
5600 With hand held air equipment, bituminous, to 6" thick	B-39	1900	.025	S.F.		.53	.10	.63	1
5700 Concrete to 6" thick, no reinforcing		1600	.030			.63	.12	.75	1.19
5800 Mesh reinforced		1400	.034			.72	.14	.86	1.36
5900 Rod reinforced	↓	765	.063	↓		1.31	.25	1.56	2.50
6000 Curbs, concrete, plain	B-6	360	.067	L.F.		1.51	.82	2.33	3.43
6100 Reinforced		275	.087			1.98	1.07	3.05	4.49
6200 Granite		360	.067			1.51	.82	2.33	3.43
6300 Bituminous	↓	528	.045	↓		1.03	.56	1.59	2.33

02 41 13.33 Minor Site Demolition

	Crew	Daily Output	Labor-Hours	Unit	Material	Labor	Equipment	Total	Total Incl O&P
0010 **MINOR SITE DEMOLITION** R024119-10									
0015 No hauling, abandon catch basin or manhole	B-6	7	3.429	Ea.		78	42	120	176
0020 Remove existing catch basin or manhole, masonry		4	6			136	73.50	209.50	310
0030 Catch basin or manhole frames and covers, stored	↓	13	1.846	↓		42	22.50	64.50	95

02 41 Demolition

02 41 13 – Selective Site Demolition

02 41 13.33 Minor Site Demolition

		Crew	Daily Output	Labor-Hours	Unit	Material	2009 Bare Costs		Total	Total Incl O&P
							Labor	Equipment		
0040	Remove and reset	B-6	7	3.429	Ea.		78	42	120	176
1000	Masonry walls, block, solid	B-5	1800	.022	C.F.		.50	.66	1.16	1.56
1200	Brick, solid		900	.044			1	1.31	2.31	3.12
1400	Stone, with mortar		900	.044			1	1.31	2.31	3.12
1500	Dry set		1500	.027			.60	.79	1.39	1.88
2900	Pipe removal, sewer/water, no excavation, 12" diameter	B-6	175	.137	L.F.		3.11	1.68	4.79	7.05
2960	21"-24" diameter		120	.200	"		4.54	2.45	6.99	10.30
4000	Sidewalk removal, bituminous, 2-1/2" thick		325	.074	S.Y.		1.67	.90	2.57	3.79
4050	Brick, set in mortar		185	.130			2.94	1.59	4.53	6.65
4100	Concrete, plain, 4"		160	.150			3.40	1.84	5.24	7.70
4200	Mesh reinforced		150	.160			3.63	1.96	5.59	8.20
4300	Slab on grade removal, plain	B-5	45	.889	C.Y.		20	26.50	46.50	62.50
4310	Mesh reinforced		33	1.212			27.50	36	63.50	85.50
4320	Rod reinforced		25	1.600			36	47.50	83.50	113
4400	For congested sites or small quantities, add up to								200%	200%
4450	For disposal on site, add	B-11A	232	.069			1.69	4.66	6.35	7.95
4500	To 5 miles, add	B-34D	76	.105			2.39	8	10.39	12.80

02 41 13.60 Selective Demolition Fencing

		Crew	Daily Output	Labor-Hours	Unit	Material	2009 Bare Costs		Total	Total Incl O&P
							Labor	Equipment		
0010	**SELECTIVE DEMOLITION FENCING** R024119-10									
1600	Fencing, barbed wire, 3 strand	2 Clab	430	.037	L.F.		.76		.76	1.29
1650	5 strand	"	280	.057			1.17		1.17	1.99
1700	Chain link, posts & fabric, 8' to 10' high, remove only	B-6	445	.054			1.22	.66	1.88	2.78
1750	Remove and reset	"	70	.343			7.80	4.20	12	17.60

02 41 16 – Structure Demolition

02 41 16.13 Building Demolition

		Crew	Daily Output	Labor-Hours	Unit	Material	2009 Bare Costs		Total	Total Incl O&P
							Labor	Equipment		
0010	**BUILDING DEMOLITION** Large urban projects, incl. 20 mi. haul R024119-10									
0500	Small bldgs, or single bldgs, no salvage included, steel	B-3	14800	.003	C.F.		.07	.14	.21	.27
0600	Concrete	"	11300	.004	"		.10	.18	.28	.36
0605	Concrete, plain	B-5	33	1.212	C.Y.		27.50	36	63.50	85.50
0610	Reinforced		25	1.600			36	47.50	83.50	113
0615	Concrete walls		34	1.176			26.50	35	61.50	82.50
0620	Elevated slabs		26	1.538			34.50	45.50	80	109
0650	Masonry	B-3	14800	.003	C.F.		.07	.14	.21	.27
0700	Wood	"	14800	.003	"		.07	.14	.21	.27
1000	Single family, one story house, wood, minimum				Ea.				2,525	2,975
1020	Maximum								4,400	5,275
1200	Two family, two story house, wood, minimum								3,300	3,950
1220	Maximum								6,375	7,700
1300	Three family, three story house, wood, minimum								4,400	5,275
1320	Maximum								7,700	9,250

02 41 16.17 Building Demolition Footings and Foundations

		Crew	Daily Output	Labor-Hours	Unit	Material	2009 Bare Costs		Total	Total Incl O&P
							Labor	Equipment		
0010	**BUILDING DEMOLITION FOOTINGS AND FOUNDATIONS** R024119-10									
0200	Floors, concrete slab on grade,									
0240	4" thick, plain concrete	B-9	500	.080	S.F.		1.68	.38	2.06	3.25
0280	Reinforced, wire mesh		470	.085			1.78	.41	2.19	3.47
0300	Rods		400	.100			2.10	.48	2.58	4.07
0400	6" thick, plain concrete		375	.107			2.23	.51	2.74	4.34
0420	Reinforced, wire mesh		340	.118			2.46	.56	3.02	4.79
0440	Rods		300	.133			2.79	.64	3.43	5.40
1000	Footings, concrete, 1' thick, 2' wide	B-5	300	.133	L.F.		3.01	3.94	6.95	9.40
1080	1'-6" thick, 2' wide		250	.160			3.61	4.73	8.34	11.25
1120	3' wide		200	.200			4.51	5.90	10.41	14.05

13

02 41 Demolition

02 41 16 – Structure Demolition

02 41 16.17 Building Demolition Footings and Foundations	Crew	Daily Output	Labor-Hours	Unit	Material	2009 Bare Costs Labor	Equipment	Total	Total Incl O&P
1200	Average reinforcing, add				L.F.				10%	10%
2000	Walls, block, 4" thick	1 Clab	180	.044	S.F.		.91		.91	1.54
2040	6" thick		170	.047			.97		.97	1.64
2080	8" thick		150	.053			1.10		1.10	1.85
2100	12" thick		150	.053			1.10		1.10	1.85
2400	Concrete, plain concrete, 6" thick	B-9	160	.250			5.25	1.20	6.45	10.15
2420	8" thick		140	.286			6	1.37	7.37	11.60
2440	10" thick		120	.333			7	1.60	8.60	13.55
2500	12" thick		100	.400			8.40	1.92	10.32	16.25
2600	For average reinforcing, add								10%	10%
4000	For congested sites or small quantities, add up to								200%	200%
4200	Add for disposal, on site	B-11A	232	.069	C.Y.		1.69	4.66	6.35	7.95
4250	To five miles	B-30	220	.109	"		2.69	8.80	11.49	14.15

02 41 19 – Selective Structure Demolition

02 41 19.13 Selective Building Demolition

0010	**SELECTIVE BUILDING DEMOLITION**									
0020	Costs related to selective demolition of specific building components									
0025	are included under Common Work Results (XX 05 00)									
0030	in the component's appropriate division.									

02 41 19.16 Selective Demolition, Cutout

		Crew	Daily Output	Labor-Hours	Unit	Material	Labor	Equipment	Total	Total Incl O&P
0010	**SELECTIVE DEMOLITION, CUTOUT** R024119-10									
0020	Concrete, elev. slab, light reinforcement, under 6 C.F.	B-9	65	.615	C.F.		12.90	2.95	15.85	25
0050	Light reinforcing, over 6 C.F.		75	.533	"		11.15	2.55	13.70	21.50
0200	Slab on grade to 6" thick, not reinforced, under 8 S.F.		85	.471	S.F.		9.85	2.25	12.10	19.15
0250	8 – 16 S.F.		175	.229	"		4.79	1.09	5.88	9.30
0255	For over 16 S.F. see Div. 02 41 16.17 0400									
0600	Walls, not reinforced, under 6 C.F.	B-9	60	.667	C.F.		13.95	3.19	17.14	27
0650	6 – 12 C.F.	"	80	.500	"		10.50	2.40	12.90	20.50
0655	For over 12 C.F. see Div. 02 41 16.17 2500									
1000	Concrete, elevated slab, bar reinforced, under 6 C.F.	B-9	45	.889	C.F.		18.60	4.26	22.86	36
1050	Bar reinforced, over 6 C.F.		50	.800	"		16.75	3.83	20.58	32.50
1200	Slab on grade to 6" thick, bar reinforced, under 8 S.F.		75	.533	S.F.		11.15	2.55	13.70	21.50
1250	8 – 16 S.F.		150	.267	"		5.60	1.28	6.88	10.85
1255	For over 16 S.F. see Div. 02 41 16.17 0440									
1400	Walls, bar reinforced, under 6 C.F.	B-9	50	.800	C.F.		16.75	3.83	20.58	32.50
1450	6 – 12 C.F.	"	70	.571	"		11.95	2.74	14.69	23.50
1455	For over 12 C.F. see Div. 02 41 16.17 2500 and 2600									
2000	Brick, to 4 S.F. opening, not including toothing									
2040	4" thick	B-9	30	1.333	Ea.		28	6.40	34.40	54
2060	8" thick		18	2.222			46.50	10.65	57.15	90
2080	12" thick		10	4			84	19.15	103.15	163
2400	Concrete block, to 4 S.F. opening, 2" thick		35	1.143			24	5.45	29.45	46.50
2420	4" thick		30	1.333			28	6.40	34.40	54
2440	8" thick		27	1.481			31	7.10	38.10	60.50
2460	12" thick		24	1.667			35	8	43	68
2600	Gypsum block, to 4 S.F. opening, 2" thick		80	.500			10.50	2.40	12.90	20.50
2620	4" thick		70	.571			11.95	2.74	14.69	23.50
2640	8" thick		55	.727			15.25	3.48	18.73	30
2800	Terra cotta, to 4 S.F. opening, 4" thick		70	.571			11.95	2.74	14.69	23.50
2840	8" thick		65	.615			12.90	2.95	15.85	25
2880	12" thick		50	.800			16.75	3.83	20.58	32.50
3000	Toothing masonry cutouts, brick, soft old mortar	1 Brhe	40	.200	V.L.F.		4.50		4.50	7.45

02 41 Demolition

02 41 19 – Selective Structure Demolition

02 41 19.16 Selective Demolition, Cutout

		Crew	Daily Output	Labor-Hours	Unit	Material	2009 Bare Costs Labor	2009 Bare Costs Equipment	Total	Total Incl O&P
3100	Hard mortar	1 Brhe	30	.267	V.L.F.		6		6	9.95
3200	Block, soft old mortar		70	.114			2.57		2.57	4.26
3400	Hard mortar	↓	50	.160	↓		3.60		3.60	5.95
6000	Walls, interior, not including re-framing,									
6010	openings to 5 S.F.									
6100	Drywall to 5/8" thick	1 Clab	24	.333	Ea.		6.85		6.85	11.60
6200	Paneling to 3/4" thick		20	.400			8.20		8.20	13.90
6300	Plaster, on gypsum lath		20	.400			8.20		8.20	13.90
6340	On wire lath	↓	14	.571	↓		11.75		11.75	19.85
7000	Wood frame, not including re-framing, openings to 5 S.F.									
7200	Floors, sheathing and flooring to 2" thick	1 Clab	5	1.600	Ea.		33		33	55.50
7310	Roofs, sheathing to 1" thick, not including roofing		6	1.333			27.50		27.50	46.50
7410	Walls, sheathing to 1" thick, not including siding	↓	7	1.143			23.50		23.50	39.50

02 41 19.19 Selective Demolition, Dump Charges

		Crew	Daily Output	Labor-Hours	Unit	Material	2009 Bare Costs Labor	2009 Bare Costs Equipment	Total	Total Incl O&P
0010	**SELECTIVE DEMOLITION, DUMP CHARGES** R024119-10									
0020	Dump charges, typical urban city, tipping fees only									
0100	Building construction materials				Ton					70
0200	Trees, brush, lumber									50
0300	Rubbish only									60
0500	Reclamation station, usual charge				↓	95			95	104.50

02 41 19.21 Selective Demolition, Gutting

		Crew	Daily Output	Labor-Hours	Unit	Material	2009 Bare Costs Labor	2009 Bare Costs Equipment	Total	Total Incl O&P
0010	**SELECTIVE DEMOLITION, GUTTING** R024119-10									
0020	Building interior, including disposal, dumpster fees not included									
0500	Residential building									
0560	Minimum	B-16	400	.080	SF Flr.		1.73	1.33	3.06	4.39
0580	Maximum	"	360	.089	"		1.92	1.48	3.40	4.87
0900	Commercial building									
1000	Minimum	B-16	350	.091	SF Flr.		1.97	1.52	3.49	5
1020	Maximum	"	250	.128	"		2.76	2.13	4.89	7

02 41 19.23 Selective Demolition, Rubbish Handling

		Crew	Daily Output	Labor-Hours	Unit	Material	2009 Bare Costs Labor	2009 Bare Costs Equipment	Total	Total Incl O&P
0010	**SELECTIVE DEMOLITION, RUBBISH HANDLING** R024119-10									
0020	The following are to be added to the demolition prices									
0400	Chute, circular, prefabricated steel, 18" diameter	B-1	40	.600	L.F.	47	12.75		59.75	73.50
0440	30" diameter	"	30	.800	"	48.50	17		65.50	82
0600	Dumpster, weekly rental, 1 dump/week, 6 C.Y. capacity (2 Tons)				Week					315
0700	10 C.Y. capacity (4 Tons)									375
0725	Dumpster, weekly rental, 1 dump/week, 20 C.Y. capacity (8 Tons) R024119-20									440
0800	30 C.Y. capacity (10 Tons)									665
0840	40 C.Y. capacity (13 Tons)				↓					805
1000	Dust partition, 6 mil polyethylene, 1" x 3" frame	2 Carp	2000	.008	S.F.	.44	.22		.66	.87
1080	2" x 4" frame	"	2000	.008	"	.27	.22		.49	.68
2000	Load, haul, and dump, 50' haul	2 Clab	24	.667	C.Y.		13.70		13.70	23
2040	100' haul		16.50	.970			19.95		19.95	33.50
2080	Over 100' haul, add per 100 L.F.		35.50	.451			9.25		9.25	15.65
2120	In elevators, per 10 floors, add	↓	140	.114			2.35		2.35	3.97
3000	Loading & trucking, including 2 mile haul, chute loaded	B-16	45	.711			15.35	11.85	27.20	39
3040	Hand loading truck, 50' haul	"	48	.667			14.40	11.10	25.50	36.50
3080	Machine loading truck	B-17	120	.267			6.05	5.15	11.20	15.85
5000	Haul, per mile, up to 8 C.Y. truck	B-34B	1165	.007			.16	.46	.62	.76
5100	Over 8 C.Y. truck	"	1550	.005	↓		.12	.34	.46	.58

02 83 19.23 Encapsulation of Lead-Based Paint	Crew	Daily Output	Labor-Hours	Unit	Material	2009 Bare Costs Labor	Equipment	Total	Total Incl O&P
0010	**ENCAPSULATION OF LEAD-BASED PAINT**									
0020	Interior, brushwork, trim, under 6"	1 Pord	240	.033	L.F.	2.46	.83		3.29	4.07
0030	6" to 12" wide		180	.044		3.28	1.11		4.39	5.45
0040	Balustrades		300	.027		1.98	.67		2.65	3.27
0050	Pipe to 4" diameter		500	.016		1.19	.40		1.59	1.97
0060	To 8" diameter		375	.021		1.57	.53		2.10	2.60
0070	To 12" diameter		250	.032		2.36	.80		3.16	3.91
0080	To 16" diameter		170	.047		3.47	1.18		4.65	5.75
0090	Cabinets, ornate design		200	.040	S.F.	2.97	1		3.97	4.91
0100	Simple design		250	.032	"	2.36	.80		3.16	3.91
0110	Doors, 3' x 7', both sides, incl. frame & trim									
0120	Flush	1 Pord	6	1.333	Ea.	30.50	33.50		64	88
0130	French, 10-15 lite		3	2.667		6.05	66.50		72.55	116
0140	Panel		4	2		36.50	50		86.50	122
0150	Louvered		2.75	2.909		33.50	72.50		106	156
0160	Windows, per interior side, per 15 S.F.									
0170	1 to 6 lite	1 Pord	14	.571	Ea.	21	14.30		35.30	46.50
0180	7 to 10 lite		7.50	1.067		23	26.50		49.50	69
0190	12 lite		5.75	1.391		31	35		66	91
0200	Radiators		8	1		73.50	25		98.50	122
0210	Grilles, vents		275	.029	S.F.	2.15	.73		2.88	3.56
0220	Walls, roller, drywall or plaster		1000	.008		.59	.20		.79	.98
0230	With spunbonded reinforcing fabric		720	.011		.68	.28		.96	1.21
0240	Wood		800	.010		.74	.25		.99	1.22
0250	Ceilings, roller, drywall or plaster		900	.009		.68	.22		.90	1.11
0260	Wood		700	.011		.84	.29		1.13	1.39
0270	Exterior, brushwork, gutters and downspouts		300	.027	L.F.	1.98	.67		2.65	3.27
0280	Columns		400	.020	S.F.	1.47	.50		1.97	2.44
0290	Spray, siding		600	.013	"	.99	.33		1.32	1.64
0300	Miscellaneous									
0310	Electrical conduit, brushwork, to 2" diameter	1 Pord	500	.016	L.F.	1.19	.40		1.59	1.97
0320	Brick, block or concrete, spray		500	.016	S.F.	1.19	.40		1.59	1.97
0330	Steel, flat surfaces and tanks to 12"		500	.016		1.19	.40		1.59	1.97
0340	Beams, brushwork		400	.020		1.47	.50		1.97	2.44
0350	Trusses		400	.020		1.47	.50		1.97	2.44

02 83 19.26 Removal of Lead-Based Paint

		Crew	Daily Output	Labor-Hours	Unit	Material	Labor	Equipment	Total	Total Incl O&P
0010	**REMOVAL OF LEAD-BASED PAINT**									
0011	By chemicals, per application									
0050	Baseboard, to 6" wide	1 Pord	64	.125	L.F.	.66	3.13		3.79	5.85
0070	To 12" wide		32	.250	"	1.32	6.25		7.57	11.70
0200	Balustrades, one side		28	.286	S.F.	1.32	7.15		8.47	13.15
1400	Cabinets, simple design		32	.250		1.32	6.25		7.57	11.70
1420	Ornate design		25	.320		1.32	8		9.32	14.55
1600	Cornice, simple design		60	.133		1.32	3.33		4.65	6.90
1620	Ornate design		20	.400		5.05	10		15.05	22
2800	Doors, one side, flush		84	.095		1	2.38		3.38	5
2820	Two panel		80	.100		1.32	2.50		3.82	5.55
2840	Four panel		45	.178		1.32	4.44		5.76	8.75
2880	For trim, one side, add		64	.125	L.F.	.66	3.13		3.79	5.85
3000	Fence, picket, one side		30	.267	S.F.	1.32	6.65		7.97	12.35
3200	Grilles, one side, simple design		30	.267		1.32	6.65		7.97	12.35
3220	Ornate design		25	.320		1.32	8		9.32	14.55

02 83 Lead Remediation

02 83 19 – Lead-Based Paint Remediation

02 83 19.26 Removal of Lead-Based Paint	Crew	Daily Output	Labor-Hours	Unit	Material	2009 Bare Costs Labor	Equipment	Total	Total Incl O&P
4400	Pipes, to 4" diameter	1 Pord	90	.089	L.F.	1	2.22		3.22	4.74
4420	To 8" diameter		50	.160		2	4		6	8.75
4440	To 12" diameter		36	.222		3	5.55		8.55	12.40
4460	To 16" diameter		20	.400		4	10		14	21
4500	For hangers, add		40	.200	Ea.	2.54	5		7.54	11
4800	Siding		90	.089	S.F.	1.17	2.22		3.39	4.93
5000	Trusses, open		55	.145	SF Face	1.85	3.64		5.49	8
6200	Windows, one side only, double hung, 1/1 light, 24" x 48" high		4	2	Ea.	25.50	50		75.50	110
6220	30" x 60" high		3	2.667		34	66.50		100.50	147
6240	36" x 72" high		2.50	3.200		41	80		121	176
6280	40" x 80" high		2	4		51	100		151	220
6400	Colonial window, 6/6 light, 24" x 48" high		2	4		51	100		151	221
6420	30" x 60" high		1.50	5.333		68	133		201	293
6440	36" x 72" high		1	8		102	200		302	440
6480	40" x 80" high		1	8		102	200		302	440
6600	8/8 light, 24" x 48" high		2	4		51	100		151	221
6620	40" x 80" high		1	8		102	200		302	440
6800	12/12 light, 24" x 48" high		1	8		102	200		302	440
6820	40" x 80" high		.75	10.667		136	267		403	585
6840	Window frame & trim items, included in pricing above									

Division 3
Concrete

03 01 Maintenance of Concrete

03 01 30 – Maintenance of Cast-In-Place Concrete

03 01 30.64 Floor Patching		Crew	Daily Output	Labor-Hours	Unit	Material	2009 Bare Costs Labor	Equipment	Total	Total Incl O&P
0010	**FLOOR PATCHING**									
0012	Floor patching, 1/4" thick, small areas, regular	1 Cefi	170	.047	S.F.	2.51	1.26		3.77	4.78
0100	Epoxy	"	100	.080	"	5.45	2.14		7.59	9.45

03 05 Common Work Results for Concrete

03 05 13 – Basic Concrete Materials

03 05 13.85 Winter Protection

03 05 13.85 Winter Protection		Crew	Daily Output	Labor-Hours	Unit	Material	2009 Bare Costs Labor	Equipment	Total	Total Incl O&P
0010	**WINTER PROTECTION**									
0012	For heated ready mix, add, minimum				C.Y.	5.25			5.25	5.80
0050	Maximum				"	6.60			6.60	7.25
0100	Temporary heat to protect concrete, 24 hours, minimum	2 Clab	50	.320	M.S.F.	520	6.60		526.60	585
0200	Temporary shelter for slab on grade, wood frame/polyethylene sheeting									
0201	Build or remove, minimum	2 Carp	10	1.600	M.S.F.	245	44.50		289.50	345
0210	Maximum	"	3	5.333	"	293	149		442	570
0710	Electrically, heated pads, 15 watts/S.F., 20 uses, minimum				S.F.	.22			.22	.24
0800	Maximum				"	.36			.36	.40

03 11 Concrete Forming

03 11 13 – Structural Cast-In-Place Concrete Forming

03 11 13.25 Forms In Place, Columns

03 11 13.25 Forms In Place, Columns			Crew	Daily Output	Labor-Hours	Unit	Material	2009 Bare Costs Labor	Equipment	Total	Total Incl O&P
0010	**FORMS IN PLACE, COLUMNS**										
1500	Round fiber tube, recycled paper, 1 use, 8" diameter	G	C-1	155	.206	L.F.	1.39	5		6.39	10.05
1550	10" diameter	G		155	.206		1.98	5		6.98	10.70
1600	12" diameter	G		150	.213		2.33	5.20		7.53	11.30
1700	16" diameter	G		140	.229		4	5.55		9.55	13.80
1720	18" diameter	G		140	.229		4.68	5.55		10.23	14.55
5000	Job-built plywood, 8" x 8" columns, 1 use			165	.194	SFCA	2.41	4.72		7.13	10.60
5500	12" x 12" columns, 1 use			180	.178	"	2.27	4.33		6.60	9.80
7400	Steel framed plywood, based on 50 uses of purchased										
7420	forms, and 4 uses of bracing lumber										
7500	8" x 8" column		C-1	340	.094	SFCA	4.53	2.29		6.82	8.85
7550	10" x 10"			350	.091		3.99	2.22		6.21	8.15
7600	12" x 12"			370	.086		3.38	2.10		5.48	7.25

03 11 13.45 Forms In Place, Footings

03 11 13.45 Forms In Place, Footings		Crew	Daily Output	Labor-Hours	Unit	Material	2009 Bare Costs Labor	Equipment	Total	Total Incl O&P
0010	**FORMS IN PLACE, FOOTINGS**									
0020	Continuous wall, plywood, 1 use	C-1	375	.085	SFCA	7.45	2.08		9.53	11.70
0150	4 use	"	485	.066	"	2.42	1.61		4.03	5.40
1500	Keyway, 4 use, tapered wood, 2" x 4"	1 Carp	530	.015	L.F.	.16	.42		.58	.88
1550	2" x 6"	"	500	.016	"	.26	.45		.71	1.05
5000	Spread footings, job-built lumber, 1 use	C-1	305	.105	SFCA	2.16	2.55		4.71	6.70
5150	4 use	"	414	.077	"	.70	1.88		2.58	3.95

03 11 13.50 Forms In Place, Grade Beam

03 11 13.50 Forms In Place, Grade Beam		Crew	Daily Output	Labor-Hours	Unit	Material	2009 Bare Costs Labor	Equipment	Total	Total Incl O&P
0010	**FORMS IN PLACE, GRADE BEAM**									
0020	Job-built plywood, 1 use	C-2	530	.091	SFCA	3.25	2.24		5.49	7.35
0150	4 use	"	605	.079	"	1.06	1.96		3.02	4.47

03 11 13.65 Forms In Place, Slab On Grade

03 11 13.65 Forms In Place, Slab On Grade		Crew	Daily Output	Labor-Hours	Unit	Material	2009 Bare Costs Labor	Equipment	Total	Total Incl O&P
0010	**FORMS IN PLACE, SLAB ON GRADE**									
1000	Bulkhead forms w/keyway, wood, 6" high, 1 use	C-1	510	.063	L.F.	.84	1.53		2.37	3.50

03 11 Concrete Forming

03 11 13 – Structural Cast-In-Place Concrete Forming

03 11 13.65 Forms In Place, Slab On Grade

			Crew	Daily Output	Labor-Hours	Unit	Material	2009 Bare Costs Labor	Equipment	Total	Total Incl O&P
1400	Bulkhead form for slab, 4-1/2" high, exp metal, incl keyway & stakes	G	C-1	1200	.027	L.F.	2.78	.65		3.43	4.16
1410	5-1/2" high	G		1100	.029		3.23	.71		3.94	4.75
1420	7-1/2" high	G		960	.033		4.26	.81		5.07	6.05
1430	9-1/2" high	G		840	.038		4.84	.93		5.77	6.85
2000	Curb forms, wood, 6" to 12" high, on grade, 1 use			215	.149	SFCA	2.83	3.62		6.45	9.20
2150	4 use			275	.116	"	.92	2.83		3.75	5.80
3000	Edge forms, wood, 4 use, on grade, to 6" high			600	.053	L.F.	.38	1.30		1.68	2.61
3050	7" to 12" high			435	.074	SFCA	.74	1.79		2.53	3.83
4000	For slab blockouts, to 12" high, 1 use			200	.160	L.F.	.63	3.89		4.52	7.30
4100	Plastic (extruded), to 6" high, multiple use, on grade			800	.040	"	5.50	.97		6.47	7.70
8760	Void form, corrugated fiberboard, 6" x 12", 10' long	G		240	.133	S.F.	.84	3.24		4.08	6.40

03 11 13.85 Forms In Place, Walls

			Crew	Daily Output	Labor-Hours	Unit	Material	2009 Bare Costs Labor	Equipment	Total	Total Incl O&P
0010	**FORMS IN PLACE, WALLS**										
0100	Box out for wall openings, to 16" thick, to 10 S.F.		C-2	24	2	Ea.	21.50	49.50		71	107
0150	Over 10 S.F. (use perimeter)		"	280	.171	L.F.	1.78	4.23		6.01	9.10
0250	Brick shelf, 4" w, add to wall forms, use wall area abv shelf										
0260	1 use		C-2	240	.200	SFCA	1.87	4.94		6.81	10.40
0350	4 use			300	.160	"	.75	3.95		4.70	7.45
0500	Bulkhead, wood with keyway, 1 use, 2 piece			265	.181	L.F.	1.68	4.47		6.15	9.40
0600	Bulkhead forms with keyway, 1 piece expanded metal, 8" wall	G	C-1	1000	.032		4.26	.78		5.04	6
0610	10" wall	G	"	800	.040		4.84	.97		5.81	6.95
2000	Wall, job-built plywood, to 8' high, 1 use		C-2	370	.130	SFCA	2.62	3.20		5.82	8.30
2050	2 use			435	.110		1.61	2.72		4.33	6.35
2100	3 use			495	.097		1.17	2.39		3.56	5.35
2150	4 use			505	.095		.95	2.35		3.30	5
2400	Over 8' to 16' high, 1 use			280	.171		9.40	4.23		13.63	17.50
2450	2 use			345	.139		1.34	3.43		4.77	7.25
2500	3 use			375	.128		.96	3.16		4.12	6.40
2550	4 use			395	.122		.78	3		3.78	5.90
3000	For architectural finish, add			1820	.026		2.72	.65		3.37	4.09
7800	Modular prefabricated plywood, based on 20 uses of purchased										
7820	forms, and 4 uses of bracing lumber										
7860	To 8' high		C-2	800	.060	SFCA	1.47	1.48		2.95	4.11
8060	Over 8' to 16' high		"	600	.080	"	1.51	1.97		3.48	5

03 11 19 – Insulating Concrete Forming

03 11 19.10 Insulating Forms, Left In Place

			Crew	Daily Output	Labor-Hours	Unit	Material	2009 Bare Costs Labor	Equipment	Total	Total Incl O&P
0010	**INSULATING FORMS, LEFT IN PLACE**										
0020	S.F. is for exterior face, but includes forms for both faces										
2000	4" wall, straight block, 16" x 48" (5.33 SF)	G	2 Carp	90	.178	Ea.	15.75	4.97		20.72	25.50
2010	90 corner block, exterior 16" x 38" x 22" (6.67 SF)	G		75	.213		18.45	5.95		24.40	30.50
2020	45 corner block, exterior 16" x 34" x 18" (5.78 SF)	G		75	.213		17.05	5.95		23	29
2100	6" wall, straight block, 16" x 48" (5.33 SF)	G		90	.178		16.05	4.97		21.02	26
2110	90 corner block, exterior 16" x 32" x 24" (6.22 SF)	G		75	.213		18.45	5.95		24.40	30.50
2120	45 corner block, exterior 16" x 26" x 18" (4.89 SF)	G		75	.213		16.85	5.95		22.80	28.50
2130	Brick ledge block, 16" x 48" (5.33 SF)	G		80	.200		19.65	5.60		25.25	31
2140	Taper top block, 16" x 48" (5.33 SF)	G		80	.200		17.55	5.60		23.15	29
2200	8" wall, straight block, 16" x 48" (5.33 SF)	G		90	.178		17.05	4.97		22.02	27
2210	90 corner block, exterior 16" x 34" x 26" (6.67 SF)	G		75	.213		19.60	5.95		25.55	31.50
2220	45 corner block, exterior 16" x 28" x 20" (5.33 SF)	G		75	.213		17.75	5.95		23.70	29.50
2230	Brick ledge block, 16" x 48" (5.33 SF)	G		80	.200		20.50	5.60		26.10	32
2240	Taper top block, 16" x 48" (5.33 SF)	G		80	.200		18.55	5.60		24.15	30

03 11 Concrete Forming

03 11 23 – Permanent Stair Forming

03 11 23.75 Forms In Place, Stairs		Crew	Daily Output	Labor-Hours	Unit	Material	2009 Bare Costs Labor	2009 Bare Costs Equipment	Total	Total Incl O&P
0010	**FORMS IN PLACE, STAIRS**									
0015	(Slant length x width), 1 use	C-2	165	.291	S.F.	5.10	7.20		12.30	17.75
0150	4 use		190	.253		2.11	6.25		8.36	12.85
2000	Stairs, cast on sloping ground (length x width), 1 use		220	.218		1.75	5.40		7.15	11
2025	2 use		232	.207		.96	5.10		6.06	9.70
2050	3 use		244	.197		.70	4.86		5.56	8.95
2100	4 use		256	.188		.57	4.63		5.20	8.40

03 15 Concrete Accessories

03 15 05 – Concrete Forming Accessories

03 15 05.02 Anchor Bolts

			Crew	Daily Output	Labor-Hours	Unit	Material	Labor	Equipment	Total	Total Incl O&P
0010	**ANCHOR BOLTS**										
0015	J-type, plain, incl. nut and washer										
0020	1/2" diameter, 6" long	G	1 Carp	90	.089	Ea.	1.01	2.48		3.49	5.30
0050	10" long	G		85	.094		1.15	2.63		3.78	5.70
0100	12" long	G		85	.094		1.26	2.63		3.89	5.85
0200	5/8" diameter, 12" long	G		80	.100		1.58	2.80		4.38	6.45
0250	18" long	G		70	.114		1.86	3.19		5.05	7.45
0300	24" long	G		60	.133		2.14	3.73		5.87	8.65
0350	3/4" diameter, 8" long	G		80	.100		1.86	2.80		4.66	6.80
0400	12" long	G		70	.114		2.33	3.19		5.52	7.95
0450	18" long	G		60	.133		3.03	3.73		6.76	9.65
0500	24" long	G		50	.160		3.96	4.47		8.43	11.90

03 15 05.12 Chamfer Strips

			Crew	Daily Output	Labor-Hours	Unit	Material	Labor	Equipment	Total	Total Incl O&P
0010	**CHAMFER STRIPS**										
5000	Wood, 1/2" wide		1 Carp	535	.015	L.F.	.22	.42		.64	.95
5200	3/4" wide			525	.015		.21	.43		.64	.95
5400	1" wide			515	.016		.34	.43		.77	1.10

03 15 05.15 Column Form Accessories

			Crew	Daily Output	Labor-Hours	Unit	Material	Labor	Equipment	Total	Total Incl O&P
0010	**COLUMN FORM ACCESSORIES**										
1000	Column clamps, adjustable to 24" x 24", buy	G				Set	117			117	129
1400	Rent per month	G				"	12.75			12.75	14

03 15 05.25 Expansion Joints

			Crew	Daily Output	Labor-Hours	Unit	Material	Labor	Equipment	Total	Total Incl O&P
0010	**EXPANSION JOINTS**										
0020	Keyed, cold, 24 ga., incl. stakes, 3-1/2" high	G	1 Carp	200	.040	L.F.	2.44	1.12		3.56	4.57
0050	4-1/2" high	G		200	.040		2.78	1.12		3.90	4.95
0100	5-1/2" high	G		195	.041		3.23	1.15		4.38	5.50
2000	Premolded, bituminous fiber, 1/2" x 6"			375	.021		.40	.60		1	1.45
2050	1" x 12"			300	.027		1.91	.75		2.66	3.36
2140	Concrete expansion joint, recycled paper and fiber, 1/2" x 6"	G		390	.021		.35	.57		.92	1.35
2150	1/2" x 12"	G		360	.022		.69	.62		1.31	1.81
2500	Neoprene sponge, closed cell, 1/2" x 6"			375	.021		1.88	.60		2.48	3.08
2550	1" x 12"			300	.027		7.05	.75		7.80	9
5000	For installation in walls, add							75%			
5250	For installation in boxouts, add							25%			

03 15 05.35 Inserts

			Crew	Daily Output	Labor-Hours	Unit	Material	Labor	Equipment	Total	Total Incl O&P
0010	**INSERTS**										
1000	Inserts, slotted nut type for 3/4" bolts, 4" long	G	1 Carp	84	.095	Ea.	25	2.66		27.66	32
2100	6" long	G		84	.095		28	2.66		30.66	35.50

03 15 Concrete Accessories

03 15 05 – Concrete Forming Accessories

03 15 05.35 Inserts

			Crew	Daily Output	Labor-Hours	Unit	Material	2009 Bare Costs Labor	Equipment	Total	Total Incl O&P
2150	8" long	G	1 Carp	84	.095	Ea.	35.50	2.66		38.16	43.50
2200	Slotted, strap type, 4" long	G	↓	84	.095		24.50	2.66		27.16	31.50
2300	8" long	G	↓	84	.095	↓	35.50	2.66		38.16	43.50

03 15 05.75 Sleeves and Chases

			Crew	Daily Output	Labor-Hours	Unit	Material	Labor	Equipment	Total	Total Incl O&P
0010	**SLEEVES AND CHASES**										
0100	Plastic, 1 use, 9" long, 2" diameter		1 Carp	100	.080	Ea.	1.41	2.24		3.65	5.35
0150	4" diameter		↓	90	.089		4.14	2.48		6.62	8.75
0200	6" diameter		↓	75	.107	↓	7.30	2.98		10.28	13.10

03 15 05.80 Snap Ties

			Crew	Daily Output	Labor-Hours	Unit	Material	Labor	Equipment	Total	Total Incl O&P
0010	**SNAP TIES**, 8-1/4" L&W (Lumber and wedge)										
0100	2250 lb., w/ flat washer, 8" wall	G				C	254			254	279
0250	16" wall	G					300			300	330
0300	18" wall	G					315			315	350
0500	With plastic cone, 8" wall	G					272			272	300
0600	12" wall	G					295			295	325
0650	16" wall	G					325			325	355
0700	18" wall	G				↓	340			340	375

03 15 05.95 Wall and Foundation Form Accessories

			Crew	Daily Output	Labor-Hours	Unit	Material	Labor	Equipment	Total	Total Incl O&P
0010	**WALL AND FOUNDATION FORM ACCESSORIES**										
0020	Coil tie system										
0700	1-1/4", 36,000 lb., to 8"	G				C	2,175			2,175	2,400
1200	1-1/4" diameter x 3" long	G				"	3,775			3,775	4,150
4200	30" long	G				Ea.	7.85			7.85	8.65
4250	36" long	G				"	8.85			8.85	9.70

03 15 13 – Waterstops

03 15 13.50 Waterstops

			Crew	Daily Output	Labor-Hours	Unit	Material	Labor	Equipment	Total	Total Incl O&P
0010	**WATERSTOPS**, PVC and Rubber										
0020	PVC, ribbed 3/16" thick, 4" wide		1 Carp	155	.052	L.F.	1.08	1.44		2.52	3.63
0050	6" wide			145	.055		1.80	1.54		3.34	4.59
0500	With center bulb, 6" wide, 3/16" thick			135	.059		1.52	1.66		3.18	4.47
0550	3/8" thick			130	.062		3.20	1.72		4.92	6.45
0600	9" wide x 3/8" thick		↓	125	.064	↓	4.50	1.79		6.29	7.95

03 21 Reinforcing Steel

03 21 10 – Uncoated Reinforcing Steel

03 21 10.60 Reinforcing In Place

			Crew	Daily Output	Labor-Hours	Unit	Material	Labor	Equipment	Total	Total Incl O&P
0015	**REINFORCING IN PLACE** A615 Grade 60, incl. access. labor										
0030	Made from recycled materials										
0502	Footings, #4 to #7	G	4 Rodm	4200	.008	Lb.	.81	.23		1.04	1.29
0550	#8 to #18	G	↓	3.60	8.889	Ton	1,400	265		1,665	2,000
0702	Walls, #3 to #7	G		6000	.005	Lb.	.81	.16		.97	1.17
0750	#8 to #18	G	↓	4	8	Ton	1,475	239		1,714	2,050
2400	Dowels, 2 feet long, deformed, #3	G	2 Rodm	520	.031	Ea.	.64	.92		1.56	2.34
2410	#4	G		480	.033		1.14	.99		2.13	3.02
2420	#5	G		435	.037		1.78	1.10		2.88	3.91
2430	#6	G	↓	360	.044	↓	2.56	1.33		3.89	5.20
2600	Dowel sleeves for CIP concrete, 2-part system										
2610	Sleeve base, plastic, for 5/8" smooth dowel sleeve, fasten to edge form		1 Rodm	200	.040	Ea.	.52	1.19		1.71	2.69
2615	Sleeve, plastic, 12" long, for 5/8" smooth dowel, snap onto base			400	.020		1.03	.60		1.63	2.19
2620	Sleeve base, for 3/4" smooth dowel sleeve		↓	175	.046	↓	.49	1.36		1.85	2.96

03 21 Reinforcing Steel

03 21 10 – Uncoated Reinforcing Steel

03 21 10.60 Reinforcing In Place	Crew	Daily Output	Labor-Hours	Unit	Material	2009 Bare Costs Labor	Equipment	Total	Total Incl O&P
2625	Sleeve, 12" long, for 3/4" smooth dowel	1 Rodm	350	.023	Ea.	1.08	.68		1.76	2.40
2630	Sleeve base, for 1" smooth dowel sleeve		150	.053		.64	1.59		2.23	3.53
2635	Sleeve, 12" long, for 1" smooth dowel	↓	300	.027		1.84	.80		2.64	3.43
2700	Dowel caps, visual warning only, plastic, #3 to #8	2 Rodm	800	.020		.51	.60		1.11	1.62
2720	#8 to #18		750	.021		1.08	.64		1.72	2.32
2750	Impalement protective, plastic, #4 to #9	↓	800	.020	↓	1.95	.60		2.55	3.21

03 22 Welded Wire Fabric Reinforcing

03 22 05 – Uncoated Welded Wire Fabric

03 22 05.50 Welded Wire Fabric

			Crew	Daily Output	Labor-Hours	Unit	Material	Labor	Equipment	Total	Total Incl O&P
0011	**WELDED WIRE FABRIC**, 6 x 6 - W1.4 x W1.4 (10 x 10)	G	2 Rodm	3500	.005	S.F.	.18	.14		.32	.44
0301	6 x 6 - W2.9 x W2.9 (6 x 6) 42 lb. per C.S.F.	G		2900	.006		.33	.16		.49	.65
0501	4 x 4 - W1.4 x W1.4 (10 x 10) 31 lb. per C.S.F.	G	↓	3100	.005	↓	.27	.15		.42	.56
0750	Rolls										
0901	2 x 2 - #12 galv. for gunite reinforcing	G	2 Rodm	650	.025	S.F.	.66	.73		1.39	2.02

03 24 Fibrous Reinforcing

03 24 05 – Reinforcing Fibers

03 24 05.30 Synthetic Fibers

						Unit	Material	Labor	Equipment	Total	Total Incl O&P
0010	**SYNTHETIC FIBERS**										
0100	Synthetic fibers, add to concrete					Lb.	4.43			4.43	4.87
0110	1-1/2 lb. per C.Y.					C.Y.	6.85			6.85	7.55

03 24 05.70 Steel Fibers

						Unit	Material	Labor	Equipment	Total	Total Incl O&P
0010	**STEEL FIBERS**										
0150	Steel fibers, add to concrete	G				Lb.	.70			.70	.77
0155	25 lb. per C.Y.	G				C.Y.	17.50			17.50	19.25
0160	50 lb. per C.Y.	G					35			35	38.50
0170	75 lb. per C.Y.	G					54			54	59.50
0180	100 lb. per C.Y.	G				↓	70			70	77

03 30 Cast-In-Place Concrete

03 30 53 – Miscellaneous Cast-In-Place Concrete

03 30 53.40 Concrete In Place

		Crew	Daily Output	Labor-Hours	Unit	Material	Labor	Equipment	Total	Total Incl O&P
0010	**CONCRETE IN PLACE**									
0020	Including forms (4 uses), reinforcing steel, concrete, placement,									
0050	and finishing unless otherwise indicated									
0500	Chimney foundations, industrial, minimum	C-14C	32.22	3.476	C.Y.	166	91	.80	257.80	335
0510	Maximum	"	23.71	4.724	"	203	124	1.09	328.09	435
3540	Equipment pad, 3' x 3' x 6" thick	C-14H	45	1.067	Ea.	58	29	.58	87.58	114
3550	4' x 4' x 6" thick		30	1.600		84	43.50	.86	128.36	167
3560	5' x 5' x 8" thick		18	2.667		140	72.50	1.44	213.94	279
3570	6' x 6' x 8" thick		14	3.429		188	93	1.85	282.85	365
3580	8' x 8' x 10" thick		8	6		395	163	3.24	561.24	715
3590	10' x 10' x 12" thick	↓	5	9.600	↓	655	261	5.20	921.20	1,175
3800	Footings, spread under 1 C.Y.	C-14C	28	4	C.Y.	209	105	.92	314.92	410
3825	1 C.Y to 5 C.Y.		43	2.605		221	68	.60	289.60	360
3850	Over 5 C.Y.	↓	75	1.493	↓	198	39	.34	237.34	284

03 30 53 – Miscellaneous Cast-In-Place Concrete

03 30 53.40 Concrete In Place	Crew	Daily Output	Labor-Hours	Unit	Material	2009 Bare Costs Labor	Equipment	Total	Total Incl O&P
3900	Footings, strip, 18" x 9", unreinforced	C-14L	40	2.400	C.Y.	120	61.50	.65	182.15	236
3920	18" x 9", reinforced	C-14C	35	3.200		157	83.50	.74	241.24	315
3925	20" x 10", unreinforced	C-14L	45	2.133		117	54.50	.58	172.08	222
3930	20" x 10", reinforced	C-14C	40	2.800		147	73.50	.64	221.14	287
3935	24" x 12", unreinforced	C-14L	55	1.745		116	44.50	.47	160.97	204
3940	24" x 12", reinforced	C-14C	48	2.333		147	61	.54	208.54	267
3945	36" x 12", unreinforced	C-14L	70	1.371		113	35	.37	148.37	183
3950	36" x 12", reinforced	C-14C	60	1.867		141	49	.43	190.43	238
4000	Foundation mat, under 10 C.Y.		38.67	2.896		255	76	.67	331.67	410
4050	Over 20 C.Y.	↓	56.40	1.986	↓	218	52	.46	270.46	330
4520	Handicap access ramp, railing both sides, 3' wide	C-14H	14.58	3.292	L.F.	300	89.50	1.78	391.28	485
4525	5' wide		12.22	3.928		310	107	2.12	419.12	525
4530	With 6" curb and rails both sides, 3' wide		8.55	5.614		310	153	3.03	466.03	600
4535	5' wide	↓	7.31	6.566	↓	315	178	3.55	496.55	650
4650	Slab on grade, not including finish, 4" thick	C-14E	60.75	1.449	C.Y.	127	38.50	.43	165.93	207
4700	6" thick	"	92	.957	"	121	25.50	.29	146.79	177
4751	Slab on grade, incl. troweled finish, not incl. forms									
4760	or reinforcing, over 10,000 S.F., 4" thick	C-14F	3425	.021	S.F.	1.35	.52	.01	1.88	2.34
4820	6" thick	"	3350	.021	"	1.97	.54	.01	2.52	3.05
5000	Slab on grade, incl. textured finish, not incl. forms									
5001	or reinforcing, 4" thick	C-14G	2873	.019	S.F.	1.31	.48	.01	1.80	2.23
5010	6" thick		2590	.022		2.05	.53	.01	2.59	3.13
5020	8" thick	↓	2320	.024	↓	2.68	.59	.01	3.28	3.91
6203	Retaining walls, gravity, 4' high				C.Y.	153			153	168
6800	Stairs, not including safety treads, free standing, 3'-6" wide	C-14H	83	.578	LF Nose	6.20	15.70	.31	22.21	33.50
6850	Cast on ground		125	.384	"	5	10.45	.21	15.66	23.50
7000	Stair landings, free standing		200	.240	S.F.	5.25	6.50	.13	11.88	16.95
7050	Cast on ground	↓	475	.101	"	4	2.75	.05	6.80	9.10

03 31 05 – Normal Weight Structural Concrete

03 31 05.35 Normal Weight Concrete, Ready Mix

	03 31 05.35	Crew	Daily Output	Labor-Hours	Unit	Material	2009 Bare Costs Labor	Equipment	Total	Total Incl O&P
0010	**NORMAL WEIGHT CONCRETE, READY MIX**, delivered									
0012	Includes local aggregate, sand, Portland cement, and water									
0015	Excludes all additives and treatments									
0020	2000 psi				C.Y.	97			97	107
0100	2500 psi					98.50			98.50	108
0150	3000 psi					101			101	111
0200	3500 psi					104			104	114
0300	4000 psi					106			106	116
0350	4500 psi					109			109	120
0400	5000 psi					111			111	122
0411	6000 psi					127			127	139
0412	8000 psi					206			206	227
0413	10,000 psi					293			293	320
0414	12,000 psi					355			355	390
1000	For high early strength cement, add					10%				
1300	For winter concrete (hot water), add					5.25			5.25	5.80
1400	For hot weather concrete (ice), add					7.20			7.20	7.90
1410	For mid-range water reducer, add					3.95			3.95	4.35
1420	For high-range water reducer/superplasticizer, add				↓	5.85			5.85	6.40

03 31 05 – Normal Weight Structural Concrete

03 31 05.35 Normal Weight Concrete, Ready Mix	Crew	Daily Output	Labor-Hours	Unit	Material	2009 Bare Costs Labor	Equipment	Total	Total Incl O&P
1430	For retarder, add				C.Y.	2			2	2.20
1440	For non-Chloride accelerator, add					4.75			4.75	5.25
1450	For Chloride accelerator, per 1%, add					2.75			2.75	3.03
1460	For fiber reinforcing, synthetic (1 Lb./C.Y.), add					6			6	6.60
1500	For Saturday delivery, add					5.50			5.50	6.05
1510	For truck holding/waiting time past 1st hour per load, add				Hr.	92			92	101
1520	For short load (less than 4 C.Y.), add per load				Ea.	100			100	110
2000	For all lightweight aggregate, add				C.Y.	45%				

03 31 05.70 Placing Concrete

		Crew	Daily Output	Labor-Hours	Unit	Material	Labor	Equipment	Total	Total Incl O&P
0010	**PLACING CONCRETE**									
0020	Includes labor and equipment to place and vibrate									
1900	Footings, continuous, shallow, direct chute	C-6	120	.400	C.Y.		8.75	.43	9.18	15.20
1950	Pumped	C-20	150	.427			9.65	5.25	14.90	22
2000	With crane and bucket	C-7	90	.800			18.30	13.30	31.60	45
2400	Footings, spread, under 1 C.Y., direct chute	C-6	55	.873			19.15	.94	20.09	33
2600	Over 5 C.Y., direct chute		120	.400			8.75	.43	9.18	15.20
2900	Foundation mats, over 20 C.Y., direct chute		350	.137			3.01	.15	3.16	5.20
4300	Slab on grade, up to 6" thick, direct chute		110	.436			9.55	.47	10.02	16.50
4350	Pumped	C-20	130	.492			11.10	6.10	17.20	25.50
4400	With crane and bucket	C-7	110	.655			15	10.90	25.90	37
4900	Walls, 8" thick, direct chute	C-6	90	.533			11.70	.58	12.28	20
4950	Pumped	C-20	100	.640			14.45	7.90	22.35	32.50
5000	With crane and bucket	C-7	80	.900			20.50	14.95	35.45	51
5050	12" thick, direct chute	C-6	100	.480			10.55	.52	11.07	18.15
5100	Pumped	C-20	110	.582			13.15	7.20	20.35	30
5200	With crane and bucket	C-7	90	.800			18.30	13.30	31.60	45
5600	Wheeled concrete dumping, add to placing costs above									
5610	Walking cart, 50' haul, add	C-18	32	.281	C.Y.		5.85	1.72	7.57	11.80
5620	150' haul, add		24	.375			7.80	2.30	10.10	15.70
5700	250' haul, add		18	.500			10.40	3.07	13.47	21
5800	Riding cart, 50' haul, add	C-19	80	.113			2.34	1.08	3.42	5.15
5810	150' haul, add		60	.150			3.12	1.44	4.56	6.85
5900	250' haul, add		45	.200			4.15	1.92	6.07	9.15

03 35 29 – Tooled Concrete Finishing

03 35 29.30 Finishing Floors

		Crew	Daily Output	Labor-Hours	Unit	Material	Labor	Equipment	Total	Total Incl O&P
0010	**FINISHING FLOORS**									
0020	Manual screed finish	C-10	4800	.005	S.F.		.12		.12	.20
0100	Manual screed and bull float		4000	.006			.15		.15	.24
0125	Manual screed, bull float, manual float		2000	.012			.30		.30	.48
0150	Manual screed, bull float, manual float & broom finish		1850	.013			.32		.32	.52
0200	Manual screed, bull float, manual float, manual steel trowel		1265	.019			.47		.47	.76
0250	Manual screed, bull float, machine float & trowel (walk-behind)	C-10C	1715	.014			.35	.02	.37	.59
0300	Power screed, bull float, machine float & trowel (walk-behind)	C-10D	2400	.010			.25	.05	.30	.45
0350	Power screed, bull float, machine float & trowel (ride-on)	C-10E	4000	.006			.15	.06	.21	.31
1600	Exposed local aggregate finish, minimum	1 Cefi	625	.013		.22	.34		.56	.80
1650	Maximum	"	465	.017		.66	.46		1.12	1.47

03 35 29.35 Control Joints, Saw Cut

0010	**CONTROL JOINTS, SAW CUT**

03 35 Concrete Finishing

03 35 29 – Tooled Concrete Finishing

03 35 29.35 Control Joints, Saw Cut

		Crew	Daily Output	Labor-Hours	Unit	Material	2009 Bare Costs Labor	Equipment	Total	Total Incl O&P
0100	Sawcut in green concrete									
0120	1" depth	C-27	2000	.008	L.F.	.07	.21	.07	.35	.49
0140	1-1/2" depth		1800	.009		.10	.24	.08	.42	.58
0160	2" depth	↓	1600	.010		.13	.27	.09	.49	.68
0200	Clean out control joint of debris	C-28	6000	.001	↓		.04		.04	.06
0300	Joint sealant									
0320	Backer rod, polyethylene, 1/4" diameter	1 Cefi	460	.017	L.F.	.04	.47		.51	.80
0340	Sealant, polyurethane									
0360	1/4" x 1/4" (308 LF/Gal)	1 Cefi	270	.030	L.F.	.18	.79		.97	1.46
0380	1/4" x 1/2" (154 LF/Gal)	"	255	.031	"	.35	.84		1.19	1.74

03 35 29.60 Finishing Walls

		Crew	Daily Output	Labor-Hours	Unit	Material	2009 Bare Costs Labor	Equipment	Total	Total Incl O&P
0010	**FINISHING WALLS**									
0020	Break ties and patch voids	1 Cefi	540	.015	S.F.	.03	.40		.43	.67
0050	Burlap rub with grout	"	450	.018		.03	.48		.51	.79
0300	Bush hammer, green concrete	B-39	1000	.048			1	.19	1.19	1.91
0350	Cured concrete	"	650	.074	↓		1.54	.29	1.83	2.93

03 35 33 – Stamped Concrete Finishing

03 35 33.50 Slab Texture Stamping

		Crew	Daily Output	Labor-Hours	Unit	Material	2009 Bare Costs Labor	Equipment	Total	Total Incl O&P
0010	**SLAB TEXTURE STAMPING**									
0020	Buy re-usable stamp, small (2 S.F.)				Ea.	82.50			82.50	91
0030	Medium (4 S.F.)					165			165	182
0120	Large (8 S.F.)				↓	330			330	365
0200	Commonly used chemicals for texture systems									
0210	Hardener, colored powder				S.F.	.53			.53	.59
0220	Release agent, colored powder					.09			.09	.10
0230	Curing & sealing compound, solvent based					.06			.06	.06
0300	Broadcasting hardener & release agent, stamping, tooling	3 Cefi	1000	.024	↓		.64		.64	1.03

03 39 Concrete Curing

03 39 13 – Water Concrete Curing

03 39 13.50 Water Curing

		Crew	Daily Output	Labor-Hours	Unit	Material	2009 Bare Costs Labor	Equipment	Total	Total Incl O&P
0011	**WATER CURING**									
0020	With burlap, 4 uses assumed, 7.5 oz.	2 Clab	5500	.003	S.F.	.09	.06		.15	.19
0101	10 oz.	"	5500	.003	"	.16	.06		.22	.27

03 39 23 – Membrane Concrete Curing

03 39 23.13 Chemical Compound Membrane Concrete Curing

		Crew	Daily Output	Labor-Hours	Unit	Material	2009 Bare Costs Labor	Equipment	Total	Total Incl O&P
0010	**CHEMICAL COMPOUND MEMBRANE CONCRETE CURING**									
0301	Sprayed membrane curing compound	2 Clab	9500	.002	S.F.	.05	.03		.08	.12

03 39 23.23 Sheet Membrane Concrete Curing

		Crew	Daily Output	Labor-Hours	Unit	Material	2009 Bare Costs Labor	Equipment	Total	Total Incl O&P
0010	**SHEET MEMBRANE CONCRETE CURING**									
0201	Curing blanket, burlap/poly, 2-ply	2 Clab	7000	.002	S.F.	.16	.05		.21	.26

03 41 Precast Structural Concrete

03 41 23 – Precast Concrete Stairs

03 41 23.50 Precast Stairs	Crew	Daily Output	Labor-Hours	Unit	Material	2009 Bare Costs Labor	Equipment	Total	Total Incl O&P
0010 **PRECAST STAIRS**									
0020 Precast concrete treads on steel stringers, 3' wide	C-12	75	.640	Riser	137	17.45	10.25	164.70	192
0300 Front entrance, 5' wide with 48" platform, 2 risers		16	3	Flight	450	82	48	580	685
0350 5 risers		12	4		705	109	64	878	1,025
0500 6' wide, 2 risers		15	3.200		505	87.50	51.50	644	760
0550 5 risers		11	4.364		785	119	70	974	1,125
0700 7' wide, 2 risers		14	3.429		610	93.50	55	758.50	890
1200 Basement entrance stairs, steel bulkhead doors, minimum	B-51	22	2.182		1,350	46	8.85	1,404.85	1,575
1250 Maximum	"	11	4.364		2,250	92	17.65	2,359.65	2,650

03 48 Precast Concrete Specialties

03 48 43 – Precast Concrete Trim

03 48 43.40 Precast Lintels

03 48 43.40 Precast Lintels	Crew	Daily Output	Labor-Hours	Unit	Material	2009 Bare Costs Labor	Equipment	Total	Total Incl O&P
0010 **PRECAST LINTELS**									
0800 Precast concrete, 4" wide, 8" high, to 5' long	D-10	28	1.143	Ea.	25.50	31.50	21.50	78.50	104
0850 5'-12' long		24	1.333		82.50	37	25	144.50	180
1000 6" wide, 8" high, to 5' long		26	1.231		39.50	34	23	96.50	126
1050 5'-12' long		22	1.455		99.50	40	27.50	167	207

03 48 43.90 Precast Window Sills

03 48 43.90 Precast Window Sills	Crew	Daily Output	Labor-Hours	Unit	Material	2009 Bare Costs Labor	Equipment	Total	Total Incl O&P
0010 **PRECAST WINDOW SILLS**									
0600 Precast concrete, 4" tapers to 3", 9" wide	D-1	70	.229	L.F.	11.75	5.80		17.55	22.50
0650 11" wide		60	.267		15.60	6.80		22.40	28.50
0700 13" wide, 3 1/2" tapers to 2 1/2", 12" wall		50	.320		15	8.15		23.15	30

03 63 Epoxy Grouting

03 63 05 – Grouting of Dowels and Fasteners

03 63 05.10 Epoxy Only

03 63 05.10 Epoxy Only	Crew	Daily Output	Labor-Hours	Unit	Material	2009 Bare Costs Labor	Equipment	Total	Total Incl O&P
0010 **EPOXY ONLY**									
1500 Chemical anchoring, epoxy cartridge, excludes layout, drilling, fastener									
1530 For fastener 3/4" diam. x 6" embedment	2 Skwk	72	.222	Ea.	3.54	6.25		9.79	14.40
1535 1" diam. x 8" embedment		66	.242		5.30	6.80		12.10	17.35
1540 1-1/4" diam. x 10" embedment		60	.267		10.60	7.50		18.10	24.50
1545 1-3/4" diam. x 12" embedment		54	.296		17.70	8.30		26	33.50
1550 14" embedment		48	.333		21	9.35		30.35	39.50
1555 2" diam. x 12" embedment		42	.381		28.50	10.70		39.20	49
1560 18" embedment		32	.500		35.50	14.05		49.55	62.50

03 82 Concrete Boring

03 82 16 – Concrete Drilling

03 82 16.10 Concrete Impact Drilling

03 82 16.10 Concrete Impact Drilling	Crew	Daily Output	Labor-Hours	Unit	Material	2009 Bare Costs Labor	Equipment	Total	Total Incl O&P
0010 **CONCRETE IMPACT DRILLING**									
0050 Up to 4" deep in conc/brick floor/wall, incl. bit & layout, no anchor									
0100 Holes, 1/4" diameter	1 Carp	75	.107	Ea.	.06	2.98		3.04	5.10
0150 For each additional inch of depth, add		430	.019		.01	.52		.53	.90
0200 3/8" diameter		63	.127		.05	3.55		3.60	6.05
0250 For each additional inch of depth, add		340	.024		.01	.66		.67	1.12
0300 1/2" diameter		50	.160		.06	4.47		4.53	7.60

03 82 Concrete Boring

03 82 16 – Concrete Drilling

03 82 16.10 Concrete Impact Drilling		Crew	Daily Output	Labor-Hours	Unit	Material	2009 Bare Costs Labor	Equipment	Total	Total Incl O&P
0350	For each additional inch of depth, add	1 Carp	250	.032	Ea.	.01	.89		.90	1.53
0400	5/8" diameter		48	.167		.08	4.66		4.74	8
0450	For each additional inch of depth, add		240	.033		.02	.93		.95	1.59
0500	3/4" diameter		45	.178		.10	4.97		5.07	8.50
0550	For each additional inch of depth, add		220	.036		.03	1.02		1.05	1.75
0600	7/8" diameter		43	.186		.12	5.20		5.32	8.95
0650	For each additional inch of depth, add		210	.038		.03	1.06		1.09	1.83
0700	1" diameter		40	.200		.14	5.60		5.74	9.60
0750	For each additional inch of depth, add		190	.042		.04	1.18		1.22	2.03
0800	1-1/4" diameter		38	.211		.20	5.90		6.10	10.15
0850	For each additional inch of depth, add		180	.044		.05	1.24		1.29	2.16
0900	1-1/2" diameter		35	.229		.29	6.40		6.69	11.10
0950	For each additional inch of depth, add		165	.048		.07	1.36		1.43	2.37
1000	For ceiling installations, add						40%			

Division 4
Masonry

04 01 Maintenance of Masonry

04 01 20 – Maintenance of Unit Masonry

04 01 20.20 Pointing Masonry

		Crew	Daily Output	Labor-Hours	Unit	Material	2009 Bare Costs Labor	Equipment	Total	Total Incl O&P
0010	**POINTING MASONRY**									
0300	Cut and repoint brick, hard mortar, running bond	1 Bric	80	.100	S.F.	.53	2.84		3.37	5.30
0320	Common bond		77	.104		.53	2.95		3.48	5.45
0360	Flemish bond		70	.114		.56	3.24		3.80	5.95
0400	English bond		65	.123		.56	3.49		4.05	6.40
0600	Soft old mortar, running bond		100	.080		.53	2.27		2.80	4.34
0620	Common bond		96	.083		.53	2.36		2.89	4.50
0640	Flemish bond		90	.089		.56	2.52		3.08	4.79
0680	English bond		82	.098	▼	.56	2.77		3.33	5.20
0700	Stonework, hard mortar		140	.057	L.F.	.70	1.62		2.32	3.46
0720	Soft old mortar		160	.050	"	.70	1.42		2.12	3.12
1000	Repoint, mask and grout method, running bond		95	.084	S.F.	.70	2.39		3.09	4.73
1020	Common bond		90	.089		.70	2.52		3.22	4.95
1040	Flemish bond		86	.093		.74	2.64		3.38	5.20
1060	English bond		77	.104		.74	2.95		3.69	5.70
2000	Scrub coat, sand grout on walls, minimum		120	.067		2.84	1.89		4.73	6.25
2020	Maximum	▼	98	.082	▼	3.94	2.31		6.25	8.20

04 01 30 – Unit Masonry Cleaning

04 01 30.60 Brick Washing

			Crew	Daily Output	Labor-Hours	Unit	Material	2009 Bare Costs Labor	Equipment	Total	Total Incl O&P
0010	**BRICK WASHING**	R040130-10									
0012	Acid cleanser, smooth brick surface		1 Bric	560	.014	S.F.	.05	.41		.46	.73
0050	Rough brick			400	.020		.07	.57		.64	1.02
0060	Stone, acid wash		▼	600	.013	▼	.08	.38		.46	.72
1000	Muriatic acid, price per gallon in 5 gallon lots					Gal.	10.50			10.50	11.55

04 05 Common Work Results for Masonry

04 05 05 – Selective Masonry Demolition

04 05 05.10 Selective Demolition

			Crew	Daily Output	Labor-Hours	Unit	Material	2009 Bare Costs Labor	Equipment	Total	Total Incl O&P
0010	**SELECTIVE DEMOLITION**	R024119-10									
0200	Bond beams, 8" block with #4 bar		2 Clab	32	.500	L.F.		10.30		10.30	17.40
0300	Concrete block walls, unreinforced, 2" thick			1200	.013	S.F.		.27		.27	.46
0310	4" thick			1150	.014			.29		.29	.48
0320	6" thick			1100	.015			.30		.30	.51
0330	8" thick			1050	.015			.31		.31	.53
0340	10" thick			1000	.016			.33		.33	.56
0360	12" thick			950	.017			.35		.35	.59
0380	Reinforced alternate courses, 2" thick			1130	.014			.29		.29	.49
0390	4" thick			1080	.015			.30		.30	.51
0400	6" thick			1035	.015			.32		.32	.54
0410	8" thick			990	.016			.33		.33	.56
0420	10" thick			940	.017			.35		.35	.59
0430	12" thick			890	.018			.37		.37	.62
0440	Reinforced alternate courses & vertically 48" OC, 4" thick			900	.018			.37		.37	.62
0450	6" thick			850	.019			.39		.39	.65
0460	8" thick			800	.020			.41		.41	.70
0480	10" thick			750	.021			.44		.44	.74
0490	12" thick		▼	700	.023	▼		.47		.47	.79
1000	Chimney, 16" x 16", soft old mortar		1 Clab	55	.145	C.F.		2.99		2.99	5.05
1020	Hard mortar			40	.200			4.11		4.11	6.95
1030	16" x 20", soft old mortar			55	.145	▼		2.99		2.99	5.05

04 05 05 – Selective Masonry Demolition

04 05 05.10 Selective Demolition		Crew	Daily Output	Labor-Hours	Unit	Material	2009 Bare Costs Labor	2009 Bare Costs Equipment	Total	Total Incl O&P
1040	Hard mortar	1 Clab	40	.200	C.F.		4.11		4.11	6.95
1050	16" x 24", soft old mortar		55	.145			2.99		2.99	5.05
1060	Hard mortar		40	.200			4.11		4.11	6.95
1080	20" x 20", soft old mortar		55	.145			2.99		2.99	5.05
1100	Hard mortar		40	.200			4.11		4.11	6.95
1110	20" x 24", soft old mortar		55	.145			2.99		2.99	5.05
1120	Hard mortar		40	.200			4.11		4.11	6.95
1140	20" x 32", soft old mortar		55	.145			2.99		2.99	5.05
1160	Hard mortar		40	.200			4.11		4.11	6.95
1200	48" x 48", soft old mortar		55	.145			2.99		2.99	5.05
1220	Hard mortar	↓	40	.200	↓		4.11		4.11	6.95
1250	Metal, high temp steel jacket, 24" diameter	E-2	130	.369	V.L.F.		11.15	13.40	24.55	36
1260	60" diameter	"	60	.800			24	29	53	77.50
1280	Flue lining, up to 12" x 12"	1 Clab	200	.040			.82		.82	1.39
1282	Up to 24" x 24"		150	.053			1.10		1.10	1.85
2000	Columns, 8" x 8", soft old mortar		48	.167			3.43		3.43	5.80
2020	Hard mortar		40	.200			4.11		4.11	6.95
2060	16" x 16", soft old mortar		16	.500			10.30		10.30	17.40
2100	Hard mortar		14	.571			11.75		11.75	19.85
2140	24" x 24", soft old mortar		8	1			20.50		20.50	35
2160	Hard mortar		6	1.333			27.50		27.50	46.50
2200	36" x 36", soft old mortar		4	2			41		41	69.50
2220	Hard mortar		3	2.667	↓		55		55	92.50
2230	Alternate pricing method, soft old mortar		30	.267	C.F.		5.50		5.50	9.25
2240	Hard mortar	↓	23	.348	"		7.15		7.15	12.10
3000	Copings, precast or masonry, to 8" wide									
3020	Soft old mortar	1 Clab	180	.044	L.F.		.91		.91	1.54
3040	Hard mortar	"	160	.050	"		1.03		1.03	1.74
3100	To 12" wide									
3120	Soft old mortar	1 Clab	160	.050	L.F.		1.03		1.03	1.74
3140	Hard mortar	"	140	.057	"		1.17		1.17	1.99
4000	Fireplace, brick, 30" x 24" opening									
4020	Soft old mortar	1 Clab	2	4	Ea.		82		82	139
4040	Hard mortar		1.25	6.400			132		132	222
4100	Stone, soft old mortar		1.50	5.333			110		110	185
4120	Hard mortar		1	8	↓		164		164	278
5000	Veneers, brick, soft old mortar		140	.057	S.F.		1.17		1.17	1.99
5020	Hard mortar		125	.064			1.32		1.32	2.22
5100	Granite and marble, 2" thick		180	.044			.91		.91	1.54
5120	4" thick		170	.047			.97		.97	1.64
5140	Stone, 4" thick		180	.044			.91		.91	1.54
5160	8" thick		175	.046	↓		.94		.94	1.59
5400	Alternate pricing method, stone, 4" thick		60	.133	C.F.		2.74		2.74	4.63
5420	8" thick	↓	85	.094	"		1.93		1.93	3.27

04 05 13 – Masonry Mortaring

04 05 13.10 Cement

		Crew	Daily Output	Labor-Hours	Unit	Material	2009 Bare Costs Labor	2009 Bare Costs Equipment	Total	Total Incl O&P
0010	**CEMENT** R040513-10									
0100	Masonry, 70 lb. bag, T.L. lots				Bag	9.05			9.05	9.95
0150	L.T.L. lots					9.60			9.60	10.55
0200	White, 70 lb. bag, T.L. lots					15.05			15.05	16.55
0250	L.T.L. lots				↓	16.05			16.05	17.65

04 05 16 – Masonry Grouting

04 05 16.30 Grouting

		Crew	Daily Output	Labor-Hours	Unit	Material	2009 Bare Costs Labor	2009 Bare Costs Equipment	Total	Total Incl O&P
0010	**GROUTING**									
0200	Concrete block cores, solid, 4" thk., by hand, 0.067 C.F./S.F. of wall	D-8	1100	.036	S.F.	.28	.95		1.23	1.88
0210	6" thick, pumped, 0.175 C.F. per S.F.	D-4	720	.056		.74	1.29	.18	2.21	3.16
0250	8" thick, pumped, 0.258 C.F. per S.F.		680	.059		1.10	1.37	.19	2.66	3.70
0300	10" thick, pumped, 0.340 C.F. per S.F.		660	.061		1.45	1.41	.19	3.05	4.15
0350	12" thick, pumped, 0.422 C.F. per S.F.		640	.063		1.79	1.46	.20	3.45	4.61

04 05 19 – Masonry Anchorage and Reinforcing

04 05 19.05 Anchor Bolts

		Crew	Daily Output	Labor-Hours	Unit	Material	2009 Bare Costs Labor	2009 Bare Costs Equipment	Total	Total Incl O&P
0010	**ANCHOR BOLTS**									
0020	Hooked, with nut and washer, 1/2" diam., 8" long	1 Bric	200	.040	Ea.	.83	1.13		1.96	2.79
0030	12" long		190	.042		1.26	1.19		2.45	3.37
0060	3/4" diameter, 8" long		160	.050		1.86	1.42		3.28	4.40
0070	12" long		150	.053		2.33	1.51		3.84	5.05

04 05 19.16 Masonry Anchors

		Crew	Daily Output	Labor-Hours	Unit	Material	2009 Bare Costs Labor	2009 Bare Costs Equipment	Total	Total Incl O&P
0010	**MASONRY ANCHORS**									
0020	For brick veneer, galv., corrugated, 7/8" x 7", 22 Ga.	1 Bric	10.50	.762	C	9.80	21.50		31.30	47
0100	24 Ga.		10.50	.762		9.15	21.50		30.65	46
0150	16 Ga.		10.50	.762		24	21.50		45.50	62.50
0200	Buck anchors, galv., corrugated, 16 gauge, 2" bend, 8" x 2"		10.50	.762		134	21.50		155.50	183
0250	8" x 3"		10.50	.762		138	21.50		159.50	188
0660	Cavity wall, Z-type, galvanized, 6" long, 1/8" diam.		10.50	.762		23	21.50		44.50	61
0670	3/16" diameter		10.50	.762		21	21.50		42.50	59
0680	1/4" diameter		10.50	.762		39.50	21.50		61	79.50
0850	8" long, 3/16" diameter		10.50	.762		24	21.50		45.50	62
0855	1/4" diameter		10.50	.762		46.50	21.50		68	87
1000	Rectangular type, galvanized, 1/4" diameter, 2" x 6"		10.50	.762		67	21.50		88.50	110
1050	4" x 6"		10.50	.762		83.50	21.50		105	128
1100	3/16" diameter, 2" x 6"		10.50	.762		37.50	21.50		59	77.50
1150	4" x 6"		10.50	.762		39	21.50		60.50	79
1500	Rigid partition anchors, plain, 8" long, 1" x 1/8"		10.50	.762		103	21.50		124.50	149
1550	1" x 1/4"		10.50	.762		170	21.50		191.50	223
1580	1-1/2" x 1/8"		10.50	.762		138	21.50		159.50	187
1600	1-1/2" x 1/4"		10.50	.762		280	21.50		301.50	345
1650	2" x 1/8"		10.50	.762		178	21.50		199.50	232
1700	2" x 1/4"		10.50	.762		325	21.50		346.50	395

04 05 19.26 Masonry Reinforcing Bars

			Crew	Daily Output	Labor-Hours	Unit	Material	2009 Bare Costs Labor	2009 Bare Costs Equipment	Total	Total Incl O&P
0010	**MASONRY REINFORCING BARS**	R040519-50									
0015	Steel bars A615, placed horiz., #3 & #4 bars		1 Bric	450	.018	Lb.	.78	.50		1.28	1.69
0050	Placed vertical, #3 & #4 bars			350	.023		.78	.65		1.43	1.92
0060	#5 & #6 bars			650	.012		.78	.35		1.13	1.43
0200	Joint reinforcing, regular truss, to 6" wide, mill std galvanized			30	.267	C.L.F.	16.20	7.55		23.75	30.50
0250	12" wide			20	.400		18	11.35		29.35	38.50
0400	Cavity truss with drip section, to 6" wide			30	.267		15.50	7.55		23.05	29.50
0450	12" wide			20	.400		15.85	11.35		27.20	36.50

04 21 Clay Unit Masonry

04 21 13 – Brick Masonry

04 21 13.13 Brick Veneer Masonry

	Crew	Daily Output	Labor-Hours	Unit	Material	2009 Bare Costs Labor	Equipment	Total	Total Incl O&P
0010 **BRICK VENEER MASONRY**, T.L. lots, excl. scaff., grout & reinforcing									
0015 Material costs incl. 3% brick and 25% mortar waste R042110-10									
2000 Standard, sel. common, 4" x 2-2/3" x 8", (6.75/S.F.) R042110-20	D-8	230	.174	S.F.	3.60	4.52		8.12	11.45
2020 Standard, red, 4" x 2-2/3" x 8", running bond (6.75/SF)		220	.182		5.65	4.73		10.38	14.10
2050 Full header every 6th course (7.88/S.F.) R042110-50		185	.216		6.60	5.60		12.20	16.55
2100 English, full header every 2nd course (10.13/S.F.)		140	.286		8.50	7.45		15.95	21.50
2150 Flemish, alternate header every course (9.00/S.F.)		150	.267		7.55	6.95		14.50	19.80
2200 Flemish, alt. header every 6th course (7.13/S.F.)		205	.195		6	5.10		11.10	15
2250 Full headers throughout (13.50/S.F.)		105	.381		11.30	9.90		21.20	29
2300 Rowlock course (13.50/S.F.)		100	.400		11.30	10.40		21.70	29.50
2350 Rowlock stretcher (4.50/S.F.)		310	.129		3.80	3.36		7.16	9.75
2400 Soldier course (6.75/S.F.)		200	.200		5.65	5.20		10.85	14.85
2450 Sailor course (4.50/S.F.)		290	.138		3.80	3.59		7.39	10.15
2600 Buff or gray face, running bond, (6.75/S.F.)		220	.182		6	4.73		10.73	14.45
2700 Glazed face brick, running bond		210	.190		11.15	4.95		16.10	20.50
2750 Full header every 6th course (7.88/S.F.)		170	.235		13	6.10		19.10	24.50
3000 Jumbo, 6" x 4" x 12" running bond (3.00/S.F.)		435	.092		4.86	2.39		7.25	9.30
3050 Norman, 4" x 2-2/3" x 12" running bond, (4.5/S.F.)		320	.125		5.65	3.25		8.90	11.65
3100 Norwegian, 4" x 3-1/5" x 12" (3.75/S.F.)		375	.107		4.14	2.77		6.91	9.15
3150 Economy, 4" x 4" x 8" (4.50/S.F.)		310	.129		4.18	3.36		7.54	10.15
3200 Engineer, 4" x 3-1/5" x 8" (5.63/S.F.)		260	.154		3.46	4		7.46	10.45
3250 Roman, 4" x 2" x 12" (6.00/S.F.)		250	.160		5.75	4.16		9.91	13.20
3300 SCR, 6" x 2-2/3" x 12" (4.50/S.F.)		310	.129		5.25	3.36		8.61	11.30
3350 Utility, 4" x 4" x 12" (3.00/S.F.)		450	.089		4.40	2.31		6.71	8.65
3400 For cavity wall construction, add						15%			
3450 For stacked bond, add						10%			
3500 For interior veneer construction, add						15%			
3550 For curved walls, add						30%			

04 21 13.15 Chimney

	Crew	Daily Output	Labor-Hours	Unit	Material	2009 Bare Costs Labor	Equipment	Total	Total Incl O&P
0010 **CHIMNEY**, excludes foundation, scaffolding, grout and reinforcing									
0100 Brick, 16" x 16", 8" flue	D-1	18.20	.879	V.L.F.	20.50	22.50		43	60
0150 16" x 20" with one 8" x 12" flue		16	1		32.50	25.50		58	78
0200 16" x 24" with two 8" x 8" flues		14	1.143		47.50	29		76.50	101
0250 20" x 20" with one 12" x 12" flue		13.70	1.168		38.50	29.50		68	91
0300 20" x 24" with two 8" x 12" flues		12	1.333		54	34		88	116
0350 20" x 32" with two 12" x 12" flues		10	1.600		67.50	40.50		108	142

04 21 13.18 Columns

	Crew	Daily Output	Labor-Hours	Unit	Material	2009 Bare Costs Labor	Equipment	Total	Total Incl O&P
0010 **COLUMNS**, solid, excludes scaffolding, grout and reinforcing									
0050 Brick, 8" x 8", 9 brick per VLF	D-1	56	.286	V.L.F.	7.35	7.25		14.60	20
0100 12" x 8", 13.5 brick per VLF		37	.432		11.05	11		22.05	30.50
0200 12" x 12", 20 brick per VLF		25	.640		16.35	16.30		32.65	45
0300 16" x 12", 27 brick per VLF		19	.842		22	21.50		43.50	60
0400 16" x 16", 36 brick per VLF		14	1.143		29.50	29		58.50	80.50
0500 20" x 16", 45 brick per VLF		11	1.455		37	37		74	102
0600 20" x 20", 56 brick per VLF		9	1.778		46	45		91	126

04 21 13.30 Oversized Brick

	Crew	Daily Output	Labor-Hours	Unit	Material	2009 Bare Costs Labor	Equipment	Total	Total Incl O&P
0010 **OVERSIZED BRICK**, excludes scaffolding, grout and reinforcing									
0100 Veneer, 4" x 2.25" x 16"	D-8	387	.103	S.F.	5.05	2.69		7.74	10
0105 4" x 2.75" x 16"		412	.097		4.55	2.53		7.08	9.20
0110 4" x 4" x 16"		460	.087		4.48	2.26		6.74	8.70
0120 4" x 8" x 16"		533	.075		5.50	1.95		7.45	9.30
0125 Loadbearing, 6" x 4" x 16", grouted and reinforced		387	.103		8.10	2.69		10.79	13.35

04 21 Clay Unit Masonry

04 21 13 – Brick Masonry

04 21 13.30 Oversized Brick

		Crew	Daily Output	Labor-Hours	Unit	Material	2009 Bare Costs Labor	Equipment	Total	Total Incl O&P
0130	8" x 4" x 16", grouted and reinforced	D-8	327	.122	S.F.	8.25	3.18		11.43	14.30
0135	6" x 8" x 16", grouted and reinforced		440	.091		7.45	2.36		9.81	12.10
0140	8" x 8" x 16", grouted and reinforced		400	.100		8.35	2.60		10.95	13.50
0145	Curtainwall / reinforced veneer, 6" x 4" x 16"		387	.103		12.05	2.69		14.74	17.70
0150	8" x 4" x 16"		327	.122		14.55	3.18		17.73	21.50
0155	6" x 8" x 16"		440	.091		12.15	2.36		14.51	17.25
0160	8" x 8" x 16"	↓	400	.100		14.65	2.60		17.25	20.50
0200	For 1 to 3 slots in face, add					25%				
0210	For 4 to 7 slots in face, add					15%				
0220	For bond beams, add					20%				
0230	For bullnose shapes, add					20%				
0240	For open end knockout, add					10%				
0250	For white or gray color group, add					10%				
0260	For 135 degree corner, add					250%				

04 21 13.35 Common Building Brick

					Unit	Material	Labor	Equipment	Total	Total Incl O&P
0010	**COMMON BUILDING BRICK**, C62, TL lots, material only R042110-20									
0020	Standard, minimum				M	340			340	375
0050	Average (select)				"	430			430	475

04 21 13.45 Face Brick

					Unit	Material	Labor	Equipment	Total	Total Incl O&P
0010	**FACE BRICK** Material Only, C216, TL lots R042110-20									
0300	Standard modular, 4" x 2-2/3" x 8", minimum				M	730			730	805
0350	Maximum					835			835	920
2170	For less than truck load lots, add					10			10	11
2180	For buff or gray brick, add					15			15	16.50

04 22 Concrete Unit Masonry

04 22 10 – Concrete Masonry Units

04 22 10.11 Autoclave Aerated Concrete Block

			Crew	Daily Output	Labor-Hours	Unit	Material	Labor	Equipment	Total	Total Incl O&P
0010	**AUTOCLAVE AERATED CONCRETE BLOCK**, excl. scaffolding, grout & reinforcing										
0050	Solid, 4" x 12" x 24", incl mortar	G	D-8	600	.067	S.F.	1.60	1.73		3.33	4.63
0060	6" x 12" x 24"	G		600	.067		2.17	1.73		3.90	5.25
0070	8" x 8" x 24"	G		575	.070		2.80	1.81		4.61	6.10
0080	10" x 12" x 24"	G		575	.070		3.59	1.81		5.40	6.95
0090	12" x 12" x 24"	G	↓	550	.073		4.23	1.89		6.12	7.80

04 22 10.14 Concrete Block, Back-Up

			Crew	Daily Output	Labor-Hours	Unit	Material	Labor	Equipment	Total	Total Incl O&P
0010	**CONCRETE BLOCK, BACK-UP**, C90, 2000 psi R042210-20										
0020	Normal weight, 8" x 16" units, tooled joint 1 side										
0050	Not-reinforced, 2000 psi, 2" thick		D-8	475	.084	S.F.	1.36	2.19		3.55	5.15
0200	4" thick			460	.087		1.52	2.26		3.78	5.40
0300	6" thick			440	.091		2.12	2.36		4.48	6.25
0350	8" thick			400	.100		2.23	2.60		4.83	6.75
0400	10" thick		↓	330	.121		3.03	3.15		6.18	8.60
0450	12" thick		D-9	310	.155		3.44	3.94		7.38	10.35
1000	Reinforced, alternate courses, 4" thick		D-8	450	.089		1.64	2.31		3.95	5.65
1100	6" thick			430	.093		2.25	2.42		4.67	6.50
1150	8" thick			395	.101		2.37	2.63		5	6.95
1200	10" thick		↓	320	.125		3.18	3.25		6.43	8.90
1250	12" thick		D-9	300	.160		3.59	4.07		7.66	10.70

04 22 10.16 Concrete Block, Bond Beam

0010	**CONCRETE BLOCK, BOND BEAM**, C90, 2000 psi										

04 22 Concrete Unit Masonry

04 22 10 – Concrete Masonry Units

04 22 10.16 Concrete Block, Bond Beam		Crew	Daily Output	Labor-Hours	Unit	Material	2009 Bare Costs Labor	2009 Bare Costs Equipment	Total	Total Incl O&P
0020	Not including grout or reinforcing									
0125	Regular block, 6" thick	D-8	584	.068	L.F.	2.21	1.78		3.99	5.40
0130	8" high, 8" thick	"	565	.071		2.51	1.84		4.35	5.80
0150	12" thick	D-9	510	.094		3.43	2.39		5.82	7.75
0525	Lightweight, 6" thick	D-8	592	.068		2.53	1.76		4.29	5.70

04 22 10.19 Concrete Block, Insulation Inserts										
0010	**CONCRETE BLOCK, INSULATION INSERTS**									
0100	Styrofoam, plant installed, add to block prices									
0200	8" x 16" units, 6" thick				S.F.	1.81			1.81	1.99
0250	8" thick					1.81			1.81	1.99
0300	10" thick					2.13			2.13	2.34
0350	12" thick					2.24			2.24	2.46

04 22 10.23 Concrete Block, Decorative										
0010	**CONCRETE BLOCK, DECORATIVE**, C90, 2000 psi									
5000	Split rib profile units, 1" deep ribs, 8 ribs									
5100	8" x 16" x 4" thick	D-8	345	.116	S.F.	3.15	3.02		6.17	8.45
5150	6" thick		325	.123		3.62	3.20		6.82	9.30
5200	8" thick		300	.133		4.13	3.47		7.60	10.30
5250	12" thick	D-9	275	.175		4.91	4.44		9.35	12.75
5400	For special deeper colors, 4" thick, add					1.12			1.12	1.23
5450	12" thick, add					.96			.96	1.06
5600	For white, 4" thick, add					1.12			1.12	1.23
5650	6" thick, add					1.12			1.12	1.23
5700	8" thick, add					1.04			1.04	1.14
5750	12" thick, add					.96			.96	1.06

04 22 10.24 Concrete Block, Exterior										
0010	**CONCRETE BLOCK, EXTERIOR**, C90, 2000 psi									
0020	Reinforced alt courses, tooled joints 2 sides									
0100	Normal weight, 8" x 16" x 6" thick	D-8	395	.101	S.F.	2.49	2.63		5.12	7.10
0200	8" thick		360	.111		3.65	2.89		6.54	8.80
0250	10" thick		290	.138		4.27	3.59		7.86	10.65
0300	12" thick	D-9	250	.192		4.52	4.88		9.40	13.05

04 22 10.26 Concrete Block Foundation Wall										
0010	**CONCRETE BLOCK FOUNDATION WALL**, C90 / C145									
0050	Normal-weight, cut joints, horiz joint reinf, no vert reinf									
0200	Hollow, 8" x 16" x 6" thick	D-8	455	.088	S.F.	2.56	2.29		4.85	6.60
0250	8" thick		425	.094		2.68	2.45		5.13	7
0300	10" thick		350	.114		3.49	2.97		6.46	8.75
0350	12" thick	D-9	300	.160		3.91	4.07		7.98	11.05
0500	Solid, 8" x 16" block, 6" thick	D-8	440	.091		2.55	2.36		4.91	6.70
0550	8" thick	"	415	.096		3.73	2.51		6.24	8.25
0600	12" thick	D-9	350	.137		5.40	3.49		8.89	11.75

04 22 10.32 Concrete Block, Lintels										
0010	**CONCRETE BLOCK, LINTELS**, C90, normal weight									
0100	Including grout and horizontal reinforcing									
0200	8" x 8" x 8", 1 #4 bar	D-4	300	.133	L.F.	4.86	3.10	.42	8.38	10.95
0250	2 #4 bars		295	.136		5.20	3.16	.43	8.79	11.45
1000	12" x 8" x 8", 1 #4 bar		275	.145		6.75	3.39	.46	10.60	13.60
1150	2 #5 bars		270	.148		7.50	3.45	.47	11.42	14.50

04 22 10.34 Concrete Block, Partitions										
0010	**CONCRETE BLOCK, PARTITIONS**, excludes scaffolding									

04 22 Concrete Unit Masonry

04 22 10 – Concrete Masonry Units

04 22 10.34 Concrete Block, Partitions	Crew	Daily Output	Labor-Hours	Unit	Material	2009 Bare Costs Labor	Equipment	Total	Total Incl O&P
1000 Lightweight block, tooled joints, 2 sides, hollow									
1100 Not reinforced, 8" x 16" x 4" thick	D-8	440	.091	S.F.	1.72	2.36		4.08	5.80
1150 6" thick		410	.098		2.33	2.54		4.87	6.75
1200 8" thick		385	.104		2.86	2.70		5.56	7.60
1250 10" thick		370	.108		3.73	2.81		6.54	8.75
1300 12" thick	D-9	350	.137		4.26	3.49		7.75	10.50
4000 Regular block, tooled joints, 2 sides, hollow									
4100 Not reinforced, 8" x 16" x 4" thick	D-8	430	.093	S.F.	1.43	2.42		3.85	5.60
4150 6" thick		400	.100		2.03	2.60		4.63	6.55
4200 8" thick		375	.107		2.14	2.77		4.91	6.95
4250 10" thick		360	.111		2.95	2.89		5.84	8.05
4300 12" thick	D-9	340	.141		3.35	3.59		6.94	9.65

04 23 Glass Unit Masonry

04 23 13 – Vertical Glass Unit Masonry

04 23 13.10 Glass Block

	Crew	Daily Output	Labor-Hours	Unit	Material	2009 Bare Costs Labor	Equipment	Total	Total Incl O&P
0010 **GLASS BLOCK**									
0150 8" x 8"	D-8	160	.250	S.F.	14.05	6.50		20.55	26.50
0160 end block		160	.250		39.50	6.50		46	54
0170 90 deg corner		160	.250		39	6.50		45.50	54
0180 45 deg corner		160	.250		19.15	6.50		25.65	32
0200 12" x 12"		175	.229		16	5.95		21.95	27.50
0210 4" x 8"		160	.250		10.25	6.50		16.75	22
0220 6" x 8"		160	.250		11	6.50		17.50	23
0700 For solar reflective blocks, add					100%				
1000 Thinline, plain, 3-1/8" thick, under 1,000 S.F., 6" x 6"	D-8	115	.348	S.F.	15.05	9.05		24.10	31.50
1050 8" x 8"		160	.250		9.35	6.50		15.85	21
1400 For cleaning block after installation (both sides), add		1000	.040		.11	1.04		1.15	1.84

04 24 Adobe Unit Masonry

04 24 16 – Manufactured Adobe Unit Masonry

04 24 16.06 Adobe Brick

		Crew	Daily Output	Labor-Hours	Unit	Material	2009 Bare Costs Labor	Equipment	Total	Total Incl O&P
0010 **ADOBE BRICK**, Semi-stabilized, with cement mortar										
0060 Brick, 10" x 4" x 14", 2.6/S.F.	G	D-8	560	.071	S.F.	3.54	1.86		5.40	7
0080 12" x 4" x 16", 2.3/S.F.	G		580	.069		4.50	1.79		6.29	7.90
0100 10" x 4" x 16", 2.3/S.F.	G		590	.068		4.27	1.76		6.03	7.60
0120 8" x 4" x 16", 2.3/S.F.	G		560	.071		3.34	1.86		5.20	6.75
0140 4" x 4" x 16", 2.3/S.F.	G		540	.074		2.55	1.93		4.48	6
0160 6" x 4" x 16", 2.3/S.F.	G		540	.074		2.58	1.93		4.51	6
0180 4" x 4" x 12", 3.0/S.F.	G		520	.077		2.29	2		4.29	5.85
0200 8" x 4" x 12", 3.0/S.F.	G		520	.077		2.52	2		4.52	6.10

04 27 Multiple-Wythe Unit Masonry

04 27 10 – Multiple-Wythe Masonry

04 27 10.10 Cornices

		Crew	Daily Output	Labor-Hours	Unit	Material	2009 Bare Costs Labor	Equipment	Total	Total Incl O&P
0010	**CORNICES**									
0110	Face bricks, 12 brick/S.F., minimum	D-1	30	.533	SF Face	5.95	13.55		19.50	29
0150	15 brick/S.F., maximum	"	23	.696	"	7.15	17.70		24.85	37.50

04 27 10.30 Brick Walls

		Crew	Daily Output	Labor-Hours	Unit	Material	Labor	Equipment	Total	Total Incl O&P
0010	**BRICK WALLS**, including mortar, excludes scaffolding									
0800	Face brick, 4" thick wall, 6.75 brick/S.F.	D-8	215	.186	S.F.	5.60	4.84		10.44	14.15
0850	Common brick, 4" thick wall, 6.75 brick/S.F.		240	.167		2.90	4.34		7.24	10.40
0900	8" thick, 13.50 bricks per S.F.		135	.296		6.05	7.70		13.75	19.45
1000	12" thick, 20.25 bricks per S.F.		95	.421		9.10	10.95		20.05	28
1050	16" thick, 27.00 bricks per S.F.		75	.533		12.35	13.85		26.20	36.50
1200	Reinforced, face brick, 4" thick wall, 6.75 brick/S.F.		210	.190		5.85	4.95		10.80	14.65
1220	Common brick, 4" thick wall, 6.75 brick/S.F.		235	.170		3.16	4.43		7.59	10.85
1250	8" thick, 13.50 bricks per S.F.		130	.308		6.55	8		14.55	20.50
1300	12" thick, 20.25 bricks per S.F.		90	.444		9.90	11.55		21.45	30
1350	16" thick, 27.00 bricks per S.F.		70	.571		13.40	14.85		28.25	39

04 41 Dry-Placed Stone

04 41 10 – Dry Placed Stone

04 41 10.10 Rough Stone Wall

			Crew	Daily Output	Labor-Hours	Unit	Material	Labor	Equipment	Total	Total Incl O&P
0011	**ROUGH STONE WALL**, Dry										
0100	Random fieldstone, under 18" thick	G	D-12	60	.533	C.F.	11.15	13.55		24.70	35
0150	Over 18" thick	G	"	63	.508	"	13.40	12.90		26.30	36.50
0500	Field stone veneer	G	D-8	120	.333	S.F.	9.60	8.65		18.25	25
0510	Valley stone veneer	G		120	.333		9.60	8.65		18.25	25
0520	River stone veneer	G		120	.333		9.60	8.65		18.25	25
0600	Rubble stone walls, in mortar bed, up to 18" thick	G	D-11	75	.320	C.F.	13.50	8.45		21.95	29

04 43 Stone Masonry

04 43 10 – Masonry with Natural and Processed Stone

04 43 10.45 Granite

		Crew	Daily Output	Labor-Hours	Unit	Material	Labor	Equipment	Total	Total Incl O&P
0010	**GRANITE**, cut to size									
2450	For radius under 5', add				L.F.	100%				
2500	Steps, copings, etc., finished on more than one surface									
2550	Minimum	D-10	50	.640	C.F.	91	17.70	12.05	120.75	143
2600	Maximum	"	50	.640	"	146	17.70	12.05	175.75	203
2800	Pavers, 4" x 4" x 4" blocks, split face and joints									
2850	Minimum	D-11	80	.300	S.F.	12.35	7.90		20.25	27
2900	Maximum	"	80	.300	"	27.50	7.90		35.40	43.50

04 43 10.55 Limestone

		Crew	Daily Output	Labor-Hours	Unit	Material	Labor	Equipment	Total	Total Incl O&P
0010	**LIMESTONE**, cut to size									
0020	Veneer facing panels									
0500	Texture finish, light stick, 4-1/2" thick, 5' x 12'	D-4	300	.133	S.F.	48	3.10	.42	51.52	58.50
0750	5" thick, 5' x 14' panels	D-10	275	.116		52.50	3.22	2.19	57.91	65
1000	Sugarcube finish, 2" Thick, 3' x 5' panels		275	.116		35	3.22	2.19	40.41	46
1050	3" Thick, 4' x 9' panels		275	.116		38	3.22	2.19	43.41	49.50
1200	4" Thick, 5' x 11' panels		275	.116		41	3.22	2.19	46.41	52.50
1400	Sugarcube, textured finish, 4-1/2" thick, 5' x 12'		275	.116		45	3.22	2.19	50.41	57
1450	5" thick, 5' x 14' panels		275	.116		53.50	3.22	2.19	58.91	66
2000	Coping, sugarcube finish, top & 2 sides		30	1.067	C.F.	58.50	29.50	20	108	136

04 43 Stone Masonry

04 43 10.55 Limestone

		Crew	Daily Output	Labor-Hours	Unit	Material	2009 Bare Costs Labor	Equipment	Total	Total Incl O&P
2100	Sills, lintels, jambs, trim, stops, sugarcube finish, average	D-10	20	1.600	C.F.	58.50	44	30	132.50	171
2150	Detailed		20	1.600		58.50	44	30	132.50	171
2300	Steps, extra hard, 14" wide, 6" rise		50	.640	L.F.	35	17.70	12.05	64.75	81.50
3000	Quoins, plain finish, 6"x12"x12"	D-12	25	1.280	Ea.	58.50	32.50		91	119
3050	6"x16"x24"	"	25	1.280	"	78	32.50		110.50	140

04 43 10.60 Marble

		Crew	Daily Output	Labor-Hours	Unit	Material	2009 Bare Costs Labor	Equipment	Total	Total Incl O&P
0011	**MARBLE**, ashlar, split face, 4" + or - thick, random									
0040	Lengths 1' to 4' & heights 2" to 7-1/2", average	D-8	175	.229	S.F.	16.90	5.95		22.85	28.50
0100	Base, polished, 3/4" or 7/8" thick, polished, 6" high	D-10	65	.492	L.F.	15.35	13.60	9.25	38.20	49.50
1000	Facing, polished finish, cut to size, 3/4" to 7/8" thick									
1050	Average	D-10	130	.246	S.F.	22.50	6.80	4.63	33.93	41.50
1100	Maximum	"	130	.246	"	52.50	6.80	4.63	63.93	74
2200	Window sills, 6" x 3/4" thick	D-1	85	.188	L.F.	8.50	4.79		13.29	17.30
2500	Flooring, polished tiles, 12" x 12" x 3/8" thick									
2510	Thin set, average	D-11	90	.267	S.F.	11.35	7.05		18.40	24
2600	Maximum		90	.267		174	7.05		181.05	204
2700	Mortar bed, average		65	.369		10.15	9.75		19.90	27.50
2740	Maximum		65	.369		174	9.75		183.75	208
2780	Travertine, 3/8" thick, average	D-10	130	.246		14.40	6.80	4.63	25.83	32
2790	Maximum	"	130	.246		35.50	6.80	4.63	46.93	55.50
3500	Thresholds, 3' long, 7/8" thick, 4" to 5" wide, plain	D-12	24	1.333	Ea.	16.30	34		50.30	74
3550	Beveled		24	1.333	"	18.95	34		52.95	77
3700	Window stools, polished, 7/8" thick, 5" wide		85	.376	L.F.	15.20	9.55		24.75	32.50

04 43 10.75 Sandstone or Brownstone

		Crew	Daily Output	Labor-Hours	Unit	Material	2009 Bare Costs Labor	Equipment	Total	Total Incl O&P
0011	**SANDSTONE OR BROWNSTONE**									
0100	Sawed face veneer, 2-1/2" thick, to 2' x 4' panels	D-10	130	.246	S.F.	17.25	6.80	4.63	28.68	35.50
0150	4' thick, to 3'-6" x 8' panels		100	.320		17.25	8.85	6	32.10	40
0300	Split face, random sizes		100	.320		12.40	8.85	6	27.25	35
0350	Cut stone trim (limestone)									
0360	Ribbon stone, 4" thick, 5' pieces	D-8	120	.333	Ea.	152	8.65		160.65	181
0370	Cove stone, 4" thick, 5' pieces		105	.381		153	9.90		162.90	184
0380	Cornice stone, 10" to 12" wide		90	.444		189	11.55		200.55	226
0390	Band stone, 4" thick, 5' pieces		145	.276		97.50	7.20		104.70	119
0410	Window and door trim, 3" to 4" wide		160	.250		82.50	6.50		89	102
0420	Key stone, 18" long		60	.667		87	17.35		104.35	125

04 43 10.80 Slate

		Crew	Daily Output	Labor-Hours	Unit	Material	2009 Bare Costs Labor	Equipment	Total	Total Incl O&P
0010	**SLATE**									
3500	Stair treads, sand finish, 1" thick x 12" wide									
3600	3 L.F. to 6 L.F.	D-10	120	.267	L.F.	24	7.35	5	36.35	43.50
3700	Ribbon, sand finish, 1" thick x 12" wide									
3750	To 6 L.F.	D-10	120	.267	L.F.	20	7.35	5	32.35	39.50

04 43 10.85 Window Sill

		Crew	Daily Output	Labor-Hours	Unit	Material	2009 Bare Costs Labor	Equipment	Total	Total Incl O&P
0010	**WINDOW SILL**									
0020	Bluestone, thermal top, 10" wide, 1-1/2" thick	D-1	85	.188	S.F.	16.05	4.79		20.84	25.50
0050	2" thick		75	.213	"	18.75	5.45		24.20	29.50
0100	Cut stone, 5" x 8" plain		48	.333	L.F.	11.90	8.50		20.40	27
0200	Face brick on edge, brick, 8" wide		80	.200		2.53	5.10		7.63	11.25
0400	Marble, 9" wide, 1" thick		85	.188		8.50	4.79		13.29	17.30
0900	Slate, colored, unfading, honed, 12" wide, 1" thick		85	.188		17.10	4.79		21.89	27
0950	2" thick		70	.229		24	5.80		29.80	35.50

04 51 Flue Liner Masonry

04 51 10 – Clay Flue Lining

04 51 10.10 Flue Lining

04 51 10.10 Flue Lining	Crew	Daily Output	Labor-Hours	Unit	Material	2009 Bare Costs Labor	Equipment	Total	Total Incl O&P
0010	**FLUE LINING**, including mortar									
0020	8" x 8"	D-1	125	.128	V.L.F.	4.43	3.26		7.69	10.25
0100	8" x 12"		103	.155		6.90	3.95		10.85	14.15
0200	12" x 12"		93	.172		8.85	4.38		13.23	16.95
0300	12" x 18"		84	.190		17.25	4.84		22.09	27
0400	18" x 18"		75	.213		22	5.45		27.45	33
0500	20" x 20"		66	.242		34	6.15		40.15	47.50
0600	24" x 24"		56	.286		43.50	7.25		50.75	60
1000	Round, 18" diameter		66	.242		30	6.15		36.15	43
1100	24" diameter		47	.340		62.50	8.65		71.15	83.50

04 57 Masonry Fireplaces

04 57 10 – Brick or Stone Fireplaces

04 57 10.10 Fireplace

04 57 10.10 Fireplace	Crew	Daily Output	Labor-Hours	Unit	Material	2009 Bare Costs Labor	Equipment	Total	Total Incl O&P
0010	**FIREPLACE**									
0100	Brick fireplace, not incl. foundations or chimneys									
0110	30" x 29" opening, incl. chamber, plain brickwork	D-1	.40	40	Ea.	535	1,025		1,560	2,275
0200	Fireplace box only (110 brick)	"	2	8	"	176	203		379	530
0300	For elaborate brickwork and details, add					35%	35%			
0400	For hearth, brick & stone, add	D-1	2	8	Ea.	197	203		400	550
0410	For steel angle, damper, cleanouts, add		4	4		138	102		240	320
0600	Plain brickwork, incl. metal circulator		.50	32		1,025	815		1,840	2,475
0800	Face brick only, standard size, 8" x 2-2/3" x 4"		.30	53.333	M	535	1,350		1,885	2,850
0900	Stone fireplace, fieldstone, add				SF Face	14.05			14.05	15.50
1000	Cut stone, add				"	15.50			15.50	17.05

04 72 Cast Stone Masonry

04 72 10 – Cast Stone Masonry Features

04 72 10.10 Coping

04 72 10.10 Coping	Crew	Daily Output	Labor-Hours	Unit	Material	2009 Bare Costs Labor	Equipment	Total	Total Incl O&P
0010	**COPING**, stock units									
0050	Precast concrete, 10" wide, 4" tapers to 3-1/2", 8" wall	D-1	75	.213	L.F.	17.40	5.45		22.85	28
0100	12" wide, 3-1/2" tapers to 3", 10" wall		70	.229		17.40	5.80		23.20	29
0150	16" wide, 4" tapers to 3-1/2", 14" wall		60	.267		17.10	6.80		23.90	30
0300	Limestone for 12" wall, 4" thick		90	.178		15.35	4.52		19.87	24.50
0350	6" thick		80	.200		23	5.10		28.10	34
0500	Marble, to 4" thick, no wash, 9" wide		90	.178		23	4.52		27.52	33
0550	12" wide		80	.200		35	5.10		40.10	47
0700	Terra cotta, 9" wide		90	.178		5.60	4.52		10.12	13.65
0800	Aluminum, for 12" wall		80	.200		13.30	5.10		18.40	23

04 72 20 – Cultured Stone Veneer

04 72 20.10 Cultured Stone Veneer Components

04 72 20.10 Cultured Stone Veneer Components	Crew	Daily Output	Labor-Hours	Unit	Material	2009 Bare Costs Labor	Equipment	Total	Total Incl O&P
0010	**CULTURED STONE VENEER COMPONENTS**									
0110	On wood frame and sheathing substrate, random sized cobbles, corner stones	D-8	70	.571	V.L.F.	12	14.85		26.85	37.50
0120	Field stones		140	.286	S.F.	8.70	7.45		16.15	22
0130	Random sized flats, corner stones		70	.571	V.L.F.	11.80	14.85		26.65	37.50
0140	Field stones		140	.286	S.F.	9.90	7.45		17.35	23
0150	Horizontal lined ledgestones, corner stones		75	.533	V.L.F.	12	13.85		25.85	36
0160	Field stones		150	.267	S.F.	8.70	6.95		15.65	21
0170	Random shaped flats, corner stones		65	.615	V.L.F.	12	16		28	39.50

04 72 Cast Stone Masonry

04 72 20 – Cultured Stone Veneer

04 72 20.10 Cultured Stone Veneer Components	Crew	Daily Output	Labor- Hours	Unit	Material	2009 Bare Costs Labor	Equipment	Total	Total Incl O&P
0180	Field stones	D-8	150	.267	S.F.	8.70	6.95		15.65	21
0190	Random shaped / textured face, corner stones		65	.615	V.L.F.	12	16		28	39.50
0200	Field stones		130	.308	S.F.	8.70	8		16.70	23
0210	Random shaped river rock, corner stones		65	.615	V.L.F.	12	16		28	39.50
0220	Field stones		130	.308	S.F.	8.70	8		16.70	23
0240	On concrete or CMU substrate, random sized cobbles, corner stones		70	.571	V.L.F.	11.25	14.85		26.10	37
0250	Field stones		140	.286	S.F.	8.30	7.45		15.75	21.50
0260	Random sized flats, corner stones		70	.571	V.L.F.	11	14.85		25.85	36.50
0270	Field stones		140	.286	S.F.	9.50	7.45		16.95	23
0280	Horizontal lined ledgestones, corner stones		75	.533	V.L.F.	11.25	13.85		25.10	35.50
0290	Field stones		150	.267	S.F.	8.30	6.95		15.25	20.50
0300	Random shaped flats, corner stones		70	.571	V.L.F.	11.25	14.85		26.10	37
0310	Field stones		140	.286	S.F.	8.30	7.45		15.75	21.50
0320	Random shaped / textured face, corner stones		65	.615	V.L.F.	11.25	16		27.25	39
0330	Field stones		130	.308	S.F.	8.30	8		16.30	22.50
0340	Random shaped river rock, corner stones		65	.615	V.L.F.	11.25	16		27.25	39
0350	Field stones		130	.308	S.F.	8.30	8		16.30	22.50
0360	Cultured stone veneer, #15 felt weather resistant barrier	1 Clab	3700	.002	Sq.	4.77	.04		4.81	5.35
0390	Water table or window sill, 18" long	1 Bric	80	.100	Ea.	47.50	2.84		50.34	56.50

Division 5
Metals

05 05 21.15 Drilling Steel

05 05 21.15 Drilling Steel	Crew	Daily Output	Labor-Hours	Unit	Material	2009 Bare Costs Labor	Equipment	Total	Total Incl O&P
0010 DRILLING STEEL									
1910 Drilling & layout for steel, up to 1/4" deep, no anchor									
1920 Holes, 1/4" diameter	1 Sswk	112	.071	Ea.	.08	2.14		2.22	4.20
1925 For each additional 1/4" depth, add		336	.024		.08	.71		.79	1.46
1930 3/8" diameter		104	.077		.09	2.30		2.39	4.53
1935 For each additional 1/4" depth, add		312	.026		.09	.77		.86	1.58
1940 1/2" diameter		96	.083		.10	2.50		2.60	4.91
1945 For each additional 1/4" depth, add		288	.028		.10	.83		.93	1.71
1950 5/8" diameter		88	.091		.13	2.72		2.85	5.40
1955 For each additional 1/4" depth, add		264	.030		.13	.91		1.04	1.89
1960 3/4" diameter		80	.100		.15	3		3.15	5.90
1965 For each additional 1/4" depth, add		240	.033		.15	1		1.15	2.09
1970 7/8" diameter		72	.111		.18	3.33		3.51	6.60
1975 For each additional 1/4" depth, add		216	.037		.18	1.11		1.29	2.33
1980 1" diameter		64	.125		.20	3.74		3.94	7.40
1985 For each additional 1/4" depth, add		192	.042		.20	1.25		1.45	2.62
1990 For drilling up, add						40%			

05 05 23 – Metal Fastenings

05 05 23.10 Bolts and Hex Nuts

05 05 23.10 Bolts and Hex Nuts		Crew	Daily Output	Labor-Hours	Unit	Material	2009 Bare Costs Labor	Equipment	Total	Total Incl O&P
0010 BOLTS & HEX NUTS, Steel, A307										
0100 1/4" diameter, 1/2" long	G	1 Sswk	140	.057	Ea.	.10	1.71		1.81	3.40
0200 1" long	G		140	.057		.12	1.71		1.83	3.42
0300 2" long	G		130	.062		.15	1.84		1.99	3.71
0400 3" long	G		130	.062		.22	1.84		2.06	3.78
0500 4" long	G		120	.067		.24	2		2.24	4.10
0600 3/8" diameter, 1" long	G		130	.062		.19	1.84		2.03	3.75
0700 2" long	G		130	.062		.24	1.84		2.08	3.80
0800 3" long	G		120	.067		.31	2		2.31	4.18
0900 4" long	G		120	.067		.39	2		2.39	4.27
1000 5" long	G		115	.070		.48	2.08		2.56	4.52
1100 1/2" diameter, 1-1/2" long	G		120	.067		.47	2		2.47	4.35
1200 2" long	G		120	.067		.52	2		2.52	4.42
1300 4" long	G		115	.070		.79	2.08		2.87	4.87
1400 6" long	G		110	.073		1.07	2.18		3.25	5.35
1500 8" long	G		105	.076		1.38	2.28		3.66	5.90
1600 5/8" diameter, 1-1/2" long	G		120	.067		1.09	2		3.09	5.05
1700 2" long	G		120	.067		1.18	2		3.18	5.15
1800 4" long	G		115	.070		1.64	2.08		3.72	5.80
1900 6" long	G		110	.073		2.07	2.18		4.25	6.45
2000 8" long	G		105	.076		3	2.28		5.28	7.70
2100 10" long	G		100	.080		3.74	2.40		6.14	8.70
2200 3/4" diameter, 2" long	G		120	.067		1.54	2		3.54	5.55
2300 4" long	G		110	.073		2.17	2.18		4.35	6.55
2400 6" long	G		105	.076		2.76	2.28		5.04	7.40
2500 8" long	G		95	.084		4.12	2.52		6.64	9.40
2600 10" long	G		85	.094		5.40	2.82		8.22	11.30
2700 12" long	G		80	.100		6.30	3		9.30	12.65
2800 1" diameter, 3" long	G		105	.076		4.49	2.28		6.77	9.30
2900 6" long	G		90	.089		7	2.66		9.66	12.80
3000 12" long	G		75	.107		13.30	3.19		16.49	21
3100 For galvanized, add						75%				
3200 For stainless, add						350%				

05 05 23.15 Chemical Anchors

		Crew	Daily Output	Labor-Hours	Unit	Material	2009 Bare Costs Labor	Equipment	Total	Total Incl O&P
0010	**CHEMICAL ANCHORS**									
0020	Includes layout & drilling									
1430	Chemical anchor, w/rod & epoxy cartridge, 3/4" diam. x 9-1/2" long	B-89A	27	.593	Ea.	7.55	14.40	4.08	26.03	37.50
1435	1" diameter x 11-3/4" long		24	.667		12.80	16.20	4.59	33.59	46.50
1440	1-1/4" diameter x 14" long		21	.762		23.50	18.50	5.25	47.25	63.50
1445	1-3/4" diameter x 15" long		20	.800		49.50	19.45	5.50	74.45	93
1450	18" long		17	.941		59	23	6.50	88.50	111
1455	2" diameter x 18" long		16	1		78.50	24.50	6.90	109.90	135
1460	24" long		15	1.067		102	26	7.35	135.35	164

05 05 23.20 Expansion Anchors

			Crew	Daily Output	Labor-Hours	Unit	Material	2009 Bare Costs Labor	Equipment	Total	Total Incl O&P
0010	**EXPANSION ANCHORS**										
0100	Anchors for concrete, brick or stone, no layout and drilling										
0200	Expansion shields, zinc, 1/4" diameter, 1-5/16" long, single	G	1 Carp	90	.089	Ea.	.28	2.48		2.76	4.51
0300	1-3/8" long, double	G		85	.094		.33	2.63		2.96	4.81
0500	2" long, double	G		80	.100		1.10	2.80		3.90	5.95
0700	2-1/2" long, double	G		75	.107		1.40	2.98		4.38	6.60
0900	2-3/4" long, double	G		70	.114		2.14	3.19		5.33	7.75
1100	3-15/16" long, double	G		65	.123		3.36	3.44		6.80	9.50
2100	Hollow wall anchors for gypsum wall board, plaster or tile										
2500	3/16" diameter, short	G	1 Carp	150	.053	Ea.	.36	1.49		1.85	2.92
3000	Toggle bolts, bright steel, 1/8" diameter, 2" long	G		85	.094		.17	2.63		2.80	4.64
3100	4" long	G		80	.100		.20	2.80		3	4.95
3200	3/16" diameter, 3" long	G		80	.100		.22	2.80		3.02	4.97
3300	6" long	G		75	.107		.31	2.98		3.29	5.40
3400	1/4" diameter, 3" long	G		75	.107		.30	2.98		3.28	5.40
3500	6" long	G		70	.114		.42	3.19		3.61	5.85
3600	3/8" diameter, 3" long	G		70	.114		.68	3.19		3.87	6.15
3700	6" long	G		60	.133		1.13	3.73		4.86	7.55
3800	1/2" diameter, 4" long	G		60	.133		1.58	3.73		5.31	8.05
3900	6" long	G		50	.160		1.86	4.47		6.33	9.60
4000	Nailing anchors										
4100	Nylon nailing anchor, 1/4" diameter, 1" long		1 Carp	3.20	2.500	C	9.75	70		79.75	129
4200	1-1/2" long			2.80	2.857		11.20	80		91.20	147
4300	2" long			2.40	3.333		14.25	93		107.25	174
4400	Metal nailing anchor, 1/4" diameter, 1" long	G		3.20	2.500		14.75	70		84.75	134
4500	1-1/2" long	G		2.80	2.857		19.55	80		99.55	157
4600	2" long	G		2.40	3.333		24.50	93		117.50	185
5000	Screw anchors for concrete, masonry,										
5100	stone & tile, no layout or drilling included										
5700	Lag screw shields, 1/4" diameter, short	G	1 Carp	90	.089	Ea.	.33	2.48		2.81	4.56
5800	Long	G		85	.094		.33	2.63		2.96	4.81
5900	3/8" diameter, short	G		85	.094		.61	2.63		3.24	5.10
6000	Long	G		80	.100		.64	2.80		3.44	5.45
6100	1/2" diameter, short	G		80	.100		.89	2.80		3.69	5.70
6200	Long	G		75	.107		1.06	2.98		4.04	6.20
6300	5/8" diameter, short	G		70	.114		1.47	3.19		4.66	7
6400	Long	G		65	.123		1.79	3.44		5.23	7.75
6600	Lead, #6 & #8, 3/4" long	G		260	.031		.10	.86		.96	1.56
6700	#10 - #14, 1-1/2" long	G		200	.040		.22	1.12		1.34	2.13
6800	#16 & #18, 1-1/2" long	G		160	.050		.28	1.40		1.68	2.67
6900	Plastic, #6 & #8, 3/4" long			260	.031		.03	.86		.89	1.48
7000	#8 & #10, 7/8" long			240	.033		.03	.93		.96	1.60

05 05 Common Work Results for Metals

05 05 23 – Metal Fastenings

05 05 23.20 Expansion Anchors		Crew	Daily Output	Labor-Hours	Unit	Material	2009 Bare Costs Labor	Equipment	Total	Total Incl O&P
7100	#10 & #12, 1" long	1 Carp	220	.036	Ea.	.04	1.02		1.06	1.76
7200	#14 & #16, 1-1/2" long		160	.050		.06	1.40		1.46	2.43
8950	Self-drilling concrete screw, hex washer head, 3/16" diam. x 1-3/4" long **G**		300	.027		.16	.75		.91	1.44
8960	2-1/4" long **G**		250	.032		.18	.89		1.07	1.71
8970	Phillips flat head, 3/16" diam. x 1-3/4" long **G**		300	.027		.15	.75		.90	1.43
8980	2-1/4" long **G**		250	.032		.17	.89		1.06	1.70

05 05 23.30 Lag Screws

		Crew	Daily Output	Labor-Hours	Unit	Material	Labor	Equipment	Total	Total Incl O&P
0010	**LAG SCREWS**									
0020	Steel, 1/4" diameter, 2" long **G**	1 Carp	200	.040	Ea.	.08	1.12		1.20	1.98
0100	3/8" diameter, 3" long **G**		150	.053		.26	1.49		1.75	2.81
0200	1/2" diameter, 3" long **G**		130	.062		.45	1.72		2.17	3.41
0300	5/8" diameter, 3" long **G**		120	.067		1.12	1.86		2.98	4.38

05 05 23.50 Powder Actuated Tools and Fasteners

		Crew	Daily Output	Labor-Hours	Unit	Material	Labor	Equipment	Total	Total Incl O&P
0010	**POWDER ACTUATED TOOLS & FASTENERS**									
0020	Stud driver, .22 caliber, buy, minimum				Ea.	224			224	246
0100	Maximum				"	360			360	395
0300	Powder charges for above, low velocity				C	8.65			8.65	9.50
0400	Standard velocity					12.30			12.30	13.55
0600	Drive pins & studs, 1/4" & 3/8" diam., to 3" long, minimum **G**	1 Carp	4.80	1.667		2.88	46.50		49.38	82
0700	Maximum **G**	"	4	2		11.25	56		67.25	107

05 05 23.55 Rivets

		Crew	Daily Output	Labor-Hours	Unit	Material	Labor	Equipment	Total	Total Incl O&P
0010	**RIVETS**									
0100	Aluminum rivet & mandrel, 1/2" grip length x 1/8" diameter **G**	1 Carp	4.80	1.667	C	6.15	46.50		52.65	86
0200	3/16" diameter **G**		4	2		9.45	56		65.45	105
0300	Aluminum rivet, steel mandrel, 1/8" diameter **G**		4.80	1.667		9.40	46.50		55.90	89.50
0400	3/16" diameter **G**		4	2		8.60	56		64.60	104
0500	Copper rivet, steel mandrel, 1/8" diameter **G**		4.80	1.667		19.90	46.50		66.40	101
0600	Monel rivet, steel mandrel, 1/8" diameter **G**		4.80	1.667		57.50	46.50		104	142
0700	3/16" diameter **G**		4	2		165	56		221	276
0800	Stainless rivet & mandrel, 1/8" diameter **G**		4.80	1.667		30	46.50		76.50	112
0900	3/16" diameter **G**		4	2		56.50	56		112.50	157
1000	Stainless rivet, steel mandrel, 1/8" diameter **G**		4.80	1.667		23.50	46.50		70	105
1100	3/16" diameter **G**		4	2		42.50	56		98.50	142
1200	Steel rivet and mandrel, 1/8" diameter **G**		4.80	1.667		10.45	46.50		56.95	90.50
1300	3/16" diameter **G**		4	2		15.70	56		71.70	112
1400	Hand riveting tool, minimum				Ea.	143			143	157
1500	Maximum					273			273	300
1600	Power riveting tool, minimum					1,025			1,025	1,125
1700	Maximum					2,600			2,600	2,850

05 12 Structural Steel Framing

05 12 23 – Structural Steel for Buildings

05 12 23.10 Ceiling Supports

		Crew	Daily Output	Labor-Hours	Unit	Material	Labor	Equipment	Total	Total Incl O&P
0010	**CEILING SUPPORTS**									
1000	Entrance door/folding partition supports, shop fabricated **G**	E-4	60	.533	L.F.	30	16.25	2.23	48.48	66.50
1100	Linear accelerator door supports **G**		14	2.286		137	69.50	9.60	216.10	295
1200	Lintels or shelf angles, hung, exterior hot dipped galv. **G**		267	.120		20.50	3.65	.50	24.65	30
1250	Two coats primer paint instead of galv. **G**		267	.120		17.75	3.65	.50	21.90	27
1400	Monitor support, ceiling hung, expansion bolted **G**		4	8	Ea.	475	244	33.50	752.50	1,025
1450	Hung from pre-set inserts **G**		6	5.333		510	162	22.50	694.50	900

05 12 23 – Structural Steel for Buildings

05 12 23.10 Ceiling Supports		Crew	Daily Output	Labor-Hours	Unit	Material	2009 Bare Costs Labor	Equipment	Total	Total Incl O&P
1600	Motor supports for overhead doors	E-4	4	8	Ea.	242	244	33.50	519.50	775
1700	Partition support for heavy folding partitions, without pocket		24	1.333	L.F.	68.50	40.50	5.60	114.60	159
1750	Supports at pocket only		12	2.667		137	81	11.15	229.15	320
2000	Rolling grilles & fire door supports		34	.941	↓	58.50	28.50	3.94	90.94	124
2100	Spider-leg light supports, expansion bolted to ceiling slab		8	4	Ea.	195	122	16.75	333.75	465
2150	Hung from pre-set inserts		12	2.667	"	210	81	11.15	302.15	400
2400	Toilet partition support		36	.889	L.F.	68.50	27	3.72	99.22	131
2500	X-ray travel gantry support	↓	12	2.667	"	234	81	11.15	326.15	425

05 12 23.15 Columns, Lightweight

	COLUMNS, LIGHTWEIGHT	Crew	Daily Output	Labor-Hours	Unit	Material	Labor	Equipment	Total	Total Incl O&P
0010	**COLUMNS, LIGHTWEIGHT**									
1000	Lightweight units (lally), 3-1/2" diameter	E-2	780	.062	L.F.	6.10	1.86	2.23	10.19	12.65
1050	4" diameter	"	900	.053	"	8.95	1.61	1.93	12.49	14.95
8000	Lally columns, to 8', 3-1/2" diameter	2 Carp	24	.667	Ea.	48.50	18.65		67.15	85
8080	4" diameter	"	20	.800	"	71.50	22.50		94	117

05 12 23.17 Columns, Structural

	COLUMNS, STRUCTURAL	Crew	Daily Output	Labor-Hours	Unit	Material	Labor	Equipment	Total	Total Incl O&P
0010	**COLUMNS, STRUCTURAL**									
0015	Made from recycled materials									
0020	Shop fab'd for 100-ton, 1-2 story project, bolted connections									
0800	Steel, concrete filled, extra strong pipe, 3-1/2" diameter	E-2	660	.073	L.F.	49.50	2.20	2.64	54.34	61.50
0830	4" diameter		780	.062		55	1.86	2.23	59.09	66.50
0890	5" diameter		1020	.047		65.50	1.42	1.71	68.63	77
0930	6" diameter		1200	.040		87	1.21	1.45	89.66	100
0940	8" diameter	↓	1100	.044	↓	87	1.32	1.58	89.90	100
1100	For galvanizing, add				Lb.	.40			.40	.44
1300	For web ties, angles, etc., add per added lb.	1 Sswk	945	.008		1.50	.25		1.75	2.14
1500	Steel pipe, extra strong, no concrete, 3" to 5" diameter	E-2	16000	.003		1.50	.09	.11	1.70	1.94
1600	6" to 12" diameter	E-2	14000	.003		1.50	.10	.12	1.72	1.98
2700	12" x 8" x 1/2" thk wall		24000	.002		1.50	.06	.07	1.63	1.84
2800	Heavy section		32000	.002		1.50	.05	.05	1.60	1.79
5100	Structural tubing, rect, 5" to 6" wide, light section	↓	8000	.006	↓	1.50	.18	.22	1.90	2.23
8090	For projects 75 to 99 tons, add				All	10%				
8092	50 to 74 tons, add					20%				
8094	25 to 49 tons, add					30%	10%			
8096	10 to 24 tons, add					50%	25%			
8098	2 to 9 tons, add					75%	50%			
8099	Less than 2 tons, add				↓	100%	100%			

05 12 23.45 Lintels

	LINTELS	Crew	Daily Output	Labor-Hours	Unit	Material	Labor	Equipment	Total	Total Incl O&P
0010	**LINTELS**									
0015	Made from recycled materials									
0020	Plain steel angles, shop fabricated, under 500 lb.	1 Bric	550	.015	Lb.	1.16	.41		1.57	1.95
0100	500 to 1000 lb.		640	.013	"	1.13	.35		1.48	1.83
2000	Steel angles, 3-1/2" x 3", 1/4" thick, 2'-6" long		47	.170	Ea.	16.20	4.83		21.03	26
2100	4'-6" long		26	.308		29	8.70		37.70	46.50
2600	4" x 3-1/2", 1/4" thick, 5'-0" long		21	.381		37	10.80		47.80	59
2700	9'-0" long	↓	12	.667	↓	67	18.90		85.90	105

05 12 23.65 Plates

	PLATES	Crew	Daily Output	Labor-Hours	Unit	Material	Labor	Equipment	Total	Total Incl O&P
0010	**PLATES**									
0015	Made from recycled materials									
0020	For connections & stiffener plates, shop fabricated									
0050	1/8" thick (5.1 Lb./S.F.)				S.F.	7.65			7.65	8.40
0100	1/4" thick (10.2 Lb./S.F.)					15.30			15.30	16.85
0300	3/8" thick (15.3 Lb./S.F.)				↓	23			23	25.50

05 12 Structural Steel Framing

05 12 23 – Structural Steel for Buildings

05 12 23.65 Plates

			Crew	Daily Output	Labor-Hours	Unit	Material	2009 Bare Costs Labor	Equipment	Total	Total Incl O&P
0400	1/2" thick (20.4 Lb./S.F.)	G				S.F.	30.50			30.50	33.50
0450	3/4" thick (30.6 Lb./S.F.)	G					46			46	50.50
0500	1" thick (40.8 Lb.S.F.)	G					61			61	67.50
2000	Steel plate, warehouse prices, no shop fabrication										
2100	1/4" thick (10.2 Lb./S.F.)	G				S.F.	11.20			11.20	12.35

05 12 23.79 Structural Steel

			Crew	Daily Output	Labor-Hours	Unit	Material	2009 Bare Costs Labor	Equipment	Total	Total Incl O&P
0010	**STRUCTURAL STEEL**										
0020	Shop fab'd for 100-ton, 1-2 story project, bolted conn's.										
0050	Beams, W 6 x 9	G	E-2	720	.067	L.F.	16.20	2.01	2.42	20.63	24
0100	W 8 x 10	G		720	.067		18	2.01	2.42	22.43	26
0200	Columns, W 6 x 15	G		540	.089		29.50	2.68	3.22	35.40	40.50
0250	W 8 x 31	G		540	.089		60.50	2.68	3.22	66.40	75
7990	For projects 75 to 99 tons, add					All	10%				
7992	50 to 75 tons, add						20%				
7994	25 to 49 tons, add						30%	10%			
7996	10 to 24 tons, add						50%	25%			
7998	2 to 9 tons, add						75%	50%			
7999	Less than 2 tons, add						100%	100%			

05 31 Steel Decking

05 31 23 – Steel Roof Decking

05 31 23.50 Roof Decking

			Crew	Daily Output	Labor-Hours	Unit	Material	2009 Bare Costs Labor	Equipment	Total	Total Incl O&P
0010	**ROOF DECKING**										
0015	Made from recycled materials										
2100	Open type, galv., 1-1/2" deep wide rib, 22 gauge, under 50 squares	G	E-4	4500	.007	S.F.	2.58	.22	.03	2.83	3.28
2600	20 gauge, under 50 squares	G		3865	.008		3.03	.25	.03	3.31	3.85
2900	18 gauge, under 50 squares	G		3800	.008		3.91	.26	.04	4.21	4.84
3050	16 gauge, under 50 squares	G		3700	.009		5.25	.26	.04	5.55	6.35

05 31 33 – Steel Form Decking

05 31 33.50 Form Decking

			Crew	Daily Output	Labor-Hours	Unit	Material	2009 Bare Costs Labor	Equipment	Total	Total Incl O&P
0010	**FORM DECKING**										
0015	Made from recycled materials										
6100	Slab form, steel, 28 gauge, 9/16" deep, uncoated	G	E-4	4000	.008	S.F.	1.72	.24	.03	1.99	2.40
6200	Galvanized	G		4000	.008		1.52	.24	.03	1.79	2.18
6220	24 gauge, 1" deep, uncoated	G		3900	.008		1.87	.25	.03	2.15	2.58
6240	Galvanized	G		3900	.008		2.20	.25	.03	2.48	2.94
6300	24 gauge, 1-5/16" deep, uncoated	G		3800	.008		1.99	.26	.04	2.29	2.72
6400	Galvanized	G		3800	.008		2.34	.26	.04	2.64	3.10
6500	22 gauge, 1-5/16" deep, uncoated	G		3700	.009		2.50	.26	.04	2.80	3.30
6600	Galvanized	G		3700	.009		2.55	.26	.04	2.85	3.36
6700	22 gauge, 2" deep uncoated	G		3600	.009		3.28	.27	.04	3.59	4.17
6800	Galvanized	G		3600	.009		3.22	.27	.04	3.53	4.10

05 41 Structural Metal Stud Framing

05 41 13 – Load-Bearing Metal Stud Framing

05 41 13.05 Bracing

		Crew	Daily Output	Labor-Hours	Unit	Material	2009 Bare Costs Labor	Equipment	Total	Total Incl O&P
0010	**BRACING**, shear wall X-bracing, per 10' x 10' bay, one face									
0015	Made of recycled materials									
0120	Metal strap, 20 ga x 4" wide	G 2 Carp	18	.889	Ea.	27	25		52	71.50
0130	6" wide	G	18	.889		42	25		67	88
0160	18 ga x 4" wide	G	16	1		39	28		67	90.50
0170	6" wide	G	16	1		57	28		85	111
0410	Continuous strap bracing, per horizontal row on both faces									
0420	Metal strap, 20 ga x 2" wide, studs 12" O.C.	G 1 Carp	7	1.143	C.L.F.	68	32		100	129
0430	16" O.C.	G	8	1		68	28		96	123
0440	24" O.C.	G	10	.800		68	22.50		90.50	113
0450	18 ga x 2" wide, studs 12" O.C.	G	6	1.333		94.50	37.50		132	167
0460	16" O.C.	G	7	1.143		94.50	32		126.50	158
0470	24" O.C.	G	8	1		94.50	28		122.50	152

05 41 13.10 Bridging

		Crew	Daily Output	Labor-Hours	Unit	Material	2009 Bare Costs Labor	Equipment	Total	Total Incl O&P
0010	**BRIDGING**, solid between studs w/ 1-1/4" leg track, per stud bay									
0015	Made from recycled materials									
0200	Studs 12" O.C., 18 ga x 2-1/2" wide	G 1 Carp	125	.064	Ea.	1.16	1.79		2.95	4.29
0210	3-5/8" wide	G	120	.067		1.40	1.86		3.26	4.69
0220	4" wide	G	120	.067		1.48	1.86		3.34	4.78
0230	6" wide	G	115	.070		1.93	1.94		3.87	5.40
0240	8" wide	G	110	.073		2.45	2.03		4.48	6.15
0300	16 ga x 2-1/2" wide	G	115	.070		1.44	1.94		3.38	4.88
0310	3-5/8" wide	G	110	.073		1.76	2.03		3.79	5.40
0320	4" wide	G	110	.073		1.88	2.03		3.91	5.50
0330	6" wide	G	105	.076		2.42	2.13		4.55	6.25
0340	8" wide	G	100	.080		3.08	2.24		5.32	7.15
1200	Studs 16" O.C., 18 ga x 2-1/2" wide	G	125	.064		1.49	1.79		3.28	4.65
1210	3-5/8" wide	G	120	.067		1.80	1.86		3.66	5.15
1220	4" wide	G	120	.067		1.90	1.86		3.76	5.25
1230	6" wide	G	115	.070		2.48	1.94		4.42	6
1240	8" wide	G	110	.073		3.14	2.03		5.17	6.90
1300	16 ga x 2-1/2" wide	G	115	.070		1.85	1.94		3.79	5.30
1310	3-5/8" wide	G	110	.073		2.26	2.03		4.29	5.95
1320	4" wide	G	110	.073		2.41	2.03		4.44	6.10
1330	6" wide	G	105	.076		3.10	2.13		5.23	7
1340	8" wide	G	100	.080		3.94	2.24		6.18	8.10
2200	Studs 24" O.C., 18 ga x 2-1/2" wide	G	125	.064		2.15	1.79		3.94	5.40
2210	3-5/8" wide	G	120	.067		2.60	1.86		4.46	6
2220	4" wide	G	120	.067		2.75	1.86		4.61	6.15
2230	6" wide	G	115	.070		3.58	1.94		5.52	7.25
2240	8" wide	G	110	.073		4.54	2.03		6.57	8.45
2300	16 ga x 2-1/2" wide	G	115	.070		2.67	1.94		4.61	6.25
2310	3-5/8" wide	G	110	.073		3.27	2.03		5.30	7.05
2320	4" wide	G	110	.073		3.49	2.03		5.52	7.25
2330	6" wide	G	105	.076		4.49	2.13		6.62	8.55
2340	8" wide	G	100	.080		5.70	2.24		7.94	10.10
3000	Continuous bridging, per row									
3100	16 ga x 1-1/2" channel thru studs 12" O.C.	G 1 Carp	6	1.333	C.L.F.	65	37.50		102.50	135
3110	16" O.C.	G	7	1.143		65	32		97	126
3120	24" O.C.	G	8.80	.909		65	25.50		90.50	115
4100	2" x 2" angle x 18 ga, studs 12" O.C.	G	7	1.143		94.50	32		126.50	158
4110	16" O.C.	G	9	.889		94.50	25		119.50	146

05 41 Structural Metal Stud Framing

05 41 13 – Load-Bearing Metal Stud Framing

05 41 13.10 Bridging

			Crew	Daily Output	Labor-Hours	Unit	Material	2009 Bare Costs Labor	Equipment	Total	Total Incl O&P
4120	24" O.C.	G	1 Carp	12	.667	C.L.F.	94.50	18.65		113.15	136
4200	16 ga, studs 12" O.C.	G		5	1.600		120	44.50		164.50	208
4210	16" O.C.	G		7	1.143		120	32		152	186
4220	24" O.C.	G		10	.800		120	22.50		142.50	170

05 41 13.25 Framing, Boxed Headers/Beams

			Crew	Daily Output	Labor-Hours	Unit	Material	2009 Bare Costs Labor	Equipment	Total	Total Incl O&P
0010	**FRAMING, BOXED HEADERS/BEAMS**										
0015	Made from recycled materials										
0200	Double, 18 ga x 6" deep	G	2 Carp	220	.073	L.F.	6.70	2.03		8.73	10.85
0210	8" deep	G		210	.076		7.45	2.13		9.58	11.80
0220	10" deep	G		200	.080		9	2.24		11.24	13.75
0230	12" deep	G		190	.084		9.90	2.35		12.25	14.90
0300	16 ga x 8" deep	G		180	.089		8.55	2.48		11.03	13.60
0310	10" deep	G		170	.094		10.30	2.63		12.93	15.75
0320	12" deep	G		160	.100		11.20	2.80		14	17.05
0400	14 ga x 10" deep	G		140	.114		11.95	3.19		15.14	18.55
0410	12" deep	G		130	.123		13.10	3.44		16.54	20.50
1210	Triple, 18 ga x 8" deep	G		170	.094		10.80	2.63		13.43	16.35
1220	10" deep	G		165	.097		13	2.71		15.71	18.90
1230	12" deep	G		160	.100		14.35	2.80		17.15	20.50
1300	16 ga x 8" deep	G		145	.110		12.45	3.08		15.53	18.90
1310	10" deep	G		140	.114		14.85	3.19		18.04	22
1320	12" deep	G		135	.119		16.20	3.31		19.51	23.50
1400	14 ga x 10" deep	G		115	.139		16.30	3.89		20.19	24.50
1410	12" deep	G		110	.145		18.10	4.07		22.17	27

05 41 13.30 Framing, Stud Walls

			Crew	Daily Output	Labor-Hours	Unit	Material	2009 Bare Costs Labor	Equipment	Total	Total Incl O&P
0010	**FRAMING, STUD WALLS** w/ top & bottom track, no openings,										
0020	Headers, beams, bridging or bracing										
0025	Made from recycled materials										
4100	8' high walls, 18 ga x 2-1/2" wide, studs 12" O.C.	G	2 Carp	54	.296	L.F.	11.25	8.30		19.55	26.50
4110	16" O.C.	G		77	.208		9	5.80		14.80	19.70
4120	24" O.C.	G		107	.150		6.75	4.18		10.93	14.45
4130	3-5/8" wide, studs 12" O.C.	G		53	.302		13.40	8.45		21.85	29
4140	16" O.C.	G		76	.211		10.70	5.90		16.60	22
4150	24" O.C.	G		105	.152		8.05	4.26		12.31	16.05
4160	4" wide, studs 12" O.C.	G		52	.308		14	8.60		22.60	30
4170	16" O.C.	G		74	.216		11.20	6.05		17.25	22.50
4180	24" O.C.	G		103	.155		8.40	4.34		12.74	16.60
4190	6" wide, studs 12" O.C.	G		51	.314		17.85	8.75		26.60	34.50
4200	16" O.C.	G		73	.219		14.30	6.15		20.45	26
4210	24" O.C.	G		101	.158		10.75	4.43		15.18	19.35
4220	8" wide, studs 12" O.C.	G		50	.320		22	8.95		30.95	39
4230	16" O.C.	G		72	.222		17.50	6.20		23.70	30
4240	24" O.C.	G		100	.160		13.20	4.47		17.67	22
4300	16 ga x 2-1/2" wide, studs 12" O.C.	G		47	.340		13.25	9.50		22.75	30.50
4310	16" O.C.	G		68	.235		10.50	6.60		17.10	22.50
4320	24" O.C.	G		94	.170		7.75	4.76		12.51	16.55
4330	3-5/8" wide, studs 12" O.C.	G		46	.348		15.85	9.70		25.55	34
4340	16" O.C.	G		66	.242		12.55	6.80		19.35	25.50
4350	24" O.C.	G		92	.174		9.25	4.86		14.11	18.40
4360	4" wide, studs 12" O.C.	G		45	.356		16.75	9.95		26.70	35
4370	16" O.C.	G		65	.246		13.25	6.90		20.15	26.50
4380	24" O.C.	G		90	.178		9.80	4.97		14.77	19.15

05 41 13.30 Framing, Stud Walls		Crew	Daily Output	Labor-Hours	Unit	Material	2009 Bare Costs Labor	Equipment	Total	Total Incl O&P
4390	6" wide, studs 12" O.C.	G	2 Carp	44	.364	L.F.	21	10.15		31.15	40
4400	16" O.C.	G		64	.250		16.70	7		23.70	30
4410	24" O.C.	G		88	.182		12.35	5.10		17.45	22
4420	8" wide, studs 12" O.C.	G		43	.372		26	10.40		36.40	46
4430	16" O.C.	G		63	.254		20.50	7.10		27.60	34.50
4440	24" O.C.	G		86	.186		15.25	5.20		20.45	25.50
5100	10' high walls, 18 ga x 2-1/2" wide, studs 12" O.C.	G		54	.296		13.50	8.30		21.80	29
5110	16" O.C.	G		77	.208		10.70	5.80		16.50	21.50
5120	24" O.C.	G		107	.150		7.85	4.18		12.03	15.70
5130	3-5/8" wide, studs 12" O.C.	G		53	.302		16.05	8.45		24.50	32
5140	16" O.C.	G		76	.211		12.70	5.90		18.60	24
5150	24" O.C.	G		105	.152		9.35	4.26		13.61	17.50
5160	4" wide, studs 12" O.C.	G		52	.308		16.80	8.60		25.40	33
5170	16" O.C.	G		74	.216		13.30	6.05		19.35	25
5180	24" O.C.	G		103	.155		9.80	4.34		14.14	18.15
5190	6" wide, studs 12" O.C.	G		51	.314		21.50	8.75		30.25	38.50
5200	16" O.C.	G		73	.219		16.95	6.15		23.10	29
5210	24" O.C.	G		101	.158		12.55	4.43		16.98	21.50
5220	8" wide, studs 12" O.C.	G		50	.320		26	8.95		34.95	43.50
5230	16" O.C.	G		72	.222		20.50	6.20		26.70	33.50
5240	24" O.C.	G		100	.160		15.35	4.47		19.82	24.50
5300	16 ga x 2-1/2" wide, studs 12" O.C.	G		47	.340		16	9.50		25.50	33.50
5310	16" O.C.	G		68	.235		12.55	6.60		19.15	25
5320	24" O.C.	G		94	.170		9.10	4.76		13.86	18.05
5330	3-5/8" wide, studs 12" O.C.	G		46	.348		19.15	9.70		28.85	37.50
5340	16" O.C.	G		66	.242		15.05	6.80		21.85	28
5350	24" O.C.	G		92	.174		10.90	4.86		15.76	20
5360	4" wide, studs 12" O.C.	G		45	.356		20	9.95		29.95	39
5370	16" O.C.	G		65	.246		15.85	6.90		22.75	29
5380	24" O.C.	G		90	.178		11.50	4.97		16.47	21
5390	6" wide, studs 12" O.C.	G		44	.364		25.50	10.15		35.65	45
5400	16" O.C.	G		64	.250		19.95	7		26.95	34
5410	24" O.C.	G		88	.182		14.55	5.10		19.65	24.50
5420	8" wide, studs 12" O.C.	G		43	.372		31	10.40		41.40	52
5430	16" O.C.	G		63	.254		24.50	7.10		31.60	39
5440	24" O.C.	G		86	.186		17.90	5.20		23.10	28.50
6190	12' high walls, 18 ga x 6" wide, studs 12" O.C.	G		41	.390		25	10.90		35.90	46
6200	16" O.C.	G		58	.276		19.60	7.70		27.30	34.50
6210	24" O.C.	G		81	.198		14.30	5.50		19.80	25
6220	8" wide, studs 12" O.C.	G		40	.400		30.50	11.20		41.70	52.50
6230	16" O.C.	G		57	.281		24	7.85		31.85	40
6240	24" O.C.	G		80	.200		17.50	5.60		23.10	28.50
6390	16 ga x 6" wide, studs 12" O.C.	G		35	.457		29.50	12.80		42.30	54
6400	16" O.C.	G		51	.314		23	8.75		31.75	40.50
6410	24" O.C.	G		70	.229		16.70	6.40		23.10	29
6420	8" wide, studs 12" O.C.	G		34	.471		36.50	13.15		49.65	62
6430	16" O.C.	G		50	.320		28.50	8.95		37.45	46.50
6440	24" O.C.	G		69	.232		20.50	6.50		27	33.50
6530	14 ga x 3-5/8" wide, studs 12" O.C.	G		34	.471		28	13.15		41.15	53
6540	16" O.C.	G		48	.333		22	9.30		31.30	40
6550	24" O.C.	G		65	.246		15.80	6.90		22.70	29
6560	4" wide, studs 12" O.C.	G		33	.485		29.50	13.55		43.05	55.50
6570	16" O.C.	G		47	.340		23	9.50		32.50	41.50

05 41 Structural Metal Stud Framing

05 41 13 – Load-Bearing Metal Stud Framing

05 41 13.30 Framing, Stud Walls

			Crew	Daily Output	Labor-Hours	Unit	Material	2009 Bare Costs Labor	Equipment	Total	Total Incl O&P
6580	24" O.C.	G	2 Carp	64	.250	L.F.	16.65	7		23.65	30
6730	12 ga x 3-5/8" wide, studs 12" O.C.	G		31	.516		39	14.45		53.45	67.50
6740	16" O.C.	G		43	.372		30	10.40		40.40	50.50
6750	24" O.C.	G		59	.271		21	7.60		28.60	36.50
6760	4" wide, studs 12" O.C.	G		30	.533		41.50	14.90		56.40	71
6770	16" O.C.	G		42	.381		32	10.65		42.65	53.50
6780	24" O.C.	G		58	.276		22.50	7.70		30.20	38
7390	16' high walls, 16 ga x 6" wide, studs 12" O.C.	G		33	.485		38.50	13.55		52.05	65
7400	16" O.C.	G		48	.333		29.50	9.30		38.80	48.50
7410	24" O.C.	G		67	.239		21	6.65		27.65	34.50
7420	8" wide, studs 12" O.C.	G		32	.500		47	14		61	75.50
7430	16" O.C.	G		47	.340		36.50	9.50		46	56
7440	24" O.C.	G		66	.242		26	6.80		32.80	40
7560	14 ga x 4" wide, studs 12" O.C.	G		31	.516		38.50	14.45		52.95	67
7570	16" O.C.	G		45	.356		29.50	9.95		39.45	49.50
7580	24" O.C.	G		61	.262		21	7.35		28.35	35.50
7590	6" wide, studs 12" O.C.	G		30	.533		48.50	14.90		63.40	78.50
7600	16" O.C.	G		44	.364		37.50	10.15		47.65	58
7610	24" O.C.	G		60	.267		26.50	7.45		33.95	41.50
7760	12 ga x 4" wide, studs 12" O.C.	G		29	.552		54.50	15.40		69.90	86
7770	16" O.C.	G		40	.400		41.50	11.20		52.70	65
7780	24" O.C.	G		55	.291		29	8.15		37.15	46
7790	6" wide, studs 12" O.C.	G		28	.571		69	15.95		84.95	103
7800	16" O.C.	G		39	.410		53	11.45		64.45	77.50
7810	24" O.C.	G		54	.296		37	8.30		45.30	54.50
8590	20' high walls, 14 ga x 6" wide, studs 12" O.C.	G		29	.552		59.50	15.40		74.90	91.50
8600	16" O.C.	G		42	.381		45.50	10.65		56.15	68.50
8610	24" O.C.	G		57	.281		32	7.85		39.85	48.50
8620	8" wide, studs 12" O.C.	G		28	.571		72.50	15.95		88.45	107
8630	16" O.C.	G		41	.390		56	10.90		66.90	80
8640	24" O.C.	G		56	.286		39.50	8		47.50	56.50
8790	12 ga x 6" wide, studs 12" O.C.	G		27	.593		85	16.55		101.55	122
8800	16" O.C.	G		37	.432		65	12.10		77.10	92
8810	24" O.C.	G		51	.314		45	8.75		53.75	64.50
8820	8" wide, studs 12" O.C.	G		26	.615		103	17.20		120.20	143
8830	16" O.C.	G		36	.444		79	12.40		91.40	108
8840	24" O.C.	G		50	.320		54.50	8.95		63.45	75

05 42 Cold-Formed Metal Joist Framing

05 42 13 – Cold-Formed Metal Floor Joist Framing

05 42 13.05 Bracing

			Crew	Daily Output	Labor-Hours	Unit	Material	2009 Bare Costs Labor	Equipment	Total	Total Incl O&P
0010	**BRACING**, continuous, per row, top & bottom										
0015	Made from recycled materials										
0120	Flat strap, 20 ga x 2" wide, joists at 12" O.C.	G	1 Carp	4.67	1.713	C.L.F.	71.50	48		119.50	160
0130	16" O.C.	G		5.33	1.501		69	42		111	147
0140	24" O.C.	G		6.66	1.201		66.50	33.50		100	130
0150	18 ga x 2" wide, joists at 12" O.C.	G		4	2		93.50	56		149.50	198
0160	16" O.C.	G		4.67	1.713		92	48		140	182
0170	24" O.C.	G		5.33	1.501		90.50	42		132.50	171

05 42 13.10 Bridging

0010	**BRIDGING**, solid between joists w/ 1-1/4" leg track, per joist bay										

05 42 Cold-Formed Metal Joist Framing

05 42 13 – Cold-Formed Metal Floor Joist Framing

05 42 13.10 Bridging

		Crew	Daily Output	Labor-Hours	Unit	Material	2009 Bare Costs Labor	Equipment	Total	Total Incl O&P
0015	Made from recycled materials									
0230	Joists 12" O.C., 18 ga track x 6" wide	G 1 Carp	80	.100	Ea.	1.93	2.80		4.73	6.85
0240	8" wide	G	75	.107		2.45	2.98		5.43	7.75
0250	10" wide	G	70	.114		3.02	3.19		6.21	8.75
0260	12" wide	G	65	.123		3.50	3.44		6.94	9.65
0330	16 ga track x 6" wide	G	70	.114		2.42	3.19		5.61	8.05
0340	8" wide	G	65	.123		3.08	3.44		6.52	9.20
0350	10" wide	G	60	.133		3.78	3.73		7.51	10.45
0360	12" wide	G	55	.145		4.36	4.07		8.43	11.65
0440	14 ga track x 8" wide	G	60	.133		3.87	3.73		7.60	10.55
0450	10" wide	G	55	.145		4.76	4.07		8.83	12.10
0460	12" wide	G	50	.160		5.50	4.47		9.97	13.60
0550	12 ga track x 10" wide	G	45	.178		7	4.97		11.97	16.10
0560	12" wide	G	40	.200		7.90	5.60		13.50	18.15
1230	16" O.C., 18 ga track x 6" wide	G	80	.100		2.48	2.80		5.28	7.45
1240	8" wide	G	75	.107		3.14	2.98		6.12	8.50
1250	10" wide	G	70	.114		3.88	3.19		7.07	9.65
1260	12" wide	G	65	.123		4.49	3.44		7.93	10.75
1330	16 ga track x 6" wide	G	70	.114		3.10	3.19		6.29	8.80
1340	8" wide	G	65	.123		3.94	3.44		7.38	10.15
1350	10" wide	G	60	.133		4.85	3.73		8.58	11.65
1360	12" wide	G	55	.145		5.60	4.07		9.67	13
1440	14 ga track x 8" wide	G	60	.133		4.97	3.73		8.70	11.75
1450	10" wide	G	55	.145		6.10	4.07		10.17	13.55
1460	12" wide	G	50	.160		7.05	4.47		11.52	15.30
1550	12 ga track x 10" wide	G	45	.178		9	4.97		13.97	18.25
1560	12" wide	G	40	.200		10.15	5.60		15.75	20.50
2230	24" O.C., 18 ga track x 6" wide	G	80	.100		3.58	2.80		6.38	8.65
2240	8" wide	G	75	.107		4.54	2.98		7.52	10.05
2250	10" wide	G	70	.114		5.60	3.19		8.79	11.55
2260	12" wide	G	65	.123		6.50	3.44		9.94	12.95
2330	16 ga track x 6" wide	G	70	.114		4.49	3.19		7.68	10.35
2340	8" wide	G	65	.123		5.70	3.44		9.14	12.10
2350	10" wide	G	60	.133		7	3.73		10.73	14
2360	12" wide	G	55	.145		8.10	4.07		12.17	15.75
2440	14 ga track x 8" wide	G	60	.133		7.20	3.73		10.93	14.20
2450	10" wide	G	55	.145		8.85	4.07		12.92	16.55
2460	12" wide	G	50	.160		10.20	4.47		14.67	18.80
2550	12 ga track x 10" wide	G	45	.178		13	4.97		17.97	22.50
2560	12" wide	G	40	.200		14.70	5.60		20.30	25.50

05 42 13.25 Framing, Band Joist

		Crew	Daily Output	Labor-Hours	Unit	Material	2009 Bare Costs Labor	Equipment	Total	Total Incl O&P
0010	**FRAMING, BAND JOIST** (track) fastened to bearing wall									
0015	Made from recycled materials									
0220	18 ga track x 6" deep	G 2 Carp	1000	.016	L.F.	1.58	.45		2.03	2.49
0230	8" deep	G	920	.017		2	.49		2.49	3.01
0240	10" deep	G	860	.019		2.47	.52		2.99	3.59
0320	16 ga track x 6" deep	G	900	.018		1.97	.50		2.47	3.01
0330	8" deep	G	840	.019		2.51	.53		3.04	3.66
0340	10" deep	G	780	.021		3.09	.57		3.66	4.37
0350	12" deep	G	740	.022		3.56	.60		4.16	4.94
0430	14 ga track x 8" deep	G	750	.021		3.16	.60		3.76	4.49
0440	10" deep	G	720	.022		3.89	.62		4.51	5.30

05 42 Cold-Formed Metal Joist Framing

05 42 13 – Cold-Formed Metal Floor Joist Framing

05 42 13.25 Framing, Band Joist

		Crew	Daily Output	Labor-Hours	Unit	Material	2009 Bare Costs Labor	Equipment	Total	Total Incl O&P
0450	12" deep **G**	2 Carp	700	.023	L.F.	4.49	.64		5.13	6
0540	12 ga track x 10" deep **G**		670	.024		5.70	.67		6.37	7.45
0550	12" deep **G**	↓	650	.025	↓	6.45	.69		7.14	8.25

05 42 13.30 Framing, Boxed Headers/Beams

		Crew	Daily Output	Labor-Hours	Unit	Material	2009 Bare Costs Labor	Equipment	Total	Total Incl O&P
0010	**FRAMING, BOXED HEADERS/BEAMS**									
0015	Made from recycled materials									
0200	Double, 18 ga x 6" deep **G**	2 Carp	220	.073	L.F.	6.70	2.03		8.73	10.85
0210	8" deep **G**		210	.076		7.45	2.13		9.58	11.80
0220	10" deep **G**		200	.080		9	2.24		11.24	13.75
0230	12" deep **G**		190	.084		9.90	2.35		12.25	14.90
0300	16 ga x 8" deep **G**		180	.089		8.55	2.48		11.03	13.60
0310	10" deep **G**		170	.094		10.30	2.63		12.93	15.75
0320	12" deep **G**		160	.100		11.20	2.80		14	17.05
0400	14 ga x 10" deep **G**		140	.114		11.95	3.19		15.14	18.55
0410	12" deep **G**		130	.123		13.10	3.44		16.54	20.50
0500	12 ga x 10" deep **G**		110	.145		15.80	4.07		19.87	24.50
0510	12" deep **G**		100	.160		17.50	4.47		21.97	27
1210	Triple, 18 ga x 8" deep **G**		170	.094		10.80	2.63		13.43	16.35
1220	10" deep **G**		165	.097		13	2.71		15.71	18.90
1230	12" deep **G**		160	.100		14.35	2.80		17.15	20.50
1300	16 ga x 8" deep **G**		145	.110		12.45	3.08		15.53	18.90
1310	10" deep **G**		140	.114		14.85	3.19		18.04	22
1320	12" deep **G**		135	.119		16.20	3.31		19.51	23.50
1400	14 ga x 10" deep **G**		115	.139		17.35	3.89		21.24	25.50
1410	12" deep **G**		110	.145		19.15	4.07		23.22	28
1500	12 ga x 10" deep **G**		90	.178		23	4.97		27.97	34
1510	12" deep **G**	↓	85	.188	↓	25.50	5.25		30.75	37.50

05 42 13.40 Framing, Joists

		Crew	Daily Output	Labor-Hours	Unit	Material	2009 Bare Costs Labor	Equipment	Total	Total Incl O&P
0010	**FRAMING, JOISTS**, no band joists (track), web stiffeners, headers,									
0020	Beams, bridging or bracing									
0025	Made from recycled materials									
0030	Joists (2" flange) and fasteners, materials only									
0220	18 ga x 6" deep **G**				L.F.	2.08			2.08	2.29
0230	8" deep **G**					2.47			2.47	2.71
0240	10" deep **G**					2.90			2.90	3.19
0320	16 ga x 6" deep **G**					2.54			2.54	2.80
0330	8" deep **G**					3.05			3.05	3.35
0340	10" deep **G**					3.56			3.56	3.92
0350	12" deep **G**					4.03			4.03	4.44
0430	14 ga x 8" deep **G**					3.84			3.84	4.23
0440	10" deep **G**					4.43			4.43	4.87
0450	12" deep **G**					5.05			5.05	5.55
0540	12 ga x 10" deep **G**					6.45			6.45	7.10
0550	12" deep **G**				↓	7.35			7.35	8.10
1010	Installation of joists to band joists, beams & headers, labor only									
1220	18 ga x 6" deep	2 Carp	110	.145	Ea.		4.07		4.07	6.85
1230	8" deep		90	.178			4.97		4.97	8.40
1240	10" deep		80	.200			5.60		5.60	9.45
1320	16 ga x 6" deep		95	.168			4.71		4.71	7.95
1330	8" deep		70	.229			6.40		6.40	10.80
1340	10" deep		60	.267			7.45		7.45	12.60
1350	12" deep	↓	55	.291	↓		8.15		8.15	13.75

05 42 Cold-Formed Metal Joist Framing

05 42 13 – Cold-Formed Metal Floor Joist Framing

05 42 13.40 Framing, Joists

	05 42 13.40 Framing, Joists		Crew	Daily Output	Labor-Hours	Unit	Material	2009 Bare Costs Labor	Equipment	Total	Total Incl O&P
1430	14 ga x 8" deep		2 Carp	65	.246	Ea.		6.90		6.90	11.65
1440	10" deep			45	.356			9.95		9.95	16.80
1450	12" deep			35	.457			12.80		12.80	21.50
1540	12 ga x 10" deep			40	.400			11.20		11.20	18.90
1550	12" deep			30	.533			14.90		14.90	25

05 42 13.45 Framing, Web Stiffeners

	05 42 13.45 Framing, Web Stiffeners		Crew	Daily Output	Labor-Hours	Unit	Material	2009 Bare Costs Labor	Equipment	Total	Total Incl O&P
0010	**FRAMING, WEB STIFFENERS** at joist bearing, fabricated from										
0020	Stud piece (1-5/8" flange) to stiffen joist (2" flange)										
0025	Made from recycled materials										
2120	For 6" deep joist, with 18 ga x 2-1/2" stud	G	1 Carp	120	.067	Ea.	2.49	1.86		4.35	5.90
2130	3-5/8" stud	G		110	.073		2.75	2.03		4.78	6.45
2140	4" stud	G		105	.076		2.66	2.13		4.79	6.55
2150	6" stud	G		100	.080		2.92	2.24		5.16	7
2160	8" stud	G		95	.084		3	2.35		5.35	7.30
2220	8" deep joist, with 2-1/2" stud	G		120	.067		2.73	1.86		4.59	6.15
2230	3-5/8" stud	G		110	.073		2.96	2.03		4.99	6.70
2240	4" stud	G		105	.076		2.91	2.13		5.04	6.80
2250	6" stud	G		100	.080		3.20	2.24		5.44	7.30
2260	8" stud	G		95	.084		3.44	2.35		5.79	7.75
2320	10" deep joist, with 2-1/2" stud	G		110	.073		3.85	2.03		5.88	7.65
2330	3-5/8" stud	G		100	.080		4.23	2.24		6.47	8.45
2340	4" stud	G		95	.084		4.18	2.35		6.53	8.60
2350	6" stud	G		90	.089		4.55	2.48		7.03	9.20
2360	8" stud	G		85	.094		4.62	2.63		7.25	9.55
2420	12" deep joist, with 2-1/2" stud	G		110	.073		4.07	2.03		6.10	7.90
2430	3-5/8" stud	G		100	.080		4.42	2.24		6.66	8.65
2440	4" stud	G		95	.084		4.34	2.35		6.69	8.75
2450	6" stud	G		90	.089		4.78	2.48		7.26	9.45
2460	8" stud	G		85	.094		5.15	2.63		7.78	10.10
3130	For 6" deep joist, with 16 ga x 3-5/8" stud	G		100	.080		2.89	2.24		5.13	6.95
3140	4" stud	G		95	.084		2.87	2.35		5.22	7.15
3150	6" stud	G		90	.089		3.15	2.48		5.63	7.65
3160	8" stud	G		85	.094		3.31	2.63		5.94	8.10
3230	8" deep joist, with 3-5/8" stud	G		100	.080		3.21	2.24		5.45	7.30
3240	4" stud	G		95	.084		3.15	2.35		5.50	7.45
3250	6" stud	G		90	.089		3.49	2.48		5.97	8.05
3260	8" stud	G		85	.094		3.73	2.63		6.36	8.55
3330	10" deep joist, with 3-5/8" stud	G		85	.094		4.38	2.63		7.01	9.25
3340	4" stud	G		80	.100		4.48	2.80		7.28	9.65
3350	6" stud	G		75	.107		4.86	2.98		7.84	10.40
3360	8" stud	G		70	.114		5.05	3.19		8.24	10.95
3430	12" deep joist, with 3-5/8" stud	G		85	.094		4.79	2.63		7.42	9.70
3440	4" stud	G		80	.100		4.70	2.80		7.50	9.90
3450	6" stud	G		75	.107		5.20	2.98		8.18	10.80
3460	8" stud	G		70	.114		5.55	3.19		8.74	11.50
4230	For 8" deep joist, with 14 ga x 3-5/8" stud	G		90	.089		4.16	2.48		6.64	8.80
4240	4" stud	G		85	.094		4.24	2.63		6.87	9.10
4250	6" stud	G		80	.100		4.59	2.80		7.39	9.80
4260	8" stud	G		75	.107		4.92	2.98		7.90	10.45
4330	10" deep joist, with 3-5/8" stud	G		75	.107		5.85	2.98		8.83	11.50
4340	4" stud	G		70	.114		5.80	3.19		8.99	11.75
4350	6" stud	G		65	.123		6.35	3.44		9.79	12.80

05 42 Cold-Formed Metal Joist Framing

05 42 13 – Cold-Formed Metal Floor Joist Framing

05 42 13.45 Framing, Web Stiffeners

			Crew	Daily Output	Labor-Hours	Unit	Material	2009 Bare Costs Labor	2009 Bare Costs Equipment	Total	Total Incl O&P
4360	8" stud	G	1 Carp	60	.133	Ea.	6.65	3.73		10.38	13.60
4430	12" deep joist, with 3-5/8" stud	G		75	.107		6.20	2.98		9.18	11.90
4440	4" stud	G		70	.114		6.30	3.19		9.49	12.35
4450	6" stud	G		65	.123		6.85	3.44		10.29	13.35
4460	8" stud	G		60	.133		7.35	3.73		11.08	14.40
5330	For 10" deep joist, with 12 ga x 3-5/8" stud	G		65	.123		6.15	3.44		9.59	12.60
5340	4" stud	G		60	.133		6.35	3.73		10.08	13.25
5350	6" stud	G		55	.145		7	4.07		11.07	14.55
5360	8" stud	G		50	.160		7.70	4.47		12.17	16
5430	12" deep joist, with 3-5/8" stud	G		65	.123		6.85	3.44		10.29	13.30
5440	4" stud	G		60	.133		6.70	3.73		10.43	13.65
5450	6" stud	G		55	.145		7.65	4.07		11.72	15.25
5460	8" stud	G		50	.160		8.80	4.47		13.27	17.20

05 42 23 – Cold-Formed Metal Roof Joist Framing

05 42 23.05 Framing, Bracing

			Crew	Daily Output	Labor-Hours	Unit	Material	2009 Bare Costs Labor	2009 Bare Costs Equipment	Total	Total Incl O&P
0010	**FRAMING, BRACING**										
0015	Made from recycled materials										
0020	Continuous bracing, per row										
0100	16 ga x 1-1/2" channel thru rafters/trusses @ 16" O.C.	G	1 Carp	4.50	1.778	C.L.F.	65	49.50		114.50	156
0120	24" O.C.	G		6	1.333		65	37.50		102.50	135
0300	2" x 2" angle x 18 ga, rafters/trusses @ 16" O.C.	G		6	1.333		94.50	37.50		132	167
0320	24" O.C.	G		8	1		94.50	28		122.50	152
0400	16 ga, rafters/trusses @ 16" O.C.	G		4.50	1.778		120	49.50		169.50	216
0420	24" O.C.	G		6.50	1.231		120	34.50		154.50	190

05 42 23.10 Framing, Bridging

			Crew	Daily Output	Labor-Hours	Unit	Material	2009 Bare Costs Labor	2009 Bare Costs Equipment	Total	Total Incl O&P
0010	**FRAMING, BRIDGING**										
0015	Made from recycled materials										
0020	Solid, between rafters w/ 1-1/4" leg track, per rafter bay										
1200	Rafters 16" O.C., 18 ga x 4" deep	G	1 Carp	60	.133	Ea.	1.90	3.73		5.63	8.40
1210	6" deep	G		57	.140		2.48	3.92		6.40	9.35
1220	8" deep	G		55	.145		3.14	4.07		7.21	10.30
1230	10" deep	G		52	.154		3.88	4.30		8.18	11.50
1240	12" deep	G		50	.160		4.49	4.47		8.96	12.50
2200	24" O.C., 18 ga x 4" deep	G		60	.133		2.75	3.73		6.48	9.30
2210	6" deep	G		57	.140		3.58	3.92		7.50	10.60
2220	8" deep	G		55	.145		4.54	4.07		8.61	11.85
2230	10" deep	G		52	.154		5.60	4.30		9.90	13.40
2240	12" deep	G		50	.160		6.50	4.47		10.97	14.70

05 42 23.50 Framing, Parapets

			Crew	Daily Output	Labor-Hours	Unit	Material	2009 Bare Costs Labor	2009 Bare Costs Equipment	Total	Total Incl O&P
0010	**FRAMING, PARAPETS**										
0015	Made from recycled materials										
0100	3' high installed on 1st story, 18 ga x 4" wide studs, 12" O.C.	G	2 Carp	100	.160	L.F.	7	4.47		11.47	15.25
0110	16" O.C.	G		150	.107		5.95	2.98		8.93	11.60
0120	24" O.C.	G		200	.080		4.92	2.24		7.16	9.20
0200	6" wide studs, 12" O.C.	G		100	.160		9	4.47		13.47	17.45
0210	16" O.C.	G		150	.107		7.65	2.98		10.63	13.45
0220	24" O.C.	G		200	.080		6.35	2.24		8.59	10.75
1100	Installed on 2nd story, 18 ga x 4" wide studs, 12" O.C.	G		95	.168		7	4.71		11.71	15.65
1110	16" O.C.	G		145	.110		5.95	3.08		9.03	11.75
1120	24" O.C.	G		190	.084		4.92	2.35		7.27	9.40
1200	6" wide studs, 12" O.C.	G		95	.168		9	4.71		13.71	17.85
1210	16" O.C.	G		145	.110		7.65	3.08		10.73	13.60

05 42 Cold-Formed Metal Joist Framing

05 42 23 – Cold-Formed Metal Roof Joist Framing

05 42 23.50 Framing, Parapets

			Crew	Daily Output	Labor-Hours	Unit	Material	2009 Bare Costs Labor	2009 Bare Costs Equipment	Total	Total Incl O&P
1220	24" O.C.	G	2 Carp	190	.084	L.F.	6.35	2.35		8.70	10.95
2100	Installed on gable, 18 ga x 4" wide studs, 12" O.C.	G		85	.188		7	5.25		12.25	16.60
2110	16" O.C.	G		130	.123		5.95	3.44		9.39	12.35
2120	24" O.C.	G		170	.094		4.92	2.63		7.55	9.85
2200	6" wide studs, 12" O.C.	G		85	.188		9	5.25		14.25	18.80
2210	16" O.C.	G		130	.123		7.65	3.44		11.09	14.20
2220	24" O.C.	G		170	.094		6.35	2.63		8.98	11.40

05 42 23.60 Framing, Roof Rafters

			Crew	Daily Output	Labor-Hours	Unit	Material	2009 Bare Costs Labor	2009 Bare Costs Equipment	Total	Total Incl O&P
0010	**FRAMING, ROOF RAFTERS**										
0015	Made from recycled materials										
0100	Boxed ridge beam, double, 18 ga x 6" deep	G	2 Carp	160	.100	L.F.	6.70	2.80		9.50	12.15
0110	8" deep	G		150	.107		7.45	2.98		10.43	13.25
0120	10" deep	G		140	.114		9	3.19		12.19	15.35
0130	12" deep	G		130	.123		9.90	3.44		13.34	16.70
0200	16 ga x 6" deep	G		150	.107		7.60	2.98		10.58	13.40
0210	8" deep	G		140	.114		8.55	3.19		11.74	14.80
0220	10" deep	G		130	.123		10.30	3.44		13.74	17.10
0230	12" deep	G		120	.133		11.20	3.73		14.93	18.60
1100	Rafters, 2" flange, material only, 18 ga x 6" deep	G					2.08			2.08	2.29
1110	8" deep	G					2.47			2.47	2.71
1120	10" deep	G					2.90			2.90	3.19
1130	12" deep	G					3.37			3.37	3.71
1200	16 ga x 6" deep	G					2.54			2.54	2.80
1210	8" deep	G					3.05			3.05	3.35
1220	10" deep	G					3.56			3.56	3.92
1230	12" deep	G					4.03			4.03	4.44
2100	Installation only, ordinary rafter to 4:12 pitch, 18 ga x 6" deep		2 Carp	35	.457	Ea.		12.80		12.80	21.50
2110	8" deep			30	.533			14.90		14.90	25
2120	10" deep			25	.640			17.90		17.90	30
2130	12" deep			20	.800			22.50		22.50	38
2200	16 ga x 6" deep			30	.533			14.90		14.90	25
2210	8" deep			25	.640			17.90		17.90	30
2220	10" deep			20	.800			22.50		22.50	38
2230	12" deep			15	1.067			30		30	50.50
8100	Add to labor, ordinary rafters on steep roofs							25%			
8110	Dormers & complex roofs							50%			
8200	Hip & valley rafters to 4:12 pitch							25%			
8210	Steep roofs							50%			
8220	Dormers & complex roofs							75%			
8300	Hip & valley jack rafters to 4:12 pitch							50%			
8310	Steep roofs							75%			
8320	Dormers & complex roofs							100%			

05 42 23.70 Framing, Soffits and Canopies

			Crew	Daily Output	Labor-Hours	Unit	Material	2009 Bare Costs Labor	2009 Bare Costs Equipment	Total	Total Incl O&P
0010	**FRAMING, SOFFITS & CANOPIES**										
0015	Made from recycled materials										
0130	Continuous ledger track @ wall, studs @ 16" O.C., 18 ga x 4" wide	G	2 Carp	535	.030	L.F.	1.27	.84		2.11	2.80
0140	6" wide	G		500	.032		1.65	.89		2.54	3.33
0150	8" wide	G		465	.034		2.09	.96		3.05	3.93
0160	10" wide	G		430	.037		2.59	1.04		3.63	4.60
0230	Studs @ 24" O.C., 18 ga x 4" wide	G		800	.020		1.21	.56		1.77	2.28
0240	6" wide	G		750	.021		1.58	.60		2.18	2.74
0250	8" wide	G		700	.023		2	.64		2.64	3.27

05 42 Cold-Formed Metal Joist Framing

05 42 23 – Cold-Formed Metal Roof Joist Framing

05 42 23.70 Framing, Soffits and Canopies		Crew	Daily Output	Labor-Hours	Unit	Material	2009 Bare Costs Labor	Equipment	Total	Total Incl O&P
0260	10" wide	G	2 Carp	650	.025	L.F.	2.47	.69		3.16	3.87
1000	Horizontal soffit and canopy members, material only										
1030	1-5/8" flange studs, 18 ga x 4" deep	G				L.F.	1.68			1.68	1.85
1040	6" deep	G					2.12			2.12	2.34
1050	8" deep	G					2.57			2.57	2.82
1140	2" flange joists, 18 ga x 6" deep	G					2.38			2.38	2.61
1150	8" deep	G					2.82			2.82	3.10
1160	10" deep	G					3.31			3.31	3.64
4030	Installation only, 18 ga, 1-5/8" flange x 4" deep		2 Carp	130	.123	Ea.		3.44		3.44	5.80
4040	6" deep			110	.145			4.07		4.07	6.85
4050	8" deep			90	.178			4.97		4.97	8.40
4140	2" flange, 18 ga x 6" deep			110	.145			4.07		4.07	6.85
4150	8" deep			90	.178			4.97		4.97	8.40
4160	10" deep			80	.200			5.60		5.60	9.45
6010	Clips to attach facia to rafter tails, 2" x 2" x 18 ga angle	G	1 Carp	120	.067		1.12	1.86		2.98	4.38
6020	16 ga angle	G	"	100	.080		1.42	2.24		3.66	5.35

05 44 Cold-Formed Metal Trusses

05 44 13 – Cold-Formed Metal Roof Trusses

05 44 13.60 Framing, Roof Trusses		Crew	Daily Output	Labor-Hours	Unit	Material	2009 Bare Costs Labor	Equipment	Total	Total Incl O&P
0010	**FRAMING, ROOF TRUSSES**										
0015	Made from recycled materials										
0020	Fabrication of trusses on ground, Fink (W) or King Post, to 4:12 pitch										
0120	18 ga x 4" chords, 16' span	G	2 Carp	12	1.333	Ea.	78.50	37.50		116	149
0130	20' span	G		11	1.455		98	40.50		138.50	177
0140	24' span	G		11	1.455		118	40.50		158.50	198
0150	28' span	G		10	1.600		137	44.50		181.50	227
0160	32' span	G		10	1.600		157	44.50		201.50	248
0250	6" chords, 28' span	G		9	1.778		173	49.50		222.50	275
0260	32' span	G		9	1.778		198	49.50		247.50	300
0270	36' span	G		8	2		223	56		279	340
0280	40' span	G		8	2		248	56		304	370
1120	5:12 to 8:12 pitch, 18 ga x 4" chords, 16' span	G		10	1.600		89.50	44.50		134	174
1130	20' span	G		9	1.778		112	49.50		161.50	207
1140	24' span	G		9	1.778		134	49.50		183.50	232
1150	28' span	G		8	2		157	56		213	267
1160	32' span	G		8	2		179	56		235	292
1250	6" chords, 28' span	G		7	2.286		198	64		262	325
1260	32' span	G		7	2.286		227	64		291	355
1270	36' span	G		6	2.667		255	74.50		329.50	405
1280	40' span	G		6	2.667		283	74.50		357.50	435
2120	9:12 to 12:12 pitch, 18 ga x 4" chords, 16' span	G		8	2		112	56		168	218
2130	20' span	G		7	2.286		140	64		204	262
2140	24' span	G		7	2.286		168	64		232	293
2150	28' span	G		6	2.667		196	74.50		270.50	340
2160	32' span	G		6	2.667		224	74.50		298.50	370
2250	6" chords, 28' span	G		5	3.200		248	89.50		337.50	425
2260	32' span	G		5	3.200		283	89.50		372.50	460
2270	36' span	G		4	4		320	112		432	540
2280	40' span	G		4	4		355	112		467	580
5120	Erection only of roof trusses, to 4:12 pitch, 16' span		F-6	48	.833			21	16	37	53

05 44 Cold-Formed Metal Trusses

05 44 13 – Cold-Formed Metal Roof Trusses

05 44 13.60 Framing, Roof Trusses		Crew	Daily Output	Labor-Hours	Unit	Material	2009 Bare Costs Labor	Equipment	Total	Total Incl O&P
5130	20' span	F-6	46	.870	Ea.		22	16.70	38.70	55.50
5140	24' span		44	.909			23	17.45	40.45	57.50
5150	28' span		42	.952			24	18.30	42.30	60.50
5160	32' span		40	1			25.50	19.20	44.70	63.50
5170	36' span		38	1.053			26.50	20	46.50	67
5180	40' span		36	1.111			28	21.50	49.50	70.50
5220	5:12 to 8:12 pitch, 16' span		42	.952			24	18.30	42.30	60.50
5230	20' span		40	1			25.50	19.20	44.70	63.50
5240	24' span		38	1.053			26.50	20	46.50	67
5250	28' span		36	1.111			28	21.50	49.50	70.50
5260	32' span		34	1.176			29.50	22.50	52	75
5270	36' span		32	1.250			31.50	24	55.50	79.50
5280	40' span		30	1.333			33.50	25.50	59	84.50
5320	9:12 to 12:12 pitch, 16' span		36	1.111			28	21.50	49.50	70.50
5330	20' span		34	1.176			29.50	22.50	52	75
5340	24' span		32	1.250			31.50	24	55.50	79.50
5350	28' span		30	1.333			33.50	25.50	59	84.50
5360	32' span		28	1.429			36	27.50	63.50	90.50
5370	36' span		26	1.538			39	29.50	68.50	98
5380	40' span		24	1.667			42	32	74	106

05 51 Metal Stairs

05 51 13 – Metal Pan Stairs

05 51 13.50 Pan Stairs			Crew	Daily Output	Labor-Hours	Unit	Material	2009 Bare Costs Labor	Equipment	Total	Total Incl O&P
0010	**PAN STAIRS**, shop fabricated, steel stringers										
0015	Made from recycled materials										
1700	Pre-erected, steel pan tread, 3'-6" wide, 2 line pipe rail	G	E-2	87	.552	Riser	620	16.65	20	656.65	735
1800	With flat bar picket rail	G	"	87	.552	"	695	16.65	20	731.65	815

05 51 23 – Metal Fire Escapes

05 51 23.50 Fire Escape Stairs			Crew	Daily Output	Labor-Hours	Unit	Material	2009 Bare Costs Labor	Equipment	Total	Total Incl O&P
0010	**FIRE ESCAPE STAIRS**, shop fabricated										
0020	One story, disappearing, stainless steel	G	2 Sswk	20	.800	V.L.F.	223	24		247	291
0100	Portable ladder					Ea.	93			93	102
1100	Fire escape, galvanized steel, 8'-0" to 10'-4" ceiling	G	2 Carp	1	16		1,525	445		1,970	2,425
1110	10'-6" to 13'-6" ceiling	G	"	1	16		2,000	445		2,445	2,950

05 52 Metal Railings

05 52 13 – Pipe and Tube Railings

05 52 13.50 Railings, Pipe			Crew	Daily Output	Labor-Hours	Unit	Material	2009 Bare Costs Labor	Equipment	Total	Total Incl O&P
0010	**RAILINGS, PIPE**, shop fab'd, 3'-6" high, posts @ 5' O.C.										
0015	Made from recycled materials										
0020	Aluminum, 2 rail, satin finish, 1-1/4" diameter	G	E-4	160	.200	L.F.	34	6.10	.84	40.94	50
0030	Clear anodized	G		160	.200		42	6.10	.84	48.94	59
0040	Dark anodized	G		160	.200		47.50	6.10	.84	54.44	65
0080	1-1/2" diameter, satin finish	G		160	.200		41	6.10	.84	47.94	57.50
0090	Clear anodized	G		160	.200		45.50	6.10	.84	52.44	62.50
0100	Dark anodized	G		160	.200		50.50	6.10	.84	57.44	68
0140	Aluminum, 3 rail, 1-1/4" diam., satin finish	G		137	.234		52	7.10	.98	60.08	72
0150	Clear anodized	G		137	.234		65.50	7.10	.98	73.58	86.50

05 52 Metal Railings

05 52 13 – Pipe and Tube Railings

05 52 13.50 Railings, Pipe

			Crew	Daily Output	Labor-Hours	Unit	Material	2009 Bare Costs Labor	Equipment	Total	Total Incl O&P
0160	Dark anodized	G	E-4	137	.234	L.F.	72.50	7.10	.98	80.58	94
0200	1-1/2" diameter, satin finish	G		137	.234		62.50	7.10	.98	70.58	83.50
0210	Clear anodized	G		137	.234		71	7.10	.98	79.08	92.50
0220	Dark anodized	G		137	.234		78	7.10	.98	86.08	100
0500	Steel, 2 rail, on stairs, primed, 1-1/4" diameter	G		160	.200		28	6.10	.84	34.94	43.50
0520	1-1/2" diameter	G		160	.200		30.50	6.10	.84	37.44	46.50
0540	Galvanized, 1-1/4" diameter	G		160	.200		39	6.10	.84	45.94	55
0560	1-1/2" diameter	G		160	.200		43.50	6.10	.84	50.44	60.50
0580	Steel, 3 rail, primed, 1-1/4" diameter	G		137	.234		41.50	7.10	.98	49.58	60.50
0600	1-1/2" diameter	G		137	.234		44.50	7.10	.98	52.58	63
0620	Galvanized, 1-1/4" diameter	G		137	.234		58.50	7.10	.98	66.58	79
0640	1-1/2" diameter	G		137	.234		69	7.10	.98	77.08	90.50
0700	Stainless steel, 2 rail, 1-1/4" diam. #4 finish	G		137	.234		101	7.10	.98	109.08	126
0720	High polish	G		137	.234		162	7.10	.98	170.08	194
0740	Mirror polish	G		137	.234		203	7.10	.98	211.08	239
0760	Stainless steel, 3 rail, 1-1/2" diam., #4 finish	G		120	.267		152	8.10	1.12	161.22	184
0770	High polish	G		120	.267		251	8.10	1.12	260.22	293
0780	Mirror finish	G		120	.267		305	8.10	1.12	314.22	350
0900	Wall rail, alum. pipe, 1-1/4" diam., satin finish	G		213	.150		19.50	4.57	.63	24.70	31
0905	Clear anodized	G		213	.150		24	4.57	.63	29.20	35.50
0910	Dark anodized	G		213	.150		29	4.57	.63	34.20	41
0915	1-1/2" diameter, satin finish	G		213	.150		21.50	4.57	.63	26.70	33
0920	Clear anodized	G		213	.150		27	4.57	.63	32.20	39.50
0925	Dark anodized	G		213	.150		33.50	4.57	.63	38.70	46.50
0930	Steel pipe, 1-1/4" diameter, primed	G		213	.150		16.95	4.57	.63	22.15	28
0935	Galvanized	G		213	.150		24.50	4.57	.63	29.70	36.50
0940	1-1/2" diameter	G		176	.182		17.45	5.55	.76	23.76	30.50
0945	Galvanized	G		213	.150		24.50	4.57	.63	29.70	36.50
0955	Stainless steel pipe, 1-1/2" diam., #4 finish	G		107	.299		80.50	9.10	1.25	90.85	107
0960	High polish	G		107	.299		164	9.10	1.25	174.35	199
0965	Mirror polish	G		107	.299		193	9.10	1.25	203.35	231
2000	Aluminum pipe & picket railing, double top rail, pickets @ 4-1/2" OC											
2010	36" high, straight & level	G	2 Sswk	80	.200	L.F.	81	6		87	101
2020	Curved & level	G		60	.267		111	8		119	137
2030	Straight & sloped	G		40	.400		93	12		105	125

05 58 Formed Metal Fabrications

05 58 25 – Formed Lamp Posts

05 58 25.40 Lamp Posts

			Crew	Daily Output	Labor-Hours	Unit	Material	Labor	Equipment	Total	Total Incl O&P
0010	**LAMP POSTS**										
0020	Aluminum, 7' high, stock units, post only	G	1 Carp	16	.500	Ea.	28.50	14		42.50	55
0100	Mild steel, plain	G	"	16	.500	"	23	14		37	49

05 71 13 – Fabricated Metal Spiral Stairs

05 71 13.50 Spiral Stairs		Crew	Daily Output	Labor-Hours	Unit	Material	2009 Bare Costs Labor	Equipment	Total	Total Incl O&P
0010	**SPIRAL STAIRS**, shop fabricated									
1810	Spiral aluminum, 5'-0" diameter, stock units	G	E-4	45	.711	Riser	680	21.50	2.98	704.48	795
1820	Custom units	G		45	.711		1,275	21.50	2.98	1,299.48	1,475
1900	Spiral, cast iron, 4'-0" diameter, ornamental, minimum	G		45	.711		605	21.50	2.98	629.48	710
1920	Maximum	G		25	1.280		825	39	5.35	869.35	990

Division 6
Wood, Plastics &
Composites

06 05 05.10 Selective Demolition Wood Framing	Crew	Daily Output	Labor-Hours	Unit	Material	2009 Bare Costs Labor	2009 Bare Costs Equipment	Total	Total Incl O&P
0010 **SELECTIVE DEMOLITION WOOD FRAMING** R024119-10									
0100 Timber connector, nailed, small	1 Clab	96	.083	Ea.		1.71		1.71	2.90
0110 Medium		60	.133			2.74		2.74	4.63
0120 Large		48	.167			3.43		3.43	5.80
0130 Bolted, small		48	.167			3.43		3.43	5.80
0140 Medium		32	.250			5.15		5.15	8.70
0150 Large		24	.333			6.85		6.85	11.60
2958 Beams, 2" x 6"	2 Clab	1100	.015	L.F.		.30		.30	.51
2960 2" x 8"		825	.019			.40		.40	.67
2965 2" x 10"		665	.024			.49		.49	.84
2970 2" x 12"		550	.029			.60		.60	1.01
2972 2" x 14"		470	.034			.70		.70	1.18
2975 4" x 8"	B-1	413	.058			1.23		1.23	2.09
2980 4" x 10"		330	.073			1.54		1.54	2.61
2985 4" x 12"		275	.087			1.85		1.85	3.13
3000 6" x 8"		275	.087			1.85		1.85	3.13
3040 6" x 10"		220	.109			2.31		2.31	3.91
3080 6" x 12"		185	.130			2.75		2.75	4.65
3120 8" x 12"		140	.171			3.64		3.64	6.15
3160 10" x 12"		110	.218			4.63		4.63	7.85
3162 Alternate pricing method		1.10	21.818	M.B.F.		465		465	785
3170 Blocking, in 16" OC wall framing, 2" x 4"	1 Clab	600	.013	L.F.		.27		.27	.46
3172 2" x 6"		400	.020			.41		.41	.70
3174 In 24" OC wall framing, 2" x 4"		600	.013			.27		.27	.46
3176 2" x 6"		400	.020			.41		.41	.70
3178 Alt method, wood blocking removal from wood framing		.40	20	M.B.F.		410		410	695
3179 Wood blocking removal from steel framing		.36	22.222	"		455		455	770
3180 Bracing, let in, 1" x 3", studs 16" OC		1050	.008	L.F.		.16		.16	.26
3181 Studs 24" OC		1080	.007			.15		.15	.26
3182 1" x 4", studs 16" OC		1050	.008			.16		.16	.26
3183 Studs 24" OC		1080	.007			.15		.15	.26
3184 1" x 6", studs 16" OC		1050	.008			.16		.16	.26
3185 Studs 24" OC		1080	.007			.15		.15	.26
3186 2" x 3", studs 16" OC		800	.010			.21		.21	.35
3187 Studs 24" OC		830	.010			.20		.20	.34
3188 2" x 4", studs 16" OC		800	.010			.21		.21	.35
3189 Studs 24" OC		830	.010			.20		.20	.34
3190 2" x 6", studs 16" OC		800	.010			.21		.21	.35
3191 Studs 24" OC		830	.010			.20		.20	.34
3192 2" x 8", studs 16" OC		800	.010			.21		.21	.35
3193 Studs 24" OC		830	.010			.20		.20	.34
3194 "T" shaped metal bracing, studs at 16" OC		1060	.008			.16		.16	.26
3195 Studs at 24" OC		1200	.007			.14		.14	.23
3196 Metal straps, studs at 16" OC		1200	.007			.14		.14	.23
3197 Studs at 24" OC		1240	.006			.13		.13	.22
3200 Columns, round, 8' to 14' tall		40	.200	Ea.		4.11		4.11	6.95
3202 Dimensional lumber sizes	2 Clab	1.10	14.545	M.B.F.		299		299	505
3250 Blocking, between joists	1 Clab	320	.025	Ea.		.51		.51	.87
3252 Bridging, metal strap, between joists		320	.025	Pr.		.51		.51	.87
3254 Wood, between joists		320	.025	"		.51		.51	.87
3260 Door buck, studs, header & access., 8' high 2" x 4" wall, 3' wide		32	.250	Ea.		5.15		5.15	8.70
3261 4' wide		32	.250			5.15		5.15	8.70
3262 5' wide		32	.250			5.15		5.15	8.70

06 05 05 – Selective Wood and Plastics Demolition

06 05 05.10 Selective Demolition Wood Framing	Crew	Daily Output	Labor-Hours	Unit	Material	2009 Bare Costs Labor	Equipment	Total	Total Incl O&P
3263	6' wide	1 Clab	32	.250	Ea.		5.15		5.15	8.70
3264	8' wide		30	.267			5.50		5.50	9.25
3265	10' wide		30	.267			5.50		5.50	9.25
3266	12' wide		30	.267			5.50		5.50	9.25
3267	2" x 6" wall, 3' wide		32	.250			5.15		5.15	8.70
3268	4' wide		32	.250			5.15		5.15	8.70
3269	5' wide		32	.250			5.15		5.15	8.70
3270	6' wide		32	.250			5.15		5.15	8.70
3271	8' wide		30	.267			5.50		5.50	9.25
3272	10' wide		30	.267			5.50		5.50	9.25
3273	12' wide		30	.267			5.50		5.50	9.25
3274	Window buck, studs, header & access, 8' high 2" x 4" wall, 2' wide		24	.333			6.85		6.85	11.60
3275	3' wide		24	.333			6.85		6.85	11.60
3276	4' wide		24	.333			6.85		6.85	11.60
3277	5' wide		24	.333			6.85		6.85	11.60
3278	6' wide		24	.333			6.85		6.85	11.60
3279	7' wide		24	.333			6.85		6.85	11.60
3280	8' wide		22	.364			7.45		7.45	12.65
3281	10' wide		22	.364			7.45		7.45	12.65
3282	12' wide		22	.364			7.45		7.45	12.65
3283	2" x 6" wall, 2' wide		24	.333			6.85		6.85	11.60
3284	3' wide		24	.333			6.85		6.85	11.60
3285	4' wide		24	.333			6.85		6.85	11.60
3286	5' wide		24	.333			6.85		6.85	11.60
3287	6' wide		24	.333			6.85		6.85	11.60
3288	7' wide		24	.333			6.85		6.85	11.60
3289	8' wide		22	.364			7.45		7.45	12.65
3290	10' wide		22	.364			7.45		7.45	12.65
3291	12' wide		22	.364			7.45		7.45	12.65
3400	Fascia boards, 1" x 6"		500	.016	L.F.		.33		.33	.56
3440	1" x 8"		450	.018			.37		.37	.62
3480	1" x 10"		400	.020			.41		.41	.70
3490	2" x 6"		450	.018			.37		.37	.62
3500	2" x 8"		400	.020			.41		.41	.70
3510	2" x 10"		350	.023			.47		.47	.79
3610	Furring, on wood walls or ceiling		4000	.002	S.F.		.04		.04	.07
3620	On masonry or concrete walls or ceiling		1200	.007	"		.14		.14	.23
3800	Headers over openings, 2 @ 2" x 6"		110	.073	L.F.		1.49		1.49	2.53
3840	2 @ 2" x 8"		100	.080			1.64		1.64	2.78
3880	2 @ 2" x 10"		90	.089			1.83		1.83	3.09
3885	Alternate pricing method		.26	30.651	M.B.F.		630		630	1,075
3920	Joists, 1" x 4"		1250	.006	L.F.		.13		.13	.22
3930	1" x 6"		1135	.007			.14		.14	.25
3950	1" x 10"		895	.009			.18		.18	.31
3960	1" x 12"		765	.010			.22		.22	.36
4200	2" x 4"	2 Clab	1000	.016			.33		.33	.56
4230	2" x 6"		970	.016			.34		.34	.57
4240	2" x 8"		940	.017			.35		.35	.59
4250	2" x 10"		910	.018			.36		.36	.61
4280	2" x 12"		880	.018			.37		.37	.63
4281	2" x 14"		850	.019			.39		.39	.65
4282	Composite joists, 9-1/2"		960	.017			.34		.34	.58
4283	11-7/8"		930	.017			.35		.35	.60

06 05 05.10 Selective Demolition Wood Framing		Crew	Daily Output	Labor-Hours	Unit	Material	2009 Bare Costs Labor	2009 Bare Costs Equipment	Total	Total Incl O&P
4284	14"	2 Clab	897	.018	L.F.		.37		.37	.62
4285	16"		865	.019	↓		.38		.38	.64
4290	Wood joists, alternate pricing method		1.50	10.667	M.B.F.		219		219	370
4500	Open web joist, 12" deep		500	.032	L.F.		.66		.66	1.11
4505	14" deep		475	.034			.69		.69	1.17
4510	16" deep		450	.036			.73		.73	1.24
4520	18" deep		425	.038			.77		.77	1.31
4530	24" deep		400	.040			.82		.82	1.39
4550	Ledger strips, 1" x 2"	1 Clab	1200	.007			.14		.14	.23
4560	1" x 3"		1200	.007			.14		.14	.23
4570	1" x 4"		1200	.007			.14		.14	.23
4580	2" x 2"		1100	.007			.15		.15	.25
4590	2" x 4"		1000	.008			.16		.16	.28
4600	2" x 6"		1000	.008			.16		.16	.28
4601	2" x 8" or 2" x 10"		800	.010			.21		.21	.35
4602	4" x 6"		600	.013			.27		.27	.46
4604	4" x 8"		450	.018			.37		.37	.62
5400	Posts, 4" x 4"	2 Clab	800	.020			.41		.41	.70
5405	4" x 6"		550	.029			.60		.60	1.01
5410	4" x 8"		440	.036			.75		.75	1.26
5425	4" x 10"		390	.041			.84		.84	1.43
5430	4" x 12"		350	.046			.94		.94	1.59
5440	6" x 6"		400	.040			.82		.82	1.39
5445	6" x 8"		350	.046			.94		.94	1.59
5450	6" x 10"		320	.050			1.03		1.03	1.74
5455	6" x 12"		290	.055			1.13		1.13	1.92
5480	8" x 8"		300	.053			1.10		1.10	1.85
5500	10" x 10"		240	.067	↓		1.37		1.37	2.32
5660	Tongue and groove floor planks		2	8	M.B.F.		164		164	278
5750	Rafters, ordinary, 16" OC, 2" x 4"		880	.018	S.F.		.37		.37	.63
5755	2" x 6"		840	.019			.39		.39	.66
5760	2" x 8"		820	.020			.40		.40	.68
5770	2" x 10"		820	.020			.40		.40	.68
5780	2" x 12"		810	.020			.41		.41	.69
5785	24" OC, 2" x 4"		1170	.014			.28		.28	.48
5786	2" x 6"		1117	.014			.29		.29	.50
5787	2" x 8"		1091	.015			.30		.30	.51
5788	2" x 10"		1091	.015			.30		.30	.51
5789	2" x 12"		1077	.015	↓		.31		.31	.52
5795	Rafters, ordinary, 2" x 4" (alternate method)		862	.019	L.F.		.38		.38	.65
5800	2" x 6" (alternate method)		850	.019			.39		.39	.65
5840	2" x 8" (alternate method)		837	.019			.39		.39	.66
5855	2" x 10" (alternate method)		825	.019			.40		.40	.67
5865	2" x 12" (alternate method)		812	.020			.40		.40	.68
5870	Sill plate, 2" x 4"	1 Clab	1170	.007			.14		.14	.24
5871	2" x 6"		780	.010			.21		.21	.36
5872	2" x 8"		586	.014	↓		.28		.28	.47
5873	Alternate pricing method		.78	10.256	M.B.F.		211		211	355
5885	Ridge board, 1" x 4"	2 Clab	900	.018	L.F.		.37		.37	.62
5886	1" x 6"		875	.018			.38		.38	.64
5887	1" x 8"		850	.019			.39		.39	.65
5888	1" x 10"		825	.019			.40		.40	.67
5889	1" x 12"		800	.020	↓		.41		.41	.70

06 05 05 – Selective Wood and Plastics Demolition

	06 05 05.10 Selective Demolition Wood Framing	Crew	Daily Output	Labor-Hours	Unit	Material	2009 Bare Costs Labor	Equipment	Total	Total Incl O&P
5890	2" x 4"	2 Clab	900	.018	L.F.		.37		.37	.62
5892	2" x 6"		875	.018			.38		.38	.64
5894	2" x 8"		850	.019			.39		.39	.65
5896	2" x 10"		825	.019			.40		.40	.67
5898	2" x 12"		800	.020			.41		.41	.70
6050	Rafter tie, 1" x 4"		1250	.013			.26		.26	.44
6052	1" x 6"		1135	.014			.29		.29	.49
6054	2" x 4"		1000	.016			.33		.33	.56
6056	2" x 6"		970	.016			.34		.34	.57
6070	Sleepers, on concrete, 1" x 2"	1 Clab	4700	.002			.03		.03	.06
6075	1" x 3"		4000	.002			.04		.04	.07
6080	2" x 4"		3000	.003			.05		.05	.09
6085	2" x 6"		2600	.003			.06		.06	.11
6086	Sheathing from roof, 5/16"	2 Clab	1600	.010	S.F.		.21		.21	.35
6088	3/8"		1525	.010			.22		.22	.36
6090	1/2"		1400	.011			.23		.23	.40
6092	5/8"		1300	.012			.25		.25	.43
6094	3/4"		1200	.013			.27		.27	.46
6096	Board sheathing from roof		1400	.011			.23		.23	.40
6100	Sheathing, from walls, 1/4"		1200	.013			.27		.27	.46
6110	5/16"		1175	.014			.28		.28	.47
6120	3/8"		1150	.014			.29		.29	.48
6130	1/2"		1125	.014			.29		.29	.49
6140	5/8"		1100	.015			.30		.30	.51
6150	3/4"		1075	.015			.31		.31	.52
6152	Board sheathing from walls		1500	.011			.22		.22	.37
6158	Subfloor, with boards		1050	.015			.31		.31	.53
6160	Plywood, 1/2" thick		768	.021			.43		.43	.72
6162	5/8" thick		760	.021			.43		.43	.73
6164	3/4" thick		750	.021			.44		.44	.74
6165	1-1/8" thick		720	.022			.46		.46	.77
6166	Underlayment, particle board, 3/8" thick	1 Clab	780	.010			.21		.21	.36
6168	1/2" thick		768	.010			.21		.21	.36
6170	5/8" thick		760	.011			.22		.22	.37
6172	3/4" thick		750	.011			.22		.22	.37
6200	Stairs and stringers, minimum	2 Clab	40	.400	Riser		8.20		8.20	13.90
6240	Maximum	"	26	.615	"		12.65		12.65	21.50
6300	Components, tread	1 Clab	110	.073	Ea.		1.49		1.49	2.53
6320	Riser		80	.100	"		2.06		2.06	3.48
6390	Stringer, 2" x 10"		260	.031	L.F.		.63		.63	1.07
6400	2" x 12"		260	.031			.63		.63	1.07
6410	3" x 10"		250	.032			.66		.66	1.11
6420	3" x 12"		250	.032			.66		.66	1.11
6590	Wood studs, 2" x 3"	2 Clab	3076	.005			.11		.11	.18
6600	2" x 4"		2000	.008			.16		.16	.28
6640	2" x 6"		1600	.010			.21		.21	.35
6720	Wall framing, including studs, plates and blocking, 2" x 4"	1 Clab	600	.013	S.F.		.27		.27	.46
6740	2" x 6"		480	.017	"		.34		.34	.58
6750	Headers, 2" x 4"		1125	.007	L.F.		.15		.15	.25
6755	2" x 6"		1125	.007			.15		.15	.25
6760	2" x 8"		1050	.008			.16		.16	.26
6765	2" x 10"		1050	.008			.16		.16	.26
6770	2" x 12"		1000	.008			.16		.16	.28

06 05 05 – Selective Wood and Plastics Demolition

06 05 05.10 Selective Demolition Wood Framing

		Crew	Daily Output	Labor-Hours	Unit	Material	2009 Bare Costs Labor	Equipment	Total	Total Incl O&P
6780	4" x 10"	1 Clab	525	.015	L.F.		.31		.31	.53
6785	4" x 12"		500	.016			.33		.33	.56
6790	6" x 8"		560	.014			.29		.29	.50
6795	6" x 10"		525	.015			.31		.31	.53
6797	6" x 12"		500	.016			.33		.33	.56
7000	Trusses									
7050	12' span	2 Clab	74	.216	Ea.		4.44		4.44	7.50
7150	24' span	F-3	66	.606			15.40	11.65	27.05	39
7200	26' span		64	.625			15.85	12	27.85	39.50
7250	28' span		62	.645			16.40	12.40	28.80	41
7300	30' span		58	.690			17.50	13.25	30.75	44
7350	32' span		56	.714			18.15	13.75	31.90	45.50
7400	34' span		54	.741			18.80	14.25	33.05	47
7450	36' span		52	.769			19.55	14.80	34.35	49.50
8000	Soffit, T & G wood	1 Clab	520	.015	S.F.		.32		.32	.53
8010	Hardboard, vinyl or aluminum	"	640	.013			.26		.26	.43
8030	Plywood	2 Carp	315	.051			1.42		1.42	2.40

06 05 05.20 Selective Demolition Millwork and Trim

		Crew	Daily Output	Labor-Hours	Unit	Material	2009 Bare Costs Labor	Equipment	Total	Total Incl O&P
0010	**SELECTIVE DEMOLITION MILLWORK AND TRIM** R024119-10									
1000	Cabinets, wood, base cabinets, per L.F.	2 Clab	80	.200	L.F.		4.11		4.11	6.95
1020	Wall cabinets, per L.F.		80	.200			4.11		4.11	6.95
1100	Steel, painted, base cabinets		60	.267			5.50		5.50	9.25
1500	Counter top, minimum		200	.080			1.64		1.64	2.78
1510	Maximum		120	.133			2.74		2.74	4.63
2000	Paneling, 4' x 8' sheets		2000	.008	S.F.		.16		.16	.28
2100	Boards, 1" x 4"		700	.023			.47		.47	.79
2120	1" x 6"		750	.021			.44		.44	.74
2140	1" x 8"		800	.020			.41		.41	.70
3000	Trim, baseboard, to 6" wide		1200	.013	L.F.		.27		.27	.46
3040	Greater than 6" and up to 12" wide		1000	.016			.33		.33	.56
3100	Ceiling trim		1000	.016			.33		.33	.56
3120	Chair rail		1200	.013			.27		.27	.46
3140	Railings with balusters		240	.067			1.37		1.37	2.32
3160	Wainscoting		700	.023	S.F.		.47		.47	.79
4000	Curtain rod	1 Clab	80	.100	L.F.		2.06		2.06	3.48
9000	Minimum labor/equipment charge	"	4	2	Job		41		41	69.50

06 05 23 – Wood, Plastic, and Composite Fastenings

06 05 23.10 Nails

		Crew	Daily Output	Labor-Hours	Unit	Material	2009 Bare Costs Labor	Equipment	Total	Total Incl O&P
0010	**NAILS**, material only, based upon 50# box purchase									
0020	Copper nails, plain				Lb.	10.20			10.20	11.20
0400	Stainless steel, plain					6.65			6.65	7.30
0500	Box, 3d to 20d, bright					.91			.91	1
0520	Galvanized					1.67			1.67	1.84
0600	Common, 3d to 60d, plain					1.11			1.11	1.22
0700	Galvanized					1.53			1.53	1.68
0800	Aluminum					4.45			4.45	4.90
1000	Annular or spiral thread, 4d to 60d, plain					1.82			1.82	2
1200	Galvanized					1.99			1.99	2.19
1400	Drywall nails, plain					.79			.79	.87
1600	Galvanized					1.59			1.59	1.75
1800	Finish nails, 4d to 10d, plain					1.11			1.11	1.22
2000	Galvanized					1.75			1.75	1.93

06 05 Common Work Results for Wood, Plastics and Composites

06 05 23 – Wood, Plastic, and Composite Fastenings

06 05 23.10 Nails

		Crew	Daily Output	Labor-Hours	Unit	Material	2009 Bare Costs Labor	Equipment	Total	Total Incl O&P
2100	Aluminum				Lb.	4.10			4.10	4.51
2300	Flooring nails, hardened steel, 2d to 10d, plain					2.04			2.04	2.24
2400	Galvanized					3.88			3.88	4.27
2500	Gypsum lath nails, 1-1/8", 13 ga. flathead, blued					2.64			2.64	2.90
2600	Masonry nails, hardened steel, 3/4" to 3" long, plain					1.97			1.97	2.17
2700	Galvanized					2.78			2.78	3.06
2900	Roofing nails, threaded, galvanized					1.38			1.38	1.52
3100	Aluminum					4.80			4.80	5.30
3300	Compressed lead head, threaded, galvanized					2.50			2.50	2.75
3600	Siding nails, plain shank, galvanized					1.50			1.50	1.65
3800	Aluminum					4.11			4.11	4.52
5000	Add to prices above for cement coating					.10			.10	.11
5200	Zinc or tin plating					.13			.13	.14
5500	Vinyl coated sinkers, 8d to 16d					.60			.60	.66

06 05 23.20 Pneumatic Nails

		Crew	Daily Output	Labor-Hours	Unit	Material	2009 Bare Costs Labor	Equipment	Total	Total Incl O&P
0010	**PNEUMATIC NAILS**									
0020	Framing, per carton of 5000, 2"				Ea.	40			40	44
0100	2-3/8"					45			45	49.50
0200	Per carton of 4000, 3"					40			40	44
0300	3-1/4"					40			40	44
0400	Per carton of 5000, 2-3/8", galv.					60			60	66
0500	Per carton of 4000, 3", galv.					65			65	71.50
0600	3-1/4", galv.					82			82	90
0700	Roofing, per carton of 7200, 1"					35.50			35.50	39
0800	1-1/4"					35			35	38.50
0900	1-1/2"					38.50			38.50	42.50
1000	1-3/4"					47.50			47.50	52.50

06 05 23.40 Sheet Metal Screws

		Crew	Daily Output	Labor-Hours	Unit	Material	2009 Bare Costs Labor	Equipment	Total	Total Incl O&P
0010	**SHEET METAL SCREWS**									
0020	Steel, standard, #8 x 3/4", plain				C	3.30			3.30	3.63
0100	Galvanized					3.39			3.39	3.73
0300	#10 x 1", plain					4.25			4.25	4.68
0400	Galvanized					4.43			4.43	4.87
1500	Self-drilling, with washers, (pinch point) #8 x 3/4", plain					8			8	8.80
1600	Galvanized					8.70			8.70	9.55
1800	#10 x 3/4", plain					10			10	11
1900	Galvanized					10.95			10.95	12.05
3000	Stainless steel w/aluminum or neoprene washers, #14 x 1", plain					32			32	35
3100	#14 x 2", plain					41			41	45

06 05 23.50 Wood Screws

		Crew	Daily Output	Labor-Hours	Unit	Material	2009 Bare Costs Labor	Equipment	Total	Total Incl O&P
0010	**WOOD SCREWS**									
0020	#8 x 1" long, steel				C	5.10			5.10	5.60
0100	Brass					11.50			11.50	12.65
0200	#8, 2" long, steel					5.30			5.30	5.85
0300	Brass					19.50			19.50	21.50
0400	#10, 1" long, steel					3.71			3.71	4.08
0500	Brass					15.65			15.65	17.20
0600	#10, 2" long, steel					5.95			5.95	6.55
0700	Brass					23.50			23.50	26
0800	#10, 3" long, steel					9.95			9.95	10.95
1000	#12, 2" long, steel					7.90			7.90	8.70
1100	Brass					31.50			31.50	34.50

06 05 23.50 Wood Screws		Crew	Daily Output	Labor-Hours	Unit	Material	2009 Bare Costs Labor	Equipment	Total	Total Incl O&P
1500	#12, 3" long, steel				C	11.45			11.45	12.60
2000	#12, 4" long, steel					25.50			25.50	28

06 05 23.60 Timber Connectors		Crew	Daily Output	Labor-Hours	Unit	Material	2009 Bare Costs Labor	Equipment	Total	Total Incl O&P
0010	**TIMBER CONNECTORS**									
0020	Add up cost of each part for total cost of connection									
0100	Connector plates, steel, with bolts, straight	2 Carp	75	.213	Ea.	25.50	5.95		31.45	38
0110	Tee, 7 gauge		50	.320		29.50	8.95		38.45	47
0120	T-Strap, 14 gauge, 12" x 8" x 2"		50	.320		29.50	8.95		38.45	47
0150	Anchor plates, 7 ga, 9" x 7"		75	.213		25.50	5.95		31.45	38
0200	Bolts, machine, sq. hd. with nut & washer, 1/2" diameter, 4" long	1 Carp	140	.057		.79	1.60		2.39	3.57
0300	7-1/2" long		130	.062		1.37	1.72		3.09	4.42
0500	3/4" diameter, 7-1/2" long		130	.062		4.12	1.72		5.84	7.45
0610	Machine bolts, w/ nut, washer, 3/4" diam., 15" L, HD's & beam hangers		95	.084		7.60	2.35		9.95	12.35
0720	Machine bolts, sq. hd. w/nut & wash		150	.053	Lb.	2.90	1.49		4.39	5.70
0800	Drilling bolt holes in timber, 1/2" diameter		450	.018	Inch		.50		.50	.84
0900	1" diameter		350	.023	"		.64		.64	1.08
1100	Framing anchor, angle, 3" x 3" x 1-1/2", 12 ga		175	.046	Ea.	2.20	1.28		3.48	4.58
1150	Framing anchors, 18 gauge, 4-1/2" x 2-3/4"		175	.046		2.20	1.28		3.48	4.58
1160	Framing anchors, 18 gauge, 4-1/2" x 3"		175	.046		2.20	1.28		3.48	4.58
1170	Clip anchors plates, 18 gauge, 12" x 1-1/8"		175	.046		2.20	1.28		3.48	4.58
1250	Holdowns, 3 gauge base, 10 gauge body		8	1		21	28		49	70.50
1260	Holdowns, 7 gauge 11-1/16" x 3-1/4"		8	1		21	28		49	70.50
1270	Holdowns, 7 gauge 14-3/8" x 3-1/8"		8	1		21	28		49	70.50
1275	Holdowns, 12 gauge 8" x 2-1/2"		8	1		21	28		49	70.50
1300	Joist and beam hangers, 18 ga. galv., for 2" x 4" joist		175	.046		.61	1.28		1.89	2.83
1400	2" x 6" to 2" x 10" joist		165	.048		1.17	1.36		2.53	3.58
1600	16 ga. galv., 3" x 6" to 3" x 10" joist		160	.050		2.49	1.40		3.89	5.10
1700	3" x 10" to 3" x 14" joist		160	.050		4.14	1.40		5.54	6.90
1800	4" x 6" to 4" x 10" joist		155	.052		2.58	1.44		4.02	5.30
1900	4" x 10" to 4" x 14" joist		155	.052		4.23	1.44		5.67	7.10
2000	Two-2" x 6" to two-2" x 10" joists		150	.053		3.52	1.49		5.01	6.40
2100	Two-2" x 10" to two-2" x 14" joists		150	.053		3.94	1.49		5.43	6.85
2300	3/16" thick, 6" x 8" joist		145	.055		54.50	1.54		56.04	62.50
2400	6" x 10" joist		140	.057		57	1.60		58.60	65
2500	6" x 12" joist		135	.059		59.50	1.66		61.16	68
2700	1/4" thick, 6" x 14" joist		130	.062		61.50	1.72		63.22	71
2900	Plywood clips, extruded aluminum H clip, for 3/4" panels					.20			.20	.22
3000	Galvanized 18 ga. back-up clip					.16			.16	.18
3200	Post framing, 16 ga. galv. for 4" x 4" base, 2 piece	1 Carp	130	.062		13.80	1.72		15.52	18.05
3300	Cap		130	.062		19.15	1.72		20.87	24
3500	Rafter anchors, 18 ga. galv., 1-1/2" wide, 5-1/4" long		145	.055		.41	1.54		1.95	3.06
3600	10-3/4" long		145	.055		1.23	1.54		2.77	3.96
3800	Shear plates, 2-5/8" diameter		120	.067		2	1.86		3.86	5.35
3900	4" diameter		115	.070		4.75	1.94		6.69	8.55
4000	Sill anchors, embedded in concrete or block, 25-1/2" long		115	.070		10.70	1.94		12.64	15.05
4100	Spike grids, 3" x 6"		120	.067		.81	1.86		2.67	4.04
4400	Split rings, 2-1/2" diameter		120	.067		1.65	1.86		3.51	4.97
4500	4" diameter		110	.073		2.50	2.03		4.53	6.20
4550	Tie plate, 20 gauge, 7" x 3 1/8"		110	.073		2.50	2.03		4.53	6.20
4560	Tie plate, 20 gauge, 5" x 4 1/8"		110	.073		2.50	2.03		4.53	6.20
4575	Twist straps, 18 gauge, 12" x 1 1/4"		110	.073		2.50	2.03		4.53	6.20
4580	Twist straps, 18 gauge, 16" x 1 1/4"		110	.073		2.50	2.03		4.53	6.20

06 05 23.60 Timber Connectors		Crew	Daily Output	Labor-Hours	Unit	Material	2009 Bare Costs Labor	Equipment	Total	Total Incl O&P
4600	Strap ties, 20 ga., 2-1/16" wide, 12 13/16" long	1 Carp	180	.044	Ea.	.80	1.24		2.04	2.98
4700	Strap ties, 16 ga., 1-3/8" wide, 12" long		180	.044		.80	1.24		2.04	2.98
4800	21-5/8" x 1-1/4"		160	.050		2.49	1.40		3.89	5.10
5000	Toothed rings, 2-5/8" or 4" diameter		90	.089		1.49	2.48		3.97	5.85
5200	Truss plates, nailed, 20 gauge, up to 32' span		17	.471	Truss	10.85	13.15		24	34
5400	Washers, 2" x 2" x 1/8"				Ea.	.34			.34	.37
5500	3" x 3" x 3/16"				"	.89			.89	.98
6000	Angles and gussets, painted									
6010	Angles, reinforcing, 16 ga., 5" x 5"									
6011	Angles, 16 ga., 2-3/8" x 7-7/8"									
6012	7 ga., 3-1/4" x 3-1/4" x 2-1/2" long	1 Carp	1.90	4.211	C	930	118		1,048	1,225
6014	3-1/4" x 3-1/4" x 5" long		1.90	4.211		1,825	118		1,943	2,200
6016	3-1/4" x 3-1/4" x 7-1/2" long		1.85	4.324		3,425	121		3,546	3,975
6018	5-3/4" x 5-3/4" x 2-1/2" long		1.85	4.324		2,225	121		2,346	2,625
6020	5-3/4" x 5-3/4" x 5" long		1.85	4.324		3,525	121		3,646	4,075
6022	5-3/4" x 5-3/4" x 7-1/2" long		1.80	4.444		5,200	124		5,324	5,925
6024	3 ga., 4-1/4" x 4-1/4" x 3" long		1.85	4.324		2,350	121		2,471	2,775
6026	4-1/4" x 4-1/4" x 6" long		1.85	4.324		5,050	121		5,171	5,750
6028	4-1/4" x 4-1/4" x 9" long		1.80	4.444		5,675	124		5,799	6,450
6030	7-1/4" x 7-1/4" x 3" long		1.80	4.444		4,075	124		4,199	4,675
6032	7-1/4" x 7-1/4" x 6" long		1.80	4.444		5,475	124		5,599	6,225
6034	7-1/4" x 7-1/4" x 9" long		1.75	4.571		12,300	128		12,428	13,700
6036	Gussets									
6038	7 ga., 8-1/8" x 8-1/8" x 2-3/4" long	1 Carp	1.80	4.444	C	3,875	124		3,999	4,475
6040	3 ga., 9-3/4" x 9-3/4" x 3-1/4" long	"	1.80	4.444	"	5,400	124		5,524	6,125
6101	Beam hangers, polymer painted									
6102	Bolted, 3 ga., (W x H x L)									
6104	3-1/4" x 9" x 12" top flange	1 Carp	1	8	C	16,500	224		16,724	18,600
6106	5-1/4" x 9" x 12" top flange		1	8		17,200	224		17,424	19,300
6108	5-1/4" x 11" x 11-3/4" top flange		1	8		19,600	224		19,824	21,900
6110	6-7/8" x 9" x 12" top flange		1	8		17,800	224		18,024	20,000
6112	6-7/8" x 11" x 13-1/2" top flange		1	8		20,600	224		20,824	23,000
6114	8-7/8" x 11" x 15-1/2" top flange		1	8		22,000	224		22,224	24,600
6116	Nailed, 3 ga., (W x H x L)									
6118	3-1/4" x 10-1/2" x 10" top flange	1 Carp	1.80	4.444	C	17,200	124		17,324	19,100
6120	3-1/4" x 10-1/2" x 12" top flange		1.80	4.444		17,200	124		17,324	19,100
6122	5-1/4" x 9-1/2" x 10" top flange		1.80	4.444		17,600	124		17,724	19,500
6124	5-1/4" x 9-1/2" x 12" top flange		1.80	4.444		17,600	124		17,724	19,500
6128	6-7/8" x 8-1/2" x 12" top flange		1.80	4.444		18,000	124		18,124	20,000
6134	Saddle hangers, glu-lam (W x H x L)									
6136	3-1/4" x 10-1/2" x 5-1/4" x 6" saddle	1 Carp	.50	16	C	12,900	445		13,345	15,000
6138	3-1/4" x 10-1/2" x 6-7/8" x 6" saddle		.50	16		13,600	445		14,045	15,700
6140	3-1/4" x 10-1/2" x 8-7/8" x 6" saddle		.50	16		14,200	445		14,645	16,500
6142	3-1/4" x 19-1/2" x 5-1/4" x 10-1/8" saddle		.40	20		12,900	560		13,460	15,100
6144	3-1/4" x 19-1/2" x 6-7/8" x 10-1/8" saddle		.40	20		13,600	560		14,160	15,800
6146	3-1/4" x 19-1/2" x 8-7/8" x 10-1/8" saddle		.40	20		14,200	560		14,760	16,600
6148	5-1/4" x 9-1/2" x 5-1/4" x 12" saddle		.50	16		15,400	445		15,845	17,700
6150	5-1/4" x 9-1/2" x 6-7/8" x 9" saddle		.50	16		16,800	445		17,245	19,300
6152	5-1/4" x 10-1/2" x spec x 12" saddle		.50	16		18,300	445		18,745	20,900
6154	5-1/4" x 18" x 5-1/4" x 12-1/8" saddle		.40	20		15,400	560		15,960	17,800
6156	5-1/4" x 18" x 6-7/8" x 12-1/8" saddle		.40	20		16,800	560		17,360	19,400
6158	5-1/4" x 18" x spec x 12-1/8" saddle		.40	20		18,300	560		18,860	21,000
6160	6-7/8" x 8-1/2" x 6-7/8" x 12" saddle		.50	16		18,400	445		18,845	21,000

06 05 23.60 Timber Connectors		Crew	Daily Output	Labor-Hours	Unit	Material	2009 Bare Costs Labor	Equipment	Total	Total Incl O&P
6162	6-7/8" x 8-1/2" x 8-7/8" x 12" saddle	1 Carp	.50	16	C	19,000	445		19,445	21,700
6164	6-7/8" x 10-1/2" x spec x 12" saddle		.50	16		18,400	445		18,845	21,000
6166	6-7/8" x 18" x 6-7/8" x 13-3/4" saddle		.40	20		18,400	560		18,960	21,100
6168	6-7/8" x 18" x 8-7/8" x 13-3/4" saddle		.40	20		19,000	560		19,560	21,800
6170	6-7/8" x 18" x spec x 13-3/4" saddle		.40	20		20,500	560		21,060	23,400
6172	8-7/8" x 18" x spec x 15-3/4" saddle		.40	20		32,000	560		32,560	36,100
6201	Beam and purlin hangers, galvanized, 12 ga.									
6202	Purlin or joist size, 3" x 8"	1 Carp	1.70	4.706	C	1,700	132		1,832	2,100
6204	3" x 10"		1.70	4.706		1,850	132		1,982	2,250
6206	3" x 12"		1.65	4.848		2,100	136		2,236	2,550
6208	3" x 14"		1.65	4.848		2,250	136		2,386	2,700
6210	3" x 16"		1.65	4.848		2,375	136		2,511	2,850
6212	4" x 8"		1.65	4.848		1,725	136		1,861	2,100
6214	4" x 10"		1.65	4.848		1,850	136		1,986	2,250
6216	4" x 12"		1.60	5		2,200	140		2,340	2,625
6218	4" x 14"		1.60	5		2,325	140		2,465	2,775
6220	4" x 16"		1.60	5		2,450	140		2,590	2,925
6222	6" x 8"		1.60	5		2,200	140		2,340	2,650
6224	6" x 10"		1.55	5.161		2,250	144		2,394	2,725
6226	6" x 12"		1.55	5.161		3,850	144		3,994	4,500
6228	6" x 14"		1.50	5.333		4,075	149		4,224	4,750
6230	6" x 16"		1.50	5.333		4,325	149		4,474	5,000
6250	Beam seats									
6252	Beam size, 5-1/4" wide									
6254	5" x 7" x 1/4"	1 Carp	1.80	4.444	C	6,500	124		6,624	7,350
6256	6" x 7" x 3/8"		1.80	4.444		7,275	124		7,399	8,200
6258	7" x 7" x 3/8"		1.80	4.444		7,775	124		7,899	8,750
6260	8" x 7" x 3/8"		1.80	4.444		9,175	124		9,299	10,300
6262	Beam size, 6-7/8" wide									
6264	5" x 9" x 1/4"	1 Carp	1.80	4.444	C	7,750	124		7,874	8,725
6266	6" x 9" x 3/8"		1.80	4.444		10,100	124		10,224	11,300
6268	7" x 9" x 3/8"		1.80	4.444		10,200	124		10,324	11,500
6270	8" x 9" x 3/8"		1.80	4.444		12,100	124		12,224	13,500
6272	Special beams, over 6-7/8" wide									
6274	5" x 10" x 3/8"	1 Carp	1.80	4.444	C	10,500	124		10,624	11,800
6276	6" x 10" x 3/8"		1.80	4.444		12,200	124		12,324	13,700
6278	7" x 10" x 3/8"		1.80	4.444		12,800	124		12,924	14,300
6280	8" x 10" x 3/8"		1.75	4.571		13,700	128		13,828	15,300
6282	5-1/4" x 12" x 5/16"		1.75	4.571		10,600	128		10,728	11,900
6284	6-1/2" x 12" x 3/8"		1.75	4.571		17,600	128		17,728	19,600
6286	5-1/4" x 16" x 5/16"		1.70	4.706		15,600	132		15,732	17,400
6288	6-1/2" x 16" x 3/8"		1.70	4.706		20,400	132		20,532	22,600
6290	5-1/4" x 20" x 5/16"		1.70	4.706		18,400	132		18,532	20,400
6292	6-1/2" x 20" x 3/8"		1.65	4.848		24,000	136		24,136	26,600
6300	Column bases									
6302	4 x 4, 16 ga.	1 Carp	1.80	4.444	C	655	124		779	930
6306	7 ga.		1.80	4.444		2,400	124		2,524	2,850
6308	4 x 6, 16 ga.		1.80	4.444		1,525	124		1,649	1,875
6312	7 ga.		1.80	4.444		2,500	124		2,624	2,950
6314	6 x 6, 16 ga.		1.75	4.571		1,725	128		1,853	2,125
6318	7 ga.		1.75	4.571		3,425	128		3,553	3,975
6320	6 x 8, 7 ga.		1.70	4.706		2,675	132		2,807	3,150
6322	6 x 10, 7 ga.		1.70	4.706		2,850	132		2,982	3,375

06 05 23 – Wood, Plastic, and Composite Fastenings

06 05 23.60 Timber Connectors		Crew	Daily Output	Labor-Hours	Unit	Material	2009 Bare Costs Labor	Equipment	Total	Total Incl O&P
6324	6 x 12, 7 ga.	1 Carp	1.70	4.706	C	3,100	132		3,232	3,625
6326	8 x 8, 7 ga.		1.65	4.848		5,225	136		5,361	5,975
6330	8 x 10, 7 ga.		1.65	4.848		6,250	136		6,386	7,100
6332	8 x 12, 7 ga.		1.60	5		6,800	140		6,940	7,700
6334	10 x 10, 3 ga.		1.60	5		6,925	140		7,065	7,850
6336	10 x 12, 3 ga.		1.60	5		7,975	140		8,115	8,975
6338	12 x 12, 3 ga.		1.55	5.161		8,650	144		8,794	9,750
6350	Column caps, painted, 3 ga.									
6352	3-1/4" x 3-5/8"	1 Carp	1.80	4.444	C	8,800	124		8,924	9,875
6354	3-1/4" x 5-1/2"		1.80	4.444		8,800	124		8,924	9,875
6356	3-5/8" x 3-5/8"		1.80	4.444		7,200	124		7,324	8,100
6358	3-5/8" x 5-1/2"		1.80	4.444		7,200	124		7,324	8,100
6360	5-1/4" x 5-1/2"		1.75	4.571		9,400	128		9,528	10,500
6362	5-1/4" x 7-1/2"		1.75	4.571		9,400	128		9,528	10,500
6364	5-1/2" x 3-5/8"		1.75	4.571		10,100	128		10,228	11,400
6366	5-1/2" x 5-1/2"		1.75	4.571		10,100	128		10,228	11,400
6368	5-1/2" x 7-1/2"		1.70	4.706		10,100	132		10,232	11,400
6370	6-7/8" x 5-1/2"		1.70	4.706		10,600	132		10,732	11,800
6372	6-7/8" x 6-7/8"		1.70	4.706		10,600	132		10,732	11,800
6374	6-7/8" x 7-1/2"		1.70	4.706		10,600	132		10,732	11,800
6376	7-1/2" x 5-1/2"		1.65	4.848		11,100	136		11,236	12,400
6378	7-1/2" x 7-1/2"		1.65	4.848		11,100	136		11,236	12,400
6380	8-7/8" x 5-1/2"		1.60	5		11,700	140		11,840	13,100
6382	8-7/8" x 7-1/2"		1.60	5		11,700	140		11,840	13,100
6384	9-1/2" x 5-1/2"		1.60	5		15,800	140		15,940	17,600
6400	Floor tie anchors, polymer paint									
6402	10 ga., 3" x 37-1/2"	1 Carp	1.80	4.444	C	4,200	124		4,324	4,825
6404	3-1/2" x 45-1/2"		1.75	4.571		4,400	128		4,528	5,075
6406	3 ga., 3-1/2" x 56"		1.70	4.706		7,775	132		7,907	8,775
6410	Girder hangers									
6412	6" wall thickness, 4" x 6"	1 Carp	1.80	4.444	C	2,375	124		2,499	2,800
6414	4" x 8"		1.80	4.444		2,650	124		2,774	3,100
6416	8" wall thickness, 4" x 6"		1.80	4.444		2,725	124		2,849	3,200
6418	4" x 8"		1.80	4.444		2,725	124		2,849	3,200
6420	Hinge connections, polymer painted									
6422	3/4" thick top plate									
6424	5-1/4" x 12" w/ 5" x 5" top	1 Carp	1	8	C	30,200	224		30,424	33,600
6426	5-1/4" x 15" w/ 6" x 6" top		.80	10		32,100	280		32,380	35,800
6428	5-1/4" x 18" w/ 7" x 7" top		.70	11.429		33,700	320		34,020	37,600
6430	5-1/4" x 26" w/ 9" x 9" top		.60	13.333		35,900	375		36,275	40,100
6432	1" thick top plate									
6434	6-7/8" x 14" w/ 5" x 5" top	1 Carp	.80	10	C	36,800	280		37,080	40,900
6436	6-7/8" x 17" w/ 6" x 6" top		.80	10		40,900	280		41,180	45,500
6438	6-7/8" x 21" w/ 7" x 7" top		.70	11.429		44,600	320		44,920	49,500
6440	6-7/8" x 31" w/ 9" x 9" top		.60	13.333		48,800	375		49,175	54,000
6442	1-1/4" thick top plate									
6444	8-7/8" x 16" w/ 5" x 5" top	1 Carp	.60	13.333	C	45,800	375		46,175	51,000
6446	8-7/8" x 21" w/ 6" x 6" top		.50	16		50,500	445		50,945	56,500
6448	8-7/8" x 26" w/ 7" x 7" top		.40	20		57,000	560		57,560	64,000
6450	8-7/8" x 39" w/ 9" x 9" top		.30	26.667		71,000	745		71,745	80,000
6460	Holddowns									
6462	Embedded along edge									
6464	26" long, 12 ga.	1 Carp	.90	8.889	C	1,150	248		1,398	1,700

06 05 23.60 Timber Connectors		Crew	Daily Output	Labor-Hours	Unit	Material	2009 Bare Costs Labor	Equipment	Total	Total Incl O&P
6466	35" long, 12 ga.	1 Carp	.85	9.412	C	1,525	263		1,788	2,125
6468	35" long, 10 ga.	↓	.85	9.412	↓	1,200	263		1,463	1,775
6470	Embedded away from edge									
6472	Medium duty, 12 ga.									
6474	18-1/2" long	1 Carp	.95	8.421	C	695	235		930	1,175
6476	23-3/4" long		.90	8.889		810	248		1,058	1,300
6478	28" long		.85	9.412		825	263		1,088	1,350
6480	35" long	↓	.85	9.412	↓	1,125	263		1,388	1,700
6482	Heavy duty, 10 ga.									
6484	28" long	1 Carp	.85	9.412	C	1,475	263		1,738	2,075
6486	35" long	"	.85	9.412	"	1,625	263		1,888	2,225
6490	Surface mounted (W x H)									
6492	2-1/2" x 5-3/4", 7 ga.	1 Carp	1	8	C	1,675	224		1,899	2,225
6494	2-1/2" x 8", 12 ga.		1	8		1,025	224		1,249	1,525
6496	2-7/8" x 6-3/8", 7 ga.		1	8		3,725	224		3,949	4,475
6498	2-7/8" x 12-1/2", 3 ga.		1	8		3,825	224		4,049	4,600
6500	3-3/16" x 9-3/8", 10 ga.		1	8		2,550	224		2,774	3,175
6502	3-1/2" x 11-5/8", 3 ga.		1	8		4,650	224		4,874	5,500
6504	3-1/2" x 14-3/4", 3 ga.		1	8		3,100	224		3,324	3,800
6506	3-1/2" x 16-1/2", 3 ga.		1	8		7,125	224		7,349	8,200
6508	3-1/2" x 20-1/2", 3 ga.		.90	8.889		7,275	248		7,523	8,450
6510	3-1/2" x 24-1/2", 3 ga.		.90	8.889		9,200	248		9,448	10,500
6512	4-1/4" x 20-3/4", 3 ga.	↓	.90	8.889	↓	6,275	248		6,523	7,325
6520	Joist hangers									
6522	Sloped, field adjustable, 18 ga.									
6524	2" x 6"	1 Carp	1.65	4.848	C	460	136		596	735
6526	2" x 8"		1.65	4.848		825	136		961	1,150
6528	2" x 10" and up		1.65	4.848		1,375	136		1,511	1,750
6530	3" x 10" and up		1.60	5		1,025	140		1,165	1,350
6532	4" x 10" and up	↓	1.55	5.161	↓	1,250	144		1,394	1,625
6536	Skewed 45°, 16 ga.									
6538	2" x 4"	1 Carp	1.75	4.571	C	735	128		863	1,025
6540	2" x 6" or 2" x 8"		1.65	4.848		745	136		881	1,050
6542	2" x 10" or 2" x 12"		1.65	4.848		855	136		991	1,175
6544	2" x 14" or 2" x 16"		1.60	5		1,525	140		1,665	1,900
6546	(2) 2" x 6" or (2) 2" x 8"		1.60	5		1,375	140		1,515	1,750
6548	(2) 2" x 10" or (2) 2" x 12"		1.55	5.161		1,475	144		1,619	1,875
6550	(2) 2" x 14" or (2) 2" x 16"		1.50	5.333		2,325	149		2,474	2,825
6552	4" x 6" or 4" x 8"		1.60	5		1,150	140		1,290	1,500
6554	4" x 10" or 4" x 12"		1.55	5.161		1,350	144		1,494	1,725
6556	4" x 14" or 4" x 16"	↓	1.55	5.161	↓	2,100	144		2,244	2,550
6560	Skewed 45°, 14 ga.									
6562	(2) 2" x 6" or (2) 2" x 8"	1 Carp	1.60	5	C	1,600	140		1,740	1,975
6564	(2) 2" x 10" or (2) 2" x 12"		1.55	5.161		2,200	144		2,344	2,675
6566	(2) 2" x 14" or (2) 2" x 16"		1.50	5.333		3,225	149		3,374	3,800
6568	4" x 6" or 4" x 8"		1.60	5		1,900	140		2,040	2,300
6570	4" x 10" or 4" x 12"		1.55	5.161		2,000	144		2,144	2,450
6572	4" x 14" or 4" x 16"	↓	1.55	5.161	↓	2,675	144		2,819	3,200
6590	Joist hangers, heavy duty 12 ga., galvanized									
6592	2" x 4"	1 Carp	1.75	4.571	C	1,075	128		1,203	1,400
6594	2" x 6"		1.65	4.848		1,175	136		1,311	1,525
6595	2" x 6", 16 gauge		1.65	4.848		1,175	136		1,311	1,525
6596	2" x 8"	↓	1.65	4.848	↓	1,775	136		1,911	2,175

06 05 23 – Wood, Plastic, and Composite Fastenings

06 05 23.60 Timber Connectors		Crew	Daily Output	Labor-Hours	Unit	Material	2009 Bare Costs Labor	Equipment	Total	Total Incl O&P
6597	2" x 8", 16 gauge	1 Carp	1.65	4.848	C	1,775	136		1,911	2,175
6598	2" x 10"		1.65	4.848		1,825	136		1,961	2,225
6600	2" x 12"		1.65	4.848		2,250	136		2,386	2,700
6602	2" x 14"		1.65	4.848		2,350	136		2,486	2,800
6604	2" x 16"		1.65	4.848		2,475	136		2,611	2,950
6606	3" x 4"		1.65	4.848		1,475	136		1,611	1,850
6608	3" x 6"		1.65	4.848		2,000	136		2,136	2,425
6610	3" x 8"		1.65	4.848		2,025	136		2,161	2,450
6612	3" x 10"		1.60	5		2,350	140		2,490	2,800
6614	3" x 12"		1.60	5		2,800	140		2,940	3,325
6616	3" x 14"		1.60	5		3,300	140		3,440	3,850
6618	3" x 16"		1.60	5		3,625	140		3,765	4,225
6620	(2) 2" x 4"		1.75	4.571		1,750	128		1,878	2,150
6622	(2) 2" x 6"		1.60	5		2,100	140		2,240	2,525
6624	(2) 2" x 8"		1.60	5		2,150	140		2,290	2,575
6626	(2) 2" x 10"		1.55	5.161		2,325	144		2,469	2,800
6628	(2) 2" x 12"		1.55	5.161		2,975	144		3,119	3,525
6630	(2) 2" x 14"		1.50	5.333		3,000	149		3,149	3,550
6632	(2) 2" x 16"		1.50	5.333		3,025	149		3,174	3,575
6634	4" x 4"		1.65	4.848		1,275	136		1,411	1,625
6636	4" x 6"		1.60	5		1,375	140		1,515	1,750
6638	4" x 8"		1.60	5		1,600	140		1,740	2,000
6640	4" x 10"		1.55	5.161		1,975	144		2,119	2,425
6642	4" x 12"		1.55	5.161		2,100	144		2,244	2,550
6644	4" x 14"		1.55	5.161		2,500	144		2,644	3,000
6646	4" x 16"		1.55	5.161		2,750	144		2,894	3,275
6648	(3) 2" x 10"		1.50	5.333		3,050	149		3,199	3,600
6650	(3) 2" x 12"		1.50	5.333		3,400	149		3,549	3,975
6652	(3) 2" x 14"		1.45	5.517		3,650	154		3,804	4,250
6654	(3) 2" x 16"		1.45	5.517		3,700	154		3,854	4,325
6656	6" x 6"		1.60	5		1,625	140		1,765	2,025
6658	6" x 8"		1.60	5		1,675	140		1,815	2,075
6660	6" x 10"		1.55	5.161		2,025	144		2,169	2,475
6662	6" x 12"		1.55	5.161		2,300	144		2,444	2,775
6664	6" x 14"		1.50	5.333		2,875	149		3,024	3,425
6666	6" x 16"	▼	1.50	5.333	▼	3,400	149		3,549	3,975
6690	Knee braces, galvanized, 12 ga.									
6692	Beam depth, 10" x 15" x 5' long	1 Carp	1.80	4.444	C	4,400	124		4,524	5,050
6694	15" x 22-1/2" x 7' long		1.70	4.706		5,050	132		5,182	5,775
6696	22-1/2" x 28-1/2" x 8' long		1.60	5		5,425	140		5,565	6,200
6698	28-1/2" x 36" x 10' long		1.55	5.161		5,650	144		5,794	6,475
6700	36" x 42" x 12' long	▼	1.50	5.333	▼	6,250	149		6,399	7,125
6710	Mudsill anchors									
6714	2" x 4" or 3" x 4"	1 Carp	115	.070	C	1,125	1.94		1,126.94	1,250
6716	2" x 6" or 3" x 6"		115	.070		1,125	1.94		1,126.94	1,250
6718	Block wall, 13-1/4" long		115	.070		780	1.94		781.94	865
6720	21-1/4" long	▼	115	.070	▼	116	1.94		117.94	131
6730	Post bases, 12 ga. galvanized									
6732	Adjustable, 3-9/16" x 3-9/16"	1 Carp	1.30	6.154	C	900	172		1,072	1,275
6734	3-9/16" x 5-1/2"		1.30	6.154		1,175	172		1,347	1,575
6736	4" x 4"		1.30	6.154		655	172		827	1,000
6738	4" x 6"		1.30	6.154		1,525	172		1,697	1,975
6740	5-1/2" x 5-1/2"	▼	1.30	6.154	▼	2,250	172		2,422	2,800

06 05 23.60 Timber Connectors		Crew	Daily Output	Labor-Hours	Unit	Material	2009 Bare Costs Labor	2009 Bare Costs Equipment	Total	Total Incl O&P
6742	6" x 6"	1 Carp	1.30	6.154	C	1,725	172		1,897	2,200
6744	Elevated, 3-9/16" x 3-1/4"		1.30	6.154		975	172		1,147	1,375
6746	5-1/2" x 3-5/16"		1.30	6.154		1,375	172		1,547	1,825
6748	5-1/2" x 5"		1.30	6.154		2,075	172		2,247	2,575
6750	Regular, 3-9/16" x 3-3/8"		1.30	6.154		745	172		917	1,100
6752	4" x 3-3/8"		1.30	6.154		1,050	172		1,222	1,450
6754	5-1/4" x 3-1/8", 18 gauge		1.30	6.154		1,100	172		1,272	1,500
6755	5-1/2" x 3-3/8"		1.30	6.154		1,100	172		1,272	1,500
6756	5-1/2" x 5-3/8"		1.30	6.154		1,575	172		1,747	2,025
6758	6" x 3-3/8"		1.30	6.154		1,900	172		2,072	2,400
6760	6" x 5-3/8"	↓	1.30	6.154	↓	2,125	172		2,297	2,625
6762	Post combination cap/bases									
6764	3-9/16" x 3-9/16"	1 Carp	1.20	6.667	C	425	186		611	785
6766	3-9/16" x 5-1/2"		1.20	6.667		965	186		1,151	1,375
6768	4" x 4"		1.20	6.667		1,925	186		2,111	2,425
6770	5-1/2" x 5-1/2"		1.20	6.667		1,075	186		1,261	1,500
6772	6" x 6"		1.20	6.667		3,500	186		3,686	4,175
6774	7-1/2" x 7-1/2"		1.20	6.667		3,850	186		4,036	4,550
6776	8" x 8"	↓	1.20	6.667	↓	3,975	186		4,161	4,700
6790	Post-beam connection caps									
6792	Beam size 3-9/16"									
6794	12 ga. post, 4" x 4"	1 Carp	1	8	C	2,425	224		2,649	3,050
6796	4" x 6"		1	8		3,250	224		3,474	3,950
6798	4" x 8"		1	8		4,800	224		5,024	5,650
6800	16 ga. post, 4" x 4"		1	8		1,100	224		1,324	1,575
6802	4" x 6"		1	8		1,825	224		2,049	2,400
6804	4" x 8"		1	8		3,100	224		3,324	3,775
6805	18 ga. post, 2- 7/8" x 3"	↓	1	8	↓	3,100	224		3,324	3,775
6806	Beam size 5-1/2"									
6808	12 ga. post, 6" x 4"	1 Carp	1	8	C	3,175	224		3,399	3,875
6810	6" x 6"		1	8		5,000	224		5,224	5,875
6812	6" x 8"		1	8		3,450	224		3,674	4,175
6816	16 ga. post, 6" x 4"		1	8		1,725	224		1,949	2,275
6818	6" x 6"	↓	1	8	↓	1,825	224		2,049	2,400
6820	Beam size 7-1/2"									
6822	12 ga. post, 8" x 4"	1 Carp	1	8	C	4,400	224		4,624	5,225
6824	8" x 6"		1	8		4,600	224		4,824	5,450
6826	8" x 8"	↓	1	8	↓	6,950	224		7,174	8,000
6840	Purlin anchors, embedded									
6842	Heavy duty, 10 ga.									
6844	Straight, 28" long	1 Carp	1.60	5	C	1,200	140		1,340	1,550
6846	35" long		1.50	5.333		1,500	149		1,649	1,900
6848	Twisted, 28" long		1.60	5		1,200	140		1,340	1,550
6850	35" long	↓	1.50	5.333	↓	1,500	149		1,649	1,900
6852	Regular duty, 12 ga.									
6854	Straight, 18-1/2" long	1 Carp	1.80	4.444	C	695	124		819	970
6856	23-3/4" long		1.70	4.706		870	132		1,002	1,175
6858	29" long		1.60	5		890	140		1,030	1,225
6860	35" long		1.50	5.333		1,225	149		1,374	1,600
6862	Twisted, 18" long		1.80	4.444		695	124		819	970
6866	28" long		1.60	5		825	140		965	1,150
6868	35" long	↓	1.50	5.333	↓	1,225	149		1,374	1,600
6870	Straight, plastic coated									

06 05 23 – Wood, Plastic, and Composite Fastenings

06 05 23.60 Timber Connectors		Crew	Daily Output	Labor-Hours	Unit	Material	2009 Bare Costs Labor	Equipment	Total	Total Incl O&P
6872	23-1/2" long	1 Carp	1.60	5	C	1,700	140		1,840	2,100
6874	26-7/8" long		1.60	5		2,000	140		2,140	2,425
6876	32-1/2" long		1.50	5.333		2,125	149		2,274	2,600
6878	35-7/8" long		1.50	5.333		2,225	149		2,374	2,675
6890	Purlin hangers, painted									
6892	12 ga., 2" x 6"	1 Carp	1.80	4.444	C	1,575	124		1,699	1,925
6894	2" x 8"		1.80	4.444		1,700	124		1,824	2,075
6896	2" x 10"		1.80	4.444		1,825	124		1,949	2,225
6898	2" x 12"		1.75	4.571		1,975	128		2,103	2,400
6900	2" x 14"		1.75	4.571		2,100	128		2,228	2,525
6902	2" x 16"		1.75	4.571		2,225	128		2,353	2,675
6904	3" x 6"		1.70	4.706		1,575	132		1,707	1,950
6906	3" x 8"		1.70	4.706		1,700	132		1,832	2,100
6908	3" x 10"		1.70	4.706		1,850	132		1,982	2,250
6910	3" x 12"		1.65	4.848		2,100	136		2,236	2,550
6912	3" x 14"		1.65	4.848		2,250	136		2,386	2,700
6914	3" x 16"		1.65	4.848		2,375	136		2,511	2,850
6916	4" x 6"		1.65	4.848		1,575	136		1,711	1,950
6918	4" x 8"		1.65	4.848		1,725	136		1,861	2,100
6920	4" x 10"		1.65	4.848		1,850	136		1,986	2,250
6922	4" x 12"		1.60	5		2,200	140		2,340	2,625
6924	4" x 14"		1.60	5		2,325	140		2,465	2,775
6926	4" x 16"		1.60	5		2,450	140		2,590	2,925
6928	6" x 6"		1.60	5		2,075	140		2,215	2,525
6930	6" x 8"		1.60	5		2,200	140		2,340	2,650
6932	6" x 10"		1.55	5.161		2,250	144		2,394	2,725
6934	double 2" x 6"		1.70	4.706		1,700	132		1,832	2,100
6936	double 2" x 8"		1.70	4.706		1,850	132		1,982	2,250
6938	double 2" x 10"		1.70	4.706		1,975	132		2,107	2,400
6940	double 2" x 12"		1.65	4.848		2,100	136		2,236	2,550
6942	double 2" x 14"		1.65	4.848		2,250	136		2,386	2,700
6944	double 2" x 16"		1.65	4.848		2,375	136		2,511	2,850
6960	11 ga., 4" x 6"		1.65	4.848		3,125	136		3,261	3,650
6962	4" x 8"		1.65	4.848		3,350	136		3,486	3,925
6964	4" x 10"		1.65	4.848		3,575	136		3,711	4,175
6966	6" x 6"		1.60	5		3,150	140		3,290	3,700
6968	6" x 8"		1.60	5		3,400	140		3,540	3,950
6970	6" x 10"		1.55	5.161		3,625	144		3,769	4,225
6972	6" x 12"		1.55	5.161		3,850	144		3,994	4,500
6974	6" x 14"		1.55	5.161		4,075	144		4,219	4,750
6976	6" x 16"		1.50	5.333		4,325	149		4,474	5,000
6978	7 ga., 8" x 6"		1.60	5		3,425	140		3,565	4,000
6980	8" x 8"		1.60	5		3,650	140		3,790	4,250
6982	8" x 10"		1.55	5.161		3,900	144		4,044	4,525
6984	8" x 12"		1.55	5.161		4,125	144		4,269	4,775
6986	8" x 14"		1.50	5.333		4,350	149		4,499	5,050
6988	8" x 16"		1.50	5.333		4,575	149		4,724	5,300
7000	Strap connectors, galvanized									
7002	12 ga., 2-1/16" x 36"	1 Carp	1.55	5.161	C	995	144		1,139	1,350
7004	2-1/16" x 47"		1.50	5.333		1,375	149		1,524	1,775
7005	10 ga., 2-1/16" x 72"		1.50	5.333		1,375	149		1,524	1,775
7006	7 ga., 2-1/16" x 34"		1.55	5.161		2,475	144		2,619	2,975
7008	2-1/16" x 45"		1.50	5.333		3,200	149		3,349	3,775

06 05 23 – Wood, Plastic, and Composite Fastenings

06 05 23.60 Timber Connectors

06 05 23.60 Timber Connectors	Crew	Daily Output	Labor-Hours	Unit	Material	2009 Bare Costs Labor	2009 Bare Costs Equipment	Total	Total Incl O&P
7010	3 ga., 3" x 32"	1 Carp	1.55	5.161	C	4,150	144		4,294	4,825
7012	3" x 41"		1.55	5.161		4,325	144		4,469	5,000
7014	3" x 50"		1.50	5.333		6,575	149		6,724	7,475
7016	3" x 59"		1.50	5.333		8,000	149		8,149	9,050
7018	3-1/2" x 68"	↓	1.45	5.517	↓	8,125	154		8,279	9,200
7030	Tension ties									
7032	19-1/8" long, 16 ga., 3/4" anchor bolt	1 Carp	1.80	4.444	C	1,150	124		1,274	1,450
7034	20" long, 12 ga., 1/2" anchor bolt		1.80	4.444		1,475	124		1,599	1,825
7036	20" long, 12 ga., 3/4" anchor bolt		1.80	4.444		1,475	124		1,599	1,825
7038	27-3/4" long, 12 ga., 3/4" anchor bolt	↓	1.75	4.571	↓	2,625	128		2,753	3,100
7050	Truss connectors, galvanized									
7052	Adjustable hanger									
7054	18 ga., 2" x 6"	1 Carp	1.65	4.848	C	445	136		581	720
7056	4" x 6"		1.65	4.848		585	136		721	875
7058	16 ga., 4" x 10"		1.60	5		860	140		1,000	1,175
7060	(2) 2" x 10"	↓	1.60	5	↓	860	140		1,000	1,175
7062	Connectors to plate									
7064	16 ga., 2" x 4" plate	1 Carp	1.80	4.444	C	440	124		564	695
7066	2" x 6" plate	"	1.80	4.444	"	570	124		694	835
7068	Hip jack connector									
7070	14 ga.	1 Carp	1.50	5.333	C	2,325	149		2,474	2,800

06 05 23.70 Rough Hardware

06 05 23.70 Rough Hardware	Crew	Daily Output	Labor-Hours	Unit	Material	2009 Bare Costs Labor	2009 Bare Costs Equipment	Total	Total Incl O&P
0010	**ROUGH HARDWARE**, average percent of carpentry material									
0020	Minimum					.50%				
0200	Maximum					1.50%				
0210	In seismic or hurricane areas, up to					10%				

06 05 23.80 Metal Bracing

06 05 23.80 Metal Bracing	Crew	Daily Output	Labor-Hours	Unit	Material	2009 Bare Costs Labor	2009 Bare Costs Equipment	Total	Total Incl O&P
0010	**METAL BRACING**									
0302	Let-in, "T" shaped, 22 ga. galv. steel, studs at 16" O.C.	1 Carp	580	.014	L.F.	.75	.39		1.14	1.48
0402	Studs at 24" O.C.		600	.013		.75	.37		1.12	1.46
0502	16 ga. galv. steel straps, studs at 16" O.C.		600	.013		.97	.37		1.34	1.70
0602	Studs at 24" O.C.	↓	620	.013	↓	.97	.36		1.33	1.68

06 11 Wood Framing

06 11 10 – Framing with Dimensional, Engineered or Composite Lumber

06 11 10.02 Blocking

06 11 10.02 Blocking	Crew	Daily Output	Labor-Hours	Unit	Material	2009 Bare Costs Labor	2009 Bare Costs Equipment	Total	Total Incl O&P
0010	**BLOCKING**									
1950	Miscellaneous, to wood construction									
2000	2" x 4"	1 Carp	250	.032	L.F.	.29	.89		1.18	1.83
2005	Pneumatic nailed		305	.026		.29	.73		1.02	1.56
2050	2" x 6"		222	.036		.52	1.01		1.53	2.27
2055	Pneumatic nailed		271	.030		.52	.83		1.35	1.96
2100	2" x 8"		200	.040		.71	1.12		1.83	2.67
2105	Pneumatic nailed		244	.033		.71	.92		1.63	2.33
2150	2" x 10"		178	.045		.99	1.26		2.25	3.21
2155	Pneumatic nailed		217	.037		.99	1.03		2.02	2.83
2200	2" x 12"		151	.053		1.39	1.48		2.87	4.02
2205	Pneumatic nailed	↓	185	.043	↓	1.39	1.21		2.60	3.56
2300	To steel construction									
2320	2" x 4"	1 Carp	208	.038	L.F.	.29	1.08		1.37	2.14

06 11 Wood Framing

06 11 10 – Framing with Dimensional, Engineered or Composite Lumber

	06 11 10.02 Blocking	Crew	Daily Output	Labor-Hours	Unit	Material	2009 Bare Costs Labor	Equipment	Total	Total Incl O&P
2340	2" x 6"	1 Carp	180	.044	L.F.	.52	1.24		1.76	2.67
2360	2" x 8"		158	.051		.71	1.42		2.13	3.17
2380	2" x 10"		136	.059		.99	1.64		2.63	3.87
2400	2" x 12"		109	.073		1.39	2.05		3.44	4.99

	06 11 10.04 Bracing									
0012	**BRACING** Let-in, with 1" x 6" boards, studs @ 16" O.C.	1 Carp	150	.053	L.F.	.64	1.49		2.13	3.22
0202	Studs @ 24" O.C.	"	230	.035	"	.64	.97		1.61	2.34

	06 11 10.06 Bridging									
0012	**BRIDGING** Wood, for joists 16" O.C., 1" x 3"	1 Carp	130	.062	Pr.	.51	1.72		2.23	3.47
0017	Pneumatic nailed		170	.047		.51	1.32		1.83	2.78
0102	2" x 3" bridging		130	.062		.47	1.72		2.19	3.43
0107	Pneumatic nailed		170	.047		.47	1.32		1.79	2.74
0302	Steel, galvanized, 18 ga., for 2" x 10" joists at 12" O.C.		130	.062		1.58	1.72		3.30	4.65
0402	24" O.C.		140	.057		1.69	1.60		3.29	4.56
0602	For 2" x 14" joists at 16" O.C.		130	.062		1.70	1.72		3.42	4.78
0902	Compression type, 16" O.C., 2" x 8" joists		200	.040		1.64	1.12		2.76	3.69
1002	2" x 12" joists		200	.040		1.64	1.12		2.76	3.69

	06 11 10.10 Beam and Girder Framing									
0010	**BEAM AND GIRDER FRAMING** R061110-30									
1000	Single, 2" x 6"	2 Carp	700	.023	L.F.	.52	.64		1.16	1.65
1005	Pneumatic nailed		812	.020		.52	.55		1.07	1.50
1020	2" x 8"		650	.025		.71	.69		1.40	1.94
1025	Pneumatic nailed		754	.021		.71	.59		1.30	1.78
1040	2" x 10"		600	.027		.99	.75		1.74	2.35
1045	Pneumatic nailed		696	.023		.99	.64		1.63	2.18
1060	2" x 12"		550	.029		1.39	.81		2.20	2.89
1065	Pneumatic nailed		638	.025		1.39	.70		2.09	2.71
1080	2" x 14"		500	.032		2.14	.89		3.03	3.86
1085	Pneumatic nailed		580	.028		2.14	.77		2.91	3.65
1100	3" x 8"		550	.029		2.10	.81		2.91	3.68
1120	3" x 10"		500	.032		2.94	.89		3.83	4.74
1140	3" x 12"		450	.036		3.54	.99		4.53	5.55
1160	3" x 14"		400	.040		4.48	1.12		5.60	6.80
1170	4" x 6"	F-3	1100	.036		2.34	.92	.70	3.96	4.89
1180	4" x 8"		1000	.040		2.79	1.02	.77	4.58	5.60
1200	4" x 10"		950	.042		3.68	1.07	.81	5.56	6.75
1220	4" x 12"		900	.044		4.74	1.13	.85	6.72	8.05
1240	4" x 14"		850	.047		5.60	1.19	.90	7.69	9.15
2000	Double, 2" x 6"	2 Carp	625	.026		1.04	.72		1.76	2.36
2005	Pneumatic nailed		725	.022		1.04	.62		1.66	2.19
2020	2" x 8"		575	.028		1.42	.78		2.20	2.88
2025	Pneumatic nailed		667	.024		1.42	.67		2.09	2.69
2040	2" x 10"		550	.029		1.98	.81		2.79	3.55
2045	Pneumatic nailed		638	.025		1.98	.70		2.68	3.37
2060	2" x 12"		525	.030		2.77	.85		3.62	4.49
2065	Pneumatic nailed		610	.026		2.77	.73		3.50	4.29
2080	2" x 14"		475	.034		4.28	.94		5.22	6.30
2085	Pneumatic nailed		551	.029		4.28	.81		5.09	6.05
3000	Triple, 2" x 6"		550	.029		1.56	.81		2.37	3.09
3005	Pneumatic nailed		638	.025		1.56	.70		2.26	2.91
3020	2" x 8"		525	.030		2.12	.85		2.97	3.77
3025	Pneumatic nailed		609	.026		2.12	.73		2.85	3.57

06 11 Wood Framing

06 11 10 – Framing with Dimensional, Engineered or Composite Lumber

06 11 10.10 Beam and Girder Framing		Crew	Daily Output	Labor-Hours	Unit	Material	2009 Bare Costs Labor	Equipment	Total	Total Incl O&P
3040	2" x 10"	2 Carp	500	.032	L.F.	2.98	.89		3.87	4.78
3045	Pneumatic nailed		580	.028		2.98	.77		3.75	4.57
3060	2" x 12"		475	.034		4.16	.94		5.10	6.15
3065	Pneumatic nailed		551	.029		4.16	.81		4.97	5.95
3080	2" x 14"		450	.036		6.40	.99		7.39	8.75
3085	Pneumatic nailed		522	.031		6.40	.86		7.26	8.50

06 11 10.12 Ceiling Framing		Crew	Daily Output	Labor-Hours	Unit	Material	2009 Bare Costs Labor	Equipment	Total	Total Incl O&P
0010	**CEILING FRAMING**									
6000	Suspended, 2" x 3"	2 Carp	1000	.016	L.F.	.32	.45		.77	1.11
6050	2" x 4"		900	.018		.29	.50		.79	1.16
6100	2" x 6"		800	.020		.52	.56		1.08	1.52
6150	2" x 8"		650	.025		.71	.69		1.40	1.94

06 11 10.14 Posts and Columns		Crew	Daily Output	Labor-Hours	Unit	Material	2009 Bare Costs Labor	Equipment	Total	Total Incl O&P
0010	**POSTS AND COLUMNS**									
0101	4" x 4"	2 Carp	390	.041	L.F.	1.56	1.15		2.71	3.66
0151	4" x 6"		275	.058		2.34	1.63		3.97	5.30
0201	4" x 8"		220	.073		2.79	2.03		4.82	6.50
0251	6" x 6"		215	.074		4.87	2.08		6.95	8.85
0301	6" x 8"		175	.091		7.60	2.56		10.16	12.65
0351	6" x 10"		150	.107		10.90	2.98		13.88	17.05

06 11 10.18 Joist Framing		Crew	Daily Output	Labor-Hours	Unit	Material	2009 Bare Costs Labor	Equipment	Total	Total Incl O&P
0010	**JOIST FRAMING**									
2002	Joists, 2" x 4"	2 Carp	1250	.013	L.F.	.29	.36		.65	.92
2007	Pneumatic nailed		1438	.011		.29	.31		.60	.85
2100	2" x 6"		1250	.013		.52	.36		.88	1.17
2105	Pneumatic nailed		1438	.011		.52	.31		.83	1.10
2152	2" x 8"		1100	.015		.71	.41		1.12	1.47
2157	Pneumatic nailed		1265	.013		.71	.35		1.06	1.38
2202	2" x 10"		900	.018		.99	.50		1.49	1.93
2207	Pneumatic nailed		1035	.015		.99	.43		1.42	1.82
2252	2" x 12"		875	.018		1.39	.51		1.90	2.38
2257	Pneumatic nailed		1006	.016		1.39	.44		1.83	2.27
2302	2" x 14"		770	.021		2.14	.58		2.72	3.33
2307	Pneumatic nailed		886	.018		2.14	.50		2.64	3.20
2352	3" x 6"		925	.017		1.52	.48		2	2.50
2402	3" x 10"		780	.021		2.94	.57		3.51	4.20
2452	3" x 12"		600	.027		3.54	.75		4.29	5.15
2502	4" x 6"		800	.020		2.34	.56		2.90	3.52
2552	4" x 10"		600	.027		3.68	.75		4.43	5.30
2602	4" x 12"		450	.036		4.74	.99		5.73	6.90
2607	Sister joist, 2" x 6"		800	.020		.52	.56		1.08	1.52
2608	Pneumatic nailed		960	.017		.52	.47		.99	1.36
3000	Composite wood joist 9-1/2" deep		.90	17.778	M.L.F.	2,000	495		2,495	3,050
3010	11-1/2" deep		.88	18.182		2,175	510		2,685	3,225
3020	14" deep		.82	19.512		2,800	545		3,345	4,000
3030	16" deep		.78	20.513		3,250	575		3,825	4,550
4000	Open web joist 12" deep		.88	18.182		3,050	510		3,560	4,200
4010	14" deep		.82	19.512		3,275	545		3,820	4,525
4020	16" deep		.78	20.513		3,250	575		3,825	4,550
4030	18" deep		.74	21.622		3,500	605		4,105	4,875
6000	Composite rim joist, 1-1/4" x 9-1/2"		90	.178		1,825	4.97		1,829.97	2,025
6010	1-1/4" x 11-1/2"		.88	18.182		2,275	510		2,785	3,350

06 11 10 – Framing with Dimensional, Engineered or Composite Lumber

06 11 10.18 Joist Framing

		Crew	Daily Output	Labor-Hours	Unit	Material	2009 Bare Costs Labor	2009 Bare Costs Equipment	Total	Total Incl O&P
6020	1-1/4" x 14-1/2"	2 Carp	.82	19.512	M.L.F.	2,625	545		3,170	3,800
6030	1-1/4" x 16-1/2"		.78	20.513		3,100	575		3,675	4,400

06 11 10.24 Miscellaneous Framing

		Crew	Daily Output	Labor-Hours	Unit	Material	Labor	Equipment	Total	Total Incl O&P
0010	**MISCELLANEOUS FRAMING**									
2002	Firestops, 2" x 4"	2 Carp	780	.021	L.F.	.29	.57		.86	1.29
2007	Pneumatic nailed		952	.017		.29	.47		.76	1.11
2102	2" x 6"		600	.027		.52	.75		1.27	1.83
2107	Pneumatic nailed		732	.022		.52	.61		1.13	1.60
5002	Nailers, treated, wood construction, 2" x 4"		800	.020		.44	.56		1	1.44
5007	Pneumatic nailed		960	.017		.44	.47		.91	1.28
5102	2" x 6"		750	.021		.69	.60		1.29	1.76
5107	Pneumatic nailed		900	.018		.69	.50		1.19	1.59
5122	2" x 8"		700	.023		1.05	.64		1.69	2.23
5127	Pneumatic nailed		840	.019		1.05	.53		1.58	2.05
5202	Steel construction, 2" x 4"		750	.021		.44	.60		1.04	1.50
5222	2" x 6"		700	.023		.69	.64		1.33	1.83
5242	2" x 8"		650	.025		1.05	.69		1.74	2.31
7002	Rough bucks, treated, for doors or windows, 2" x 6"		400	.040		.69	1.12		1.81	2.64
7007	Pneumatic nailed		480	.033		.69	.93		1.62	2.32
7102	2" x 8"		380	.042		1.05	1.18		2.23	3.14
7107	Pneumatic nailed		456	.035		1.05	.98		2.03	2.81
8001	Stair stringers, 2" x 10"		130	.123		.99	3.44		4.43	6.90
8101	2" x 12"		130	.123		1.39	3.44		4.83	7.30
8151	3" x 10"		125	.128		2.94	3.58		6.52	9.30
8201	3" x 12"		125	.128		3.54	3.58		7.12	9.95
8870	Laminated structural lumber, 1-1/4" x 11-1/2"		130	.123		2.27	3.44		5.71	8.30
8880	1-1/4" x 14-1/2"		130	.123		2.62	3.44		6.06	8.70

06 11 10.26 Partitions

		Crew	Daily Output	Labor-Hours	Unit	Material	Labor	Equipment	Total	Total Incl O&P
0010	**PARTITIONS**									
0020	Single bottom and double top plate, no waste, std. & better lumber									
0182	2" x 4" studs, 8' high, studs 12" O.C.	2 Carp	80	.200	L.F.	3.51	5.60		9.11	13.30
0187	12" O.C., pneumatic nailed		96	.167		3.51	4.66		8.17	11.75
0202	16" O.C.		100	.160		2.87	4.47		7.34	10.70
0207	16" O.C., pneumatic nailed		120	.133		2.87	3.73		6.60	9.45
0302	24" O.C.		125	.128		2.23	3.58		5.81	8.50
0307	24" O.C., pneumatic nailed		150	.107		2.23	2.98		5.21	7.50
0382	10' high, studs 12" O.C.		80	.200		4.15	5.60		9.75	14
0387	12" O.C., pneumatic nailed		96	.167		4.15	4.66		8.81	12.45
0402	16" O.C.		100	.160		3.35	4.47		7.82	11.25
0407	16" O.C., pneumatic nailed		120	.133		3.35	3.73		7.08	10
0502	24" O.C.		125	.128		2.55	3.58		6.13	8.85
0507	24" O.C., pneumatic nailed		150	.107		2.55	2.98		5.53	7.85
0582	12' high, studs 12" O.C.		65	.246		4.78	6.90		11.68	16.90
0587	12" O.C., pneumatic nailed		78	.205		4.78	5.75		10.53	14.95
0602	16" O.C.		80	.200		3.83	5.60		9.43	13.65
0607	16" O.C., pneumatic nailed		96	.167		3.83	4.66		8.49	12.10
0700	24" O.C.		100	.160		2.61	4.47		7.08	10.40
0705	24" O.C., pneumatic nailed		120	.133		2.61	3.73		6.34	9.15
0782	2" x 6" studs, 8' high, studs 12" O.C.		70	.229		6.30	6.40		12.70	17.75
0787	12" O.C., pneumatic nailed		84	.190		6.30	5.30		11.60	15.95
0802	16" O.C.		90	.178		5.15	4.97		10.12	14.05
0807	16" O.C., pneumatic nailed		108	.148		5.15	4.14		9.29	12.65

06 11 Wood Framing

06 11 10 – Framing with Dimensional, Engineered or Composite Lumber

06 11 10.26 Partitions

		Crew	Daily Output	Labor-Hours	Unit	Material	2009 Bare Costs Labor	Equipment	Total	Total Incl O&P
0902	24" O.C.	2 Carp	115	.139	L.F.	4.01	3.89		7.90	10.95
0907	24" O.C., pneumatic nailed		138	.116		4.01	3.24		7.25	9.90
0982	10' high, studs 12" O.C.		70	.229		7.45	6.40		13.85	19
0987	12" O.C., pneumatic nailed		84	.190		7.45	5.30		12.75	17.20
1002	16" O.C.		90	.178		6	4.97		10.97	15
1007	16" O.C., pneumatic nailed		108	.148		6	4.14		10.14	13.60
1102	24" O.C.		115	.139		4.58	3.89		8.47	11.60
1107	24" O.C., pneumatic nailed		138	.116		4.58	3.24		7.82	10.55
1182	12' high, studs 12" O.C.		55	.291		8.60	8.15		16.75	23
1187	12" O.C., pneumatic nailed		66	.242		8.60	6.80		15.40	21
1202	16" O.C.		70	.229		6.90	6.40		13.30	18.35
1207	16" O.C., pneumatic nailed		84	.190		6.90	5.30		12.20	16.55
1302	24" O.C.		90	.178		5.15	4.97		10.12	14.05
1307	24" O.C., pneumatic nailed		108	.148		5.15	4.14		9.29	12.65
1402	For horizontal blocking, 2" x 4", add		600	.027		.32	.75		1.07	1.61
1502	2" x 6", add		600	.027		.57	.75		1.32	1.89
1600	For openings, add	▼	250	.064	▼		1.79		1.79	3.02
1702	Headers for above openings, material only, add				B.F.	.58			.58	.64

06 11 10.28 Porch or Deck Framing

		Crew	Daily Output	Labor-Hours	Unit	Material	2009 Bare Costs Labor	Equipment	Total	Total Incl O&P
0010	**PORCH OR DECK FRAMING**									
0100	Treated lumber, posts or columns, 4" x 4"	2 Carp	390	.041	L.F.	1.52	1.15		2.67	3.61
0110	4" x 6"		275	.058		2.24	1.63		3.87	5.20
0120	4" x 8"		220	.073		3.21	2.03		5.24	6.95
0130	Girder, single, 4" x 4"		675	.024		1.52	.66		2.18	2.79
0140	4" x 6"		600	.027		2.24	.75		2.99	3.72
0150	4" x 8"		525	.030		3.21	.85		4.06	4.97
0160	Double, 2" x 4"		625	.026		.91	.72		1.63	2.21
0170	2" x 6"		600	.027		1.42	.75		2.17	2.82
0180	2" x 8"		575	.028		2.16	.78		2.94	3.70
0190	2" x 10"		550	.029		2.63	.81		3.44	4.26
0200	2" x 12"		525	.030		4.15	.85		5	6
0210	Triple, 2" x 4"		575	.028		1.37	.78		2.15	2.83
0220	2" x 6"		550	.029		2.13	.81		2.94	3.71
0230	2" x 8"		525	.030		3.24	.85		4.09	5
0240	2" x 10"		500	.032		3.95	.89		4.84	5.85
0250	2" x 12"		475	.034		6.20	.94		7.14	8.45
0260	Ledger, bolted 4' O.C., 2" x 4"		400	.040		.55	1.12		1.67	2.50
0270	2" x 6"		395	.041		.80	1.13		1.93	2.79
0280	2" x 8"		390	.041		1.16	1.15		2.31	3.22
0300	2" x 12"		380	.042		2.14	1.18		3.32	4.35
0310	Joists, 2" x 4"		1250	.013		.46	.36		.82	1.10
0320	2" x 6"		1250	.013		.71	.36		1.07	1.38
0330	2" x 8"		1100	.015		1.08	.41		1.49	1.88
0340	2" x 10"		900	.018		1.32	.50		1.82	2.29
0350	2" x 12"		875	.018		1.58	.51		2.09	2.60
0360	Railings and trim , 1" x 4"	1 Carp	300	.027		.69	.75		1.44	2.02
0370	2" x 2"		300	.027		.33	.75		1.08	1.62
0380	2" x 4"		300	.027		.44	.75		1.19	1.75
0390	2" x 6"		300	.027	▼	.69	.75		1.44	2.01
0400	Decking, 1" x 4"		275	.029	S.F.	1.55	.81		2.36	3.08
0410	2" x 4"		300	.027		1.51	.75		2.26	2.92
0420	2" x 6"	▼	320	.025	▼	1.49	.70		2.19	2.82

06 11 10 – Framing with Dimensional, Engineered or Composite Lumber

06 11 10.28 Porch or Deck Framing

		Crew	Daily Output	Labor-Hours	Unit	Material	2009 Bare Costs Labor	2009 Bare Costs Equipment	Total	Total Incl O&P
0430	5/4" x 6"	1 Carp	320	.025	S.F.	2.35	.70		3.05	3.77
0440	Balusters, square, 2" x 2"	2 Carp	660	.024	L.F.	.34	.68		1.02	1.52
0450	Turned, 2" x 2"		420	.038		.45	1.06		1.51	2.29
0460	Stair stringer, 2" x 10"		130	.123		1.32	3.44		4.76	7.25
0470	2" x 12"		130	.123		1.58	3.44		5.02	7.55
0480	Stair treads, 1" x 4"		140	.114		.52	3.19		3.71	6
0490	2" x 4"		140	.114		.46	3.19		3.65	5.90
0500	2" x 6"		160	.100		.69	2.80		3.49	5.50
0510	5/4" x 6"		160	.100	▼	1.10	2.80		3.90	5.95
0520	Turned handrail post, 4" x 4"		64	.250	Ea.	31.50	7		38.50	47
0530	Lattice panel, 4' x 8'		1600	.010	S.F.	.68	.28		.96	1.22
0540	Cedar, posts or columns, 4" x 4"		390	.041	L.F.	3.21	1.15		4.36	5.45
0550	4" x 6"		275	.058		4.86	1.63		6.49	8.10
0560	4" x 8"		220	.073		7.30	2.03		9.33	11.45
0800	Decking, 1" x 4"		550	.029		1.76	.81		2.57	3.30
0810	2" x 4"		600	.027		3.63	.75		4.38	5.25
0820	2" x 6"		640	.025		6.95	.70		7.65	8.85
0830	5/4" x 6"		640	.025		4.60	.70		5.30	6.25
0840	Railings and trim, 1" x 4"		600	.027		1.76	.75		2.51	3.19
0860	2" x 4"		600	.027		3.63	.75		4.38	5.25
0870	2" x 6"		600	.027		6.95	.75		7.70	8.90
0920	Stair treads, 1" x 4"		140	.114		1.76	3.19		4.95	7.35
0930	2" x 4"		140	.114		3.63	3.19		6.82	9.40
0940	2" x 6"		160	.100		6.95	2.80		9.75	12.40
0950	5/4" x 6"		160	.100		4.60	2.80		7.40	9.80
0980	Redwood, posts or columns, 4" x 4"		390	.041		6.45	1.15		7.60	9.05
0990	4" x 6"		275	.058		11.20	1.63		12.83	15.05
1000	4" x 8"	▼	220	.073	▼	20.50	2.03		22.53	26
1240	Redwood decking, 1" x 4"	1 Carp	275	.029	S.F.	5.70	.81		6.51	7.60
1260	2" x 6"		340	.024		10.70	.66		11.36	12.85
1270	5/4" x 6"	▼	320	.025	▼	8.25	.70		8.95	10.30
1280	Railings and trim, 1" x 4"	2 Carp	600	.027	L.F.	1.67	.75		2.42	3.10
1310	2" x 6"		600	.027		4.93	.75		5.68	6.65
1420	Alternative decking, wood / plastic composite, 5/4" x 6" ☐G		640	.025		3.01	.70		3.71	4.50
1430	Vinyl, 1-1/2" x 5-1/2"		640	.025		4.15	.70		4.85	5.75
1440	1" x 4" square edge fir		550	.029		1.57	.81		2.38	3.09
1450	1" x 4" tongue and groove fir		450	.036		1.41	.99		2.40	3.23
1460	1" x 4" mahogany		550	.029		2.51	.81		3.32	4.13
1462	5/4" x 6" PVC	▼	550	.029	▼	2.72	.81		3.53	4.36
1470	Accessories, joist hangers, 2" x 4"	1 Carp	160	.050	Ea.	.61	1.40		2.01	3.03
1480	2" x 6" through 2" x 12"	"	150	.053		1.17	1.49		2.66	3.81
1530	Post footing, incl excav, backfill, tube form & concrete, 4' deep, 8" dia	F-7	12	2.667		10.60	64.50		75.10	121
1540	10" diameter		11	2.909		16	70.50		86.50	137
1550	12" diameter	▼	10	3.200	▼	21.50	77.50		99	155

06 11 10.30 Roof Framing

		Crew	Daily Output	Labor-Hours	Unit	Material	2009 Bare Costs Labor	2009 Bare Costs Equipment	Total	Total Incl O&P
0010	**ROOF FRAMING**									
2001	Fascia boards, 2" x 8"	2 Carp	225	.071	L.F.	.71	1.99		2.70	4.14
2101	2" x 10"		180	.089		.99	2.48		3.47	5.30
5002	Rafters, to 4 in 12 pitch, 2" x 6", ordinary		1000	.016		.52	.45		.97	1.33
5021	On steep roofs		800	.020		.52	.56		1.08	1.52
5041	On dormers or complex roofs		590	.027		.52	.76		1.28	1.85
5062	2" x 8", ordinary	▼	950	.017	▼	.71	.47		1.18	1.58

06 11 Wood Framing

06 11 10 – Framing with Dimensional, Engineered or Composite Lumber

06 11 10.30 Roof Framing	Crew	Daily Output	Labor-Hours	Unit	Material	2009 Bare Costs Labor	Equipment	Total	Total Incl O&P
5081	On steep roofs	2 Carp	750	.021	L.F.	.71	.60		1.31	1.79
5101	On dormers or complex roofs		540	.030		.71	.83		1.54	2.18
5122	2" x 10", ordinary		630	.025		.99	.71		1.70	2.29
5141	On steep roofs		495	.032		.99	.90		1.89	2.62
5161	On dormers or complex roofs		425	.038		.99	1.05		2.04	2.87
5182	2" x 12", ordinary		575	.028		1.39	.78		2.17	2.84
5201	On steep roofs		455	.035		1.39	.98		2.37	3.18
5221	On dormers or complex roofs		395	.041		1.39	1.13		2.52	3.43
5250	Composite rafter, 9-1/2" deep		575	.028		2	.78		2.78	3.52
5260	11-1/2" deep		575	.028		2.17	.78		2.95	3.70
5301	Hip and valley rafters, 2" x 6", ordinary		760	.021		.52	.59		1.11	1.56
5321	On steep roofs		585	.027		.52	.76		1.28	1.86
5341	On dormers or complex roofs		510	.031		.52	.88		1.40	2.05
5361	2" x 8", ordinary		720	.022		.71	.62		1.33	1.83
5381	On steep roofs		545	.029		.71	.82		1.53	2.17
5401	On dormers or complex roofs		470	.034		.71	.95		1.66	2.39
5421	2" x 10", ordinary		570	.028		.99	.78		1.77	2.42
5441	On steep roofs		440	.036		.99	1.02		2.01	2.81
5461	On dormers or complex roofs	▼	380	.042	▼	.99	1.18		2.17	3.08
5470										
5481	Hip and valley rafters, 2" x 12", ordinary	2 Carp	525	.030	L.F.	1.39	.85		2.24	2.96
5501	On steep roofs		410	.039		1.39	1.09		2.48	3.36
5521	On dormers or complex roofs		355	.045		1.39	1.26		2.65	3.65
5541	Hip and valley jacks, 2" x 6", ordinary		600	.027		.52	.75		1.27	1.83
5561	On steep roofs		475	.034		.52	.94		1.46	2.16
5581	On dormers or complex roofs		410	.039		.52	1.09		1.61	2.41
5601	2" x 8", ordinary		490	.033		.71	.91		1.62	2.32
5621	On steep roofs		385	.042		.71	1.16		1.87	2.74
5641	On dormers or complex roofs		335	.048		.71	1.33		2.04	3.04
5661	2" x 10", ordinary		450	.036		.99	.99		1.98	2.77
5681	On steep roofs		350	.046		.99	1.28		2.27	3.25
5701	On dormers or complex roofs		305	.052		.99	1.47		2.46	3.57
5721	2" x 12", ordinary		375	.043		1.39	1.19		2.58	3.54
5741	On steep roofs		295	.054		1.39	1.52		2.91	4.08
5762	On dormers or complex roofs		255	.063		1.39	1.75		3.14	4.48
5781	Rafter tie, 1" x 4", #3		800	.020		.44	.56		1	1.43
5791	2" x 4", #3		800	.020		.29	.56		.85	1.27
5801	Ridge board, #2 or better, 1" x 6"		600	.027		.64	.75		1.39	1.96
5821	1" x 8"		550	.029		.84	.81		1.65	2.29
5841	1" x 10"		500	.032		1.23	.89		2.12	2.86
5861	2" x 6"		500	.032		.52	.89		1.41	2.08
5881	2" x 8"		450	.036		.71	.99		1.70	2.46
5901	2" x 10"		400	.040		.99	1.12		2.11	2.98
5921	Roof cants, split, 4" x 4"		650	.025		1.56	.69		2.25	2.88
5941	6" x 6"		600	.027		4.87	.75		5.62	6.60
5961	Roof curbs, untreated, 2" x 6"		520	.031		.52	.86		1.38	2.02
5981	2" x 12"		400	.040		1.39	1.12		2.51	3.41
6001	Sister rafters, 2" x 6"		800	.020		.52	.56		1.08	1.52
6021	2" x 8"		640	.025		.71	.70		1.41	1.96
6041	2" x 10"		535	.030		.99	.84		1.83	2.50
6061	2" x 12"	▼	455	.035	▼	1.39	.98		2.37	3.18

06 11 10 – Framing with Dimensional, Engineered or Composite Lumber

06 11 10.32 Sill and Ledger Framing

06 11 10.32 Sill and Ledger Framing	Crew	Daily Output	Labor-Hours	Unit	Material	2009 Bare Costs Labor	Equipment	Total	Total Incl O&P
0010 **SILL AND LEDGER FRAMING**									
2002 Ledgers, nailed, 2" x 4"	2 Carp	755	.021	L.F.	.29	.59		.88	1.32
2052 2" x 6"		600	.027		.52	.75		1.27	1.83
2102 Bolted, not including bolts, 3" x 6"		325	.049		1.52	1.38		2.90	4.01
2152 3" x 12"		233	.069		3.54	1.92		5.46	7.15
2602 Mud sills, redwood, construction grade, 2" x 4"		895	.018		2.27	.50		2.77	3.33
2622 2" x 6"		780	.021		3.40	.57		3.97	4.71
4002 Sills, 2" x 4"		600	.027		.29	.75		1.04	1.58
4052 2" x 6"		550	.029		.52	.81		1.33	1.94
4082 2" x 8"		500	.032		.71	.89		1.60	2.29
4101 2" x 10"		450	.036		.99	.99		1.98	2.77
4121 2" x 12"		400	.040		1.39	1.12		2.51	3.41
4202 Treated, 2" x 4"		550	.029		.44	.81		1.25	1.86
4222 2" x 6"		500	.032		.69	.89		1.58	2.26
4242 2" x 8"		450	.036		1.05	.99		2.04	2.83
4261 2" x 10"		400	.040		1.28	1.12		2.40	3.30
4281 2" x 12"		350	.046		2.03	1.28		3.31	4.39
4402 4" x 4"		450	.036		1.49	.99		2.48	3.32
4422 4" x 6"		350	.046		2.19	1.28		3.47	4.57
4462 4" x 8"		300	.053		3.14	1.49		4.63	6
4480 4" x 10"		260	.062		4.10	1.72		5.82	7.40

06 11 10.34 Sleepers

06 11 10.34 Sleepers	Crew	Daily Output	Labor-Hours	Unit	Material	2009 Bare Costs Labor	Equipment	Total	Total Incl O&P
0010 **SLEEPERS**									
0100 On concrete, treated, 1" x 2"	2 Carp	2350	.007	L.F.	.38	.19		.57	.74
0150 1" x 3"		2000	.008		.52	.22		.74	.95
0200 2" x 4"		1500	.011		.44	.30		.74	.99
0250 2" x 6"		1300	.012		.69	.34		1.03	1.33

06 11 10.36 Soffit and Canopy Framing

06 11 10.36 Soffit and Canopy Framing	Crew	Daily Output	Labor-Hours	Unit	Material	2009 Bare Costs Labor	Equipment	Total	Total Incl O&P
0010 **SOFFIT AND CANOPY FRAMING**									
1002 Canopy or soffit framing , 1" x 4"	2 Carp	900	.018	L.F.	.44	.50		.94	1.32
1021 1" x 6"		850	.019		.64	.53		1.17	1.59
1042 1" x 8"		750	.021		.84	.60		1.44	1.93
1102 2" x 4"		620	.026		.29	.72		1.01	1.54
1121 2" x 6"		560	.029		.52	.80		1.32	1.92
1142 2" x 8"		500	.032		.71	.89		1.60	2.29
1202 3" x 4"		500	.032		.93	.89		1.82	2.53
1221 3" x 6"		400	.040		1.52	1.12		2.64	3.57
1242 3" x 10"		300	.053		2.94	1.49		4.43	5.75
1250									

06 11 10.38 Treated Lumber Framing Material

06 11 10.38 Treated Lumber Framing Material	Crew	Daily Output	Labor-Hours	Unit	Material	2009 Bare Costs Labor	Equipment	Total	Total Incl O&P
0010 **TREATED LUMBER FRAMING MATERIAL**									
0100 2" x 4"				M.B.F.	660			660	730
0110 2" x 6"					685			685	755
0120 2" x 8"					785			785	865
0130 2" x 10"					765			765	845
0140 2" x 12"					1,025			1,025	1,125
0200 4" x 4"					1,125			1,125	1,225
0210 4" x 6"					1,100			1,100	1,200
0220 4" x 8"					1,175			1,175	1,300

06 11 10.40 Wall Framing

06 11 10.40 Wall Framing									
0010 **WALL FRAMING** R061110-30									

06 11 10.40 **Wall Framing**	Crew	Daily Output	Labor-Hours	Unit	Material	2009 Bare Costs Labor	Equipment	Total	Total Incl O&P
0100	Door buck, studs, header, access, 8' H, 2"x4" wall, 3' W	1 Carp	32	.250	Ea.	12.65	7		19.65	26
0110	4' wide		32	.250		13.70	7		20.70	27
0120	5' wide		32	.250		16.70	7		23.70	30
0130	6' wide		32	.250		18.10	7		25.10	32
0140	8' wide		30	.267		25.50	7.45		32.95	40.50
0150	10' wide		30	.267		37.50	7.45		44.95	54
0160	12' wide		30	.267		61.50	7.45		68.95	80.50
0170	2" x 6" wall, 3' wide		32	.250		20	7		27	34
0180	4' wide		32	.250		21	7		28	35
0190	5' wide		32	.250		24	7		31	38.50
0200	6' wide		32	.250		25.50	7		32.50	40
0210	8' wide		30	.267		33	7.45		40.45	49
0220	10' wide		30	.267		45	7.45		52.45	62
0230	12' wide		30	.267		69	7.45		76.45	88.50
0240	Window buck, studs, header & access, 8' high 2" x 4" wall, 2' wide		24	.333		12.90	9.30		22.20	30
0250	3' wide		24	.333		15.55	9.30		24.85	33
0260	4' wide		24	.333		17.20	9.30		26.50	34.50
0270	5' wide		24	.333		20	9.30		29.30	38
0280	6' wide		24	.333		22.50	9.30		31.80	40.50
0300	8' wide		22	.364		31.50	10.15		41.65	52
0310	10' wide		22	.364		44.50	10.15		54.65	66
0320	12' wide		22	.364		70.50	10.15		80.65	94.50
0330	2" x 6" wall, 2' wide		24	.333		22	9.30		31.30	40.50
0340	3' wide		24	.333		25.50	9.30		34.80	44
0350	4' wide		24	.333		27	9.30		36.30	46
0360	5' wide		24	.333		30.50	9.30		39.80	49.50
0370	6' wide		24	.333		33.50	9.30		42.80	52.50
0380	7' wide		24	.333		40.50	9.30		49.80	60.50
0390	8' wide		22	.364		44	10.15		54.15	65.50
0400	10' wide		22	.364		57.50	10.15		67.65	80.50
0410	12' wide		22	.364		84.50	10.15		94.65	110
2002	Headers over openings, 2" x 6"	2 Carp	360	.044	L.F.	.52	1.24		1.76	2.67
2007	2" x 6", pneumatic nailed		432	.037		.52	1.04		1.56	2.32
2052	2" x 8"		340	.047		.71	1.32		2.03	3
2057	2" x 8", pneumatic nailed		408	.039		.71	1.10		1.81	2.63
2101	2" x 10"		320	.050		.99	1.40		2.39	3.45
2106	2" x 10", pneumatic nailed		384	.042		.99	1.16		2.15	3.06
2152	2" x 12"		300	.053		1.39	1.49		2.88	4.04
2157	2" x 12", pneumatic nailed		360	.044		1.39	1.24		2.63	3.62
2191	4" x 10"		240	.067		3.68	1.86		5.54	7.20
2196	4" x 10", pneumatic nailed		288	.056		3.68	1.55		5.23	6.70
2202	4" x 12"		190	.084		4.74	2.35		7.09	9.20
2207	4" x 12", pneumatic nailed		228	.070		4.74	1.96		6.70	8.50
2241	6" x 10"		165	.097		10.90	2.71		13.61	16.60
2246	6" x 10", pneumatic nailed		198	.081		10.90	2.26		13.16	15.80
2251	6" x 12"		140	.114		9.15	3.19		12.34	15.50
2256	6" x 12", pneumatic nailed		168	.095		9.15	2.66		11.81	14.60
5002	Plates, untreated, 2" x 3"		850	.019		.32	.53		.85	1.24
5007	2" x 3", pneumatic nailed		1020	.016		.32	.44		.76	1.09
5022	2" x 4"		800	.020		.29	.56		.85	1.27
5027	2" x 4", pneumatic nailed		960	.017		.29	.47		.76	1.11
5040	2" x 6"		750	.021		.52	.60		1.12	1.58
5045	2" x 6", pneumatic nailed		900	.018		.52	.50		1.02	1.41

06 11 Wood Framing

06 11 10 – Framing with Dimensional, Engineered or Composite Lumber

06 11 10.40 Wall Framing	Crew	Daily Output	Labor-Hours	Unit	Material	2009 Bare Costs Labor	Equipment	Total	Total Incl O&P
5061	Treated, 2" x 3"	2 Carp	850	.019	L.F.	.48	.53		1.01	1.42
5066	2" x 3", treated, pneumatic nailed		1020	.016		.48	.44		.92	1.27
5081	2" x 4"		800	.020		.44	.56		1	1.44
5086	2" x 4", treated, pneumatic nailed		960	.017		.44	.47		.91	1.28
5101	2" x 6"		750	.021		.69	.60		1.29	1.76
5106	2" x 6", treated, pneumatic nailed		900	.018		.69	.50		1.19	1.59
5122	Studs, 8' high wall, 2" x 3"		1200	.013		.32	.37		.69	.98
5127	2" x 3", pneumatic nailed		1440	.011		.32	.31		.63	.87
5142	2" x 4"		1100	.015		.29	.41		.70	1.01
5147	2" x 4", pneumatic nailed		1320	.012		.29	.34		.63	.89
5162	2" x 6"		1000	.016		.52	.45		.97	1.33
5167	2" x 6", pneumatic nailed		1200	.013		.52	.37		.89	1.20
5182	3" x 4"		800	.020		.93	.56		1.49	1.97
5187	3" x 4", pneumatic nailed		960	.017		.93	.47		1.40	1.81
5201	Installed on second story, 2" x 3"		1170	.014		.32	.38		.70	1
5206	2" x 3", pneumatic nailed		1200	.013		.32	.37		.69	.98
5221	2" x 4"		1015	.016		.29	.44		.73	1.06
5226	2" x 4", pneumatic nailed		1080	.015		.29	.41		.70	1.02
5241	2" x 6"		890	.018		.52	.50		1.02	1.42
5246	2" x 6", pneumatic nailed		1020	.016		.52	.44		.96	1.31
5261	3" x 4"		800	.020		.93	.56		1.49	1.97
5266	3" x 4", pneumatic nailed		960	.017		.93	.47		1.40	1.81
5281	Installed on dormer or gable, 2" x 3"		1045	.015		.32	.43		.75	1.07
5286	2" x 3", pneumatic nailed		1254	.013		.32	.36		.68	.95
5301	2" x 4"		905	.018		.29	.49		.78	1.16
5306	2" x 4", pneumatic nailed		1086	.015		.29	.41		.70	1.02
5321	2" x 6"		800	.020		.52	.56		1.08	1.52
5326	2" x 6", pneumatic nailed		960	.017		.52	.47		.99	1.36
5341	3" x 4"		700	.023		.93	.64		1.57	2.10
5346	3" x 4", pneumatic nailed		840	.019		.93	.53		1.46	1.92
5361	6' high wall, 2" x 3"		970	.016		.32	.46		.78	1.13
5366	2" x 3", pneumatic nailed		1164	.014		.32	.38		.70	1
5381	2" x 4"		850	.019		.29	.53		.82	1.21
5386	2" x 4", pneumatic nailed		1020	.016		.29	.44		.73	1.06
5401	2" x 6"		740	.022		.52	.60		1.12	1.59
5406	2" x 6", pneumatic nailed		888	.018		.52	.50		1.02	1.42
5421	3" x 4"		600	.027		.93	.75		1.68	2.28
5426	3" x 4", pneumatic nailed		720	.022		.93	.62		1.55	2.07
5441	Installed on second story, 2" x 3"		950	.017		.32	.47		.79	1.15
5446	2" x 3", pneumatic nailed		1140	.014		.32	.39		.71	1.01
5461	2" x 4"		810	.020		.29	.55		.84	1.25
5466	2" x 4", pneumatic nailed		972	.016		.29	.46		.75	1.10
5481	2" x 6"		700	.023		.52	.64		1.16	1.65
5486	2" x 6", pneumatic nailed		840	.019		.52	.53		1.05	1.47
5501	3" x 4"		550	.029		.93	.81		1.74	2.39
5506	3" x 4", pneumatic nailed		660	.024		.93	.68		1.61	2.17
5521	Installed on dormer or gable, 2" x 3"		850	.019		.32	.53		.85	1.24
5526	2" x 3", pneumatic nailed		1020	.016		.32	.44		.76	1.09
5541	2" x 4"		720	.022		.29	.62		.91	1.37
5546	2" x 4", pneumatic nailed		864	.019		.29	.52		.81	1.20
5561	2" x 6"		620	.026		.52	.72		1.24	1.79
5566	2" x 6", pneumatic nailed		744	.022		.52	.60		1.12	1.59
5581	3" x 4"		480	.033		.93	.93		1.86	2.59

06 11 Wood Framing

06 11 10 – Framing with Dimensional, Engineered or Composite Lumber

06 11 10.40 Wall Framing

		Crew	Daily Output	Labor-Hours	Unit	Material	2009 Bare Costs Labor	Equipment	Total	Total Incl O&P
5586	3" x 4", pneumatic nailed	2 Carp	576	.028	L.F.	.93	.78		1.71	2.33
5601	3' high wall, 2" x 3"		740	.022		.32	.60		.92	1.37
5606	2" x 3", pneumatic nailed		888	.018		.32	.50		.82	1.20
5621	2" x 4"		640	.025		.29	.70		.99	1.50
5626	2" x 4", pneumatic nailed		768	.021		.29	.58		.87	1.30
5641	2" x 6"		550	.029		.52	.81		1.33	1.94
5646	2" x 6", pneumatic nailed		660	.024		.52	.68		1.20	1.72
5661	3" x 4"		440	.036		.93	1.02		1.95	2.74
5666	3" x 4", pneumatic nailed		528	.030		.93	.85		1.78	2.45
5681	Installed on second story, 2" x 3"		700	.023		.32	.64		.96	1.43
5686	2" x 3", pneumatic nailed		840	.019		.32	.53		.85	1.25
5701	2" x 4"		610	.026		.29	.73		1.02	1.56
5706	2" x 4", pneumatic nailed		732	.022		.29	.61		.90	1.35
5721	2" x 6"		520	.031		.52	.86		1.38	2.02
5726	2" x 6", pneumatic nailed		624	.026		.52	.72		1.24	1.78
5741	3" x 4"		430	.037		.93	1.04		1.97	2.78
5746	3" x 4", pneumatic nailed		516	.031		.93	.87		1.80	2.49
5761	Installed on dormer or gable, 2" x 3"		625	.026		.32	.72		1.04	1.56
5766	2" x 3", pneumatic nailed		750	.021		.32	.60		.92	1.36
5781	2" x 4"		545	.029		.29	.82		1.11	1.71
5786	2" x 4", pneumatic nailed		654	.024		.29	.68		.97	1.48
5801	2" x 6"		465	.034		.52	.96		1.48	2.20
5806	2" x 6", pneumatic nailed		558	.029		.52	.80		1.32	1.92
5821	3" x 4"		380	.042		.93	1.18		2.11	3.01
5826	3" x 4", pneumatic nailed		456	.035		.93	.98		1.91	2.68
8250	For second story & above, add						5%			
8300	For dormer & gable, add						15%			

06 11 10.42 Furring

		Crew	Daily Output	Labor-Hours	Unit	Material	2009 Bare Costs Labor	Equipment	Total	Total Incl O&P
0010	**FURRING**									
0012	Wood strips, 1" x 2", on walls, on wood	1 Carp	550	.015	L.F.	.25	.41		.66	.97
0017	Pneumatic nailed		710	.011		.25	.32		.57	.81
0300	On masonry		495	.016		.25	.45		.70	1.04
0400	On concrete		260	.031		.25	.86		1.11	1.73
0600	1" x 3", on walls, on wood		550	.015		.34	.41		.75	1.07
0607	Pneumatic nailed		710	.011		.34	.32		.66	.91
0700	On masonry		495	.016		.34	.45		.79	1.14
0800	On concrete		260	.031		.34	.86		1.20	1.83
0850	On ceilings, on wood		350	.023		.34	.64		.98	1.46
0857	Pneumatic nailed		450	.018		.34	.50		.84	1.22
0900	On masonry		320	.025		.34	.70		1.04	1.56
0950	On concrete		210	.038		.34	1.06		1.40	2.18

06 11 10.44 Grounds

		Crew	Daily Output	Labor-Hours	Unit	Material	2009 Bare Costs Labor	Equipment	Total	Total Incl O&P
0010	**GROUNDS**									
0020	For casework, 1" x 2" wood strips, on wood	1 Carp	330	.024	L.F.	.25	.68		.93	1.43
0100	On masonry		285	.028		.25	.78		1.03	1.61
0200	On concrete		250	.032		.25	.89		1.14	1.79
0400	For plaster, 3/4" deep, on wood		450	.018		.25	.50		.75	1.12
0500	On masonry		225	.036		.25	.99		1.24	1.96
0600	On concrete		175	.046		.25	1.28		1.53	2.44
0700	On metal lath		200	.040		.25	1.12		1.37	2.17

06 12 Structural Panels

06 12 10 – Structural Insulated Panels

06 12 10.10 OSB Faced Panels		Crew	Daily Output	Labor-Hours	Unit	Material	2009 Bare Costs Labor	Equipment	Total	Total Incl O&P
0010	**OSB FACED PANELS**									
0100	Structural insul. panels, 7/16" OSB both faces, EPS insul, 3-5/8" T	G	F-3	2075	.019	S.F.	2.84	.49	.37	3.70	4.35
0110	5-5/8" thick	G		1725	.023		3.37	.59	.45	4.41	5.20
0120	7-3/8" thick	G		1425	.028		3.84	.71	.54	5.09	6
0130	9-3/8" thick	G		1125	.036		4.40	.90	.68	5.98	7.10
0140	7/16" OSB one face, EPS insul, 3-5/8" thick	G		2175	.018		1.65	.47	.35	2.47	2.99
0150	5-5/8" thick	G		1825	.022		2.06	.56	.42	3.04	3.66
0160	7-3/8" thick	G		1525	.026		2.40	.67	.50	3.57	4.31
0170	9-3/8" thick	G		1225	.033		2.81	.83	.63	4.27	5.15
0190	7/16" OSB - 1/2" GWB faces , EPS insul, 3-5/8" T	G		2075	.019		3.03	.49	.37	3.89	4.56
0200	5-5/8" thick	G		1725	.023		3.34	.59	.45	4.38	5.15
0210	7-3/8" thick	G		1425	.028		3.65	.71	.54	4.90	5.80
0220	9-3/8" thick	G		1125	.036		4	.90	.68	5.58	6.65
0240	7/16" OSB - 1/2" MRGWB faces , EPS insul, 3-5/8" T	G		2075	.019		3.03	.49	.37	3.89	4.56
0250	5-5/8" thick	G		1725	.023		3.34	.59	.45	4.38	5.15
0260	7-3/8" thick			1425	.028		3.65	.71	.54	4.90	5.80
0270	9-3/8" thick		▼	1125	.036		4	.90	.68	5.58	6.65
0300	For 1/2" GWB added to OSB skin, add	G					.60			.60	.66
0310	For 1/2" MRGWB added to OSB skin, add	G					.75			.75	.83
0320	For one T1-11 skin, add to OSB-OSB	G					.74			.74	.81
0330	For one 19/32" CDX skin, add to OSB-OSB	G				▼	.69			.69	.76
0500	Structural insulated panel, 7/16" OSB both sides, straw core										
0510	4-3/8" T, walls (w/sill, splines, plates)	G	F-6	2400	.017	S.F.	7.60	.42	.32	8.34	9.40
0520	Floors (w/ splines)	G		2400	.017		7.60	.42	.32	8.34	9.40
0530	Roof (w/ splines)	G		2400	.017		7.60	.42	.32	8.34	9.40
0550	7-7/8" T, walls (w/sill, splines, plates)	G		2400	.017		10.95	.42	.32	11.69	13.10
0560	Floors (w/ splines)	G		2400	.017		10.95	.42	.32	11.69	13.10
0570	Roof (w/ splines)	G	▼	2400	.017	▼	10.95	.42	.32	11.69	13.10

06 13 Heavy Timber

06 13 13 – Log Construction

06 13 13.10 Log Structures

		Crew	Daily Output	Labor-Hours	Unit	Material	2009 Bare Costs Labor	Equipment	Total	Total Incl O&P
0010	**LOG STRUCTURES**									
0020	Exterior walls, pine, D logs, with double T & G, 6" x 6"	2 Carp	500	.032	L.F.	3.98	.89		4.87	5.90
0030	6" x 8"		375	.043		4.78	1.19		5.97	7.25
0040	8" x 6"		375	.043		3.98	1.19		5.17	6.40
0050	8" x 7"		322	.050		3.98	1.39		5.37	6.75
0060	Square/rectangular logs, with double T & G, 6" x 6"		500	.032		3.98	.89		4.87	5.90
0160	Upper floor framing, pine, posts/columns, 4" x 6"		750	.021		2.34	.60		2.94	3.58
0180	4" x 8"		562	.028		2.79	.80		3.59	4.42
0190	6" x 6"		500	.032		4.87	.89		5.76	6.85
0200	6" x 8"		375	.043		7.60	1.19		8.79	10.35
0210	8" x 8"		281	.057		7.80	1.59		9.39	11.30
0220	8" x 10"		225	.071		9.75	1.99		11.74	14.10
0230	Beams, 4" x 8"		562	.028		2.79	.80		3.59	4.42
0240	4" x 10"		449	.036		3.68	1		4.68	5.75
0250	4" x 12"		375	.043		4.74	1.19		5.93	7.20
0260	6" x 8"		375	.043		7.60	1.19		8.79	10.35
0270	6" x 10"		300	.053		10.90	1.49		12.39	14.50
0280	6" x 12"		250	.064		9.15	1.79		10.94	13.10
0290	8" x 10"	▼	225	.071	▼	9.75	1.99		11.74	14.10

06 13 Heavy Timber

06 13 13 – Log Construction

06 13 13.10 Log Structures

06 13 13.10 Log Structures		Crew	Daily Output	Labor-Hours	Unit	Material	2009 Bare Costs Labor	Equipment	Total	Total Incl O&P
0300	8" x 12"	2 Carp	188	.085	L.F.	9.05	2.38		11.43	13.95
0310	Joists, 4" x 8"		562	.028		2.79	.80		3.59	4.42
0320	4" x 10"		449	.036		3.68	1		4.68	5.75
0330	4" x 12"		375	.043		4.74	1.19		5.93	7.20
0340	6" x 8"		375	.043		7.60	1.19		8.79	10.35
0350	6" x 10"		300	.053	▼	10.90	1.49		12.39	14.50
0390	Decking, 1" x 6" T & G		964	.017	S.F.	1.53	.46		1.99	2.46
0400	1" x 8" T & G		700	.023	"	1.39	.64		2.03	2.61
0420	Gable end roof framing, rafters, 4" x 8"	▼	562	.028	L.F.	2.79	.80		3.59	4.42

06 13 23 – Heavy Timber Construction

06 13 23.10 Heavy Framing

06 13 23.10 Heavy Framing		Crew	Daily Output	Labor-Hours	Unit	Material	2009 Bare Costs Labor	Equipment	Total	Total Incl O&P
0010	**HEAVY FRAMING**									
0020	Beams, single 6" x 10"	2 Carp	1.10	14.545	M.B.F.	1,825	405		2,230	2,675
0100	Single 8" x 16"		1.20	13.333	"	2,900	375		3,275	3,825
0202	Built from 2" lumber, multiple 2" x 14"		900	.018	B.F.	.92	.50		1.42	1.85
0212	Built from 3" lumber, multiple 3" x 6"		700	.023		1.02	.64		1.66	2.20
0222	Multiple 3" x 8"		800	.020		1.05	.56		1.61	2.11
0232	Multiple 3" x 10"		900	.018		1.18	.50		1.68	2.13
0242	Multiple 3" x 12"		1000	.016		1.18	.45		1.63	2.06
0252	Built from 4" lumber, multiple 4" x 6"		800	.020		1.17	.56		1.73	2.24
0262	Multiple 4" x 8"		900	.018		1.05	.50		1.55	1.99
0272	Multiple 4" x 10"		1000	.016		1.10	.45		1.55	1.97
0282	Multiple 4" x 12"		1100	.015	▼	1.18	.41		1.59	1.99
0292	Columns, structural grade, 1500f, 4" x 4"		450	.036	L.F.	1.75	.99		2.74	3.61
0302	6" x 6"		225	.071		4.85	1.99		6.84	8.70
0402	8" x 8"		240	.067		10.75	1.86		12.61	15
0502	10" x 10"		90	.178		19	4.97		23.97	29.50
0602	12" x 12"		70	.229	▼	28	6.40		34.40	41.50
0802	Floor planks, 2" thick, T & G, 2" x 6"		1050	.015	B.F.	1.54	.43		1.97	2.42
0902	2" x 10"		1100	.015		1.54	.41		1.95	2.39
1102	3" thick, 3" x 6"		1050	.015		1.30	.43		1.73	2.15
1202	3" x 10"		1100	.015		1.30	.41		1.71	2.12
1402	Girders, structural grade, 12" x 12"		800	.020		2.32	.56		2.88	3.51
1502	10" x 16"		1000	.016		2.28	.45		2.73	3.27
2302	Roof purlins, 4" thick, structural grade		1050	.015		1.46	.43		1.89	2.33
2502	Roof trusses, add timber connectors, Division 06 05 23.60	▼	450	.036	▼	1.34	.99		2.33	3.16

06 15 Wood Decking

06 15 16 – Wood Roof Decking

06 15 16.10 Solid Wood Roof Decking

06 15 16.10 Solid Wood Roof Decking		Crew	Daily Output	Labor-Hours	Unit	Material	2009 Bare Costs Labor	Equipment	Total	Total Incl O&P
0010	**SOLID WOOD ROOF DECKING**									
0400	Cedar planks, 3" thick	2 Carp	320	.050	S.F.	6.20	1.40		7.60	9.20
0500	4" thick		250	.064		8.35	1.79		10.14	12.20
0702	Douglas fir, 3" thick		320	.050		2.33	1.40		3.73	4.92
0802	4" thick		250	.064		3.12	1.79		4.91	6.45
1002	Hemlock, 3" thick		320	.050		2.33	1.40		3.73	4.92
1102	4" thick		250	.064		3.11	1.79		4.90	6.45
1302	Western white spruce, 3" thick		320	.050		2.24	1.40		3.64	4.82
1402	4" thick	▼	250	.064	▼	2.99	1.79		4.78	6.30

06 15 Wood Decking

06 15 23 – Laminated Wood Decking

06 15 23.10 Laminated Roof Deck

	06 15 23.10 Laminated Roof Deck	Crew	Daily Output	Labor-Hours	Unit	Material	2009 Bare Costs Labor	Equipment	Total	Total Incl O&P
0010	**LAMINATED ROOF DECK**									
0020	Pine or hemlock, 3" thick	2 Carp	425	.038	S.F.	3.36	1.05		4.41	5.50
0100	4" thick		325	.049		4.47	1.38		5.85	7.25
0300	Cedar, 3" thick		425	.038		4.11	1.05		5.16	6.30
0400	4" thick		325	.049		5.25	1.38		6.63	8.10
0600	Fir, 3" thick		425	.038		3.43	1.05		4.48	5.55
0700	4" thick		325	.049		4.29	1.38		5.67	7.05

06 16 Sheathing

06 16 23 – Subflooring

06 16 23.10 Subfloor

	06 16 23.10 Subfloor	Crew	Daily Output	Labor-Hours	Unit	Material	2009 Bare Costs Labor	Equipment	Total	Total Incl O&P
0010	**SUBFLOOR**									
0011	Plywood, CDX, 1/2" thick	2 Carp	1500	.011	SF Flr.	.46	.30		.76	1.01
0015	Pneumatic nailed		1860	.009		.46	.24		.70	.92
0017	Pneumatic nailed		1860	.009		.46	.24		.70	.92
0102	5/8" thick		1350	.012		.67	.33		1	1.30
0107	Pneumatic nailed		1674	.010		.67	.27		.94	1.19
0202	3/4" thick		1250	.013		.83	.36		1.19	1.51
0207	Pneumatic nailed		1550	.010		.83	.29		1.12	1.40
0302	1-1/8" thick, 2-4-1 including underlayment		1050	.015		1.63	.43		2.06	2.51
0440	With boards, 1" x 6", S4S, laid regular		900	.018		1.40	.50		1.90	2.37
0452	1" x 8", laid regular		1000	.016		1.34	.45		1.79	2.23
0462	Laid diagonal		850	.019		1.34	.53		1.87	2.36
0502	1" x 10", laid regular		1100	.015		1.54	.41		1.95	2.39
0602	Laid diagonal		900	.018		1.54	.50		2.04	2.54
1500	OSB, 5/8" thick		1330	.012	S.F.	.54	.34		.88	1.16
1600	3/4" thick		1230	.013	"	.65	.36		1.01	1.33
8990	Subfloor adhesive, 3/8" bead	1 Carp	2300	.003	L.F.	.09	.10		.19	.26

06 16 26 – Underlayment

06 16 26.10 Wood Product Underlayment

	06 16 26.10 Wood Product Underlayment	Crew	Daily Output	Labor-Hours	Unit	Material	2009 Bare Costs Labor	Equipment	Total	Total Incl O&P
0010	**WOOD PRODUCT UNDERLAYMENT**									
0030	Plywood, underlayment grade, 3/8" thick	2 Carp	1500	.011	SF Flr.	1.03	.30		1.33	1.63
0080	Pneumatic nailed		1860	.009		1.03	.24		1.27	1.54
0102	1/2" thick		1450	.011		1	.31		1.31	1.62
0107	Pneumatic nailed		1798	.009		1	.25		1.25	1.52
0202	5/8" thick		1400	.011		1.29	.32		1.61	1.96
0207	Pneumatic nailed		1736	.009		1.29	.26		1.55	1.86
0302	3/4" thick		1300	.012		1.33	.34		1.67	2.04
0306	Pneumatic nailed		1612	.010		1.33	.28		1.61	1.93
0502	Particle board, 3/8" thick		1500	.011		.41	.30		.71	.95
0507	Pneumatic nailed		1860	.009		.41	.24		.65	.86
0602	1/2" thick		1450	.011		.45	.31		.76	1.02
0607	Pneumatic nailed		1798	.009		.45	.25		.70	.92
0802	5/8" thick		1400	.011		.54	.32		.86	1.13
0807	Pneumatic nailed		1736	.009		.54	.26		.80	1.03
0902	3/4" thick		1300	.012		.65	.34		.99	1.30
0907	Pneumatic nailed		1612	.010		.65	.28		.93	1.19
0955	Particleboard, 100% recycled straw/wheat, 4' x 8' x 1/4" **G**		1450	.011	S.F.	.26	.31		.57	.81
0960	4' x 8' x 3/8" **G**		1450	.011		.39	.31		.70	.95
0965	4' x 8' x 1/2" **G**		1350	.012		.53	.33		.86	1.14

06 16 Sheathing

06 16 26 – Underlayment

06 16 26.10 Wood Product Underlayment		Crew	Daily Output	Labor-Hours	Unit	Material	2009 Bare Costs Labor	Equipment	Total	Total Incl O&P
0970	4' x 8' x 5/8" G	2 Carp	1300	.012	S.F.	.63	.34		.97	1.27
0975	4' x 8' x 3/4" G		1250	.013		.73	.36		1.09	1.40
0980	4' x 8' x 1" G		1150	.014		.98	.39		1.37	1.74
0985	4' x 8' x 1-1/4" G		1100	.015		1.15	.41		1.56	1.96
1102	Hardboard, underlayment grade, 4' x 4', .215" thick		1500	.011	SF Flr.	.50	.30		.80	1.05

06 16 36 – Wood Panel Product Sheathing

06 16 36.10 Sheathing

06 16 36.10 Sheathing		Crew	Daily Output	Labor-Hours	Unit	Material	2009 Bare Costs Labor	Equipment	Total	Total Incl O&P
0010	**SHEATHING**, plywood on roofs R061636-20									
0012	Plywood on roofs, CDX									
0032	5/16" thick	2 Carp	1600	.010	S.F.	.53	.28		.81	1.05
0037	Pneumatic nailed		1952	.008		.53	.23		.76	.97
0052	3/8" thick		1525	.010		.45	.29		.74	.99
0057	Pneumatic nailed		1860	.009		.45	.24		.69	.90
0102	1/2" thick		1400	.011		.46	.32		.78	1.05
0103	Pneumatic nailed		1708	.009		.46	.26		.72	.95
0202	5/8" thick		1300	.012		.67	.34		1.01	1.32
0207	Pneumatic nailed		1586	.010		.67	.28		.95	1.22
0302	3/4" thick		1200	.013		.83	.37		1.20	1.54
0307	Pneumatic nailed		1464	.011		.83	.31		1.14	1.43
0502	Plywood on walls with exterior CDX, 3/8" thick		1200	.013		.45	.37		.82	1.12
0507	Pneumatic nailed		1488	.011		.45	.30		.75	1
0603	1/2" thick		1125	.014		.46	.40		.86	1.18
0608	Pneumatic nailed		1395	.011		.46	.32		.78	1.05
0702	5/8" thick		1050	.015		.67	.43		1.10	1.46
0707	Pneumatic nailed		1302	.012		.67	.34		1.01	1.32
0803	3/4" thick		975	.016		.83	.46		1.29	1.69
0808	Pneumatic nailed		1209	.013		.83	.37		1.20	1.54
0840	Oriented strand board, 7/16" thick G		1400	.011		.45	.32		.77	1.04
0845	Pneumatic nailed G		1736	.009		.45	.26		.71	.94
0846	1/2" thick G		1325	.012		.22	.34		.56	.81
0847	Pneumatic nailed G		1643	.010		.22	.27		.49	.70
0852	5/8" thick G		1250	.013		.35	.36		.71	.99
0857	Pneumatic nailed G		1550	.010		.35	.29		.64	.88
1000	For shear wall construction, add						20%			
1200	For structural 1 exterior plywood, add				S.F.	10%				
1402	With boards, on roof 1" x 6" boards, laid horizontal	2 Carp	725	.022		1.40	.62		2.02	2.57
1502	Laid diagonal		650	.025		1.40	.69		2.09	2.69
1702	1" x 8" boards, laid horizontal		875	.018		1.34	.51		1.85	2.33
1802	Laid diagonal		725	.022		1.34	.62		1.96	2.51
2000	For steep roofs, add						40%			
2200	For dormers, hips and valleys, add					5%	50%			
2402	Boards on walls, 1" x 6" boards, laid regular	2 Carp	650	.025		1.40	.69		2.09	2.69
2502	Laid diagonal		585	.027		1.40	.76		2.16	2.82
2702	1" x 8" boards, laid regular		765	.021		1.34	.58		1.92	2.46
2802	Laid diagonal		650	.025		1.34	.69		2.03	2.63
2852	Gypsum, weatherproof, 1/2" thick		1050	.015		.61	.43		1.04	1.39
2900	With embedded glass mats		1100	.015		.49	.41		.90	1.23
3000	Wood fiber, regular, no vapor barrier, 1/2" thick		1200	.013		.56	.37		.93	1.25
3100	5/8" thick		1200	.013		.71	.37		1.08	1.41
3300	No vapor barrier, in colors, 1/2" thick		1200	.013		.75	.37		1.12	1.46
3400	5/8" thick		1200	.013		.80	.37		1.17	1.51
3600	With vapor barrier one side, white, 1/2" thick		1200	.013		.55	.37		.92	1.24

06 16 Sheathing

06 16 36 – Wood Panel Product Sheathing

06 16 36.10 Sheathing

	06 16 36.10 Sheathing	Crew	Daily Output	Labor-Hours	Unit	Material	2009 Bare Costs Labor	Equipment	Total	Total Incl O&P
3700	Vapor barrier 2 sides, 1/2" thick	2 Carp	1200	.013	S.F.	.77	.37		1.14	1.48
3800	Asphalt impregnated, 25/32" thick		1200	.013		.28	.37		.65	.94
3850	Intermediate, 1/2" thick		1200	.013		.18	.37		.55	.83
4000	OSB on roof, 1/2" thick G		1455	.011		.22	.31		.53	.76
4100	5/8" thick G		1330	.012		.35	.34		.69	.96

06 17 Shop-Fabricated Structural Wood

06 17 33 – Wood I-Joists

06 17 33.10 Wood and Composite I-Joists

		Crew	Daily Output	Labor-Hours	Unit	Material	2009 Bare Costs Labor	Equipment	Total	Total Incl O&P
0010	**WOOD AND COMPOSITE I-JOISTS**									
0100	Plywood webs, incl. bridging & blocking, panels 24" O.C.									
1200	15' to 24' span, 50 psf live load	F-5	2400	.013	SF Flr.	2.28	.33		2.61	3.06
1300	55 psf live load		2250	.014		2.47	.35		2.82	3.31
1400	24' to 30' span, 45 psf live load		2600	.012		3.18	.30		3.48	4.01
1500	55 psf live load		2400	.013		3.70	.33		4.03	4.62
1600	Tubular steel open webs, 45 psf, 24" O.C., 40' span	F-3	6250	.006		2.24	.16	.12	2.52	2.87
1700	55' span		7750	.005		2.17	.13	.10	2.40	2.72
1800	70' span		9250	.004		2.82	.11	.08	3.01	3.37
1900	85 psf live load, 26' span		2300	.017		2.63	.44	.33	3.40	4

06 17 53 – Shop-Fabricated Wood Trusses

06 17 53.10 Roof Trusses

		Crew	Daily Output	Labor-Hours	Unit	Material	2009 Bare Costs Labor	Equipment	Total	Total Incl O&P
0010	**ROOF TRUSSES**									
5000	Common wood, 2" x 4" metal plate connected, 24" O.C., 4/12 slope									
5010	1' overhang, 12' span	F-5	55	.582	Ea.	47	14.20		61.20	75.50
5050	20' span	F-6	62	.645		72	16.30	12.40	100.70	121
5100	24' span		60	.667		84	16.85	12.80	113.65	135
5150	26' span		57	.702		87	17.75	13.50	118.25	141
5200	28' span		53	.755		77.50	19.05	14.50	111.05	133
5240	30' span		51	.784		108	19.80	15.05	142.85	169
5250	32' span		50	.800		112	20	15.40	147.40	174
5280	34' span		48	.833		136	21	16	173	203
5350	8/12 pitch, 1' overhang, 20' span		57	.702		76.50	17.75	13.50	107.75	129
5400	24' span		55	.727		97	18.40	14	129.40	153
5450	26' span		52	.769		101	19.45	14.80	135.25	160
5500	28' span		49	.816		111	20.50	15.70	147.20	175
5550	32' span		45	.889		134	22.50	17.10	173.60	203
5600	36' span		41	.976		167	24.50	18.75	210.25	245
5650	38' span		40	1		182	25.50	19.20	226.70	264
5700	40' span		40	1		188	25.50	19.20	232.70	270

06 18 Glued-Laminated Construction

06 18 13 – Glued-Laminated Beams

06 18 13.10 Laminated Beams

		Crew	Daily Output	Labor-Hours	Unit	Material	2009 Bare Costs Labor	Equipment	Total	Total Incl O&P
0010	**LAMINATED BEAMS**									
0050	3-1/2" x 18"	F-3	480	.083	L.F.	25.50	2.12	1.60	29.22	33.50
0100	5-1/4" x 11-7/8"		450	.089		25	2.26	1.71	28.97	33
0150	5-1/4" x 16"		360	.111		33.50	2.82	2.14	38.46	44
0200	5-1/4" x 18"		290	.138		38	3.50	2.65	44.15	51
0250	5-1/4" x 24"		220	.182		43.50	4.62	3.49	51.61	59.50
0300	7" x 11-7/8"		320	.125		36	3.17	2.40	41.57	47.50
0350	7" x 16"		260	.154		49	3.91	2.96	55.87	64
0400	7" x 18"		210	.190		56.50	4.84	3.66	65	74
0500	For premium appearance, add to S.F. prices					5%				
0550	For industrial type, deduct					15%				
0600	For stain and varnish, add					5%				
0650	For 3/4" laminations, add					25%				

06 18 13.20 Laminated Framing

		Crew	Daily Output	Labor-Hours	Unit	Material	2009 Bare Costs Labor	Equipment	Total	Total Incl O&P
0010	**LAMINATED FRAMING**									
0020	30 lb., short term live load, 15 lb. dead load									
0200	Straight roof beams, 20' clear span, beams 8' O.C.	F-3	2560	.016	SF Flr.	1.98	.40	.30	2.68	3.18
0300	Beams 16' O.C.		3200	.013		1.43	.32	.24	1.99	2.36
0500	40' clear span, beams 8' O.C.		3200	.013		3.78	.32	.24	4.34	4.95
0600	Beams 16' O.C.		3840	.010		3.09	.26	.20	3.55	4.06
0800	60' clear span, beams 8' O.C.	F-4	2880	.014		6.50	.35	.39	7.24	8.15
0900	Beams 16' O.C.	"	3840	.010		4.84	.26	.29	5.39	6.05
1100	Tudor arches, 30' to 40' clear span, frames 8' O.C.	F-3	1680	.024		8.45	.60	.46	9.51	10.80
1200	Frames 16' O.C.	"	2240	.018		6.65	.45	.34	7.44	8.45
1400	50' to 60' clear span, frames 8' O.C.	F-4	2200	.018		9.15	.46	.51	10.12	11.40
1500	Frames 16' O.C.		2640	.015		7.80	.38	.43	8.61	9.65
1700	Radial arches, 60' clear span, frames 8' O.C.		1920	.021		8.55	.53	.59	9.67	10.95
1800	Frames 16' O.C.		2880	.014		6.55	.35	.39	7.29	8.25
2000	100' clear span, frames 8' O.C.		1600	.025		8.85	.63	.70	10.18	11.60
2100	Frames 16' O.C.		2400	.017		7.80	.42	.47	8.69	9.80
2300	120' clear span, frames 8' O.C.		1440	.028		11.75	.71	.78	13.24	15
2400	Frames 16' O.C.		1920	.021		10.75	.53	.59	11.87	13.35
2600	Bowstring trusses, 20' O.C., 40' clear span	F-3	2400	.017		5.30	.42	.32	6.04	6.90
2700	60' clear span	F-4	3600	.011		4.76	.28	.31	5.35	6.05
2800	100' clear span		4000	.010		6.75	.25	.28	7.28	8.15
2900	120' clear span		3600	.011		7.25	.28	.31	7.84	8.75
3100	For premium appearance, add to S.F. prices					5%				
3300	For industrial type, deduct					15%				
3500	For stain and varnish, add					5%				
3900	For 3/4" laminations, add to straight					25%				
4100	Add to curved					15%				
4300	Alternate pricing method: (use nominal footage of									
4310	components). Straight beams, camber less than 6"	F-3	3.50	11.429	M.B.F.	2,925	290	220	3,435	3,950
4400	Columns, including hardware		2	20		3,150	510	385	4,045	4,750
4600	Curved members, radius over 32'		2.50	16		3,225	405	310	3,940	4,575
4700	Radius 10' to 32'		3	13.333		3,200	340	256	3,796	4,375
4900	For complicated shapes, add maximum					100%				
5100	For pressure treating, add to straight					35%				
5200	Add to curved					45%				
6000	Laminated veneer members, southern pine or western species									
6050	1-3/4" wide x 5-1/2" deep	2 Carp	480	.033	L.F.	2.99	.93		3.92	4.86
6100	9-1/2" deep		480	.033		4.79	.93		5.72	6.80

06 18 Glued-Laminated Construction

06 18 13 – Glued-Laminated Beams

06 18 13.20 Laminated Framing		Crew	Daily Output	Labor-Hours	Unit	Material	2009 Bare Costs Labor	Equipment	Total	Total Incl O&P
6150	14" deep	2 Carp	450	.036	L.F.	7	.99		7.99	9.40
6200	18" deep	↓	450	.036	↓	9.05	.99		10.04	11.70
6300	Parallel strand members, southern pine or western species									
6350	1-3/4" wide x 9-1/4" deep	2 Carp	480	.033	L.F.	4.79	.93		5.72	6.80
6400	11-1/4" deep		450	.036		5.95	.99		6.94	8.20
6450	14" deep		400	.040		7	1.12		8.12	9.60
6500	3-1/2" wide x 9-1/4" deep		480	.033		15.85	.93		16.78	19
6550	11-1/4" deep		450	.036		19.15	.99		20.14	22.50
6600	14" deep		400	.040		23.50	1.12		24.62	27.50
6650	7" wide x 9-1/4" deep		450	.036		33	.99		33.99	38
6700	11-1/4" deep		420	.038		39.50	1.06		40.56	45.50
6750	14" deep	↓	400	.040	↓	46.50	1.12		47.62	53.50
8000	Straight beams									
8102	20' span									
8104	3-1/8" x 9"	F-3	30	1.333	Ea.	138	34	25.50	197.50	236
8106	X 10-1/2"		30	1.333		160	34	25.50	219.50	261
8108	X 12"		30	1.333		183	34	25.50	242.50	287
8110	X 13-1/2"		30	1.333		206	34	25.50	265.50	310
8112	X 15"		29	1.379		229	35	26.50	290.50	340
8114	5-1/8" x 10-1/2"		30	1.333		263	34	25.50	322.50	375
8116	X 12"		30	1.333		300	34	25.50	359.50	415
8118	X 13-1/2"		30	1.333		340	34	25.50	399.50	455
8120	X 15"		29	1.379		375	35	26.50	436.50	505
8122	X 16-1/2"		29	1.379		415	35	26.50	476.50	545
8124	X 18"		29	1.379		450	35	26.50	511.50	585
8126	X 19-1/2"		29	1.379		490	35	26.50	551.50	625
8128	X 21"		28	1.429		525	36.50	27.50	589	670
8130	X 22-1/2"		28	1.429		565	36.50	27.50	629	710
8132	X 24"		28	1.429		600	36.50	27.50	664	750
8134	6-3/4" x 12"		29	1.379		395	35	26.50	456.50	525
8136	X 13-1/2"		29	1.379		445	35	26.50	506.50	580
8138	X 15"		29	1.379		495	35	26.50	556.50	635
8140	X 16-1/2"		28	1.429		545	36.50	27.50	609	690
8142	X 18"		28	1.429		595	36.50	27.50	659	745
8144	X 19-1/2"		28	1.429		645	36.50	27.50	709	800
8146	X 21"		27	1.481		695	37.50	28.50	761	855
8148	X 22-1/2"		27	1.481		745	37.50	28.50	811	910
8150	X 24"		27	1.481		790	37.50	28.50	856	965
8152	X 25-1/2"		27	1.481		840	37.50	28.50	906	1,025
8154	X 27"		26	1.538		890	39	29.50	958.50	1,075
8156	X 28-1/2"		26	1.538		940	39	29.50	1,008.50	1,125
8158	X 30"	↓	26	1.538	↓	990	39	29.50	1,058.50	1,200
8200	30' span									
8250	3-1/8" x 9"	F-3	30	1.333	Ea.	206	34	25.50	265.50	310
8252	X 10-1/2"		30	1.333		241	34	25.50	300.50	350
8254	X 12"		30	1.333		275	34	25.50	334.50	390
8256	X 13-1/2"		30	1.333		310	34	25.50	369.50	425
8258	X 15"		29	1.379		345	35	26.50	406.50	470
8260	5-1/8" x 10-1/2"		30	1.333		395	34	25.50	454.50	520
8262	X 12"		30	1.333		450	34	25.50	509.50	580
8264	X 13-1/2"		30	1.333		505	34	25.50	564.50	645
8266	X 15"		29	1.379		565	35	26.50	626.50	710
8268	X 16-1/2"	↓	29	1.379	↓	620	35	26.50	681.50	770

06 18 13.20 Laminated Framing	Crew	Daily Output	Labor-Hours	Unit	Material	2009 Bare Costs Labor	Equipment	Total	Total Incl O&P
8270	X 18"	F-3	29	1.379	Ea.	675	35	26.50	736.50	835
8272	X 19-1/2"		29	1.379		735	35	26.50	796.50	895
8274	X 21"		28	1.429		790	36.50	27.50	854	960
8276	X 22-1/2"		28	1.429		845	36.50	27.50	909	1,025
8278	X 24"		28	1.429		900	36.50	27.50	964	1,075
8280	6-3/4" x 12"		29	1.379		595	35	26.50	656.50	745
8282	X 13-1/2"		29	1.379		670	35	26.50	731.50	825
8284	X 15"		29	1.379		745	35	26.50	806.50	905
8286	X 16-1/2"		28	1.429		815	36.50	27.50	879	990
8288	X 18"		28	1.429		890	36.50	27.50	954	1,075
8290	X 19-1/2"		28	1.429		965	36.50	27.50	1,029	1,150
8292	X 21"		27	1.481		1,050	37.50	28.50	1,116	1,250
8294	X 22-1/2"		27	1.481		1,125	37.50	28.50	1,191	1,325
8296	X 24"		27	1.481		1,200	37.50	28.50	1,266	1,400
8298	X 25-1/2"		27	1.481		1,250	37.50	28.50	1,316	1,500
8300	X 27"		26	1.538		1,325	39	29.50	1,393.50	1,575
8302	X 28-1/2"		26	1.538		1,400	39	29.50	1,468.50	1,650
8304	X 30"	↓	26	1.538	↓	1,475	39	29.50	1,543.50	1,725
8400	40' span									
8402	3-1/8" x 9"	F-3	30	1.333	Ea.	275	34	25.50	334.50	390
8404	X 10-1/2"		30	1.333		320	34	25.50	379.50	440
8406	X 12"		30	1.333		365	34	25.50	424.50	490
8408	X 13-1/2"		30	1.333		415	34	25.50	474.50	540
8410	X 15"		29	1.379		460	35	26.50	521.50	595
8412	5-1/8" x 10-1/2"		30	1.333		525	34	25.50	584.50	665
8414	X 12"		30	1.333		600	34	25.50	659.50	745
8416	X 13-1/2"		30	1.333		675	34	25.50	734.50	830
8418	X 15"		29	1.379		750	35	26.50	811.50	915
8420	X 16-1/2"		29	1.379		825	35	26.50	886.50	1,000
8422	X 18"		29	1.379		900	35	26.50	961.50	1,075
8424	X 19-1/2"		29	1.379		975	35	26.50	1,036.50	1,175
8426	X 21"		28	1.429		1,050	36.50	27.50	1,114	1,250
8428	X 22-1/2"		28	1.429		1,125	36.50	27.50	1,189	1,350
8430	X 24"		28	1.429		1,200	36.50	27.50	1,264	1,425
8432	6-3/4" x 12"		29	1.379		790	35	26.50	851.50	960
8434	X 13-1/2"		29	1.379		890	35	26.50	951.50	1,075
8436	X 15"		29	1.379		990	35	26.50	1,051.50	1,200
8438	X 16-1/2"		28	1.429		1,100	36.50	27.50	1,164	1,300
8440	X 18"		28	1.429		1,200	36.50	27.50	1,264	1,400
8442	X 19-1/2"		28	1.429		1,275	36.50	27.50	1,339	1,525
8444	X 21"		27	1.481		1,375	37.50	28.50	1,441	1,625
8446	X 22-1/2"		27	1.481		1,475	37.50	28.50	1,541	1,725
8448	X 24"		27	1.481		1,575	37.50	28.50	1,641	1,850
8450	X 25-1/2"		27	1.481		1,675	37.50	28.50	1,741	1,950
8452	X 27"		26	1.538		1,775	39	29.50	1,843.50	2,050
8454	X 28-1/2"		26	1.538		1,875	39	29.50	1,943.50	2,175
8456	X 30"	↓	26	1.538	↓	1,975	39	29.50	2,043.50	2,275

06 22 Millwork

06 22 13 – Standard Pattern Wood Trim

06 22 13.10 Millwork

06 22 13.10 Millwork		Crew	Daily Output	Labor-Hours	Unit	Material	2009 Bare Costs Labor	Equipment	Total	Total Incl O&P
0010	**MILLWORK**									
1020	1" x 12", custom birch				L.F.	3.22			3.22	3.54
1040	Cedar					2.69			2.69	2.96
1060	Oak					3.66			3.66	4.03
1080	Redwood					4.68			4.68	5.15
1100	Southern yellow pine					2.57			2.57	2.83
1120	Sugar pine					2.51			2.51	2.76
1140	Teak					18.50			18.50	20.50
1160	Walnut					5.60			5.60	6.15
1180	White pine				▼	3.07			3.07	3.38

06 22 13.15 Moldings, Base

	06 22 13.15 Moldings, Base	Crew	Daily Output	Labor-Hours	Unit	Material	Labor	Equipment	Total	Total Incl O&P
0010	**MOLDINGS, BASE**									
0500	Base, stock pine, 9/16" x 3-1/2"	1 Carp	240	.033	L.F.	2.22	.93		3.15	4.01
0501	Oak or birch, 9/16" x 3-1/2"		240	.033		3.38	.93		4.31	5.30
0550	9/16" x 4-1/2"		200	.040		2.47	1.12		3.59	4.61
0561	Base shoe, oak, 3/4" x 1"		240	.033		1.13	.93		2.06	2.81
0570	Base, prefinished, 2-1/2" x 9/16"		242	.033		1.05	.92		1.97	2.72
0580	Shoe, prefinished, 3/8" x 5/8"		266	.030		.45	.84		1.29	1.92
0585	Flooring cant strip, 3/4" x 1/2"	▼	500	.016	▼	.46	.45		.91	1.27

06 22 13.30 Moldings, Casings

	06 22 13.30 Moldings, Casings	Crew	Daily Output	Labor-Hours	Unit	Material	Labor	Equipment	Total	Total Incl O&P
0010	**MOLDINGS, CASINGS**									
0090	Apron, stock pine, 5/8" x 2"	1 Carp	250	.032	L.F.	1.24	.89		2.13	2.87
0110	5/8" x 3-1/2"		220	.036		1.88	1.02		2.90	3.79
0300	Band, stock pine, 11/16" x 1-1/8"		270	.030		.82	.83		1.65	2.30
0350	11/16" x 1-3/4"		250	.032		1.10	.89		1.99	2.72
0700	Casing, stock pine, 11/16" x 2-1/2"		240	.033		1.33	.93		2.26	3.03
0701	Oak or birch		240	.033		2.58	.93		3.51	4.41
0750	11/16" x 3-1/2"		215	.037		1.89	1.04		2.93	3.84
0760	Door & window casing, exterior, 1-1/4" x 2"		200	.040		2.14	1.12		3.26	4.24
0770	Finger jointed, 1-1/4" x 2"		200	.040		.99	1.12		2.11	2.98
4600	Mullion casing, stock pine, 5/16" x 2"		200	.040		.84	1.12		1.96	2.81
4601	Oak or birch, 9/16" x 2-1/2"		200	.040		2.79	1.12		3.91	4.96
4700	Teak, custom, nominal 1" x 1"		215	.037		1.16	1.04		2.20	3.04
4800	Nominal 1" x 3"	▼	200	.040	▼	3.50	1.12		4.62	5.75

06 22 13.35 Moldings, Ceilings

	06 22 13.35 Moldings, Ceilings	Crew	Daily Output	Labor-Hours	Unit	Material	Labor	Equipment	Total	Total Incl O&P
0010	**MOLDINGS, CEILINGS**									
0600	Bed, stock pine, 9/16" x 1-3/4"	1 Carp	270	.030	L.F.	.97	.83		1.80	2.47
0650	9/16" x 2"		240	.033		1.23	.93		2.16	2.92
1200	Cornice molding, stock pine, 9/16" x 1-3/4"		330	.024		1.02	.68		1.70	2.27
1300	9/16" x 2-1/4"		300	.027		1.24	.75		1.99	2.62
2400	Cove scotia, stock pine, 9/16" x 1-3/4"		270	.030		.93	.83		1.76	2.42
2401	Oak or birch, 9/16" x 1-3/4"		270	.030		.83	.83		1.66	2.31
2500	11/16" x 2-3/4"		255	.031		1.35	.88		2.23	2.97
2600	Crown, stock pine, 9/16" x 3-5/8"		250	.032		1.92	.89		2.81	3.62
2700	11/16" x 4-5/8"	▼	220	.036	▼	3.22	1.02		4.24	5.25

06 22 13.40 Moldings, Exterior

	06 22 13.40 Moldings, Exterior	Crew	Daily Output	Labor-Hours	Unit	Material	Labor	Equipment	Total	Total Incl O&P
0010	**MOLDINGS, EXTERIOR**									
1500	Cornice, boards, pine, 1" x 2"	1 Carp	330	.024	L.F.	.27	.68		.95	1.45
1600	1" x 4"		250	.032		.73	.89		1.62	2.31
1700	1" x 6"		250	.032		1.15	.89		2.04	2.78
1800	1" x 8"		200	.040		1.34	1.12		2.46	3.36
1900	1" x 10"	▼	180	.044	▼	1.78	1.24		3.02	4.06

06 22 13 – Standard Pattern Wood Trim

06 22 13.40 Moldings, Exterior

		Crew	Daily Output	Labor-Hours	Unit	Material	2009 Bare Costs Labor	Equipment	Total	Total Incl O&P
2000	1" x 12"	1 Carp	180	.044	L.F.	2.31	1.24		3.55	4.64
2200	Three piece, built-up, pine, minimum		80	.100		2.15	2.80		4.95	7.10
2300	Maximum		65	.123		5.45	3.44		8.89	11.75
3000	Corner board, sterling pine, 1" x 4"		200	.040		.69	1.12		1.81	2.65
3100	1" x 6"		200	.040		1.05	1.12		2.17	3.05
3200	2" x 6"		165	.048		1.98	1.36		3.34	4.47
3300	2" x 8"		165	.048		2.63	1.36		3.99	5.20
3350	Fascia, sterling pine, 1" x 6"		250	.032		1.05	.89		1.94	2.67
3370	1" x 8"		225	.036		1.43	.99		2.42	3.25
3372	2" x 6"		225	.036		1.98	.99		2.97	3.86
3374	2" x 8"		200	.040		2.63	1.12		3.75	4.78
3376	2" x 10"		180	.044		3.29	1.24		4.53	5.70
3395	Grounds, 1" x 1" redwood		300	.027		.25	.75		1	1.54
3400	Trim, back band, 11/16" x 1-1/16"		250	.032		.89	.89		1.78	2.49
3500	Casing, 11/16" x 4-1/4"		250	.032		2.19	.89		3.08	3.92
3600	Crown, 11/16" x 4-1/4"		250	.032		3.65	.89		4.54	5.55
3700	1" x 4" and 1" x 6" railing w/ balusters 4" O.C.		22	.364		22.50	10.15		32.65	42
3800	Insect screen framing stock, 1-1/16" x 1-3/4"		395	.020		2.03	.57		2.60	3.19
4100	Verge board, sterling pine, 1" x 4"		200	.040		.69	1.12		1.81	2.65
4200	1" x 6"		200	.040		1.05	1.12		2.17	3.05
4300	2" x 6"		165	.048		1.98	1.36		3.34	4.47
4400	2" x 8"		165	.048		2.63	1.36		3.99	5.20
4700	For redwood trim, add					200%				
5000	Casing/fascia, rough-sawn cedar									
5100	1" x 2"	1 Carp	275	.029	L.F.	.32	.81		1.13	1.72
5200	1" x 6"		250	.032		1.07	.89		1.96	2.69
5300	1" x 8"		230	.035		1.74	.97		2.71	3.55
5400	2" x 4"		220	.036		2.34	1.02		3.36	4.29
5500	2" x 6"		220	.036		3.50	1.02		4.52	5.55
5600	2" x 8"		200	.040		4.69	1.12		5.81	7.05
5700	2" x 10"		180	.044		5.85	1.24		7.09	8.55
5800	2" x 12"		170	.047		4.68	1.32		6	7.35

06 22 13.45 Moldings, Trim

		Crew	Daily Output	Labor-Hours	Unit	Material	2009 Bare Costs Labor	Equipment	Total	Total Incl O&P
0010	**MOLDINGS, TRIM**									
0200	Astragal, stock pine, 11/16" x 1-3/4"	1 Carp	255	.031	L.F.	1.19	.88		2.07	2.79
0250	1-5/16" x 2-3/16"		240	.033		2.57	.93		3.50	4.40
0800	Chair rail, stock pine, 5/8" x 2-1/2"		270	.030		1.45	.83		2.28	3
0900	5/8" x 3-1/2"		240	.033		2.27	.93		3.20	4.07
1000	Closet pole, stock pine, 1-1/8" diameter		200	.040		1.01	1.12		2.13	3
1100	Fir, 1-5/8" diameter		200	.040		1.67	1.12		2.79	3.73
1150	Corner, inside, 5/16" x 1"		225	.036		.33	.99		1.32	2.04
1160	Outside, 1-1/16" x 1-1/16"		240	.033		1.24	.93		2.17	2.93
1161	1-5/16" x 1-5/16"		240	.033		1.71	.93		2.64	3.45
3300	Half round, stock pine, 1/4" x 1/2"		270	.030		.24	.83		1.07	1.66
3350	1/2" x 1"		255	.031		.61	.88		1.49	2.15
3400	Handrail, fir, single piece, stock, hardware not included									
3450	1-1/2" x 1-3/4"	1 Carp	80	.100	L.F.	1.85	2.80		4.65	6.75
3470	Pine, 1-1/2" x 1-3/4"		80	.100		1.85	2.80		4.65	6.75
3500	1-1/2" x 2-1/2"		76	.105		2.30	2.94		5.24	7.50
3600	Lattice, stock pine, 1/4" x 1-1/8"		270	.030		.42	.83		1.25	1.86
3700	1/4" x 1-3/4"		250	.032		.66	.89		1.55	2.24
3800	Miscellaneous, custom, pine, 1" x 1"		270	.030		.34	.83		1.17	1.77

06 22 Millwork

06 22 13 – Standard Pattern Wood Trim

06 22 13.45 Moldings, Trim

		Crew	Daily Output	Labor-Hours	Unit	Material	2009 Bare Costs Labor	2009 Bare Costs Equipment	Total	Total Incl O&P
3900	1" x 3"	1 Carp	240	.033	L.F.	.69	.93		1.62	2.33
4100	Birch or oak, nominal 1" x 1"		240	.033		.53	.93		1.46	2.15
4200	Nominal 1" x 3"		215	.037		1.81	1.04		2.85	3.75
4400	Walnut, nominal 1" x 1"		215	.037		.87	1.04		1.91	2.72
4500	Nominal 1" x 3"		200	.040		2.61	1.12		3.73	4.76
4700	Teak, nominal 1" x 1"		215	.037		1.23	1.04		2.27	3.11
4800	Nominal 1" x 3"		200	.040		3.52	1.12		4.64	5.75
4900	Quarter round, stock pine, 1/4" x 1/4"		275	.029		.22	.81		1.03	1.61
4950	3/4" x 3/4"		255	.031		.59	.88		1.47	2.13
5600	Wainscot moldings, 1-1/8" x 9/16", 2' high, minimum		76	.105	S.F.	11.05	2.94		13.99	17.10
5700	Maximum		65	.123	"	20	3.44		23.44	28.50

06 22 13.50 Moldings, Window and Door

		Crew	Daily Output	Labor-Hours	Unit	Material	2009 Bare Costs Labor	2009 Bare Costs Equipment	Total	Total Incl O&P
0010	**MOLDINGS, WINDOW AND DOOR**									
2800	Door moldings, stock, decorative, 1-1/8" wide, plain	1 Carp	17	.471	Set	45	13.15		58.15	71.50
2900	Detailed		17	.471	"	87.50	13.15		100.65	119
2960	Clear pine door jamb, no stops, 11/16" x 4-9/16"		240	.033	L.F.	5.05	.93		5.98	7.10
3150	Door trim set, 1 head and 2 sides, pine, 2-1/2 wide		5.90	1.356	Opng.	22.50	38		60.50	89
3170	3-1/2" wide		5.30	1.509	"	32	42		74	107
3250	Glass beads, stock pine, 3/8" x 1/2"		275	.029	L.F.	.31	.81		1.12	1.71
3270	3/8" x 7/8"		270	.030		.41	.83		1.24	1.85
4850	Parting bead, stock pine, 3/8" x 3/4"		275	.029		.36	.81		1.17	1.77
4870	1/2" x 3/4"		255	.031		.45	.88		1.33	1.98
5000	Stool caps, stock pine, 11/16" x 3-1/2"		200	.040		2.27	1.12		3.39	4.39
5100	1-1/16" x 3-1/4"		150	.053		3.59	1.49		5.08	6.45
5300	Threshold, oak, 3' long, inside, 5/8" x 3-5/8"		32	.250	Ea.	9	7		16	21.50
5400	Outside, 1-1/2" x 7-5/8"		16	.500	"	37	14		51	64
5900	Window trim sets, including casings, header, stops,									
5910	stool and apron, 2-1/2" wide, minimum	1 Carp	13	.615	Opng.	32.50	17.20		49.70	64.50
5950	Average		10	.800		38	22.50		60.50	80
6000	Maximum		6	1.333		62.50	37.50		100	132

06 22 13.60 Moldings, Soffits

		Crew	Daily Output	Labor-Hours	Unit	Material	2009 Bare Costs Labor	2009 Bare Costs Equipment	Total	Total Incl O&P
0010	**MOLDINGS, SOFFITS**									
0200	Soffits, pine, 1" x 4"	2 Carp	420	.038	L.F.	.44	1.06		1.50	2.28
0210	1" x 6"		420	.038		.64	1.06		1.70	2.50
0220	1" x 8"		420	.038		.84	1.06		1.90	2.72
0230	1" x 10"		400	.040		1.23	1.12		2.35	3.24
0240	1" x 12"		400	.040		1.60	1.12		2.72	3.65
0250	STK cedar, 1" x 4"		420	.038		.65	1.06		1.71	2.52
0260	1" x 6"		420	.038		1.09	1.06		2.15	3
0270	1" x 8"		420	.038		1.52	1.06		2.58	3.47
0280	1" x 10"		400	.040		2.09	1.12		3.21	4.19
0290	1" x 12"		400	.040		2.70	1.12		3.82	4.86
1000	Exterior AC plywood, 1/4" thick		420	.038	S.F.	.85	1.06		1.91	2.74
1050	3/8" thick		420	.038		1.03	1.06		2.09	2.93
1100	1/2" thick		420	.038		1	1.06		2.06	2.90
1150	Polyvinyl chloride, white, solid	1 Carp	230	.035		1.09	.97		2.06	2.84
1160	Perforated	"	230	.035		1.09	.97		2.06	2.84
1170	Accessories, "J" channel 5/8"	2 Carp	700	.023	L.F.	.39	.64		1.03	1.51

06 25 Prefinished Paneling

06 25 13 – Prefinished Hardboard Paneling

06 25 13.10 Paneling, Hardboard

		Crew	Daily Output	Labor-Hours	Unit	Material	2009 Bare Costs Labor	2009 Bare Costs Equipment	Total	Total Incl O&P
0010	**PANELING, HARDBOARD**									
0050	Not incl. furring or trim, hardboard, tempered, 1/8" thick Ⓖ	2 Carp	500	.032	S.F.	.38	.89		1.27	1.93
0100	1/4" thick Ⓖ		500	.032		.53	.89		1.42	2.09
0300	Tempered pegboard, 1/8" thick Ⓖ		500	.032		.47	.89		1.36	2.03
0400	1/4" thick Ⓖ		500	.032		.65	.89		1.54	2.23
0600	Untempered hardboard, natural finish, 1/8" thick Ⓖ		500	.032		.39	.89		1.28	1.94
0700	1/4" thick Ⓖ		500	.032		.42	.89		1.31	1.97
0900	Untempered pegboard, 1/8" thick Ⓖ		500	.032		.38	.89		1.27	1.93
1000	1/4" thick Ⓖ		500	.032		.42	.89		1.31	1.97
1200	Plastic faced hardboard, 1/8" thick Ⓖ		500	.032		.64	.89		1.53	2.21
1300	1/4" thick Ⓖ		500	.032		.85	.89		1.74	2.45
1500	Plastic faced pegboard, 1/8" thick Ⓖ		500	.032		.61	.89		1.50	2.18
1600	1/4" thick Ⓖ		500	.032		.75	.89		1.64	2.34
1800	Wood grained, plain or grooved, 1/4" thick, minimum Ⓖ		500	.032		.57	.89		1.46	2.14
1900	Maximum Ⓖ		425	.038		1.19	1.05		2.24	3.09
2100	Moldings for hardboard, wood or aluminum, minimum		500	.032	L.F.	.38	.89		1.27	1.93
2200	Maximum		425	.038	"	1.08	1.05		2.13	2.97

06 25 16 – Prefinished Plywood Paneling

06 25 16.10 Paneling, Plywood

		Crew	Daily Output	Labor-Hours	Unit	Material	2009 Bare Costs Labor	2009 Bare Costs Equipment	Total	Total Incl O&P
0010	**PANELING, PLYWOOD**									
2400	Plywood, prefinished, 1/4" thick, 4' x 8' sheets									
2410	with vertical grooves. Birch faced, minimum	2 Carp	500	.032	S.F.	.89	.89		1.78	2.49
2420	Average		420	.038		1.35	1.06		2.41	3.29
2430	Maximum		350	.046		1.97	1.28		3.25	4.33
2600	Mahogany, African		400	.040		2.51	1.12		3.63	4.65
2700	Philippine (Lauan)		500	.032		1.08	.89		1.97	2.70
2900	Oak or Cherry, minimum		500	.032		2.10	.89		2.99	3.82
3000	Maximum		400	.040		3.22	1.12		4.34	5.45
3200	Rosewood		320	.050		4.58	1.40		5.98	7.40
3400	Teak		400	.040		3.22	1.12		4.34	5.45
3600	Chestnut		375	.043		4.77	1.19		5.96	7.25
3800	Pecan		400	.040		2.06	1.12		3.18	4.16
3900	Walnut, minimum		500	.032		2.75	.89		3.64	4.54
3950	Maximum		400	.040		5.20	1.12		6.32	7.65
4000	Plywood, prefinished, 3/4" thick, stock grades, minimum		320	.050		1.24	1.40		2.64	3.72
4100	Maximum		224	.071		5.40	2		7.40	9.30
4300	Architectural grade, minimum		224	.071		3.98	2		5.98	7.75
4400	Maximum		160	.100		6.10	2.80		8.90	11.45
4600	Plywood, "A" face, birch, V.C., 1/2" thick, natural		450	.036		1.89	.99		2.88	3.76
4700	Select		450	.036		2.06	.99		3.05	3.95
4900	Veneer core, 3/4" thick, natural		320	.050		1.99	1.40		3.39	4.55
5000	Select		320	.050		2.23	1.40		3.63	4.81
5200	Lumber core, 3/4" thick, natural		320	.050		2.99	1.40		4.39	5.65
5500	Plywood, knotty pine, 1/4" thick, A2 grade		450	.036		1.63	.99		2.62	3.47
5600	A3 grade		450	.036		2.06	.99		3.05	3.95
5800	3/4" thick, veneer core, A2 grade		320	.050		2.11	1.40		3.51	4.68
5900	A3 grade		320	.050		2.38	1.40		3.78	4.98
6100	Aromatic cedar, 1/4" thick, plywood		400	.040		2.08	1.12		3.20	4.18
6200	1/4" thick, particle board		400	.040		1.01	1.12		2.13	3

06 25 26 – Panel System

06 25 26.10 Panel Systems

0010	**PANEL SYSTEMS**									

06 25 Prefinished Paneling

06 25 26 – Panel System

06 25 26.10 Panel Systems

		Crew	Daily Output	Labor-Hours	Unit	Material	2009 Bare Costs Labor	Equipment	Total	Total Incl O&P
0100	Raised panel, eng. wood core w/ wood veneer, std., paint grade	2 Carp	300	.053	S.F.	11.90	1.49		13.39	15.60
0110	Oak veneer		300	.053		19.70	1.49		21.19	24
0120	Maple veneer		300	.053		25	1.49		26.49	30
0130	Cherry veneer		300	.053		32	1.49		33.49	37.50
0300	Class I fire rated, paint grade		300	.053		14.25	1.49		15.74	18.15
0310	Oak veneer		300	.053		23.50	1.49		24.99	28.50
0320	Maple veneer		300	.053		30	1.49		31.49	35.50
0330	Cherry veneer		300	.053		38.50	1.49		39.99	44.50
0510	Beadboard, 5/8" MDF, standard, primed		300	.053		8.40	1.49		9.89	11.75
0520	Oak veneer, unfinished		300	.053		14.60	1.49		16.09	18.55
0530	Maple veneer, unfinished		300	.053		17.25	1.49		18.74	21.50
0610	Rustic paneling, 5/8" MDF, standard, maple veneer, unfinished	↓	300	.053	↓	22	1.49		23.49	27

06 26 Board Paneling

06 26 13 – Profile Board Paneling

06 26 13.10 Paneling, Boards

		Crew	Daily Output	Labor-Hours	Unit	Material	2009 Bare Costs Labor	Equipment	Total	Total Incl O&P
0010	**PANELING, BOARDS**									
6400	Wood board paneling, 3/4" thick, knotty pine	2 Carp	300	.053	S.F.	1.44	1.49		2.93	4.10
6500	Rough sawn cedar		300	.053		1.85	1.49		3.34	4.56
6700	Redwood, clear, 1" x 4" boards		300	.053		4.30	1.49		5.79	7.25
6900	Aromatic cedar, closet lining, boards	↓	275	.058	↓	3.32	1.63		4.95	6.40

06 43 Wood Stairs and Railings

06 43 13 – Wood Stairs

06 43 13.20 Prefabricated Wood Stairs

		Crew	Daily Output	Labor-Hours	Unit	Material	2009 Bare Costs Labor	Equipment	Total	Total Incl O&P
0010	**PREFABRICATED WOOD STAIRS**									
0100	Box stairs, prefabricated, 3'-0" wide									
0110	Oak treads, up to 14 risers	2 Carp	39	.410	Riser	85	11.45		96.45	113
0600	With pine treads for carpet, up to 14 risers	"	39	.410	"	55	11.45		66.45	80
1100	For 4' wide stairs, add				Flight	25%				
1550	Stairs, prefabricated stair handrail with balusters	1 Carp	30	.267	L.F.	75	7.45		82.45	94.50
1700	Basement stairs, prefabricated, pine treads									
1710	Pine risers, 3' wide, up to 14 risers	2 Carp	52	.308	Riser	55	8.60		63.60	75
4000	Residential, wood, oak treads, prefabricated		1.50	10.667	Flight	1,100	298		1,398	1,700
4200	Built in place	↓	.44	36.364	"	1,575	1,025		2,600	3,450
4400	Spiral, oak, 4'-6" diameter, unfinished, prefabricated,									
4500	incl. railing, 9' high	2 Carp	1.50	10.667	Flight	3,625	298		3,923	4,475

06 43 13.40 Wood Stair Parts

		Crew	Daily Output	Labor-Hours	Unit	Material	2009 Bare Costs Labor	Equipment	Total	Total Incl O&P
0010	**WOOD STAIR PARTS**									
0020	Pin top balusters, 1-1/4", oak, 34"	1 Carp	96	.083	Ea.	9.15	2.33		11.48	14
0030	36"		96	.083		9.60	2.33		11.93	14.50
0040	42"		96	.083		11.65	2.33		13.98	16.75
0050	Poplar, 34"		96	.083		6.10	2.33		8.43	10.70
0060	36"		96	.083		6.50	2.33		8.83	11.10
0070	42"		96	.083		14.20	2.33		16.53	19.55
0080	Maple, 31"		96	.083		11.65	2.33		13.98	16.75
0090	34"		96	.083		12.95	2.33		15.28	18.20
0100	36"		96	.083		13.50	2.33		15.83	18.80
0110	39"	↓	96	.083	↓	15.45	2.33		17.78	21

06 43 Wood Stairs and Railings

06 43 13 – Wood Stairs

06 43 13.40 Wood Stair Parts	Crew	Daily Output	Labor-Hours	Unit	Material	2009 Bare Costs Labor	Equipment	Total	Total Incl O&P
0120	41"	1 Carp	96	.083	Ea.	17.35	2.33		19.68	23
0130	Primed, 31"		96	.083		4.03	2.33		6.36	8.35
0140	34"		96	.083		4.33	2.33		6.66	8.70
0150	36"		96	.083		4.51	2.33		6.84	8.90
0160	39"		96	.083		5.05	2.33		7.38	9.50
0170	41"		96	.083		5.55	2.33		7.88	10.10
0180	Box top balusters, 1-1/4", oak, 34"		60	.133		9.40	3.73		13.13	16.65
0190	36"		60	.133		10.25	3.73		13.98	17.55
0200	42"		60	.133		11.45	3.73		15.18	18.85
0210	Poplar, 34"		60	.133		6	3.73		9.73	12.90
0220	36"		60	.133		6.75	3.73		10.48	13.70
0230	42"		60	.133		7.50	3.73		11.23	14.55
0240	Maple, 31"		60	.133		13.45	3.73		17.18	21
0250	34"		60	.133		14.15	3.73		17.88	22
0260	36"		60	.133		16.35	3.73		20.08	24.50
0270	39"		60	.133		16.60	3.73		20.33	24.50
0280	41"		60	.133		17.15	3.73		20.88	25
0290	Primed, 31"		60	.133		4.46	3.73		8.19	11.20
0300	34"		60	.133		4.79	3.73		8.52	11.55
0310	36"		60	.133		4.97	3.73		8.70	11.75
0320	39"		60	.133		5.60	3.73		9.33	12.50
0330	41"		60	.133		6.25	3.73		9.98	13.20
0340	Square balusters, cut from lineal stock, pine, 1-1/16" x 1-1/16"		180	.044	L.F.	1.32	1.24		2.56	3.55
0350	1-5/16" x 1-5/16"		180	.044		1.85	1.24		3.09	4.14
0360	1-5/8" x 1-5/8"		180	.044		3.03	1.24		4.27	5.45
0370	Turned newel, oak, 3-1/2" square, 50" high		8	1	Ea.	119	28		147	178
0380	65" high		8	1		154	28		182	218
0390	Poplar, 3-1/2" square, 50" high		8	1		99	28		127	157
0400	65" high		8	1		128	28		156	188
0410	Maple, 3-1/2" square, 50" high		8	1		138	28		166	199
0420	65" high		8	1		160	28		188	223
0430	Square newel, oak, 3-1/2" square, 50" high		8	1		226	28		254	296
0440	65" high		8	1		242	28		270	315
0450	Poplar, 3-1/2" square, 50" high		8	1		215	28		243	284
0460	65" high		8	1		220	28		248	290
0470	Maple, 3" square, 50" high		8	1		243	28		271	315
0480	65" high		8	1		265	28		293	340
0490	Railings, oak, minimum		96	.083	L.F.	7.50	2.33		9.83	12.20
0500	Average		96	.083		9.20	2.33		11.53	14.05
0510	Maximum		96	.083		9.35	2.33		11.68	14.25
0520	Maple, minimum		96	.083		8.80	2.33		11.13	13.65
0530	Average		96	.083		11	2.33		13.33	16.05
0540	Maximum		96	.083		13.20	2.33		15.53	18.50
0550	Oak, for bending rail, minimum		48	.167		21	4.66		25.66	31
0560	Average		48	.167		30	4.66		34.66	41
0570	Maximum		48	.167		32	4.66		36.66	43.50
0580	Maple, for bending rail, minimum		48	.167		32	4.66		36.66	43.50
0590	Average		48	.167		35	4.66		39.66	46.50
0600	Maximum		48	.167		37.50	4.66		42.16	49
0610	Risers, oak, 3/4" x 8", 36" long		80	.100	Ea.	22	2.80		24.80	28.50
0620	42" long		80	.100		25.50	2.80		28.30	32.50
0630	48" long		80	.100		29.50	2.80		32.30	36.50
0640	54" long		80	.100		33	2.80		35.80	40.50

06 43 13 – Wood Stairs

06 43 13.40 Wood Stair Parts

		Crew	Daily Output	Labor-Hours	Unit	Material	2009 Bare Costs Labor	Equipment	Total	Total Incl O&P
0650	60" long	1 Carp	80	.100	Ea.	36.50	2.80		39.30	45
0660	72" long		80	.100		44	2.80		46.80	53
0670	Poplar, 3/4" x 8", 36" long		80	.100		14.35	2.80		17.15	20.50
0680	42" long		80	.100		16.75	2.80		19.55	23
0690	48" long		80	.100		19.15	2.80		21.95	25.50
0700	54" long		80	.100		21.50	2.80		24.30	28
0710	60" long		80	.100		24	2.80		26.80	31
0720	72" long		80	.100		28.50	2.80		31.30	36
0730	Pine, 1" x 8", 36" long		80	.100		2.51	2.80		5.31	7.50
0740	42" long		80	.100		2.93	2.80		5.73	7.95
0750	48" long		80	.100		3.34	2.80		6.14	8.40
0760	54" long		80	.100		3.76	2.80		6.56	8.85
0770	60" long		80	.100		4.18	2.80		6.98	9.35
0780	72" long		80	.100		5	2.80		7.80	10.25
0790	Treads, oak, no returns, 1-1/32" x 11-1/2" x 36" long		32	.250		43.50	7		50.50	60
0800	42" long		32	.250		50.50	7		57.50	68
0810	48" long		32	.250		58	7		65	75.50
0820	54" long		32	.250		65	7		72	83.50
0830	60" long		32	.250		72.50	7		79.50	91.50
0840	72" long		32	.250		87	7		94	107
0850	Mitred return one end, 1-1/32" x 11-1/2" x 36" long		24	.333		65	9.30		74.30	87.50
0860	42" long		24	.333		76	9.30		85.30	99.50
0870	48" long		24	.333		86.50	9.30		95.80	111
0880	54" long		24	.333		97.50	9.30		106.80	123
0890	60" long		24	.333		108	9.30		117.30	135
0900	72" long		24	.333		130	9.30		139.30	159
0910	Mitred return two ends, 1-1/32" x 11-1/2" x 36" long		12	.667		81	18.65		99.65	121
0920	42" long		12	.667		95	18.65		113.65	136
0930	48" long		12	.667		108	18.65		126.65	151
0940	54" long		12	.667		122	18.65		140.65	166
0950	60" long		12	.667		135	18.65		153.65	181
0960	72" long		12	.667		162	18.65		180.65	211
0970	Starting step, oak, 48", bullnose		8	1		163	28		191	228
0980	Double end bullnose		8	1		193	28		221	260
0990	Half circle		8	1		216	28		244	285
1010	Double end half circle		8	1		286	28		314	365
1030	Skirt board, pine, 1" x 10"		55	.145	L.F.	1.23	4.07		5.30	8.20
1040	1" x 12"		52	.154	"	1.60	4.30		5.90	9
1050	Oak landing tread, 1-1/16" thick		54	.148	S.F.	8.25	4.14		12.39	16.10
1060	Oak cove molding		96	.083	L.F.	2.75	2.33		5.08	6.95
1070	Oak stringer molding		96	.083		2.75	2.33		5.08	6.95
1080	Oak shoe molding		96	.083		2.75	2.33		5.08	6.95
1090	Rail bolt, 5/16" x 3-1/2"		48	.167	Ea.	1.65	4.66		6.31	9.70
1100	5/16" x 4-1/2"		48	.167		2.20	4.66		6.86	10.30
1120	Newel post anchor		16	.500		27.50	14		41.50	54
1130	Tapered plug, 1/2"		240	.033		.28	.93		1.21	1.88
1140	1"		240	.033		.39	.93		1.32	2
1150	Rail bolt kit		24	.333		11	9.30		20.30	28

06 43 16 – Wood Railings

06 43 16.10 Wood Handrails and Railings

		Crew	Daily Output	Labor-Hours	Unit	Material	2009 Bare Costs Labor	Equipment	Total	Total Incl O&P
0010	**WOOD HANDRAILS AND RAILINGS**									
0020	Custom design, architectural grade, hardwood, minimum	1 Carp	38	.211	L.F.	9.50	5.90		15.40	20.50

06 43 Wood Stairs and Railings

06 43 16 – Wood Railings

06 43 16.10 Wood Handrails and Railings	Crew	Daily Output	Labor-Hours	Unit	Material	2009 Bare Costs Labor	Equipment	Total	Total Incl O&P
0100 Maximum	1 Carp	30	.267	L.F.	51.50	7.45		58.95	69
0300 Stock interior railing with spindles 4" O.C., 4' long		40	.200		31.50	5.60		37.10	44.50
0400 8' long		48	.167		31.50	4.66		36.16	43

06 44 Ornamental Woodwork

06 44 19 – Wood Grilles

06 44 19.10 Grilles

	Crew	Daily Output	Labor-Hours	Unit	Material	2009 Bare Costs Labor	Equipment	Total	Total Incl O&P
0010 **GRILLES** and panels, hardwood, sanded									
0020 2' x 4' to 4' x 8', custom designs, unfinished, minimum	1 Carp	38	.211	S.F.	14.65	5.90		20.55	26
0050 Average		30	.267		31.50	7.45		38.95	47.50
0100 Maximum		19	.421		49	11.75		60.75	73.50
0300 As above, but prefinished, minimum		38	.211		14.65	5.90		20.55	26
0400 Maximum		19	.421		55	11.75		66.75	80.50

06 44 33 – Wood Mantels

06 44 33.10 Fireplace Mantels

	Crew	Daily Output	Labor-Hours	Unit	Material	2009 Bare Costs Labor	Equipment	Total	Total Incl O&P
0010 **FIREPLACE MANTELS**									
0015 6" molding, 6' x 3'-6" opening, minimum	1 Carp	5	1.600	Opng.	228	44.50		272.50	325
0100 Maximum		5	1.600		410	44.50		454.50	530
0300 Prefabricated pine, colonial type, stock, deluxe		2	4		1,250	112		1,362	1,600
0400 Economy		3	2.667		530	74.50		604.50	710

06 44 33.20 Fireplace Mantel Beam

	Crew	Daily Output	Labor-Hours	Unit	Material	2009 Bare Costs Labor	Equipment	Total	Total Incl O&P
0010 **FIREPLACE MANTEL BEAM**									
0020 Rough texture wood, 4" x 8"	1 Carp	36	.222	L.F.	6	6.20		12.20	17.10
0100 4" x 10"		35	.229	"	7	6.40		13.40	18.50
0300 Laminated hardwood, 2-1/4" x 10-1/2" wide, 6' long		5	1.600	Ea.	110	44.50		154.50	197
0400 8' long		5	1.600	"	150	44.50		194.50	241
0600 Brackets for above, rough sawn		12	.667	Pr.	10	18.65		28.65	42.50
0700 Laminated		12	.667	"	15	18.65		33.65	48

06 44 39 – Wood Posts and Columns

06 44 39.10 Decorative Beams

	Crew	Daily Output	Labor-Hours	Unit	Material	2009 Bare Costs Labor	Equipment	Total	Total Incl O&P
0010 **DECORATIVE BEAMS**									
0020 Rough sawn cedar, non-load bearing, 4" x 4"	2 Carp	180	.089	L.F.	1.47	2.48		3.95	5.80
0100 4" x 6"		170	.094		2.21	2.63		4.84	6.90
0200 4" x 8"		160	.100		2.95	2.80		5.75	8
0300 4" x 10"		150	.107		3.68	2.98		6.66	9.10
0400 4" x 12"		140	.114		4.42	3.19		7.61	10.25
0500 8" x 8"		130	.123		7.50	3.44		10.94	14.05
0600 Plastic beam, "hewn finish", 6" x 2"		240	.067		2.89	1.86		4.75	6.35
0601 6" x 4"		220	.073		3.37	2.03		5.40	7.15

06 44 39.20 Columns

	Crew	Daily Output	Labor-Hours	Unit	Material	2009 Bare Costs Labor	Equipment	Total	Total Incl O&P
0010 **COLUMNS**									
0050 Aluminum, round colonial, 6" diameter	2 Carp	80	.200	V.L.F.	16	5.60		21.60	27
0100 8" diameter		62.25	.257		17	7.20		24.20	31
0200 10" diameter		55	.291		19.10	8.15		27.25	35
0250 Fir, stock units, hollow round, 6" diameter		80	.200		17.20	5.60		22.80	28.50
0300 8" diameter		80	.200		20.50	5.60		26.10	32
0350 10" diameter		70	.229		25.50	6.40		31.90	39.50
0400 Solid turned, to 8' high, 3-1/2" diameter		80	.200		8.40	5.60		14	18.70
0500 4-1/2" diameter		75	.213		12.80	5.95		18.75	24

06 44 Ornamental Woodwork

06 44 39 – Wood Posts and Columns

06 44 39.20 Columns		Crew	Daily Output	Labor-Hours	Unit	Material	2009 Bare Costs Labor	2009 Bare Costs Equipment	Total	Total Incl O&P
0600	5-1/2" diameter	2 Carp	70	.229	V.L.F.	16.40	6.40		22.80	29
0800	Square columns, built-up, 5" x 5"		65	.246		10.20	6.90		17.10	23
0900	Solid, 3-1/2" x 3-1/2"		130	.123		6.70	3.44		10.14	13.15
1600	Hemlock, tapered, T & G, 12" diam, 10' high		100	.160		31.50	4.47		35.97	42
1700	16' high		65	.246		46.50	6.90		53.40	62.50
1900	10' high, 14" diameter		100	.160		78.50	4.47		82.97	94
2000	18' high		65	.246		63.50	6.90		70.40	81.50
2200	18" diameter, 12' high		65	.246		114	6.90		120.90	138
2300	20' high		50	.320		80	8.95		88.95	103
2500	20" diameter, 14' high		40	.400		129	11.20		140.20	161
2600	20' high		35	.457		120	12.80		132.80	154
2800	For flat pilasters, deduct					33%				
3000	For splitting into halves, add				Ea.	116			116	128
4000	Rough sawn cedar posts, 4" x 4"	2 Carp	250	.064	V.L.F.	6.30	1.79		8.09	9.95
4100	4" x 6"		235	.068		15	1.90		16.90	19.70
4200	6" x 6"		220	.073		26.50	2.03		28.53	32.50
4300	8" x 8"		200	.080		60	2.24		62.24	70

06 48 Wood Frames

06 48 13 – Exterior Wood Door Frames

06 48 13.10 Exterior Wood Door Frames and Accessories

06 48 13.10		Crew	Daily Output	Labor-Hours	Unit	Material	2009 Bare Costs Labor	2009 Bare Costs Equipment	Total	Total Incl O&P
0010	**EXTERIOR WOOD DOOR FRAMES AND ACCESSORIES**									
0400	Exterior frame, incl. ext. trim, pine, 5/4 x 4-9/16" deep	2 Carp	375	.043	L.F.	5.55	1.19		6.74	8.10
0420	5-3/16" deep		375	.043		11.30	1.19		12.49	14.45
0440	6-9/16" deep		375	.043		8.50	1.19		9.69	11.35
0600	Oak, 5/4 x 4-9/16" deep		350	.046		12	1.28		13.28	15.35
0620	5-3/16" deep		350	.046		13.50	1.28		14.78	17
0640	6-9/16" deep		350	.046		15.30	1.28		16.58	19
0800	Walnut, 5/4 x 4-9/16" deep		350	.046		14	1.28		15.28	17.55
0820	5-3/16" deep		350	.046		18	1.28		19.28	22
0840	6-9/16" deep		350	.046		20	1.28		21.28	24
1000	Sills, 8/4 x 8" deep, oak, no horns		100	.160		19	4.47		23.47	28.50
1020	2" horns		100	.160		19	4.47		23.47	28.50
1040	3" horns		100	.160		19.95	4.47		24.42	29.50
1100	8/4 x 10" deep, oak, no horns		90	.178		21.50	4.97		26.47	32.50
1120	2" horns		90	.178		19.80	4.97		24.77	30.50
1140	3" horns		90	.178		20.50	4.97		25.47	31.50
2000	Exterior, colonial, frame & trim, 3' opng., in-swing, minimum		22	.727	Ea.	315	20.50		335.50	385
2010	Average		21	.762		470	21.50		491.50	550
2020	Maximum		20	.800		1,075	22.50		1,097.50	1,225
2100	5'-4" opening, in-swing, minimum		17	.941		360	26.50		386.50	440
2120	Maximum		15	1.067		1,075	30		1,105	1,225
2140	Out-swing, minimum		17	.941		370	26.50		396.50	450
2160	Maximum		15	1.067		1,100	30		1,130	1,275
2400	6'-0" opening, in-swing, minimum		16	1		345	28		373	430
2420	Maximum		10	1.600		1,100	44.50		1,144.50	1,300
2460	Out-swing, minimum		16	1		375	28		403	465
2480	Maximum		10	1.600		1,300	44.50		1,344.50	1,500
2600	For two sidelights, add, minimum		30	.533	Opng.	355	14.90		369.90	415
2620	Maximum		20	.800	"	1,125	22.50		1,147.50	1,300
2700	Custom birch frame, 3'-0" opening		16	1	Ea.	205	28		233	274

06 48 Wood Frames

06 48 13 – Exterior Wood Door Frames

06 48 13.10 Exterior Wood Door Frames and Accessories	Crew	Daily Output	Labor-Hours	Unit	Material	2009 Bare Costs Labor	Equipment	Total	Total Incl O&P
2750	6'-0" opening	2 Carp	16	1	Ea.	305	28		333	385
2900	Exterior, modern, plain trim, 3' opng., in-swing, minimum		26	.615		35.50	17.20		52.70	68
2920	Average		24	.667		42	18.65		60.65	77.50
2940	Maximum		22	.727		50.50	20.50		71	90

06 48 16 – Interior Wood Door Frames

06 48 16.10 Interior Wood Door Jamb and Frames

	06 48 16.10 Interior Wood Door Jamb and Frames	Crew	Daily Output	Labor-Hours	Unit	Material	Labor	Equipment	Total	Total Incl O&P
0010	**INTERIOR WOOD DOOR JAMB AND FRAMES**									
3000	Interior frame, pine, 11/16" x 3-5/8" deep	2 Carp	375	.043	L.F.	5.50	1.19		6.69	8.05
3020	4-9/16" deep		375	.043		6.45	1.19		7.64	9.10
3040	5-3/16" deep		375	.043		4.28	1.19		5.47	6.75
3200	Oak, 11/16" x 3-5/8" deep		350	.046		4.35	1.28		5.63	6.95
3220	4-9/16" deep		350	.046		4.50	1.28		5.78	7.10
3240	5-3/16" deep		350	.046		4.58	1.28		5.86	7.20
3400	Walnut, 11/16" x 3-5/8" deep		350	.046		7.30	1.28		8.58	10.20
3420	4-9/16" deep		350	.046		7.50	1.28		8.78	10.40
3440	5-3/16" deep		350	.046		7.90	1.28		9.18	10.80
3800	Threshold, oak, 5/8" x 3-5/8" deep		200	.080		3.83	2.24		6.07	8
3820	4-5/8" deep		190	.084		4.58	2.35		6.93	9.05
3840	5-5/8" deep		180	.089		7.15	2.48		9.63	12.05

06 49 Wood Screens and Exterior Wood Shutters

06 49 19 – Exterior Wood Shutters

06 49 19.10 Shutters, Exterior

	06 49 19.10 Shutters, Exterior	Crew	Daily Output	Labor-Hours	Unit	Material	Labor	Equipment	Total	Total Incl O&P
0010	**SHUTTERS, EXTERIOR**									
0012	Aluminum, louvered, 1'-4" wide, 3'-0" long	1 Carp	10	.800	Pr.	55.50	22.50		78	99
0200	4'-0" long		10	.800		66.50	22.50		89	111
0300	5'-4" long		10	.800		87.50	22.50		110	135
0400	6'-8" long		9	.889		112	25		137	165
1000	Pine, louvered, primed, each 1'-2" wide, 3'-3" long		10	.800		96.50	22.50		119	144
1100	4'-7" long		10	.800		130	22.50		152.50	181
1250	Each 1'-4" wide, 3'-0" long		10	.800		98.50	22.50		121	146
1350	5'-3" long		10	.800		148	22.50		170.50	201
1500	Each 1'-6" wide, 3'-3" long		10	.800		102	22.50		124.50	151
1600	4'-7" long		10	.800		144	22.50		166.50	196
1620	Hemlock, louvered, 1'-2" wide, 5'-7" long		10	.800		158	22.50		180.50	212
1630	Each 1'-4" wide, 2'-2" long		10	.800		98.50	22.50		121	146
1640	3'-0" long		10	.800		98.50	22.50		121	146
1650	3'-3" long		10	.800		106	22.50		128.50	155
1660	3'-11" long		10	.800		121	22.50		143.50	171
1670	4'-3" long		10	.800		118	22.50		140.50	168
1680	5'-3" long		10	.800		147	22.50		169.50	199
1690	5'-11" long		10	.800		166	22.50		188.50	221
1700	Door blinds, 6'-9" long, each 1'-3" wide		9	.889		167	25		192	226
1710	1'-6" wide		9	.889		180	25		205	240
1720	Hemlock, solid raised panel, each 1'-4" wide, 3'-3" long		10	.800		159	22.50		181.50	213
1730	3'-11" long		10	.800		186	22.50		208.50	243
1740	4'-3" long		10	.800		202	22.50		224.50	260
1750	4'-7" long		10	.800		217	22.50		239.50	277
1760	4'-11" long		10	.800		237	22.50		259.50	298
1770	5'-11" long		10	.800		269	22.50		291.50	335

06 49 Wood Screens and Exterior Wood Shutters

06 49 19 – Exterior Wood Shutters

06 49 19.10 Shutters, Exterior	Crew	Daily Output	Labor-Hours	Unit	Material	2009 Bare Costs Labor	Equipment	Total	Total Incl O&P
1800	Door blinds, 6'-9" long, each 1'-3" wide	1 Carp	9	.889	Pr.	305	25		330	375
1900	1'-6" wide		9	.889		330	25		355	405
2500	Polystyrene, solid raised panel, each 1'-4" wide, 3'-3" long		10	.800		40.50	22.50		63	82.50
2600	3'-11" long		10	.800		46.50	22.50		69	89.50
2700	4'-7" long		10	.800		51	22.50		73.50	94
2800	5'-3" long		10	.800		55	22.50		77.50	98.50
2900	6'-8" long		9	.889		66	25		91	115
4500	Polystyrene, louvered, each 1'-2" wide, 3'-3" long		10	.800		27.50	22.50		50	68.50
4600	4'-7" long		10	.800		34	22.50		56.50	75.50
4750	5'-3" long		10	.800		39	22.50		61.50	81
4850	6'-8" long		9	.889		46.50	25		71.50	93
6000	Vinyl, louvered, each 1'-2" x 4'-7" long		10	.800		34	22.50		56.50	75.50
6200	Each 1'-4" x 6'-8" long		9	.889		47.50	25		72.50	94.50
8000	PVC exterior rolling shutters									
8100	including crank control	1 Carp	8	1	Ea.	475	28		503	570
8500	Insulative - 6' x 6'8" stock unit	"	8	1	"	675	28		703	790

06 51 Structural Plastic Shapes and Plates

06 51 13 – Plastic Lumber

06 51 13.10 Recycled Plastic Lumber

			Crew	Daily Output	Labor-Hours	Unit	Material	2009 Bare Costs Labor	Equipment	Total	Total Incl O&P
0010	**RECYCLED PLASTIC LUMBER**										
4000	Sheeting, recycled plastic, black or white, 4' x 8' x 1/8"	G	2 Carp	1100	.015	S.F.	1.39	.41		1.80	2.22
4010	4' x 8' x 3/16"	G		1100	.015		2.10	.41		2.51	3
4020	4' x 8' x 1/4"	G		950	.017		2.53	.47		3	3.58
4030	4' x 8' x 3/8"	G		950	.017		3.53	.47		4	4.68
4040	4' x 8' x 1/2"	G		900	.018		4.56	.50		5.06	5.85
4050	4' x 8' x 5/8"	G		900	.018		6	.50		6.50	7.45
4060	4' x 8' x 3/4"	G		850	.019		6.30	.53		6.83	7.85
4070	Add for colors					Ea.	5%			5%	6%
8500	100% recycled plastic, var colors, NLB, 2" x 2"	G				L.F.	2.62			2.62	2.88
8510	2" x 4"	G					2.80			2.80	3.08
8520	2" x 6"	G					4.37			4.37	4.81
8530	2" x 8"	G					6.05			6.05	6.65
8540	2" x 10"	G					8.90			8.90	9.80
8550	5/4" x 4"	G					2.20			2.20	2.42
8560	5/4" x 6"	G					3.08			3.08	3.39
8570	1" x 6"	G					2.08			2.08	2.29
8580	1/2" x 8"	G					2.30			2.30	2.53
8590	2" x 10" T & G	G					6.35			6.35	7
8600	3" x 10" T & G	G					10.15			10.15	11.15
8610	Add for premium colors						20%			20%	22%
8620	Mailbox post, recycled plastic, black/green , 4"x 4"x 6' H, incl. box	G	1 Clab	3	2.667	Ea.	58	55		113	156

06 63 Plastic Railings

06 63 10 – Plastic (PVC) Railings

06 63 10.10 Plastic Railings	Crew	Daily Output	Labor-Hours	Unit	Material	2009 Bare Costs Labor	Equipment	Total	Total Incl O&P
0010	**PLASTIC RAILINGS**									
0100	Horizontal PVC handrail with balusters, 3-1/2" wide, 36" high	1 Carp	96	.083	L.F.	23.50	2.33		25.83	29.50
0150	42" high		96	.083		26.50	2.33		28.83	33
0200	Angled PVC handrail with balusters, 3-1/2" wide, 36" high		72	.111		27	3.11		30.11	35
0250	42" high		72	.111		30	3.11		33.11	38.50
0300	Post sleeve for 4 x 4 post		96	.083		12	2.33		14.33	17.15
0400	Post cap for 4 x 4 post, flat profile		48	.167	Ea.	11.35	4.66		16.01	20.50
0450	Newel post style profile		48	.167		21	4.66		25.66	31
0500	Raised corbeled profile		48	.167		22	4.66		26.66	32.50
0550	Post base trim for 4 x 4 post		96	.083		12.80	2.33		15.13	18

06 65 Plastic Simulated Wood Trim

06 65 10 – PVC Trim

06 65 10.10 PVC Trim, Exterior

06 65 10.10 PVC Trim, Exterior	Crew	Daily Output	Labor-Hours	Unit	Material	2009 Bare Costs Labor	Equipment	Total	Total Incl O&P
0010	**PVC TRIM, EXTERIOR**									
0100	Cornerboards, 5/4" x 6" x 6"	1 Carp	240	.033	L.F.	5.70	.93		6.63	7.85
0110	Door / window casing, 1" x 4"		200	.040		1.28	1.12		2.40	3.30
0120	1" x 6"		200	.040		2.02	1.12		3.14	4.11
0130	1" x 8"		195	.041		2.66	1.15		3.81	4.87
0140	1" x 10"		195	.041		3.39	1.15		4.54	5.65
0150	1" x 12"		190	.042		4.12	1.18		5.30	6.50
0160	5/4" x 4"		195	.041		1.63	1.15		2.78	3.73
0170	5/4" x 6"		195	.041		2.57	1.15		3.72	4.77
0180	5/4" x 8"		190	.042		3.38	1.18		4.56	5.70
0190	5/4" x 10"		190	.042		4.32	1.18		5.50	6.75
0200	5/4" x 12"		185	.043		5.25	1.21		6.46	7.85
0210	Fascia, 1" x 4"		250	.032		1.28	.89		2.17	2.92
0220	1" x 6"		250	.032		2.02	.89		2.91	3.73
0230	1" x 8"		225	.036		2.66	.99		3.65	4.61
0240	1" x 10"		225	.036		3.39	.99		4.38	5.40
0250	1" x 12"		200	.040		4.12	1.12		5.24	6.40
0260	5/4" x 4"		240	.033		1.63	.93		2.56	3.36
0270	5/4" x 6"		240	.033		2.57	.93		3.50	4.40
0280	5/4" x 8"		215	.037		3.38	1.04		4.42	5.50
0290	5/4" x 10"		215	.037		4.32	1.04		5.36	6.50
0300	5/4" x 12"		190	.042		5.25	1.18		6.43	7.80
0310	Frieze, 1" x 4"		250	.032		1.28	.89		2.17	2.92
0320	1" x 6"		250	.032		2.02	.89		2.91	3.73
0330	1" x 8"		225	.036		2.66	.99		3.65	4.61
0340	1" x 10"		225	.036		3.39	.99		4.38	5.40
0350	1" x 12"		200	.040		4.12	1.12		5.24	6.40
0360	5/4" x 4"		240	.033		1.63	.93		2.56	3.36
0370	5/4" x 6"		240	.033		2.57	.93		3.50	4.40
0380	5/4" x 8"		215	.037		3.38	1.04		4.42	5.50
0390	5/4" x 10"		215	.037		4.32	1.04		5.36	6.50
0400	5/4" x 12"		190	.042		5.25	1.18		6.43	7.80
0410	Rake, 1" x 4"		200	.040		1.28	1.12		2.40	3.30
0420	1" x 6"		200	.040		2.02	1.12		3.14	4.11
0430	1" x 8"		190	.042		2.66	1.18		3.84	4.92
0440	1" x 10"		190	.042		3.39	1.18		4.57	5.70
0450	1" x 12"		180	.044		4.12	1.24		5.36	6.65

06 65 Plastic Simulated Wood Trim

06 65 10 – PVC Trim

06 65 10.10 PVC Trim, Exterior		Crew	Daily Output	Labor-Hours	Unit	Material	2009 Bare Costs Labor	Equipment	Total	Total Incl O&P
0460	5/4" x 4"	1 Carp	195	.041	L.F.	1.63	1.15		2.78	3.73
0470	5/4" x 6"		195	.041		2.57	1.15		3.72	4.77
0480	5/4" x 8"		185	.043		3.38	1.21		4.59	5.75
0490	5/4" x 10"		185	.043		4.32	1.21		5.53	6.80
0500	5/4" x 12"		175	.046		5.25	1.28		6.53	7.95
0510	Rake trim, 1" x 4"		225	.036		1.28	.99		2.27	3.09
0520	1" x 6"		225	.036		2.02	.99		3.01	3.90
0560	5/4" x 4"		220	.036		1.63	1.02		2.65	3.51
0570	5/4" x 6"		220	.036		2.57	1.02		3.59	4.55
0610	Soffit, 1" x 4"	2 Carp	420	.038		1.28	1.06		2.34	3.21
0620	1" x 6"		420	.038		2.02	1.06		3.08	4.02
0630	1" x 8"		420	.038		2.66	1.06		3.72	4.73
0640	1" x 10"		400	.040		3.39	1.12		4.51	5.60
0650	1" x 12"		400	.040		4.12	1.12		5.24	6.40
0660	5/4" x 4"		410	.039		1.63	1.09		2.72	3.63
0670	5/4" x 6"		410	.039		2.57	1.09		3.66	4.67
0680	5/4" x 8"		410	.039		3.38	1.09		4.47	5.55
0690	5/4" x 10"		390	.041		4.32	1.15		5.47	6.70
0700	5/4" x 12"		390	.041		5.25	1.15		6.40	7.75

Division 7
Thermal and Moisture
Protection

07 01 Operation and Maint. of Thermal and Moisture Protection

07 01 50 – Maintenance of Membrane Roofing

07 01 50.10 Roof Coatings		Crew	Daily Output	Labor-Hours	Unit	Material	2009 Bare Costs Labor	2009 Bare Costs Equipment	Total	Total Incl O&P
0010	**ROOF COATINGS**									
0012	Asphalt, brush grade, material only				Gal.	8.25			8.25	9.10
0800	Glass fibered roof & patching cement, 5 gallon					7.35			7.35	8.05
1100	Roof patch & flashing cement, 5 gallon					11.70			11.70	12.85

07 05 Common Work Results for Thermal and Moisture Protection

07 05 05 – Selective Demolition

07 05 05.10 Selective Demo., Thermal and Moist. Protection

	07 05 05.10 Selective Demo., Thermal and Moist. Protection		Crew	Daily Output	Labor-Hours	Unit	Material	2009 Bare Costs Labor	2009 Bare Costs Equipment	Total	Total Incl O&P
0010	**SELECTIVE DEMOLITION, THERMAL AND MOISTURE PROTECTION**										
0020	Caulking / sealant, to 1" x 1" joint	R024119-10	1 Clab	600	.013	L.F.		.27		.27	.46
0120	Downspouts, including hangers			350	.023	"		.47		.47	.79
0220	Flashing, sheet metal			290	.028	S.F.		.57		.57	.96
0420	Gutters, aluminum or wood, edge hung			240	.033	L.F.		.68		.68	1.16
0520	Built-in			100	.080	"		1.64		1.64	2.78
0620	Insulation, air / vapor barrier			3500	.002	S.F.		.05		.05	.08
0670	Batts or blankets			1400	.006	C.F.		.12		.12	.20
0720	Foamed or sprayed in place		2 Clab	1000	.016	B.F.		.33		.33	.56
0770	Loose fitting		1 Clab	3000	.003	C.F.		.05		.05	.09
0870	Rigid board			3450	.002	B.F.		.05		.05	.08
1120	Roll roofing, cold adhesive			12	.667	Sq.		13.70		13.70	23
1170	Roof accessories, adjustable metal chimney flashing			9	.889	Ea.		18.25		18.25	31
1325	Plumbing vent flashing			32	.250	"		5.15		5.15	8.70
1375	Ridge vent strip, aluminum			310	.026	L.F.		.53		.53	.90
1620	Skylight to 10 S.F.			8	1	Ea.		20.50		20.50	35
2120	Roof edge, aluminum soffit and fascia			570	.014	L.F.		.29		.29	.49
2170	Concrete coping, up to 12" wide		2 Clab	160	.100			2.06		2.06	3.48
2220	Drip edge		1 Clab	1650	.005			.10		.10	.17
2270	Gravel stop			1650	.005			.10		.10	.17
2370	Sheet metal coping, up to 12" wide			240	.033			.68		.68	1.16
2470	Roof insulation board, over 2" thick		B-2	7800	.005	B.F.		.11		.11	.18
2520	Up to 2" thick		"	3900	.010	S.F.		.21		.21	.36
2620	Roof ventilation, louvered gable vent		1 Clab	16	.500	Ea.		10.30		10.30	17.40
2675	Rafter vents			960	.008	"		.17		.17	.29
2720	Soffit vent and/or fascia vent			575	.014	L.F.		.29		.29	.48
2775	Soffit vent strip, aluminum, 3" to 4" wide			160	.050			1.03		1.03	1.74
2820	Roofing accessories, shingle moulding, to 1" x 4"			1600	.005			.10		.10	.17
2870	Cant strip		B-2	2000	.020			.42		.42	.71
2920	Concrete block walkway		1 Clab	230	.035			.71		.71	1.21
3070	Roofing, felt paper, 15#			70	.114	Sq.		2.35		2.35	3.97
3125	#30 felt			30	.267	"		5.50		5.50	9.25
3170	Asphalt shingles, 1 layer		B-2	3500	.011	S.F.		.24		.24	.41
3370	Modified bitumen			26	1.538	Sq.		32		32	54.50
3420	Built-up, no gravel, 3 ply			25	1.600			33.50		33.50	56.50
3470	4 ply			21	1.905			40		40	67.50
3620	5 ply			1600	.025	S.F.		.52		.52	.89
3725	Gravel removal, minimum			5000	.008			.17		.17	.28
3730	Maximum			2000	.020			.42		.42	.71
3870	Fiberglass sheet			1200	.033			.70		.70	1.18
4120	Slate shingles			1900	.021			.44		.44	.75
4170	Ridge shingles, clay or slate			2000	.020	L.F.		.42		.42	.71
4320	Single ply membrane, attached at seams			52	.769	Sq.		16.10		16.10	27.50

07 05 Common Work Results for Thermal and Moisture Protection

07 05 05 – Selective Demolition

07 05 05.10 Selective Demo., Thermal and Moist. Protection	Crew	Daily Output	Labor-Hours	Unit	Material	2009 Bare Costs Labor	Equipment	Total	Total Incl O&P
4370	Ballasted	B-2	75	.533	Sq.		11.15		11.15	18.90
4420	Fully adhered	↓	39	1.026	↓		21.50		21.50	36.50
4550	Roof hatch, 2'-6" x 3'-0"	1 Clab	10	.800	Ea.		16.45		16.45	28
4670	Wood shingles	B-2	2200	.018	S.F.		.38		.38	.64
4820	Sheet metal roofing	"	2150	.019			.39		.39	.66
4970	Siding, horizontal wood clapboards	1 Clab	380	.021			.43		.43	.73
5025	Exterior insulation finish system	"	120	.067			1.37		1.37	2.32
5070	Tempered hardboard, remove and reset	1 Carp	380	.021			.59		.59	.99
5120	Tempered hardboard sheet siding	"	375	.021	↓		.60		.60	1.01
5170	Metal, corner strips	1 Clab	850	.009	L.F.		.19		.19	.33
5225	Horizontal strips		444	.018	S.F.		.37		.37	.63
5320	Vertical strips		400	.020			.41		.41	.70
5520	Wood shingles		350	.023			.47		.47	.79
5620	Stucco siding		360	.022			.46		.46	.77
5670	Textured plywood		725	.011			.23		.23	.38
5720	Vinyl siding		510	.016	↓		.32		.32	.55
5770	Corner strips		900	.009	L.F.		.18		.18	.31
5870	Wood, boards, vertical	↓	400	.020	S.F.		.41		.41	.70
5920	Waterproofing, protection / drain board	2 Clab	3900	.004	B.F.		.08		.08	.14
5970	Over 1/2" thick		1750	.009	S.F.		.19		.19	.32
6020	To 1/2" thick	↓	2000	.008	"		.16		.16	.28

07 11 Dampproofing

07 11 13 – Bituminous Dampproofing

07 11 13.10 Bituminous Asphalt Coating

0010	BITUMINOUS ASPHALT COATING									
0030	Brushed on, below grade, 1 coat	1 Rofc	665	.012	S.F.	.16	.28		.44	.70
0100	2 coat		500	.016	↓	.32	.38		.70	1.05
0300	Sprayed on, below grade, 1 coat, 25.6 S.F./gal.		830	.010		.16	.23		.39	.60
0400	2 coat, 20.5 S.F./gal.		500	.016		.32	.38		.70	1.04
0600	Troweled on, asphalt with fibers, 1/16" thick		500	.016		.37	.38		.75	1.10
0700	1/8" thick		400	.020		.65	.47		1.12	1.58
1000	1/2" thick	↓	350	.023	↓	2.12	.54		2.66	3.32

07 11 16 – Cementitious Dampproofing

07 11 16.20 Cementitious Parging

0010	CEMENTITIOUS PARGING									
0020	Portland cement, 2 coats, 1/2" thick	D-1	250	.064	S.F.	.28	1.63		1.91	3.01
0100	Waterproofed Portland cement, 1/2" thick, 2 coats	"	250	.064	"	2.62	1.63		4.25	5.60

07 19 Water Repellents

07 19 19 – Silicone Water Repellents

07 19 19.10 Silicone Based Water Repellents		Crew	Daily Output	Labor-Hours	Unit	Material	2009 Bare Costs Labor	Equipment	Total	Total Incl O&P
0010	**SILICONE BASED WATER REPELLENTS**									
0020	Water base liquid, roller applied	2 Rofc	7000	.002	S.F.	.65	.05		.70	.82
0200	Silicone or stearate, sprayed on CMU, 1 coat	1 Rofc	4000	.002		.36	.05		.41	.49
0300	2 coats	"	3000	.003		.73	.06		.79	.92

07 21 Thermal Insulation

07 21 13 – Board Insulation

07 21 13.10 Rigid Insulation

			Crew	Daily Output	Labor-Hours	Unit	Material	Labor	Equipment	Total	Total Incl O&P
0010	**RIGID INSULATION**, for walls										
0040	Fiberglass, 1.5#/CF, unfaced, 1" thick, R4.1	G	1 Carp	1000	.008	S.F.	.44	.22		.66	.86
0060	1-1/2" thick, R6.2	G		1000	.008		.64	.22		.86	1.08
0080	2" thick, R8.3	G		1000	.008		.69	.22		.91	1.14
0120	3" thick, R12.4	G		800	.010		.80	.28		1.08	1.35
0370	3#/CF, unfaced, 1" thick, R4.3	G		1000	.008		.49	.22		.71	.92
0390	1-1/2" thick, R6.5	G		1000	.008		.95	.22		1.17	1.43
0400	2" thick, R8.7	G		890	.009		1.15	.25		1.40	1.69
0420	2-1/2" thick, R10.9	G		800	.010		1.01	.28		1.29	1.58
0440	3" thick, R13	G		800	.010		1.01	.28		1.29	1.58
0520	Foil faced, 1" thick, R4.3	G		1000	.008		.92	.22		1.14	1.39
0540	1-1/2" thick, R6.5	G		1000	.008		1.36	.22		1.58	1.88
0560	2" thick, R8.7	G		890	.009		1.71	.25		1.96	2.30
0580	2-1/2" thick, R10.9	G		800	.010		2.02	.28		2.30	2.69
0600	3" thick, R13	G		800	.010		2.19	.28		2.47	2.88
0670	6#/CF, unfaced, 1" thick, R4.3	G		1000	.008		.98	.22		1.20	1.46
0690	1-1/2" thick, R6.5	G		890	.009		1.50	.25		1.75	2.07
0700	2" thick, R8.7	G		800	.010		2.12	.28		2.40	2.80
0721	2-1/2" thick, R10.9	G		800	.010		2.32	.28		2.60	3.02
0741	3" thick, R13	G		730	.011		2.78	.31		3.09	3.58
0821	Foil faced, 1" thick, R4.3	G		1000	.008		1.38	.22		1.60	1.90
0840	1-1/2" thick, R6.5	G		890	.009		1.98	.25		2.23	2.60
0850	2" thick, R8.7	G		800	.010		2.59	.28		2.87	3.32
0880	2-1/2" thick, R10.9	G		800	.010		3.11	.28		3.39	3.89
0900	3" thick, R13	G		730	.011		3.72	.31		4.03	4.61
1500	Foamglass, 1-1/2" thick, R4.5	G		800	.010		1.37	.28		1.65	1.98
1550	3" thick, R9	G		730	.011		3.29	.31		3.60	4.14
1600	Isocyanurate, 4' x 8' sheet, foil faced, both sides										
1610	1/2" thick, R3.9	G	1 Carp	800	.010	S.F.	.30	.28		.58	.80
1620	5/8" thick, R4.5	G		800	.010		.51	.28		.79	1.03
1630	3/4" thick, R5.4	G		800	.010		.35	.28		.63	.86
1640	1" thick, R7.2	G		800	.010		.55	.28		.83	1.08
1650	1-1/2" thick, R10.8	G		730	.011		.64	.31		.95	1.22
1660	2" thick, R14.4	G		730	.011		.81	.31		1.12	1.41
1670	3" thick, R21.6	G		730	.011		1.90	.31		2.21	2.61
1680	4" thick, R28.8	G		730	.011		2.14	.31		2.45	2.87
1700	Perlite, 1" thick, R2.77	G		800	.010		.30	.28		.58	.80
1750	2" thick, R5.55	G		730	.011		.60	.31		.91	1.18
1900	Extruded polystyrene, 25 PSI compressive strength, 1" thick, R5	G		800	.010		.51	.28		.79	1.03
1940	2" thick R10	G		730	.011		1.04	.31		1.35	1.66
1960	3" thick, R15	G		730	.011		1.40	.31		1.71	2.06
2100	Expanded polystyrene, 1" thick, R3.85	G		800	.010		.24	.28		.52	.73
2120	2" thick, R7.69	G		730	.011		.61	.31		.92	1.19

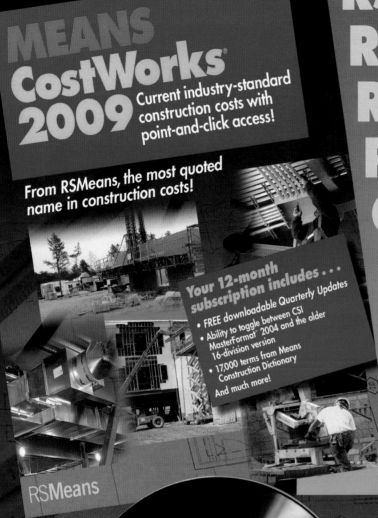

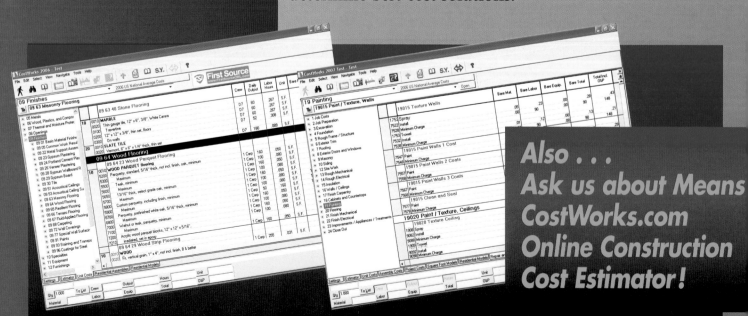

07 21 Thermal Insulation

07 21 13 – Board Insulation

07 21 13.10 Rigid Insulation		Crew	Daily Output	Labor-Hours	Unit	Material	2009 Bare Costs Labor	Equipment	Total	Total Incl O&P
2140	3" thick, R11.49	G	1 Carp	730	.011	S.F.	.78	.31		1.09	1.38

07 21 13.13 Foam Board Insulation

0010	**FOAM BOARD INSULATION**										
0600	Polystyrene, expanded, 1" thick, R4	G	1 Carp	680	.012	S.F.	.31	.33		.64	.90
0700	2" thick, R8	G	"	675	.012	"	.62	.33		.95	1.24

07 21 16 – Blanket Insulation

07 21 16.10 Blanket Insulation for Floors

0010	**BLANKET INSULATION FOR FLOORS**										
0020	Including spring type wire fasteners										
2000	Fiberglass, blankets or batts, paper or foil backing										
2100	1 side, 3-1/2" thick, R11	G	1 Carp	700	.011	S.F.	.42	.32		.74	1
2150	6" thick, R19	G		600	.013		.52	.37		.89	1.20
2200	8-1/2" thick, R30	G		550	.015		.74	.41		1.15	1.50

07 21 16.20 Blanket Insulation for Walls

0010	**BLANKET INSULATION FOR WALLS**										
0020	Kraft faced fiberglass, 3-1/2" thick, R11, 15" wide	G	1 Carp	1350	.006	S.F.	.34	.17		.51	.65
0030	23" wide	G		1600	.005		.34	.14		.48	.61
0061	3-1/2" thick, R13, 11" wide	G		1350	.006		.34	.17		.51	.65
0080	15" wide	G		1350	.006		.34	.17		.51	.65
0110	R-15, 11" wide	G		1150	.007		.41	.19		.60	.78
0120	15" wide	G		1350	.006		.41	.17		.58	.73
0141	6" thick, R19, 11" wide	G		1350	.006		.44	.17		.61	.76
0201	9" thick, R30, 15" wide	G		1350	.006		.66	.17		.83	1.01
0241	12" thick, R38, 15" wide	G		1350	.006		.95	.17		1.12	1.33
0410	Foil faced fiberglass, 3-1/2" thick, R13, 11" wide	G		1150	.007		.57	.19		.76	.96
0420	15" wide	G		1350	.006		.57	.17		.74	.91
0442	R15, 11" wide	G		1150	.007		.56	.19		.75	.95
0444	15" wide	G		1350	.006		.56	.17		.73	.90
0461	6" thick, R19, 15" wide	G		1600	.005		.77	.14		.91	1.09
0482	R-21, 11" wide	G		1150	.007		.82	.19		1.01	1.23
0501	9" thick, R30, 15" wide	G		1350	.006		.89	.17		1.06	1.26
0620	Unfaced fiberglass, 3-1/2" thick, R-13, 11" wide	G		1150	.007		.29	.19		.48	.65
0821	3-1/2" thick, R13, 15" wide	G		1600	.005		.29	.14		.43	.56
0832	R15, 11" wide	G		1150	.007		.37	.19		.56	.74
0834	15" wide	G		1350	.006		.37	.17		.54	.69
0861	6" thick, R19, 15" wide	G		1350	.006		.42	.17		.59	.74
0901	9" thick, R30, 15" wide	G		1150	.007		.75	.19		.94	1.16
0941	12" thick, R38, 15" wide	G		1150	.007		1.28	.19		1.47	1.74
1300	Mineral fiber batts, kraft faced										
1320	3-1/2" thick, R12	G	1 Carp	1600	.005	S.F.	.38	.14		.52	.66
1340	6" thick, R19	G		1600	.005		.48	.14		.62	.77
1380	10" thick, R30	G		1350	.006		.73	.17		.90	1.08
1850	Friction fit wire insulation supports, 16" O.C.			960	.008	Ea.	.08	.23		.31	.48

07 21 23 – Loose-Fill Insulation

07 21 23.10 Poured Loose-Fill Insulation

0010	**POURED LOOSE-FILL INSULATION**										
0020	Cellulose fiber, R3.8 per inch	G	1 Carp	200	.040	C.F.	.60	1.12		1.72	2.55
0080	Fiberglass wool, R4 per inch	G		200	.040		.46	1.12		1.58	2.40
0100	Mineral wool, R3 per inch	G		200	.040		.39	1.12		1.51	2.32
0300	Polystyrene, R4 per inch	G		200	.040		3.09	1.12		4.21	5.30
0400	Vermiculite or perlite, R2.7 per inch	G		200	.040		1.75	1.12		2.87	3.82

07 21 Thermal Insulation

07 21 23 – Loose-Fill Insulation

07 21 23.20 Masonry Loose-Fill Insulation

		Crew	Daily Output	Labor-Hours	Unit	Material	2009 Bare Costs Labor	Equipment	Total	Total Incl O&P
0010	**MASONRY LOOSE-FILL INSULATION**, vermiculite or perlite	G								
0100	In cores of concrete block, 4" thick wall, .115 CF/SF	G	D-1	4800	.003	S.F.	.20	.08		.28	.36
0700	Foamed in place, urethane in 2-5/8" cavity	G	G-2A	1035	.023		.42	.48	.58	1.48	1.95
0800	For each 1" added thickness, add	G	"	2372	.010	↓	.12	.21	.26	.59	.78

07 21 26 – Blown Insulation

07 21 26.10 Blown Insulation

		Crew	Daily Output	Labor-Hours	Unit	Material	Labor	Equipment	Total	Total Incl O&P
0010	**BLOWN INSULATION** Ceilings, with open access	G	G-4	5000	.005	S.F.	.19	.10	.07	.36	.45
0020	Cellulose, 3-1/2" thick, R13	G									
0030	5-3/16" thick, R19	G		3800	.006		.28	.13	.09	.50	.64
0050	6-1/2" thick, R22	G		3000	.008		.36	.17	.11	.64	.81
1000	Fiberglass, 5.5" thick, R11	G		3800	.006		.19	.13	.09	.41	.54
1050	6" thick, R12	G		3000	.008		.23	.17	.11	.51	.66
1100	8.8" thick, R19	G		2200	.011		.33	.23	.15	.71	.92
1300	11.5" thick, R26	G		1500	.016		.46	.34	.22	1.02	1.32
1350	13" thick, R30	G		1400	.017		.50	.36	.24	1.10	1.43
1450	16" thick, R38	G		1145	.021		.61	.44	.29	1.34	1.74
1500	20" thick, R49	G		920	.026		.77	.55	.36	1.68	2.18
2000	Mineral wool, 4" thick, R12	G		3500	.007		.23	.15	.10	.48	.60
2050	6" thick, R17	G		2500	.010		.26	.20	.13	.59	.78
2100	9" thick, R23	G	↓	1750	.014	↓	.34	.29	.19	.82	1.07
2500	Wall installation, incl. drilling & patching from outside, two 1"										
2510	diam. holes @ 16" O.C., top & mid-point of wall, add to above										
2700	For masonry	G	G-4	415	.058	S.F.	.06	1.23	.80	2.09	3.02
2800	For wood siding	G		840	.029		.06	.61	.40	1.07	1.54
2900	For stucco/plaster	G	↓	665	.036	↓	.06	.77	.50	1.33	1.91

07 21 27 – Reflective Insulation

07 21 27.10 Reflective Insulation Options

		Crew	Daily Output	Labor-Hours	Unit	Material	Labor	Equipment	Total	Total Incl O&P
0010	**REFLECTIVE INSULATION OPTIONS**										
0020	Aluminum foil on reinforced scrim	G	1 Carp	19	.421	C.S.F.	14.20	11.75		25.95	35.50
0100	Reinforced with woven polyolefin	G		19	.421		17.20	11.75		28.95	39
0500	With single bubble air space, R8.8	G		15	.533		25.50	14.90		40.40	53
0600	With double bubble air space, R9.8	G	↓	15	.533	↓	26.50	14.90		41.40	54

07 21 29 – Sprayed Insulation

07 21 29.10 Sprayed-On Insulation

		Crew	Daily Output	Labor-Hours	Unit	Material	Labor	Equipment	Total	Total Incl O&P
0010	**SPRAYED-ON INSULATION**										
0300	Closed cell, spray polyurethane foam, 2 pounds per cubic foot density										
0310	1" thick	G	G-2A	6000	.004	S.F.	.70	.08	.10	.88	1.03
0320	2" thick	G		3000	.008		1.40	.16	.20	1.76	2.05
0330	3" thick	G		2000	.012		2.11	.25	.30	2.66	3.09
0335	3-1/2" thick	G		1715	.014		2.45	.29	.35	3.09	3.59
0340	4" thick	G		1500	.016		2.80	.33	.40	3.53	4.11
0350	5" thick	G		1200	.020		3.50	.41	.50	4.41	5.15
0355	5-1/2" thick	G		1090	.022		3.85	.45	.56	4.86	5.65
0360	6" thick	G	↓	1000	.024	↓	4.22	.49	.61	5.32	6.20

07 22 16 – Roof Board Insulation

07 22 16.10 Roof Deck Insulation		Crew	Daily Output	Labor-Hours	Unit	Material	2009 Bare Costs Labor	Equipment	Total	Total Incl O&P
0010	**ROOF DECK INSULATION**									
0020	Fiberboard low density, 1/2" thick R1.39	G	1 Rofc	1000	.008	S.F.	.24	.19		.43	.61
0030	1" thick R2.78	G		800	.010		.42	.24		.66	.89
0080	1 1/2" thick R4.17	G		800	.010		.64	.24		.88	1.13
0100	2" thick R5.56	G		800	.010		.85	.24		1.09	1.37
0110	Fiberboard high density, 1/2" thick R1.3	G		1000	.008		.22	.19		.41	.59
0120	1" thick R2.5	G		800	.010		.44	.24		.68	.91
0130	1-1/2" thick R3.8	G		800	.010		.66	.24		.90	1.16
0200	Fiberglass, 3/4" thick R2.78	G		1000	.008		.56	.19		.75	.97
0400	15/16" thick R3.70	G		1000	.008		.74	.19		.93	1.16
0460	1-1/16" thick R4.17	G		1000	.008		.93	.19		1.12	1.37
0600	1-5/16" thick R5.26	G		1000	.008		1.26	.19		1.45	1.74
0650	2-1/16" thick R8.33	G		800	.010		1.35	.24		1.59	1.92
0700	2-7/16" thick R10	G		800	.010		1.55	.24		1.79	2.14
1650	Perlite, 1/2" thick R1.32	G		1050	.008		.34	.18		.52	.70
1655	3/4" thick R2.08	G		800	.010		.37	.24		.61	.84
1660	1" thick R2.78	G		800	.010		.46	.24		.70	.94
1670	1-1/2" thick R4.17	G		800	.010		.48	.24		.72	.96
1680	2" thick R5.56	G		700	.011		.80	.27		1.07	1.37
1685	2-1/2" thick R6.67	G		700	.011		.95	.27		1.22	1.54
1700	Polyisocyanurate, 2#/CF density, 3/4" thick, R5.1	G		1500	.005		.44	.13		.57	.71
1705	1" thick R7.14	G		1400	.006		.50	.14		.64	.80
1715	1-1/2" thick R10.87	G		1250	.006		.63	.15		.78	.97
1725	2" thick R14.29	G		1100	.007		.81	.17		.98	1.20
1735	2-1/2" thick R16.67	G		1050	.008		1.01	.18		1.19	1.44
1745	3" thick R21.74	G		1000	.008		1.26	.19		1.45	1.74
1755	3-1/2" thick R25	G		1000	.008	▼	1.93	.19		2.12	2.47
1765	Tapered for drainage	G		1400	.006	B.F.	1.93	.14		2.07	2.37
1900	Extruded Polystyrene									
1910	15 PSI compressive strength, 1" thick, R5	G	1 Rofc	1500	.005	S.F.	.49	.13		.62	.77
1920	2" thick, R10	G		1250	.006		.62	.15		.77	.96
1930	3" thick R15	G		1000	.008		1.27	.19		1.46	1.75
1932	4" thick R20	G		1000	.008	▼	1.64	.19		1.83	2.15
1934	Tapered for drainage	G		1500	.005	B.F.	.53	.13		.66	.81
1940	25 PSI compressive strength, 1" thick R5	G		1500	.005	S.F.	.67	.13		.80	.97
1942	2" thick R10	G		1250	.006		1.27	.15		1.42	1.68
1944	3" thick R15	G		1000	.008		1.93	.19		2.12	2.47
1946	4" thick R20	G		1000	.008	▼	2.72	.19		2.91	3.34
1948	Tapered for drainage	G		1500	.005	B.F.	.56	.13		.69	.85
1950	40 psi compressive strength, 1" thick R5	G		1500	.005	S.F.	.51	.13		.64	.79
1952	2" thick R10	G		1250	.006		.96	.15		1.11	1.34
1954	3" thick R15	G		1000	.008		1.40	.19		1.59	1.89
1956	4" thick R20	G		1000	.008	▼	1.88	.19		2.07	2.42
1958	Tapered for drainage	G		1400	.006	B.F.	.71	.14		.85	1.03
1960	60 PSI compressive strength, 1" thick R5	G		1450	.006	S.F.	.59	.13		.72	.89
1962	2" thick R10	G		1200	.007		1.05	.16		1.21	1.45
1964	3" thick R15	G		975	.008		1.55	.19		1.74	2.06
1966	4" thick R20	G		950	.008	▼	2.17	.20		2.37	2.75
1968	Tapered for drainage	G		1400	.006	B.F.	.85	.14		.99	1.19
2010	Expanded polystyrene, 1#/CF density, 3/4" thick R2.89	G		1500	.005	S.F.	.31	.13		.44	.57
2020	1" thick R3.85	G		1500	.005		.31	.13		.44	.57
2100	2" thick R7.69	G		1250	.006		.62	.15		.77	.96
2110	3" thick R11.49	G		1250	.006	▼	.91	.15		1.06	1.28

07 22 Roof and Deck Insulation

07 22 16 – Roof Board Insulation

07 22 16.10 Roof Deck Insulation		Crew	Daily Output	Labor-Hours	Unit	Material	2009 Bare Costs Labor	Equipment	Total	Total Incl O&P
2120	4" thick R15.38	G 1 Rofc	1200	.007	S.F.	.98	.16		1.14	1.37
2130	5" thick R19.23	G	1150	.007		1.05	.16		1.21	1.46
2140	6" thick R23.26	G	1150	.007		1.23	.16		1.39	1.65
2150	Tapered for drainage	G	1500	.005	B.F.	.49	.13		.62	.77
2400	Composites with 2" EPS									
2410	1" fiberboard	G 1 Rofc	950	.008	S.F.	1.07	.20		1.27	1.54
2420	7/16" oriented strand board	G	800	.010		1.26	.24		1.50	1.82
2430	1/2" plywood	G	800	.010		1.37	.24		1.61	1.94
2440	1" perlite	G	800	.010		1.12	.24		1.36	1.66
2450	Composites with 1-1/2" polyisocyanurate									
2460	1" fiberboard	G 1 Rofc	800	.010	S.F.	1.46	.24		1.70	2.04
2470	1" perlite	G	850	.009		1.53	.22		1.75	2.09
2480	7/16" oriented strand board	G	800	.010		1.77	.24		2.01	2.38

07 24 Exterior Insulation and Finish Systems

07 24 13 – Polymer Based Exterior Insulation and Finish Systems

07 24 13.10 Exterior Insulation and Finish Systems

		Crew	Daily Output	Labor-Hours	Unit	Material	2009 Bare Costs Labor	Equipment	Total	Total Incl O&P
0010	**EXTERIOR INSULATION AND FINISH SYSTEMS**									
0095	Field applied, 1" EPS insulation	G J-1	295	.136	S.F.	2.18	3.28	.43	5.89	8.25
0100	With 1/2" cement board sheathing	G	220	.182		2.90	4.40	.58	7.88	11.05
0105	2" EPS insulation	G	295	.136		2.55	3.28	.43	6.26	8.70
0110	With 1/2" cement board sheathing	G	220	.182		3.27	4.40	.58	8.25	11.50
0115	3" EPS insulation	G	295	.136		2.72	3.28	.43	6.43	8.85
0120	With 1/2" cement board sheathing	G	220	.182		3.44	4.40	.58	8.42	11.65
0125	4" EPS insulation	G	295	.136		3.16	3.28	.43	6.87	9.35
0130	With 1/2" cement board sheathing	G	220	.182		4.60	4.40	.58	9.58	12.95
0140	Premium finish add		1265	.032		.31	.77	.10	1.18	1.71
0150	Heavy duty reinforcement add		914	.044		1.08	1.06	.14	2.28	3.09
0160	2.5#/S.Y. metal lath substrate add	1 Lath	75	.107	S.Y.	2.37	2.66		5.03	6.90
0170	3.4#/S.Y. metal lath substrate add	"	75	.107	"	2.58	2.66		5.24	7.15
0180	Color or texture change,	J-1	1265	.032	S.F.	.83	.77	.10	1.70	2.28
0190	With substrate leveling base coat	1 Plas	530	.015		.83	.38		1.21	1.54
0210	With substrate sealing base coat	1 Pord	1224	.007		.08	.16		.24	.36
0370	V groove shape in panel face				L.F.	.60			.60	.66
0380	U groove shape in panel face				"	.79			.79	.87

07 26 Vapor Retarders

07 26 10 – Vapor Retarders Made of Natural and Synthetic Materials

07 26 10.10 Building Paper

		Crew	Daily Output	Labor-Hours	Unit	Material	2009 Bare Costs Labor	Equipment	Total	Total Incl O&P
0011	**BUILDING PAPER** Aluminum and kraft laminated, foil 1 side	1 Carp	3700	.002	S.F.	.05	.06		.11	.16
0101	Foil 2 sides	G	3700	.002		.09	.06		.15	.20
0301	Asphalt, two ply, 30#, for subfloors		1900	.004		.19	.12		.31	.41
0401	Asphalt felt sheathing paper, 15#		3700	.002		.05	.06		.11	.15
0450	Housewrap, exterior, spun bonded polypropylene									
0470	Small roll	G 1 Carp	3800	.002	S.F.	.23	.06		.29	.35
0480	Large roll	G "	4000	.002	"	.12	.06		.18	.22
0500	Material only, 3' x 111.1' roll	G			Ea.	75			75	82.50
0520	9' x 111.1' roll	G			"	120			120	132
0601	Polyethylene vapor barrier, standard, .002" thick	G 1 Carp	3700	.002	S.F.	.01	.06		.07	.11

07 26 Vapor Retarders

07 26 10 – Vapor Retarders Made of Natural and Synthetic Materials

07 26 10.10 Building Paper		Crew	Daily Output	Labor-Hours	Unit	Material	2009 Bare Costs Labor	Equipment	Total	Total Incl O&P
0701	.004" thick	G	1 Carp	3700	.002	S.F.	.04	.06		.10	.14
0901	.006" thick	G		3700	.002		.06	.06		.12	.16
1201	.010" thick	G		3700	.002		.07	.06		.13	.17
1501	Red rosin paper, 5 sq rolls, 4 lb per square			3700	.002		.02	.06		.08	.12
1601	5 lbs. per square			3700	.002		.03	.06		.09	.13
1801	Reinf. waterproof, .002" polyethylene backing, 1 side			3700	.002		.06	.06		.12	.16
1901	2 sides			3700	.002		.07	.06		.13	.18
3000	Building wrap, spunbonded polyethylene		2 Carp	8000	.002		.13	.06		.19	.23

07 31 Shingles and Shakes

07 31 13 – Asphalt Shingles

07 31 13.10 Asphalt Roof Shingles

		Crew	Daily Output	Labor-Hours	Unit	Material	2009 Bare Costs Labor	Equipment	Total	Total Incl O&P
0010	**ASPHALT ROOF SHINGLES**									
0100	Standard strip shingles									
0150	Inorganic, class A, 210-235 lb/sq	1 Rofc	5.50	1.455	Sq.	50	34.50		84.50	118
0155	Pneumatic nailed		7	1.143		50	27		77	105
0200	Organic, class C, 235-240 lb/sq		5	1.600		51	38		89	125
0205	Pneumatic nailed		6.25	1.280		51	30.50		81.50	111
0250	Standard, laminated multi-layered shingles									
0300	Class A, 240-260 lb/sq	1 Rofc	4.50	1.778	Sq.	61.50	42		103.50	144
0305	Pneumatic nailed		5.63	1.422		61.50	33.50		95	129
0350	Class C, 260-300 lb/square, 4 bundles/square		4	2		67	47.50		114.50	161
0355	Pneumatic nailed		5	1.600		67	38		105	143
0400	Premium, laminated multi-layered shingles									
0450	Class A, 260-300 lb, 4 bundles/sq	1 Rofc	3.50	2.286	Sq.	80	54		134	187
0455	Pneumatic nailed		4.37	1.831		80	43.50		123.50	167
0500	Class C, 300-385 lb/square, 5 bundles/square		3	2.667		106	63		169	232
0505	Pneumatic nailed		3.75	2.133		106	50.50		156.50	209
0800	#15 felt underlayment		64	.125		4.77	2.96		7.73	10.65
0825	#30 felt underlayment		58	.138		9.55	3.26		12.81	16.45
0850	Self adhering polyethylene and rubberized asphalt underlayment		22	.364		55.50	8.60		64.10	76.50
0900	Ridge shingles		330	.024	L.F.	1.49	.57		2.06	2.69
0905	Pneumatic nailed		412.50	.019	"	1.49	.46		1.95	2.48
1000	For steep roofs (7 to 12 pitch or greater), add						50%			

07 31 19 – Mineral-Fiber Cement Shingles

07 31 19.10 Fiber Cement Shingles

		Crew	Daily Output	Labor-Hours	Unit	Material	2009 Bare Costs Labor	Equipment	Total	Total Incl O&P
0010	**FIBER CEMENT SHINGLES**									
0012	Field shingles, 16" x 9.35", 500 lb per square	1 Rofc	2.20	3.636	Sq.	335	86		421	525
0200	Shakes, 16" x 9.35", 550 lb per square		2.20	3.636	"	305	86		391	490
0301	Hip & ridge, 4.75 x 14"		100	.080	L.F.	8.25	1.89		10.14	12.55
0400	Hexagonal, 16" x 16"		3	2.667	Sq.	227	63		290	365
0500	Square, 16" x 16"		3	2.667		203	63		266	340
2000	For steep roofs (7/12 pitch or greater), add						50%			

07 31 26 – Slate Shingles

07 31 26.10 Slate Roof Shingles

			Crew	Daily Output	Labor-Hours	Unit	Material	2009 Bare Costs Labor	Equipment	Total	Total Incl O&P
0010	**SLATE ROOF SHINGLES**	R073126-20									
0100	Buckingham Virginia black, 3/16" - 1/4" thick	G	1 Rots	1.75	4.571	Sq.	475	108		583	720
0200	1/4" thick	G		1.75	4.571		475	108		583	720
0900	Pennsylvania black, Bangor, #1 clear	G		1.75	4.571		490	108		598	735
1200	Vermont, unfading, green, mottled green	G		1.75	4.571		480	108		588	720

07 31 Shingles and Shakes

07 31 26 – Slate Shingles

07 31 26.10 Slate Roof Shingles

		Crew	Daily Output	Labor-Hours	Unit	Material	2009 Bare Costs Labor	Equipment	Total	Total Incl O&P
1300	Semi-weathering green & gray	G 1 Rots	1.75	4.571	Sq.	350	108		458	575
1400	Purple	G	1.75	4.571		425	108		533	660
1500	Black or gray	G	1.75	4.571		460	108		568	700
2700	Ridge shingles, slate		200	.040	L.F.	9.30	.94		10.24	11.90

07 31 29 – Wood Shingles and Shakes

07 31 29.13 Wood Shingles

		Crew	Daily Output	Labor-Hours	Unit	Material	2009 Bare Costs Labor	Equipment	Total	Total Incl O&P
0010	**WOOD SHINGLES**									
0012	16" No. 1 red cedar shingles, 5" exposure, on roof	1 Carp	2.50	3.200	Sq.	330	89.50		419.50	515
0015	Pneumatic nailed		3.25	2.462		330	69		399	480
0200	7-1/2" exposure, on walls		2.05	3.902		221	109		330	425
0205	Pneumatic nailed		2.67	2.996		221	84		305	385
0300	18" No. 1 red cedar perfections, 5-1/2" exposure, on roof		2.75	2.909		330	81.50		411.50	495
0305	Pneumatic nailed		3.57	2.241		330	62.50		392.50	465
0500	7-1/2" exposure, on walls		2.25	3.556		242	99.50		341.50	435
0505	Pneumatic nailed		2.92	2.740		242	76.50		318.50	395
0600	Resquared, and rebutted, 5-1/2" exposure, on roof		3	2.667		365	74.50		439.50	530
0605	Pneumatic nailed		3.90	2.051		365	57.50		422.50	500
0900	7-1/2" exposure, on walls		2.45	3.265		269	91.50		360.50	450
0905	Pneumatic nailed		3.18	2.516		269	70.50		339.50	415
1000	Add to above for fire retardant shingles, 16" long					63			63	69.50
1050	18" long					63			63	69.50
1060	Preformed ridge shingles	1 Carp	400	.020	L.F.	3.50	.56		4.06	4.80
2000	White cedar shingles, 16" long, extras, 5" exposure, on roof		2.40	3.333	Sq.	178	93		271	355
2005	Pneumatic nailed		3.12	2.564		178	71.50		249.50	315
2050	5" exposure on walls		2	4		178	112		290	385
2055	Pneumatic nailed		2.60	3.077		178	86		264	340
2100	7-1/2" exposure, on walls		2	4		127	112		239	330
2105	Pneumatic nailed		2.60	3.077		127	86		213	285
2150	"B" grade, 5" exposure on walls		2	4		150	112		262	355
2155	Pneumatic nailed		2.60	3.077		150	86		236	310
2300	For 15# organic felt underlayment on roof, 1 layer, add		64	.125		4.77	3.49		8.26	11.15
2400	2 layers, add		32	.250		9.55	7		16.55	22.50
2600	For steep roofs (7/12 pitch or greater), add to above						50%			
3000	Ridge shakes or shingle wood	1 Carp	280	.029	L.F.	3.60	.80		4.40	5.30

07 31 29.16 Wood Shakes

		Crew	Daily Output	Labor-Hours	Unit	Material	2009 Bare Costs Labor	Equipment	Total	Total Incl O&P
0010	**WOOD SHAKES**									
1100	Hand-split red cedar shakes, 1/2" thick x 24" long, 10" exp. on roof	1 Carp	2.50	3.200	Sq.	252	89.50		341.50	430
1105	Pneumatic nailed		3.25	2.462		252	69		321	395
1110	3/4" thick x 24" long, 10" exp. on roof		2.25	3.556		252	99.50		351.50	445
1115	Pneumatic nailed		2.92	2.740		252	76.50		328.50	405
1200	1/2" thick, 18" long, 8-1/2" exp. on roof		2	4		239	112		351	450
1205	Pneumatic nailed		2.60	3.077		239	86		325	410
1210	3/4" thick x 18" long, 8 1/2" exp. on roof		1.80	4.444		239	124		363	475
1215	Pneumatic nailed		2.34	3.419		239	95.50		334.50	425
1255	10" exp. on walls		2	4		231	112		343	445
1260	10" exposure on walls, pneumatic nailed		2.60	3.077		231	86		317	400
1700	Add to above for fire retardant shakes, 24" long					63			63	69.50
1800	18" long					63			63	69.50
1810	Ridge shakes	1 Carp	350	.023	L.F.	3.50	.64		4.14	4.93

07 32 Roof Tiles

07 32 13 – Clay Roof Tiles

07 32 13.10 Clay Tiles		Crew	Daily Output	Labor-Hours	Unit	Material	2009 Bare Costs Labor	Equipment	Total	Total Incl O&P
0010	**CLAY TILES**									
0200	Lanai tile or Classic tile, 158 pc per sq	G	1 Rots	1.65	4.848	Sq.	500	114		614	760
0300	Americana, 158 pc per sq, most colors	G		1.65	4.848		660	114		774	940
0350	Green, gray or brown	G		1.65	4.848		625	114		739	895
0400	Blue	G		1.65	4.848		625	114		739	895
0600	Spanish tile, 171 pc per sq, red	G		1.80	4.444		325	105		430	545
0800	Buff, green, gray, brown	G		1.80	4.444		600	105		705	850
0900	Glazed white	G		1.80	4.444		665	105		770	920
1100	Mission tile, 192 pc per sq, machine scored finish, red	G		1.15	6.957		740	164		904	1,125
1700	French tile, 133 pc per sq, smooth finish, red	G		1.35	5.926		675	140		815	995
1750	Blue or green	G		1.35	5.926		870	140		1,010	1,200
1800	Norman black 317 pc per sq	G		1	8		1,050	188		1,238	1,500
2200	Williamsburg tile, 158 pc per sq, aged cedar	G		1.35	5.926		730	140		870	1,050
2250	Gray or green	G		1.35	5.926	↓	595	140		735	910
2350	Ridge shingles, clay tile	G		200	.040	L.F.	10.70	.94		11.64	13.50
2510	One piece mission tile, natural red, 75 pc per square	G		1.65	4.848	Sq.	231	114		345	460
2530	Mission Tile, 134 pc per square	G	↓	1.15	6.957		263	164		427	590
3000	For steep roofs (7/12 pitch or greater), add to above							50%			
3010	Clay tile, #15 felt underlayment		1 Rofc	64	.125		4.77	2.96		7.73	10.65
3020	Clay tile, #30 felt underlayment			58	.138		9.55	3.26		12.81	16.45
3040	Clay tile, polyethylene and rubberized asph. underlayment		↓	22	.364	↓	55.50	8.60		64.10	76.50

07 32 16 – Concrete Roof Tiles

07 32 16.10 Concrete Tiles		Crew	Daily Output	Labor-Hours	Unit	Material	2009 Bare Costs Labor	Equipment	Total	Total Incl O&P
0010	**CONCRETE TILES**									
0020	Corrugated, 13" x 16-1/2", 90 per sq, 950 lb per sq									
0050	Earthtone colors, nailed to wood deck		1 Rots	1.35	5.926	Sq.	96.50	140		236.50	360
0150	Blues			1.35	5.926		97.50	140		237.50	360
0200	Greens			1.35	5.926		97.50	140		237.50	360
0250	Premium colors		↓	1.35	5.926	↓	214	140		354	490
0500	Shakes, 13" x 16-1/2", 90 per sq, 950 lb per sq									
0600	All colors, nailed to wood deck		1 Rots	1.50	5.333	Sq.	252	126		378	505
1500	Accessory pieces, ridge & hip, 10" x 16-1/2", 8 lbs. each					Ea.	3.40			3.40	3.74
1700	Rake, 6-1/2" x 16-3/4", 9 lbs. each						3.40			3.40	3.74
1800	Mansard hip, 10" x 16-1/2", 9.2 lbs. each						3.40			3.40	3.74
1900	Hip starter, 10" x 16-1/2", 10.5 lbs. each						10.70			10.70	11.80
2000	3 or 4 way apex, 10" each side, 11.5 lbs. each					↓	12.35			12.35	13.60

07 33 Natural Roof Coverings

07 33 63 – Vegetated Roofing

07 33 63.10 Green Roof Systems		Crew	Daily Output	Labor-Hours	Unit	Material	2009 Bare Costs Labor	Equipment	Total	Total Incl O&P
0010	**GREEN ROOF SYSTEMS**									
0020	Soil mixture for green roof 30% sand, 55% gravel, 15% soil									
0100	Hoist and spread soil mixture 4 inch depth up to five stories tall roof	G	B-13B	4000	.014	S.F.	.25	.32	.28	.85	1.12
0150	6 inch depth	G		2667	.021		.38	.48	.42	1.28	1.68
0200	8 inch depth	G		2000	.028		.50	.64	.56	1.70	2.25
0250	10 inch depth	G		1600	.035		.63	.80	.70	2.13	2.80
0300	12 inch depth	G	↓	1335	.042	↓	.76	.96	.84	2.56	3.36
0350	Mobilization 55 ton crane to site	G	1 Eqhv	3.60	2.222	Ea.		65		65	107
0355	Hoisting cost to five stories per day (Avg. 28 picks per day)	G	B-13B	1	56	Day		1,275	1,125	2,400	3,375
0360	Mobilization or demobilization, 100 ton crane to site driver & escort	G	A-3E	2.50	6.400	Ea.		167	51.50	218.50	335

07 33 Natural Roof Coverings

07 33 63 – Vegetated Roofing

07 33 63.10 Green Roof Systems

			Crew	Daily Output	Labor-Hours	Unit	Material	2009 Bare Costs Labor	Equipment	Total	Total Incl O&P
0365	Hoisting cost six to ten stories per day (Avg. 21 picks per day)	G	B-13C	1	56	Day		1,275	1,600	2,875	3,900
0370	Hoist and spread soil mixture 4 inch depth six to ten stories tall roof	G		4000	.014	S.F.	.25	.32	.40	.97	1.25
0375	6 inch depth	G		2667	.021		.38	.48	.60	1.46	1.88
0380	8 inch depth	G		2000	.028		.50	.64	.81	1.95	2.52
0385	10 inch depth	G		1600	.035		.63	.80	1.01	2.44	3.14
0390	12 inch depth	G		1335	.042		.76	.96	1.21	2.93	3.76
0400	Green roof edging treated lumber 4" x4" no hoisting included	G	2 Carp	400	.040	L.F.	1.52	1.12		2.64	3.56
0410	4" x 6"	G		400	.040		2.24	1.12		3.36	4.35
0420	4" x 8"	G		360	.044		3.21	1.24		4.45	5.65
0430	4" x 6" double stacked	G		300	.053		4.47	1.49		5.96	7.45
0500	Green roof edging redwood lumber 4" x4" no hoisting included	G		400	.040		6.45	1.12		7.57	9
0510	4" x 6"	G		400	.040		11.20	1.12		12.32	14.20
0520	4" x 8"	G		360	.044		20.50	1.24		21.74	24.50
0530	4" x 6" double stacked	G		300	.053		22.50	1.49		23.99	27
0600	Planting sedum, light soil, potted, 2-1/4" diameter, two per SF	G	1 Clab	420	.019	S.F.	7	.39		7.39	8.35
0610	one per SF	G	"	840	.010		3.50	.20		3.70	4.18
0630	Planting sedum mat per SF including shipping (4000 SF Minimum)	G	4 Clab	4000	.008		8.40	.16		8.56	9.55
0640	Installation sedum mat system (no soil required) per SF (4000 SF minimum)	G	"	4000	.008		11.70	.16		11.86	13.15
0645	Note: pricing of sedum mats shipped in full truck loads (4000-5000 SF)											

07 41 Roof Panels

07 41 13 – Metal Roof Panels

07 41 13.10 Aluminum Roof Panels

			Crew	Daily Output	Labor-Hours	Unit	Material	Labor	Equipment	Total	Total Incl O&P
0010	**ALUMINUM ROOF PANELS**										
0020	Corrugated or ribbed, .0155" thick, natural		G-3	1200	.027	S.F.	.87	.69		1.56	2.12
0300	Painted		"	1200	.027	"	1.27	.69		1.96	2.56

07 41 33 – Plastic Roof Panels

07 41 33.10 Fiberglass Panels

			Crew	Daily Output	Labor-Hours	Unit	Material	Labor	Equipment	Total	Total Incl O&P
0010	**FIBERGLASS PANELS**										
0012	Corrugated panels, roofing, 8 oz per S.F.		G-3	1000	.032	S.F.	1.58	.83		2.41	3.13
0100	12 oz per S.F.			1000	.032		3.43	.83		4.26	5.15
0300	Corrugated siding, 6 oz per S.F.			880	.036		1.58	.94		2.52	3.32
0400	8 oz per S.F.			880	.036		1.58	.94		2.52	3.32
0600	12 oz. siding, textured			880	.036		3.32	.94		4.26	5.25
0900	Flat panels, 6 oz per S.F., clear or colors			880	.036		1.83	.94		2.77	3.59
1300	8 oz per S.F., clear or colors			880	.036		2.38	.94		3.32	4.20

07 42 Wall Panels

07 42 13 – Metal Wall Panels

07 42 13.20 Aluminum Siding

			Crew	Daily Output	Labor-Hours	Unit	Material	Labor	Equipment	Total	Total Incl O&P
0011	**ALUMINUM SIDING**										
6040	.024 thick smooth white single 8" wide		2 Carp	515	.031	S.F.	1.98	.87		2.85	3.65
6060	Double 4" pattern			515	.031		1.41	.87		2.28	3.02
6080	Double 5" pattern			550	.029		1.45	.81		2.26	2.97
6120	Embossed white, 8" wide			515	.031		1.98	.87		2.85	3.65
6140	Double 4" pattern			515	.031		2.09	.87		2.96	3.77
6160	Double 5" pattern			550	.029		2.09	.81		2.90	3.67
6170	Vertical, embossed white, 12" wide			590	.027		2.09	.76		2.85	3.58
6320	.019 thick, insulated, smooth white, 8" wide			515	.031		1.85	.87		2.72	3.51

07 42 Wall Panels

07 42 13 – Metal Wall Panels

07 42 13.20 Aluminum Siding	Crew	Daily Output	Labor-Hours	Unit	Material	2009 Bare Costs Labor	Equipment	Total	Total Incl O&P
6340	Double 4" pattern	2 Carp	515	.031	S.F.	1.83	.87		2.70	3.48
6360	Double 5" pattern		550	.029		1.83	.81		2.64	3.38
6400	Embossed white, 8" wide		515	.031		2.14	.87		3.01	3.82
6420	Double 4" pattern		515	.031		2.17	.87		3.04	3.86
6440	Double 5" pattern		550	.029		2.17	.81		2.98	3.76
6500	Shake finish 10" wide white		550	.029		2.30	.81		3.11	3.90
6600	Vertical pattern, 12" wide, white		590	.027		1.92	.76		2.68	3.39
6640	For colors add					.12			.12	.13
6700	Accessories, white									
6720	Starter strip 2-1/8"	2 Carp	610	.026	L.F.	.32	.73		1.05	1.59
6740	Sill trim		450	.036		.54	.99		1.53	2.27
6760	Inside corner		610	.026		1.36	.73		2.09	2.74
6780	Outside corner post		610	.026		3	.73		3.73	4.54
6800	Door & window trim		440	.036		.52	1.02		1.54	2.29
6820	For colors add					.10			.10	.11
6900	Soffit & fascia 1' overhang solid	2 Carp	110	.145		3	4.07		7.07	10.15
6920	Vented		110	.145		3	4.07		7.07	10.15
6940	2' overhang solid		100	.160		4.92	4.47		9.39	12.95
6960	Vented		100	.160		4.92	4.47		9.39	12.95

07 42 13.30 Steel Siding

07 42 13.30 Steel Siding	Crew	Daily Output	Labor-Hours	Unit	Material	2009 Bare Costs Labor	Equipment	Total	Total Incl O&P
0010	**STEEL SIDING**									
0020	Beveled, vinyl coated, 8" wide	1 Carp	265	.030	S.F.	1.81	.84		2.65	3.42
0050	10" wide	"	275	.029		1.94	.81		2.75	3.50
0081	Galv., corrugated or ribbed, on steel frame, 30 gauge	G-3	775	.041		1.21	1.07		2.28	3.13
0101	28 gauge		775	.041		1.27	1.07		2.34	3.20
0301	26 gauge		775	.041		1.78	1.07		2.85	3.76
0401	24 gauge		775	.041		1.79	1.07		2.86	3.77
0601	22 gauge		775	.041		2.06	1.07		3.13	4.07
0701	Colored, corrugated/ribbed, on steel frame, 10 yr fnsh, 28 ga.		775	.041		1.88	1.07		2.95	3.87
0901	26 gauge		775	.041		1.96	1.07		3.03	3.96
1001	24 gauge		775	.041		2.29	1.07		3.36	4.32

07 46 Siding

07 46 23 – Wood Siding

07 46 23.10 Wood Board Siding

07 46 23.10 Wood Board Siding	Crew	Daily Output	Labor-Hours	Unit	Material	2009 Bare Costs Labor	Equipment	Total	Total Incl O&P
0010	**WOOD BOARD SIDING**									
2000	Board & batten, cedar, "B" grade, 1" x 10"	1 Carp	375	.021	S.F.	2.09	.60		2.69	3.31
2200	Redwood, clear, vertical grain, 1" x 10"		375	.021		6.20	.60		6.80	7.80
2400	White pine, #2 & better, 1" x 10"		375	.021		2	.60		2.60	3.21
2410	Board & batten siding, white pine #2, 1" x 12"		420	.019		2	.53		2.53	3.10
3200	Wood, cedar bevel, A grade, 1/2" x 6"		295	.027		4.62	.76		5.38	6.40
3300	1/2" x 8"		330	.024		4.97	.68		5.65	6.60
3500	3/4" x 10", clear grade		375	.021		6.70	.60		7.30	8.35
3600	"B" grade		375	.021		6.25	.60		6.85	7.90
3800	Cedar, rough sawn, 1" x 4", A grade, natural		220	.036		3.77	1.02		4.79	5.85
3900	Stained		220	.036		4.17	1.02		5.19	6.30
4100	1" x 12", board & batten, #3 & Btr., natural		420	.019		3.94	.53		4.47	5.25
4200	Stained		420	.019		4.20	.53		4.73	5.50
4400	1" x 8" channel siding, #3 & Btr., natural		330	.024		2.34	.68		3.02	3.72
4500	Stained		330	.024		2.50	.68		3.18	3.90
4700	Redwood, clear, beveled, vertical grain, 1/2" x 4"		220	.036		3.10	1.02		4.12	5.15

07 46 Siding

07 46 23 – Wood Siding

07 46 23.10 Wood Board Siding

		Crew	Daily Output	Labor-Hours	Unit	Material	2009 Bare Costs Labor	Equipment	Total	Total Incl O&P
4750	1/2" x 6"	1 Carp	295	.027	S.F.	3.10	.76		3.86	4.69
4800	1/2" x 8"		330	.024		3.24	.68		3.92	4.71
5000	3/4" x 10"		375	.021		3.32	.60		3.92	4.66
5200	Channel siding, 1" x 10", B grade		375	.021		3.32	.60		3.92	4.66
5250	Redwood, T&G boards, B grade, 1" x 4"		220	.036		3.10	1.02		4.12	5.15
5270	1" x 8"		330	.024		3.24	.68		3.92	4.71
5400	White pine, rough sawn, 1" x 8", natural		330	.024		2.20	.68		2.88	3.57
5500	Stained		330	.024		2.60	.68		3.28	4.01
5600	Tongue and groove, 1" x 8"		330	.024		2.20	.68		2.88	3.57

07 46 29 – Plywood Siding

07 46 29.10 Plywood Siding Options

		Crew	Daily Output	Labor-Hours	Unit	Material	2009 Bare Costs Labor	Equipment	Total	Total Incl O&P
0010	**PLYWOOD SIDING OPTIONS**									
0900	Plywood, medium density overlaid, 3/8" thick	2 Carp	750	.021	S.F.	.96	.60		1.56	2.07
1000	1/2" thick		700	.023		1.27	.64		1.91	2.48
1100	3/4" thick		650	.025		1.31	.69		2	2.60
1600	Texture 1-11, cedar, 5/8" thick, natural		675	.024		2.45	.66		3.11	3.82
1700	Factory stained		675	.024		1.96	.66		2.62	3.28
1900	Texture 1-11, fir, 5/8" thick, natural		675	.024		1.13	.66		1.79	2.36
2000	Factory stained		675	.024		1.73	.66		2.39	3.02
2050	Texture 1-11, S.Y.P., 5/8" thick, natural		675	.024		1.21	.66		1.87	2.45
2100	Factory stained		675	.024		1.25	.66		1.91	2.50
2200	Rough sawn cedar, 3/8" thick, natural		675	.024		1.21	.66		1.87	2.45
2300	Factory stained		675	.024		.97	.66		1.63	2.19
2500	Rough sawn fir, 3/8" thick, natural		675	.024		.77	.66		1.43	1.97
2600	Factory stained		675	.024		.97	.66		1.63	2.19
2800	Redwood, textured siding, 5/8" thick		675	.024		1.94	.66		2.60	3.25

07 46 33 – Plastic Siding

07 46 33.10 Vinyl Siding

		Crew	Daily Output	Labor-Hours	Unit	Material	2009 Bare Costs Labor	Equipment	Total	Total Incl O&P
0010	**VINYL SIDING**									
3995	Clapboard profile, woodgrain texture, .048 thick, double 4	2 Carp	495	.032	S.F.	.92	.90		1.82	2.54
4000	Double 5		550	.029		.92	.81		1.73	2.38
4005	Single 8		495	.032		.96	.90		1.86	2.59
4010	Single 10		550	.029		1.15	.81		1.96	2.64
4015	.044 thick, double 4		495	.032		.84	.90		1.74	2.45
4020	Double 5		550	.029		.84	.81		1.65	2.29
4025	.042 thick, double 4		495	.032		.84	.90		1.74	2.45
4030	Double 5		550	.029		.84	.81		1.65	2.29
4035	Cross sawn texture, .040 thick, double 4		495	.032		.66	.90		1.56	2.26
4040	Double 5		550	.029		.66	.81		1.47	2.10
4045	Smooth texture, .042 thick, double 4		495	.032		.75	.90		1.65	2.36
4050	Double 5		550	.029		.75	.81		1.56	2.20
4055	Single 8		495	.032		.96	.90		1.86	2.59
4060	Cedar texture, .044 thick, double 4		495	.032		.92	.90		1.82	2.54
4065	Double 6		600	.027		.92	.75		1.67	2.27
4070	Dutch lap profile, woodgrain texture, .048 thick, double 5		550	.029		.92	.81		1.73	2.38
4075	.044 thick, double 4.5		525	.030		.92	.85		1.77	2.45
4080	.042 thick, double 4.5		525	.030		.75	.85		1.60	2.27
4085	.040 thick, double 4.5		525	.030		.66	.85		1.51	2.17
4090	Shingle profile, random grooves, double 7		400	.040		2.57	1.12		3.69	4.72
4095	Triple 5		400	.040		2.57	1.12		3.69	4.72
4100	Shake profile, 10" wide		400	.040		2.19	1.12		3.31	4.30
4105	Vertical pattern, .046 thick, double 5		550	.029		1.22	.81		2.03	2.71

07 46 Siding

07 46 33 – Plastic Siding

07 46 33.10 Vinyl Siding

		Crew	Daily Output	Labor-Hours	Unit	Material	2009 Bare Costs Labor	Equipment	Total	Total Incl O&P
4110	.044 thick, triple 3	2 Carp	550	.029	S.F.	1.37	.81		2.18	2.88
4115	.040 thick, triple 4		550	.029		1.22	.81		2.03	2.71
4120	.040 thick, triple 2.66		550	.029		1.47	.81		2.28	2.99
4125	Insulation, fan folded extruded polystyrene, 1/4"		2000	.008		.25	.22		.47	.66
4130	3/8"		2000	.008		.27	.22		.49	.68
4135	Accessories, J channel, 5/8" pocket		700	.023	L.F.	.40	.64		1.04	1.51
4140	3/4" pocket		695	.023		.44	.64		1.08	1.57
4145	1-1/4" pocket		680	.024		.56	.66		1.22	1.72
4150	Flexible, 3/4" pocket		600	.027		1.57	.75		2.32	2.98
4155	Under sill finish trim		500	.032		.39	.89		1.28	1.93
4160	Vinyl starter strip		700	.023		.31	.64		.95	1.42
4165	Aluminum starter strip		700	.023		.22	.64		.86	1.32
4170	Window casing, 2-1/2" wide, 3/4" pocket		510	.031		.99	.88		1.87	2.56
4175	Outside corner, woodgrain finish, 4" face, 3/4" pocket		700	.023		1.65	.64		2.29	2.90
4180	5/8" pocket		700	.023		1.67	.64		2.31	2.92
4185	Smooth finish, 4" face, 3/4" pocket		700	.023		1.73	.64		2.37	2.98
4190	7/8" pocket		690	.023		1.78	.65		2.43	3.06
4195	1-1/4" pocket		700	.023		1.16	.64		1.80	2.36
4200	Soffit and fascia, 1' overhang, solid		120	.133		3.31	3.73		7.04	9.95
4205	Vented		120	.133		3.31	3.73		7.04	9.95
4210	2' overhang, solid		110	.145		4.52	4.07		8.59	11.80
4215	Vented		110	.145		4.52	4.07		8.59	11.80
4220	Colors for siding and soffits, add				S.F.	.16			.16	.18
4225	Colors for accessories and trim, add				L.F.	.32			.32	.35

07 46 46 -- Mineral-Fiber Cement Siding

07 46 46.10 Fiber Cement Siding

		Crew	Daily Output	Labor-Hours	Unit	Material	2009 Bare Costs Labor	Equipment	Total	Total Incl O&P
0010	**FIBER CEMENT SIDING**									
0020	Lap siding, 5/16" thick, 6" wide, 4-3/4" exposure, smooth texture	2 Carp	415	.039	S.F.	1.34	1.08		2.42	3.29
0025	Woodgrain texture		415	.039		1.34	1.08		2.42	3.29
0030	7-1/2" wide, 6-1/4" exposure, smooth texture		425	.038		1.36	1.05		2.41	3.28
0035	Woodgrain texture		425	.038		1.36	1.05		2.41	3.28
0040	8" wide, 6-3/4" exposure, smooth texture		425	.038		1.51	1.05		2.56	3.44
0045	Roughsawn texture		425	.038		1.51	1.05		2.56	3.44
0050	9-1/2" wide, 8-1/4" exposure, smooth texture		440	.036		1.45	1.02		2.47	3.32
0055	Woodgrain texture		440	.036		1.45	1.02		2.47	3.32
0060	12" wide, 10-3/8" exposure, smooth texture		455	.035		1.34	.98		2.32	3.13
0065	Woodgrain texture		455	.035		1.34	.98		2.32	3.13
0070	Panel siding, 5/16" thick, smooth texture		750	.021		1.12	.60		1.72	2.25
0075	Stucco texture		750	.021		1.12	.60		1.72	2.25
0080	Grooved woodgrain texture		750	.021		1.12	.60		1.72	2.25
0085	V - grooved woodgrain texture		750	.021		1.12	.60		1.72	2.25
0090	Wood starter strip		400	.040	L.F.	.52	1.12		1.64	2.46

07 46 73 – Soffit

07 46 73.10 Soffit Options

		Crew	Daily Output	Labor-Hours	Unit	Material	2009 Bare Costs Labor	Equipment	Total	Total Incl O&P
0010	**SOFFIT OPTIONS**									
0012	Aluminum, residential, .020" thick	1 Carp	210	.038	S.F.	1.62	1.06		2.68	3.58
0100	Baked enamel on steel, 16 or 18 gauge		105	.076		5.35	2.13		7.48	9.50
0300	Polyvinyl chloride, white, solid		230	.035		1.09	.97		2.06	2.84
0400	Perforated		230	.035		1.09	.97		2.06	2.84
0500	For colors, add					.13			.13	.14

07 51 Built-Up Bituminous Roofing

07 51 13 – Built-Up Asphalt Roofing

07 51 13.10 Built-Up Roofing Components		Crew	Daily Output	Labor-Hours	Unit	Material	2009 Bare Costs Labor	Equipment	Total	Total Incl O&P
0010	**BUILT-UP ROOFING COMPONENTS**									
0012	Asphalt saturated felt, #30, 2 square per roll	1 Rofc	58	.138	Sq.	9.55	3.26		12.81	16.45
0200	#15, 4 sq per roll, plain or perforated, not mopped		58	.138		4.77	3.26		8.03	11.20
0250	Perforated		58	.138		4.77	3.26		8.03	11.20
0300	Roll roofing, smooth, #65		15	.533		9.30	12.60		21.90	33
0500	#90		15	.533		32	12.60		44.60	58
0520	Mineralized		15	.533		33.50	12.60		46.10	59.50
0540	D.C. (Double coverage), 19" selvage edge		10	.800		49.50	18.90		68.40	89
0580	Adhesive (lap cement)				Gal.	4.50			4.50	4.95

07 51 13.20 Built-Up Roofing Systems		Crew	Daily Output	Labor-Hours	Unit	Material	Labor	Equipment	Total	Total Incl O&P
0010	**BUILT-UP ROOFING SYSTEMS**									
0120	Asphalt flood coat with gravel/slag surfacing, not including									
0140	Insulation, flashing or wood nailers									
0200	Asphalt base sheet, 3 plies #15 asphalt felt, mopped	G-1	22	2.545	Sq.	84	56.50	20	160.50	218
0350	On nailable decks		21	2.667		88.50	59	21	168.50	229
0500	4 plies #15 asphalt felt, mopped		20	2.800		117	62	22	201	265
0550	On nailable decks		19	2.947		104	65.50	23	192.50	259
2000	Asphalt flood coat, smooth surface									
2200	Asphalt base sheet & 3 plies #15 asphalt felt, mopped	G-1	24	2.333	Sq.	89.50	52	18.35	159.85	213
2400	On nailable decks		23	2.435		83	54	19.15	156.15	211
2600	4 plies #15 asphalt felt, mopped		24	2.333		105	52	18.35	175.35	230
2700	On nailable decks		23	2.435		98.50	54	19.15	171.65	228
4500	Coal tar pitch with gravel/slag surfacing									
4600	4 plies #15 tarred felt, mopped	G-1	21	2.667	Sq.	156	59	21	236	305
4800	3 plies glass fiber felt (type IV), mopped	"	19	2.947	"	127	65.50	23	215.50	285

07 51 13.30 Cants		Crew	Daily Output	Labor-Hours	Unit	Material	Labor	Equipment	Total	Total Incl O&P
0010	**CANTS**									
0012	Lumber, treated, 4" x 4" cut diagonally	1 Rofc	325	.025	L.F.	1.52	.58		2.10	2.73
0100	Foamglass		325	.025		2.23	.58		2.81	3.51
0300	Mineral or fiber, trapezoidal, 1"x 4" x 48"		325	.025		.19	.58		.77	1.27
0400	1-1/2" x 5-5/8" x 48"		325	.025		.31	.58		.89	1.40

07 52 Modified Bituminous Membrane Roofing

07 52 13 – Atactic-Polypropylene-Modified Bituminous Membrane Roofing

07 52 13.10 APP Modified Bituminous Membrane		Crew	Daily Output	Labor-Hours	Unit	Material	Labor	Equipment	Total	Total Incl O&P
0010	**APP MODIFIED BITUMINOUS MEMBRANE** R075213-30									
0020	Base sheet, #15 glass fiber felt, nailed to deck	1 Rofc	58	.138	Sq.	6.75	3.26		10.01	13.40
0030	Spot mopped to deck	G-1	295	.190		10.55	4.21	1.49	16.25	21
0040	Fully mopped to deck	"	192	.292		15.10	6.45	2.29	23.84	31
0050	#15 organic felt, nailed to deck	1 Rofc	58	.138		5.55	3.26		8.81	12.05
0060	Spot mopped to deck	G-1	295	.190		9.30	4.21	1.49	15	19.60
0070	Fully mopped to deck	"	192	.292		13.85	6.45	2.29	22.59	29.50
2100	APP mod., smooth surf. cap sheet, poly. reinf., torched, 160 mils	G-5	2100	.019	S.F.	.48	.41	.08	.97	1.37
2150	170 mils		2100	.019		.55	.41	.08	1.04	1.45
2200	Granule surface cap sheet, poly. reinf., torched, 180 mils		2000	.020		.59	.43	.09	1.11	1.54
2250	Smooth surface flashing, torched, 160 mils		1260	.032		.48	.69	.14	1.31	1.93
2300	170 mils		1260	.032		.55	.69	.14	1.38	2.01
2350	Granule surface flashing, torched, 180 mils		1260	.032		.59	.69	.14	1.42	2.05
2400	Fibrated aluminum coating	1 Rofc	3800	.002		.12	.05		.17	.22

07 52 Modified Bituminous Membrane Roofing

07 52 16 – Styrene-Butadiene-Styrene Modified Bituminous Membrane Roofing

07 52 16.10 SBS Modified Bituminous Membrane	Crew	Daily Output	Labor-Hours	Unit	Material	2009 Bare Costs Labor	Equipment	Total	Total Incl O&P
0010 **SBS MODIFIED BITUMINOUS MEMBRANE**									
0080 SBS modified, granule surf cap sheet, polyester rein., mopped									
1500 Glass fiber reinforced, mopped, 160 mils	G-1	2000	.028	S.F.	.49	.62	.22	1.33	1.91
1600 Smooth surface cap sheet, mopped, 145 mils		2100	.027		.49	.59	.21	1.29	1.85
1700 Smooth surface flashing, 145 mils		1260	.044		.49	.99	.35	1.83	2.72
1800 150 mils		1260	.044		.48	.99	.35	1.82	2.71
1900 Granular surface flashing, 150 mils		1260	.044		.53	.99	.35	1.87	2.76
2000 160 mils		1260	.044		.77	.99	.35	2.11	3.03

07 57 Coated Foamed Roofing

07 57 13 – Sprayed Polyurethane Foam Roofing

07 57 13.10 Sprayed Polyurethane Foam Roofing (S.P.F.)

	Crew	Daily Output	Labor-Hours	Unit	Material	2009 Bare Costs Labor	Equipment	Total	Total Incl O&P
0010 **SPRAYED POLYURETHANE FOAM ROOFING (S.P.F.)**									
0100 Primer for metal substrate (when required)	G-2A	3000	.008	S.F.	.42	.16	.20	.78	.97
0200 Primer for non-metal substrate (when required)		3000	.008		.16	.16	.20	.52	.69
0300 Closed cell spray, polyureathane foam, 3 lbs per CF density, 1", R6.7		15000	.002		.70	.03	.04	.77	.87
0400 2", R13.4		13125	.002		1.40	.04	.05	1.49	1.66
0500 3", R20.1		11485	.002		2.10	.04	.05	2.19	2.45
0700 Spray-on silicone coating		2500	.010		.92	.20	.24	1.36	1.63
0800 Warranty 5-20 year manufacturer's									.15
0900 Warranty 20 year, no dollar limit									.20

07 58 Roll Roofing

07 58 10 – Asphalt Roll Roofing

07 58 10.10 Roll Roofing

	Crew	Daily Output	Labor-Hours	Unit	Material	2009 Bare Costs Labor	Equipment	Total	Total Incl O&P
0010 **ROLL ROOFING**									
0100 Asphalt, mineral surface									
0200 1 ply #15 organic felt, 1 ply mineral surfaced									
0300 Selvage roofing, lap 19", nailed & mopped	G-1	27	2.074	Sq.	67.50	46	16.30	129.80	176
0400 3 plies glass fiber felt (type IV), 1 ply mineral surfaced									
0500 Selvage roofing, lapped 19", mopped	G-1	25	2.240	Sq.	104	49.50	17.60	171.10	224
0600 Coated glass fiber base sheet, 2 plies of glass fiber									
0700 Felt (type IV), 1 ply mineral surfaced selvage									
0800 Roofing, lapped 19", mopped	G-1	25	2.240	Sq.	112	49.50	17.60	179.10	233
0900 On nailable decks	"	24	2.333	"	102	52	18.35	172.35	228
1000 3 plies glass fiber felt (type III), 1 ply mineral surfaced									
1100 Selvage roofing, lapped 19", mopped	G-1	25	2.240	Sq.	104	49.50	17.60	171.10	224

07 61 Sheet Metal Roofing

07 61 13 – Standing Seam Sheet Metal Roofing

07 61 13.10 Standing Seam Sheet Metal Roofing, Field Fab.	Crew	Daily Output	Labor-Hours	Unit	Material	2009 Bare Costs Labor	Equipment	Total	Total Incl O&P
0010 **STANDING SEAM SHEET METAL ROOFING, FIELD FABRICATED**									
0400 Copper standing seam roofing, over 10 squares, 16 oz, 125 lb per sq	1 Shee	1.30	6.154	Sq.	1,075	192		1,267	1,500
0600 18 oz, 140 lb per sq	"	1.20	6.667		1,200	208		1,408	1,650
1200 For abnormal conditions or small areas, add					25%	100%			
1300 For lead-coated copper, add				↓	25%				

07 61 16 – Batten Seam Sheet Metal Roofing

07 61 16.10 Batten Seam Sheet Metal Roofing, Field Fabricated

	Crew	Daily Output	Labor-Hours	Unit	Material	2009 Bare Costs Labor	Equipment	Total	Total Incl O&P
0009 **BATTEN SEAM SHEET METAL ROOFING, FIELD FABRICATED**									
0012 Copper batten seam roofing, over 10 sq, 16 oz, 130 lb per sq	1 Shee	1.10	7.273	Sq.	1,075	227		1,302	1,550
0100 Zinc / copper alloy batten seam roofing, .020 thick		1.20	6.667		1,050	208		1,258	1,525
0200 Copper roofing, batten seam, over 10 sq, 18 oz, 145 lb per sq		1	8		1,200	249		1,449	1,725
0800 Zinc, copper alloy roofing, batten seam, .027" thick		1.15	6.957		1,500	217		1,717	2,050
0900 .032" thick		1.10	7.273		1,800	227		2,027	2,350
1000 .040" thick	↓	1.05	7.619	↓	2,150	237		2,387	2,775

07 61 19 – Flat Seam Sheet Metal Roofing

07 61 19.10 Flat Seam Sheet Metal Roofing, Field Fabricated

	Crew	Daily Output	Labor-Hours	Unit	Material	2009 Bare Costs Labor	Equipment	Total	Total Incl O&P
0010 **FLAT SEAM SHEET METAL ROOFING, FIELD FABRICATED**									
0900 Copper flat seam roofing, over 10 squares, 16 oz, 115 lb per sq	1 Shee	1.20	6.667	Sq.	1,075	208		1,283	1,525

07 62 Sheet Metal Flashing and Trim

07 62 10 – Sheet Metal Trim

07 62 10.10 Sheet Metal Cladding

	Crew	Daily Output	Labor-Hours	Unit	Material	2009 Bare Costs Labor	Equipment	Total	Total Incl O&P
0010 **SHEET METAL CLADDING**									
0100 Aluminum, up to 6 bends, .032" thick, window casing	1 Carp	180	.044	S.F.	.87	1.24		2.11	3.06
0200 Window sill		72	.111	L.F.	.87	3.11		3.98	6.20
0300 Door casing		180	.044	S.F.	.87	1.24		2.11	3.06
0400 Fascia		250	.032		.87	.89		1.76	2.47
0500 Rake trim		225	.036		.87	.99		1.86	2.64
0700 .024" thick, window casing		180	.044	↓	1.55	1.24		2.79	3.81
0800 Window sill		72	.111	L.F.	1.55	3.11		4.66	6.95
0900 Door casing		180	.044	S.F.	1.55	1.24		2.79	3.81
1000 Fascia		250	.032		1.55	.89		2.44	3.22
1100 Rake trim		225	.036		1.55	.99		2.54	3.39
1200 Vinyl coated aluminum, up to 6 bends, window casing		180	.044	↓	1.03	1.24		2.27	3.23
1300 Window sill		72	.111	L.F.	1.03	3.11		4.14	6.40
1400 Door casing		180	.044	S.F.	1.03	1.24		2.27	3.23
1500 Fascia		250	.032		1.03	.89		1.92	2.64
1600 Rake trim	↓	225	.036	↓	1.03	.99		2.02	2.81

07 65 Flexible Flashing

07 65 10 – Sheet Metal Flashing

07 65 10.10 Sheet Metal Flashing and Counter Flashing	Crew	Daily Output	Labor-Hours	Unit	Material	2009 Bare Costs Labor	Equipment	Total	Total Incl O&P
0010 **SHEET METAL FLASHING AND COUNTER FLASHING**									
0011 Including up to 4 bends									
0020 Aluminum, mill finish, .013" thick	1 Rofc	145	.055	S.F.	.67	1.30		1.97	3.12
0030 .016" thick		145	.055		.79	1.30		2.09	3.25
0060 .019" thick		145	.055		1.02	1.30		2.32	3.50
0100 .032" thick		145	.055		1.48	1.30		2.78	4.01
0200 .040" thick		145	.055		2.21	1.30		3.51	4.81
0300 .050" thick		145	.055		2.52	1.30		3.82	5.15
0325 Mill finish 5" x 7" step flashing, .016" thick		1920	.004	Ea.	.16	.10		.26	.36
0350 Mill finish 12" x 12" step flashing, .016" thick		1600	.005	"	.61	.12		.73	.89
0400 Painted finish, add				S.F.	.37			.37	.41
1600 Copper, 16 oz, sheets, under 1000 lbs.	1 Rofc	115	.070		6.80	1.65		8.45	10.45
1900 20 oz sheets, under 1000 lbs.		110	.073		7.90	1.72		9.62	11.85
2200 24 oz sheets, under 1000 lbs.		105	.076		9.60	1.80		11.40	13.90
2500 32 oz sheets, under 1000 lbs.		100	.080		12.80	1.89		14.69	17.55
2700 W shape for valleys, 16 oz, 24" wide		100	.080	L.F.	14.65	1.89		16.54	19.55
5800 Lead, 2.5 lb. per SF, up to 12" wide		135	.059	S.F.	4.14	1.40		5.54	7.10
5900 Over 12" wide		135	.059		4.14	1.40		5.54	7.10
8650 Copper, 16 oz		100	.080		5.25	1.89		7.14	9.20
8900 Stainless steel sheets, 32 ga, .010" thick		155	.052		3.79	1.22		5.01	6.40
9000 28 ga, .015" thick		155	.052		4.70	1.22		5.92	7.40
9100 26 ga, .018" thick		155	.052		5.70	1.22		6.92	8.50
9200 24 ga, .025" thick		155	.052		7.40	1.22		8.62	10.40
9290 For mechanically keyed flashing, add					40%				
9320 Steel sheets, galvanized, 20 gauge	1 Rofc	130	.062	S.F.	1.19	1.46		2.65	3.97
9340 30 gauge		160	.050		.50	1.18		1.68	2.71
9400 Terne coated stainless steel, .015" thick, 28 ga		155	.052		7.35	1.22		8.57	10.35
9500 .018" thick, 26 ga		155	.052		8.30	1.22		9.52	11.35
9600 Zinc and copper alloy (brass), .020" thick		155	.052		5.15	1.22		6.37	7.90
9700 .027" thick		155	.052		6.90	1.22		8.12	9.85
9800 .032" thick		155	.052		8.05	1.22		9.27	11.10
9900 .040" thick		155	.052		9.85	1.22		11.07	13.05

07 65 13 – Laminated Sheet Flashing

07 65 13.10 Laminated Sheet Flashing

	Crew	Daily Output	Labor-Hours	Unit	Material	2009 Bare Costs Labor	Equipment	Total	Total Incl O&P
0010 **LAMINATED SHEET FLASHING**, Including up to 4 bends									
2800 Copper, paperbacked 1 side, 2 oz	1 Rofc	330	.024	S.F.	1.42	.57		1.99	2.61
2900 3 oz		330	.024		1.85	.57		2.42	3.09
3100 Paperbacked 2 sides, 2 oz		330	.024		1.43	.57		2	2.62
3150 3 oz		330	.024		1.84	.57		2.41	3.07
3200 5 oz		330	.024		2.76	.57		3.33	4.09
6100 Lead-coated copper, fabric-backed, 2 oz		330	.024		2.42	.57		2.99	3.71
6200 5 oz		330	.024		2.78	.57		3.35	4.11
6400 Mastic-backed 2 sides, 2 oz		330	.024		1.89	.57		2.46	3.13
6500 5 oz		330	.024		2.34	.57		2.91	3.62
6700 Paperbacked 1 side, 2 oz		330	.024		1.63	.57		2.20	2.84
6800 3 oz		330	.024		1.92	.57		2.49	3.16
7000 Paperbacked 2 sides, 2 oz		330	.024		1.69	.57		2.26	2.91
7100 5 oz		330	.024		2.75	.57		3.32	4.08
8550 3 ply copper and fabric, 3 oz		155	.052		2.73	1.22		3.95	5.25
8600 7 oz		155	.052		5.70	1.22		6.92	8.50
8700 Lead on copper and fabric, 5 oz		155	.052		2.78	1.22		4	5.30
8800 7 oz		155	.052		4.94	1.22		6.16	7.70

07 65 Flexible Flashing

07 65 13 – Laminated Sheet Flashing

07 65 13.10 Laminated Sheet Flashing	Crew	Daily Output	Labor-Hours	Unit	Material	2009 Bare Costs Labor	Equipment	Total	Total Incl O&P
9300 Stainless steel, paperbacked 2 sides, .005" thick	1 Rofc	330	.024	S.F.	3.34	.57		3.91	4.72

07 71 Roof Specialties

07 71 19 – Manufactured Gravel Stops and Fascias

07 71 19.10 Gravel Stop

		Crew	Daily Output	Labor-Hours	Unit	Material	Labor	Equipment	Total	Total Incl O&P
0010	**GRAVEL STOP**									
0020	Aluminum, .050" thick, 4" face height, mill finish	1 Shee	145	.055	L.F.	5.95	1.72		7.67	9.45
0080	Duranodic finish		145	.055		5.75	1.72		7.47	9.20
0100	Painted		145	.055		6.65	1.72		8.37	10.20
1350	Galv steel, 24 ga., 4" leg, plain, with continuous cleat, 4" face		145	.055		2.34	1.72		4.06	5.45
1500	Polyvinyl chloride, 6" face height		135	.059		4.32	1.85		6.17	7.85
1800	Stainless steel, 24 ga., 6" face height		135	.059		11.15	1.85		13	15.35

07 71 19.30 Fascia

		Crew	Daily Output	Labor-Hours	Unit	Material	Labor	Equipment	Total	Total Incl O&P
0010	**FASCIA**									
0100	Aluminum, reverse board and batten, .032" thick, colored, no furring incl	1 Shee	145	.055	S.F.	5.75	1.72		7.47	9.20
0200	Residential type, aluminum	1 Carp	200	.040	L.F.	1.51	1.12		2.63	3.55
0300	Steel, galv and enameled, stock, no furring, long panels	1 Shee	145	.055	S.F.	3.76	1.72		5.48	7.05
0600	Short panels	"	115	.070	"	5	2.17		7.17	9.15

07 71 23 – Manufactured Gutters and Downspouts

07 71 23.10 Downspouts

		Crew	Daily Output	Labor-Hours	Unit	Material	Labor	Equipment	Total	Total Incl O&P
0010	**DOWNSPOUTS**									
0020	Aluminum 2" x 3", .020" thick, embossed	1 Shee	190	.042	L.F.	1.04	1.31		2.35	3.34
0100	Enameled		190	.042		1.59	1.31		2.90	3.95
0300	Enameled, .024" thick, 2" x 3"		180	.044		1.84	1.38		3.22	4.34
0400	3" x 4"		140	.057		2.55	1.78		4.33	5.80
0600	Round, corrugated aluminum, 3" diameter, .020" thick		190	.042		1.42	1.31		2.73	3.76
0700	4" diameter, .025" thick		140	.057		2.39	1.78		4.17	5.60
0900	Wire strainer, round, 2" diameter		155	.052	Ea.	3.32	1.61		4.93	6.35
1000	4" diameter		155	.052		6.10	1.61		7.71	9.45
1200	Rectangular, perforated, 2" x 3"		145	.055		2.30	1.72		4.02	5.40
1300	3" x 4"		145	.055		3.32	1.72		5.04	6.55
1500	Copper, round, 16 oz., stock, 2" diameter		190	.042	L.F.	9	1.31		10.31	12.10
1600	3" diameter		190	.042		8.60	1.31		9.91	11.65
1800	4" diameter		145	.055		9	1.72		10.72	12.80
1900	5" diameter		130	.062		14.75	1.92		16.67	19.45
2100	Rectangular, corrugated copper, stock, 2" x 3"		190	.042		7.15	1.31		8.46	10.10
2200	3" x 4"		145	.055		9.30	1.72		11.02	13.10
2400	Rectangular, plain copper, stock, 2" x 3"		190	.042		7	1.31		8.31	9.90
2500	3" x 4"		145	.055		9.10	1.72		10.82	12.90
2700	Wire strainers, rectangular, 2" x 3"		145	.055	Ea.	5.50	1.72		7.22	8.95
2800	3" x 4"		145	.055		6.60	1.72		8.32	10.15
3000	Round, 2" diameter		145	.055		4.40	1.72		6.12	7.75
3100	3" diameter		145	.055		6.10	1.72		7.82	9.65
3300	4" diameter		145	.055		11.30	1.72		13.02	15.30
3400	5" diameter		115	.070		16.60	2.17		18.77	22
3600	Lead-coated copper, round, stock, 2" diameter		190	.042	L.F.	11.45	1.31		12.76	14.80
3700	3" diameter		190	.042		12.60	1.31		13.91	16.05
3900	4" diameter		145	.055		14.70	1.72		16.42	19.10
4300	Rectangular, corrugated, stock, 2" x 3"		190	.042		12.75	1.31		14.06	16.25
4500	Plain, stock, 2" x 3"		190	.042		9.75	1.31		11.06	12.95

07 71 Roof Specialties

07 71 23 – Manufactured Gutters and Downspouts

07 71 23.10 Downspouts

		Crew	Daily Output	Labor-Hours	Unit	Material	2009 Bare Costs Labor	Equipment	Total	Total Incl O&P
4600	3" x 4"	1 Shee	145	.055	L.F.	10	1.72		11.72	13.90
4800	Steel, galvanized, round, corrugated, 2" or 3" diameter, 28 gauge		190	.042		1.74	1.31		3.05	4.11
4900	4" diameter, 28 gauge		145	.055		2.48	1.72		4.20	5.60
5700	Rectangular, corrugated, 28 gauge, 2" x 3"		190	.042		1.60	1.31		2.91	3.96
5800	3" x 4"		145	.055		1.60	1.72		3.32	4.65
6000	Rectangular, plain, 28 gauge, galvanized, 2" x 3"		190	.042		1.74	1.31		3.05	4.11
6100	3" x 4"		145	.055		2.51	1.72		4.23	5.65
6300	Epoxy painted, 24 gauge, corrugated, 2" x 3"		190	.042		1.90	1.31		3.21	4.29
6400	3" x 4"		145	.055	↓	2.12	1.72		3.84	5.20
6600	Wire strainers, rectangular, 2" x 3"		145	.055	Ea.	5.90	1.72		7.62	9.40
6700	3" x 4"		145	.055		11.75	1.72		13.47	15.85
6900	Round strainers, 2" or 3" diameter		145	.055		2.84	1.72		4.56	6
7000	4" diameter		145	.055	↓	11.75	1.72		13.47	15.85
8200	Vinyl, rectangular, 2" x 3"		210	.038	L.F.	.75	1.19		1.94	2.82
8300	Round, 2-1/2"	↓	220	.036	"	.75	1.13		1.88	2.73

07 71 23.20 Downspout Elbows

		Crew	Daily Output	Labor-Hours	Unit	Material	2009 Bare Costs Labor	Equipment	Total	Total Incl O&P
0010	**DOWNSPOUT ELBOWS**									
0020	Aluminum, 2" x 3", embossed	1 Shee	100	.080	Ea.	1.17	2.49		3.66	5.45
0100	Enameled		100	.080		1.17	2.49		3.66	5.45
0200	3" x 4", .025" thick, embossed		100	.080		3.99	2.49		6.48	8.55
0300	Enameled		100	.080		5.15	2.49		7.64	9.85
0400	Round corrugated, 3", embossed, .020" thick		100	.080		2.38	2.49		4.87	6.80
0500	4", .025" thick		100	.080		6.50	2.49		8.99	11.35
0600	Copper, 16 oz. round, 2" diameter		100	.080		13.40	2.49		15.89	18.95
0700	3" diameter		100	.080		7.70	2.49		10.19	12.65
0800	4" diameter		100	.080		12.85	2.49		15.34	18.30
1000	2" x 3" corrugated		100	.080		7.10	2.49		9.59	12
1100	3" x 4" corrugated		100	.080		12	2.49		14.49	17.40
1300	Vinyl, 2-1/2" diameter, 45° or 75°		100	.080		3.50	2.49		5.99	8.05
1400	Tee Y junction	↓	75	.107	↓	12	3.32		15.32	18.80

07 71 23.30 Gutters

		Crew	Daily Output	Labor-Hours	Unit	Material	2009 Bare Costs Labor	Equipment	Total	Total Incl O&P
0010	**GUTTERS**									
0012	Aluminum, stock units, 5" K type, .027" thick, plain	1 Shee	120	.067	L.F.	2.32	2.08		4.40	6.05
0020	Inside corner		25	.320	Ea.	7.45	9.95		17.40	25
0030	Outside corner		25	.320	"	7.45	9.95		17.40	25
0100	Enameled		120	.067	L.F.	2.45	2.08		4.53	6.20
0110	Inside corner		25	.320	Ea.	7.45	9.95		17.40	25
0120	Outside corner		25	.320	"	7.45	9.95		17.40	25
0300	5" K type type, .032" thick, plain		120	.067	L.F.	3.27	2.08		5.35	7.10
0310	Inside corner		25	.320	Ea.	7.45	9.95		17.40	25
0320	Outside corner		25	.320	"	7.45	9.95		17.40	25
0400	Enameled		120	.067	L.F.	2.92	2.08		5	6.70
0410	Inside corner		25	.320	Ea.	7.45	9.95		17.40	25
0420	Outside corner		25	.320	"	7.45	9.95		17.40	25
0700	Copper, half round, 16 oz, stock units, 4" wide		120	.067	L.F.	6.25	2.08		8.33	10.40
0900	5" wide		120	.067		6.30	2.08		8.38	10.45
1000	6" wide		115	.070		8.90	2.17		11.07	13.45
1200	K type, 16 oz, stock, 4" wide		120	.067		6.15	2.08		8.23	10.25
1300	5" wide		120	.067		7.30	2.08		9.38	11.55
1500	Lead coated copper, half round, stock, 4" wide		120	.067		10.50	2.08		12.58	15.05
1600	6" wide		115	.070		21	2.17		23.17	26.50
1800	K type, stock, 4" wide	↓	120	.067		11.35	2.08		13.43	16

07 71 Roof Specialties

07 71 23 – Manufactured Gutters and Downspouts

07 71 23.30 Gutters

		Crew	Daily Output	Labor-Hours	Unit	Material	2009 Bare Costs Labor	Equipment	Total	Total Incl O&P
1900	5" wide	1 Shee	120	.067	L.F.	14.25	2.08		16.33	19.15
2100	Stainless steel, half round or box, stock, 4" wide		120	.067		5.90	2.08		7.98	9.95
2200	5" wide		120	.067		6.35	2.08		8.43	10.45
2400	Steel, galv, half round or box, 28 ga, 5" wide, plain		120	.067		1.46	2.08		3.54	5.10
2500	Enameled		120	.067		1.51	2.08		3.59	5.15
2700	26 ga, stock, 5" wide		120	.067		1.59	2.08		3.67	5.25
2800	6" wide		120	.067		2.30	2.08		4.38	6
3000	Vinyl, O.G., 4" wide	1 Carp	110	.073		1.14	2.03		3.17	4.69
3100	5" wide		110	.073		1.35	2.03		3.38	4.93
3200	4" half round, stock units		110	.073		.91	2.03		2.94	4.44
3250	Joint connectors				Ea.	1.82			1.82	2
3300	Wood, clear treated cedar, fir or hemlock, 3" x 4"	1 Carp	100	.080	L.F.	7	2.24		9.24	11.50
3400	4" x 5"	"	100	.080	"	8	2.24		10.24	12.60

07 71 23.35 Gutter Guard

		Crew	Daily Output	Labor-Hours	Unit	Material	2009 Bare Costs Labor	Equipment	Total	Total Incl O&P
0010	**GUTTER GUARD**									
0020	6" wide strip, aluminum mesh	1 Carp	500	.016	L.F.	2.18	.45		2.63	3.16
0100	Vinyl mesh	"	500	.016	"	1.45	.45		1.90	2.36

07 71 43 – Drip Edge

07 71 43.10 Drip Edge, Rake Edge, Ice Belts

		Crew	Daily Output	Labor-Hours	Unit	Material	2009 Bare Costs Labor	Equipment	Total	Total Incl O&P
0010	**DRIP EDGE, RAKE EDGE, ICE BELTS**									
0020	Aluminum, .016" thick, 5" wide, mill finish	1 Carp	400	.020	L.F.	.42	.56		.98	1.41
0100	White finish		400	.020		.46	.56		1.02	1.46
0200	8" wide, mill finish		400	.020		.56	.56		1.12	1.57
0300	Ice belt, 28" wide, mill finish		100	.080		3.80	2.24		6.04	7.95
0310	Vented, mill finish		400	.020		2.63	.56		3.19	3.84
0320	Painted finish		400	.020		2.72	.56		3.28	3.94
0400	Galvanized, 5" wide		400	.020		.54	.56		1.10	1.54
0500	8" wide, mill finish		400	.020		.78	.56		1.34	1.81
0510	Rake edge, aluminum, 1-1/2" x 1-1/2"		400	.020		.24	.56		.80	1.21
0520	3-1/2" x 1-1/2"		400	.020		.46	.56		1.02	1.46

07 72 Roof Accessories

07 72 23 – Relief Vents

07 72 23.20 Vents

		Crew	Daily Output	Labor-Hours	Unit	Material	2009 Bare Costs Labor	Equipment	Total	Total Incl O&P
0010	**VENTS**									
0100	Soffit or eave, aluminum, mill finish, strips, 2-1/2" wide	1 Carp	200	.040	L.F.	.38	1.12		1.50	2.31
0200	3" wide		200	.040		.40	1.12		1.52	2.33
0300	Enamel finish, 3" wide		200	.040		.50	1.12		1.62	2.44
0400	Mill finish, rectangular, 4" x 16"		72	.111	Ea.	1.43	3.11		4.54	6.80
0500	8" x 16"		72	.111		1.84	3.11		4.95	7.25
2420	Roof ventilator	Q-9	16	1		48	28		76	100

07 72 26 – Ridge Vents

07 72 26.10 Ridge Vents and Accessories

		Crew	Daily Output	Labor-Hours	Unit	Material	2009 Bare Costs Labor	Equipment	Total	Total Incl O&P
0010	**RIDGE VENTS AND ACCESSORIES**									
0100	Aluminum strips, mill finish	1 Rofc	160	.050	L.F.	2.50	1.18		3.68	4.91
0150	Painted finish		160	.050	"	3.61	1.18		4.79	6.15
0200	Connectors		48	.167	Ea.	3.50	3.94		7.44	11.05
0300	End caps		48	.167	"	1.30	3.94		5.24	8.65
0400	Galvanized strips		160	.050	L.F.	2.14	1.18		3.32	4.51
0430	Molded polyethylene, shingles not included		160	.050	"	3.05	1.18		4.23	5.50

07 72 Roof Accessories

07 72 26 – Ridge Vents

07 72 26.10 Ridge Vents and Accessories	Crew	Daily Output	Labor-Hours	Unit	Material	2009 Bare Costs Labor	Equipment	Total	Total Incl O&P
0440	End plugs	1 Rofc	48	.167	Ea.	1.30	3.94		5.24	8.65
0450	Flexible roll, shingles not included	↓	160	.050	L.F.	2.48	1.18		3.66	4.88
2300	Ridge vent strip, mill finish	1 Shee	155	.052	"	2.76	1.61		4.37	5.75

07 72 53 – Snow Guards

07 72 53.10 Snow Guard Options

		Crew	Daily Output	Labor-Hours	Unit	Material	2009 Bare Costs Labor	Equipment	Total	Total Incl O&P
0010	**SNOW GUARD OPTIONS**									
0100	Slate & asphalt shingle roofs	1 Rofc	160	.050	Ea.	9.40	1.18		10.58	12.50
0200	Standing seam metal roofs		48	.167		14.60	3.94		18.54	23.50
0300	Surface mount for metal roofs		48	.167	↓	7.70	3.94		11.64	15.65
0400	Double rail pipe type, including pipe	↓	130	.062	L.F.	22.50	1.46		23.96	27.50

07 72 80 – Vents

07 72 80.30 Vent Options

		Crew	Daily Output	Labor-Hours	Unit	Material	2009 Bare Costs Labor	Equipment	Total	Total Incl O&P
0010	**VENT OPTIONS**									
0800	Polystyrene baffles, 12" wide for 16" O.C. rafter spacing	1 Carp	90	.089	Ea.	.40	2.48		2.88	4.64

07 92 Joint Sealants

07 92 10 – Caulking and Sealants

07 92 10.10 Caulking and Sealant Options

		Crew	Daily Output	Labor-Hours	Unit	Material	2009 Bare Costs Labor	Equipment	Total	Total Incl O&P
0010	**CAULKING AND SEALANT OPTIONS**									
0020	Acoustical sealant, elastomeric, cartridges				Ea.	4.60			4.60	5.05
0100	Acrylic latex caulk, white									
0200	11 fl. oz cartridge				Ea.	2.14			2.14	2.35
0500	1/4" x 1/2"	1 Bric	248	.032	L.F.	.17	.91		1.08	1.71
0600	1/2" x 1/2"		250	.032		.35	.91		1.26	1.88
0800	3/4" x 3/4"		230	.035		.79	.99		1.78	2.50
0900	3/4" x 1"		200	.040		1.05	1.13		2.18	3.03
1000	1" x 1"	↓	180	.044	↓	1.31	1.26		2.57	3.53
1400	Butyl based, bulk				Gal.	26			26	28.50
1500	Cartridges				"	31.50			31.50	34.50
1700	Bulk, in place 1/4" x 1/2", 154 L.F./gal.	1 Bric	230	.035	L.F.	.17	.99		1.16	1.82
1800	1/2" x 1/2", 77 L.F./gal.	"	180	.044	"	.34	1.26		1.60	2.46
2000	Latex acrylic based, bulk				Gal.	27			27	30
2100	Cartridges				"	33.50			33.50	36.50
2200	Bulk in place, 1/4" x 1/2", 154 L.F./gal.	1 Bric	230	.035	L.F.	.18	.99		1.17	1.82
2300	Polysulfide compounds, 1 component, bulk				Gal.	51			51	56
2400	Cartridges				"	54.50			54.50	60
2600	1 or 2 component, in place, 1/4" x 1/4", 308 L.F./gal.	1 Bric	145	.055	L.F.	.17	1.56		1.73	2.77
2700	1/2" x 1/4", 154 L.F./gal.		135	.059		.33	1.68		2.01	3.16
2900	3/4" x 3/8", 68 L.F./gal.		130	.062		.75	1.74		2.49	3.72
3000	1" x 1/2", 38 L.F./gal.	↓	130	.062	↓	1.35	1.74		3.09	4.37
3200	Polyurethane, 1 or 2 component				Gal.	54			54	59.50
3300	Cartridges				"	56			56	61.50
3500	Bulk, in place, 1/4" x 1/4"	1 Bric	150	.053	L.F.	.18	1.51		1.69	2.70
3600	1/2" x 1/4"		145	.055		.35	1.56		1.91	2.98
3800	3/4" x 3/8", 68 L.F./gal.		130	.062		.79	1.74		2.53	3.76
3900	1" x 1/2"	↓	110	.073	↓	1.40	2.06		3.46	4.96
4100	Silicone rubber, bulk				Gal.	41			41	45
4200	Cartridges				"	41			41	45

Division 8
Openings

08 01 53.81 Solid Vinyl Replacement Windows		Crew	Daily Output	Labor-Hours	Unit	Material	2009 Bare Costs Labor	Equipment	Total	Total Incl O&P
0010	**SOLID VINYL REPLACEMENT WINDOWS** R085313-20									
0020	Double hung, insulated glass, up to 83 united inches	G	2 Carp	8	2	Ea.	330	56		386	455
0040	84 to 93	G		8	2		360	56		416	490
0060	94 to 101	G		6	2.667		360	74.50		434.50	520
0080	102 to 111	G		6	2.667		380	74.50		454.50	540
0100	112 to 120	G		6	2.667		410	74.50		484.50	580
0120	For each united inch over 120 , add	G		800	.020	Inch	5.20	.56		5.76	6.65
0140	Casement windows, one operating sash , 42 to 60 united inches	G		8	2	Ea.	191	56		247	305
0160	61 to 70	G		8	2		217	56		273	335
0180	71 to 80	G		8	2		236	56		292	355
0200	81 to 96	G		8	2		252	56		308	370
0220	Two operating sash, 58 to 78 united inches	G		8	2		380	56		436	515
0240	79 to 88	G		8	2		405	56		461	545
0260	89 to 98	G		8	2		445	56		501	585
0280	99 to 108	G		6	2.667		465	74.50		539.50	640
0300	109 to 121	G		6	2.667		500	74.50		574.50	680
0320	Three operating sash, 73 to 108 united inches	G		8	2		600	56		656	755
0340	109 to 118	G		8	2		635	56		691	795
0360	119 to 128	G		6	2.667		655	74.50		729.50	845
0380	129 to 138	G		6	2.667		700	74.50		774.50	895
0400	139 to 156	G		6	2.667		740	74.50		814.50	940
0420	Four operating sash, 98 to 118 united inches	G		8	2		865	56		921	1,050
0440	119 to 128	G		8	2		930	56		986	1,125
0460	129 to 138	G		6	2.667		985	74.50		1,059.50	1,200
0480	139 to 148	G		6	2.667		1,050	74.50		1,124.50	1,275
0500	149 to 168	G		6	2.667		1,100	74.50		1,174.50	1,350
0520	169 to 178	G		6	2.667		1,200	74.50		1,274.50	1,450
0540	For venting unit to fixed unit, deduct	G					16			16	17.60
0560	Fixed picture window, up to 63 united inches	G	2 Carp	8	2		136	56		192	245
0580	64 to 83	G		8	2		160	56		216	271
0600	84 to 101	G		8	2		207	56		263	325
0620	For each united inch over 101, add	G		900	.018	Inch	2.29	.50		2.79	3.36
0640	Picture window opt., low E glazing, up to 101 united inches	G				Ea.	19.50			19.50	21.50
0660	102 to 124	G					26			26	28.50
0680	124 and over	G					38			38	42
0700	Options, low E glazing, up to w/gas 101 united inches	G					31			31	34
0720	102 to 124	G					31			31	34
0740	124 and over	G					31			31	34
0760	Muntins, between glazing, square, per lite						3			3	3.30
0780	Diamond shape, per full or partial diamond						4			4	4.40
0800	Cellulose fiber insulation, poured into sash balance cavity	G	1 Carp	36	.222	C.F.	.60	6.20		6.80	11.15
0820	Silicone caulking at perimeter	G	"	800	.010	L.F.	.13	.28		.41	.62

08 05 Common Work Results for Openings

08 05 05 – Selective Windows and Doors Demolition

08 05 05.10 Selective Demolition Doors

		Crew	Daily Output	Labor-Hours	Unit	Material	2009 Bare Costs Labor	2009 Bare Costs Equipment	Total	Total Incl O&P
0010	**SELECTIVE DEMOLITION DOORS** R024119-10									
0200	Doors, exterior, 1-3/4" thick, single, 3' x 7' high	1 Clab	16	.500	Ea.		10.30		10.30	17.40
0220	Double, 6' x 7' high		12	.667			13.70		13.70	23
0500	Interior, 1-3/8" thick, single, 3' x 7' high		20	.400			8.20		8.20	13.90
0520	Double, 6' x 7' high		16	.500			10.30		10.30	17.40
0700	Bi-folding, 3' x 6'-8" high		20	.400			8.20		8.20	13.90
0720	6' x 6'-8" high		18	.444			9.15		9.15	15.45
0900	Bi-passing, 3' x 6'-8" high		16	.500			10.30		10.30	17.40
0940	6' x 6'-8" high		14	.571			11.75		11.75	19.85
1500	Remove and reset, minimum	1 Carp	8	1			28		28	47.50
1520	Maximum		6	1.333			37.50		37.50	63
2000	Frames, including trim, metal		8	1			28		28	47.50
2200	Wood	2 Carp	32	.500			14		14	23.50
2201	Alternate pricing method	1 Carp	200	.040	L.F.		1.12		1.12	1.89
3000	Special doors, counter doors	2 Carp	6	2.667	Ea.		74.50		74.50	126
3300	Glass, sliding, including frames		12	1.333			37.50		37.50	63
3400	Overhead, commercial, 12' x 12' high		4	4			112		112	189
3500	Residential, 9' x 7' high		8	2			56		56	94.50
3540	16' x 7' high		7	2.286			64		64	108
3600	Remove and reset, minimum		4	4			112		112	189
3620	Maximum		2.50	6.400			179		179	300
3660	Remove and reset elec. garage door opener	1 Carp	8	1			28		28	47.50
4000	Residential lockset, exterior		30	.267			7.45		7.45	12.60
4200	Deadbolt lock		32	.250			7		7	11.80
9000	Minimum labor/equipment charge		4	2	Job		56		56	94.50

08 05 05.20 Selective Demolition of Windows

		Crew	Daily Output	Labor-Hours	Unit	Material	2009 Bare Costs Labor	2009 Bare Costs Equipment	Total	Total Incl O&P
0010	**SELECTIVE DEMOLITION OF WINDOWS** R024119-10									
0200	Aluminum, including trim, to 12 S.F.	1 Clab	16	.500	Ea.		10.30		10.30	17.40
0240	To 25 S.F.		11	.727			14.95		14.95	25.50
0280	To 50 S.F.		5	1.600			33		33	55.50
0320	Storm windows/screens, to 12 S.F.		27	.296			6.10		6.10	10.30
0360	To 25 S.F.		21	.381			7.85		7.85	13.25
0400	To 50 S.F.		16	.500			10.30		10.30	17.40
0500	Screens, incl. aluminum frame, small		20	.400			8.20		8.20	13.90
0510	Large		16	.500			10.30		10.30	17.40
0600	Glass, minimum		200	.040	S.F.		.82		.82	1.39
0620	Maximum		150	.053	"		1.10		1.10	1.85
2000	Wood, including trim, to 12 S.F.		22	.364	Ea.		7.45		7.45	12.65
2020	To 25 S.F.		18	.444			9.15		9.15	15.45
2060	To 50 S.F.		13	.615			12.65		12.65	21.50
2065	To 180 S.F.		8	1			20.50		20.50	35
4300	Remove bay/bow window	2 Carp	6	2.667			74.50		74.50	126
4410	Remove skylight, plstc domes, flush/curb mtd	G-3	395	.081	S.F.		2.09		2.09	3.53
5020	Remove and reset window, minimum	1 Carp	6	1.333	Ea.		37.50		37.50	63
5040	Average		4	2			56		56	94.50
5080	Maximum		2	4			112		112	189
9100	Window awning, residential	1 Clab	80	.100	L.F.		2.06		2.06	3.48

08 11 Metal Doors and Frames

08 11 63 – Metal Screen and Storm Doors and Frames

08 11 63.23 Aluminum Screen and Storm Doors and Frames	Crew	Daily Output	Labor-Hours	Unit	Material	2009 Bare Costs Labor	Equipment	Total	Total Incl O&P
0010	**ALUMINUM SCREEN AND STORM DOORS AND FRAMES**								
0020	Combination storm and screen								
0400	Clear anodic coating, 6'-8" x 2'-6" wide	2 Carp	15	1.067	Ea.	172	30		202	240
0420	2'-8" wide		14	1.143		156	32		188	225
0440	3'-0" wide		14	1.143		156	32		188	225
0500	For 7' door height, add					5%				
1000	Mill finish, 6'-8" x 2'-6" wide	2 Carp	15	1.067	Ea.	225	30		255	299
1020	2'-8" wide		14	1.143		225	32		257	300
1040	3'-0" wide		14	1.143		246	32		278	325
1100	For 7'-0" door, add					5%				
1500	White painted, 6'-8" x 2'-6" wide	2 Carp	15	1.067		305	30		335	385
1520	2'-8" wide		14	1.143		260	32		292	340
1540	3'-0" wide		14	1.143		305	32		337	390
1541	Storm door, painted, alum., insul., 6'-8" x 2'-6" wide		14	1.143		262	32		294	340
1545	2'-8" wide		14	1.143		262	32		294	340
1600	For 7'-0" door, add					5%				
1800	Aluminum screen door, minimum, 6'-8" x 2'-8" wide	2 Carp	14	1.143		100	32		132	164
1810	3'-0" wide		14	1.143		240	32		272	320
1820	Average, 6'-8" x 2'-8" wide		14	1.143		164	32		196	234
1830	3'-0" wide		14	1.143		240	32		272	320
1840	Maximum, 6'-8" x 2'-8" wide		14	1.143		325	32		357	415
1850	3'-0" wide		14	1.143		262	32		294	340
2000	Wood door & screen, see Div. 08 14 33.20									

08 12 Metal Frames

08 12 13 – Hollow Metal Frames

08 12 13.13 Standard Hollow Metal Frames

		Crew	Daily Output	Labor-Hours	Unit	Material	2009 Bare Costs Labor	Equipment	Total	Total Incl O&P
0010	**STANDARD HOLLOW METAL FRAMES**									
0020	16 ga., up to 5-3/4" jamb depth									
0025	6'-8" high, 3'-0" wide, single	2 Carp	16	1	Ea.	123	28		151	184
0028	3'-6" wide, single		16	1		116	28		144	175
0030	4'-0" wide, single		16	1		116	28		144	175
0040	6'-0" wide, double		14	1.143		153	32		185	223
0045	8'-0" wide, double		14	1.143		159	32		191	229
0100	7'-0" high, 3'-0" wide, single		16	1		110	28		138	169
0110	3'-6" wide, single		16	1		119	28		147	179
0112	4'-0" wide, single		16	1		151	28		179	215
0140	6'-0" wide, double		14	1.143		160	32		192	230
0145	8'-0" wide, double		14	1.143		150	32		182	219
1000	16 ga., up to 4-7/8" deep, 7'-0" H, 3'-0" W, single		16	1		130	28		158	191
1140	6'-0" wide, double		14	1.143		158	32		190	227
2800	14 ga., up to 3-7/8" deep, 7'-0" high, 3'-0" wide, single		16	1		134	28		162	195
2840	6'-0" wide, double		14	1.143		161	32		193	231
3000	14 ga., up to 5-3/4" deep, 6'-8" high, 3'-0" wide, single		16	1		140	28		168	202
3002	3'-6" wide, single		16	1		140	28		168	202
3005	4'-0" wide, single		16	1		140	28		168	202
3600	up to 5-3/4" jamb depth, 7'-0" high, 4'-0" wide, single		15	1.067		111	30		141	173
3620	6'-0" wide, double		12	1.333		174	37.50		211.50	254
3640	8'-0" wide, double		12	1.333		164	37.50		201.50	243
3700	8'-0" high, 4'-0" wide, single		15	1.067		146	30		176	211
3740	8'-0" wide, double		12	1.333		181	37.50		218.50	262

08 12 Metal Frames

08 12 13 – Hollow Metal Frames

08 12 13.13 Standard Hollow Metal Frames

		Crew	Daily Output	Labor-Hours	Unit	Material	2009 Bare Costs Labor	2009 Bare Costs Equipment	Total	Total Incl O&P
4000	6-3/4" deep, 7'-0" high, 4'-0" wide, single	2 Carp	15	1.067	Ea.	172	30		202	240
4020	6'-0" wide, double		12	1.333		204	37.50		241.50	287
4040	8'-0", wide double		12	1.333		222	37.50		259.50	305
4100	8'-0" high, 4'-0" wide, single		15	1.067		174	30		204	242
4140	8'-0" wide, double		12	1.333		202	37.50		239.50	285
4400	8-3/4" deep, 7'-0" high, 4'-0" wide, single		15	1.067		163	30		193	230
4440	8'-0" wide, double		12	1.333		231	37.50		268.50	320
4500	8'-0" high, 4'-0" wide, single		15	1.067		174	30		204	242
4540	8'-0" wide, double		12	1.333		203	37.50		240.50	286
4900	For welded frames, add					47.50			47.50	52.50
5400	14 ga., "B" label, up to 5-3/4" deep, 7'-0" high, 4'-0" wide, single	2 Carp	15	1.067		170	30		200	238
5440	8'-0" wide, double		12	1.333		208	37.50		245.50	292
5800	6-3/4" deep, 7'-0" high, 4'-0" wide, single		15	1.067		150	30		180	216
5840	8'-0" wide, double		12	1.333		249	37.50		286.50	335
6200	8-3/4" deep, 7'-0" high, 4'-0" wide, single		15	1.067		220	30		250	294
6240	8'-0" wide, double		12	1.333		245	37.50		282.50	335
6300	For "A" label use same price as "B" label									
6400	For baked enamel finish, add					30%	15%			
6500	For galvanizing, add					15%				
6600	For hospital stop, add				Ea.	282			282	310
7900	Transom lite frames, fixed, add	2 Carp	155	.103	S.F.	45.50	2.89		48.39	55
8000	Movable, add	"	130	.123	"	55	3.44		58.44	66.50

08 13 Metal Doors

08 13 13 – Hollow Metal Doors

08 13 13.15 Metal Fire Doors

			Crew	Daily Output	Labor-Hours	Unit	Material	2009 Bare Costs Labor	2009 Bare Costs Equipment	Total	Total Incl O&P
0010	**METAL FIRE DOORS**	R081313-20									
0015	Steel, flush, "B" label, 90 minute										
0020	Full panel, 20 ga., 2'-0" x 6'-8"		2 Carp	20	.800	Ea.	340	22.50		362.50	415
0040	2'-8" x 6'-8"			18	.889		345	25		370	420
0060	3'-0" x 6'-8"			17	.941		345	26.50		371.50	425
0080	3'-0" x 7'-0"			17	.941		360	26.50		386.50	440
0140	18 ga., 3'-0" x 6'-8"			16	1		395	28		423	485
0160	2'-8" x 7'-0"			17	.941		415	26.50		441.50	500
0180	3'-0" x 7'-0"			16	1		395	28		423	485
0200	4'-0" x 7'-0"			15	1.067		520	30		550	620
0220	For "A" label, 3 hour, 18 ga., use same price as "B" label										
0240	For vision lite, add					Ea.	112			112	124
0520	Flush, "B" label 90 min., composite, 20 ga., 2'-0" x 6'-8"		2 Carp	18	.889		430	25		455	515
0540	2'-8" x 6'-8"			17	.941		435	26.50		461.50	525
0560	3'-0" x 6'-8"			16	1		440	28		468	530
0580	3'-0" x 7'-0"			16	1		450	28		478	545
0640	Flush, "A" label 3 hour, composite, 18 ga., 3'-0" x 6'-8"			15	1.067		495	30		525	595
0660	2'-8" x 7'-0"			16	1		515	28		543	620
0680	3'-0" x 7'-0"			15	1.067		505	30		535	605
0700	4'-0" x 7'-0"			14	1.143		610	32		642	730

08 13 13.20 Residential Steel Doors

			Crew	Daily Output	Labor-Hours	Unit	Material	2009 Bare Costs Labor	2009 Bare Costs Equipment	Total	Total Incl O&P
0010	**RESIDENTIAL STEEL DOORS**										
0020	Prehung, insulated, exterior										
0030	Embossed, full panel, 2'-8" x 6'-8"	G	2 Carp	17	.941	Ea.	277	26.50		303.50	350
0040	3'-0" x 6'-8"	G		15	1.067		249	30		279	325

08 13 13 – Hollow Metal Doors

08 13 13.20 Residential Steel Doors		Crew	Daily Output	Labor-Hours	Unit	Material	2009 Bare Costs Labor	Equipment	Total	Total Incl O&P
0060	3'-0" x 7'-0" **G**	2 Carp	15	1.067	Ea.	291	30		321	370
0070	5'-4" x 6'-8", double **G**		8	2		575	56		631	730
0220	Half glass, 2'-8" x 6'-8"		17	.941		278	26.50		304.50	350
0240	3'-0" x 6'-8"		16	1		278	28		306	355
0260	3'-0" x 7'-0"		16	1		335	28		363	420
0270	5'-4" x 6'-8", double		8	2		570	56		626	725
0720	Raised plastic face, full panel, 2'-8" x 6'-8"		16	1		278	28		306	355
0740	3'-0" x 6'-8"		15	1.067		280	30		310	360
0760	3'-0" x 7'-0"		15	1.067		284	30		314	360
0780	5'-4" x 6'-8", double		8	2		525	56		581	675
0820	Half glass, 2'-8" x 6'-8"		17	.941		310	26.50		336.50	390
0840	3'-0" x 6'-8"		16	1		315	28		343	395
0860	3'-0" x 7'-0"		16	1		345	28		373	430
0880	5'-4" x 6'-8", double		8	2		670	56		726	830
1320	Flush face, full panel, 2'-8" x 6'-8"		16	1		230	28		258	300
1340	3'-0" x 6'-8"		15	1.067		230	30		260	305
1360	3'-0" x 7'-0"		15	1.067		290	30		320	370
1380	5'-4" x 6'-8", double		8	2		470	56		526	610
1420	Half glass, 2'-8" x 6'-8"		17	.941		289	26.50		315.50	365
1440	3'-0" x 6'-8"		16	1		289	28		317	370
1460	3'-0" x 7'-0"		16	1		335	28		363	420
1480	5'-4" x 6'-8", double		8	2		555	56		611	705
1500	Sidelight, full lite, 1'-0" x 6'-8" with grille					228			228	251
1510	1'-0" x 6'-8", low e					248			248	273
1520	1'-0" x 6'-8", half lite					254			254	280
1530	1'-0" x 6'-8", half lite, low e					262			262	289
2300	Interior, residential, closet, bi-fold, 6'-8" x 2'-0" wide	2 Carp	16	1		148	28		176	211
2330	3'-0" wide		16	1		165	28		193	230
2360	4'-0" wide		15	1.067		251	30		281	325
2400	5'-0" wide		14	1.143		290	32		322	375
2420	6'-0" wide		13	1.231		325	34.50		359.50	420
2510	Bi-passing closet, incl. hardware, no frame or trim incl.									
2511	Mirrored, metal frame, 6'-8" x 4'-0" wide	2 Carp	10	1.600	Opng.	176	44.50		220.50	270
2512	5'-0" wide		10	1.600		204	44.50		248.50	300
2513	6'-0" wide		10	1.600		233	44.50		277.50	330
2514	7'-0" wide		9	1.778		259	49.50		308.50	370
2515	8'-0" wide		9	1.778		490	49.50		539.50	625
2611	Mirrored, metal, 8'-0" x 4'-0" wide		10	1.600		300	44.50		344.50	405
2612	5'-0" wide		10	1.600		315	44.50		359.50	420
2613	6'-0" wide		10	1.600		340	44.50		384.50	450
2614	7'-0" wide		9	1.778		380	49.50		429.50	500
2615	8'-0" wide		9	1.778		425	49.50		474.50	550

08 14 Wood Doors

08 14 13 – Carved Wood Doors

08 14 13.10 Types of Wood Doors, Carved	Crew	Daily Output	Labor-Hours	Unit	Material	2009 Bare Costs Labor	2009 Bare Costs Equipment	Total	Total Incl O&P
0010 **TYPES OF WOOD DOORS, CARVED**									
3000 Solid wood, 1-3/4" thick stile and rail									
3020 Mahogany, 3'-0" x 7'-0", minimum	2 Carp	14	1.143	Ea.	850	32		882	990
3030 Maximum		10	1.600		1,350	44.50		1,394.50	1,550
3040 3'-6" x 8'-0", minimum		10	1.600		980	44.50		1,024.50	1,150
3050 Maximum		8	2		1,800	56		1,856	2,075
3100 Pine, 3'-0" x 7'-0", minimum		14	1.143		415	32		447	510
3110 Maximum		10	1.600		705	44.50		749.50	850
3120 3'-6" x 8'-0", minimum		10	1.600		750	44.50		794.50	900
3130 Maximum		8	2		1,200	56		1,256	1,425
3200 Red oak, 3'-0" x 7'-0", minimum		14	1.143		1,600	32		1,632	1,825
3210 Maximum		10	1.600		1,925	44.50		1,969.50	2,200
3220 3'-6" x 8'-0", minimum		10	1.600		1,775	44.50		1,819.50	2,025
3230 Maximum		8	2		3,150	56		3,206	3,575
4000 Hand carved door, mahogany									
4020 3'-0" x 7'-0", minimum	2 Carp	14	1.143	Ea.	1,600	32		1,632	1,800
4030 Maximum		11	1.455		3,125	40.50		3,165.50	3,525
4040 3'-6" x 8'-0", minimum		10	1.600		2,000	44.50		2,044.50	2,275
4050 Maximum		8	2		3,075	56		3,131	3,475
4200 Rose wood, 3'-0" x 7'-0", minimum		14	1.143		4,850	32		4,882	5,400
4210 Maximum		11	1.455		13,300	40.50		13,340.50	14,700
4220 3'-6" x 8'-0", minimum		10	1.600		5,475	44.50		5,519.50	6,100
4280 For 6'-8" high door, deduct from 7'-0" door					34			34	37.50
4400 For custom finish, add					370			370	405
4600 Side light, mahogany, 7'-0" x 1'-6" wide, minimum	2 Carp	18	.889		895	25		920	1,025
4610 Maximum		14	1.143		2,625	32		2,657	2,950
4620 8'-0" x 1'-6" wide, minimum		14	1.143		1,675	32		1,707	1,900
4630 Maximum		10	1.600		1,900	44.50		1,944.50	2,175
4640 Side light, oak, 7'-0" x 1'-6" wide, minimum		18	.889		1,025	25		1,050	1,175
4650 Maximum		14	1.143		1,825	32		1,857	2,075
4660 8'-0" x 1'-6" wide, minimum		14	1.143		970	32		1,002	1,125
4670 Maximum		10	1.600		1,825	44.50		1,869.50	2,100

08 14 16 – Flush Wood Doors

08 14 16.09 Smooth Wood Doors

	Crew	Daily Output	Labor-Hours	Unit	Material	Labor	Equipment	Total	Total Incl O&P
0010 **SMOOTH WOOD DOORS**									
4160 Walnut faced, 1-3/4" x 6'-8" x 3'-0" wide	1 Carp	17	.471	Ea.	274	13.15		287.15	320

08 14 33 – Stile and Rail Wood Doors

08 14 33.10 Wood Doors Paneled

	Crew	Daily Output	Labor-Hours	Unit	Material	Labor	Equipment	Total	Total Incl O&P
0010 **WOOD DOORS PANELED**									
0020 Interior, six panel, hollow core, 1-3/8" thick									
0040 Molded hardboard, 2'-0" x 6'-8"	2 Carp	17	.941	Ea.	55	26.50		81.50	105
0060 2'-6" x 6'-8"		17	.941		59.50	26.50		86	110
0070 2'-8" x 6'-8"		17	.941		62	26.50		88.50	113
0080 3'-0" x 6'-8"		17	.941		65.50	26.50		92	117
0140 Embossed print, molded hardboard, 2'-0" x 6'-8"		17	.941		59.50	26.50		86	110
0160 2'-6" x 6'-8"		17	.941		59.50	26.50		86	110
0180 3'-0" x 6'-8"		17	.941		65.50	26.50		92	117
0540 Six panel, solid, 1-3/8" thick, pine, 2'-0" x 6'-8"		15	1.067		144	30		174	209
0560 2'-6" x 6'-8"		14	1.143		162	32		194	232
0580 3'-0" x 6'-8"		13	1.231		186	34.50		220.50	263
1020 Two panel, bored rail, solid, 1-3/8" thick, pine, 1'-6" x 6'-8"		16	1		263	28		291	335
1040 2'-0" x 6'-8"		15	1.067		345	30		375	430

08 14 33 – Stile and Rail Wood Doors

08 14 33.10 Wood Doors Paneled		Crew	Daily Output	Labor-Hours	Unit	Material	2009 Bare Costs Labor	Equipment	Total	Total Incl O&P
1060	2'-6" x 6'-8"	2 Carp	14	1.143	Ea.	395	32		427	490
1340	Two panel, solid, 1-3/8" thick, fir, 2'-0" x 6'-8"		15	1.067		144	30		174	209
1360	2'-6" x 6'-8"		14	1.143		162	32		194	232
1380	3'-0" x 6'-8"		13	1.231		395	34.50		429.50	495
1740	Five panel, solid, 1-3/8" thick, fir, 2'-0" x 6'-8"		15	1.067		258	30		288	335
1760	2'-6" x 6'-8"		14	1.143		415	32		447	510
1780	3'-0" x 6'-8"		13	1.231		415	34.50		449.50	515

08 14 33.20 Wood Doors Residential		Crew	Daily Output	Labor-Hours	Unit	Material	2009 Bare Costs Labor	Equipment	Total	Total Incl O&P
0010	**WOOD DOORS RESIDENTIAL**									
0200	Exterior, combination storm & screen, pine									
0260	2'-8" wide	2 Carp	10	1.600	Ea.	276	44.50		320.50	380
0280	3'-0" wide		9	1.778		282	49.50		331.50	395
0300	7'-1" x 3'-0" wide		9	1.778		325	49.50		374.50	440
0400	Full lite, 6'-9" x 2'-6" wide		11	1.455		296	40.50		336.50	395
0420	2'-8" wide		10	1.600		296	44.50		340.50	400
0440	3'-0" wide		9	1.778		300	49.50		349.50	415
0500	7'-1" x 3'-0" wide		9	1.778		325	49.50		374.50	440
0604	Door, screen, plain full		12	1.333		137	37.50		174.50	213
0614	Divided		12	1.333		157	37.50		194.50	236
0634	Decor full		12	1.333		315	37.50		352.50	410
0700	Dutch door, pine, 1-3/4" x 6'-8" x 2'-8" wide, minimum		12	1.333		675	37.50		712.50	810
0720	Maximum		10	1.600		825	44.50		869.50	985
0800	3'-0" wide, minimum		12	1.333		690	37.50		727.50	820
0820	Maximum		10	1.600		900	44.50		944.50	1,075
1000	Entrance door, colonial, 1-3/4" x 6'-8" x 2'-8" wide		16	1		450	28		478	545
1020	6 panel pine, 3'-0" wide		15	1.067		430	30		460	525
1100	8 panel pine, 2'-8" wide		16	1		570	28		598	675
1120	3'-0" wide		15	1.067		555	30		585	660
1200	For tempered safety glass lites, (min of 2) add					65.50			65.50	72
1300	Flush, birch, solid core, 1-3/4" x 6'-8" x 2'-8" wide	2 Carp	16	1		102	28		130	161
1320	3'-0" wide		15	1.067		114	30		144	176
1350	7'-0" x 2'-8" wide		16	1		112	28		140	171
1360	3'-0" wide		15	1.067		119	30		149	182
1740	Mahogany, 2'-8" x 6'-8"		15	1.067		810	30		840	940
1760	3'-0" x 6'-8"		15	1.067		830	30		860	965
2700	Interior, closet, bi-fold, w/hardware, no frame or trim incl.									
2720	Flush, birch, 6'-6" or 6'-8" x 2'-6" wide	2 Carp	13	1.231	Ea.	51	34.50		85.50	114
2740	3'-0" wide		13	1.231		55	34.50		89.50	118
2760	4'-0" wide		12	1.333		97.50	37.50		135	170
2780	5'-0" wide		11	1.455		100	40.50		140.50	179
2800	6'-0" wide		10	1.600		109	44.50		153.50	196
3000	Raised panel pine, 6'-6" or 6'-8" x 2'-6" wide		13	1.231		162	34.50		196.50	237
3020	3'-0" wide		13	1.231		254	34.50		288.50	335
3040	4'-0" wide		12	1.333		305	37.50		342.50	400
3060	5'-0" wide		11	1.455		365	40.50		405.50	470
3080	6'-0" wide		10	1.600		400	44.50		444.50	515
3200	Louvered, pine 6'-6" or 6'-8" x 2'-6" wide		13	1.231		106	34.50		140.50	174
3220	3'-0" wide		13	1.231		165	34.50		199.50	239
3240	4'-0" wide		12	1.333		198	37.50		235.50	281
3260	5'-0" wide		11	1.455		224	40.50		264.50	315
3280	6'-0" wide		10	1.600		248	44.50		292.50	350
4400	Bi-passing closet, incl. hardware and frame, no trim incl.									

08 14 Wood Doors

08 14 33 – Stile and Rail Wood Doors

08 14 33.20 Wood Doors Residential	Crew	Daily Output	Labor-Hours	Unit	Material	2009 Bare Costs Labor	Equipment	Total	Total Incl O&P
4420	Flush, lauan, 6'-8" x 4'-0" wide	2 Carp	12	1.333	Opng.	169	37.50		206.50	248
4440	5'-0" wide		11	1.455		184	40.50		224.50	271
4460	6'-0" wide		10	1.600		198	44.50		242.50	294
4600	Flush, birch, 6'-8" x 4'-0" wide		12	1.333		213	37.50		250.50	297
4620	5'-0" wide		11	1.455		209	40.50		249.50	299
4640	6'-0" wide		10	1.600		256	44.50		300.50	355
4800	Louvered, pine, 6'-8" x 4'-0" wide		12	1.333		420	37.50		457.50	525
4820	5'-0" wide		11	1.455		390	40.50		430.50	495
4840	6'-0" wide		10	1.600	↓	515	44.50		559.50	640
4900	Mirrored, 6'-8" x 4'-0" wide		12	1.333	Ea.	300	37.50		337.50	395
5000	Paneled, pine, 6'-8" x 4'-0" wide		12	1.333	Opng.	495	37.50		532.50	610
5020	5'-0" wide		11	1.455		405	40.50		445.50	515
5040	6'-0" wide		10	1.600		590	44.50		634.50	725
5061	Hardboard, 6'-8" x 4'-0" wide		10	1.600		186	44.50		230.50	281
5062	5'-0" wide		10	1.600		194	44.50		238.50	289
5063	6'-0" wide	↓	10	1.600	↓	213	44.50		257.50	310
6100	Folding accordion, closet, including track and frame									
6200	Rigid PVC	2 Carp	10	1.600	Ea.	32	44.50		76.50	111
6220	For custom partition, add					25%	10%			
7310	Passage doors, flush, no frame included									
7320	Hardboard, hollow core, 1-3/8" x 6'-8" x 1'-6" wide	2 Carp	18	.889	Ea.	42.50	25		67.50	88.50
7330	2'-0" wide		18	.889		42	25		67	88
7340	2'-6" wide		18	.889		46.50	25		71.50	93
7350	2'-8" wide		18	.889		48	25		73	95
7360	3'-0" wide		17	.941		51	26.50		77.50	101
7420	Lauan, hollow core, 1-3/8" x 6'-8" x 1'-6" wide		18	.889		30.50	25		55.50	75.50
7440	2'-0" wide		18	.889		28	25		53	72.50
7450	2'-4" wide		18	.889		31.50	25		56.50	76.50
7460	2'-6" wide		18	.889		31.50	25		56.50	76.50
7480	2'-8" wide		18	.889		33	25		58	78.50
7500	3'-0" wide		17	.941		34.50	26.50		61	82.50
7700	Birch, hollow core, 1-3/8" x 6'-8" x 1'-6" wide		18	.889		37	25		62	83
7720	2'-0" wide		18	.889		41	25		66	87
7740	2'-6" wide		18	.889		49	25		74	95.50
7760	2'-8" wide		18	.889		49.50	25		74.50	96.50
7780	3'-0" wide		17	.941		51	26.50		77.50	101
8000	Pine louvered, 1-3/8" x 6'-8" x 1'-6" wide		19	.842		101	23.50		124.50	151
8020	2'-0" wide		18	.889		128	25		153	182
8040	2'-6" wide		18	.889		139	25		164	195
8060	2'-8" wide		18	.889		146	25		171	203
8080	3'-0" wide		17	.941		157	26.50		183.50	217
8300	Pine paneled, 1-3/8" x 6'-8" x 1'-6" wide		19	.842		111	23.50		134.50	162
8320	2'-0" wide		18	.889		128	25		153	182
8330	2'-4" wide		18	.889		150	25		175	207
8340	2'-6" wide		18	.889		158	25		183	215
8360	2'-8" wide		18	.889		161	25		186	219
8380	3'-0" wide		17	.941		165	26.50		191.50	227
8450	French door, pine, 15 lites, 1-3/8"x6'-8"x2'-6" wide		18	.889		178	25		203	238
8470	2'-8" wide		18	.889		228	25		253	293
8490	3'-0" wide	↓	17	.941	↓	252	26.50		278.50	320
8550	For over 20 doors, deduct					15%				
8804	Pocket door, 6 panel pine, 2'-6" x 6'-8"	2 Carp	10.50	1.524	Ea.	218	42.50		260.50	310
8814	2'-8" x 6'-8"	↓	10.50	1.524	↓	230	42.50		272.50	325

08 14 Wood Doors

08 14 33 – Stile and Rail Wood Doors

08 14 33.20 Wood Doors Residential	Crew	Daily Output	Labor-Hours	Unit	Material	2009 Bare Costs Labor	Equipment	Total	Total Incl O&P
8824	3'-0" x 6'-8"	2 Carp	10.50	1.524	Ea.	239	42.50		281.50	335

08 14 40 – Interior Cafe Doors

08 14 40.10 Cafe Style Doors

		Crew	Daily Output	Labor-Hours	Unit	Material	2009 Bare Costs Labor	Equipment	Total	Total Incl O&P
0010	**CAFE STYLE DOORS**									
6520	Interior cafe doors, 2'-6" opening, stock, panel pine	2 Carp	16	1	Ea.	188	28		216	254
6540	3'-0" opening	"	16	1	"	195	28		223	263
6550	Louvered pine									
6560	2'-6" opening	2 Carp	16	1	Ea.	164	28		192	228
8000	3'-0" opening		16	1		165	28		193	230
8010	2'-6" opening, hardwood		16	1		283	28		311	360
8020	3'-0" opening		16	1		315	28		343	395

08 16 Composite Doors

08 16 13 – Fiberglass Doors

08 16 13.10 Entrance Doors, Fiberous Glass

		Crew	Daily Output	Labor-Hours	Unit	Material	2009 Bare Costs Labor	Equipment	Total	Total Incl O&P
0010	**ENTRANCE DOORS, FIBEROUS GLASS**									
0020	Exterior, fiberglass, door, 2'-8" wide x 6'-8" high	2 Carp	15	1.067	Ea.	248	30		278	325
0040	3'-0" wide x 6'-8" high		15	1.067		248	30		278	325
0060	3'-0" wide x 7'-0" high		15	1.067		445	30		475	540
0080	3'-0" wide x 6'-8" high, with two lites		15	1.067		278	30		308	355
0100	3'-0" wide x 8'-0" high, with two lites		15	1.067		450	30		480	545
0110	glass, 3'-0" wide x 6'-8" high		15	1.067		287	30		317	365
0120	3'-0" wide x 6'-8" high, low e		15	1.067		315	30		345	400
0130	3'-0" wide x 8'-0" high		15	1.067		570	30		600	675
0140	3'-0" wide x 8'-0" high, low e		15	1.067		630	30		660	740
0150	Side lights, 1'-0" wide x 6'-8" high,					232			232	255
0160	1'-0" wide x 6'-8" high, low e					246			246	271
0180	1'-0" wide x 6'-8" high, full glass					275			275	305
0190	1'-0" wide x 6'-8" high, low e					284			284	315

08 17 Integrated Door Opening Assemblies

08 17 23 – Integrated Wood Door Opening Assemblies

08 17 23.10 Pre-Hung Doors

		Crew	Daily Output	Labor-Hours	Unit	Material	2009 Bare Costs Labor	Equipment	Total	Total Incl O&P
0010	**PRE-HUNG DOORS**									
0300	Exterior, wood, comb. storm & screen, 6'-9" x 2'-6" wide	2 Carp	15	1.067	Ea.	289	30		319	370
0320	2'-8" wide		15	1.067		289	30		319	370
0340	3'-0" wide		15	1.067		296	30		326	375
0360	For 7'-0" high door, add					22.50			22.50	25
1600	Entrance door, flush, birch, solid core									
1620	4-5/8" solid jamb, 1-3/4" x 6'-8" x 2'-8" wide	2 Carp	16	1	Ea.	285	28		313	365
1640	3'-0" wide		16	1		350	28		378	435
1642	5-5/8" jamb		16	1		325	28		353	405
1680	For 7'-0" high door, add					20.50			20.50	22.50
2000	Entrance door, colonial, 6 panel pine									
2020	4-5/8" solid jamb, 1-3/4" x 6'-8" x 2'-8" wide	2 Carp	16	1	Ea.	530	28		558	630
2040	3'-0" wide	"	16	1		530	28		558	630
2060	For 7'-0" high door, add					52.50			52.50	58
2200	For 5-5/8" solid jamb, add					40.50			40.50	44.50
2250	French door, 6'-8" x 4'-0" wide, 1/2" insul. glass and grille	2 Carp	7	2.286	Pr.	1,100	64		1,164	1,300

08 17 Integrated Door Opening Assemblies

08 17 23 – Integrated Wood Door Opening Assemblies

08 17 23.10 Pre-Hung Doors

		Crew	Daily Output	Labor-Hours	Unit	Material	2009 Bare Costs Labor	Equipment	Total	Total Incl O&P
2260	5'-0" wide	2 Carp	7	2.286	Pr.	1,200	64		1,264	1,425
2270	Pine french door, 8 panels/leaf, 6'-8" x 5'-0" wide, including frame	↓	7	2.286	↓	1,250	64		1,314	1,475
2500	Exterior, metal face, insulated, incl. jamb, brickmold and									
2520	threshold, flush, 2'-8" x 6'-8"	2 Carp	16	1	Ea.	242	28		270	315
2550	3'-0" x 6'-8"		16	1		242	28		270	315
3500	Embossed, 6 panel, 2'-8" x 6'-8"		16	1		224	28		252	294
3550	3'-0" x 6'-8"		16	1		236	28		264	310
3600	2 narrow lites, 2'-8" x 6'-8"		16	1		260	28		288	335
3650	3'-0" x 6'-8"		16	1		265	28		293	340
3700	Half glass, 2'-8" x 6'-8"		16	1		269	28		297	345
3750	3'-0" x 6'-8"		16	1		270	28		298	345
3800	2 top lites, 2'-8" x 6'-8"		16	1		261	28		289	335
3850	3'-0" x 6'-8"	↓	16	1	↓	261	28		289	335
4000	Interior, passage door, 4-5/8" solid jamb									
4400	Lauan, flush, solid core, 1-3/8" x 6'-8" x 2'-6" wide	2 Carp	17	.941	Ea.	182	26.50		208.50	245
4420	2'-8" wide		17	.941		182	26.50		208.50	245
4440	3'-0" wide		16	1		195	28		223	263
4600	Hollow core, 1-3/8" x 6'-8" x 2'-6" wide		17	.941		123	26.50		149.50	180
4620	2'-8" wide		17	.941		123	26.50		149.50	180
4640	3'-0" wide	↓	16	1		124	28		152	184
4700	For 7'-0" high door, add					24			24	26.50
5000	Birch, flush, solid core, 1-3/8" x 6'-8" x 2'-6" wide	2 Carp	17	.941		170	26.50		196.50	232
5020	2'-8" wide		17	.941		195	26.50		221.50	260
5040	3'-0" wide		16	1		205	28		233	274
5200	Hollow core, 1-3/8" x 6'-8" x 2'-6" wide		17	.941		141	26.50		167.50	200
5220	2'-8" wide		17	.941		147	26.50		173.50	206
5240	3'-0" wide	↓	16	1		148	28		176	210
5280	For 7'-0" high door, add					21			21	23
5500	Hardboard paneled, 1-3/8" x 6'-8" x 2'-6" wide	2 Carp	17	.941		142	26.50		168.50	201
5520	2'-8" wide		17	.941		149	26.50		175.50	208
5540	3'-0" wide		16	1		146	28		174	209
6000	Pine paneled, 1-3/8" x 6'-8" x 2'-6" wide		17	.941		247	26.50		273.50	315
6020	2'-8" wide		17	.941		267	26.50		293.50	340
6040	3'-0" wide	↓	16	1		271	28		299	345
6500	For 5-5/8" solid jamb, add					14.60			14.60	16.05
6520	For split jamb, deduct					16.45			16.45	18.10
7600	Oak, 6 panel, 1-3/4" x 6'-8" x 3'-0"	1 Carp	17	.471	↓	635	13.15		648.15	715

08 31 Access Doors and Panels

08 31 13 – Access Doors and Frames

08 31 13.20 Bulkhead/Cellar Doors

		Crew	Daily Output	Labor-Hours	Unit	Material	2009 Bare Costs Labor	Equipment	Total	Total Incl O&P
0010	**BULKHEAD/CELLAR DOORS**									
0020	Steel, not incl. sides, 44" x 62"	1 Carp	5.50	1.455	Ea.	350	40.50		390.50	455
0100	52" x 73"		5.10	1.569		365	44		409	475
0500	With sides and foundation plates, 57" x 45" x 24"		4.70	1.702		650	47.50		697.50	795
0600	42" x 49" x 51"	↓	4.30	1.860	↓	755	52		807	920

08 32 Sliding Glass Doors

08 32 19 – Sliding Wood-Framed Glass Doors

08 32 19.10 Sliding Wood Doors

		Crew	Daily Output	Labor-Hours	Unit	Material	2009 Bare Costs Labor	Equipment	Total	Total Incl O&P
0010	**SLIDING WOOD DOORS**									
0020	Wood, 5/8" tempered insul. glass, 6' wide, premium	2 Carp	4	4	Ea.	1,200	112		1,312	1,525
0100	Economy		4	4		850	112		962	1,125
0150	8' wide, wood, premium		3	5.333		1,425	149		1,574	1,825
0200	Economy		3	5.333		935	149		1,084	1,275
0250	12' wide, wood, vinyl clad		2.50	6.400		2,900	179		3,079	3,475
0300	Economy		2.50	6.400		2,150	179		2,329	2,650

08 32 19.15 Sliding Glass Vinyl-Clad Wood Doors

			Crew	Daily Output	Labor-Hours	Unit	Material	2009 Bare Costs Labor	Equipment	Total	Total Incl O&P
0010	**SLIDING GLASS VINYL-CLAD WOOD DOORS**										
0012	Vinyl clad, 1" insul. glass, 6'-0" x 6'-10" high	G	2 Carp	4	4	Opng.	1,350	112		1,462	1,675
0030	6'-0" x 8'-0" high	G		4	4	Ea.	1,950	112		2,062	2,350
0100	8'-0" x 6'-10" high	G		4	4	Opng.	2,000	112		2,112	2,400
0104	8'-0" x 6'-8" double	G		4	4	Ea.	1,875	112		1,987	2,275
0500	3 leaf, 9'-0" x 6'-10" high	G		3	5.333	Opng.	2,400	149		2,549	2,875
0600	12'-0" x 6'-10" high	G		3	5.333	"	3,175	149		3,324	3,750

08 36 Panel Doors

08 36 13 – Sectional Doors

08 36 13.20 Residential Garage Doors

		Crew	Daily Output	Labor-Hours	Unit	Material	2009 Bare Costs Labor	Equipment	Total	Total Incl O&P
0010	**RESIDENTIAL GARAGE DOORS**									
0050	Hinged, wood, custom, double door, 9' x 7'	2 Carp	4	4	Ea.	540	112		652	780
0070	16' x 7'		3	5.333		955	149		1,104	1,300
0200	Overhead, sectional, incl. hardware, fiberglass, 9' x 7', standard		5.28	3.030		715	84.50		799.50	930
0220	Deluxe		5.28	3.030		900	84.50		984.50	1,125
0300	16' x 7', standard		6	2.667		1,550	74.50		1,624.50	1,825
0320	Deluxe		6	2.667		1,925	74.50		1,999.50	2,225
0500	Hardboard, 9' x 7', standard		8	2		540	56		596	690
0520	Deluxe		8	2		765	56		821	935
0600	16' x 7', standard		6	2.667		1,125	74.50		1,199.50	1,350
0620	Deluxe		6	2.667		1,300	74.50		1,374.50	1,550
0700	Metal, 9' x 7', standard		5.28	3.030		555	84.50		639.50	760
0720	Deluxe		8	2		805	56		861	980
0800	16' x 7', standard		3	5.333		705	149		854	1,025
0820	Deluxe		6	2.667		1,375	74.50		1,449.50	1,625
0900	Wood, 9' x 7', standard		8	2		705	56		761	870
0920	Deluxe		8	2		2,025	56		2,081	2,325
1000	16' x 7', standard		6	2.667		1,425	74.50		1,499.50	1,675
1020	Deluxe		6	2.667		2,950	74.50		3,024.50	3,375
1800	Door hardware, sectional	1 Carp	4	2		310	56		366	435
1810	Door tracks only		4	2		139	56		195	248
1820	One side only		7	1.143		100	32		132	164
3000	Swing-up, including hardware, fiberglass, 9' x 7', standard	2 Carp	8	2		970	56		1,026	1,175
3020	Deluxe		8	2		790	56		846	965
3100	16' x 7', standard		6	2.667		1,225	74.50		1,299.50	1,475
3120	Deluxe		6	2.667		980	74.50		1,054.50	1,200
3200	Hardboard, 9' x 7', standard		8	2		345	56		401	475
3220	Deluxe		8	2		455	56		511	600
3300	16' x 7', standard		6	2.667		480	74.50		554.50	655
3320	Deluxe		6	2.667		715	74.50		789.50	910
3400	Metal, 9' x 7', standard		8	2		380	56		436	515
3420	Deluxe		8	2		800	56		856	975

08 36 Panel Doors

08 36 13 – Sectional Doors

08 36 13.20 Residential Garage Doors		Crew	Daily Output	Labor-Hours	Unit	Material	2009 Bare Costs Labor	Equipment	Total	Total Incl O&P
3500	16' x 7', standard	2 Carp	6	2.667	Ea.	600	74.50		674.50	785
3520	Deluxe		6	2.667		960	74.50		1,034.50	1,175
3600	Wood, 9' x 7', standard		8	2		450	56		506	590
3620	Deluxe		8	2		850	56		906	1,025
3700	16' x 7', standard		6	2.667		725	74.50		799.50	925
3720	Deluxe		6	2.667		1,025	74.50		1,099.50	1,250
3900	Door hardware only, swing up	1 Carp	4	2		142	56		198	251
3920	One side only		7	1.143		69	32		101	130
4000	For electric operator, economy, add		8	1		405	28		433	495
4100	Deluxe, including remote control		8	1		595	28		623	700
4500	For transmitter/receiver control , add to operator				Total	97			97	107
4600	Transmitters, additional				"	45.50			45.50	50
6000	Replace section, on sectional door, fiberglass, 9' x 7'	1 Carp	4	2	Ea.	440	56		496	575
6020	16' x 7'		3.50	2.286		505	64		569	665
6200	Hardboard, 9' x 7'		4	2		110	56		166	216
6220	16' x 7'		3.50	2.286		211	64		275	340
6300	Metal, 9' x 7'		4	2		179	56		235	292
6320	16' x 7'		3.50	2.286		294	64		358	435
6500	Wood, 9' x 7'		4	2		109	56		165	215
6520	16' x 7'		3.50	2.286		211	64		275	340

08 51 Metal Windows

08 51 13 – Aluminum Windows

08 51 13.20 Aluminum Windows

08 51 13.20 Aluminum Windows		Crew	Daily Output	Labor-Hours	Unit	Material	2009 Bare Costs Labor	Equipment	Total	Total Incl O&P
0010	**ALUMINUM WINDOWS**, incl. frame and glazing, commercial grade									
1000	Stock units, casement, 3'-1" x 3'-2" opening	2 Sswk	10	1.600	Ea.	355	48		403	480
1040	Insulating glass	"	10	1.600		425	48		473	555
1050	Add for storms					115			115	126
1600	Projected, with screen, 3'-1" x 3'-2" opening	2 Sswk	10	1.600		335	48		383	460
1650	Insulating glass	"	10	1.600		260	48		308	380
1700	Add for storms					112			112	123
2000	4'-5" x 5'-3" opening	2 Sswk	8	2		380	60		440	535
2050	Insulating glass	"	8	2		400	60		460	555
2100	Add for storms					120			120	132
2500	Enamel finish windows, 3'-1" x 3'-2"	2 Sswk	10	1.600		340	48		388	465
2550	Insulating glass		10	1.600		230	48		278	345
2600	4'-5" x 5'-3"		8	2		385	60		445	540
2700	Insulating glass		8	2		315	60		375	460
3000	Single hung, 2' x 3' opening, enameled, standard glazed		10	1.600		198	48		246	310
3100	Insulating glass		10	1.600		240	48		288	355
3300	2'-8" x 6'-8" opening, standard glazed		8	2		350	60		410	500
3400	Insulating glass		8	2		450	60		510	610
3700	3'-4" x 5'-0" opening, standard glazed		9	1.778		286	53		339	415
3800	Insulating glass		9	1.778		315	53		368	450
4000	Sliding aluminum, 3' x 2' opening, standard glazed		10	1.600		209	48		257	320
4100	Insulating glass		10	1.600		224	48		272	340
4300	5' x 3' opening, standard glazed		9	1.778		320	53		373	450
4400	Insulating glass		9	1.778		370	53		423	510
4600	8' x 4' opening, standard glazed		6	2.667		335	80		415	525
4700	Insulating glass		6	2.667		545	80		625	750
5000	9' x 5' opening, standard glazed		4	4		510	120		630	790

08 51 Metal Windows

08 51 13 – Aluminum Windows

08 51 13.20 Aluminum Windows	Crew	Daily Output	Labor-Hours	Unit	Material	2009 Bare Costs Labor	Equipment	Total	Total Incl O&P
5100	Insulating glass	2 Sswk	4	4	Ea.	820	120		940	1,125
5500	Sliding, with thermal barrier and screen, 6' x 4', 2 track		8	2		695	60		755	880
5700	4 track	↓	8	2		880	60		940	1,075
6000	For above units with bronze finish, add					12%				
6200	For installation in concrete openings, add				↓	5%				

08 51 23 – Steel Windows

08 51 23.10 Steel Sash

		Crew	Daily Output	Labor-Hours	Unit	Material	Labor	Equipment	Total	Total Incl O&P
0010	**STEEL SASH** Custom units, glazing and trim not included									
1200	Basement/utility window, 1'-11" x 2'-8"	1 Carp	14	.571	Ea.	135	15.95		150.95	176

08 51 23.40 Basement Utility Windows

		Crew	Daily Output	Labor-Hours	Unit	Material	Labor	Equipment	Total	Total Incl O&P
0010	**BASEMENT UTILITY WINDOWS**									
0015	1'-3" x 2'-8"	1 Carp	16	.500	Ea.	119	14		133	155
1100	1'-7" x 2'-8"	"	16	.500	"	133	14		147	170

08 51 66 – Metal Window Screens

08 51 66.10 Screens

		Crew	Daily Output	Labor-Hours	Unit	Material	Labor	Equipment	Total	Total Incl O&P
0010	**SCREENS**									
0020	For metal sash, aluminum or bronze mesh, flat screen	2 Sswk	1200	.013	S.F.	4.10	.40		4.50	5.30
0500	Wicket screen, inside window	"	1000	.016	"	6.30	.48		6.78	7.85
0600	Residential, aluminum mesh and frame, 2' x 3'	2 Carp	32	.500	Ea.	13.20	14		27.20	38
0610	Rescreen		50	.320		11	8.95		19.95	27
0620	3' x 5'		32	.500		43.50	14		57.50	71.50
0630	Rescreen		45	.356		27	9.95		36.95	46.50
0640	4' x 8'		25	.640		69	17.90		86.90	106
0650	Rescreen		40	.400		43	11.20		54.20	66.50
0660	Patio door		25	.640	↓	124	17.90		141.90	167
0680	Rescreening	↓	1600	.010	S.F.	1.70	.28		1.98	2.34
1000	For solar louvers, add	2 Sswk	160	.100	"	23.50	3		26.50	31.50

08 52 Wood Windows

08 52 10 – Plain Wood Windows

08 52 10.10 Wood Windows

		Crew	Daily Output	Labor-Hours	Unit	Material	Labor	Equipment	Total	Total Incl O&P
0010	**WOOD WINDOWS**, including frame, screens and grilles									
0020	Residential, stock units									
0050	Awning type, double insulated glass, 2'-10" x 1'-9" opening	2 Carp	12	1.333	Opng.	215	37.50		252.50	299
0100	2'-10" x 6'-0" opening	1 Carp	8	1		525	28		553	625
0200	4'-0" x 3'-6" single pane		10	.800	↓	355	22.50		377.50	430
0300	6' x 5' single pane	↓	8	1	Ea.	490	28		518	585
1000	Casement, 2'-0" x 3'-4" high	2 Carp	20	.800		233	22.50		255.50	294
1020	2'-0" x 4'-0"		18	.889		255	25		280	325
1040	2'-0" x 5'-0"		17	.941		291	26.50		317.50	365
1060	2'-0" x 6'-0"		16	1		287	28		315	365
1080	4'-0" x 3'-4"		15	1.067		575	30		605	685
1100	4'-0" x 4'-0"		15	1.067		645	30		675	760
1120	4'-0" x 5'-0"		14	1.143		730	32		762	860
1140	4'-0" x 6'-0"		12	1.333	↓	825	37.50		862.50	970
1600	Casement units, 8' x 5', with screens, double insulated glass		2.50	6.400	Opng.	1,375	179		1,554	1,825
1700	Low E glass		2.50	6.400		1,575	179		1,754	2,025
2300	Casements, including screens, 2'-0" x 3'-4", dbl insulated glass		11	1.455		262	40.50		302.50	355
2400	Low E glass		11	1.455		262	40.50		302.50	355
2600	2 lite, 4'-0" x 4'-0", double insulated glass		9	1.778	↓	480	49.50		529.50	610

08 52 Wood Windows

08 52 10 – Plain Wood Windows

08 52 10.10 Wood Windows

		Crew	Daily Output	Labor-Hours	Unit	Material	2009 Bare Costs Labor	Equipment	Total	Total Incl O&P
2700	Low E glass	2 Carp	9	1.778	Opng.	495	49.50		544.50	625
2900	3 lite, 5'-2" x 5'-0", double insulated glass		7	2.286		730	64		794	915
3000	Low E glass		7	2.286		775	64		839	965
3200	4 lite, 7'-0" x 5'-0", double insulated glass		6	2.667		1,050	74.50		1,124.50	1,275
3300	Low E glass		6	2.667		1,100	74.50		1,174.50	1,350
3500	5 lite, 8'-6" x 5'-0", double insulated glass		5	3.200		1,375	89.50		1,464.50	1,675
3600	Low E glass		5	3.200		1,475	89.50		1,564.50	1,750
3800	For removable wood grilles, diamond pattern, add				Leaf	34			34	37.50
3900	Rectangular pattern, add				"	34			34	37.50
4000	Bow, fixed lites, 8' x 5', double insulated glass	2 Carp	3	5.333	Opng.	1,375	149		1,524	1,775
4100	Low E glass	"	3	5.333	"	1,875	149		2,024	2,325
4150	6'-0" x 5'-0"	1 Carp	8	1	Ea.	1,250	28		1,278	1,425
4300	Fixed lites, 9'-9" x 5'-0", double insulated glass	2 Carp	2	8	Opng.	950	224		1,174	1,425
4400	Low E glass	"	2	8	"	1,050	224		1,274	1,525
5000	Bow, casement, 8'-1" x 4'-8" high	3 Carp	8	3	Ea.	1,550	84		1,634	1,875
5020	9'-6" x 4'-8"		8	3		1,650	84		1,734	1,975
5040	8'-1" x 5'-1"		8	3		1,875	84		1,959	2,225
5060	9'-6" x 5'-1"		6	4		1,900	112		2,012	2,275
5080	8'-1" x 6'-0"		6	4		1,875	112		1,987	2,275
5100	9'-6" x 6'-0"		6	4		1,975	112		2,087	2,375
5800	Skylights, hatches, vents, and sky roofs, see Div. 08 62 13.00									
9000	Minimum labor/equipment charge	2 Carp	2	8	Job		224		224	380

08 52 10.20 Awning Window

		Crew	Daily Output	Labor-Hours	Unit	Material	2009 Bare Costs Labor	Equipment	Total	Total Incl O&P
0010	**AWNING WINDOW**, Including frame, screens and grilles									
0100	Average quality, builders model, 34" x 22", double insulated glass	1 Carp	10	.800	Ea.	249	22.50		271.50	310
0200	Low E glass		10	.800		258	22.50		280.50	320
0300	40" x 28", double insulated glass		9	.889		310	25		335	380
0400	Low E Glass		9	.889		330	25		355	405
0500	48" x 36", double insulated glass		8	1		450	28		478	545
0600	Low E glass		8	1		475	28		503	575
0800	Vinyl clad, premium, double insulated glass, 24" x 17"		12	.667		207	18.65		225.65	260
0840	24" x 28"		11	.727		251	20.50		271.50	310
0860	36" x 17"		11	.727		250	20.50		270.50	310
0900	36" x 40"		10	.800		385	22.50		407.50	465
1100	40" x 22"		10	.800		283	22.50		305.50	350
1200	36" x 28"		9	.889		300	25		325	370
1300	36" x 36"		9	.889		335	25		360	410
1400	48" x 28"		8	1		360	28		388	450
1500	60" x 36"		8	1		505	28		533	605
2200	Metal clad, 36" x 25"		9	.889		269	25		294	340
2300	40" x 30"		9	.889		335	25		360	410
2400	48" x 28"		8	1		345	28		373	430
2500	60" x 36"		8	1		365	28		393	455
4000	Impact windows, minimum, add					60%				
4010	Impact windows, maximum, add					160%				

08 52 10.30 Wood Windows

		Crew	Daily Output	Labor-Hours	Unit	Material	2009 Bare Costs Labor	Equipment	Total	Total Incl O&P
0010	**WOOD WINDOWS**, double hung									
0020	Including frame, double insulated glass, screens and grilles									
0040	Double hung, 2'-2" x 3'-4" high	2 Carp	15	1.067	Ea.	194	30		224	264
0060	2'-2" x 4'-4"		14	1.143		217	32		249	293
0080	2'-6" x 3'-4"		13	1.231		203	34.50		237.50	281
0100	2'-6" x 4'-0"		12	1.333		220	37.50		257.50	305

08 52 Wood Windows

08 52 10 – Plain Wood Windows

08 52 10.30 Wood Windows		Crew	Daily Output	Labor-Hours	Unit	Material	2009 Bare Costs Labor	Equipment	Total	Total Incl O&P
0120	2'-6" x 4'-8"	2 Carp	12	1.333	Ea.	242	37.50		279.50	330
0140	2'-10" x 3'-4"		10	1.600		213	44.50		257.50	310
0160	2'-10" x 4'-0"		10	1.600		238	44.50		282.50	335
0180	3'-7" x 3'-4"		9	1.778		242	49.50		291.50	350
0200	3'-7" x 5'-4"		9	1.778		310	49.50		359.50	425
0220	3'-10" x 5'-4"	▼	8	2	▼	490	56		546	630
3200	20 S.F. and over	1 Carp	106	.075	S.F.	16.65	2.11		18.76	22
3800	Triple glazing for above, add				"	8.10			8.10	8.90
4300	Double hung, 2'-4" x 3'-2", double insulated glass	2 Carp	6.80	2.353	Ea.	177	66		243	305
4400	Low E glass		6.80	2.353		192	66		258	320
4600	3'-0" x 4'-6" opening, double insulated glass		6	2.667		238	74.50		312.50	390
4700	Low E glass		6	2.667		260	74.50		334.50	410
5000	Picture window, wood, double insulated glass, 4'-6" x 4'-6"		5	3.200		460	89.50		549.50	655
5100	5'-8" x 4'-6"	▼	4.50	3.556		515	99.50		614.50	740
5300	Add for snap-in muntins, 4'-6" x 4'-6"					66.50			66.50	73.50
5400	5'-8" x 4'-6"				▼	64.50			64.50	71
6000	Replacement sash, double hung, double glazing, to 12 S.F.	1 Carp	64	.125	S.F.	24	3.49		27.49	32.50
6100	12 S.F. to 20 S.F.		94	.085	"	19.05	2.38		21.43	25
7000	Sash, single lite, 2'-0" x 2'-0" high		20	.400	Ea.	45	11.20		56.20	68.50
7050	2'-6" x 2'-0" high		19	.421		56.50	11.75		68.25	82
7100	2'-6" x 2'-6" high		18	.444		67.50	12.40		79.90	95.50
7150	3'-0" x 2'-0" high	▼	17	.471	▼	88	13.15		101.15	119

08 52 10.40 Casement Window

			Crew	Daily Output	Labor-Hours	Unit	Material	2009 Bare Costs Labor	Equipment	Total	Total Incl O&P
0010	**CASEMENT WINDOW**, including frame, screen and grilles	R085216-10									
0100	Avg. quality, bldrs. model, 2'-0" x 3'-0" H, dbl. insulated glass	G	1 Carp	10	.800	Ea.	276	22.50		298.50	345
0150	Low E glass	G		10	.800		225	22.50		247.50	286
0200	2'-0" x 4'-6" high, double insulated glass	G		9	.889		345	25		370	420
0250	Low E glass	G		9	.889		263	25		288	330
0300	2'-4" x 6'-0" high, double insulated glass	G		8	1		380	28		408	470
0350	Low E glass	G		8	1		465	28		493	565
0522	Vinyl clad, premium, double insulated glass, 2'-0" x 3'-0"	G		10	.800		266	22.50		288.50	330
0524	2'-0" x 4'-0"	G		9	.889		310	25		335	380
0525	2'-0" x 5'-0"	G		8	1		355	28		383	440
0528	2'-0" x 6'-0"	G	▼	8	1	▼	355	28		383	440
2000	Bay, casement units, 8' x 5', w/screens, dbl Insul glass		2 Carp	2.50	6.400	Opng.	1,575	179		1,754	2,025
2100	Low E glass		"	2.50	6.400	"	1,650	179		1,829	2,100
3020	Vinyl clad, premium, double insulated glass, multiple leaf units										
3080	Single unit, 1'-6" x 5'-0"	G	2 Carp	20	.800	Ea.	310	22.50		332.50	380
3100	2'-0" x 2'-0"	G		20	.800		206	22.50		228.50	265
3140	2'-0" x 2'-6"	G		20	.800		266	22.50		288.50	330
3220	2'-0" x 3'-6"	G		20	.800		267	22.50		289.50	330
3260	2'-0" x 4'-0"	G		19	.842		310	23.50		333.50	380
3300	2'-0" x 4'-6"	G		19	.842		310	23.50		333.50	380
3340	2'-0" x 5'-0"	G		18	.889		355	25		380	430
3460	2'-4" x 3'-0"	G		20	.800		269	22.50		291.50	335
3500	2'-4" x 4'-0"	G		19	.842		335	23.50		358.50	405
3540	2'-4" x 5'-0"	G		18	.889		400	25		425	475
3700	Double unit, 2'-8" x 5'-0"	G		18	.889		560	25		585	655
3740	2'-8" x 6'-0"	G		17	.941		650	26.50		676.50	760
3840	3'-0" x 4'-6"	G		18	.889		505	25		530	595
3860	3'-0" x 5'-0"	G		17	.941		650	26.50		676.50	760
3880	3'-0" x 6'-0"	G	▼	17	.941	▼	700	26.50		726.50	815

08 52 10.40 Casement Window

		Crew	Daily Output	Labor-Hours	Unit	Material	2009 Bare Costs Labor	Equipment	Total	Total Incl O&P
3980	3'-4" x 2'-6"	G 2 Carp	19	.842	Ea.	430	23.50		453.50	510
4000	3'-4" x 3'-0"	G	12	1.333		430	37.50		467.50	535
4030	3'-4" x 4'-0"	G	18	.889		520	25		545	615
4050	3'-4" x 5'-0"	G	12	1.333		650	37.50		687.50	780
4100	3'-4" x 6'-0"	G	11	1.455		700	40.50		740.50	840
4200	3'-6" x 3'-0"	G	18	.889		510	25		535	600
4340	4'-0" x 3'-0"	G	18	.889		480	25		505	570
4380	4'-0" x 3'-6"	G	17	.941		520	26.50		546.50	615
4420	4'-0" x 4'-0"	G	16	1		610	28		638	725
4460	4'-0" x 4'-4"	G	16	1		610	28		638	720
4540	4'-0" x 5'-0"	G	16	1		665	28		693	780
4580	4'-0" x 6'-0"	G	15	1.067		755	30		785	880
4740	4'-8" x 3'-0"	G	18	.889		545	25		570	640
4780	4'-8" x 3'-6"	G	17	.941		580	26.50		606.50	685
4820	4'-8" x 4'-0"	G	16	1		675	28		703	795
4860	4'-8" x 5'-0"	G	15	1.067		770	30		800	900
4900	4'-8" x 6'-0"	G	15	1.067		870	30		900	1,000
5060	Triple unit, 5'-0" x 5'-0"	G	15	1.067		1,025	30		1,055	1,175
5100	5'-6" x 3'-0"	G	17	.941		700	26.50		726.50	815
5140	5'-6" x 3'-6"	G	16	1		760	28		788	885
5180	5'-6" x 4'-6"	G	15	1.067		855	30		885	990
5220	5'-6" x 5'-6"	G	15	1.067		1,125	30		1,155	1,275
5300	6'-0" x 4'-6"	G	15	1.067		835	30		865	965
5850	5'-0" x 3'-0"	G	12	1.333		590	37.50		627.50	715
5900	5'-0" x 4'-0"	G	11	1.455		895	40.50		935.50	1,050
6100	5'-0" x 5'-6"	G	10	1.600		1,050	44.50		1,094.50	1,225
6150	5'-0" x 6'-0"	G	10	1.600		1,150	44.50		1,194.50	1,350
6200	6'-0" x 3'-0"	G	12	1.333		1,150	37.50		1,187.50	1,350
6250	6'-0" x 3'-4"	G	12	1.333		740	37.50		777.50	880
6300	6'-0" x 4'-0"	G	11	1.455		825	40.50		865.50	980
6350	6'-0" x 5'-0"	G	10	1.600		900	44.50		944.50	1,075
6400	6'-0" x 6'-0"	G	10	1.600		1,075	44.50		1,119.50	1,250
6500	Quadruple unit, 7'-0" x 4'-0"	G	9	1.778		1,075	49.50		1,124.50	1,250
6700	8'-0" x 4'-6"	G	9	1.778		1,350	49.50		1,399.50	1,575
6950	6'-8" x 4'-0"	G	10	1.600		1,050	44.50		1,094.50	1,225
7000	6'-8" x 6'-0"	G	10	1.600		1,400	44.50		1,444.50	1,625
8100	Metal clad, deluxe, dbl. insul. glass, 2'-0" x 3'-0" high	G 1 Carp	10	.800		226	22.50		248.50	287
8120	2'-0" x 4'-0" high	G	9	.889		273	25		298	340
8140	2'-0" x 5'-0" high	G	8	1		310	28		338	390
8160	2'-0" x 6'-0" high	G	8	1		355	28		383	440
8190	For installation, add per leaf						15%			
8200	For multiple leaf units, deduct for stationary sash									
8220	2' high				Ea.	22			22	24
8240	4'-6" high					25			25	28
8260	6' high					33.50			33.50	37
8300	Impact windows, minimum, add					60%				
8310	Impact windows, maximum, add					160%				

08 52 10.50 Double Hung

		Crew	Daily Output	Labor-Hours	Unit	Material	2009 Bare Costs Labor	Equipment	Total	Total Incl O&P
0010	**DOUBLE HUNG**, Including frame, screens and grilles									
0100	Avg. quality, bldrs. model, 2'-0" x 3'-0" high, dbl insul. glass	G 1 Carp	10	.800	Ea.	184	22.50		206.50	240
0150	Low E glass	G	10	.800		208	22.50		230.50	267
0200	3'-0" x 4'-0" high, double insulated glass	G	9	.889		257	25		282	325

08 52 Wood Windows

08 52 10 – Plain Wood Windows

08 52 10.50 Double Hung

		Crew	Daily Output	Labor-Hours	Unit	Material	2009 Bare Costs Labor	2009 Bare Costs Equipment	Total	Total Incl O&P
0250	Low E glass	1 Carp	9	.889	Ea.	275	25		300	345
0300	4'-0" x 4'-6" high, double insulated glass		8	1		315	28		343	395
0350	Low E glass		8	1		340	28		368	425
1000	Vinyl clad, premium, double insulated glass, 2'-6" x 3'-0"		10	.800		225	22.50		247.50	286
1100	3'-0" x 3'-6"		10	.800		261	22.50		283.50	325
1200	3'-0" x 4'-0"		9	.889		370	25		395	445
1300	3'-0" x 4'-6"		9	.889		340	25		365	410
1400	3'-0" x 5'-0"		8	1		315	28		343	395
1500	3'-6" x 6'-0"		8	1		370	28		398	455
2000	Metal clad, deluxe, dbl. insul. glass, 2'-6" x 3'-0" high		10	.800		262	22.50		284.50	325
2100	3'-0" x 3'-6" high		10	.800		299	22.50		321.50	370
2200	3'-0" x 4'-0" high		9	.889		315	25		340	385
2300	3'-0" x 4'-6" high		9	.889		330	25		355	405
2400	3'-0" x 5'-0" high		8	1		360	28		388	445
2500	3'-6" x 6'-0" high		8	1		435	28		463	525
8000	Impact windows, minimum, add					60%				
8010	Impact windows, maximum, add					160%				

08 52 10.55 Picture Window

		Crew	Daily Output	Labor-Hours	Unit	Material	2009 Bare Costs Labor	2009 Bare Costs Equipment	Total	Total Incl O&P
0010	**PICTURE WINDOW**, Including frame and grilles									
0100	Average quality, bldrs. model, 3'-6" x 4'-0" high, dbl insulated glass	2 Carp	12	1.333	Ea.	420	37.50		457.50	530
0150	Low E glass		12	1.333		465	37.50		502.50	575
0200	4'-0" x 4'-6" high, double insulated glass		11	1.455		475	40.50		515.50	595
0250	Low E glass		11	1.455		500	40.50		540.50	620
0300	5'-0" x 4'-0" high, double insulated glass		11	1.455		555	40.50		595.50	680
0350	Low E glass		11	1.455		575	40.50		615.50	705
0400	6'-0" x 4'-6" high, double insulated glass		10	1.600		605	44.50		649.50	740
0450	Low E glass		10	1.600		610	44.50		654.50	750

08 52 10.65 Wood Sash

		Crew	Daily Output	Labor-Hours	Unit	Material	2009 Bare Costs Labor	2009 Bare Costs Equipment	Total	Total Incl O&P
0010	**WOOD SASH**, Including glazing but not trim									
0050	Custom, 5'-0" x 4'-0", 1" dbl. glazed, 3/16" thick lites	2 Carp	3.20	5	Ea.	167	140		307	420
0100	1/4" thick lites		5	3.200		172	89.50		261.50	340
0200	1" thick, triple glazed		5	3.200		395	89.50		484.50	580
0300	7'-0" x 4'-6" high, 1" double glazed, 3/16" thick lites		4.30	3.721		400	104		504	615
0400	1/4" thick lites		4.30	3.721		450	104		554	670
0500	1" thick, triple glazed		4.30	3.721		515	104		619	740
0600	8'-6" x 5'-0" high, 1" double glazed, 3/16" thick lites		3.50	4.571		540	128		668	805
0700	1/4" thick lites		3.50	4.571		590	128		718	865
0800	1" thick, triple glazed		3.50	4.571		595	128		723	870
0900	Window frames only, based on perimeter length				L.F.	3.68			3.68	4.05

08 52 10.70 Sliding Windows

		Crew	Daily Output	Labor-Hours	Unit	Material	2009 Bare Costs Labor	2009 Bare Costs Equipment	Total	Total Incl O&P
0010	**SLIDING WINDOWS**									
0100	Average quality, bldrs. model, 3'-0" x 3'-0" high, double insulated	1 Carp	10	.800	Ea.	244	22.50		266.50	305
0120	Low E glass		10	.800		264	22.50		286.50	330
0200	4'-0" x 3'-6" high, double insulated		9	.889		282	25		307	350
0220	Low E glass		9	.889		305	25		330	375
0300	6'-0" x 5'-0" high, double insulated		8	1		420	28		448	510
0320	Low E glass		8	1		460	28		488	555
6000	Sliding, insulating glass, including screens,									
6100	3'-0" x 3'-0"	2 Carp	6.50	2.462	Ea.	286	69		355	430
6200	4'-0" x 3'-6"		6.30	2.540		325	71		396	475
6300	5'-0" x 4'-0"		6	2.667		400	74.50		474.50	565

08 52 Wood Windows

08 52 13 – Metal-Clad Wood Windows

08 52 13.10 Awning Windows, Metal-Clad

		Crew	Daily Output	Labor-Hours	Unit	Material	2009 Bare Costs Labor	2009 Bare Costs Equipment	Total	Total Incl O&P
0010	**AWNING WINDOWS, METAL-CLAD**									
2000	Metal clad, awning deluxe, double insulated glass, 34" x 22"	1 Carp	10	.800	Ea.	242	22.50		264.50	305
2100	40" x 22"	"	10	.800	"	285	22.50		307.50	355

08 52 13.35 Picture and Sliding Windows Metal-Clad

		Crew	Daily Output	Labor-Hours	Unit	Material	2009 Bare Costs Labor	2009 Bare Costs Equipment	Total	Total Incl O&P
0010	**PICTURE AND SLIDING WINDOWS METAL-CLAD**									
2000	Metal clad, dlx picture, dbl. insul. glass, 4'-0" x 4'-0" high	2 Carp	12	1.333	Ea.	360	37.50		397.50	465
2100	4'-0" x 6'-0" high		11	1.455		535	40.50		575.50	660
2200	5'-0" x 6'-0" high		10	1.600		590	44.50		634.50	720
2300	6'-0" x 6'-0" high		10	1.600		675	44.50		719.50	820
2400	Metal clad, dlx sliding, double insulated glass, 3'-0" x 3'-0" high [G]	1 Carp	10	.800		320	22.50		342.50	395
2420	4'-0" x 3'-6" high [G]		9	.889		395	25		420	475
2440	5'-0" x 4'-0" high [G]		9	.889		475	25		500	560
2460	6'-0" x 5'-0" high [G]		8	1		700	28		728	820

08 52 16 – Plastic-Clad Wood Windows

08 52 16.10 Bow Window

		Crew	Daily Output	Labor-Hours	Unit	Material	2009 Bare Costs Labor	2009 Bare Costs Equipment	Total	Total Incl O&P
0010	**BOW WINDOW** Including frames, screens, and grilles									
0020	End panels operable									
1000	Bow type, casement, wood, bldrs mdl, 8' x 5' dbl insltd glass, 4 panel	2 Carp	10	1.600	Ea.	1,500	44.50		1,544.50	1,725
1050	Low E glass		10	1.600		1,300	44.50		1,344.50	1,525
1100	10'-0" x 5'-0", double insulated glass, 6 panels		6	2.667		1,350	74.50		1,424.50	1,625
1200	Low E glass, 6 panels		6	2.667		1,450	74.50		1,524.50	1,725
1300	Vinyl clad, bldrs model, double insulated glass, 6'-0" x 4'-0", 3 panel		10	1.600		1,475	44.50		1,519.50	1,700
1340	9'-0" x 4'-0", 4 panel		8	2		1,450	56		1,506	1,700
1380	10'-0" x 6'-0", 5 panels		7	2.286		2,250	64		2,314	2,575
1420	12'-0" x 6'-0", 6 panels		6	2.667		2,925	74.50		2,999.50	3,350
1600	Metal clad, casement, bldrs mdl, 6'-0" x 4'-0", dbl insltd gls, 3 panels		10	1.600		920	44.50		964.50	1,075
1640	9'-0" x 4'-0", 4 panels		8	2		1,250	56		1,306	1,475
1680	10'-0" x 5'-0", 5 panels		7	2.286		1,725	64		1,789	2,000
1720	12'-0" x 6'-0", 6 panels		6	2.667		2,400	74.50		2,474.50	2,775
2000	Bay window, builders model, 8' x 5' dbl insul glass,		10	1.600		1,825	44.50		1,869.50	2,100
2050	Low E glass,		10	1.600		2,225	44.50		2,269.50	2,525
2100	12'-0" x 6'-0", double insulated glass, 6 panels		6	2.667		2,300	74.50		2,374.50	2,650
2200	Low E glass		6	2.667		2,175	74.50		2,249.50	2,525
2280	6'-0" x 4'-0"		11	1.455		1,100	40.50		1,140.50	1,300
2300	Vinyl clad, premium, double insulated glass, 8'-0" x 5'-0"		10	1.600		1,300	44.50		1,344.50	1,500
2340	10'-0" x 5'-0"		8	2		1,850	56		1,906	2,125
2380	10'-0" x 6'-0"		7	2.286		1,925	64		1,989	2,225
2420	12'-0" x 6'-0"		6	2.667		2,300	74.50		2,374.50	2,650
2430	14'-0" x 3'-0"		7	2.286		1,575	64		1,639	1,850
2440	14'-0" x 6'-0"		5	3.200		2,500	89.50		2,589.50	2,900
2600	Metal clad, deluxe, dbl insul. glass, 8'-0" x 5'-0" high, 4 panels		10	1.600		1,625	44.50		1,669.50	1,850
2640	10'-0" x 5'-0" high, 5 panels		8	2		1,750	56		1,806	2,025
2680	10'-0" x 6'-0" high, 5 panels		7	2.286		2,050	64		2,114	2,375
2720	12'-0" x 6'-0" high, 6 panels		6	2.667		2,850	74.50		2,924.50	3,275
3000	Double hung, bldrs. model, bay, 8' x 4' high, dbl insulated glass		10	1.600		1,300	44.50		1,344.50	1,500
3050	Low E glass		10	1.600		1,375	44.50		1,419.50	1,600
3100	9'-0" x 5'-0" high, double insulated glass		6	2.667		1,400	74.50		1,474.50	1,650
3200	Low E glass		6	2.667		1,475	74.50		1,549.50	1,750
3300	Vinyl clad, premium, double insulated glass, 7'-0" x 4'-6"		10	1.600		1,325	44.50		1,369.50	1,550
3340	8'-0" x 4'-6"		8	2		1,375	56		1,431	1,600
3380	8'-0" x 5'-0"		7	2.286		1,425	64		1,489	1,675

08 52 Wood Windows

08 52 16 – Plastic-Clad Wood Windows

08 52 16.10 Bow Window		Crew	Daily Output	Labor-Hours	Unit	Material	2009 Bare Costs Labor	2009 Bare Costs Equipment	Total	Total Incl O&P
3420	9'-0" x 5'-0"	2 Carp	6	2.667	Ea.	1,475	74.50		1,549.50	1,750
3600	Metal clad, deluxe, dbl insul. glass, 7'-0" x 4'-0" high		10	1.600		1,225	44.50		1,269.50	1,425
3640	8'-0" x 4'-0" high		8	2		1,275	56		1,331	1,500
3680	8'-0" x 5'-0" high		7	2.286		1,325	64		1,389	1,550
3720	9'-0" x 5'-0" high	↓	6	2.667	↓	1,400	74.50		1,474.50	1,675

08 52 16.20 Half Round, Vinyl Clad		Crew	Daily Output	Labor-Hours	Unit	Material	2009 Bare Costs Labor	2009 Bare Costs Equipment	Total	Total Incl O&P
0010	**HALF ROUND, VINYL CLAD**, double insulated glass, incl. grille									
0800	14" height x 24" base	2 Carp	9	1.778	Ea.	405	49.50		454.50	530
1040	15" height x 25" base		8	2		375	56		431	510
1060	16" height x 28" base		7	2.286		405	64		469	555
1080	17" height x 29" base	↓	7	2.286		420	64		484	570
2000	19" height x 33" base	1 Carp	6	1.333		450	37.50		487.50	560
2100	20" height x 35" base		6	1.333		450	37.50		487.50	560
2200	21" height x 37" base	↓	6	1.333		475	37.50		512.50	590
2250	23" height x 41" base	2 Carp	6	2.667		520	74.50		594.50	695
2300	26" height x 48" base		6	2.667		535	74.50		609.50	710
2350	30" height x 56" base	↓	6	2.667		615	74.50		689.50	800
3000	36" height x 67"base	1 Carp	4	2		1,050	56		1,106	1,250
3040	38" height x 71" base	2 Carp	5	3.200		975	89.50		1,064.50	1,225
3050	40" height x 75" base	"	5	3.200		1,275	89.50		1,364.50	1,550
5000	Elliptical, 71" x 16"	1 Carp	11	.727		975	20.50		995.50	1,100
5100	95" x 21"	"	10	.800	↓	1,375	22.50		1,397.50	1,575

08 52 16.30 Palladian Windows		Crew	Daily Output	Labor-Hours	Unit	Material	2009 Bare Costs Labor	2009 Bare Costs Equipment	Total	Total Incl O&P
0010	**PALLADIAN WINDOWS**									
0020	Vinyl clad, double insulated glass, including frame and grilles									
0040	3'-2" x 2'-6" high	2 Carp	11	1.455	Ea.	1,275	40.50		1,315.50	1,475
0060	3'-2" x 4'-10"		11	1.455		1,775	40.50		1,815.50	2,025
0080	3'-2" x 6'-4"		10	1.600		1,750	44.50		1,794.50	2,000
0100	4'-0" x 4'-0"	↓	10	1.600		1,575	44.50		1,619.50	1,800
0120	4'-0" x 5'-4"	3 Carp	10	2.400		1,875	67		1,942	2,175
0140	4'-0" x 6'-0"		9	2.667		1,950	74.50		2,024.50	2,275
0160	4'-0" x 7'-4"		9	2.667		2,125	74.50		2,199.50	2,450
0180	5'-5" x 4'-10"		9	2.667		2,275	74.50		2,349.50	2,625
0200	5'-5" x 6'-10"		9	2.667		2,575	74.50		2,649.50	2,975
0220	5'-5" x 7'-9"		9	2.667		2,800	74.50		2,874.50	3,200
0240	6'-0" x 7'-11"		8	3		3,475	84		3,559	3,975
0260	8'-0" x 6'-0"	↓	8	3		3,075	84		3,159	3,525

08 52 16.40 Transom Windows		Crew	Daily Output	Labor-Hours	Unit	Material	2009 Bare Costs Labor	2009 Bare Costs Equipment	Total	Total Incl O&P
0010	**TRANSOM WINDOWS**									
0050	Vinyl clad, premium, double insulated glass, 32" x 8"	1 Carp	16	.500	Ea.	173	14		187	214
0100	36" x 8"		16	.500		184	14		198	226
0110	36" x 12"	↓	16	.500		196	14		210	240
1000	Vinyl clad, premium, dbl. insul. glass, 4'-0" x 4'-0"	2 Carp	12	1.333		485	37.50		522.50	595
1100	4'-0" x 6'-0"		11	1.455		910	40.50		950.50	1,075
1200	5'-0" x 6'-0"		10	1.600		1,000	44.50		1,044.50	1,200
1300	6'-0" x 6'-0"	↓	10	1.600	↓	1,025	44.50		1,069.50	1,200

08 52 16.45 Trapezoid Windows		Crew	Daily Output	Labor-Hours	Unit	Material	2009 Bare Costs Labor	2009 Bare Costs Equipment	Total	Total Incl O&P
0010	**TRAPEZOID WINDOWS**									
0100	Vinyl clad, including frame and exterior trim									
0900	20" base x 44" leg x 53" leg	2 Carp	13	1.231	Ea.	375	34.50		409.50	475
1000	24" base x 90" leg x 102" leg		8	2		640	56		696	800
3000	36" base x 40" leg x 22" leg	↓	12	1.333	↓	410	37.50		447.50	515

08 52 16 – Plastic-Clad Wood Windows

08 52 16.45 Trapezoid Windows

		Crew	Daily Output	Labor-Hours	Unit	Material	2009 Bare Costs Labor	2009 Bare Costs Equipment	Total	Total Incl O&P
3010	36" base x 44" leg x 25" leg	2 Carp	13	1.231	Ea.	430	34.50		464.50	535
3050	36" base x 26" leg x 48" leg		9	1.778		440	49.50		489.50	570
3100	36" base x 42" legs, 50" peak		9	1.778		520	49.50		569.50	655
3200	36" base x 60" leg x 81" leg		11	1.455		680	40.50		720.50	820
4320	44" base x 23" leg x 56" leg		11	1.455		535	40.50		575.50	660
4350	44" base x 59" leg x 92" leg		10	1.600		800	44.50		844.50	955
4500	46" base x 15" leg x 46" leg		8	2		405	56		461	540
4550	46" base x 16" leg x 48" leg		8	2		430	56		486	570
4600	46" base x 50" leg x 80" leg		7	2.286		650	64		714	825
6600	66" base x 12" leg x 42" leg		8	2		550	56		606	700
6650	66" base x 12" legs, 28" peak		9	1.778		440	49.50		489.50	570
6700	68" base x 23" legs, 31" peak		8	2		550	56		606	700

08 52 16.70 Vinyl Clad, Premium, Dbl. Insulated Glass

			Crew	Daily Output	Labor-Hours	Unit	Material	2009 Bare Costs Labor	2009 Bare Costs Equipment	Total	Total Incl O&P
0010	**VINYL CLAD, PREMIUM, DBL. INSULATED GLASS**										
1000	Sliding, 3'-0" x 3'-0"	G	1 Carp	10	.800	Ea.	575	22.50		597.50	670
1020	4'-0" x 1'-11"	G		11	.727		555	20.50		575.50	645
1040	4'-0" x 3'-0"	G		10	.800		665	22.50		687.50	770
1050	4'-0" x 3'-6"	G		9	.889		670	25		695	775
1090	4'-0" x 5'-0"	G		9	.889		870	25		895	995
1100	5'-0" x 4'-0"	G		9	.889		860	25		885	985
1120	5'-0" x 5'-0"	G		8	1		960	28		988	1,100
1140	6'-0" x 4'-0"	G		8	1		970	28		998	1,125
1150	6'-0" x 5'-0"	G		8	1		1,075	28		1,103	1,250

08 52 50 – Window Accessories

08 52 50.10 Window Grille or Muntin

		Crew	Daily Output	Labor-Hours	Unit	Material	2009 Bare Costs Labor	2009 Bare Costs Equipment	Total	Total Incl O&P
0010	**WINDOW GRILLE OR MUNTIN**, snap in type									
0020	Standard pattern interior grilles									
2000	Wood, awning window, glass size 28" x 16" high	1 Carp	30	.267	Ea.	22.50	7.45		29.95	37.50
2060	44" x 24" high		32	.250		32.50	7		39.50	47.50
2100	Casement, glass size, 20" x 36" high		30	.267		27.50	7.45		34.95	43
2180	20" x 56" high		32	.250		39.50	7		46.50	55.50
2200	Double hung, glass size, 16" x 24" high		24	.333	Set	48	9.30		57.30	68.50
2280	32" x 32" high		34	.235	"	128	6.60		134.60	152
2500	Picture, glass size, 48" x 48" high		30	.267	Ea.	112	7.45		119.45	136
2580	60" x 68" high		28	.286	"	160	8		168	190
2600	Sliding, glass size, 14" x 36" high		24	.333	Set	25.50	9.30		34.80	44
2680	36" x 36" high		22	.364	"	39	10.15		49.15	60

08 52 66 – Wood Window Screens

08 52 66.10 Wood Screens

		Crew	Daily Output	Labor-Hours	Unit	Material	2009 Bare Costs Labor	2009 Bare Costs Equipment	Total	Total Incl O&P
0010	**WOOD SCREENS**									
0020	Over 3 S.F., 3/4" frames	2 Carp	375	.043	S.F.	4.28	1.19		5.47	6.75
0100	1-1/8" frames	"	375	.043	"	7.60	1.19		8.79	10.35

08 52 69 – Wood Storm Windows

08 52 69.10 Storm Windows

		Crew	Daily Output	Labor-Hours	Unit	Material	2009 Bare Costs Labor	2009 Bare Costs Equipment	Total	Total Incl O&P
0010	**STORM WINDOWS**, aluminum residential									
0300	Basement, mill finish, incl. fiberglass screen									
0320	1'-10" x 1'-0" high	2 Carp	30	.533	Ea.	32	14.90		46.90	60
0340	2'-9" x 1'-6" high		30	.533		34	14.90		48.90	62.50
0360	3'-4" x 2'-0" high		30	.533		42	14.90		56.90	71
1600	Double-hung, combination, storm & screen									
1700	Custom, clear anodic coating, 2'-0" x 3'-5" high	2 Carp	30	.533	Ea.	82	14.90		96.90	115

08 52 69.10 Storm Windows	Crew	Daily Output	Labor-Hours	Unit	Material	2009 Bare Costs Labor	Equipment	Total	Total Incl O&P
1720	2'-6" x 5'-0" high	2 Carp	28	.571	Ea.	108	15.95		123.95	146
1740	4'-0" x 6'-0" high		25	.640		226	17.90		243.90	279
1800	White painted, 2'-0" x 3'-5" high		30	.533		96	14.90		110.90	131
1820	2'-6" x 5'-0" high		28	.571		154	15.95		169.95	196
1840	4'-0" x 6'-0" high		25	.640		273	17.90		290.90	330
2000	Average quality, clear anodic coating, 2'-0" x 3'-5" high		30	.533		82	14.90		96.90	115
2020	2'-6" x 5'-0" high		28	.571		102	15.95		117.95	139
2040	4'-0" x 6'-0" high		25	.640		120	17.90		137.90	162
2400	White painted, 2'-0" x 3'-5" high		30	.533		81	14.90		95.90	114
2420	2'-6" x 5'-0" high		28	.571		88	15.95		103.95	124
2440	4'-0" x 6'-0" high		25	.640		96	17.90		113.90	136
2600	Mill finish, 2'-0" x 3'-5" high		30	.533		73	14.90		87.90	106
2620	2'-6" x 5'-0" high		28	.571		82	15.95		97.95	117
2640	4'-0" x 6'-8" high	↓	25	.640	↓	92	17.90		109.90	131
4000	Picture window, storm, 1 lite, white or bronze finish									
4020	4'-6" x 4'-6" high	2 Carp	25	.640	Ea.	122	17.90		139.90	164
4040	5'-8" x 4'-6" high		20	.800		138	22.50		160.50	190
4400	Mill finish, 4'-6" x 4'-6" high		25	.640		122	17.90		139.90	164
4420	5'-8" x 4'-6" high	↓	20	.800	↓	138	22.50		160.50	190
4600	3 lite, white or bronze finish									
4620	4'-6" x 4'-6" high	2 Carp	25	.640	Ea.	148	17.90		165.90	193
4640	5'-8" x 4'-6" high		20	.800		164	22.50		186.50	218
4800	Mill finish, 4'-6" x 4'-6" high		25	.640		131	17.90		148.90	174
4820	5'-8" x 4'-6" high	↓	20	.800		148	22.50		170.50	201
5000	Sliding glass door, storm 6' x 6'-8", standard	1 Glaz	2	4		745	110		855	995
5100	Economy	"	2	4	↓	450	110		560	675
6000	Sliding window, storm, 2 lite, white or bronze finish									
6020	3'-4" x 2'-7" high	2 Carp	28	.571	Ea.	110	15.95		125.95	148
6040	4'-4" x 3'-3" high		25	.640		150	17.90		167.90	195
6060	5'-4" x 6'-0" high	↓	20	.800	↓	240	22.50		262.50	300
9000	Magnetic interior storm window									
9100	3/16" plate glass	1 Glaz	107	.075	S.F.	4.95	2.05		7	8.85

08 53 13.10 Solid Vinyl Windows

		Crew	Daily Output	Labor-Hours	Unit	Material	Labor	Equipment	Total	Total Incl O&P
0010	**SOLID VINYL WINDOWS**									
0020	Double hung, including frame and screen, 2'-0" x 2'-6"	2 Carp	15	1.067	Ea.	178	30		208	247
0040	2'-0" x 3'-6"		14	1.143		184	32		216	256
0060	2'-6" x 4'-6"		13	1.231		219	34.50		253.50	299
0080	3'-0" x 4'-0"		10	1.600		192	44.50		236.50	287
0100	3'-0" x 4'-6"		9	1.778		246	49.50		295.50	355
0120	4'-0" x 4'-6"		8	2		310	56		366	435
0140	4'-0" x 6'-0"		7	2.286	↓	345	64		409	490
9100	Minimum labor/equipment charge	↓	2	8	Job		224		224	380

08 53 13.20 Vinyl Single Hung Windows

			Crew	Daily Output	Labor-Hours	Unit	Material	Labor	Equipment	Total	Total Incl O&P
0010	**VINYL SINGLE HUNG WINDOWS**										
0100	Grids, low E, J fin, ext. jambs, 21" x 53"	G	2 Carp	18	.889	Ea.	151	25		176	208
0110	21" x 57"	G		17	.941		155	26.50		181.50	215
0120	21" x 65"	G		16	1		161	28		189	225
0130	25" x 41"	G	↓	20	.800	↓	142	22.50		164.50	194

08 53 13.20 Vinyl Single Hung Windows

		Crew	Daily Output	Labor-Hours	Unit	Material	2009 Bare Costs Labor	2009 Bare Costs Equipment	Total	Total Incl O&P
0140	25" x 49" G	2 Carp	18	.889	Ea.	157	25		182	215
0150	25" x 57" G		17	.941		161	26.50		187.50	222
0160	25" x 65" G		16	1		168	28		196	233
0170	29" x 41" G		18	.889		151	25		176	209
0180	29" x 53" G		18	.889		162	25		187	221
0190	29" x 57" G		17	.941		166	26.50		192.50	228
0200	29" x 65" G		16	1		173	28		201	238
0210	33" x 41" G		20	.800		157	22.50		179.50	210
0220	33" x 53" G		18	.889		169	25		194	228
0230	33" x 57" G		17	.941		173	26.50		199.50	235
0240	33" x 65" G		16	1		180	28		208	246
0250	37" x 41" G		20	.800		165	22.50		187.50	220
0260	37" x 53" G		18	.889		178	25		203	237
0270	37" x 57" G		17	.941		181	26.50		207.50	244
0280	37" x 65" G		16	1		189	28		217	256
0500	Vinyl clad, premium, double insulated glass, circle, 24" diameter	1 Carp	6	1.333		415	37.50		452.50	520
1000	2'-4" diameter		6	1.333		515	37.50		552.50	630
1500	2'-11" diameter		6	1.333		600	37.50		637.50	725

08 53 13.30 Vinyl Double Hung Windows

		Crew	Daily Output	Labor-Hours	Unit	Material	2009 Bare Costs Labor	2009 Bare Costs Equipment	Total	Total Incl O&P
0010	**VINYL DOUBLE HUNG WINDOWS**									
0100	Grids, low E, J fin, ext. jambs, 21" x 53" G	2 Carp	18	.889	Ea.	173	25		198	232
0102	21" x 37" G		18	.889		154	25		179	211
0104	21" x 41" G		18	.889		157	25		182	215
0106	21" x 49" G		18	.889		165	25		190	224
0110	21" x 57" G		17	.941		176	26.50		202.50	239
0120	21" x 65" G		16	1		184	28		212	250
0128	25" x 37" G		20	.800		162	22.50		184.50	217
0130	25" x 41" G		20	.800		166	22.50		188.50	220
0140	25" x 49" G		18	.889		171	25		196	230
0145	25" x 53" G		18	.889		178	25		203	238
0150	25" x 57" G		17	.941		179	26.50		205.50	242
0160	25" x 65" G		16	1		190	28		218	257
0162	25" x 69" G		16	1		198	28		226	266
0164	25" x 77" G		16	1		209	28		237	278
0168	29" x 37" G		18	.889		168	25		193	226
0170	29" x 41" G		18	.889		171	25		196	230
0172	29" x 49" G		18	.889		180	25		205	240
0180	29" x 53" G		18	.889		184	25		209	244
0190	29" x 57" G		17	.941		188	26.50		214.50	251
0200	29" x 65" G		16	1		195	28		223	263
0202	29" x 69" G		16	1		202	28		230	271
0205	29" x 77" G		16	1		214	28		242	284
0208	33" x 37" G		20	.800		172	22.50		194.50	227
0210	33" x 41" G		20	.800		176	22.50		198.50	232
0215	33" x 49" G		20	.800		186	22.50		208.50	242
0220	33" x 53" G		18	.889		190	25		215	251
0230	33" x 57" G		17	.941		194	26.50		220.50	258
0240	33" x 65" G		16	1		199	28		227	267
0242	33" x 69" G		16	1		211	28		239	280
0246	33" x 77" G		16	1		222	28		250	292
0250	37" x 41" G		20	.800		180	22.50		202.50	236
0255	37" x 49" G		20	.800		190	22.50		212.50	247

08 53 13.30 Vinyl Double Hung Windows		Crew	Daily Output	Labor-Hours	Unit	Material	2009 Bare Costs Labor	Equipment	Total	Total Incl O&P
0260	37" x 53" G	2 Carp	18	.889	Ea.	197	25		222	259
0270	37" x 57" G		17	.941		202	26.50		228.50	267
0280	37" x 65" G		16	1		207	28		235	276
0282	37" x 69" G		16	1		212	28		240	281
0286	37" x 77" G		16	1		231	28		259	300
0300	Solid vinyl, average quality, double insulated glass, 2'-0" x 3'-0" G	1 Carp	10	.800		279	22.50		301.50	345
0310	3'-0" x 4'-0" G		9	.889		175	25		200	235
0320	4'-0" x 4'-6" G		8	1		286	28		314	365
0330	Premium, double insulated glass, 2'-6" x 3'-0" G		10	.800		194	22.50		216.50	251
0340	3'-0" x 3'-6" G		9	.889		225	25		250	290
0350	3'-0" x 4'-0" G		9	.889		238	25		263	305
0360	3'-0" x 4'-6" G		9	.889		243	25		268	310
0370	3'-0" x 5'-0" G		8	1		261	28		289	335
0380	3'-6" x 6'-0" G		8	1		300	28		328	380

08 53 13.40 Vinyl Casement Windows		Crew	Daily Output	Labor-Hours	Unit	Material	2009 Bare Costs Labor	Equipment	Total	Total Incl O&P
0010	**VINYL CASEMENT WINDOWS**									
0100	Grids, low E, J fin, ext. jambs, 1 lt, 21" x 41" G	2 Carp	20	.800	Ea.	225	22.50		247.50	286
0110	21" x 47" G		20	.800		245	22.50		267.50	310
0120	21" x 53" G		20	.800		265	22.50		287.50	330
0128	24" x 35" G		19	.842		216	23.50		239.50	278
0130	24" x 41" G		19	.842		235	23.50		258.50	298
0140	24" x 47" G		19	.842		255	23.50		278.50	320
0150	24" x 53" G		19	.842		275	23.50		298.50	340
0158	28" x 35" G		19	.842		230	23.50		253.50	294
0160	28" x 41" G		19	.842		249	23.50		272.50	315
0170	28" x 47" G		19	.842		269	23.50		292.50	335
0180	28" x 53" G		19	.842		297	23.50		320.50	365
0184	28" x 59" G		19	.842		300	23.50		323.50	375
0188	Two lites, 33" x 35" G		18	.889		370	25		395	445
0190	33" x 41" G		18	.889		395	25		420	475
0200	33" x 47" G		18	.889		425	25		450	510
0210	33" x 53" G		18	.889		455	25		480	540
0212	33" x 59" G		18	.889		485	25		510	570
0215	33" x 72" G		18	.889		500	25		525	590
0220	41" x 41" G		18	.889		435	25		460	515
0230	41" x 47" G		18	.889		465	25		490	550
0240	41" x 53" G		17	.941		490	26.50		516.50	585
0242	41" x 59" G		17	.941		520	26.50		546.50	615
0246	41" x 72" G		17	.941		545	26.50		571.50	640
0250	47" x 41" G		17	.941		440	26.50		466.50	525
0260	47" x 47" G		17	.941		465	26.50		491.50	560
0270	47" x 53" G		17	.941		495	26.50		521.50	590
0272	47" x 59" G		17	.941		540	26.50		566.50	640
0280	56" x 41" G		15	1.067		470	30		500	565
0290	56" x 47" G		15	1.067		495	30		525	595
0300	56" x 53" G		15	1.067		540	30		570	645
0302	56" x 59" G		15	1.067		565	30		595	670
0310	56" x 72" G		15	1.067		610	30		640	725
0340	Solid vinyl, premium, double insulated glass, 2'-0" x 3'-0" high G	1 Carp	10	.800		252	22.50		274.50	315
0360	2'-0" x 4'-0" high G		9	.889		280	25		305	350
0380	2'-0" x 5'-0" high G		8	1		276	28		304	355

08 53 Plastic Windows

08 53 13 – Vinyl Windows

08 53 13.50 Vinyl Picture Windows

		Crew	Daily Output	Labor-Hours	Unit	Material	2009 Bare Costs Labor	Equipment	Total	Total Incl O&P
0010	**VINYL PICTURE WINDOWS**									
0100	Grids, low E, J fin, ext. jambs, 33" x 47"	2 Carp	12	1.333	Ea.	221	37.50		258.50	305
0110	35" x 71"		12	1.333		230	37.50		267.50	315
0120	41" x 47"		12	1.333		252	37.50		289.50	340
0130	41" x 71"		12	1.333		273	37.50		310.50	365
0140	47" x 47"		12	1.333		285	37.50		322.50	380
0150	47" x 71"		11	1.455		299	40.50		339.50	400
0160	53" x 47"		11	1.455		280	40.50		320.50	380
0170	53" x 71"		11	1.455		293	40.50		333.50	395
0180	59" x 47"		11	1.455		320	40.50		360.50	420
0190	59" x 71"		11	1.455		340	40.50		380.50	445
0200	71" x 47"		10	1.600		350	44.50		394.50	465
0210	71" x 71"		10	1.600		370	44.50		414.50	485

08 53 13.60 Vinyl Half Round Windows

		Crew	Daily Output	Labor-Hours	Unit	Material	2009 Bare Costs Labor	Equipment	Total	Total Incl O&P
0010	**VINYL HALF ROUND WINDOWS**, Including grille, j fin low E, ext. jambs									
0100	10" height x 20" base	2 Carp	8	2	Ea.	475	56		531	615
0110	15" height x 30" base		8	2		415	56		471	555
0120	17" height x 34" base		7	2.286		265	64		329	400
0130	19" height x 38" base		7	2.286		286	64		350	425
0140	19" height x 33" base		7	2.286		555	64		619	725
0150	24" height x 48" base	1 Carp	6	1.333		297	37.50		334.50	390
0160	25" height x 50" base	"	6	1.333		755	37.50		792.50	895
0170	30" height x 60" base	2 Carp	6	2.667		620	74.50		694.50	810

08 54 Composite Windows

08 54 13 – Fiberglass Windows

08 54 13.10 Fiberglass Single Hung Windows

			Crew	Daily Output	Labor-Hours	Unit	Material	2009 Bare Costs Labor	Equipment	Total	Total Incl O&P
0010	**FIBERGLASS SINGLE HUNG WINDOWS**										
0100	Grids, low E, 18" x 24"	G	2 Carp	18	.889	Ea.	255	25		280	325
0110	18" x 40"	G		17	.941		275	26.50		301.50	350
0130	24" x 40"	G		20	.800		295	22.50		317.50	365
0230	36" x 36"	G		17	.941		290	26.50		316.50	365
0250	36" x 48"	G		20	.800		330	22.50		352.50	405
0260	36" x 60"	G		18	.889		350	25		375	425
0280	36" x 72"	G		16	1		360	28		388	445
0290	48" x 40"	G		16	1		360	28		388	445

08 61 Roof Windows

08 61 13 – Metal Roof Windows

08 61 13.10 Roof Windows

		Crew	Daily Output	Labor-Hours	Unit	Material	2009 Bare Costs Labor	Equipment	Total	Total Incl O&P
0010	**ROOF WINDOWS**, fixed high perf tmpd glazing, metallic framed									
0020	46" x 21-1/2"	1 Carp	8	1	Ea.	251	28		279	325
0100	46" x 28"		8	1		285	28		313	365
0125	57" x 44"		6	1.333		365	37.50		402.50	470
0130	72" x 28"		7	1.143		365	32		397	460
0150	Venting, high performance tempered glazing, 46" x 21-1/2"		8	1		380	28		408	465
0175	46" x 28"		8	1		410	28		438	500
0200	57" x 44"		6	1.333		495	37.50		532.50	610

08 61 Roof Windows

08 61 13 – Metal Roof Windows

08 61 13.10 Roof Windows

		Crew	Daily Output	Labor-Hours	Unit	Material	2009 Bare Costs Labor	Equipment	Total	Total Incl O&P
0500	Flashing set for shingled roof, 46" x 21-1/2"	1 Carp	5	1.600	Ea.	45	44.50		89.50	125
0525	46" x 28"		5	1.600		45	44.50		89.50	125
0550	57" x 44"		5	1.600		43.50	44.50		88	124
0560	72" x 28"		6	1.333		43.50	37.50		81	111
0575	Flashing set for low pitched roof, 46" x 21-1/2"		5	1.600		184	44.50		228.50	278
0600	46" x 28"		5	1.600		188	44.50		232.50	282
0625	57" x 44"		5	1.600		218	44.50		262.50	315
0650	Flashing set for tile roof 46" x 21-1/2"		5	1.600		107	44.50		151.50	194
0675	46" x 28"		5	1.600		111	44.50		155.50	198
0700	57" x 44"		5	1.600		127	44.50		171.50	215

08 61 16 – Wood Roof Windows

08 61 16.16 Roof Windows, Wood Framed

		Crew	Daily Output	Labor-Hours	Unit	Material	2009 Bare Costs Labor	Equipment	Total	Total Incl O&P
0010	**ROOF WINDOWS, WOOD FRAMED**									
5600	Roof window incl. frame, flashing, double insulated glass & screens,									
5610	complete unit, 22" x 38"	2 Carp	3	5.333	Ea.	229	149		378	505
5650	2'-5" x 3'-8"		3.20	5		293	140		433	555
5700	3'-5" x 4'-9"		3.40	4.706		375	132		507	635

08 62 Unit Skylights

08 62 13 – Domed Unit Skylights

08 62 13.20 Skylights

			Crew	Daily Output	Labor-Hours	Unit	Material	2009 Bare Costs Labor	Equipment	Total	Total Incl O&P
0010	**SKYLIGHTS**, Plastic domes, flush or curb mounted ten or										
0100	more units										
0300	Nominal size under 10 S.F., double	G	G-3	130	.246	S.F.	26.50	6.35		32.85	39.50
0400	Single			160	.200		23.50	5.15		28.65	34
0600	10 S.F. to 20 S.F., double	G		315	.102		22.50	2.63		25.13	29
0700	Single			395	.081		23	2.09		25.09	28.50
0900	20 S.F. to 30 S.F., double	G		395	.081		21	2.09		23.09	26.50
1000	Single			465	.069		19.05	1.78		20.83	24
1200	30 S.F. to 65 S.F., double	G		465	.069		21	1.78		22.78	26
1300	Single			610	.052		16.50	1.36		17.86	20.50
1500	For insulated 4" curbs, double, add						25%				
1600	Single, add						30%				
1800	For integral insulated 9" curbs, double, add						30%				
1900	Single, add						40%				
2120	Ventilating insulated plexiglass dome with										
2130	curb mounting, 36" x 36"	G	G-3	12	2.667	Ea.	380	69		449	535
2150	52" x 52"	G		12	2.667		570	69		639	745
2160	28" x 52"	G		10	3.200		455	82.50		537.50	640
2170	36" x 52"	G		10	3.200		495	82.50		577.50	685
2180	For electric opening system, add	G					292			292	320
2210	Operating skylight, with thermopane glass, 24" x 48"	G	G-3	10	3.200		560	82.50		642.50	760
2220	32" x 48"	G	"	9	3.556		590	92		682	800
2310	Non venting insulated plexiglass dome skylight with										
2320	Flush mount 22" x 46"	G	G-3	15.23	2.101	Ea.	320	54.50		374.50	440
2330	30" x 30"	G		16	2		292	51.50		343.50	405
2340	46" x 46"	G		13.91	2.301		540	59.50		599.50	695
2350	Curb mount 22" x 46"	G		15.23	2.101		279	54.50		333.50	395
2360	30" x 30"	G		16	2		266	51.50		317.50	380
2370	46" x 46"	G		13.91	2.301		500	59.50		559.50	650

08 62 Unit Skylights

08 62 13 – Domed Unit Skylights

08 62 13.20 Skylights

		Crew	Daily Output	Labor-Hours	Unit	Material	2009 Bare Costs Labor	Equipment	Total	Total Incl O&P
2381	Non-insulated flush mount 22" x 46"	G-3	15.23	2.101	Ea.	215	54.50		269.50	330
2382	30" x 30"		16	2		194	51.50		245.50	300
2383	46" x 46"		13.91	2.301		365	59.50		424.50	500
2384	Curb mount 22" x 46"		15.23	2.101		181	54.50		235.50	291
2385	30" x 30"		16	2		175	51.50		226.50	280
4000	Skylight, solar tube kit, incl dome, flashing, diffuser, 1 pipe, 9" dia. [G]	1 Carp	2	4		220	112		332	430
4010	13" dia. [G]		2	4		290	112		402	510
4020	21" dia. [G]		2	4		490	112		602	730
4030	Accessories for, 1' long x 9" dia pipe [G]		24	.333		29	9.30		38.30	48
4040	2' long x 9" dia pipe [G]		24	.333		44	9.30		53.30	64.50
4050	4' long x 9" dia pipe [G]		20	.400		72	11.20		83.20	98
4060	1' long x 13" dia pipe [G]		24	.333		36	9.30		45.30	55.50
4070	2' long x 13" dia pipe [G]		24	.333		57	9.30		66.30	78.50
4080	4' long x 13" dia pipe [G]		20	.400		99	11.20		110.20	128
4090	2' long x 21" dia pipe [G]		16	.500		96	14		110	130
4100	4' long x 21" dia pipe [G]		12	.667		162	18.65		180.65	210
4110	45 degree elbow, 9" [G]		16	.500		67	14		81	97
4120	13" [G]		16	.500		83	14		97	115
4130	Interior decorative ring, 9" [G]		20	.400		19	11.20		30.20	40
4140	13" [G]		20	.400		21	11.20		32.20	42

08 71 Door Hardware

08 71 20 – Hardware

08 71 20.15 Hardware

		Crew	Daily Output	Labor-Hours	Unit	Material	2009 Bare Costs Labor	Equipment	Total	Total Incl O&P
0009	**HARDWARE**									
0010	Average percentage for hardware, total job cost									
0025	Minimum									.75%
0050	Maximum									3.50%
0500	Total hardware for building, average distribution					85%	15%			
1000	Door hardware, apartment, interior				Door Ea.	140			140	154
2100	Pocket door					134			134	147
4000	Door knocker, bright brass	1 Carp	32	.250		47	7		54	63.50
4100	Mail slot, bright brass, 2" x 11"	"	25	.320		59	8.95		67.95	80
4200	Peep hole, add to price of door					10			10	11

08 71 20.40 Lockset

		Crew	Daily Output	Labor-Hours	Unit	Material	2009 Bare Costs Labor	Equipment	Total	Total Incl O&P
0010	**LOCKSET**, Standard duty									
0020	Non-keyed, passage	1 Carp	12	.667	Ea.	49	18.65		67.65	85
0100	Privacy		12	.667		61	18.65		79.65	98.50
0400	Keyed, single cylinder function		10	.800		83.50	22.50		106	130
0500	Lever handled, keyed, single cylinder function		10	.800		148	22.50		170.50	201
1700	Residential, interior door, minimum		16	.500		17	14		31	42
1720	Maximum		8	1		43.50	28		71.50	95.50
1800	Exterior, minimum		14	.571		36.50	15.95		52.45	67
1810	Average		8	1		69	28		97	123
1820	Maximum		8	1		156	28		184	220

08 71 20.50 Door Stops

		Crew	Daily Output	Labor-Hours	Unit	Material	2009 Bare Costs Labor	Equipment	Total	Total Incl O&P
0010	**DOOR STOPS**									
0020	Holder & bumper, floor or wall	1 Carp	32	.250	Ea.	33.50	7		40.50	49
1300	Wall bumper, 4" diameter, with rubber pad, aluminum		32	.250		10.25	7		17.25	23
1600	Door bumper, floor type, aluminum		32	.250		5.85	7		12.85	18.25

08 71 Door Hardware

08 71 20 – Hardware

08 71 20.50 Door Stops	Crew	Daily Output	Labor-Hours	Unit	Material	2009 Bare Costs Labor	Equipment	Total	Total Incl O&P
1900	Plunger type, door mounted	1 Carp	32	.250	Ea.	27	7		34	41.50

08 71 20.60 Entrance Locks

		Crew	Daily Output	Labor-Hours	Unit	Material	Labor	Equipment	Total	Total Incl O&P
0010	**ENTRANCE LOCKS**									
0015	Cylinder, grip handle deadlocking latch	1 Carp	9	.889	Ea.	127	25		152	182
0020	Deadbolt		8	1		154	28		182	217
0100	Push and pull plate, dead bolt		8	1		146	28		174	209
0900	For handicapped lever, add					161			161	177

08 71 20.65 Thresholds

		Crew	Daily Output	Labor-Hours	Unit	Material	Labor	Equipment	Total	Total Incl O&P
0010	**THRESHOLDS**									
0011	Threshold 3' long saddles aluminum	1 Carp	48	.167	L.F.	4.02	4.66		8.68	12.30
0100	Aluminum, 8" wide, 1/2" thick		12	.667	Ea.	38.50	18.65		57.15	73.50
0500	Bronze		60	.133	L.F.	38.50	3.73		42.23	49
0600	Bronze, panic threshold, 5" wide, 1/2" thick		12	.667	Ea.	65.50	18.65		84.15	104
0700	Rubber, 1/2" thick, 5-1/2" wide		20	.400		37.50	11.20		48.70	60.50
0800	2-3/4" wide		20	.400		16	11.20		27.20	36.50

08 71 20.75 Door Hardware Accessories

		Crew	Daily Output	Labor-Hours	Unit	Material	Labor	Equipment	Total	Total Incl O&P
0010	**DOOR HARDWARE ACCESSORIES**									
1000	Knockers, brass, standard	1 Carp	16	.500	Ea.	39.50	14		53.50	67
1100	Deluxe		10	.800		123	22.50		145.50	173
4100	Deluxe		18	.444		44	12.40		56.40	69.50
4500	Rubber door silencers		540	.015		.26	.41		.67	.99

08 71 20.90 Hinges

			Crew	Daily Output	Labor-Hours	Unit	Material	Labor	Equipment	Total	Total Incl O&P
0010	**HINGES**	R087120-10									
0012	Full mortise, avg. freq., steel base, USP, 4-1/2" x 4-1/2"					Pr.	24.50			24.50	27
0100	5" x 5", USP						41			41	45
0200	6" x 6", USP						87			87	95.50
0400	Brass base, 4-1/2" x 4-1/2", US10						50			50	55
0500	5" x 5", US10						73.50			73.50	81
0600	6" x 6", US10						125			125	137
0800	Stainless steel base, 4-1/2" x 4-1/2", US32						74.50			74.50	82
0900	For non removable pin, add (security item)					Ea.	4.24			4.24	4.66
0910	For floating pin, driven tips, add						3.09			3.09	3.40
0930	For hospital type tip on pin, add						13.35			13.35	14.70
0940	For steeple type tip on pin, add						11.70			11.70	12.85
0950	Full mortise, high frequency, steel base, 3-1/2" x 3-1/2", US26D					Pr.	25.50			25.50	28.50
1000	4-1/2" x 4-1/2", USP						58.50			58.50	64.50
1100	5" x 5", USP						54.50			54.50	60
1200	6" x 6", USP						134			134	147
1400	Brass base, 3-1/2" x 3-1/2", US4						45			45	49.50
1430	4-1/2" x 4-1/2", US10						78.50			78.50	86.50
1500	5" x 5", US10						118			118	130
1600	6" x 6", US10						170			170	187
1800	Stainless steel base, 4-1/2" x 4-1/2", US32						125			125	137
1810	5" x 4-1/2", US32						175			175	193
1930	For hospital type tip on pin, add					Ea.	11.65			11.65	12.80
1950	Full mortise, low frequency, steel base, 3-1/2" x 3-1/2", US26D					Pr.	10.55			10.55	11.60
2000	4-1/2" x 4-1/2", USP						12.05			12.05	13.25
2100	5" x 5", USP						30			30	33
2200	6" x 6", USP						59.50			59.50	65.50
2300	4-1/2" x 4-1/2", US3						17.40			17.40	19.15
2310	5" x 5", US3						43			43	47.50
2400	Brass bass, 4-1/2" x 4-1/2", US10						42			42	46

08 71 Door Hardware

08 71 20 – Hardware

08 71 20.90 Hinges

		Crew	Daily Output	Labor-Hours	Unit	Material	2009 Bare Costs Labor	Equipment	Total	Total Incl O&P
2500	5" x 5", US10				Pr.	63.50			63.50	70
2800	Stainless steel base, 4-1/2" x 4-1/2", US32				↓	72			72	79.50

08 71 20.92 Mortised Hinges

		Crew	Daily Output	Labor-Hours	Unit	Material	2009 Bare Costs Labor	Equipment	Total	Total Incl O&P
0010	**MORTISED HINGES**									
0200	Average frequency, steel plated, ball bearing, 3-1/2" x 3-1/2"				Pr.	25			25	27.50
0300	Bronze, ball bearing					31.50			31.50	34.50
0900	High frequency, steel plated, ball bearing					78.50			78.50	86.50
1100	Bronze, ball bearing					80.50			80.50	88.50
1300	Average frequency, steel plated, ball bearing, 4-1/2" x 4-1/2"					31			31	34
1500	Bronze, ball bearing, to 36" wide					33			33	36.50
1700	Low frequency, steel, plated, plain bearing					15.65			15.65	17.20
1900	Bronze, plain bearing				↓	17.55			17.55	19.35

08 71 20.95 Kick Plates

		Crew	Daily Output	Labor-Hours	Unit	Material	2009 Bare Costs Labor	Equipment	Total	Total Incl O&P
0010	**KICK PLATES**									
0020	Stainless steel, 6" high, for 3' door	1 Carp	15	.533	Ea.	31	14.90		45.90	59
0500	Bronze 6" high, for 3' door	"	15	.533	"	43.50	14.90		58.40	73

08 71 21 – Astragals

08 71 21.10 Exterior Mouldings, Astragals

		Crew	Daily Output	Labor-Hours	Unit	Material	2009 Bare Costs Labor	Equipment	Total	Total Incl O&P
0010	**EXTERIOR MOULDINGS, ASTRAGALS**									
4170	Astragal for double doors, aluminum	1 Carp	4	2	Opng.	23	56		79	120
4174	Bronze	"	4	2	"	35	56		91	133

08 71 25 – Weatherstripping

08 71 25.10 Mechanical Seals, Weatherstripping

		Crew	Daily Output	Labor-Hours	Unit	Material	2009 Bare Costs Labor	Equipment	Total	Total Incl O&P
0010	**MECHANICAL SEALS, WEATHERSTRIPPING**									
1000	Doors, wood frame, interlocking, for 3' x 7' door, zinc	1 Carp	3	2.667	Opng.	15.50	74.50		90	143
1100	Bronze		3	2.667		24	74.50		98.50	153
1300	6' x 7' opening, zinc		2	4		16.90	112		128.90	208
1400	Bronze		2	4	↓	32	112		144	224
1500	Vinyl V strip	↓	6.40	1.250	Ea.	7.75	35		42.75	67.50
1700	Wood frame, spring type, bronze									
1800	3' x 7' door	1 Carp	7.60	1.053	Opng.	21.50	29.50		51	73
1900	6' x 7' door		7	1.143		26.50	32		58.50	83
1920	Felt, 3' x 7' door		14	.571		2.35	15.95		18.30	29.50
1930	6' x 7' door		13	.615		2.53	17.20		19.73	32
1950	Rubber, 3' x 7' door		7.60	1.053		4.80	29.50		34.30	55
1960	6' x 7' door	↓	7	1.143	↓	5.55	32		37.55	60
2200	Metal frame, spring type, bronze									
2300	3' x 7' door	1 Carp	3	2.667	Opng.	35.50	74.50		110	165
2400	6' x 7' door	"	2.50	3.200	"	46	89.50		135.50	202
2500	For stainless steel, spring type, add					133%				
2700	Metal frame, extruded sections, 3' x 7' door, aluminum	1 Carp	2	4	Opng.	45	112		157	239
2800	Bronze		2	4		114	112		226	315
3100	6' x 7' door, aluminum		1.20	6.667		57.50	186		243.50	380
3200	Bronze	↓	1.20	6.667	↓	134	186		320	460
3500	Threshold weatherstripping									
3650	Door sweep, flush mounted, aluminum	1 Carp	25	.320	Ea.	13.30	8.95		22.25	30
3700	Vinyl		25	.320		15.70	8.95		24.65	32.50
5000	Garage door bottom weatherstrip, 12' aluminum, clear		14	.571		21.50	15.95		37.45	50.50
5010	Bronze		14	.571		81.50	15.95		97.45	117
5050	Bottom protection, Rubber		14	.571		22	15.95		37.95	51.50
5100	Threshold	↓	14	.571	↓	88.50	15.95		104.45	125

08 75 Window Hardware

08 75 10 – Window Handles and Latches

08 75 10.10 Handles and Latches	Crew	Daily Output	Labor-Hours	Unit	Material	2009 Bare Costs Labor	Equipment	Total	Total Incl O&P
0010 **HANDLES AND LATCHES**									
1000 Handles, surface mounted, aluminum	1 Carp	24	.333	Ea.	3.40	9.30		12.70	19.50
1020 Brass		24	.333		5.75	9.30		15.05	22
1040 Chrome		24	.333		5.65	9.30		14.95	22
1500 Recessed, aluminum		12	.667		1.95	18.65		20.60	33.50
1520 Brass		12	.667		2.05	18.65		20.70	34
1540 Chrome		12	.667		2	18.65		20.65	33.50
2000 Latches, aluminum		20	.400		1.96	11.20		13.16	21
2020 Brass		20	.400		2.21	11.20		13.41	21.50
2040 Chrome		20	.400		2.11	11.20		13.31	21

08 75 30 – Weatherstripping

08 75 30.10 Mechanical Weather Seals	Crew	Daily Output	Labor-Hours	Unit	Material	2009 Bare Costs Labor	Equipment	Total	Total Incl O&P
0010 **MECHANICAL WEATHER SEALS**, Window, double hung, 3' X 5'									
0020 Zinc	1 Carp	7.20	1.111	Opng.	12.30	31		43.30	66
0100 Bronze		7.20	1.111		25.50	31		56.50	80.50
0200 Vinyl V strip		7	1.143		4.55	32		36.55	59
0500 As above but heavy duty, zinc		4.60	1.739		17.30	48.50		65.80	101
0600 Bronze		4.60	1.739		29	48.50		77.50	114
9000 Minimum labor/equipment charge	1 Clab	4.60	1.739	Job		35.50		35.50	60.50

08 79 Hardware Accessories

08 79 20 – Door Accessories

08 79 20.10 Door Hardware Accessories	Crew	Daily Output	Labor-Hours	Unit	Material	2009 Bare Costs Labor	Equipment	Total	Total Incl O&P
0010 **DOOR HARDWARE ACCESSORIES**									
0140 Door bolt, surface, 4"	1 Carp	32	.250	Ea.	8.65	7		15.65	21.50
0160 Door latch	"	12	.667	"	8.05	18.65		26.70	40.50
0200 Sliding closet door									
0220 Track and hanger, single	1 Carp	10	.800	Ea.	49.50	22.50		72	92.50
0240 Double		8	1		72.50	28		100.50	127
0260 Door guide, single		48	.167		23.50	4.66		28.16	34
0280 Double		48	.167		32	4.66		36.66	43.50
0600 Deadbolt and lock cover plate, brass or stainless steel		30	.267		26.50	7.45		33.95	42
0620 Hole cover plate, brass or chrome		35	.229		6.95	6.40		13.35	18.45
2240 Mortise lockset, passage, lever handle		9	.889		194	25		219	255
4000 Security chain, standard		18	.444		7.40	12.40		19.80	29

08 81 Glass Glazing

08 81 10 – Float Glass

08 81 10.10 Various Types and Thickness of Float Glass	Crew	Daily Output	Labor-Hours	Unit	Material	2009 Bare Costs Labor	Equipment	Total	Total Incl O&P
0010 **VARIOUS TYPES AND THICKNESS OF FLOAT GLASS**									
0020 3/16" Plain	2 Glaz	130	.123	S.F.	4.48	3.37		7.85	10.50
0200 Tempered, clear		130	.123		6.05	3.37		9.42	12.20
0300 Tinted		130	.123		7.05	3.37		10.42	13.30
0600 1/4" thick, clear, plain		120	.133		5.25	3.65		8.90	11.85
0700 Tinted		120	.133		7.40	3.65		11.05	14.20
0800 Tempered, clear		120	.133		7.45	3.65		11.10	14.25
0900 Tinted		120	.133		9.75	3.65		13.40	16.75
1600 3/8" thick, clear, plain		75	.213		9.10	5.85		14.95	19.70
1700 Tinted		75	.213		14.80	5.85		20.65	26

08 81 Glass Glazing

08 81 10 – Float Glass

08 81 10.10 Various Types and Thickness of Float Glass

		Crew	Daily Output	Labor-Hours	Unit	Material	2009 Bare Costs Labor	Equipment	Total	Total Incl O&P
1800	Tempered, clear	2 Glaz	75	.213	S.F.	15.10	5.85		20.95	26.50
1900	Tinted		75	.213		16.85	5.85		22.70	28
2200	1/2" thick, clear, plain		55	.291		18.15	7.95		26.10	33
2300	Tinted		55	.291		26	7.95		33.95	41.50
2400	Tempered, clear		55	.291		22	7.95		29.95	37.50
2500	Tinted		55	.291		26	7.95		33.95	41.50
2800	5/8" thick, clear, plain		45	.356		26	9.75		35.75	44.50
2900	Tempered, clear		45	.356		29.50	9.75		39.25	48.50
8900	For low emissivity coating for 3/16" & 1/4" only, add to above					15%				

08 81 25 – Glazing Variables

08 81 25.10 Applications of Glazing

		Crew	Daily Output	Labor-Hours	Unit	Material	2009 Bare Costs Labor	Equipment	Total	Total Incl O&P
0010	**APPLICATIONS OF GLAZING**									
0600	For glass replacement, add				S.F.		100%			
0700	For gasket settings, add				L.F.	5.55			5.55	6.10
0900	For sloped glazing, add				S.F.		25%			
2000	Fabrication, polished edges, 1/4" thick				Inch	.48			.48	.53
2100	1/2" thick					1.21			1.21	1.33
2500	Mitered edges, 1/4" thick					1.21			1.21	1.33
2600	1/2" thick					1.96			1.96	2.16

08 81 30 – Insulating Glass

08 81 30.10 Reduce Heat Transfer Glass

			Crew	Daily Output	Labor-Hours	Unit	Material	2009 Bare Costs Labor	Equipment	Total	Total Incl O&P
0010	**REDUCE HEAT TRANSFER GLASS**, 2 lites 1/8" float, 1/2" thk under 15 S.F.										
0100	Tinted	G	2 Glaz	95	.168	S.F.	12.55	4.61		17.16	21.50
0280	Double glazed, 5/8" thk unit, 3/16" float, 15-30 S.F., clear			90	.178		12.95	4.87		17.82	22.50
0400	1" thk, dbl. glazed, 1/4" float, 30-70 S.F., clear	G		75	.213		15.20	5.85		21.05	26.50
0500	Tinted	G		75	.213		22	5.85		27.85	34
2000	Both lites, light & heat reflective	G		85	.188		29.50	5.15		34.65	41
2500	Heat reflective, film inside, 1" thick unit, clear	G		85	.188		26	5.15		31.15	37
2600	Tinted	G		85	.188		28	5.15		33.15	39.50
3000	Film on weatherside, clear, 1/2" thick unit	G		95	.168		18.50	4.61		23.11	28
3100	5/8" thick unit	G		90	.178		18.10	4.87		22.97	28
3200	1" thick unit	G		85	.188		25.50	5.15		30.65	36.50

08 81 40 – Plate Glass

08 81 40.10 Plate Glass

		Crew	Daily Output	Labor-Hours	Unit	Material	2009 Bare Costs Labor	Equipment	Total	Total Incl O&P
0010	**PLATE GLASS** Twin ground, polished,									
0015	3/16" thick, material				S.F.	4.73			4.73	5.20
0020	3/16" thick	2 Glaz	100	.160		4.73	4.38		9.11	12.45
0100	1/4" thick		94	.170		6.50	4.66		11.16	14.85
0200	3/8" thick		60	.267		11.15	7.30		18.45	24.50
0300	1/2" thick		40	.400		21.50	10.95		32.45	41.50

08 81 55 – Window Glass

08 81 55.10 Sheet Glass

		Crew	Daily Output	Labor-Hours	Unit	Material	2009 Bare Costs Labor	Equipment	Total	Total Incl O&P
0010	**SHEET GLASS** (window), clear float, stops, putty bed									
0015	1/8" thick, clear float	2 Glaz	480	.033	S.F.	4.27	.91		5.18	6.20
0500	3/16" thick, clear		480	.033		5.25	.91		6.16	7.30
0600	Tinted		480	.033		7.15	.91		8.06	9.35
0700	Tempered		480	.033		8.65	.91		9.56	11

08 83 Mirrors

08 83 13 – Mirrored Glass Glazing

08 83 13.10 Mirrors	Crew	Daily Output	Labor-Hours	Unit	Material	2009 Bare Costs Labor	Equipment	Total	Total Incl O&P
0010 **MIRRORS**, No frames, wall type, 1/4" plate glass, polished edge									
0100 Up to 5 S.F.	2 Glaz	125	.128	S.F.	9.10	3.51		12.61	15.80
0200 Over 5 S.F.		160	.100		8.80	2.74		11.54	14.25
0500 Door type, 1/4" plate glass, up to 12 S.F.		160	.100		7.85	2.74		10.59	13.20
1000 Float glass, up to 10 S.F., 1/8" thick		160	.100		5.05	2.74		7.79	10.15
1100 3/16" thick		150	.107		6.15	2.92		9.07	11.65
1500 12" x 12" wall tiles, square edge, clear		195	.082		1.78	2.25		4.03	5.65
1600 Veined		195	.082		4.69	2.25		6.94	8.85
2010 Bathroom, unframed, laminated	▼	160	.100	▼	13.20	2.74		15.94	19.05

08 87 Glazing Surface Films

08 87 53 – Security Films

08 87 53.10 Security Film

	Crew	Daily Output	Labor-Hours	Unit	Material	2009 Bare Costs Labor	Equipment	Total	Total Incl O&P
0010 **SECURITY FILM**, clear, 32000psi tensile strength, adhered to glass									
0100 .002" thick, daylight installation	H-2	950	.025	S.F.	1.45	.63		2.08	2.65
0150 .004" thick, daylight installation		800	.030		1.60	.75		2.35	3.01
0200 .006" thick, daylight installation		700	.034		1.70	.86		2.56	3.30
0210 Install for anchorage		600	.040		1.89	1		2.89	3.75
0400 .007" thick, daylight installation		600	.040		1.80	1		2.80	3.65
0410 Install for anchorage		500	.048		2	1.21		3.21	4.20
0500 .008" thick, daylight installation		500	.048		2.10	1.21		3.31	4.31
0510 Install for anchorage		500	.048		2.33	1.21		3.54	4.57
0600 .015" thick, daylight installation		400	.060		3.50	1.51		5.01	6.35
0610 Install for anchorage	▼	400	.060	▼	2.33	1.51		3.84	5.10
0900 Security film anchorage, mechanical attachment and cover plate	H-3	370	.043	L.F.	8.70	1.04		9.74	11.35
0950 Security film anchorage, wet glaze structural caulking	1 Glaz	225	.036	"	.83	.97		1.80	2.52
1000 Adhered security film removal	1 Clab	275	.029	S.F.		.60		.60	1.01

08 91 Louvers

08 91 19 – Fixed Louvers

08 91 19.10 Aluminum Louvers

	Crew	Daily Output	Labor-Hours	Unit	Material	2009 Bare Costs Labor	Equipment	Total	Total Incl O&P
0010 **ALUMINUM LOUVERS**									
0020 Aluminum with screen, residential, 8" x 8"	1 Carp	38	.211	Ea.	10.50	5.90		16.40	21.50
0100 12" x 12"		38	.211		11.85	5.90		17.75	23
0200 12" x 18"		35	.229		13.50	6.40		19.90	25.50
0250 14" x 24"		30	.267		17	7.45		24.45	31.50
0300 18" x 24"		27	.296		24.50	8.30		32.80	41
0500 24" x 30"		24	.333		34	9.30		43.30	53.50
0700 Triangle, adjustable, small		20	.400		29	11.20		40.20	51
0800 Large		15	.533		50	14.90		64.90	80.50
2100 Midget, aluminum, 3/4" deep, 1" diameter		85	.094		.77	2.63		3.40	5.30
2150 3" diameter		60	.133		1.62	3.73		5.35	8.10
2200 4" diameter		50	.160		3.07	4.47		7.54	10.95
2250 6" diameter	▼	30	.267	▼	3.63	7.45		11.08	16.60

08 95 Vents

08 95 13 – Soffit Vents

	08 95 13.10 Wall Louvers	Crew	Daily Output	Labor-Hours	Unit	Material	2009 Bare Costs Labor	2009 Bare Costs Equipment	Total	Total Incl O&P
0010	**WALL LOUVERS**									
2400	Under eaves vent, aluminum, mill finish, 16" x 4"	1 Carp	48	.167	Ea.	1.29	4.66		5.95	9.30
2500	16" x 8"	"	48	.167	"	1.49	4.66		6.15	9.55

08 95 16 – Wall Vents

	08 95 16.10 Louvers	Crew	Daily Output	Labor-Hours	Unit	Material	2009 Bare Costs Labor	2009 Bare Costs Equipment	Total	Total Incl O&P
0010	**LOUVERS**									
0020	Redwood, 2'-0" diameter, full circle	1 Carp	16	.500	Ea.	159	14		173	198
0100	Half circle		16	.500		152	14		166	191
0200	Octagonal		16	.500		121	14		135	157
0300	Triangular, 5/12 pitch, 5'-0" at base		16	.500		257	14		271	305
7000	Vinyl gable vent, 8" x 8"		38	.211		11.20	5.90		17.10	22.50
7020	12" x 12"		38	.211		23.50	5.90		29.40	36
7080	12" x 18"		35	.229		30	6.40		36.40	44
7200	18" x 24"		30	.267		36.50	7.45		43.95	53

Division 9
Finishes

09 01 Maintenance of Finishes

09 01 70 – Maintenance of Wall Finishes

09 01 70.10 Gypsum Wallboard Repairs

		Crew	Daily Output	Labor-Hours	Unit	Material	2009 Bare Costs Labor	Equipment	Total	Total Incl O&P
0010	**GYPSUM WALLBOARD REPAIRS**									
0100	Fill and sand, pin / nail holes	1 Carp	960	.008	Ea.		.23		.23	.39
0110	Screw head pops		480	.017			.47		.47	.79
0120	Dents, up to 2" square		48	.167		.01	4.66		4.67	7.90
0130	2" to 4" square		24	.333		.03	9.30		9.33	15.80
0140	Cut square, patch, sand and finish, holes, up to 2" square		12	.667		.04	18.65		18.69	31.50
0150	2" to 4" square		11	.727		.10	20.50		20.60	34.50
0160	4" to 8" square		10	.800		.25	22.50		22.75	38.50
0170	8" to 12" square		8	1		.48	28		28.48	48
0180	12" to 32" square		6	1.333		3.04	37.50		40.54	66.50
0210	16" by 48"		5	1.600		2.31	44.50		46.81	78
0220	32" by 48"		4	2		4.13	56		60.13	99
0230	48" square		3.50	2.286		5.85	64		69.85	114
0240	60" square		3.20	2.500		9.80	70		79.80	129
0500	Skim coat surface with joint compound		1600	.005	S.F.	.03	.14		.17	.27

09 05 Common Work Results for Finishes

09 05 05 – Selective Finishes Demolition

09 05 05.10 Selective Demolition, Ceilings

		Crew	Daily Output	Labor-Hours	Unit	Material	2009 Bare Costs Labor	Equipment	Total	Total Incl O&P
0010	**SELECTIVE DEMOLITION, CEILINGS** R024119-10									
0200	Ceiling, drywall, furred and nailed or screwed	2 Clab	800	.020	S.F.		.41		.41	.70
1000	Plaster, lime and horse hair, on wood lath, incl. lath		700	.023			.47		.47	.79
1200	Suspended ceiling, mineral fiber, 2' x 2' or 2' x 4'		1500	.011			.22		.22	.37
1250	On suspension system, incl. system		1200	.013			.27		.27	.46
1500	Tile, wood fiber, 12" x 12", glued		900	.018			.37		.37	.62
1540	Stapled		1500	.011			.22		.22	.37
2000	Wood, tongue and groove, 1" x 4"		1000	.016			.33		.33	.56
2040	1" x 8"		1100	.015			.30		.30	.51
2400	Plywood or wood fiberboard, 4' x 8' sheets		1200	.013			.27		.27	.46

09 05 05.20 Selective Demolition, Flooring

		Crew	Daily Output	Labor-Hours	Unit	Material	2009 Bare Costs Labor	Equipment	Total	Total Incl O&P
0010	**SELECTIVE DEMOLITION, FLOORING** R024119-10									
0200	Brick with mortar	2 Clab	475	.034	S.F.		.69		.69	1.17
0400	Carpet, bonded, including surface scraping		2000	.008			.16		.16	.28
0480	Tackless		9000	.002			.04		.04	.06
0550	Carpet tile, releasable adhesive		5000	.003			.07		.07	.11
0560	Permanent adhesive		1850	.009			.18		.18	.30
0800	Resilient, sheet goods		1400	.011			.23		.23	.40
0850	Vinyl or rubber cove base	1 Clab	1000	.008	L.F.		.16		.16	.28
0860	Vinyl or rubber cove base, molded corner	"	1000	.008	Ea.		.16		.16	.28
0870	For glued and caulked installation, add to labor						50%			
0900	Vinyl composition tile, 12" x 12"	2 Clab	1000	.016	S.F.		.33		.33	.56
2000	Tile, ceramic, thin set		675	.024			.49		.49	.82
2020	Mud set		625	.026			.53		.53	.89
3000	Wood, block, on end	1 Carp	400	.020			.56		.56	.95
3200	Parquet		450	.018			.50		.50	.84
3400	Strip flooring, interior, 2-1/4" x 25/32" thick		325	.025			.69		.69	1.16
3500	Exterior, porch flooring, 1" x 4"		220	.036			1.02		1.02	1.72
3800	Subfloor, tongue and groove, 1" x 6"		325	.025			.69		.69	1.16
3820	1" x 8"		430	.019			.52		.52	.88
3840	1" x 10"		520	.015			.43		.43	.73
4000	Plywood, nailed		600	.013			.37		.37	.63

09 05 Common Work Results for Finishes

09 05 05 – Selective Finishes Demolition

09 05 05.20 Selective Demolition, Flooring

		Crew	Daily Output	Labor-Hours	Unit	Material	2009 Bare Costs Labor	Equipment	Total	Total Incl O&P
4100	Glued and nailed	1 Carp	400	.020	S.F.		.56		.56	.95
4200	Hardboard, 1/4" thick	↓	760	.011	↓		.29		.29	.50

09 05 05.30 Selective Demolition, Walls and Partitions

		Crew	Daily Output	Labor-Hours	Unit	Material	2009 Bare Costs Labor	Equipment	Total	Total Incl O&P
0010	**SELECTIVE DEMOLITION, WALLS AND PARTITIONS** R024119-10									
0020	Walls, concrete, reinforced	B-39	120	.400	C.F.		8.35	1.60	9.95	13.90
0025	Plain	"	160	.300	"		6.25	1.20	7.45	11.90
1000	Drywall, nailed or screwed	1 Clab	1000	.008	S.F.		.16		.16	.28
1010	2 layers		400	.020			.41		.41	.70
1500	Fiberboard, nailed		900	.009			.18		.18	.31
1568	Plenum barrier, sheet lead	↓	300	.027			.55		.55	.93
2200	Metal or wood studs, finish 2 sides, fiberboard	B-1	520	.046			.98		.98	1.66
2250	Lath and plaster		260	.092			1.96		1.96	3.31
2300	Plasterboard (drywall)		520	.046			.98		.98	1.66
2350	Plywood	↓	450	.053			1.13		1.13	1.91
2800	Paneling, 4' x 8' sheets	1 Clab	475	.017			.35		.35	.59
3000	Plaster, lime and horsehair, on wood lath		400	.020			.41		.41	.70
3020	On metal lath		335	.024	↓		.49		.49	.83
3450	Plaster, interior gypsum, acoustic, or cement		60	.133	S.Y.		2.74		2.74	4.63
3500	Stucco, on masonry		145	.055			1.13		1.13	1.92
3510	Commercial 3-coat		80	.100			2.06		2.06	3.48
3520	Interior stucco		25	.320	↓		6.60		6.60	11.10
3760	Tile, ceramic, on walls, thin set		300	.027	S.F.		.55		.55	.93
3765	Mud set	↓	250	.032	"		.66		.66	1.11

09 22 Supports for Plaster and Gypsum Board

09 22 03 – Fastening Methods for Finishes

09 22 03.20 Drilling Plaster/Drywall

		Crew	Daily Output	Labor-Hours	Unit	Material	2009 Bare Costs Labor	Equipment	Total	Total Incl O&P
0010	**DRILLING PLASTER/DRYWALL**									
1100	Drilling & layout for drywall/plaster walls, up to 1" deep, no anchor									
1200	Holes, 1/4" diameter	1 Carp	150	.053	Ea.	.01	1.49		1.50	2.53
1300	3/8" diameter		140	.057		.01	1.60		1.61	2.71
1400	1/2" diameter		130	.062		.01	1.72		1.73	2.92
1500	3/4" diameter		120	.067		.01	1.86		1.87	3.16
1600	1" diameter		110	.073		.02	2.03		2.05	3.46
1700	1-1/4" diameter		100	.080		.03	2.24		2.27	3.81
1800	1-1/2" diameter	↓	90	.089		.04	2.48		2.52	4.24
1900	For ceiling installations, add				↓		40%			

09 22 13 – Metal Furring

09 22 13.13 Metal Channel Furring

		Crew	Daily Output	Labor-Hours	Unit	Material	2009 Bare Costs Labor	Equipment	Total	Total Incl O&P
0010	**METAL CHANNEL FURRING**									
0030	Beams and columns, 7/8" channels, galvanized, 12" O.C.	1 Lath	155	.052	S.F.	.38	1.29		1.67	2.49
0050	16" O.C.		170	.047		.31	1.17		1.48	2.24
0070	24" O.C.		185	.043		.21	1.08		1.29	1.97
0100	Ceilings, on steel, 7/8" channels, galvanized, 12" O.C.		210	.038		.34	.95		1.29	1.92
0300	16" O.C.		290	.028		.31	.69		1	1.45
0400	24" O.C.		420	.019		.21	.47		.68	1
0600	1-5/8" channels, galvanized, 12" O.C.		190	.042		.46	1.05		1.51	2.21
0700	16" O.C.		260	.031		.41	.77		1.18	1.70
0900	24" O.C.		390	.021		.28	.51		.79	1.13
0930	7/8" channels with sound isolation clips, 12" O.C.	↓	120	.067	↓	1.53	1.66		3.19	4.37

09 22 Supports for Plaster and Gypsum Board

09 22 13 – Metal Furring

09 22 13.13 Metal Channel Furring		Crew	Daily Output	Labor-Hours	Unit	Material	2009 Bare Costs Labor	Equipment	Total	Total Incl O&P
0940	16" O.C.	1 Lath	100	.080	S.F.	2.12	1.99		4.11	5.55
0950	24" O.C.		165	.048		1.39	1.21		2.60	3.48
0960	1-5/8" channels, galvanized, 12" O.C.		110	.073		1.64	1.81		3.45	4.74
0970	16" O.C.		100	.080		2.22	1.99		4.21	5.65
0980	24" O.C.		155	.052		1.46	1.29		2.75	3.69
1000	Walls, 7/8" channels, galvanized, 12" O.C.		235	.034		.34	.85		1.19	1.75
1200	16" O.C.		265	.030		.31	.75		1.06	1.56
1300	24" O.C.		350	.023		.21	.57		.78	1.15
1500	1-5/8" channels, galvanized, 12" O.C.		210	.038		.46	.95		1.41	2.05
1600	16" O.C.		240	.033		.41	.83		1.24	1.80
1800	24" O.C.		305	.026		.28	.65		.93	1.36
1920	7/8" channels with sound isolation clips, 12" O.C.		125	.064		1.53	1.59		3.12	4.26
1940	16" O.C.		100	.080		2.12	1.99		4.11	5.55
1950	24" O.C.		150	.053		1.39	1.33		2.72	3.68
1960	1-5/8" channels, galvanized, 12" O.C.		115	.070		1.64	1.73		3.37	4.61
1970	16" O.C.		95	.084		2.22	2.10		4.32	5.85
1980	24" O.C.		140	.057		1.46	1.42		2.88	3.91

09 22 16 – Non-Structural Metal Framing

09 22 16.13 Metal Studs and Track		Crew	Daily Output	Labor-Hours	Unit	Material	2009 Bare Costs Labor	Equipment	Total	Total Incl O&P
0010	**METAL STUDS AND TRACK**									
1600	Non-load bearing, galv, 8' high, 25 ga. 1-5/8" wide, 16" O.C.	1 Carp	619	.013	S.F.	.33	.36		.69	.97
1610	24" O.C.		950	.008		.25	.24		.49	.67
1620	2-1/2" wide, 16" O.C.		613	.013		.42	.36		.78	1.08
1630	24" O.C.		938	.009		.31	.24		.55	.74
1640	3-5/8" wide, 16" O.C.		600	.013		.48	.37		.85	1.15
1650	24" O.C.		925	.009		.36	.24		.60	.80
1660	4" wide, 16" O.C.		594	.013		.51	.38		.89	1.20
1670	24" O.C.		925	.009		.38	.24		.62	.83
1680	6" wide, 16" O.C.		588	.014		.72	.38		1.10	1.43
1690	24" O.C.		906	.009		.54	.25		.79	1.01
1700	20 ga. studs, 1-5/8" wide, 16" O.C.		494	.016		.53	.45		.98	1.35
1710	24" O.C.		763	.010		.40	.29		.69	.94
1720	2-1/2" wide, 16" O.C.		488	.016		.62	.46		1.08	1.45
1730	24" O.C.		750	.011		.47	.30		.77	1.01
1740	3-5/8" wide, 16" O.C.		481	.017		.67	.46		1.13	1.53
1750	24" O.C.		738	.011		.50	.30		.80	1.06
1760	4" wide, 16" O.C.		475	.017		.80	.47		1.27	1.67
1770	24" O.C.		738	.011		.60	.30		.90	1.17
1780	6" wide, 16" O.C.		469	.017		.94	.48		1.42	1.84
1790	24" O.C.		725	.011		.70	.31		1.01	1.29
2000	Non-load bearing, galv, 10' high, 25 ga. 1-5/8" wide, 16" O.C.		495	.016		.31	.45		.76	1.10
2100	24" O.C.		760	.011		.23	.29		.52	.75
2200	2-1/2" wide, 16" O.C.		490	.016		.39	.46		.85	1.20
2250	24" O.C.		750	.011		.29	.30		.59	.82
2300	3-5/8" wide, 16" O.C.		480	.017		.45	.47		.92	1.29
2350	24" O.C.		740	.011		.33	.30		.63	.87
2400	4" wide, 16" O.C.		475	.017		.49	.47		.96	1.33
2450	24" O.C.		740	.011		.36	.30		.66	.90
2500	6" wide, 16" O.C.		470	.017		.68	.48		1.16	1.55
2550	24" O.C.		725	.011		.50	.31		.81	1.07
2600	20 ga. studs, 1-5/8" wide, 16" O.C.		395	.020		.50	.57		1.07	1.51
2650	24" O.C.		610	.013		.37	.37		.74	1.03

09 22 16 – Non-Structural Metal Framing

09 22 16.13 Metal Studs and Track

		Crew	Daily Output	Labor-Hours	Unit	Material	2009 Bare Costs Labor	Equipment	Total	Total Incl O&P
2700	2-1/2" wide, 16" O.C.	1 Carp	390	.021	S.F.	.59	.57		1.16	1.62
2750	24" O.C.		600	.013		.43	.37		.80	1.11
2800	3-5/8" wide, 16" OC		385	.021		.64	.58		1.22	1.68
2850	24" O.C.		590	.014		.47	.38		.85	1.16
2900	4" wide, 16" O.C.		380	.021		.75	.59		1.34	1.82
2950	24" O.C.		590	.014		.56	.38		.94	1.25
3000	6" wide, 16" O.C.		375	.021		.89	.60		1.49	1.99
3050	24" O.C.		580	.014		.65	.39		1.04	1.37
3060	Non-load bearing, galv, 12' high, 25 ga. 1-5/8" wide, 16" O.C.		413	.019		.30	.54		.84	1.25
3070	24" O.C.		633	.013		.22	.35		.57	.84
3080	2-1/2" wide, 16" O.C.		408	.020		.38	.55		.93	1.35
3090	24" O.C.		625	.013		.27	.36		.63	.90
3100	3-5/8" wide, 16" O.C.		400	.020		.43	.56		.99	1.42
3110	24" O.C.		617	.013		.31	.36		.67	.95
3120	4" wide, 16" O.C.		396	.020		.47	.56		1.03	1.46
3130	24" O.C.		617	.013		.34	.36		.70	.98
3140	6" wide, 16" O.C.		392	.020		.65	.57		1.22	1.68
3150	24" O.C.		604	.013		.47	.37		.84	1.15
3160	20 ga. studs, 1-5/8" wide, 16" O.C.		329	.024		.48	.68		1.16	1.68
3170	24" O.C.		508	.016		.35	.44		.79	1.12
3180	2-1/2" wide, 16" O.C.		325	.025		.56	.69		1.25	1.78
3190	24" O.C.		500	.016		.41	.45		.86	1.21
3200	3-5/8" wide, 16" O.C.		321	.025		.61	.70		1.31	1.85
3210	24" O.C.		492	.016		.44	.45		.89	1.26
3220	4" wide, 16" O.C.		317	.025		.72	.71		1.43	1.98
3230	24" O.C.		492	.016		.52	.45		.97	1.35
3240	6" wide, 16" O.C.		313	.026		.85	.71		1.56	2.15
3250	24" O.C.	▼	483	.017	▼	.62	.46		1.08	1.46
5000	Load bearing studs, see Div. 05 41 13.30									

09 22 26 – Suspension Systems

09 22 26.13 Ceiling Suspension Systems

		Crew	Daily Output	Labor-Hours	Unit	Material	2009 Bare Costs Labor	Equipment	Total	Total Incl O&P
0010	**CEILING SUSPENSION SYSTEMS** For gypsum board or plaster									
8000	Suspended ceilings, including carriers									
8200	1-1/2" carriers, 24" O.C. with:									
8300	7/8" channels, 16" O.C.	1 Lath	165	.048	S.F.	.58	1.21		1.79	2.59
8320	24" O.C.		200	.040		.48	1		1.48	2.14
8400	1-5/8" channels, 16" O.C.		155	.052		.69	1.29		1.98	2.84
8420	24" O.C.	▼	190	.042	▼	.55	1.05		1.60	2.31
8600	2" carriers, 24" O.C. with:									
8700	7/8" channels, 16" O.C.	1 Lath	155	.052	S.F.	.56	1.29		1.85	2.69
8720	24" O.C.		190	.042		.46	1.05		1.51	2.20
8800	1-5/8" channels, 16" O.C.		145	.055		.66	1.37		2.03	2.95
8820	24" O.C.	▼	180	.044	▼	.53	1.11		1.64	2.37

09 22 36 – Lath

09 22 36.13 Gypsum Lath

		Crew	Daily Output	Labor-Hours	Unit	Material	2009 Bare Costs Labor	Equipment	Total	Total Incl O&P
0011	**GYPSUM LATH** Plain or perforated, nailed, 3/8" thick	1 Lath	765	.010	S.F.	.61	.26		.87	1.09
0101	1/2" thick, nailed		720	.011		.65	.28		.93	1.17
0301	Clipped to steel studs, 3/8" thick		675	.012		.61	.30		.91	1.15
0401	1/2" thick		630	.013		.65	.32		.97	1.23
0601	Firestop gypsum base, to steel studs, 3/8" thick		630	.013		.32	.32		.64	.86
0701	1/2" thick		585	.014		.36	.34		.70	.95
0901	Foil back, to steel studs, 3/8" thick	▼	675	.012		.39	.30		.69	.91

09 22 Supports for Plaster and Gypsum Board

09 22 36 – Lath

09 22 36.13 Gypsum Lath

		Crew	Daily Output	Labor-Hours	Unit	Material	2009 Bare Costs Labor	2009 Bare Costs Equipment	Total	Total Incl O&P
1001	1/2" thick	1 Lath	630	.013	S.F.	.42	.32		.74	.97
1501	For ceiling installations, add		1950	.004			.10		.10	.17
1601	For columns and beams, add		1550	.005			.13		.13	.21

09 22 36.23 Metal Lath

			Crew	Daily Output	Labor-Hours	Unit	Material	2009 Bare Costs Labor	2009 Bare Costs Equipment	Total	Total Incl O&P
0010	**METAL LATH**	R092000-50									
3601	2.5 lb. diamond painted, on wood framing, on walls		1 Lath	765	.010	S.F.	.33	.26		.59	.79
3701	On ceilings			675	.012		.33	.30		.63	.85
4201	3.4 lb. diamond painted, wired to steel framing, on walls			675	.012		.46	.30		.76	.99
4301	On ceilings			540	.015		.46	.37		.83	1.11
5101	Rib lath, painted, wired to steel, on walls, 2.75 lb.			675	.012		.36	.30		.66	.88
5201	3.4 lb.			630	.013		.56	.32		.88	1.12
5701	Suspended ceiling system, incl. 3.4 lb. diamond lath, painted			135	.059		.50	1.48		1.98	2.94
5801	Galvanized			135	.059		.50	1.48		1.98	2.94

09 22 36.83 Accessories, Plaster

		Crew	Daily Output	Labor-Hours	Unit	Material	2009 Bare Costs Labor	2009 Bare Costs Equipment	Total	Total Incl O&P
0010	**ACCESSORIES, PLASTER**									
0020	Casing bead, expanded flange, galvanized	1 Lath	2.70	2.963	C.L.F.	46.50	74		120.50	171
0900	Channels, cold rolled, 16 ga., 3/4" deep, galvanized					34			34	37.50
1620	Corner bead, expanded bullnose, 3/4" radius, #10, galvanized	1 Lath	2.60	3.077		34	76.50		110.50	162
1650	#1, galvanized		2.55	3.137		55	78		133	187
1670	Expanded wing, 2-3/4" wide, #1, galvanized		2.65	3.019		34.50	75		109.50	160
1700	Inside corner (corner rite), 3" x 3", painted		2.60	3.077		30	76.50		106.50	157
1750	Strip-ex, 4" wide, painted		2.55	3.137		30	78		108	159
1800	Expansion joint, 3/4" grounds, limited expansion, galv., 1 piece		2.70	2.963		109	74		183	239
2100	Extreme expansion, galvanized, 2 piece		2.60	3.077		173	76.50		249.50	315

09 23 Gypsum Plastering

09 23 13 – Acoustical Gypsum Plastering

09 23 13.10 Perlite or Vermiculite Plaster

			Crew	Daily Output	Labor-Hours	Unit	Material	2009 Bare Costs Labor	2009 Bare Costs Equipment	Total	Total Incl O&P
0010	**PERLITE OR VERMICULITE PLASTER**	R092000-50									
0020	In 100 lb. bags, under 200 bags					Bag	14.40			14.40	15.85
0301	2 coats, no lath included, on walls		J-1	830	.048	S.F.	.45	1.17	.15	1.77	2.59
0401	On ceilings			710	.056		.45	1.36	.18	1.99	2.95
0901	3 coats, no lath included, on walls			665	.060		.70	1.46	.19	2.35	3.38
1001	On ceilings			565	.071		.70	1.71	.22	2.63	3.85
1700	For irregular or curved surfaces, add to above					S.Y.		30%			
1800	For columns and beams, add to above							50%			
1900	For soffits, add to ceiling prices							40%			

09 23 20 – Gypsum Plaster

09 23 20.10 Gypsum Plaster On Walls and Ceilings

			Crew	Daily Output	Labor-Hours	Unit	Material	2009 Bare Costs Labor	2009 Bare Costs Equipment	Total	Total Incl O&P
0010	**GYPSUM PLASTER ON WALLS AND CEILINGS**	R092000-50									
0020	80# bag, less than 1 ton					Bag	15.20			15.20	16.70
0302	2 coats, no lath included, on walls		J-1	750	.053	S.F.	.42	1.29	.17	1.88	2.78
0402	On ceilings			660	.061		.42	1.47	.19	2.08	3.09
0903	3 coats, no lath included, on walls			620	.065		.60	1.56	.20	2.36	3.46
1002	On ceilings			560	.071		.97	1.73	.23	2.93	4.17
1600	For irregular or curved surfaces, add							30%			
1800	For columns & beams, add							50%			

09 24 Portland Cement Plastering

09 24 23 – Portland Cement Stucco

09 24 23.40 Stucco

		Crew	Daily Output	Labor-Hours	Unit	Material	2009 Bare Costs Labor	Equipment	Total	Total Incl O&P
0010	**STUCCO** R092000-50									
0011	3 coats 1" thick, float finish, with mesh, on wood frame	J-2	470	.102	S.F.	.88	2.48	.27	3.63	5.35
0101	On masonry construction	J-1	495	.081		.24	1.96	.26	2.46	3.78
0151	2 coats, 3/4" thick, float finish, no lath incl.	"	980	.041		.25	.99	.13	1.37	2.05
0301	For trowel finish, add	1 Plas	1530	.005			.13		.13	.22
0600	For coloring and special finish, add, minimum	J-1	685	.058	S.Y.	.40	1.41	.19	2	2.97
0700	Maximum		200	.200	"	1.39	4.84	.63	6.86	10.25
1001	Exterior stucco, with bonding agent, 3 coats, on walls		1800	.022	S.F.	.36	.54	.07	.97	1.37
1201	Ceilings		1620	.025		.36	.60	.08	1.04	1.48
1301	Beams		720	.056		.36	1.35	.18	1.89	2.81
1501	Columns		900	.044		.36	1.08	.14	1.58	2.32
1601	Mesh, painted, nailed to wood, 1.8 lb.	1 Lath	540	.015		.64	.37		1.01	1.30
1801	3.6 lb.		495	.016		.48	.40		.88	1.18
1901	Wired to steel, painted, 1.8 lb.		477	.017		.64	.42		1.06	1.38
2101	3.6 lb.		450	.018		.48	.44		.92	1.25

09 26 Veneer Plastering

09 26 13 – Gypsum Veneer Plastering

09 26 13.20 Blueboard

		Crew	Daily Output	Labor-Hours	Unit	Material	2009 Bare Costs Labor	Equipment	Total	Total Incl O&P
0010	**BLUEBOARD** For use with thin coat									
0100	plaster application (see Div. 09 26 13.80)									
1000	3/8" thick, on walls or ceilings, standard, no finish included	2 Carp	1900	.008	S.F.	.28	.24		.52	.71
1100	With thin coat plaster finish		875	.018		.36	.51		.87	1.26
1400	On beams, columns, or soffits, standard, no finish included		675	.024		.32	.66		.98	1.47
1450	With thin coat plaster finish		475	.034		.40	.94		1.34	2.03
3000	1/2" thick, on walls or ceilings, standard, no finish included		1900	.008		.33	.24		.57	.76
3100	With thin coat plaster finish		875	.018		.41	.51		.92	1.31
3300	Fire resistant, no finish included		1900	.008		.33	.24		.57	.76
3400	With thin coat plaster finish		875	.018		.41	.51		.92	1.31
3450	On beams, columns, or soffits, standard, no finish included		675	.024		.38	.66		1.04	1.54
3500	With thin coat plaster finish		475	.034		.46	.94		1.40	2.10
3700	Fire resistant, no finish included		675	.024		.38	.66		1.04	1.54
3800	With thin coat plaster finish		475	.034		.46	.94		1.40	2.10
5000	5/8" thick, on walls or ceilings, fire resistant, no finish included		1900	.008		.34	.24		.58	.77
5100	With thin coat plaster finish		875	.018		.42	.51		.93	1.32
5500	On beams, columns, or soffits, no finish included		675	.024		.39	.66		1.05	1.55
5600	With thin coat plaster finish		475	.034		.47	.94		1.41	2.11
6000	For high ceilings, over 8' high, add		3060	.005			.15		.15	.25
6500	For over 3 stories high, add per story		6100	.003			.07		.07	.12

09 26 13.80 Thin Coat Plaster

		Crew	Daily Output	Labor-Hours	Unit	Material	2009 Bare Costs Labor	Equipment	Total	Total Incl O&P
0010	**THIN COAT PLASTER** R092000-50									
0012	1 coat veneer, not incl. lath	J-1	3600	.011	S.F.	.08	.27	.04	.39	.57
1000	In 50 lb. bags				Bag	11.10			11.10	12.20

09 28 Backing Boards and Underlayments

09 28 13 – Cementitious Backing Boards

09 28 13.10 Cementitious Backerboard	Crew	Daily Output	Labor-Hours	Unit	Material	2009 Bare Costs Labor	Equipment	Total	Total Incl O&P
0010	**CEMENTITIOUS BACKERBOARD**									
0070	Cementitious backerboard, on floor, 3' x 4'x 1/2" sheets	2 Carp	525	.030	S.F.	.79	.85		1.64	2.30
0080	3' x 5' x 1/2" sheets		525	.030		.78	.85		1.63	2.29
0090	3' x 6' x 1/2" sheets		525	.030		.72	.85		1.57	2.23
0100	3' x 4'x 5/8" sheets		525	.030		1.04	.85		1.89	2.58
0110	3' x 5' x 5/8" sheets		525	.030		1.03	.85		1.88	2.57
0120	3' x 6' x 5/8" sheets		525	.030		.96	.85		1.81	2.50
0150	On wall, 3' x 4'x 1/2" sheets		350	.046		.79	1.28		2.07	3.02
0160	3' x 5' x 1/2" sheets		350	.046		.78	1.28		2.06	3.01
0170	3' x 6' x 1/2" sheets		350	.046		.72	1.28		2	2.95
0180	3' x 4'x 5/8" sheets		350	.046		1.04	1.28		2.32	3.30
0190	3' x 5' x 5/8" sheets		350	.046		1.03	1.28		2.31	3.29
0200	3' x 6' x 5/8" sheets		350	.046		.96	1.28		2.24	3.22
0250	On counter, 3' x 4'x 1/2" sheets		180	.089		.79	2.48		3.27	5.05
0260	3' x 5' x 1/2" sheets		180	.089		.78	2.48		3.26	5.05
0270	3' x 6' x 1/2" sheets		180	.089		.72	2.48		3.20	4.99
0300	3' x 4'x 5/8" sheets		180	.089		1.04	2.48		3.52	5.35
0310	3' x 5' x 5/8" sheets		180	.089		1.03	2.48		3.51	5.35
0320	3' x 6' x 5/8" sheets		180	.089		.96	2.48		3.44	5.25

09 29 Gypsum Board

09 29 10 – Gypsum Board Panels

09 29 10.20 Taping and Finishing

		Crew	Daily Output	Labor-Hours	Unit	Material	2009 Bare Costs Labor	Equipment	Total	Total Incl O&P
0010	**TAPING AND FINISHING**									
3600	For taping and finishing joints, add	2 Carp	2000	.008	S.F.	.04	.22		.26	.43
4500	For thin coat plaster instead of taping, add	J-1	3600	.011	"	.08	.27	.04	.39	.57

09 29 10.30 Gypsum Board

		Crew	Daily Output	Labor-Hours	Unit	Material	2009 Bare Costs Labor	Equipment	Total	Total Incl O&P
0010	**GYPSUM BOARD** on walls & ceilings R092910-10									
0100	Nailed or screwed to studs unless otherwise noted									
0150	3/8" thick, on walls, standard, no finish included	2 Carp	2000	.008	S.F.	.31	.22		.53	.72
0200	On ceilings, standard, no finish included		1800	.009		.31	.25		.56	.76
0250	On beams, columns, or soffits, no finish included		675	.024		.31	.66		.97	1.46
0300	1/2" thick, on walls, standard, no finish included		2000	.008		.31	.22		.53	.72
0350	Taped and finished (level 4 finish)		965	.017		.35	.46		.81	1.17
0390	With compound skim coat (level 5 finish)		775	.021		.40	.58		.98	1.42
0400	Fire resistant, no finish included		2000	.008		.32	.22		.54	.73
0450	Taped and finished (level 4 finish)		965	.017		.36	.46		.82	1.18
0490	With compound skim coat (level 5 finish)		775	.021		.41	.58		.99	1.43
0500	Water resistant, no finish included		2000	.008		.48	.22		.70	.91
0550	Taped and finished (level 4 finish)		965	.017		.52	.46		.98	1.35
0590	With compound skim coat (level 5 finish)		775	.021		.57	.58		1.15	1.60
0600	Prefinished, vinyl, clipped to studs		900	.018		.83	.50		1.33	1.75
0700	Mold resistant, no finish included		2000	.008		.49	.22		.71	.92
0710	Taped and finished (level 4 finish)		965	.017		.53	.46		.99	1.37
0720	With compound skim coat (level 5 finish)		775	.021		.58	.58		1.16	1.61
1000	On ceilings, standard, no finish included		1800	.009		.31	.25		.56	.76
1050	Taped and finished (level 4 finish)		765	.021		.35	.58		.93	1.38
1090	With compound skim coat (level 5 finish)		610	.026		.40	.73		1.13	1.68
1100	Fire resistant, no finish included		1800	.009		.32	.25		.57	.77
1150	Taped and finished (level 4 finish)		765	.021		.36	.58		.94	1.39
1195	With compound skim coat (level 5 finish)		610	.026		.41	.73		1.14	1.69

09 29 Gypsum Board

09 29 10 – Gypsum Board Panels

09 29 10.30 Gypsum Board

	Crew	Daily Output	Labor-Hours	Unit	Material	2009 Bare Costs Labor	Equipment	Total	Total Incl O&P
1200	Water resistant, no finish included	2 Carp	1800	.009	S.F.	.48	.25		.73	.95
1250	Taped and finished (level 4 finish)		765	.021		.52	.58		1.10	1.56
1290	With compound skim coat (level 5 finish)		610	.026		.57	.73		1.30	1.86
1310	Mold resistant, no finish included		1800	.009		.49	.25		.74	.96
1320	Taped and finished (level 4 finish)		765	.021		.53	.58		1.11	1.58
1330	With compound skim coat (level 5 finish)		610	.026		.58	.73		1.31	1.87
1500	On beams, columns, or soffits, standard, no finish included		675	.024		.36	.66		1.02	1.51
1550	Taped and finished (level 4 finish)		475	.034		.35	.94		1.29	1.98
1590	With compound skim coat (level 5 finish)		540	.030		.40	.83		1.23	1.84
1600	Fire resistant, no finish included		675	.024		.32	.66		.98	1.47
1650	Taped and finished (level 4 finish)		475	.034		.36	.94		1.30	1.99
1690	With compound skim coat (level 5 finish)		540	.030		.41	.83		1.24	1.85
1700	Water resistant, no finish included		675	.024		.55	.66		1.21	1.73
1750	Taped and finished (level 4 finish)		475	.034		.52	.94		1.46	2.16
1790	With compound skim coat (level 5 finish)		540	.030		.57	.83		1.40	2.02
1800	Mold resistant, no finish included		675	.024		.56	.66		1.22	1.74
1810	Taped and finished (level 4 finish)		475	.034		.53	.94		1.47	2.18
1820	With compound skim coat (level 5 finish)		540	.030		.58	.83		1.41	2.03
2000	5/8" thick, on walls, standard, no finish included		2000	.008		.34	.22		.56	.75
2050	Taped and finished (level 4 finish)		965	.017		.38	.46		.84	1.20
2090	With compound skim coat (level 5 finish)		775	.021		.43	.58		1.01	1.45
2100	Fire resistant, no finish included		2000	.008		.36	.22		.58	.78
2150	Taped and finished (level 4 finish)		965	.017		.40	.46		.86	1.22
2195	With compound skim coat (level 5 finish)		775	.021		.45	.58		1.03	1.47
2200	Water resistant, no finish included		2000	.008		.47	.22		.69	.90
2250	Taped and finished (level 4 finish)		965	.017		.51	.46		.97	1.34
2290	With compound skim coat (level 5 finish)		775	.021		.56	.58		1.14	1.59
2300	Prefinished, vinyl, clipped to studs		900	.018		1.02	.50		1.52	1.96
2510	Mold resistant, no finish included		2000	.008		.60	.22		.82	1.04
2520	Taped and finished (level 4 finish)		965	.017		.64	.46		1.10	1.49
2530	With compound skim coat (level 5 finish)		775	.021		.69	.58		1.27	1.74
3000	On ceilings, standard, no finish included		1800	.009		.34	.25		.59	.79
3050	Taped and finished (level 4 finish)		765	.021		.38	.58		.96	1.41
3090	With compound skim coat (level 5 finish)		615	.026		.43	.73		1.16	1.70
3100	Fire resistant, no finish included		1800	.009		.36	.25		.61	.82
3150	Taped and finished (level 4 finish)		765	.021		.40	.58		.98	1.43
3190	With compound skim coat (level 5 finish)		615	.026		.45	.73		1.18	1.72
3200	Water resistant, no finish included		1800	.009		.47	.25		.72	.94
3250	Taped and finished (level 4 finish)		765	.021		.51	.58		1.09	1.55
3290	With compound skim coat (level 5 finish)		615	.026		.56	.73		1.29	1.84
3300	Mold resistant, no finish included		1800	.009		.60	.25		.85	1.08
3310	Taped and finished (level 4 finish)		765	.021		.64	.58		1.22	1.70
3320	With compound skim coat (level 5 finish)		615	.026		.69	.73		1.42	1.99
3500	On beams, columns, or soffits, no finish included		675	.024		.39	.66		1.05	1.55
3550	Taped and finished (level 4 finish)		475	.034		.44	.94		1.38	2.07
3590	With compound skim coat (level 5 finish)		380	.042		.49	1.18		1.67	2.53
3600	Fire resistant, no finish included		675	.024		.41	.66		1.07	1.58
3650	Taped and finished (level 4 finish)		475	.034		.46	.94		1.40	2.10
3690	With compound skim coat (level 5 finish)		380	.042		.45	1.18		1.63	2.48
3700	Water resistant, no finish included		675	.024		.54	.66		1.20	1.71
3750	Taped and finished (level 4 finish)		475	.034		.59	.94		1.53	2.24
3790	With compound skim coat (level 5 finish)		380	.042		.56	1.18		1.74	2.60
3800	Mold resistant, no finish included		675	.024		.69	.66		1.35	1.88

09 29 Gypsum Board

09 29 10 – Gypsum Board Panels

09 29 10.30 Gypsum Board

		Crew	Daily Output	Labor-Hours	Unit	Material	2009 Bare Costs Labor	Equipment	Total	Total Incl O&P
3810	Taped and finished (level 4 finish)	2 Carp	475	.034	S.F.	.74	.94		1.68	2.40
3820	With compound skim coat (level 5 finish)		380	.042		.69	1.18		1.87	2.75
4000	Fireproofing, beams or columns, 2 layers, 1/2" thick, incl finish		330	.048		.68	1.36		2.04	3.04
4010	Mold resistant		330	.048		1.02	1.36		2.38	3.41
4050	5/8" thick		300	.053		.80	1.49		2.29	3.41
4060	Mold resistant		300	.053		1.28	1.49		2.77	3.93
4100	3 layers, 1/2" thick		225	.071		1	1.99		2.99	4.46
4110	Mold resistant		225	.071		1.51	1.99		3.50	5
4150	5/8" thick		210	.076		1.21	2.13		3.34	4.93
4160	Mold resistant		210	.076		1.93	2.13		4.06	5.70
5200	For work over 8' high, add		3060	.005			.15		.15	.25
5270	For textured spray, add	2 Lath	1600	.010		.08	.25		.33	.49
5350	For finishing inner corners, add	2 Carp	950	.017	L.F.	.09	.47		.56	.90
5355	For finishing outer corners, add	"	1250	.013		.21	.36		.57	.83
5500	For acoustical sealant, add per bead	1 Carp	500	.016		.04	.45		.49	.81
5550	Sealant, 1 quart tube				Ea.	6.85			6.85	7.55
5600	Sound deadening board, 1/4" gypsum	2 Carp	1800	.009	S.F.	.37	.25		.62	.83
5650	1/2" wood fiber	"	1800	.009	"	.28	.25		.53	.73

09 29 15 – Gypsum Board Accessories

09 29 15.10 Accessories, Drywall

		Crew	Daily Output	Labor-Hours	Unit	Material	2009 Bare Costs Labor	Equipment	Total	Total Incl O&P
0011	**ACCESSORIES, DRYWALL** Casing bead, galvanized steel	1 Carp	290	.028	L.F.	.22	.77		.99	1.54
0101	Vinyl		290	.028		.22	.77		.99	1.55
0401	Corner bead, galvanized steel, 1-1/4" x 1-1/4"		350	.023		.18	.64		.82	1.28
0411	1-1/4" x 1-1/4", 10' long		35	.229	Ea.	1.82	6.40		8.22	12.80
0601	Vinyl corner bead		400	.020	L.F.	.18	.56		.74	1.15
0901	Furring channel, galv. steel, 7/8" deep, standard		260	.031		.28	.86		1.14	1.75
1001	Resilient		260	.031		.27	.86		1.13	1.74
1101	J trim, galvanized steel, 1/2" wide		300	.027			.75		.75	1.26
1121	5/8" wide		300	.027		.25	.75		1	1.53
1160	Screws #6 x 1" A				M	7.20			7.20	7.90
1170	#6 x 1-5/8" A				"	10.55			10.55	11.60
1501	Z stud, galvanized steel, 1-1/2" wide	1 Carp	260	.031	L.F.	.37	.86		1.23	1.86

09 30 Tiling

09 30 13 – Ceramic Tiling

09 30 13.10 Ceramic Tile

		Crew	Daily Output	Labor-Hours	Unit	Material	2009 Bare Costs Labor	Equipment	Total	Total Incl O&P
0010	**CERAMIC TILE**									
0050	Base, using 1' x 4" high pc. with 1" x 1" tiles, mud set	D-7	82	.195	L.F.	4.95	4.65		9.60	12.90
0100	Thin set	"	128	.125		4.53	2.98		7.51	9.75
0300	For 6" high base, 1" x 1" tile face, add					.72			.72	.79
0400	For 2" x 2" tile face, add to above					.38			.38	.42
0600	Cove base, 4-1/4" x 4-1/4" high, mud set	D-7	91	.176		3.93	4.19		8.12	11.10
0700	Thin set		128	.125		4.07	2.98		7.05	9.25
0900	6" x 4-1/4" high, mud set		100	.160		3.67	3.82		7.49	10.15
1000	Thin set		137	.117		3.47	2.79		6.26	8.30
1200	Sanitary cove base, 6" x 4-1/4" high, mud set		93	.172		4	4.10		8.10	11
1300	Thin set		124	.129		3.84	3.08		6.92	9.15
1500	6" x 6" high, mud set		84	.190		4.94	4.54		9.48	12.75
1600	Thin set		117	.137		4.74	3.26		8	10.45
1800	Bathroom accessories, average		82	.195	Ea.	10.70	4.65		15.35	19.20

09 30 13 – Ceramic Tiling

09 30 13.10 Ceramic Tile		Crew	Daily Output	Labor-Hours	Unit	Material	2009 Bare Costs Labor	Equipment	Total	Total Incl O&P
1900	Bathtub, 5', rec. 4-1/4" x 4-1/4" tile wainscot, adhesive set 6' high	D-7	2.90	5.517	Ea.	156	132		288	385
2100	7' high wainscot		2.50	6.400		179	153		332	440
2200	8' high wainscot		2.20	7.273		190	173		363	485
2400	Bullnose trim, 4-1/4" x 4-1/4", mud set		82	.195	L.F.	3.68	4.65		8.33	11.50
2500	Thin set		128	.125		3.30	2.98		6.28	8.40
2700	6" x 4-1/4" bullnose trim, mud set		84	.190		2.88	4.54		7.42	10.45
2800	Thin set		124	.129		2.74	3.08		5.82	7.95
3000	Floors, natural clay, random or uniform, thin set, color group 1		183	.087	S.F.	4.34	2.09		6.43	8.10
3100	Color group 2		183	.087		4.66	2.09		6.75	8.50
3255	Floors, glazed, thin set, 6" x 6", color group 1		200	.080		3.73	1.91		5.64	7.15
3260	8" x 8" tile		250	.064		3.73	1.53		5.26	6.55
3270	12" x 12" tile		325	.049		4.62	1.17		5.79	7
3280	16" x 16" tile		550	.029		6.30	.69		6.99	8.05
3285	Border, 6" x 12" tile		275	.058		11.70	1.39		13.09	15.10
3290	3" x 12" tile		200	.080		34	1.91		35.91	40.50
3300	Porcelain type, 1 color, color group 2, 1" x 1"		183	.087		4.71	2.09		6.80	8.55
3310	2" x 2" or 2" x 1", thin set		190	.084		5.40	2.01		7.41	9.15
3350	For random blend, 2 colors, add					.88			.88	.97
3360	4 colors, add					1.24			1.24	1.36
4300	Specialty tile, 4-1/4" x 4-1/4" x 1/2", decorator finish	D-7	183	.087		10	2.09		12.09	14.35
4500	Add for epoxy grout, 1/16" joint, 1" x 1" tile		800	.020		.62	.48		1.10	1.45
4600	2" x 2" tile		820	.020		.59	.47		1.06	1.40
4800	Pregrouted sheets, walls, 4-1/4" x 4-1/4", 6" x 4-1/4"									
4810	and 8-1/2" x 4-1/4", 4 S.F. sheets, silicone grout	D-7	240	.067	S.F.	4.73	1.59		6.32	7.75
5100	Floors, unglazed, 2 S.F. sheets,									
5110	Urethane adhesive	D-7	180	.089	S.F.	4.71	2.12		6.83	8.60
5400	Walls, interior, thin set, 4-1/4" x 4-1/4" tile		190	.084		2.49	2.01		4.50	5.95
5500	6" x 4-1/4" tile		190	.084		2.86	2.01		4.87	6.35
5700	8-1/2" x 4-1/4" tile		190	.084		3.93	2.01		5.94	7.55
5800	6" x 6" tile		200	.080		3.34	1.91		5.25	6.75
5810	8" x 8" tile		225	.071		4.36	1.70		6.06	7.50
5820	12" x 12" tile		300	.053		3.62	1.27		4.89	6
5830	16" x 16" tile		500	.032		3.90	.76		4.66	5.50
6000	Decorated wall tile, 4-1/4" x 4-1/4", minimum		270	.059		3.66	1.41		5.07	6.30
6100	Maximum		180	.089		42	2.12		44.12	49.50
6600	Crystalline glazed, 4-1/4" x 4-1/4", mud set, plain		100	.160		4.43	3.82		8.25	10.95
6700	4-1/4" x 4-1/4", scored tile		100	.160		5.25	3.82		9.07	11.85
6900	6"x 6" plain		93	.172		5.65	4.10		9.75	12.80
7000	For epoxy grout, 1/16" joints, 4-1/4" tile, add		800	.020		.39	.48		.87	1.20
7200	For tile set in dry mortar, add		1735	.009			.22		.22	.35
7300	For tile set in Portland cement mortar, add		290	.055		.27	1.32		1.59	2.41
9300	Ceramic tiles, recycled glass, standard colors, 2" x 2" thru 6" x 6" [G]		190	.084		16.70	2.01		18.71	21.50
9310	6" x 6" [G]		200	.080		16.70	1.91		18.61	21.50
9320	8" x 8" [G]		225	.071		18.10	1.70		19.80	22.50
9330	12"x12" [G]		300	.053		18.10	1.27		19.37	22
9340	Earthtones, 2" x 2" to 4" x 8" [G]		190	.084		21	2.01		23.01	26
9350	6" x 6" [G]		200	.080		21	1.91		22.91	26
9360	8" x 8" [G]		225	.071		22.50	1.70		24.20	27
9370	12" x 12" [G]		300	.053		22.50	1.27		23.77	26.50
9380	Deep colors, 2" x 2" to 4" x 8" [G]		190	.084		30	2.01		32.01	36
9390	6" x 6" [G]		200	.080		30	1.91		31.91	36
9400	8" x 8" [G]		225	.071		31.50	1.70		33.20	37.50
9410	12" x1 2" [G]		300	.053		31.50	1.27		32.77	37

09 30 Tiling

09 30 16 – Quarry Tiling

09 30 16.10 Quarry Tile

		Crew	Daily Output	Labor-Hours	Unit	Material	2009 Bare Costs Labor	Equipment	Total	Total Incl O&P
0010	**QUARRY TILE**									
0100	Base, cove or sanitary, mud set, to 5" high, 1/2" thick	D-7	110	.145	L.F.	5.60	3.47		9.07	11.70
0300	Bullnose trim, red, mud set, 6" x 6" x 1/2" thick		120	.133		4.36	3.18		7.54	9.90
0400	4" x 4" x 1/2" thick		110	.145		4.79	3.47		8.26	10.80
0600	4" x 8" x 1/2" thick, using 8" as edge		130	.123		4.25	2.94		7.19	9.40
0700	Floors, mud set, 1,000 S.F. lots, red, 4" x 4" x 1/2" thick		120	.133	S.F.	7.50	3.18		10.68	13.35
0900	6" x 6" x 1/2" thick		140	.114		7.05	2.73		9.78	12.10
1000	4" x 8" x 1/2" thick		130	.123		7.50	2.94		10.44	12.95
1300	For waxed coating, add					.67			.67	.74
1500	For non-standard colors, add					.40			.40	.44
1600	For abrasive surface, add					.47			.47	.52
1800	Brown tile, imported, 6" x 6" x 3/4"	D-7	120	.133		8.45	3.18		11.63	14.35
1900	8" x 8" x 1"		110	.145		9.10	3.47		12.57	15.55
2100	For thin set mortar application, deduct		700	.023			.55		.55	.87
2700	Stair tread, 6" x 6" x 3/4", plain		50	.320		5.25	7.65		12.90	18.05
2800	Abrasive		47	.340		5.30	8.10		13.40	18.85
3000	Wainscot, 6" x 6" x 1/2", thin set, red		105	.152		4.24	3.63		7.87	10.50
3100	Non-standard colors		105	.152		4.69	3.63		8.32	11
3300	Window sill, 6" wide, 3/4" thick		90	.178	L.F.	5.25	4.24		9.49	12.60
3400	Corners		80	.200	Ea.	5.85	4.77		10.62	14.10

09 30 29 – Metal Tiling

09 30 29.10 Metal Tile

		Crew	Daily Output	Labor-Hours	Unit	Material	2009 Bare Costs Labor	Equipment	Total	Total Incl O&P
0010	**METAL TILE** 4' x 4' sheet, 24 ga., tile pattern, nailed									
0200	Stainless steel	2 Carp	512	.031	S.F.	24.50	.87		25.37	28.50
0400	Aluminized steel	"	512	.031	"	13.15	.87		14.02	15.95

09 51 Acoustical Ceilings

09 51 23 – Acoustical Tile Ceilings

09 51 23.10 Suspended Acoustic Ceiling Tiles

		Crew	Daily Output	Labor-Hours	Unit	Material	2009 Bare Costs Labor	Equipment	Total	Total Incl O&P
0010	**SUSPENDED ACOUSTIC CEILING TILES**, not including									
0100	suspension system									
0300	Fiberglass boards, film faced, 2' x 2' or 2' x 4', 5/8" thick	1 Carp	625	.013	S.F.	.73	.36		1.09	1.40
0400	3/4" thick		600	.013		1.69	.37		2.06	2.49
0500	3" thick, thermal, R11		450	.018		1.53	.50		2.03	2.52
0600	Glass cloth faced fiberglass, 3/4" thick		500	.016		3.04	.45		3.49	4.10
0700	1" thick		485	.016		2.16	.46		2.62	3.16
0820	1-1/2" thick, nubby face		475	.017		2.66	.47		3.13	3.73
1110	Mineral fiber tile, lay-in, 2' x 2' or 2' x 4', 5/8" thick, fine texture		625	.013		.57	.36		.93	1.23
1115	Rough textured		625	.013		1.26	.36		1.62	1.99
1125	3/4" thick, fine textured		600	.013		1.56	.37		1.93	2.35
1130	Rough textured		600	.013		1.88	.37		2.25	2.70
1135	Fissured		600	.013		2.45	.37		2.82	3.33
1150	Tegular, 5/8" thick, fine textured		470	.017		1.45	.48		1.93	2.40
1155	Rough textured		470	.017		1.90	.48		2.38	2.89
1165	3/4" thick, fine textured		450	.018		2.08	.50		2.58	3.13
1170	Rough textured		450	.018		2.35	.50		2.85	3.43
1175	Fissured		450	.018		3.63	.50		4.13	4.83
1180	For aluminum face, add					6.05			6.05	6.65
1185	For plastic film face, add					.94			.94	1.03
1190	For fire rating, add					.45			.45	.50

09 51 Acoustical Ceilings

09 51 23 - Acoustical Tile Ceilings

09 51 23.10 Suspended Acoustic Ceiling Tiles

		Crew	Daily Output	Labor-Hours	Unit	Material	2009 Bare Costs Labor	Equipment	Total	Total Incl O&P
1300	Mirror faced panels, 15/16" thick, 2' x 2'	1 Carp	500	.016	S.F.	10.45	.45		10.90	12.25
1900	Eggcrate, acrylic, 1/2" x 1/2" x 1/2" cubes		500	.016		1.82	.45		2.27	2.76
2100	Polystyrene eggcrate, 3/8" x 3/8" x 1/2" cubes		510	.016		1.53	.44		1.97	2.42
2200	1/2" x 1/2" x 1/2" cubes		500	.016		2.04	.45		2.49	3
2400	Luminous panels, prismatic, acrylic		400	.020		2.22	.56		2.78	3.39
2500	Polystyrene		400	.020		1.14	.56		1.70	2.20
2700	Flat white acrylic		400	.020		3.86	.56		4.42	5.20
2800	Polystyrene		400	.020		2.65	.56		3.21	3.87
3000	Drop pan, white, acrylic		400	.020		5.65	.56		6.21	7.20
3100	Polystyrene		400	.020		4.73	.56		5.29	6.15
3600	Perforated aluminum sheets, .024" thick, corrugated, painted		490	.016		2.25	.46		2.71	3.25
3700	Plain		500	.016		3.76	.45		4.21	4.90
3750	Wood fiber in cementitious binder, 2' x 2' or 4', painted, 1" thick		600	.013		1.54	.37		1.91	2.32
3760	2" thick		550	.015		2.53	.41		2.94	3.47
3770	2-1/2" thick		500	.016		3.47	.45		3.92	4.58
3780	3" thick		450	.018		3.90	.50		4.40	5.15

09 51 23.30 Suspended Ceilings, Complete

		Crew	Daily Output	Labor-Hours	Unit	Material	2009 Bare Costs Labor	Equipment	Total	Total Incl O&P
0010	**SUSPENDED CEILINGS, COMPLETE** Including standard									
0100	suspension system but not incl. 1-1/2" carrier channels									
0600	Fiberglass ceiling board, 2' x 4' x 5/8", plain faced	1 Carp	500	.016	S.F.	1.36	.45		1.81	2.26
0700	Offices, 2' x 4' x 3/4"		380	.021		2.32	.59		2.91	3.54
1800	Tile, Z bar suspension, 5/8" mineral fiber tile		150	.053		2.16	1.49		3.65	4.90
1900	3/4" mineral fiber tile		150	.053		2.31	1.49		3.80	5.05

09 51 53 - Direct-Applied Acoustical Ceilings

09 51 53.10 Ceiling Tile

		Crew	Daily Output	Labor-Hours	Unit	Material	2009 Bare Costs Labor	Equipment	Total	Total Incl O&P
0010	**CEILING TILE**, Stapled or cemented									
0100	12" x 12" or 12" x 24", not including furring									
0600	Mineral fiber, vinyl coated, 5/8" thick	1 Carp	300	.027	S.F.	1.85	.75		2.60	3.30
0700	3/4" thick		300	.027		1.87	.75		2.62	3.32
0900	Fire rated, 3/4" thick, plain faced		300	.027		1.40	.75		2.15	2.80
1000	Plastic coated face		300	.027		1.18	.75		1.93	2.56
1200	Aluminum faced, 5/8" thick, plain		300	.027		1.51	.75		2.26	2.92
3300	For flameproofing, add					.10			.10	.11
3400	For sculptured 3 dimensional, add					.29			.29	.32
3900	For ceiling primer, add					.13			.13	.14
4000	For ceiling cement, add					.37			.37	.41

09 53 Acoustical Ceiling Suspension Assemblies

09 53 23 - Metal Acoustical Ceiling Suspension Assemblies

09 53 23.30 Ceiling Suspension Systems

		Crew	Daily Output	Labor-Hours	Unit	Material	2009 Bare Costs Labor	Equipment	Total	Total Incl O&P
0010	**CEILING SUSPENSION SYSTEMS** for boards and tile									
0050	Class A suspension system, 15/16" T bar, 2' x 4' grid	1 Carp	800	.010	S.F.	.63	.28		.91	1.16
0300	2' x 2' grid	"	650	.012		.79	.34		1.13	1.44
0350	For 9/16" grid, add					.16			.16	.18
0360	For fire rated grid, add					.09			.09	.10
0370	For colored grid, add					.21			.21	.23
0400	Concealed Z bar suspension system, 12" module	1 Carp	520	.015		.65	.43		1.08	1.45
0600	1-1/2" carrier channels, 4' O.C., add		470	.017		.15	.48		.63	.97
0650	1-1/2" x 3-1/2" channels		470	.017		.38	.48		.86	1.22
0700	Carrier channels for ceilings with									

09 53 Acoustical Ceiling Suspension Assemblies

09 53 23 – Metal Acoustical Ceiling Suspension Assemblies

09 53 23.30 Ceiling Suspension Systems	Crew	Daily Output	Labor-Hours	Unit	Material	2009 Bare Costs Labor	Equipment	Total	Total Incl O&P
0900	recessed lighting fixtures, add	1 Carp	460	.017	S.F.	.28	.49		.77	1.12
5000	Wire hangers, #12 wire	"	300	.027	Ea.	1.23	.75		1.98	2.61

09 63 Masonry Flooring

09 63 13 – Brick Flooring

09 63 13.10 Miscellaneous Brick Flooring

		Crew	Daily Output	Labor-Hours	Unit	Material	Labor	Equipment	Total	Total Incl O&P
0010	**MISCELLANEOUS BRICK FLOORING**									
0020	Acid proof shales, red, 8" x 3-3/4" x 1-1/4" thick	D-7	.43	37.209	M	820	885		1,705	2,325
0050	2-1/4" thick	D-1	.40	40		950	1,025		1,975	2,725
0200	Acid proof clay brick, 8" x 3-3/4" x 2-1/4" thick	"	.40	40		870	1,025		1,895	2,625
0260	Cast ceramic, pressed, 4" x 8" x 1/2", unglazed	D-7	100	.160	S.F.	5.95	3.82		9.77	12.65
0270	Glazed		100	.160		7.95	3.82		11.77	14.80
0280	Hand molded flooring, 4" x 8" x 3/4", unglazed		95	.168		7.85	4.02		11.87	15.10
0290	Glazed		95	.168		9.85	4.02		13.87	17.30
0300	8" hexagonal, 3/4" thick, unglazed		85	.188		8.60	4.49		13.09	16.65
0310	Glazed		85	.188		15.55	4.49		20.04	24.50
0450	Acid proof joints, 1/4" wide	D-1	65	.246		1.37	6.25		7.62	11.90
0500	Pavers, 8" x 4", 1" to 1-1/4" thick, red	D-7	95	.168		3.47	4.02		7.49	10.25
0510	Ironspot	"	95	.168		4.89	4.02		8.91	11.85
0540	1-3/8" to 1-3/4" thick, red	D-1	95	.168		3.34	4.28		7.62	10.75
0560	Ironspot		95	.168		4.84	4.28		9.12	12.40
0580	2-1/4" thick, red		90	.178		3.40	4.52		7.92	11.25
0590	Ironspot		90	.178		5.30	4.52		9.82	13.30
0700	Paver, adobe brick, 6" x 12", 1/2" joint		42	.381		.90	9.70		10.60	17.05
0710	Mexican red, 12" x 12"	1 Tilf	48	.167		1.61	4.44		6.05	8.90
0720	Saltillo, 12" x 12"	"	48	.167		1.13	4.44		5.57	8.40
0800	For sidewalks and patios with pavers, see Div. 32 14 16.10									
0870	For epoxy joints, add	D-1	600	.027	S.F.	2.59	.68		3.27	3.97
0880	For Furan underlayment, add	"	600	.027		2.14	.68		2.82	3.47
0890	For waxed surface, steam cleaned, add	A-1H	1000	.008		.19	.16	.05	.40	.55

09 63 40 – Stone Flooring

09 63 40.10 Marble

		Crew	Daily Output	Labor-Hours	Unit	Material	Labor	Equipment	Total	Total Incl O&P
0010	**MARBLE**									
0020	Thin gauge tile, 12" x 6", 3/8", white Carara	D-7	60	.267	S.F.	11.15	6.35		17.50	22.50
0100	Travertine		60	.267		12.20	6.35		18.55	23.50
0200	12" x 12" x 3/8", thin set, floors		60	.267		6.05	6.35		12.40	16.85
0300	On walls		52	.308		9.35	7.35		16.70	22
1000	Marble threshold, 4" wide x 36" long x 5/8" thick, white		60	.267	Ea.	6.05	6.35		12.40	16.85

09 63 40.20 Slate Tile

		Crew	Daily Output	Labor-Hours	Unit	Material	Labor	Equipment	Total	Total Incl O&P
0010	**SLATE TILE**									
0020	Vermont, 6" x 6" x 1/4" thick, thin set	D-7	180	.089	S.F.	6.80	2.12		8.92	10.85

09 64 23 – Wood Parquet Flooring

09 64 23.10 Wood Parquet		Crew	Daily Output	Labor-Hours	Unit	Material	2009 Bare Costs Labor	Equipment	Total	Total Incl O&P
0010	**WOOD PARQUET** flooring									
5200	Parquetry, standard, 5/16" thick, not incl. finish, oak, minimum	1 Carp	160	.050	S.F.	4.41	1.40		5.81	7.20
5300	Maximum		100	.080		4.95	2.24		7.19	9.25
5500	Teak, minimum		160	.050		4.79	1.40		6.19	7.60
5600	Maximum		100	.080		8.35	2.24		10.59	13
5650	13/16" thick, select grade oak, minimum		160	.050		9.20	1.40		10.60	12.50
5700	Maximum		100	.080		14	2.24		16.24	19.20
5800	Custom parquetry, including finish, minimum		100	.080		15.45	2.24		17.69	21
5900	Maximum		50	.160		20.50	4.47		24.97	30
6700	Parquetry, prefinished white oak, 5/16" thick, minimum		160	.050		3.61	1.40		5.01	6.35
6800	Maximum		100	.080		7.30	2.24		9.54	11.85
7000	Walnut or teak, parquetry, minimum		160	.050		5.35	1.40		6.75	8.25
7100	Maximum	▼	100	.080	▼	9.35	2.24		11.59	14.10
7200	Acrylic wood parquet blocks, 12" x 12" x 5/16",									
7210	Irradiated, set in epoxy	1 Carp	160	.050	S.F.	7.80	1.40		9.20	10.95

09 64 29 – Wood Strip and Plank Flooring

09 64 29.10 Wood		Crew	Daily Output	Labor-Hours	Unit	Material	2009 Bare Costs Labor	Equipment	Total	Total Incl O&P
0010	**WOOD**									
0020	Fir, vertical grain, 1" x 4", not incl. finish, grade B & better	1 Carp	255	.031	S.F.	2.74	.88		3.62	4.49
0100	C grade & better		255	.031		2.58	.88		3.46	4.32
0300	Flat grain, 1" x 4", not incl. finish, B & better		255	.031		3.13	.88		4.01	4.92
0400	C & better		255	.031		3.01	.88		3.89	4.79
4000	Maple, strip, 25/32" x 2-1/4", not incl. finish, select		170	.047		5.25	1.32		6.57	8
4100	#2 & better		170	.047		3.32	1.32		4.64	5.85
4300	33/32" x 3-1/4", not incl. finish, #1 grade		170	.047		4.19	1.32		5.51	6.85
4400	#2 & better	▼	170	.047	▼	3.73	1.32		5.05	6.30
4600	Oak, white or red, 25/32" x 2-1/4", not incl. finish									
4700	#1 common	1 Carp	170	.047	S.F.	3.18	1.32		4.50	5.70
4900	Select quartered, 2-1/4" wide		170	.047		2.94	1.32		4.26	5.45
5000	Clear		170	.047		4.01	1.32		5.33	6.65
6100	Prefinished, white oak, prime grade, 2-1/4" wide		170	.047		6.40	1.32		7.72	9.25
6200	3-1/4" wide		185	.043		8.20	1.21		9.41	11.05
6400	Ranch plank		145	.055		7.90	1.54		9.44	11.30
6500	Hardwood blocks, 9" x 9", 25/32" thick		160	.050		5.75	1.40		7.15	8.70
7400	Yellow pine, 3/4" x 3-1/8", T & G, C & better, not incl. finish	▼	200	.040		2.35	1.12		3.47	4.48
7500	Refinish wood floor, sand, 2 cts poly, wax, soft wood, min.	1 Clab	400	.020		.78	.41		1.19	1.56
7600	Hard wood, max		130	.062		1.16	1.26		2.42	3.42
7800	Sanding and finishing, 2 coats polyurethane	▼	295	.027	▼	.78	.56		1.34	1.80
7900	Subfloor and underlayment, see Div. 06 16									
8015	Transition molding, 2 1/4" wide, 5' long	1 Carp	19.20	.417	Ea.	15	11.65		26.65	36
8300	Floating floor, wood composition strip, complete	1 Clab	133	.060	S.F.	4.23	1.24		5.47	6.75
8310	Components, T & G wood composite strips					3.66			3.66	4.03
8320	Film					.15			.15	.16
8330	Foam					.30			.30	.33
8340	Adhesive					.17			.17	.18
8350	Installation kit				▼	.17			.17	.19
8360	Trim, 2" wide x 3' long				L.F.	2.49			2.49	2.74
8370	Reducer moulding				"	4.29			4.29	4.72
8600	Flooring, wood, bamboo strips, unfinished, 5/8" x 4" x 3' [G]	1 Carp	255	.031	S.F.	4.44	.88		5.32	6.35
8610	5/8" x 4" x 4' [G]		275	.029		4.61	.81		5.42	6.40
8620	5/8" x 4" x 6' [G]		295	.027		5.05	.76		5.81	6.85
8630	Finished, 5/8" x 4" x 3' [G]	▼	255	.031	▼	4.89	.88		5.77	6.90

09 64 Wood Flooring

09 64 29 – Wood Strip and Plank Flooring

09 64 29.10 Wood		Crew	Daily Output	Labor-Hours	Unit	Material	2009 Bare Costs Labor	Equipment	Total	Total Incl O&P
8640	5/8" x 4" x 4'	G 1 Carp	275	.029	S.F.	5.10	.81		5.91	7
8650	5/8" x 4" x 6'	G	295	.027	↓	5.15	.76		5.91	7
8660	Stair treads, unfinished, 1-1/16" x 11-1/2" x 4'	G	18	.444	Ea.	42.50	12.40		54.90	68
8670	Finished, 1-1/16" x 11-1/2" x 4'	G	18	.444		65	12.40		77.40	92.50
8680	Stair risers, unfinished, 5/8" x 7-1/2" x 4'	G	18	.444		15.75	12.40		28.15	38.50
8690	Finished, 5/8" x 7-1/2" x 4'	G	18	.444		30	12.40		42.40	54
8700	Stair nosing, unfinished, 6' long	G	16	.500		35	14		49	62
8710	Finished, 6' long	G	16	.500	↓	44	14		58	71.50

09 65 Resilient Flooring

09 65 10 – Resilient Tile Underlayment

09 65 10.10 Latex Underlayment

		Crew	Daily Output	Labor-Hours	Unit	Material	2009 Bare Costs Labor	Equipment	Total	Total Incl O&P
0010	**LATEX UNDERLAYMENT**									
3600	Latex underlayment, 1/8" thk., cementitious for resilient flooring	1 Tilf	160	.050	S.F.	1.72	1.33		3.05	4.03
4000	Liquid, fortified				Gal.	38.50			38.50	42

09 65 13 – Resilient Base and Accessories

09 65 13.13 Resilient Base

		Crew	Daily Output	Labor-Hours	Unit	Material	2009 Bare Costs Labor	Equipment	Total	Total Incl O&P
0010	**RESILIENT BASE**									
0800	Base, cove, rubber or vinyl									
1100	Standard colors, 0.080" thick, 2-1/2" high	1 Tilf	315	.025	L.F.	.75	.68		1.43	1.92
1150	4" high		315	.025		.93	.68		1.61	2.11
1200	6" high		315	.025		1.25	.68		1.93	2.47
1450	1/8" thick, 2-1/2" high		315	.025		.75	.68		1.43	1.92
1500	4" high		315	.025		.71	.68		1.39	1.87
1550	6" high		315	.025	↓	1.30	.68		1.98	2.52
1600	Corners, 2-1/2" high		315	.025	Ea.	1.45	.68		2.13	2.69
1630	4" high		315	.025		2.75	.68		3.43	4.12
1660	6" high		315	.025	↓	2.85	.68		3.53	4.23

09 65 16 – Resilient Sheet Flooring

09 65 16.10 Rubber and Vinyl Sheet Flooring

		Crew	Daily Output	Labor-Hours	Unit	Material	2009 Bare Costs Labor	Equipment	Total	Total Incl O&P
0010	**RUBBER AND VINYL SHEET FLOORING**									
5500	Linoleum, sheet goods	1 Tilf	360	.022	S.F.	4.55	.59		5.14	5.95
5900	Rubber, sheet goods, 36" wide, 1/8" thick		120	.067		6.25	1.78		8.03	9.75
5950	3/16" thick		100	.080		8.50	2.13		10.63	12.75
6000	1/4" thick		90	.089		10	2.37		12.37	14.80
8000	Vinyl sheet goods, backed, .065" thick, minimum		250	.032		3.50	.85		4.35	5.20
8050	Maximum		200	.040		3.39	1.07		4.46	5.45
8100	.080" thick, minimum		230	.035		3.50	.93		4.43	5.35
8150	Maximum		200	.040		4.95	1.07		6.02	7.15
8200	.125" thick, minimum		230	.035		4	.93		4.93	5.90
8250	Maximum		200	.040	↓	5.95	1.07		7.02	8.25
8700	Adhesive cement, 1 gallon per 200 to 300 S.F.				Gal.	24.50			24.50	26.50
8800	Asphalt primer, 1 gallon per 300 S.F.					11.65			11.65	12.80
8900	Emulsion, 1 gallon per 140 S.F.				↓	15			15	16.50

09 65 19 – Resilient Tile Flooring

09 65 19.10 Miscellaneous Resilient Tile Flooring

		Crew	Daily Output	Labor-Hours	Unit	Material	2009 Bare Costs Labor	Equipment	Total	Total Incl O&P
0010	**MISCELLANEOUS RESILIENT TILE FLOORING**									
2200	Cork tile, standard finish, 1/8" thick	G 1 Tilf	315	.025	S.F.	5.75	.68		6.43	7.45
2250	3/16" thick	G ↓	315	.025	↓	5.75	.68		6.43	7.45

09 65 Resilient Flooring

09 65 19 – Resilient Tile Flooring

09 65 19.10 Miscellaneous Resilient Tile Flooring		Crew	Daily Output	Labor-Hours	Unit	Material	2009 Bare Costs Labor	Equipment	Total	Total Incl O&P
2300	5/16" thick	G	1 Tilf	315	.025	S.F.	7.15	.68		7.83	8.95
2350	1/2" thick	G		315	.025		8.45	.68		9.13	10.40
2500	Urethane finish, 1/8" thick	G		315	.025		6.85	.68		7.53	8.65
2550	3/16" thick	G		315	.025		7.20	.68		7.88	9
2600	5/16" thick	G		315	.025		9.20	.68		9.88	11.25
2650	1/2" thick	G		315	.025		12.40	.68		13.08	14.75
6050	Rubber tile, marbleized colors, 12" x 12", 1/8" thick			400	.020		6.20	.53		6.73	7.65
6100	3/16" thick			400	.020		9.25	.53		9.78	11.05
6300	Special tile, plain colors, 1/8" thick			400	.020		9.25	.53		9.78	11.05
6350	3/16" thick			400	.020		9.50	.53		10.03	11.30
7000	Vinyl composition tile, 12" x 12", 1/16" thick			500	.016		.87	.43		1.30	1.64
7050	Embossed			500	.016		2.07	.43		2.50	2.96
7100	Marbleized			500	.016		2.07	.43		2.50	2.96
7150	Solid			500	.016		2.66	.43		3.09	3.61
7200	3/32" thick, embossed			500	.016		1.30	.43		1.73	2.11
7250	Marbleized			500	.016		2.37	.43		2.80	3.29
7300	Solid			500	.016		2.20	.43		2.63	3.10
7350	1/8" thick, marbleized			500	.016		1.70	.43		2.13	2.55
7400	Solid			500	.016		2.14	.43		2.57	3.03
7450	Conductive			500	.016		5.35	.43		5.78	6.60
7500	Vinyl tile, 12" x 12", .050" thick, minimum			500	.016		2.95	.43		3.38	3.93
7550	Maximum			500	.016		5.90	.43		6.33	7.20
7600	1/8" thick, minimum			500	.016		4.20	.43		4.63	5.30
7650	Solid colors			500	.016		5.85	.43		6.28	7.15
7700	Marbleized or Travertine pattern			500	.016		5.65	.43		6.08	6.90
7750	Florentine pattern			500	.016		5.70	.43		6.13	7
7800	Maximum			500	.016		12.35	.43		12.78	14.30

09 65 33 – Conductive Resilient Flooring

09 65 33.10 Conductive Rubber and Vinyl Flooring

		Crew	Daily Output	Labor-Hours	Unit	Material	Labor	Equipment	Total	Total Incl O&P
0010	**CONDUCTIVE RUBBER AND VINYL FLOORING**									
1700	Conductive flooring, rubber tile, 1/8" thick	1 Tilf	315	.025	S.F.	5	.68		5.68	6.60
1800	Homogeneous vinyl tile, 1/8" thick	"	315	.025	"	6.95	.68		7.63	8.75

09 66 Terrazzo Flooring

09 66 13 – Portland Cement Terrazzo Flooring

09 66 13.10 Portland Cement Terrazzo

		Crew	Daily Output	Labor-Hours	Unit	Material	Labor	Equipment	Total	Total Incl O&P
0010	**PORTLAND CEMENT TERRAZZO**, cast-in-place									
4300	Stone chips, onyx gemstone, per 50 lb. bag				Bag	16.30			16.30	17.95

09 66 13.30 Terrazzo, Precast

		Crew	Daily Output	Labor-Hours	Unit	Material	Labor	Equipment	Total	Total Incl O&P
0010	**TERRAZZO, PRECAST**									
0020	Base, 6" high, straight	1 Mstz	70	.114	L.F.	10.40	3.02		13.42	16.30
0100	Cove		60	.133		12.15	3.52		15.67	19
0300	8" high, straight		60	.133		10.85	3.52		14.37	17.55
0400	Cove		50	.160		15.95	4.22		20.17	24.50
0600	For white cement, add					.43			.43	.47
0700	For 16 ga. zinc toe strip, add					1.60			1.60	1.76
0900	Curbs, 4" x 4" high	1 Mstz	40	.200		30.50	5.30		35.80	42
1000	8" x 8" high	"	30	.267		34.50	7.05		41.55	49.50
1200	Floor tiles, non-slip, 1" thick, 12" x 12"	D-1	60	.267	S.F.	17.70	6.80		24.50	30.50
1300	1-1/4" thick, 12" x 12"		60	.267		19.60	6.80		26.40	33

185

09 66 Terrazzo Flooring

09 66 13 - Portland Cement Terrazzo Flooring

09 66 13.30 Terrazzo, Precast		Crew	Daily Output	Labor-Hours	Unit	Material	2009 Bare Costs Labor	2009 Bare Costs Equipment	Total	Total Incl O&P
1500	16" x 16"	D-1	50	.320	S.F.	21.50	8.15		29.65	37
1600	1-1/2" thick, 16" x 16"	↓	45	.356		19.45	9.05		28.50	36.50
4800	Wainscot, 12" x 12" x 1" tiles	1 Mstz	12	.667		6.50	17.60		24.10	35
4900	16" x 16" x 1-1/2" tiles	"	8	1	↓	13.70	26.50		40.20	57.50

09 68 Carpeting

09 68 05 - Carpet Accessories

09 68 05.11 Flooring Transition Strip

		Crew	Daily Output	Labor-Hours	Unit	Material	2009 Bare Costs Labor	2009 Bare Costs Equipment	Total	Total Incl O&P
0010	**FLOORING TRANSITION STRIP**									
0107	Clamp down brass divider, 12' strip, vinyl to carpet	1 Tilf	31.25	.256	Ea.	26.50	6.80		33.30	40
0117	Vinyl to hard surface	"	31.25	.256	"	26.50	6.80		33.30	40

09 68 10 - Carpet Pad

09 68 10.10 Commercial Grade Carpet Pad

		Crew	Daily Output	Labor-Hours	Unit	Material	2009 Bare Costs Labor	2009 Bare Costs Equipment	Total	Total Incl O&P
0010	**COMMERCIAL GRADE CARPET PAD**									
9001	Sponge rubber pad, minimum	1 Tilf	1350	.006	S.F.	.47	.16		.63	.76
9101	Maximum		1350	.006		1.11	.16		1.27	1.47
9201	Felt pad, minimum		1350	.006		.45	.16		.61	.74
9301	Maximum		1350	.006		.78	.16		.94	1.11
9401	Bonded urethane pad, minimum		1350	.006		.51	.16		.67	.81
9501	Maximum		1350	.006		.87	.16		1.03	1.20
9601	Prime urethane pad, minimum		1350	.006		.29	.16		.45	.57
9701	Maximum	↓	1350	.006	↓	.54	.16		.70	.84

09 68 13 - Tile Carpeting

09 68 13.10 Carpet Tile

		Crew	Daily Output	Labor-Hours	Unit	Material	2009 Bare Costs Labor	2009 Bare Costs Equipment	Total	Total Incl O&P
0010	**CARPET TILE**									
0100	Tufted nylon, 18" x 18", hard back, 20 oz.	1 Tilf	150	.053	S.Y.	22	1.42		23.42	27
0110	26 oz.		150	.053		38	1.42		39.42	44
0200	Cushion back, 20 oz.		150	.053		28	1.42		29.42	33
0210	26 oz.	↓	150	.053	↓	43.50	1.42		44.92	50.50

09 68 16 - Sheet Carpeting

09 68 16.10 Sheet Carpet

		Crew	Daily Output	Labor-Hours	Unit	Material	2009 Bare Costs Labor	2009 Bare Costs Equipment	Total	Total Incl O&P
0010	**SHEET CARPET**									
0701	Nylon, level loop, 26 oz., light to medium traffic	1 Tilf	445	.018	S.F.	2.89	.48		3.37	3.95
0901	32 oz., medium traffic		445	.018		3.58	.48		4.06	4.71
1101	40 oz., medium to heavy traffic		445	.018		5.35	.48		5.83	6.60
2101	Nylon, plush, 20 oz., light traffic		445	.018		1.81	.48		2.29	2.76
2801	24 oz., light to medium traffic		445	.018		1.81	.48		2.29	2.76
2901	30 oz., medium traffic		445	.018		2.67	.48		3.15	3.70
3001	36 oz., medium traffic		445	.018		3.56	.48		4.04	4.68
3101	42 oz., medium to heavy traffic		370	.022		4	.58		4.58	5.30
3201	46 oz., medium to heavy traffic		370	.022		4.67	.58		5.25	6.05
3301	54 oz., heavy traffic		370	.022		5.10	.58		5.68	6.50
3501	Olefin, 15 oz., light traffic		445	.018		.96	.48		1.44	1.83
3651	22 oz., light traffic		445	.018		1.14	.48		1.62	2.02
4501	50 oz., medium to heavy traffic, level loop		445	.018		12.20	.48		12.68	14.20
4701	32 oz., medium to heavy traffic, patterned		400	.020		12.10	.53		12.63	14.15
4901	48 oz., heavy traffic, patterned	↓	400	.020	↓	12.45	.53		12.98	14.55
5000	For less than full roll (approx. 1500 S.F.), add					25%				
5100	For small rooms, less than 12' wide, add						25%			

09 68 Carpeting

09 68 16 – Sheet Carpeting

09 68 16.10 Sheet Carpet

		Crew	Daily Output	Labor-Hours	Unit	Material	2009 Bare Costs		Total	Total Incl O&P
							Labor	Equipment		
5200	For large open areas (no cuts), deduct						25%			
5600	For bound carpet baseboard, add	1 Tilf	300	.027	L.F.	2.25	.71		2.96	3.62
5610	For stairs, not incl. price of carpet, add	"	30	.267	Riser		7.10		7.10	11.40
8950	For tackless, stretched installation, add padding to above									
9850	For "branded" fiber, add				S.Y.	25%				

09 68 20 – Athletic Carpet

09 68 20.10 Indoor Athletic Carpet

		Crew	Daily Output	Labor-Hours	Unit	Material	Labor	Equipment	Total	Total Incl O&P
0010	**INDOOR ATHLETIC CARPET**									
3700	Polyethylene, in rolls, no base incl., landscape surfaces	1 Tilf	275	.029	S.F.	3.05	.78		3.83	4.60
3800	Nylon action surface, 1/8" thick		275	.029		3.20	.78		3.98	4.76
3900	1/4" thick		275	.029		4.61	.78		5.39	6.30
4000	3/8" thick		275	.029		5.80	.78		6.58	7.60

09 72 Wall Coverings

09 72 23 – Wallpapering

09 72 23.10 Wallpaper

		Crew	Daily Output	Labor-Hours	Unit	Material	Labor	Equipment	Total	Total Incl O&P
0010	**WALLPAPER** including sizing; add 10-30 percent waste @ takeoff R097223-10									
0050	Aluminum foil	1 Pape	275	.029	S.F.	.97	.73		1.70	2.27
0100	Copper sheets, .025" thick, vinyl backing		240	.033		5.20	.84		6.04	7.05
0300	Phenolic backing		240	.033		6.75	.84		7.59	8.75
0600	Cork tiles, light or dark, 12" x 12" x 3/16"		240	.033		4.25	.84		5.09	6.05
0700	5/16" thick		235	.034		3.62	.86		4.48	5.40
0900	1/4" basketweave		240	.033		5.55	.84		6.39	7.50
1000	1/2" natural, non-directional pattern		240	.033		6.70	.84		7.54	8.70
1100	3/4" natural, non-directional pattern		240	.033		10.90	.84		11.74	13.35
1200	Granular surface, 12" x 36", 1/2" thick		385	.021		1.20	.52		1.72	2.18
1300	1" thick		370	.022		1.55	.54		2.09	2.60
1500	Polyurethane coated, 12" x 12" x 3/16" thick		240	.033		3.76	.84		4.60	5.50
1600	5/16" thick		235	.034		5.35	.86		6.21	7.30
1800	Cork wallpaper, paperbacked, natural		480	.017		1.90	.42		2.32	2.78
1900	Colors		480	.017		2.65	.42		3.07	3.61
2100	Flexible wood veneer, 1/32" thick, plain woods		100	.080		2.25	2.01		4.26	5.80
2200	Exotic woods		95	.084		3.41	2.12		5.53	7.20
2400	Gypsum-based, fabric-backed, fire resistant									
2500	for masonry walls, minimum, 21 oz./S.Y.	1 Pape	800	.010	S.F.	.80	.25		1.05	1.29
2600	Average		720	.011		1.17	.28		1.45	1.75
2700	Maximum (small quantities)		640	.013		1.30	.31		1.61	1.95
2750	Acrylic, modified, semi-rigid PVC, .028" thick	2 Carp	330	.048		1.08	1.36		2.44	3.48
2800	.040" thick	"	320	.050		1.42	1.40		2.82	3.92
3000	Vinyl wall covering, fabric-backed, lightweight (12-15 oz./S.Y.)	1 Pape	640	.013		.65	.31		.96	1.24
3300	Medium weight, type 2 (20-24 oz./S.Y.)		480	.017		.77	.42		1.19	1.54
3400	Heavy weight, type 3 (28 oz./S.Y.)		435	.018		1.59	.46		2.05	2.51
3600	Adhesive, 5 gal. lots (18SY/Gal.)				Gal.	10.15			10.15	11.15
3700	Wallpaper, average workmanship, solid pattern, low cost paper	1 Pape	640	.013	S.F.	.35	.31		.66	.91
3900	basic patterns (matching required), avg. cost paper		535	.015		.63	.38		1.01	1.31
4000	Paper at $85 per double roll, quality workmanship		435	.018		2.35	.46		2.81	3.35
4100	Linen wall covering, paper backed									
4150	Flame treatment, minimum				S.F.	.85			.85	.94
4180	Maximum					1.48			1.48	1.63
4200	Grass cloths with lining paper, minimum	1 Pape	400	.020		.74	.50		1.24	1.63

187

09 72 Wall Coverings

09 72 23 – Wallpapering

09 72 23.10 Wallpaper		Crew	Daily Output	Labor-Hours	Unit	Material	2009 Bare Costs Labor	Equipment	Total	Total Incl O&P
4300	Maximum	1 Pape	350	.023	S.F.	2.37	.57		2.94	3.55

09 91 Painting

09 91 03 – Paint Restoration

09 91 03.20 Sanding

		Crew	Daily Output	Labor-Hours	Unit	Material	2009 Bare Costs Labor	Equipment	Total	Total Incl O&P
0010	**SANDING** and puttying interior trim, compared to									
0100	Painting 1 coat, on quality work				L.F.		100%			
0300	Medium work						50%			
0400	Industrial grade						25%			
0500	Surface protection, placement and removal									
0510	Basic drop cloths	1 Pord	6400	.001	S.F.		.03		.03	.05
0520	Masking with paper		800	.010		.06	.25		.31	.48
0530	Volume cover up (using plastic sheathing, or building paper)		16000	.001			.01		.01	.02

09 91 03.30 Exterior Surface Preparation

		Crew	Daily Output	Labor-Hours	Unit	Material	2009 Bare Costs Labor	Equipment	Total	Total Incl O&P
0010	**EXTERIOR SURFACE PREPARATION**									
0015	Doors, per side, not incl. frames or trim									
0020	Scrape & sand									
0030	Wood, flush	1 Pord	616	.013	S.F.		.32		.32	.53
0040	Wood, detail		496	.016			.40		.40	.66
0050	Wood, louvered		280	.029			.71		.71	1.17
0060	Wood, overhead		616	.013			.32		.32	.53
0070	Wire brush									
0080	Metal, flush	1 Pord	640	.013	S.F.		.31		.31	.51
0090	Metal, detail		520	.015			.38		.38	.63
0100	Metal, louvered		360	.022			.56		.56	.91
0110	Metal or fibr., overhead		640	.013			.31		.31	.51
0120	Metal, roll up		560	.014			.36		.36	.59
0130	Metal, bulkhead		640	.013			.31		.31	.51
0140	Power wash, based on 2500 lb. operating pressure									
0150	Metal, flush	A-1H	2240	.004	S.F.		.07	.02	.09	.15
0160	Metal, detail		2120	.004			.08	.02	.10	.16
0170	Metal, louvered		2000	.004			.08	.03	.11	.17
0180	Metal or fibr., overhead		2400	.003			.07	.02	.09	.14
0190	Metal, roll up		2400	.003			.07	.02	.09	.14
0200	Metal, bulkhead		2200	.004			.07	.02	.09	.16
0400	Windows, per side, not incl. trim									
0410	Scrape & sand									
0420	Wood, 1-2 lite	1 Pord	320	.025	S.F.		.63		.63	1.02
0430	Wood, 3-6 lite		280	.029			.71		.71	1.17
0440	Wood, 7-10 lite		240	.033			.83		.83	1.36
0450	Wood, 12 lite		200	.040			1		1	1.64
0460	Wood, Bay / Bow		320	.025			.63		.63	1.02
0470	Wire brush									
0480	Metal, 1-2 lite	1 Pord	480	.017	S.F.		.42		.42	.68
0490	Metal, 3-6 lite		400	.020			.50		.50	.82
0500	Metal, Bay / Bow		480	.017			.42		.42	.68
0510	Power wash, based on 2500 lb. operating pressure									
0520	1-2 lite	A-1H	4400	.002	S.F.		.04	.01	.05	.07
0530	3-6 lite		4320	.002			.04	.01	.05	.07
0540	7-10 lite		4240	.002			.04	.01	.05	.08
0550	12 lite		4160	.002			.04	.01	.05	.08

09 91 Painting

09 91 03 – Paint Restoration

09 91 03.30 Exterior Surface Preparation

09 91 03.30 Exterior Surface Preparation		Crew	Daily Output	Labor-Hours	Unit	Material	2009 Bare Costs Labor	Equipment	Total	Total Incl O&P
0560	Bay / Bow	A-1H	4400	.002	S.F.		.04	.01	.05	.07
0600	Siding, scrape and sand, light=10-30%, med.=30-70%									
0610	Heavy=70-100% of surface to sand									
0650	Texture 1-11, light	1 Pord	480	.017	S.F.		.42		.42	.68
0660	Med.		440	.018			.45		.45	.74
0670	Heavy		360	.022			.56		.56	.91
0680	Wood shingles, shakes, light		440	.018			.45		.45	.74
0690	Med.		360	.022			.56		.56	.91
0700	Heavy		280	.029			.71		.71	1.17
0710	Clapboard, light		520	.015			.38		.38	.63
0720	Med.		480	.017			.42		.42	.68
0730	Heavy	▼	400	.020	▼		.50		.50	.82
0740	Wire brush									
0750	Aluminum, light	1 Pord	600	.013	S.F.		.33		.33	.55
0760	Med.		520	.015			.38		.38	.63
0770	Heavy	▼	440	.018	▼		.45		.45	.74
0780	Pressure wash, based on 2500 lb. operating pressure									
0790	Stucco	A-1H	3080	.003	S.F.		.05	.02	.07	.11
0800	Aluminum or vinyl		3200	.003			.05	.02	.07	.11
0810	Siding, masonry, brick & block	▼	2400	.003	▼		.07	.02	.09	.14
1300	Miscellaneous, wire brush									
1310	Metal, pedestrian gate	1 Pord	100	.080	S.F.		2		2	3.28

09 91 03.40 Interior Surface Preparation

09 91 03.40 Interior Surface Preparation		Crew	Daily Output	Labor-Hours	Unit	Material	2009 Bare Costs Labor	Equipment	Total	Total Incl O&P
0010	**INTERIOR SURFACE PREPARATION**									
0020	Doors, per side, not incl. frames or trim									
0030	Scrape & sand									
0040	Wood, flush	1 Pord	616	.013	S.F.		.32		.32	.53
0050	Wood, detail		496	.016			.40		.40	.66
0060	Wood, louvered	▼	280	.029	▼		.71		.71	1.17
0070	Wire brush									
0080	Metal, flush	1 Pord	640	.013	S.F.		.31		.31	.51
0090	Metal, detail		520	.015			.38		.38	.63
0100	Metal, louvered	▼	360	.022	▼		.56		.56	.91
0110	Hand wash									
0120	Wood, flush	1 Pord	2160	.004	S.F.		.09		.09	.15
0130	Wood, detailed		2000	.004			.10		.10	.16
0140	Wood, louvered		1360	.006			.15		.15	.24
0150	Metal, flush		2160	.004			.09		.09	.15
0160	Metal, detail		2000	.004			.10		.10	.16
0170	Metal, louvered	▼	1360	.006	▼		.15		.15	.24
0400	Windows, per side, not incl. trim									
0410	Scrape & sand									
0420	Wood, 1-2 lite	1 Pord	360	.022	S.F.		.56		.56	.91
0430	Wood, 3-6 lite		320	.025			.63		.63	1.02
0440	Wood, 7-10 lite		280	.029			.71		.71	1.17
0450	Wood, 12 lite		240	.033			.83		.83	1.36
0460	Wood, Bay / Bow	▼	360	.022	▼		.56		.56	.91
0470	Wire brush									
0480	Metal, 1-2 lite	1 Pord	520	.015	S.F.		.38		.38	.63
0490	Metal, 3-6 lite		440	.018			.45		.45	.74
0500	Metal, Bay / Bow	▼	520	.015	▼		.38		.38	.63
0600	Walls, sanding, light=10-30%, medium - 30-70%,									

09 91 Painting

09 91 03 – Paint Restoration

09 91 03.40 Interior Surface Preparation

		Crew	Daily Output	Labor-Hours	Unit	Material	2009 Bare Costs Labor	Equipment	Total	Total Incl O&P
0610	heavy=70-100% of surface to sand									
0650	Walls, sand									
0660	Drywall, gypsum, plaster, light	1 Pord	3077	.003	S.F.		.07		.07	.11
0670	Drywall, gypsum, plaster, med.		2160	.004			.09		.09	.15
0680	Drywall, gypsum, plaster, heavy		923	.009			.22		.22	.36
0690	Wood, T&G, light		2400	.003			.08		.08	.14
0700	Wood, T&G, med.		1600	.005			.13		.13	.20
0710	Wood, T&G, heavy		800	.010			.25		.25	.41
0720	Walls, wash									
0730	Drywall, gypsum, plaster	1 Pord	3200	.003	S.F.		.06		.06	.10
0740	Wood, T&G		3200	.003			.06		.06	.10
0750	Masonry, brick & block, smooth		2800	.003			.07		.07	.12
0760	Masonry, brick & block, coarse		2000	.004			.10		.10	.16
8000	For chemical washing, see Div. 04 01 30.60									

09 91 03.41 Scrape After Fire Damage

		Crew	Daily Output	Labor-Hours	Unit	Material	2009 Bare Costs Labor	Equipment	Total	Total Incl O&P
0010	**SCRAPE AFTER FIRE DAMAGE**									
0050	Boards, 1" x 4"	1 Pord	336	.024	L.F.		.60		.60	.98
0060	1" x 6"		260	.031			.77		.77	1.26
0070	1" x 8"		207	.039			.97		.97	1.58
0080	1" x 10"		174	.046			1.15		1.15	1.88
0500	Framing, 2" x 4"		265	.030			.75		.75	1.24
0510	2" x 6"		221	.036			.91		.91	1.48
0520	2" x 8"		190	.042			1.05		1.05	1.72
0530	2" x 10"		165	.048			1.21		1.21	1.99
0540	2" x 12"		144	.056			1.39		1.39	2.28
1000	Heavy framing, 3" x 4"		226	.035			.89		.89	1.45
1010	4" x 4"		210	.038			.95		.95	1.56
1020	4" x 6"		191	.042			1.05		1.05	1.72
1030	4" x 8"		165	.048			1.21		1.21	1.99
1040	4" x 10"		144	.056			1.39		1.39	2.28
1060	4" x 12"		131	.061			1.53		1.53	2.50
2900	For sealing, minimum		825	.010	S.F.	.14	.24		.38	.55
2920	Maximum		460	.017	"	.28	.43		.71	1.02

09 91 13 – Exterior Painting

09 91 13.30 Fences

		Crew	Daily Output	Labor-Hours	Unit	Material	2009 Bare Costs Labor	Equipment	Total	Total Incl O&P
0010	**FENCES** R099100-20									
0100	Chain link or wire metal, one side, water base									
0110	Roll & brush, first coat	1 Pord	960	.008	S.F.	.05	.21		.26	.40
0120	Second coat		1280	.006		.05	.16		.21	.32
0130	Spray, first coat		2275	.004		.05	.09		.14	.20
0140	Second coat		2600	.003		.05	.08		.13	.19
0150	Picket, water base									
0160	Roll & brush, first coat	1 Pord	865	.009	S.F.	.06	.23		.29	.44
0170	Second coat		1050	.008		.06	.19		.25	.37
0180	Spray, first coat		2275	.004		.06	.09		.15	.20
0190	Second coat		2600	.003		.06	.08		.14	.19
0200	Stockade, water base									
0210	Roll & brush, first coat	1 Pord	1040	.008	S.F.	.06	.19		.25	.37
0220	Second coat		1200	.007		.06	.17		.23	.33
0230	Spray, first coat		2275	.004		.06	.09		.15	.20
0240	Second coat		2600	.003		.06	.08		.14	.19

09 91 13 – Exterior Painting

09 91 13.42 Miscellaneous, Exterior	Crew	Daily Output	Labor-Hours	Unit	Material	2009 Bare Costs Labor	Equipment	Total	Total Incl O&P
0010 **MISCELLANEOUS, EXTERIOR** R099100-20									
0100 Railing, ext., decorative wood, incl. cap & baluster									
0110 Newels & spindles @ 12" O.C.									
0120 Brushwork, stain, sand, seal & varnish									
0130 First coat	1 Pord	90	.089	L.F.	.61	2.22		2.83	4.31
0140 Second coat	"	120	.067	"	.61	1.67		2.28	3.40
0150 Rough sawn wood, 42" high, 2" x 2" verticals, 6" O.C.									
0160 Brushwork, stain, each coat	1 Pord	90	.089	L.F.	.18	2.22		2.40	3.84
0170 Wrought iron, 1" rail, 1/2" sq. verticals									
0180 Brushwork, zinc chromate, 60" high, bars 6" O.C.									
0190 Primer	1 Pord	130	.062	L.F.	.60	1.54		2.14	3.18
0200 Finish coat		130	.062		.19	1.54		1.73	2.73
0210 Additional coat	↓	190	.042	↓	.22	1.05		1.27	1.96
0220 Shutters or blinds, single panel, 2' x 4', paint all sides									
0230 Brushwork, primer	1 Pord	20	.400	Ea.	.59	10		10.59	17.05
0240 Finish coat, exterior latex		20	.400		.44	10		10.44	16.90
0250 Primer & 1 coat, exterior latex		13	.615		.89	15.40		16.29	26
0260 Spray, primer		35	.229		.85	5.70		6.55	10.30
0270 Finish coat, exterior latex		35	.229		.93	5.70		6.63	10.35
0280 Primer & 1 coat, exterior latex	↓	20	.400	↓	.91	10		10.91	17.40
0290 For louvered shutters, add				S.F.	10%				
0300 Stair stringers, exterior, metal									
0310 Roll & brush, zinc chromate, to 14", each coat	1 Pord	320	.025	L.F.	.06	.63		.69	1.09
0320 Rough sawn wood, 4" x 12"									
0330 Roll & brush, exterior latex, each coat	1 Pord	215	.037	L.F.	.06	.93		.99	1.59
0340 Trellis/lattice, 2" x 2" @ 3" O.C. with 2" x 8" supports									
0350 Spray, latex, per side, each coat	1 Pord	475	.017	S.F.	.06	.42		.48	.76
0450 Decking, ext., sealer, alkyd, brushwork, sealer coat		1140	.007		.09	.18		.27	.39
0460 1st coat		1140	.007		.07	.18		.25	.37
0470 2nd coat		1300	.006		.05	.15		.20	.31
0500 Paint, alkyd, brushwork, primer coat		1140	.007		.07	.18		.25	.37
0510 1st coat		1140	.007		.07	.18		.25	.37
0520 2nd coat		1300	.006		.05	.15		.20	.31
0600 Sand paint, alkyd, brushwork, 1 coat	↓	150	.053	↓	.11	1.33		1.44	2.30

09 91 13.60 Siding Exterior

	Crew	Daily Output	Labor-Hours	Unit	Material	Labor	Equipment	Total	Total Incl O&P
0010 **SIDING EXTERIOR**, Alkyd (oil base)									
0450 Steel siding, oil base, paint 1 coat, brushwork	2 Pord	2015	.008	S.F.	.06	.20		.26	.39
0500 Spray		4550	.004		.09	.09		.18	.24
0800 Paint 2 coats, brushwork		1300	.012		.11	.31		.42	.63
1000 Spray		2750	.006		.17	.15		.32	.43
1200 Stucco, rough, oil base, paint 2 coats, brushwork		1300	.012		.11	.31		.42	.63
1400 Roller		1625	.010		.12	.25		.37	.53
1600 Spray		2925	.005		.13	.14		.27	.36
1800 Texture 1-11 or clapboard, oil base, primer coat, brushwork		1300	.012		.09	.31		.40	.60
2000 Spray		4550	.004		.09	.09		.18	.24
2100 Paint 1 coat, brushwork		1300	.012		.08	.31		.39	.59
2200 Spray		4550	.004		.08	.09		.17	.23
2400 Paint 2 coats, brushwork		810	.020		.17	.49		.66	.99
2600 Spray		2600	.006		.19	.15		.34	.45
3000 Stain 1 coat, brushwork		1520	.011		.06	.26		.32	.50
3200 Spray		5320	.003		.07	.08		.15	.20
3400 Stain 2 coats, brushwork	↓	950	.017	↓	.12	.42		.54	.83

191

09 91 13.60 Siding Exterior

		Crew	Daily Output	Labor-Hours	Unit	Material	2009 Bare Costs Labor	Equipment	Total	Total Incl O&P
4000	Spray	2 Pord	3050	.005	S.F.	.14	.13		.27	.37
4200	Wood shingles, oil base primer coat, brushwork		1300	.012		.08	.31		.39	.59
4400	Spray		3900	.004		.08	.10		.18	.26
4600	Paint 1 coat, brushwork		1300	.012		.07	.31		.38	.58
4800	Spray		3900	.004		.09	.10		.19	.27
5000	Paint 2 coats, brushwork		810	.020		.14	.49		.63	.96
5200	Spray		2275	.007		.13	.18		.31	.44
5800	Stain 1 coat, brushwork		1500	.011		.06	.27		.33	.51
6000	Spray		3900	.004		.06	.10		.16	.24
6500	Stain 2 coats, brushwork		950	.017		.12	.42		.54	.83
7000	Spray		2660	.006		.17	.15		.32	.44
8000	For latex paint, deduct					10%				
8100	For work over 12' H, from pipe scaffolding, add						15%			
8200	For work over 12' H, from extension ladder, add						25%			
8300	For work over 12' H, from swing staging, add						35%			

09 91 13.62 Siding, Misc.

		Crew	Daily Output	Labor-Hours	Unit	Material	2009 Bare Costs Labor	Equipment	Total	Total Incl O&P
0010	**SIDING, MISC.**, latex paint R099100-10									
0100	Aluminum siding									
0110	Brushwork, primer	2 Pord	2275	.007	S.F.	.05	.18		.23	.35
0120	Finish coat, exterior latex		2275	.007		.04	.18		.22	.34
0130	Primer & 1 coat exterior latex		1300	.012		.10	.31		.41	.61
0140	Primer & 2 coats exterior latex		975	.016		.15	.41		.56	.83
0150	Mineral fiber shingles									
0160	Brushwork, primer	2 Pord	1495	.011	S.F.	.09	.27		.36	.54
0170	Finish coat, industrial enamel		1495	.011		.13	.27		.40	.58
0180	Primer & 1 coat enamel		810	.020		.22	.49		.71	1.05
0190	Primer & 2 coats enamel		540	.030		.35	.74		1.09	1.60
0200	Roll, primer		1625	.010		.10	.25		.35	.51
0210	Finish coat, industrial enamel		1625	.010		.14	.25		.39	.56
0220	Primer & 1 coat enamel		975	.016		.24	.41		.65	.94
0230	Primer & 2 coats enamel		650	.025		.38	.62		1	1.43
0240	Spray, primer		3900	.004		.08	.10		.18	.26
0250	Finish coat, industrial enamel		3900	.004		.12	.10		.22	.30
0260	Primer & 1 coat enamel		2275	.007		.20	.18		.38	.50
0270	Primer & 2 coats enamel		1625	.010		.31	.25		.56	.74
0280	Waterproof sealer, first coat		4485	.004		.08	.09		.17	.24
0290	Second coat		5235	.003		.07	.08		.15	.21
0300	Rough wood incl. shingles, shakes or rough sawn siding									
0310	Brushwork, primer	2 Pord	1280	.013	S.F.	.12	.31		.43	.64
0320	Finish coat, exterior latex		1280	.013		.07	.31		.38	.59
0330	Primer & 1 coat exterior latex		960	.017		.19	.42		.61	.89
0340	Primer & 2 coats exterior latex		700	.023		.27	.57		.84	1.23
0350	Roll, primer		2925	.005		.16	.14		.30	.39
0360	Finish coat, exterior latex		2925	.005		.09	.14		.23	.32
0370	Primer & 1 coat exterior latex		1790	.009		.25	.22		.47	.64
0380	Primer & 2 coats exterior latex		1300	.012		.34	.31		.65	.87
0390	Spray, primer		3900	.004		.13	.10		.23	.32
0400	Finish coat, exterior latex		3900	.004		.07	.10		.17	.25
0410	Primer & 1 coat exterior latex		2600	.006		.20	.15		.35	.47
0420	Primer & 2 coats exterior latex		2080	.008		.27	.19		.46	.61
0430	Waterproof sealer, first coat		4485	.004		.14	.09		.23	.30
0440	Second coat		4485	.004		.08	.09		.17	.24

09 91 Painting

09 91 13.62 Siding, Misc.

		Crew	Daily Output	Labor-Hours	Unit	Material	2009 Bare Costs Labor	Equipment	Total	Total Incl O&P
0450	Smooth wood incl. butt, T&G, beveled, drop or B&B siding									
0460	Brushwork, primer	2 Pord	2325	.007	S.F.	.09	.17		.26	.37
0470	Finish coat, exterior latex		1280	.013		.07	.31		.38	.59
0480	Primer & 1 coat exterior latex		800	.020		.16	.50		.66	1
0490	Primer & 2 coats exterior latex		630	.025		.23	.64		.87	1.30
0500	Roll, primer		2275	.007		.09	.18		.27	.39
0510	Finish coat, exterior latex		2275	.007		.08	.18		.26	.38
0520	Primer & 1 coat exterior latex		1300	.012		.18	.31		.49	.69
0530	Primer & 2 coats exterior latex		975	.016		.26	.41		.67	.95
0540	Spray, primer		4550	.004		.07	.09		.16	.22
0550	Finish coat, exterior latex		4550	.004		.07	.09		.16	.22
0560	Primer & 1 coat exterior latex		2600	.006		.14	.15		.29	.41
0570	Primer & 2 coats exterior latex		1950	.008		.21	.21		.42	.57
0580	Waterproof sealer, first coat		5230	.003		.08	.08		.16	.22
0590	Second coat		5980	.003		.08	.07		.15	.20
0600	For oil base paint, add					10%				

09 91 13.70 Doors and Windows, Exterior

		Crew	Daily Output	Labor-Hours	Unit	Material	2009 Bare Costs Labor	Equipment	Total	Total Incl O&P
0010	**DOORS AND WINDOWS, EXTERIOR** R099100-10									
0100	Door frames & trim, only									
0110	Brushwork, primer R099100-20	1 Pord	512	.016	L.F.	.05	.39		.44	.70
0120	Finish coat, exterior latex		512	.016		.05	.39		.44	.70
0130	Primer & 1 coat, exterior latex		300	.027		.11	.67		.78	1.21
0140	Primer & 2 coats, exterior latex		265	.030		.16	.75		.91	1.42
0150	Doors, flush, both sides, incl. frame & trim									
0160	Roll & brush, primer	1 Pord	10	.800	Ea.	4.05	20		24.05	37.50
0170	Finish coat, exterior latex		10	.800		4.15	20		24.15	37.50
0180	Primer & 1 coat, exterior latex		7	1.143		8.20	28.50		36.70	56
0190	Primer & 2 coats, exterior latex		5	1.600		12.35	40		52.35	79
0200	Brushwork, stain, sealer & 2 coats polyurethane		4	2		21	50		71	105
0210	Doors, French, both sides, 10-15 lite, incl. frame & trim									
0220	Brushwork, primer	1 Pord	6	1.333	Ea.	2.03	33.50		35.53	56.50
0230	Finish coat, exterior latex		6	1.333		2.07	33.50		35.57	57
0240	Primer & 1 coat, exterior latex		3	2.667		4.10	66.50		70.60	114
0250	Primer & 2 coats, exterior latex		2	4		6.05	100		106.05	171
0260	Brushwork, stain, sealer & 2 coats polyurethane		2.50	3.200		7.70	80		87.70	140
0270	Doors, louvered, both sides, incl. frame & trim									
0280	Brushwork, primer	1 Pord	7	1.143	Ea.	4.05	28.50		32.55	51.50
0290	Finish coat, exterior latex		7	1.143		4.15	28.50		32.65	51.50
0300	Primer & 1 coat, exterior latex		4	2		8.20	50		58.20	91
0310	Primer & 2 coats, exterior latex		3	2.667		12.10	66.50		78.60	122
0320	Brushwork, stain, sealer & 2 coats polyurethane		4.50	1.778		21	44.50		65.50	96
0330	Doors, panel, both sides, incl. frame & trim									
0340	Roll & brush, primer	1 Pord	6	1.333	Ea.	4.05	33.50		37.55	59
0350	Finish coat, exterior latex		6	1.333		4.15	33.50		37.65	59
0360	Primer & 1 coat, exterior latex		3	2.667		8.20	66.50		74.70	118
0370	Primer & 2 coats, exterior latex		2.50	3.200		12.10	80		92.10	144
0380	Brushwork, stain, sealer & 2 coats polyurethane		3	2.667		21	66.50		87.50	132
0400	Windows, per ext. side, based on 15 SF									
0410	1 to 6 lite									
0420	Brushwork, primer	1 Pord	13	.615	Ea.	.80	15.40		16.20	26
0430	Finish coat, exterior latex		13	.615		.82	15.40		16.22	26
0440	Primer & 1 coat, exterior latex		8	1		1.62	25		26.62	43

09 91 Painting

09 91 13 – Exterior Painting

09 91 13.70 Doors and Windows, Exterior

	Crew	Daily Output	Labor-Hours	Unit	Material	2009 Bare Costs Labor	Equipment	Total	Total Incl O&P
0450 Primer & 2 coats, exterior latex	1 Pord	6	1.333	Ea.	2.39	33.50		35.89	57
0460 Stain, sealer & 1 coat varnish	↓	7	1.143	↓	3.04	28.50		31.54	50.50
0470 7 to 10 lite									
0480 Brushwork, primer	1 Pord	11	.727	Ea.	.80	18.20		19	31
0490 Finish coat, exterior latex		11	.727		.82	18.20		19.02	31
0500 Primer & 1 coat, exterior latex		7	1.143		1.62	28.50		30.12	49
0510 Primer & 2 coats, exterior latex		5	1.600		2.39	40		42.39	68
0520 Stain, sealer & 1 coat varnish	↓	6	1.333	↓	3.04	33.50		36.54	58
0530 12 lite									
0540 Brushwork, primer	1 Pord	10	.800	Ea.	.80	20		20.80	34
0550 Finish coat, exterior latex		10	.800		.82	20		20.82	34
0560 Primer & 1 coat, exterior latex		6	1.333		1.62	33.50		35.12	56.50
0570 Primer & 2 coats, exterior latex		5	1.600		2.39	40		42.39	68
0580 Stain, sealer & 1 coat varnish	↓	6	1.333		3.06	33.50		36.56	58
0590 For oil base paint, add				↓	10%				

09 91 13.80 Trim, Exterior

	Crew	Daily Output	Labor-Hours	Unit	Material	2009 Bare Costs Labor	Equipment	Total	Total Incl O&P
0010 **TRIM, EXTERIOR** R099100-10									
0100 Door frames & trim (see Doors, interior or exterior)									
0110 Fascia, latex paint, one coat coverage									
0120 1" x 4", brushwork	1 Pord	640	.013	L.F.	.02	.31		.33	.53
0130 Roll		1280	.006		.02	.16		.18	.28
0140 Spray		2080	.004		.01	.10		.11	.18
0150 1" x 6" to 1" x 10", brushwork		640	.013		.06	.31		.37	.57
0160 Roll		1230	.007		.06	.16		.22	.34
0170 Spray		2100	.004		.05	.10		.15	.21
0180 1" x 12", brushwork		640	.013		.06	.31		.37	.57
0190 Roll		1050	.008		.06	.19		.25	.38
0200 Spray	↓	2200	.004	↓	.05	.09		.14	.20
0210 Gutters & downspouts, metal, zinc chromate paint									
0220 Brushwork, gutters, 5", first coat	1 Pord	640	.013	L.F.	.07	.31		.38	.58
0230 Second coat		960	.008		.06	.21		.27	.41
0240 Third coat		1280	.006		.05	.16		.21	.32
0250 Downspouts, 4", first coat		640	.013		.07	.31		.38	.58
0260 Second coat		960	.008		.06	.21		.27	.41
0270 Third coat	↓	1280	.006	↓	.05	.16		.21	.32
0280 Gutters & downspouts, wood									
0290 Brushwork, gutters, 5", primer	1 Pord	640	.013	L.F.	.05	.31		.36	.57
0300 Finish coat, exterior latex		640	.013		.05	.31		.36	.56
0310 Primer & 1 coat exterior latex		400	.020		.11	.50		.61	.94
0320 Primer & 2 coats exterior latex		325	.025		.16	.62		.78	1.19
0330 Downspouts, 4", primer		640	.013		.05	.31		.36	.57
0340 Finish coat, exterior latex		640	.013		.05	.31		.36	.56
0350 Primer & 1 coat exterior latex		400	.020		.11	.50		.61	.94
0360 Primer & 2 coats exterior latex	↓	325	.025	↓	.08	.62		.70	1.10
0370 Molding, exterior, up to 14" wide									
0380 Brushwork, primer	1 Pord	640	.013	L.F.	.06	.31		.37	.58
0390 Finish coat, exterior latex		640	.013		.06	.31		.37	.57
0400 Primer & 1 coat exterior latex		400	.020		.13	.50		.63	.96
0410 Primer & 2 coats exterior latex		315	.025		.13	.64		.77	1.18
0420 Stain & fill		1050	.008		.07	.19		.26	.39
0430 Shellac		1850	.004		.08	.11		.19	.27
0440 Varnish	↓	1275	.006	↓	.09	.16		.25	.36

09 91 Painting

09 91 13 – Exterior Painting

09 91 13.90 Walls, Masonry (CMU), Exterior

		Daily Output	Labor-Hours	Unit	Material	2009 Bare Costs Labor	Equipment	Total	Total Incl O&P
						Crew			
0350	**WALLS, MASONRY (CMU), EXTERIOR**								
0360	Concrete masonry units (CMU), smooth surface								
0370	Brushwork, latex, first coat	1 Pord	640	.013	S.F.	.04	.31		.35	.56
0380	Second coat		960	.008		.04	.21		.25	.38
0390	Waterproof sealer, first coat		736	.011		.23	.27		.50	.71
0400	Second coat		1104	.007		.23	.18		.41	.56
0410	Roll, latex, paint, first coat		1465	.005		.05	.14		.19	.28
0420	Second coat		1790	.004		.04	.11		.15	.22
0430	Waterproof sealer, first coat		1680	.005		.23	.12		.35	.45
0440	Second coat		2060	.004		.23	.10		.33	.42
0450	Spray, latex, paint, first coat		1950	.004		.04	.10		.14	.22
0460	Second coat		2600	.003		.03	.08		.11	.17
0470	Waterproof sealer, first coat		2245	.004		.23	.09		.32	.41
0480	Second coat		2990	.003		.23	.07		.30	.37
0490	Concrete masonry unit (CMU), porous								
0500	Brushwork, latex, first coat	1 Pord	640	.013	S.F.	.09	.31		.40	.61
0510	Second coat		960	.008		.04	.21		.25	.39
0520	Waterproof sealer, first coat		736	.011		.23	.27		.50	.71
0530	Second coat		1104	.007		.23	.18		.41	.56
0540	Roll latex, first coat		1465	.005		.07	.14		.21	.29
0550	Second coat		1790	.004		.04	.11		.15	.23
0560	Waterproof sealer, first coat		1680	.005		.23	.12		.35	.45
0570	Second coat		2060	.004		.23	.10		.33	.42
0580	Spray latex, first coat		1950	.004		.05	.10		.15	.22
0590	Second coat		2600	.003		.03	.08		.11	.17
0600	Waterproof sealer, first coat		2245	.004		.23	.09		.32	.41
0610	Second coat		2990	.003		.23	.07		.30	.37

09 91 23 – Interior Painting

09 91 23.20 Cabinets and Casework

		Crew	Daily Output	Labor-Hours	Unit	Material	2009 Bare Costs Labor	Equipment	Total	Total Incl O&P
0010	**CABINETS AND CASEWORK**									
1000	Primer coat, oil base, brushwork	1 Pord	650	.012	S.F.	.05	.31		.36	.56
2000	Paint, oil base, brushwork, 1 coat		650	.012		.06	.31		.37	.57
2500	2 coats		400	.020		.12	.50		.62	.95
3000	Stain, brushwork, wipe off		650	.012		.06	.31		.37	.57
4000	Shellac, 1 coat, brushwork		650	.012		.07	.31		.38	.58
4500	Varnish, 3 coats, brushwork, sand after 1st coat		325	.025		.23	.62		.85	1.26
5000	For latex paint, deduct					10%				

09 91 23.33 Doors and Windows, Interior Alkyd (Oil Base)

		Crew	Daily Output	Labor-Hours	Unit	Material	2009 Bare Costs Labor	Equipment	Total	Total Incl O&P
0010	**DOORS AND WINDOWS, INTERIOR ALKYD (OIL BASE)**									
0500	Flush door & frame, 3' x 7', oil, primer, brushwork	1 Pord	10	.800	Ea.	2.34	20		22.34	35.50
1000	Paint, 1 coat		10	.800		2.25	20		22.25	35.50
1200	2 coats		6	1.333		3.70	33.50		37.20	58.50
1400	Stain, brushwork, wipe off		18	.444		1.29	11.10		12.39	19.60
1600	Shellac, 1 coat, brushwork		25	.320		1.48	8		9.48	14.70
1800	Varnish, 3 coats, brushwork, sand after 1st coat		9	.889		4.77	22		26.77	42
2000	Panel door & frame, 3' x 7', oil, primer, brushwork		6	1.333		1.96	33.50		35.46	56.50
2200	Paint, 1 coat		6	1.333		2.25	33.50		35.75	57
2400	2 coats		3	2.667		6.45	66.50		72.95	116
2600	Stain, brushwork, panel door, 3' x 7', not incl. frame		16	.500		1.29	12.50		13.79	22
2800	Shellac, 1 coat, brushwork		22	.364		1.48	9.10		10.58	16.50
3000	Varnish, 3 coats, brushwork, sand after 1st coat		7.50	1.067		4.77	26.50		31.27	49

09 91 23.33 Doors and Windows, Interior Alkyd (Oil Base)	Crew	Daily Output	Labor-Hours	Unit	Material	2009 Bare Costs Labor	2009 Bare Costs Equipment	Total	Total Incl O&P
3020	French door, incl. 3' x 7', 6 lites, frame & trim									
3022	Paint, 1 coat, over existing paint	1 Pord	5	1.600	Ea.	4.50	40		44.50	70.50
3024	2 coats, over existing paint		5	1.600		8.75	40		48.75	75
3026	Primer & 1 coat		3.50	2.286		8.45	57		65.45	103
3028	Primer & 2 coats		3	2.667		12.95	66.50		79.45	123
3032	Varnish or polyurethane, 1 coat		5	1.600		5.35	40		45.35	71.50
3034	2 coats, sanding between		3	2.667		10.75	66.50		77.25	121
4400	Windows, including frame and trim, per side									
4600	Colonial type, 6/6 lites, 2' x 3', oil, primer, brushwork	1 Pord	14	.571	Ea.	.31	14.30		14.61	24
5800	Paint, 1 coat		14	.571		.36	14.30		14.66	24
6000	2 coats		9	.889		.69	22		22.69	37.50
6200	3' x 5' opening, 6/6 lites, primer coat, brushwork		12	.667		.78	16.65		17.43	28.50
6400	Paint, 1 coat		12	.667		.89	16.65		17.54	28.50
6600	2 coats		7	1.143		1.73	28.50		30.23	49
6800	4' x 8' opening, 6/6 lites, primer coat, brushwork		8	1		1.65	25		26.65	43
7000	Paint, 1 coat		8	1		1.90	25		26.90	43
7200	2 coats		5	1.600		3.68	40		43.68	69.50
8000	Single lite type, 2' x 3', oil base, primer coat, brushwork		33	.242		.31	6.05		6.36	10.30
8200	Paint, 1 coat		33	.242		.36	6.05		6.41	10.35
8400	2 coats		20	.400		.69	10		10.69	17.15
8600	3' x 5' opening, primer coat, brushwork		20	.400		.78	10		10.78	17.25
8800	Paint, 1 coat		20	.400		.89	10		10.89	17.40
8900	2 coats		13	.615		1.73	15.40		17.13	27
9200	4' x 8' opening, primer coat, brushwork		14	.571		1.65	14.30		15.95	25.50
9400	Paint, 1 coat		14	.571		1.90	14.30		16.20	25.50
9600	2 coats		8	1		3.68	25		28.68	45

09 91 23.35 Doors and Windows, Interior Latex

		Crew	Daily Output	Labor-Hours	Unit	Material	2009 Bare Costs Labor	2009 Bare Costs Equipment	Total	Total Incl O&P
0010	**DOORS & WINDOWS, INTERIOR LATEX** R099100-10									
0100	Doors, flush, both sides, incl. frame & trim									
0110	Roll & brush, primer	1 Pord	10	.800	Ea.	3.64	20		23.64	37
0120	Finish coat, latex		10	.800		4.21	20		24.21	37.50
0130	Primer & 1 coat latex		7	1.143		7.85	28.50		36.35	55.50
0140	Primer & 2 coats latex		5	1.600		11.80	40		51.80	78.50
0160	Spray, both sides, primer		20	.400		3.83	10		13.83	20.50
0170	Finish coat, latex		20	.400		4.42	10		14.42	21.50
0180	Primer & 1 coat latex		11	.727		8.30	18.20		26.50	39
0190	Primer & 2 coats latex		8	1		12.50	25		37.50	55
0200	Doors, French, both sides, 10-15 lite, incl. frame & trim									
0210	Roll & brush, primer	1 Pord	6	1.333	Ea.	1.82	33.50		35.32	56.50
0220	Finish coat, latex		6	1.333		2.10	33.50		35.60	57
0230	Primer & 1 coat latex		3	2.667		3.92	66.50		70.42	113
0240	Primer & 2 coats latex		2	4		5.90	100		105.90	171
0260	Doors, louvered, both sides, incl. frame & trim									
0270	Roll & brush, primer	1 Pord	7	1.143	Ea.	3.64	28.50		32.14	51
0280	Finish coat, latex		7	1.143		4.21	28.50		32.71	51.50
0290	Primer & 1 coat, latex		4	2		7.65	50		57.65	90.50
0300	Primer & 2 coats, latex		3	2.667		12.05	66.50		78.55	122
0320	Spray, both sides, primer		20	.400		3.83	10		13.83	20.50
0330	Finish coat, latex		20	.400		4.42	10		14.42	21.50
0340	Primer & 1 coat, latex		11	.727		8.30	18.20		26.50	39
0350	Primer & 2 coats, latex		8	1		12.75	25		37.75	55
0360	Doors, panel, both sides, incl. frame & trim									

09 91 Painting

09 91 23 – Interior Painting

09 91 23.35 Doors and Windows, Interior Latex

09 91 23.35 Doors and Windows, Interior Latex		Crew	Daily Output	Labor-Hours	Unit	Material	2009 Bare Costs Labor	Equipment	Total	Total Incl O&P
0370	Roll & brush, primer	1 Pord	6	1.333	Ea.	3.83	33.50		37.33	58.50
0380	Finish coat, latex		6	1.333		4.21	33.50		37.71	59
0390	Primer & 1 coat, latex		3	2.667		7.85	66.50		74.35	118
0400	Primer & 2 coats, latex		2.50	3.200		12.05	80		92.05	144
0420	Spray, both sides, primer		10	.800		3.83	20		23.83	37
0430	Finish coat, latex		10	.800		4.42	20		24.42	38
0440	Primer & 1 coat, latex		5	1.600		8.30	40		48.30	74.50
0450	Primer & 2 coats, latex	▼	4	2	▼	12.75	50		62.75	96
0460	Windows, per interior side, based on 15 SF									
0470	1 to 6 lite									
0480	Brushwork, primer	1 Pord	13	.615	Ea.	.72	15.40		16.12	26
0490	Finish coat, enamel		13	.615		.83	15.40		16.23	26
0500	Primer & 1 coat enamel		8	1		1.55	25		26.55	42.50
0510	Primer & 2 coats enamel	▼	6	1.333	▼	2.38	33.50		35.88	57
0530	7 to 10 lite									
0540	Brushwork, primer	1 Pord	11	.727	Ea.	.72	18.20		18.92	31
0550	Finish coat, enamel		11	.727		.83	18.20		19.03	31
0560	Primer & 1 coat enamel		7	1.143		1.55	28.50		30.05	48.50
0570	Primer & 2 coats enamel	▼	5	1.600	▼	2.38	40		42.38	68
0590	12 lite									
0600	Brushwork, primer	1 Pord	10	.800	Ea.	.72	20		20.72	34
0610	Finish coat, enamel		10	.800		.83	20		20.83	34
0620	Primer & 1 coat enamel		6	1.333		1.55	33.50		35.05	56
0630	Primer & 2 coats enamel	▼	5	1.600	▼	2.38	40		42.38	68
0650	For oil base paint, add					10%				

09 91 23.39 Doors and Windows, Interior Latex, Zero Voc

09 91 23.39 Doors and Windows, Interior Latex, Zero Voc			Crew	Daily Output	Labor-Hours	Unit	Material	2009 Bare Costs Labor	Equipment	Total	Total Incl O&P
0010	**DOORS & WINDOWS, INTERIOR LATEX, ZERO VOC**										
0100	Doors flush, both sides, incl. frame & trim										
0110	Roll & brush, primer	G	1 Pord	10	.800	Ea.	4.75	20		24.75	38.50
0120	Finish coat, latex	G		10	.800		4.82	20		24.82	38.50
0130	Primer & 1 coat latex	G		7	1.143		9.55	28.50		38.05	57.50
0140	Primer & 2 coats latex	G		5	1.600		14.10	40		54.10	81
0160	Spray, both sides, primer	G		20	.400		5	10		15	22
0170	Finish coat, latex	G		20	.400		5.05	10		15.05	22
0180	Primer & 1 coat latex	G		11	.727		10.10	18.20		28.30	41
0190	Primer & 2 coats latex	G	▼	8	1	▼	14.95	25		39.95	57.50
0200	Doors, French, both sides, 10-15 lite, incl. frame & trim										
0210	Roll & brush, primer	G	1 Pord	6	1.333	Ea.	2.38	33.50		35.88	57
0220	Finish coat, latex	G		6	1.333		2.41	33.50		35.91	57
0230	Primer & 1 coat latex	G		3	2.667		4.79	66.50		71.29	114
0240	Primer & 2 coats latex	G	▼	2	4	▼	7.05	100		107.05	172
0360	Doors, panel, both sides, incl. frame & trim										
0370	Roll & brush, primer	G	1 Pord	6	1.333	Ea.	5	33.50		38.50	60
0380	Finish coat, latex	G		6	1.333		4.82	33.50		38.32	60
0390	Primer & 1 coat, latex	G		3	2.667		9.55	66.50		76.05	120
0400	Primer & 2 coats, latex	G		2.50	3.200		14.40	80		94.40	147
0420	Spray, both sides, primer	G		10	.800		5	20		25	38.50
0430	Finish coat, latex	G		10	.800		5.05	20		25.05	38.50
0440	Primer & 1 coat, latex	G		5	1.600		10.10	40		50.10	76.50
0450	Primer & 2 coats, latex	G	▼	4	2	▼	15.25	50		65.25	99
0460	Windows, per interior side, based on 15 SF										
0470	1 to 6 lite										

09 91 Painting

09 91 23 – Interior Painting

09 91 23.39 Doors and Windows, Interior Latex, Zero Voc		Crew	Daily Output	Labor-Hours	Unit	Material	2009 Bare Costs Labor	Equipment	Total	Total Incl O&P
0480	Brushwork, primer	G 1 Pord	13	.615	Ea.	.94	15.40		16.34	26
0490	Finish coat, enamel	G	13	.615		.95	15.40		16.35	26
0500	Primer & 1 coat enamel	G	8	1		1.89	25		26.89	43
0510	Primer & 2 coats enamel	G	6	1.333		2.84	33.50		36.34	57.50

09 91 23.40 Floors, Interior

		Crew	Daily Output	Labor-Hours	Unit	Material	Labor	Equipment	Total	Total Incl O&P
0010	**FLOORS, INTERIOR**									
0100	Concrete paint, latex									
0110	Brushwork									
0120	1st coat	1 Pord	975	.008	S.F.	.14	.21		.35	.50
0130	2nd coat		1150	.007		.10	.17		.27	.40
0140	3rd coat		1300	.006		.08	.15		.23	.33
0150	Roll									
0160	1st coat	1 Pord	2600	.003	S.F.	.19	.08		.27	.34
0170	2nd coat		3250	.002		.12	.06		.18	.23
0180	3rd coat		3900	.002		.09	.05		.14	.17
0190	Spray									
0200	1st coat	1 Pord	2600	.003	S.F.	.17	.08		.25	.31
0210	2nd coat		3250	.002		.09	.06		.15	.20
0220	3rd coat		3900	.002		.07	.05		.12	.16
0300	Acid stain and sealer									
0310	Stain, one coat	1 Pord	650	.012	S.F.	.11	.31		.42	.62
0320	Two coats		570	.014		.22	.35		.57	.81
0330	Acrylic sealer, one coat		2600	.003		.15	.08		.23	.30
0340	Two coats		1400	.006		.30	.14		.44	.56

09 91 23.52 Miscellaneous, Interior

		Crew	Daily Output	Labor-Hours	Unit	Material	Labor	Equipment	Total	Total Incl O&P
0010	**MISCELLANEOUS, INTERIOR**									
2400	Floors, conc./wood, oil base, primer/sealer coat, brushwork	2 Pord	1950	.008	S.F.	.06	.21		.27	.40
2450	Roller		5200	.003		.06	.08		.14	.19
2600	Spray		6000	.003		.06	.07		.13	.17
2650	Paint 1 coat, brushwork		1950	.008		.08	.21		.29	.43
2800	Roller		5200	.003		.09	.08		.17	.23
2850	Spray		6000	.003		.09	.07		.16	.21
3000	Stain, wood floor, brushwork, 1 coat		4550	.004		.06	.09		.15	.21
3200	Roller		5200	.003		.06	.08		.14	.20
3250	Spray		6000	.003		.06	.07		.13	.18
3400	Varnish, wood floor, brushwork		4550	.004		.08	.09		.17	.22
3450	Roller		5200	.003		.08	.08		.16	.22
3600	Spray		6000	.003		.08	.07		.15	.20
3800	Grilles, per side, oil base, primer coat, brushwork	1 Pord	520	.015		.10	.38		.48	.74
3850	Spray		1140	.007		.11	.18		.29	.41
3880	Paint 1 coat, brushwork		520	.015		.12	.38		.50	.76
3900	Spray		1140	.007		.13	.18		.31	.43
3920	Paint 2 coats, brushwork		325	.025		.23	.62		.85	1.26
3940	Spray		650	.012		.26	.31		.57	.79
4500	Louvers, one side, primer, brushwork		524	.015		.06	.38		.44	.69
4520	Paint one coat, brushwork		520	.015		.05	.38		.43	.69
4530	Spray		1140	.007		.06	.18		.24	.36
4540	Paint two coats, brushwork		325	.025		.11	.62		.73	1.13
4550	Spray		650	.012		.12	.31		.43	.63
4560	Paint three coats, brushwork		270	.030		.16	.74		.90	1.38
4570	Spray		500	.016		.18	.40		.58	.85
5000	Pipe, 1 – 4" diameter, primer or sealer coat, oil base, brushwork	2 Pord	1250	.013	L.F.	.06	.32		.38	.58

09 91 Painting

09 91 23 – Interior Painting

09 91 23.52 Miscellaneous, Interior

		Crew	Daily Output	Labor-Hours	Unit	Material	2009 Bare Costs Labor	Equipment	Total	Total Incl O&P
5100	Spray	2 Pord	2165	.007	L.F.	.06	.18		.24	.36
5200	Paint 1 coat, brushwork		1250	.013		.06	.32		.38	.59
5300	Spray		2165	.007		.06	.18		.24	.36
5350	Paint 2 coats, brushwork		775	.021		.11	.52		.63	.97
5400	Spray		1240	.013		.12	.32		.44	.67
5450	5" - 8" diameter, primer or sealer coat, brushwork		620	.026		.12	.65		.77	1.19
5500	Spray		1085	.015		.19	.37		.56	.81
5550	Paint 1 coat, brushwork		620	.026		.17	.65		.82	1.25
5600	Spray		1085	.015		.19	.37		.56	.81
5650	Paint 2 coats, brushwork		385	.042		.22	1.04		1.26	1.95
5700	Spray		620	.026		.25	.65		.90	1.33
6600	Radiators, per side, primer, brushwork	1 Pord	520	.015	S.F.	.06	.38		.44	.69
6620	Paint, one coat		520	.015		.05	.38		.43	.69
6640	Two coats		340	.024		.11	.59		.70	1.08
6660	Three coats		283	.028		.16	.71		.87	1.33
7000	Trim, wood, incl. puttying, under 6" wide									
7200	Primer coat, oil base, brushwork	1 Pord	650	.012	L.F.	.03	.31		.34	.53
7250	Paint, 1 coat, brushwork		650	.012		.03	.31		.34	.53
7400	2 coats		400	.020		.06	.50		.56	.88
7450	3 coats		325	.025		.09	.62		.71	1.10
7500	Over 6" wide, primer coat, brushwork		650	.012		.05	.31		.36	.56
7550	Paint, 1 coat, brushwork		650	.012		.06	.31		.37	.57
7600	2 coats		400	.020		.12	.50		.62	.95
7650	3 coats		325	.025		.17	.62		.79	1.20
8000	Cornice, simple design, primer coat, oil base, brushwork		650	.012	S.F.	.05	.31		.36	.56
8250	Paint, 1 coat		650	.012		.06	.31		.37	.57
8300	2 coats		400	.020		.12	.50		.62	.95
8350	Ornate design, primer coat		350	.023		.05	.57		.62	1
8400	Paint, 1 coat		350	.023		.06	.57		.63	1.01
8450	2 coats		400	.020		.12	.50		.62	.95
8600	Balustrades, primer coat, oil base, brushwork		520	.015		.05	.38		.43	.69
8650	Paint, 1 coat		520	.015		.06	.38		.44	.70
8700	2 coats		325	.025		.12	.62		.74	1.14
8900	Trusses and wood frames, primer coat, oil base, brushwork		800	.010		.05	.25		.30	.47
8950	Spray		1200	.007		.05	.17		.22	.33
9000	Paint 1 coat, brushwork		750	.011		.06	.27		.33	.51
9200	Spray		1200	.007		.07	.17		.24	.34
9220	Paint 2 coats, brushwork		500	.016		.12	.40		.52	.79
9240	Spray		600	.013		.13	.33		.46	.69
9260	Stain, brushwork, wipe off		600	.013		.06	.33		.39	.62
9280	Varnish, 3 coats, brushwork		275	.029		.23	.73		.96	1.44
9350	For latex paint, deduct					10%				

09 91 23.72 Walls and Ceilings, Interior

		Crew	Daily Output	Labor-Hours	Unit	Material	2009 Bare Costs Labor	Equipment	Total	Total Incl O&P
0010	**WALLS AND CEILINGS, INTERIOR**									
0100	Concrete, drywall or plaster, oil base, primer or sealer coat									
0200	Smooth finish, brushwork	1 Pord	1150	.007	S.F.	.05	.17		.22	.35
0240	Roller		1350	.006		.05	.15		.20	.29
0280	Spray		2750	.003		.04	.07		.11	.16
0300	Sand finish, brushwork		975	.008		.05	.21		.26	.39
0340	Roller		1150	.007		.05	.17		.22	.35
0380	Spray		2275	.004		.04	.09		.13	.19
0400	Paint 1 coat, smooth finish, brushwork		1200	.007		.06	.17		.23	.33

09 91 23.72 Walls and Ceilings, Interior

		Crew	Daily Output	Labor-Hours	Unit	Material	2009 Bare Costs Labor	Equipment	Total	Total Incl O&P
0440	Roller	1 Pord	1300	.006	S.F.	.06	.15		.21	.31
0480	Spray		2275	.004		.05	.09		.14	.20
0500	Sand finish, brushwork		1050	.008		.06	.19		.25	.37
0540	Roller		1600	.005		.06	.13		.19	.26
0580	Spray		2100	.004		.05	.10		.15	.22
0800	Paint 2 coats, smooth finish, brushwork		680	.012		.11	.29		.40	.61
0840	Roller		800	.010		.12	.25		.37	.54
0880	Spray		1625	.005		.10	.12		.22	.31
0900	Sand finish, brushwork		605	.013		.11	.33		.44	.66
0940	Roller		1020	.008		.12	.20		.32	.45
0980	Spray		1700	.005		.10	.12		.22	.30
1200	Paint 3 coats, smooth finish, brushwork		510	.016		.17	.39		.56	.82
1240	Roller		650	.012		.18	.31		.49	.69
1280	Spray		1625	.005		.15	.12		.27	.37
1300	Sand finish, brushwork		454	.018		.17	.44		.61	.90
1340	Roller		680	.012		.18	.29		.47	.67
1380	Spray		1133	.007		.15	.18		.33	.46
1600	Glaze coating, 2 coats, spray, clear		1200	.007		.49	.17		.66	.81
1640	Multicolor		1200	.007		.95	.17		1.12	1.32
1660	Painting walls, complete, including surface prep, primer &									
1670	2 coats finish, on drywall or plaster, with roller	1 Pord	325	.025	S.F.	.18	.62		.80	1.21
1700	For latex paint, deduct					10%				
1800	For ceiling installations, add						25%			
2000	Masonry or concrete block, oil base, primer or sealer coat									
2100	Smooth finish, brushwork	1 Pord	1224	.007	S.F.	.06	.16		.22	.34
2180	Spray		2400	.003		.09	.08		.17	.24
2200	Sand finish, brushwork		1089	.007		.10	.18		.28	.41
2280	Spray		2400	.003		.09	.08		.17	.24
2400	Paint 1 coat, smooth finish, brushwork		1100	.007		.10	.18		.28	.41
2480	Spray		2400	.003		.09	.08		.17	.24
2500	Sand finish, brushwork		979	.008		.10	.20		.30	.44
2580	Spray		2400	.003		.09	.08		.17	.24
2800	Paint 2 coats, smooth finish, brushwork		756	.011		.20	.26		.46	.65
2880	Spray		1360	.006		.18	.15		.33	.44
2900	Sand finish, brushwork		672	.012		.20	.30		.50	.71
2980	Spray		1360	.006		.18	.15		.33	.44
3200	Paint 3 coats, smooth finish, brushwork		560	.014		.30	.36		.66	.92
3280	Spray		1088	.007		.28	.18		.46	.60
3300	Sand finish, brushwork		498	.016		.30	.40		.70	.99
3380	Spray		1088	.007		.28	.18		.46	.60
3600	Glaze coating, 3 coats, spray, clear		900	.009		.70	.22		.92	1.13
3620	Multicolor		900	.009		1.10	.22		1.32	1.57
4000	Block filler, 1 coat, brushwork		425	.019		.13	.47		.60	.92
4100	Silicone, water repellent, 2 coats, spray		2000	.004		.30	.10		.40	.49
4120	For latex paint, deduct					10%				
8200	For work 8 – 15' H, add						10%			
8300	For work over 15' H, add						20%			
8400	For light textured surfaces, add						10%			
8410	Heavy textured, add						25%			

09 91 23.74 Walls and Ceilings, Interior, Zero VOC Latex

0010	**WALLS AND CEILINGS, INTERIOR, ZERO VOC LATEX**									
0100	Concrete, dry wall or plaster, latex, primer or sealer coat									

09 91 Painting

09 91 23 – Interior Painting

09 91 23.74 Walls and Ceilings, Interior, Zero VOC Latex

			Crew	Daily Output	Labor-Hours	Unit	Material	2009 Bare Costs Labor	Equipment	Total	Total Incl O&P
0200	Smooth finish, brushwork	G	1 Pord	1150	.007	S.F.	.06	.17		.23	.36
0240	Roller	G		1350	.006		.06	.15		.21	.30
0280	Spray	G		2750	.003		.05	.07		.12	.17
0300	Sand finish, brushwork	G		975	.008		.06	.21		.27	.40
0340	Roller	G		1150	.007		.07	.17		.24	.36
0380	Spray	G		2275	.004		.05	.09		.14	.20
0400	Paint 1 coat, smooth finish, brushwork	G		1200	.007		.06	.17		.23	.34
0440	Roller	G		1300	.006		.06	.15		.21	.32
0480	Spray	G		2275	.004		.05	.09		.14	.20
0500	Sand finish, brushwork	G		1050	.008		.06	.19		.25	.37
0540	Roller	G		1600	.005		.06	.13		.19	.27
0580	Spray	G		2100	.004		.05	.10		.15	.22
0800	Paint 2 coats, smooth finish, brushwork	G		680	.012		.12	.29		.41	.61
0840	Roller	G		800	.010		.12	.25		.37	.54
0880	Spray	G		1625	.005		.10	.12		.22	.31
0900	Sand finish, brushwork	G		605	.013		.11	.33		.44	.67
0940	Roller	G		1020	.008		.12	.20		.32	.45
0980	Spray	G		1700	.005		.10	.12		.22	.30
1200	Paint 3 coats, smooth finish, brushwork	G		510	.016		.17	.39		.56	.83
1240	Roller	G		650	.012		.18	.31		.49	.70
1280	Spray	G		1625	.005		.16	.12		.28	.37
1800	For ceiling installations, add	G						25%			
8200	For work 8 – 15' H, add							10%			
8300	For work over 15' H, add							20%			

09 91 23.75 Dry Fall Painting

			Crew	Daily Output	Labor-Hours	Unit	Material	2009 Bare Costs Labor	Equipment	Total	Total Incl O&P
0010	**DRY FALL PAINTING**										
0100	Walls										
0200	Wallboard and smooth plaster, one coat, brush		1 Pord	910	.009	S.F.	.05	.22		.27	.41
0210	Roll			1560	.005		.05	.13		.18	.26
0220	Spray			2600	.003		.05	.08		.13	.18
0230	Two coats, brush			520	.015		.10	.38		.48	.74
0240	Roll			877	.009		.10	.23		.33	.48
0250	Spray			1560	.005		.10	.13		.23	.32
0260	Concrete or textured plaster, one coat, brush			747	.011		.05	.27		.32	.49
0270	Roll			1300	.006		.05	.15		.20	.30
0280	Spray			1560	.005		.05	.13		.18	.26
0290	Two coats, brush			422	.019		.10	.47		.57	.89
0300	Roll			747	.011		.10	.27		.37	.55
0310	Spray			1300	.006		.10	.15		.25	.36
0320	Concrete block, one coat, brush			747	.011		.05	.27		.32	.49
0330	Roll			1300	.006		.05	.15		.20	.30
0340	Spray			1560	.005		.05	.13		.18	.26
0350	Two coats, brush			422	.019		.10	.47		.57	.89
0360	Roll			747	.011		.10	.27		.37	.55
0370	Spray			1300	.006		.10	.15		.25	.36
0380	Wood, one coat, brush			747	.011		.05	.27		.32	.49
0390	Roll			1300	.006		.05	.15		.20	.30
0400	Spray			877	.009		.05	.23		.28	.42
0410	Two coats, brush			487	.016		.10	.41		.51	.78
0420	Roll			747	.011		.10	.27		.37	.55
0430	Spray			650	.012		.10	.31		.41	.61
0440	Ceilings										

09 91 Painting

09 91 23 – Interior Painting

09 91 23.75 Dry Fall Painting		Crew	Daily Output	Labor-Hours	Unit	Material	2009 Bare Costs Labor	Equipment	Total	Total Incl O&P
0450	Wallboard and smooth plaster, one coat, brush	1 Pord	600	.013	S.F.	.05	.33		.38	.60
0460	Roll		1040	.008		.05	.19		.24	.36
0470	Spray		1560	.005		.05	.13		.18	.26
0480	Two coats, brush		341	.023		.10	.59		.69	1.07
0490	Roll		650	.012		.10	.31		.41	.61
0500	Spray		1300	.006		.10	.15		.25	.36
0510	Concrete or textured plaster, one coat, brush		487	.016		.05	.41		.46	.72
0520	Roll		877	.009		.05	.23		.28	.42
0530	Spray		1560	.005		.05	.13		.18	.26
0540	Two coats, brush		276	.029		.10	.72		.82	1.30
0550	Roll		520	.015		.10	.38		.48	.74
0560	Spray		1300	.006		.10	.15		.25	.36
0570	Structural steel, bar joists or metal deck, one coat, spray		1560	.005		.05	.13		.18	.26
0580	Two coats, spray		1040	.008		.10	.19		.29	.42

09 93 Staining and Transparent Finishing

09 93 23 – Interior Staining and Finishing

09 93 23.10 Varnish

		Crew	Daily Output	Labor-Hours	Unit	Material	2009 Bare Costs Labor	Equipment	Total	Total Incl O&P
0010	**VARNISH**									
0012	1 coat + sealer, on wood trim, brush, no sanding included	1 Pord	400	.020	S.F.	.07	.50		.57	.90
0100	Hardwood floors, 2 coats, no sanding included, roller	"	1890	.004	"	.15	.11		.26	.34

09 96 High-Performance Coatings

09 96 56 – Epoxy Coatings

09 96 56.20 Wall Coatings

		Crew	Daily Output	Labor-Hours	Unit	Material	2009 Bare Costs Labor	Equipment	Total	Total Incl O&P
0010	**WALL COATINGS**									
0100	Acrylic glazed coatings, minimum	1 Pord	525	.015	S.F.	.28	.38		.66	.93
0200	Maximum		305	.026		.60	.66		1.26	1.73
0300	Epoxy coatings, minimum		525	.015		.37	.38		.75	1.03
0400	Maximum		170	.047		1.12	1.18		2.30	3.16
0600	Exposed aggregate, troweled on, 1/16" to 1/4", minimum		235	.034		.56	.85		1.41	2.01
0700	Maximum (epoxy or polyacrylate)		130	.062		1.20	1.54		2.74	3.84
0900	1/2" to 5/8" aggregate, minimum		130	.062		1.11	1.54		2.65	3.74
1000	Maximum		80	.100		1.90	2.50		4.40	6.20
1200	1" aggregate size, minimum		90	.089		1.93	2.22		4.15	5.75
1300	Maximum		55	.145		2.95	3.64		6.59	9.20
1500	Exposed aggregate, sprayed on, 1/8" aggregate, minimum		295	.027		.51	.68		1.19	1.67
1600	Maximum		145	.055		.96	1.38		2.34	3.32

Division 10
Specialties

10 21 16.10 Partitions, Shower	Crew	Daily Output	Labor-Hours	Unit	Material	2009 Bare Costs Labor	Equipment	Total	Total Incl O&P
0010 **PARTITIONS, SHOWER** Floor mounted, no plumbing									
0100　Cabinet, incl. base, no door, painted steel, 1" thick walls	2 Shee	5	3.200	Ea.	905	99.50		1,004.50	1,175
0300　　With door, fiberglass		4.50	3.556		705	111		816	960
0600　　　Galvanized and painted steel, 1" thick walls		5	3.200		955	99.50		1,054.50	1,225
0800　Stall, 1" thick wall, no base, enameled steel		5	3.200		980	99.50		1,079.50	1,250
1500　Circular fiberglass, cabinet 36" diameter,		4	4		725	125		850	1,000
1700　　One piece, 36" diameter, less door		4	4		610	125		735	880
1800　　　With door		3.50	4.571		1,000	142		1,142	1,350
2400　Glass stalls, with doors, no receptors, chrome on brass		3	5.333		1,475	166		1,641	1,900
2700　　Anodized aluminum		4	4		1,025	125		1,150	1,325
3200　Receptors, precast terrazzo, 32" x 32"	2 Marb	14	1.143		242	31		273	315
3300　　48" x 34"		9.50	1.684		330	45.50		375.50	435
3500　　Plastic, simulated terrazzo receptor, 32" x 32"		14	1.143		100	31		131	161
3600　　　32" x 48"		12	1.333		190	36		226	269
3800　　Precast concrete, colors, 32" x 32"		14	1.143		187	31		218	257
3900　　　48" x 48"		8	2		390	54		444	520
4100　Shower doors, economy plastic, 24" wide	1 Shee	9	.889		128	27.50		155.50	187
4200　　Tempered glass door, economy		8	1		169	31		200	239
4400　　　Folding, tempered glass, aluminum frame		6	1.333		385	41.50		426.50	495
4700　　　Deluxe, tempered glass, chrome on brass frame, minimum		8	1		223	31		254	298
4800　　　　Maximum		1	8		880	249		1,129	1,400
4850　　　On anodized aluminum frame, minimum		2	4		144	125		269	365
4900　　　　Maximum		1	8		520	249		769	990
5100　Shower enclosure, tempered glass, anodized alum. frame									
5120　　2 panel & door, corner unit, 32" x 32"	1 Shee	2	4	Ea.	505	125		630	770
5140　　Neo-angle corner unit, 16" x 24" x 16"	"	2	4		930	125		1,055	1,225
5200　Shower surround, 3 wall, polypropylene, 32" x 32"	1 Carp	4	2		275	56		331	400
5220　　PVC, 32" x 32"		4	2		340	56		396	470
5240　　Fiberglass		4	2		370	56		426	505
5250　　2 wall, polypropylene, 32" x 32"		4	2		282	56		338	405
5270　　PVC		4	2		350	56		406	480
5290　　Fiberglass		4	2		355	56		411	485
5300　Tub doors, tempered glass & frame, minimum	1 Shee	8	1		204	31		235	277
5400　　Maximum		6	1.333		475	41.50		516.50	595
5600　　Chrome plated, brass frame, minimum		8	1		270	31		301	350
5700　　Maximum		6	1.333		660	41.50		701.50	795
5900　Tub/shower enclosure, temp. glass, alum. frame, minimum		2	4		365	125		490	610
6200　　Maximum		1.50	5.333		760	166		926	1,125
6500　　On chrome-plated brass frame, minimum		2	4		505	125		630	765
6600　　Maximum		1.50	5.333		1,075	166		1,241	1,450
6800　Tub surround, 3 wall, polypropylene	1 Carp	4	2		228	56		284	345
6900　　PVC		4	2		345	56		401	475
7000　　Fiberglass, minimum		4	2		360	56		416	490
7100　　Maximum		3	2.667		605	74.50		679.50	790

10 28 Toilet, Bath, and Laundry Accessories

10 28 13 - Toilet Accessories

10 28 13.13 Commercial Toilet Accessories

		Crew	Daily Output	Labor-Hours	Unit	Material	2009 Bare Costs Labor	2009 Bare Costs Equipment	Total	Total Incl O&P
0010	**COMMERCIAL TOILET ACCESSORIES**									
0200	Curtain rod, stainless steel, 5' long, 1" diameter	1 Carp	13	.615	Ea.	41	17.20		58.20	74.50
0300	1-1/4" diameter		13	.615		36.50	17.20		53.70	69
0800	Grab bar, straight, 1-1/4" diameter, stainless steel, 18" long		24	.333		29.50	9.30		38.80	48.50
1100	36" long		20	.400		36	11.20		47.20	58.50
1105	42" long		20	.400		42	11.20		53.20	65
3000	Mirror, with stainless steel 3/4" square frame, 18" x 24"		20	.400		54.50	11.20		65.70	79
3100	36" x 24"		15	.533		137	14.90		151.90	176
3300	72" x 24"		6	1.333		330	37.50		367.50	430
4300	Robe hook, single, regular		36	.222		15.50	6.20		21.70	27.50
4400	Heavy duty, concealed mounting		36	.222		15.55	6.20		21.75	27.50
6400	Towel bar, stainless steel, 18" long		23	.348		39.50	9.70		49.20	60
6500	30" long		21	.381		101	10.65		111.65	129
7400	Tumbler holder, tumbler only		30	.267		38.50	7.45		45.95	55
7500	Soap, tumbler & toothbrush		30	.267		25	7.45		32.45	40

10 28 16 - Bath Accessories

10 28 16.20 Medicine Cabinets

		Crew	Daily Output	Labor-Hours	Unit	Material	2009 Bare Costs Labor	2009 Bare Costs Equipment	Total	Total Incl O&P
0010	**MEDICINE CABINETS**									
0020	With mirror, st. st. frame, 16" x 22", unlighted	1 Carp	14	.571	Ea.	81.50	15.95		97.45	117
0100	Wood frame		14	.571		113	15.95		128.95	151
0300	Sliding mirror doors, 20" x 16" x 4-3/4", unlighted		7	1.143		101	32		133	165
0400	24" x 19" x 8-1/2", lighted		5	1.600		159	44.50		203.50	251
0600	Triple door, 30" x 32", unlighted, plywood body		7	1.143		239	32		271	315
0700	Steel body		7	1.143		315	32		347	400
0900	Oak door, wood body, beveled mirror, single door		7	1.143		154	32		186	223
1000	Double door		6	1.333		375	37.50		412.50	475

10 28 23 - Laundry Accessories

10 28 23.13 Built-In Ironing Boards

		Crew	Daily Output	Labor-Hours	Unit	Material	2009 Bare Costs Labor	2009 Bare Costs Equipment	Total	Total Incl O&P
0010	**BUILT-IN IRONING BOARDS**									
0020	Including cabinet, board & light, minimum	1 Carp	2	4	Ea.	299	112		411	520

10 31 Manufactured Fireplaces

10 31 13 - Manufactured Fireplace Chimneys

10 31 13.10 Fireplace Chimneys

		Crew	Daily Output	Labor-Hours	Unit	Material	2009 Bare Costs Labor	2009 Bare Costs Equipment	Total	Total Incl O&P
0010	**FIREPLACE CHIMNEYS**									
0500	Chimney dbl. wall, all stainless, over 8'-6", 7" diam., add to fireplace	1 Carp	33	.242	V.L.F.	70	6.80		76.80	88.50
0600	10" diameter, add to fireplace		32	.250		93	7		100	114
0700	12" diameter, add to fireplace		31	.258		111	7.20		118.20	134
0800	14" diameter, add to fireplace		30	.267		180	7.45		187.45	211
1000	Simulated brick chimney top, 4' high, 16" x 16"		10	.800	Ea.	244	22.50		266.50	305
1100	24" x 24"		7	1.143	"	455	32		487	555

10 31 13.20 Chimney Accessories

		Crew	Daily Output	Labor-Hours	Unit	Material	2009 Bare Costs Labor	2009 Bare Costs Equipment	Total	Total Incl O&P
0010	**CHIMNEY ACCESSORIES**									
0020	Chimney screens, galv., 13" x 13" flue	1 Bric	8	1	Ea.	52	28.50		80.50	105
0050	24" x 24" flue		5	1.600		121	45.50		166.50	208
0200	Stainless steel, 13" x 13" flue		8	1		320	28.50		348.50	395
0250	20" x 20" flue		5	1.600		435	45.50		480.50	555
2400	Squirrel and bird screens, galvanized, 8" x 8" flue		16	.500		45.50	14.20		59.70	73.50
2450	13" x 13" flue		12	.667		49	18.90		67.90	85.50

10 31 Manufactured Fireplaces

10 31 16 – Manufactured Fireplace Forms

10 31 16.10 Fireplace Forms		Crew	Daily Output	Labor-Hours	Unit	Material	2009 Bare Costs Labor	Equipment	Total	Total Incl O&P
0010	**FIREPLACE FORMS**									
1800	Fireplace forms, no accessories, 32" opening	1 Bric	3	2.667	Ea.	620	75.50		695.50	805
1900	36" opening		2.50	3.200		785	90.50		875.50	1,025
2000	40" opening		2	4		1,050	113		1,163	1,350
2100	78" opening		1.50	5.333		1,525	151		1,676	1,925

10 31 23 – Prefabricated Fireplaces

10 31 23.10 Fireplace, Prefabricated

		Crew	Daily Output	Labor-Hours	Unit	Material	2009 Bare Costs Labor	Equipment	Total	Total Incl O&P
0010	**FIREPLACE, PREFABRICATED**, free standing or wall hung									
0100	With hood & screen, minimum	1 Carp	1.30	6.154	Ea.	1,175	172		1,347	1,575
0150	Average		1	8		1,625	224		1,849	2,175
0200	Maximum		.90	8.889		3,125	248		3,373	3,850
1500	Simulated logs, gas fired, 40,000 BTU, 2' long, minimum		7	1.143	Set	400	32		432	495
1600	Maximum		6	1.333		940	37.50		977.50	1,100
1700	Electric, 1,500 BTU, 1'-6" long, minimum		7	1.143		166	32		198	237
1800	11,500 BTU, maximum		6	1.333		370	37.50		407.50	475
2000	Fireplace, built-in, 36" hearth, radiant		1.30	6.154	Ea.	560	172		732	910
2100	Recirculating, small fan		1	8		875	224		1,099	1,350
2150	Large fan		.90	8.889		1,825	248		2,073	2,425
2200	42" hearth, radiant		1.20	6.667		790	186		976	1,175
2300	Recirculating, small fan		.90	8.889		1,025	248		1,273	1,550
2350	Large fan		.80	10		1,825	280		2,105	2,475
2400	48" hearth, radiant		1.10	7.273		1,700	203		1,903	2,225
2500	Recirculating, small fan		.80	10		2,125	280		2,405	2,800
2550	Large fan		.70	11.429		3,275	320		3,595	4,150
3000	See through, including doors		.80	10		2,775	280		3,055	3,525
3200	Corner (2 wall)		1	8		2,775	224		2,999	3,450

10 32 Fireplace Specialties

10 32 13 – Fireplace Dampers

10 32 13.10 Dampers

		Crew	Daily Output	Labor-Hours	Unit	Material	2009 Bare Costs Labor	Equipment	Total	Total Incl O&P
0010	**DAMPERS**									
0800	Damper, rotary control, steel, 30" opening	1 Bric	6	1.333	Ea.	77.50	38		115.50	148
0850	Cast iron, 30" opening		6	1.333		86	38		124	157
1200	Steel plate, poker control, 60" opening		8	1		273	28.50		301.50	345
1250	84" opening, special order		5	1.600		495	45.50		540.50	620
1400	"Universal" type, chain operated, 32" x 20" opening		8	1		212	28.50		240.50	280
1450	48" x 24" opening		5	1.600		315	45.50		360.50	425

10 32 23 – Fireplace Doors

10 32 23.10 Doors

		Crew	Daily Output	Labor-Hours	Unit	Material	2009 Bare Costs Labor	Equipment	Total	Total Incl O&P
0010	**DOORS**									
0400	Cleanout doors and frames, cast iron, 8" x 8"	1 Bric	12	.667	Ea.	36.50	18.90		55.40	71.50
0450	12" x 12"		10	.800		44	22.50		66.50	86
0500	18" x 24"		8	1		127	28.50		155.50	187
0550	Cast iron frame, steel door, 24" x 30"		5	1.600		275	45.50		320.50	375
1600	Dutch Oven door and frame, cast iron, 12" x 15" opening		13	.615		111	17.45		128.45	151
1650	Copper plated, 12" x 15" opening		13	.615		218	17.45		235.45	268

10 35 Stoves

10 35 13 – Heating Stoves

10 35 13.10 Woodburning Stoves	Crew	Daily Output	Labor-Hours	Unit	Material	2009 Bare Costs Labor	Equipment	Total	Total Incl O&P
0010 **WOODBURNING STOVES**									
0015 Cast iron, minimum	2 Carp	1.30	12.308	Ea.	1,125	345		1,470	1,825
0020 Average		1	16		1,525	445		1,970	2,425
0030 Maximum	↓	.80	20		2,750	560		3,310	3,975
0050 For gas log lighter, add				↓	37.50			37.50	41.50

10 44 Fire Protection Specialties

10 44 16 – Fire Extinguishers

10 44 16.13 Portable Fire Extinguishers

	Crew	Daily Output	Labor-Hours	Unit	Material	2009 Bare Costs Labor	Equipment	Total	Total Incl O&P
0010 **PORTABLE FIRE EXTINGUISHERS**									
0120 CO_2, portable with swivel horn, 5 lb.				Ea.	159			159	175
0140 With hose and "H" horn, 10 lb.				"	215			215	237
1000 Dry chemical, pressurized									
1040 Standard type, portable, painted, 2-1/2 lb.				Ea.	35			35	38.50
1080 10 lb.					91			91	100
1100 20 lb.					135			135	149
1120 30 lb.					295			295	325
2000 ABC all purpose type, portable, 2-1/2 lb.					34			34	37
2080 9-1/2 lb.				↓	75			75	82.50

10 55 Postal Specialties

10 55 23 – Mail Boxes

10 55 23.10 Mail Boxes

	Crew	Daily Output	Labor-Hours	Unit	Material	2009 Bare Costs Labor	Equipment	Total	Total Incl O&P
0011 **MAIL BOXES**									
1900 Letter slot, residential	1 Carp	20	.400	Ea.	86	11.20		97.20	113
2400 Residential, galv. steel, small 20" x 7" x 9"	1 Clab	16	.500		153	10.30		163.30	185
2410 With galv. steel post, 54" long		6	1.333		209	27.50		236.50	277
2420 Large, 24" x 12" x 15"		16	.500		134	10.30		144.30	164
2430 With galv. steel post, 54" long		6	1.333		241	27.50		268.50	310
2440 Decorative, polyethylene, 22" x 10" x 10"		16	.500		43	10.30		53.30	64.50
2450 With alum. post, decorative, 54" long	↓	6	1.333	↓	100	27.50		127.50	157

10 56 Storage Assemblies

10 56 13 – Metal Storage Shelving

10 56 13.10 Shelving

	Crew	Daily Output	Labor-Hours	Unit	Material	2009 Bare Costs Labor	Equipment	Total	Total Incl O&P
0010 **SHELVING**									
0020 Metal, industrial, cross-braced, 3' wide, 12" deep	1 Sswk	175	.046	SF Shlf	6.55	1.37		7.92	9.85
0100 24" deep		330	.024		4.70	.73		5.43	6.55
2200 Wide span, 1600 lb. capacity per shelf, 6' wide, 24" deep		380	.021		7	.63		7.63	8.90
2400 36" deep	↓	440	.018	↓	8.30	.54		8.84	10.15
3000 Residential, vinyl covered wire, wardrobe, 12" deep	1 Carp	195	.041	L.F.	5.90	1.15		7.05	8.45
3100 16" deep		195	.041		8	1.15		9.15	10.75
3200 Standard, 6" deep		195	.041		3.23	1.15		4.38	5.50
3300 9" deep		195	.041		4.86	1.15		6.01	7.30
3400 12" deep		195	.041		6.45	1.15		7.60	9.05
3500 16" deep		195	.041		9.90	1.15		11.05	12.85
3600 20" deep	↓	195	.041	↓	13.15	1.15		14.30	16.40

10 56 Storage Assemblies

10 56 13 – Metal Storage Shelving

10 56 13.10 Shelving	Crew	Daily Output	Labor-Hours	Unit	Material	2009 Bare Costs Labor	Equipment	Total	Total Incl O&P
3700	Support bracket	1 Carp	80	.100	Ea.	4.10	2.80		6.90	9.25

10 57 Wardrobe and Closet Specialties

10 57 23 – Closet and Utility Shelving

10 57 23.19 Wood Closet and Utility Shelving

		Crew	Daily Output	Labor-Hours	Unit	Material	2009 Bare Costs Labor	Equipment	Total	Total Incl O&P
0010	**WOOD CLOSET AND UTILITY SHELVING**									
0020	Pine, clear grade, no edge band, 1" x 8"	1 Carp	115	.070	L.F.	1.98	1.94		3.92	5.45
0100	1" x 10"		110	.073		2.22	2.03		4.25	5.90
0200	1" x 12"		105	.076		2.64	2.13		4.77	6.50
0450	1" x 18"		95	.084		3.96	2.35		6.31	8.35
0460	1" x 24"		85	.094		5.30	2.63		7.93	10.25
0600	Plywood, 3/4" thick with lumber edge, 12" wide		75	.107		1.67	2.98		4.65	6.90
0700	24" wide		70	.114		3	3.19		6.19	8.70
0900	Bookcase, clear grade pine, shelves 12" O.C., 8" deep, /SF shelf		70	.114	S.F.	6.45	3.19		9.64	12.50
1000	12" deep shelves		65	.123	"	8.60	3.44		12.04	15.25
1200	Adjustable closet rod and shelf, 12" wide, 3' long		20	.400	Ea.	8.50	11.20		19.70	28.50
1300	8' long		15	.533	"	24	14.90		38.90	51.50
1500	Prefinished shelves with supports, stock, 8" wide		75	.107	L.F.	3.80	2.98		6.78	9.25
1600	10" wide		70	.114	"	4.21	3.19		7.40	10.05

10 73 Protective Covers

10 73 16 – Canopies

10 73 16.10 Canopies, Residential

		Crew	Daily Output	Labor-Hours	Unit	Material	2009 Bare Costs Labor	Equipment	Total	Total Incl O&P
0010	**CANOPIES, RESIDENTIAL** Prefabricated									
0500	Carport, free standing, baked enamel, alum., .032", 40 psf									
0520	16' x 8', 4 posts	2 Carp	3	5.333	Ea.	4,075	149		4,224	4,725
0600	20' x 10', 6 posts		2	8		4,250	224		4,474	5,050
0605	30' x 10', 8 posts		2	8		6,375	224		6,599	7,375
1000	Door canopies, extruded alum., .032", 42" projection, 4' wide	1 Carp	8	1		285	28		313	365
1020	6' wide	"	6	1.333		360	37.50		397.50	460
1040	8' wide	2 Carp	9	1.778		455	49.50		504.50	585
1060	10' wide		7	2.286		670	64		734	845
1080	12' wide		5	3.200		775	89.50		864.50	1,000
1200	54" projection, 4' wide	1 Carp	8	1		335	28		363	420
1220	6' wide	"	6	1.333		420	37.50		457.50	525
1240	8' wide	2 Carp	9	1.778		485	49.50		534.50	620
1260	10' wide		7	2.286		725	64		789	905
1280	12' wide		5	3.200		775	89.50		864.50	1,000
1300	Painted, add					20%				
1310	Bronze anodized, add					50%				
3000	Window awnings, aluminum, window 3' high, 4' wide	1 Carp	10	.800		325	22.50		347.50	400
3020	6' wide	"	8	1		330	28		358	415
3040	9' wide	2 Carp	9	1.778		310	49.50		359.50	425
3060	12' wide	"	5	3.200		360	89.50		449.50	545
3100	Window, 4' high, 4' wide	1 Carp	10	.800		162	22.50		184.50	216
3120	6' wide	"	8	1		230	28		258	300
3140	9' wide	2 Carp	9	1.778		385	49.50		434.50	505
3160	12' wide	"	5	3.200		410	89.50		499.50	605
3200	Window, 6' high, 4' wide	1 Carp	10	.800		241	22.50		263.50	305

10 73 Protective Covers

10 73 16 – Canopies

10 73 16.10 Canopies, Residential		Crew	Daily Output	Labor-Hours	Unit	Material	2009 Bare Costs Labor	Equipment	Total	Total Incl O&P
3220	6' wide	1 Carp	8	1	Ea.	370	28		398	455
3240	9' wide	2 Carp	9	1.778		875	49.50		924.50	1,050
3260	12' wide	"	5	3.200		1,350	89.50		1,439.50	1,625
3400	Roll-up aluminum, 2'-6" wide	1 Carp	14	.571		185	15.95		200.95	231
3420	3' wide		12	.667		196	18.65		214.65	247
3440	4' wide		10	.800		226	22.50		248.50	286
3460	6' wide		8	1		278	28		306	355
3480	9' wide	2 Carp	9	1.778		360	49.50		409.50	480
3500	12' wide	"	5	3.200		465	89.50		554.50	660
3600	Window awnings, canvas, 24" drop, 3' wide	1 Carp	30	.267	L.F.	52.50	7.45		59.95	70.50
3620	4' wide		40	.200		45.50	5.60		51.10	59.50
3700	30" drop, 3' wide		30	.267		79.50	7.45		86.95	99.50
3720	4' wide		40	.200		66	5.60		71.60	82.50
3740	5' wide		45	.178		58.50	4.97		63.47	73
3760	6' wide		48	.167		53	4.66		57.66	66.50
3780	8' wide		48	.167		43.50	4.66		48.16	55.50
3800	10' wide		50	.160		41	4.47		45.47	53

10 74 Manufactured Exterior Specialties

10 74 23 – Cupolas

10 74 23.10 Wood Cupolas		Crew	Daily Output	Labor-Hours	Unit	Material	2009 Bare Costs Labor	Equipment	Total	Total Incl O&P
0010	**WOOD CUPOLAS**									
0020	Stock units, pine, painted, 18" sq., 28" high, alum. roof	1 Carp	4.10	1.951	Ea.	162	54.50		216.50	270
0100	Copper roof		3.80	2.105		189	59		248	310
0300	23" square, 33" high, aluminum roof		3.70	2.162		355	60.50		415.50	490
0400	Copper roof		3.30	2.424		480	68		548	640
0600	30" square, 37" high, aluminum roof		3.70	2.162		495	60.50		555.50	645
0700	Copper roof		3.30	2.424		535	68		603	705
0900	Hexagonal, 31" wide, 46" high, copper roof		4	2		660	56		716	820
1000	36" wide, 50" high, copper roof		3.50	2.286		1,150	64		1,214	1,375
1200	For deluxe stock units, add to above					25%				
1400	For custom built units, add to above					50%	50%			

10 74 33 – Weathervanes

10 74 33.10 Residential Weathervanes		Crew	Daily Output	Labor-Hours	Unit	Material	2009 Bare Costs Labor	Equipment	Total	Total Incl O&P
0010	**RESIDENTIAL WEATHERVANES**									
0020	Residential types, minimum	1 Carp	8	1	Ea.	183	28		211	250
0100	Maximum	"	2	4	"	1,750	112		1,862	2,125

10 75 Flagpoles

10 75 16 – Ground-Set Flagpoles

10 75 16.10 Flagpoles		Crew	Daily Output	Labor-Hours	Unit	Material	2009 Bare Costs Labor	Equipment	Total	Total Incl O&P
0010	**FLAGPOLES**, Ground set									
0050	Not including base or foundation									
0100	Aluminum, tapered, ground set 20' high	K-1	2	8	Ea.	970	200	121	1,291	1,550
0200	25' high		1.70	9.412		1,150	235	143	1,528	1,800
0300	30' high		1.50	10.667		1,250	266	162	1,678	2,000
0500	40' high		1.20	13.333		2,275	335	202	2,812	3,300

Division 11
Equipment

11 24 Maintenance Equipment

11 24 19 – Vacuum Cleaning Systems

11 24 19.10 Vacuum Cleaning

		Crew	Daily Output	Labor-Hours	Unit	Material	2009 Bare Costs Labor	Equipment	Total	Total Incl O&P
0010	**VACUUM CLEANING**									
0020	Central, 3 inlet, residential	1 Skwk	.90	8.889	Total	925	249		1,174	1,450
0400	5 inlet system, residential		.50	16		1,225	450		1,675	2,100
0600	7 inlet system, commercial		.40	20		1,400	560		1,960	2,475
0800	9 inlet system, residential	↓	.30	26.667		3,000	750		3,750	4,575
4010	Rule of thumb: First 1200 S.F., installed									1,180
4020	For each additional S.F., add				S.F.					.19

11 26 Unit Kitchens

11 26 13 – Metal Unit Kitchens

11 26 13.10 Commercial Unit Kitchens

		Crew	Daily Output	Labor-Hours	Unit	Material	2009 Bare Costs Labor	Equipment	Total	Total Incl O&P
0010	**COMMERCIAL UNIT KITCHENS**									
1500	Combination range, refrigerator and sink, 30" wide, minimum	L-1	2	5	Ea.	900	160		1,060	1,250
1550	Maximum		1	10		3,175	320		3,495	4,025
1570	60" wide, average		1.40	7.143		2,725	229		2,954	3,350
1590	72" wide, average	↓	1.20	8.333	↓	3,700	267		3,967	4,525

11 31 Residential Appliances

11 31 13 – Residential Kitchen Appliances

11 31 13.13 Cooking Equipment

		Crew	Daily Output	Labor-Hours	Unit	Material	2009 Bare Costs Labor	Equipment	Total	Total Incl O&P
0010	**COOKING EQUIPMENT**									
0020	Cooking range, 30" free standing, 1 oven, minimum	2 Clab	10	1.600	Ea.	320	33		353	405
0050	Maximum		4	4		1,775	82		1,857	2,100
0150	2 oven, minimum		10	1.600		1,450	33		1,483	1,650
0200	Maximum	↓	10	1.600		1,625	33		1,658	1,850
0350	Built-in, 30" wide, 1 oven, minimum	1 Elec	6	1.333		690	42		732	830
0400	Maximum	2 Carp	2	8		1,375	224		1,599	1,900
0500	2 oven, conventional, minimum		4	4		1,350	112		1,462	1,675
0550	1 conventional, 1 microwave, maximum	↓	2	8		1,000	224		1,224	1,475
0700	Free-standing, 1 oven, 21" wide range, minimum	2 Clab	10	1.600		315	33		348	400
0750	21" wide, maximum	"	4	4		325	82		407	500
0900	Countertop cooktops, 4 burner, standard, minimum	1 Elec	6	1.333		244	42		286	335
0950	Maximum		3	2.667		585	84		669	780
1050	As above, but with grill and griddle attachment, minimum		6	1.333		1,675	42		1,717	1,925
1100	Maximum		3	2.667		3,375	84		3,459	3,825
1250	Microwave oven, minimum		4	2		104	63		167	217
1300	Maximum		2	4		450	126		576	700
5380	Oven, built in, standard		4	2		440	63		503	585
5390	Deluxe	↓	2	4	↓	1,900	126		2,026	2,300

11 31 13.23 Refrigeration Equipment

		Crew	Daily Output	Labor-Hours	Unit	Material	2009 Bare Costs Labor	Equipment	Total	Total Incl O&P
0010	**REFRIGERATION EQUIPMENT**									
2000	Deep freeze, 15 to 23 C.F., minimum	2 Clab	10	1.600	Ea.	440	33		473	540
2050	Maximum		5	3.200		615	66		681	785
2200	30 C.F., minimum		8	2		805	41		846	955
2250	Maximum	↓	3	5.333		915	110		1,025	1,175
5200	Icemaker, automatic, 20 lb. per day	1 Plum	7	1.143		850	37		887	995
5350	51 lb. per day	"	2	4		1,175	129		1,304	1,500
5450	Refrigerator, no frost, 6 C.F.	2 Clab	15	1.067		238	22		260	298
5500	Refrigerator, no frost, 10 C.F. to 12 C.F. minimum	↓	10	1.600	↓	445	33		478	545

11 31 Residential Appliances

11 31 13 – Residential Kitchen Appliances

11 31 13.23 Refrigeration Equipment

11 31 13.23 Refrigeration Equipment		Crew	Daily Output	Labor-Hours	Unit	Material	2009 Bare Costs Labor	Equipment	Total	Total Incl O&P
5600	Maximum	2 Clab	6	2.667	Ea.	520	55		575	670
5750	14 C.F. to 16 C.F., minimum		9	1.778		460	36.50		496.50	565
5800	Maximum		5	3.200		620	66		686	795
5950	18 C.F. to 20 C.F., minimum		8	2		575	41		616	705
6000	Maximum		4	4		1,000	82		1,082	1,250
6150	21 C.F. to 29 C.F., minimum		7	2.286		855	47		902	1,025
6200	Maximum		3	5.333		2,950	110		3,060	3,425
6790	Refrigerator, energy-star qualified, 18 CF, minimum [G]	2 Carp	4	4		525	112		637	770
6795	Maximum [G]		2	8		1,050	224		1,274	1,525
6797	21.7 CF, minimum [G]		4	4		875	112		987	1,150
6799	Maximum [G]		4	4		1,150	112		1,262	1,475

11 31 13.33 Kitchen Cleaning Equipment

11 31 13.33 Kitchen Cleaning Equipment		Crew	Daily Output	Labor-Hours	Unit	Material	Labor	Equipment	Total	Total Incl O&P
0010	KITCHEN CLEANING EQUIPMENT									
2750	Dishwasher, built-in, 2 cycles, minimum	L-1	4	2.500	Ea.	206	80		286	360
2800	Maximum		2	5		310	160		470	605
2950	4 or more cycles, minimum		4	2.500		310	80		390	470
2960	Average		4	2.500		410	80		490	580
3000	Maximum		2	5		1,000	160		1,160	1,400
3100	Energy-star qualified, minimum [G]		4	2.500		315	80		395	475
3110	Maximum [G]		2	5		1,000	160		1,160	1,375

11 31 13.43 Waste Disposal Equipment

11 31 13.43 Waste Disposal Equipment		Crew	Daily Output	Labor-Hours	Unit	Material	Labor	Equipment	Total	Total Incl O&P
0010	WASTE DISPOSAL EQUIPMENT									
1750	Compactor, residential size, 4 to 1 compaction, minimum	1 Carp	5	1.600	Ea.	485	44.50		529.50	605
1800	Maximum	"	3	2.667		515	74.50		589.50	690
3300	Garbage disposal, sink type, minimum	L-1	10	1		66.50	32		98.50	126
3350	Maximum	"	10	1		192	32		224	264

11 31 13.53 Kitchen Ventilation Equipment

11 31 13.53 Kitchen Ventilation Equipment		Crew	Daily Output	Labor-Hours	Unit	Material	Labor	Equipment	Total	Total Incl O&P
0010	KITCHEN VENTILATION EQUIPMENT									
4150	Hood for range, 2 speed, vented, 30" wide, minimum	L-3	5	2	Ea.	52.50	57.50		110	154
4200	Maximum		3	3.333		810	95.50		905.50	1,050
4300	42" wide, minimum		5	2		250	57.50		307.50	370
4330	Custom		5	2		1,400	57.50		1,457.50	1,650
4350	Maximum		3	3.333		1,725	95.50		1,820.50	2,050
4500	For ventless hood, 2 speed, add					16.55			16.55	18.20
4650	For vented 1 speed, deduct from maximum					44			44	48.50

11 31 23 – Residential Laundry Appliances

11 31 23.13 Washers

11 31 23.13 Washers		Crew	Daily Output	Labor-Hours	Unit	Material	Labor	Equipment	Total	Total Incl O&P
0010	WASHERS									
6650	Washing machine, automatic, minimum	1 Plum	3	2.667	Ea.	315	86		401	485
6700	Maximum		1	8		1,100	258		1,358	1,625
6750	Energy star qualified, front loading, minimum [G]		3	2.667		510	86		596	700
6760	Maximum [G]		1	8		1,600	258		1,858	2,175
6764	Top loading, minimum [G]		3	2.667		485	86		571	675
6766	Maximum [G]		3	2.667		990	86		1,076	1,250

11 31 23.23 Dryers

11 31 23.23 Dryers		Crew	Daily Output	Labor-Hours	Unit	Material	Labor	Equipment	Total	Total Incl O&P
0010	DRYERS									
6770	Electric, front loading, energy-star qualified, minimum [G]	L-2	3	5.333	Ea.	255	130		385	500
6780	Maximum [G]	"	2	8		1,425	195		1,620	1,875
7450	Vent kits for dryers	1 Carp	10	.800		31.50	22.50		54	73

11 31 Residential Appliances

11 31 33 – Miscellaneous Residential Appliances

11 31 33.13 Sump Pumps	Crew	Daily Output	Labor-Hours	Unit	Material	2009 Bare Costs Labor	Equipment	Total	Total Incl O&P
0010 **SUMP PUMPS**									
6400 Cellar drainer, pedestal, 1/3 H.P., molded PVC base	1 Plum	3	2.667	Ea.	119	86		205	272
6450 Solid brass	"	2	4	"	263	129		392	500
6460 Sump pump, see also Div. 22 14 29.16									

11 31 33.23 Water Heaters

	Crew	Daily Output	Labor-Hours	Unit	Material	2009 Bare Costs Labor	Equipment	Total	Total Incl O&P
0010 **WATER HEATERS**									
6900 Electric, glass lined, 30 gallon, minimum	L-1	5	2	Ea.	420	64		484	570
6950 Maximum		3	3.333		585	107		692	820
7100 80 gallon, minimum		2	5		760	160		920	1,100
7150 Maximum	▼	1	10		1,050	320		1,370	1,675
7180 Gas, glass lined, 30 gallon, minimum	2 Plum	5	3.200		565	103		668	790
7220 Maximum		3	5.333		785	172		957	1,150
7260 50 gallon, minimum		2.50	6.400		845	206		1,051	1,275
7300 Maximum	▼	1.50	10.667	▼	1,175	345		1,520	1,850
7310 Water heater, see also Div. 22 33 30.13									

11 31 33.43 Air Quality

	Crew	Daily Output	Labor-Hours	Unit	Material	2009 Bare Costs Labor	Equipment	Total	Total Incl O&P
0010 **AIR QUALITY**									
2450 Dehumidifier, portable, automatic, 15 pint				Ea.	150			150	165
2550 40 pint					179			179	197
3550 Heater, electric, built-in, 1250 watt, ceiling type, minimum	1 Elec	4	2		96	63		159	208
3600 Maximum		3	2.667		156	84		240	310
3700 Wall type, minimum		4	2		143	63		206	260
3750 Maximum		3	2.667		154	84		238	305
3900 1500 watt wall type, with blower		4	2		143	63		206	260
3950 3000 watt	▼	3	2.667		292	84		376	455
4850 Humidifier, portable, 8 gallons per day					160			160	176
5000 15 gallons per day				▼	193			193	212

11 33 Retractable Stairs

11 33 10 – Disappearing Stairs

11 33 10.10 Disappearing Stairway

	Crew	Daily Output	Labor-Hours	Unit	Material	2009 Bare Costs Labor	Equipment	Total	Total Incl O&P
0010 **DISAPPEARING STAIRWAY** No trim included									
0020 One piece, yellow pine, 8'-0" ceiling	2 Carp	4	4	Ea.	288	112		400	505
0030 9'-0" ceiling		4	4		330	112		442	555
0040 10'-0" ceiling		3	5.333		300	149		449	580
0050 11'-0" ceiling		3	5.333		545	149		694	850
0060 12'-0" ceiling	▼	3	5.333		545	149		694	850
0100 Custom grade, pine, 8'-6" ceiling, minimum	1 Carp	4	2		202	56		258	315
0150 Average		3.50	2.286		215	64		279	345
0200 Maximum		3	2.667		275	74.50		349.50	430
0500 Heavy duty, pivoted, from 7'-7" to 12'-10" floor to floor		3	2.667		650	74.50		724.50	840
0600 16'-0" ceiling		2	4		1,325	112		1,437	1,650
0800 Economy folding, pine, 8'-6" ceiling		4	2		195	56		251	310
0900 9'-6" ceiling	▼	4	2		225	56		281	345
1100 Automatic electric, aluminum, floor to floor height, 8' to 9'	2 Carp	1	16	▼	8,250	445		8,695	9,825

11 41 Food Storage Equipment

11 41 13 – Refrigerated Food Storage Cases

11 41 13.30 Wine Cellar	Crew	Daily Output	Labor-Hours	Unit	Material	2009 Bare Costs Labor	Equipment	Total	Total Incl O&P
0010 **WINE CELLAR**, refrigerated, Redwood interior, carpeted, walk-in type									
0020 6'-8" high, including racks									
0200 80" W x 48" D for 900 bottles	2 Carp	1.50	10.667	Ea.	3,350	298		3,648	4,175
0250 80" W x 72" D for 1300 bottles		1.33	12.030		4,425	335		4,760	5,425
0300 80" W x 94" D for 1900 bottles		1.17	13.675		5,725	380		6,105	6,950

Division 12
Furnishings

12 21 Window Blinds

12 21 13 – Horizontal Louver Blinds

12 21 13.13 Metal Horizontal Louver Blinds		Crew	Daily Output	Labor-Hours	Unit	Material	2009 Bare Costs Labor	Equipment	Total	Total Incl O&P
0010	**METAL HORIZONTAL LOUVER BLINDS**									
0020	Horizontal, 1" aluminum slats, solid color, stock	1 Carp	590	.014	S.F.	4.75	.38		5.13	5.90
0090	Custom, minimum		590	.014		5.25	.38		5.63	6.45
0100	Maximum		440	.018		7.15	.51		7.66	8.70
0450	Stock, minimum		590	.014		4.86	.38		5.24	6
0500	Maximum	▼	440	.018	▼	7.90	.51		8.41	9.55

12 22 Curtains and Drapes

12 22 16 – Drapery Track and Accessories

12 22 16.10 Drapery Hardware

		Crew	Daily Output	Labor-Hours	Unit	Material	2009 Bare Costs Labor	Equipment	Total	Total Incl O&P
0010	**DRAPERY HARDWARE**									
0030	Standard traverse, per foot, minimum	1 Carp	59	.136	L.F.	3.35	3.79		7.14	10.10
0100	Maximum		51	.157	"	7.90	4.38		12.28	16.05
0200	Decorative traverse, 28"-48", minimum		22	.364	Ea.	25	10.15		35.15	44.50
0220	Maximum		21	.381		43	10.65		53.65	65.50
0300	48"-84", minimum		20	.400		20.50	11.20		31.70	41.50
0320	Maximum		19	.421		58.50	11.75		70.25	84
0400	66"-120", minimum		18	.444		62	12.40		74.40	89
0420	Maximum		17	.471		78	13.15		91.15	108
0500	84"-156", minimum		16	.500		65	14		79	95
0520	Maximum		15	.533		98.50	14.90		113.40	134
0600	130"-240", minimum		14	.571		31	15.95		46.95	61
0620	Maximum	▼	13	.615		134	17.20		151.20	176
0700	Slide rings, each, minimum					1.41			1.41	1.55
0720	Maximum					2.32			2.32	2.55
3000	Ripplefold, snap-a-pleat system, 3' or less, minimum	1 Carp	15	.533		52	14.90		66.90	82.50
3020	Maximum	"	14	.571	▼	70.50	15.95		86.45	105
3200	Each additional foot, add, minimum				L.F.	2.19			2.19	2.41
3220	Maximum				"	6.65			6.65	7.30
4000	Traverse rods, adjustable, 28" to 48"	1 Carp	22	.364	Ea.	20.50	10.15		30.65	39.50
4020	48" to 84"		20	.400		29	11.20		40.20	51
4040	66" to 120"		18	.444		40	12.40		52.40	65
4060	84" to 156"		16	.500		41	14		55	68.50
4080	100" to 180"		14	.571		42	15.95		57.95	73
4100	228" to 312"		13	.615		58.50	17.20		75.70	93
4500	Curtain rod, 28" to 48", single		22	.364		5	10.15		15.15	22.50
4510	Double		22	.364		8.55	10.15		18.70	26.50
4520	48" to 86", single		20	.400		8.60	11.20		19.80	28.50
4530	Double		20	.400		14.30	11.20		25.50	34.50
4540	66" to 120", single		18	.444		14.40	12.40		26.80	37
4550	Double	▼	18	.444		22.50	12.40		34.90	45.50
4600	Valance, pinch pleated fabric, 12" deep, up to 54" long, minimum					36			36	39.50
4610	Maximum					90			90	99
4620	Up to 77" long, minimum					55			55	60.50
4630	Maximum					145			145	160
5000	Stationary rods, first 2'				▼	7.45			7.45	8.20

12 23 Interior Shutters

12 23 10 – Wood Interior Shutters

12 23 10.10 Wood Interior Shutters	Crew	Daily Output	Labor-Hours	Unit	Material	2009 Bare Costs Labor	Equipment	Total	Total Incl O&P
0010 **WOOD INTERIOR SHUTTERS**, louvered									
0200 Two panel, 27" wide, 36" high	1 Carp	5	1.600	Set	114	44.50		158.50	201
0300 33" wide, 36" high		5	1.600		149	44.50		193.50	240
0500 47" wide, 36" high		5	1.600		199	44.50		243.50	294
1000 Four panel, 27" wide, 36" high		5	1.600		134	44.50		178.50	224
1100 33" wide, 36" high		5	1.600		165	44.50		209.50	257
1300 47" wide, 36" high		5	1.600		234	44.50		278.50	335

12 23 10.13 Wood Panels

	Crew	Daily Output	Labor-Hours	Unit	Material	Labor	Equipment	Total	Total Incl O&P
0010 **WOOD PANELS**									
3000 Wood folding panels with movable louvers, 7" x 20" each	1 Carp	17	.471	Pr.	46	13.15		59.15	72.50
3300 8" x 28" each		17	.471		60	13.15		73.15	88
3450 9" x 36" each		17	.471		74	13.15		87.15	104
3600 10" x 40" each		17	.471		80	13.15		93.15	110
4000 Fixed louver type, stock units, 8" x 20" each		17	.471		79.50	13.15		92.65	110
4150 10" x 28" each		17	.471		88	13.15		101.15	119
4300 12" x 36" each		17	.471		104	13.15		117.15	136
4450 18" x 40" each		17	.471		168	13.15		181.15	207
5000 Insert panel type, stock, 7" x 20" each		17	.471		19.25	13.15		32.40	43
5150 8" x 28" each		17	.471		35.50	13.15		48.65	61
5300 9" x 36" each		17	.471		45	13.15		58.15	71.50
5450 10" x 40" each		17	.471		48	13.15		61.15	75
5600 Raised panel type, stock, 10" x 24" each		17	.471		134	13.15		147.15	170
5650 12" x 26" each		17	.471		156	13.15		169.15	194
5700 14" x 30" each		17	.471		210	13.15		223.15	253
5750 16" x 36" each		17	.471		288	13.15		301.15	335
6000 For custom built pine, add					22%				
6500 For custom built hardwood blinds, add					42%				

12 24 Window Shades

12 24 13 – Roller Window Shades

12 24 13.10 Shades

	Crew	Daily Output	Labor-Hours	Unit	Material	Labor	Equipment	Total	Total Incl O&P
0010 **SHADES**									
0020 Basswood, roll-up, stain finish, 3/8" slats	1 Carp	300	.027	S.F.	13.65	.75		14.40	16.30
5011 Insulative shades **G**		125	.064		10.75	1.79		12.54	14.85
6011 Solar screening, fiberglass **G**		85	.094		4.60	2.63		7.23	9.50
8011 Interior insulative shutter									
8111 Stock unit, 15" x 60" **G**	1 Carp	17	.471	Pr.	10.75	13.15		23.90	34

12 32 Manufactured Wood Casework

12 32 16 – Manufactured Plastic-Laminate-Clad Casework

12 32 16.20 Plastic Laminate Casework Doors

	Crew	Daily Output	Labor-Hours	Unit	Material	Labor	Equipment	Total	Total Incl O&P
0010 **PLASTIC LAMINATE CASEWORK DOORS**									
1000 For casework frames, see Div. 12 32 23.15									
1100 For casework hardware, see Div. 12 32 23.35									
6000 Plastic laminate on particle board									
6100 12" wide, 18" high	1 Carp	25	.320	Ea.	10.50	8.95		19.45	26.50
6120 24" high		24	.333		14	9.30		23.30	31
6140 30" high		23	.348		17.50	9.70		27.20	35.50
6160 36" high		21	.381		21	10.65		31.65	41

12 32 Manufactured Wood Casework

12 32 16 – Manufactured Plastic-Laminate-Clad Casework

12 32 16.20 Plastic Laminate Casework Doors		Crew	Daily Output	Labor-Hours	Unit	Material	2009 Bare Costs Labor	Equipment	Total	Total Incl O&P
6200	48" high	1 Carp	16	.500	Ea.	28	14		42	54.50
6250	60" high		13	.615		35	17.20		52.20	67.50
6300	72" high		12	.667		42	18.65		60.65	77.50
6320	15" wide, 18" high		24.50	.327		13.10	9.15		22.25	30
6340	24" high		23.50	.340		17.50	9.50		27	35.50
6360	30" high		22.50	.356		22	9.95		31.95	41
6380	36" high		20.50	.390		26.50	10.90		37.40	47.50
6400	48" high		15.50	.516		35	14.45		49.45	63
6450	60" high		12.50	.640		44	17.90		61.90	78.50
6480	72" high		11.50	.696		52.50	19.45		71.95	91
6500	18" wide, 18" high		24	.333		15.75	9.30		25.05	33
6550	24" high		23	.348		21	9.70		30.70	39.50
6600	30" high		22	.364		26.50	10.15		36.65	46
6650	36" high		20	.400		31.50	11.20		42.70	53.50
6700	48" high		15	.533		42	14.90		56.90	71
6750	60" high		12	.667		52.50	18.65		71.15	89.50
6800	72" high		11	.727		63	20.50		83.50	104

12 32 16.25 Plastic Laminate Drawer Fronts

		Crew	Daily Output	Labor-Hours	Unit	Material	2009 Bare Costs Labor	Equipment	Total	Total Incl O&P
0010	**PLASTIC LAMINATE DRAWER FRONTS**									
2800	Plastic laminate on particle board front									
3000	4" high, 12" wide	1 Carp	17	.471	Ea.	4.62	13.15		17.77	27
3200	18" wide		16	.500		6.95	14		20.95	31
3600	24" wide		15	.533		9.25	14.90		24.15	35
3800	6" high, 12" wide		16	.500		7	14		21	31
4000	18" wide		15	.533		10.50	14.90		25.40	36.50
4500	24" wide		14	.571		14	15.95		29.95	42.50
4800	9" high, 12" wide		15	.533		10.50	14.90		25.40	36.50
5000	18" wide		14	.571		15.75	15.95		31.70	44.50
5200	24" wide		13	.615		21	17.20		38.20	52

12 32 23 – Hardwood Casework

12 32 23.10 Manufactured Wood Casework, Stock Units

		Crew	Daily Output	Labor-Hours	Unit	Material	2009 Bare Costs Labor	Equipment	Total	Total Incl O&P
0010	**MANUFACTURED WOOD CASEWORK, STOCK UNITS**									
0700	Kitchen base cabinets, hardwood, not incl. counter tops,									
0710	24" deep, 35" high, prefinished									
0800	One top drawer, one door below, 12" wide	2 Carp	24.80	.645	Ea.	227	18.05		245.05	281
0820	15" wide		24	.667		272	18.65		290.65	330
0840	18" wide		23.30	.687		297	19.20		316.20	360
0860	21" wide		22.70	.705		310	19.70		329.70	375
0880	24" wide		22.30	.717		360	20		380	435
1000	Four drawers, 12" wide		24.80	.645		335	18.05		353.05	395
1020	15" wide		24	.667		365	18.65		383.65	430
1040	18" wide		23.30	.687		385	19.20		404.20	455
1060	24" wide		22.30	.717		410	20		430	485
1200	Two top drawers, two doors below, 27" wide		22	.727		370	20.50		390.50	445
1220	30" wide		21.40	.748		370	21		391	445
1240	33" wide		20.90	.766		375	21.50		396.50	450
1260	36" wide		20.30	.788		390	22		412	465
1280	42" wide		19.80	.808		420	22.50		442.50	500
1300	48" wide		18.90	.847		450	23.50		473.50	535
1500	Range or sink base, two doors below, 30" wide		21.40	.748		300	21		321	365
1520	33" wide		20.90	.766		320	21.50		341.50	390
1540	36" wide		20.30	.788		340	22		362	405

12 32 23 – Hardwood Casework

12 32 23.10 Manufactured Wood Casework, Stock Units

		Crew	Daily Output	Labor-Hours	Unit	Material	2009 Bare Costs Labor	Equipment	Total	Total Incl O&P
1560	42" wide	2 Carp	19.80	.808	Ea.	360	22.50		382.50	440
1580	48" wide	↓	18.90	.847		380	23.50		403.50	455
1800	For sink front units, deduct					60			60	66
2000	Corner base cabinets, 36" wide, standard	2 Carp	18	.889		630	25		655	730
2100	Lazy Susan with revolving door	"	16.50	.970	↓	540	27		567	640
4000	Kitchen wall cabinets, hardwood, 12" deep with two doors									
4050	12" high, 30" wide	2 Carp	24.80	.645	Ea.	197	18.05		215.05	248
4100	36" wide		24	.667		228	18.65		246.65	283
4400	15" high, 30" wide		24	.667		198	18.65		216.65	250
4420	33" wide		23.30	.687		223	19.20		242.20	278
4440	36" wide		22.70	.705		243	19.70		262.70	300
4450	42" wide		22.70	.705		284	19.70		303.70	345
4700	24" high, 30" wide		23.30	.687		310	19.20		329.20	375
4720	36" wide		22.70	.705		345	19.70		364.70	415
4740	42" wide		22.30	.717		335	20		355	400
5000	30" high, one door, 12" wide		22	.727		191	20.50		211.50	246
5020	15" wide		21.40	.748		210	21		231	267
5040	18" wide		20.90	.766		234	21.50		255.50	294
5060	24" wide		20.30	.788		271	22		293	335
5300	Two doors, 27" wide		19.80	.808		289	22.50		311.50	360
5320	30" wide		19.30	.829		320	23		343	390
5340	36" wide		18.80	.851		360	24		384	435
5360	42" wide		18.50	.865		385	24		409	465
5380	48" wide		18.40	.870		400	24.50		424.50	480
6000	Corner wall, 30" high, 24" wide		18	.889		209	25		234	272
6050	30" wide		17.20	.930		260	26		286	330
6100	36" wide		16.50	.970		290	27		317	365
6500	Revolving Lazy Susan		15.20	1.053		390	29.50		419.50	480
7000	Broom cabinet, 84" high, 24" deep, 18" wide		10	1.600		515	44.50		559.50	640
7500	Oven cabinets, 84" high, 24" deep, 27" wide		8	2	↓	785	56		841	955
7750	Valance board trim	↓	396	.040	L.F.	12.25	1.13		13.38	15.40
9000	For deluxe models of all cabinets, add					40%				
9500	For custom built in place, add					25%	10%			
9550	Rule of thumb, kitchen cabinets not including									
9560	appliances & counter top, minimum	2 Carp	30	.533	L.F.	130	14.90		144.90	168
9600	Maximum	"	25	.640	"	310	17.90		327.90	375

12 32 23.15 Manufactured Wood Casework Frames

		Crew	Daily Output	Labor-Hours	Unit	Material	2009 Bare Costs Labor	Equipment	Total	Total Incl O&P
0010	**MANUFACTURED WOOD CASEWORK FRAMES**									
0050	Base cabinets, counter storage, 36" high									
0100	One bay, 18" wide	1 Carp	2.70	2.963	Ea.	131	83		214	284
0400	Two bay, 36" wide		2.20	3.636		200	102		302	390
1100	Three bay, 54" wide		1.50	5.333		238	149		387	515
2800	Bookcases, one bay, 7' high, 18" wide		2.40	3.333		154	93		247	325
3500	Two bay, 36" wide		1.60	5		224	140		364	480
4100	Three bay, 54" wide		1.20	6.667		370	186		556	720
6100	Wall mounted cabinet, one bay, 24" high, 18" wide		3.60	2.222		85	62		147	199
6800	Two bay, 36" wide		2.20	3.636		124	102		226	310
7400	Three bay, 54" wide		1.70	4.706		154	132		286	390
8400	30" high, one bay, 18" wide		3.60	2.222		92	62		154	206
9000	Two bay, 36" wide		2.15	3.721		122	104		226	310
9400	Three bay, 54" wide		1.60	5		153	140		293	405
9800	Wardrobe, 7' high, single, 24" wide	↓	2.70	2.963		170	83		253	325

12 32 Manufactured Wood Casework

12 32 23 – Hardwood Casework

12 32 23.15 Manufactured Wood Casework Frames

		Crew	Daily Output	Labor-Hours	Unit	Material	2009 Bare Costs Labor	Equipment	Total	Total Incl O&P
9950	Partition, adjustable shelves & drawers, 48" wide	1 Carp	1.40	5.714	Ea.	325	160		485	625

12 32 23.20 Manufactured Hardwood Casework Doors

		Crew	Daily Output	Labor-Hours	Unit	Material	2009 Bare Costs Labor	Equipment	Total	Total Incl O&P
0010	**MANUFACTURED HARDWOOD CASEWORK DOORS**									
2000	Glass panel, hardwood frame									
2200	12" wide, 18" high	1 Carp	34	.235	Ea.	18.75	6.60		25.35	31.50
2400	24" high		33	.242		25	6.80		31.80	39
2600	30" high		32	.250		31.50	7		38.50	46.50
2800	36" high		30	.267		37.50	7.45		44.95	54
3000	48" high		23	.348		50	9.70		59.70	71.50
3200	60" high		17	.471		62.50	13.15		75.65	91
3400	72" high		15	.533		75	14.90		89.90	108
3600	15" wide, 18" high		33	.242		23.50	6.80		30.30	37.50
3800	24" high		32	.250		31.50	7		38.50	46.50
4000	30" high		30	.267		39	7.45		46.45	55.50
4250	36" high		28	.286		47	8		55	65
4300	48" high		22	.364		62.50	10.15		72.65	86
4350	60" high		16	.500		78	14		92	110
4400	72" high		14	.571		94	15.95		109.95	130
4450	18" wide, 18" high		32	.250		28	7		35	43
4500	24" high		30	.267		37.50	7.45		44.95	54
4550	30" high		29	.276		47	7.70		54.70	64.50
4600	36" high		27	.296		56.50	8.30		64.80	76
4650	48" high		21	.381		75	10.65		85.65	101
4700	60" high		15	.533		94	14.90		108.90	128
4750	72" high		13	.615		113	17.20		130.20	153
5000	Hardwood, raised panel									
5100	12" wide, 18" high	1 Carp	16	.500	Ea.	27	14		41	53
5150	24" high		15.50	.516		36	14.45		50.45	64
5200	30" high		15	.533		45	14.90		59.90	74.50
5250	36" high		14	.571		54	15.95		69.95	86.50
5300	48" high		11	.727		72	20.50		92.50	114
5320	60" high		8	1		90	28		118	147
5340	72" high		7	1.143		108	32		140	173
5360	15" wide, 18" high		15.50	.516		34	14.45		48.45	61.50
5380	24" high		15	.533		45	14.90		59.90	74.50
5400	30" high		14.50	.552		56.50	15.40		71.90	88
5420	36" high		13.50	.593		67.50	16.55		84.05	103
5440	48" high		10.50	.762		90	21.50		111.50	135
5460	60" high		7.50	1.067		113	30		143	175
5480	72" high		6.50	1.231		135	34.50		169.50	207
5500	18" wide, 18" high		15	.533		40.50	14.90		55.40	69.50
5550	24" high		14.50	.552		54	15.40		69.40	85.50
5600	30" high		14	.571		67.50	15.95		83.45	102
5650	36" high		13	.615		81	17.20		98.20	118
5700	48" high		10	.800		108	22.50		130.50	157
5750	60" high		7	1.143		135	32		167	203
5800	72" high		6	1.333		162	37.50		199.50	241

12 32 23.25 Manufactured Wood Casework Drawer Fronts

		Crew	Daily Output	Labor-Hours	Unit	Material	2009 Bare Costs Labor	Equipment	Total	Total Incl O&P
0010	**MANUFACTURED WOOD CASEWORK DRAWER FRONTS**									
0100	Solid hardwood front									
1000	4" high, 12" wide	1 Carp	17	.471	Ea.	2.66	13.15		15.81	25
1200	18" wide		16	.500		4	14		18	28

12 32 Manufactured Wood Casework

12 32 23 – Hardwood Casework

12 32 23.25 Manufactured Wood Casework Drawer Fronts

		Crew	Daily Output	Labor-Hours	Unit	Material	2009 Bare Costs Labor	Equipment	Total	Total Incl O&P
1400	24" wide	1 Carp	15	.533	Ea.	5.30	14.90		20.20	31
1600	6" high, 12" wide		16	.500		4	14		18	28
1800	18" wide		15	.533		6	14.90		20.90	31.50
2000	24" wide		14	.571		8	15.95		23.95	36
2200	9" high, 12" wide		15	.533		6	14.90		20.90	31.50
2400	18" wide		14	.571		9	15.95		24.95	37
2600	24" wide		13	.615		12	17.20		29.20	42

12 32 23.30 Manufactured Wood Casework Vanities

		Crew	Daily Output	Labor-Hours	Unit	Material	2009 Bare Costs Labor	Equipment	Total	Total Incl O&P
0010	**MANUFACTURED WOOD CASEWORK VANITIES**									
8000	Vanity bases, 2 doors, 30" high, 21" deep, 24" wide	2 Carp	20	.800	Ea.	275	22.50		297.50	345
8050	30" wide		16	1		315	28		343	400
8100	36" wide		13.33	1.200		345	33.50		378.50	435
8150	48" wide		11.43	1.400		425	39		464	535
9000	For deluxe models of all vanities, add to above					40%				
9500	For custom built in place, add to above					25%	10%			

12 32 23.35 Manufactured Wood Casework Hardware

		Crew	Daily Output	Labor-Hours	Unit	Material	2009 Bare Costs Labor	Equipment	Total	Total Incl O&P
0010	**MANUFACTURED WOOD CASEWORK HARDWARE**									
1000	Catches, minimum	1 Carp	235	.034	Ea.	1	.95		1.95	2.71
1020	Average		119.40	.067		3.29	1.87		5.16	6.80
1040	Maximum		80	.100		6.25	2.80		9.05	11.60
2000	Door/drawer pulls, handles									
2200	Handles and pulls, projecting, metal, minimum	1 Carp	160	.050	Ea.	4.43	1.40		5.83	7.25
2220	Average		95.24	.084		6.90	2.35		9.25	11.50
2240	Maximum		68	.118		9.40	3.29		12.69	15.85
2300	Wood, minimum		160	.050		4.67	1.40		6.07	7.50
2320	Average		95.24	.084		6.25	2.35		8.60	10.80
2340	Maximum		68	.118		8.55	3.29		11.84	15
2600	Flush, metal, minimum		160	.050		4.67	1.40		6.07	7.50
2620	Average		95.24	.084		6.25	2.35		8.60	10.80
2640	Maximum		68	.118		8.55	3.29		11.84	15
3000	Drawer tracks/glides, minimum		48	.167	Pr.	7.90	4.66		12.56	16.60
3020	Average		32	.250		13.40	7		20.40	26.50
3040	Maximum		24	.333		23	9.30		32.30	41.50
4000	Cabinet hinges, minimum		160	.050		2.69	1.40		4.09	5.30
4020	Average		95.24	.084		6	2.35		8.35	10.55
4040	Maximum		68	.118		10.15	3.29		13.44	16.75

12 34 Manufactured Plastic Casework

12 34 16 – Manufactured Solid-Plastic Casework

12 34 16.10 Outdoor Casework

		Crew	Daily Output	Labor-Hours	Unit	Material	2009 Bare Costs Labor	Equipment	Total	Total Incl O&P
0010	**OUTDOOR CASEWORK**									
0020	Cabinet, base, sink/range, 36"	2 Carp	20.30	.788	Ea.	1,675	22		1,697	1,875
0100	Base, 36"		20.30	.788		2,325	22		2,347	2,575
0200	Filler strip, 1" x 30"		158	.101		31	2.83		33.83	39

12 36 Countertops

12 36 16 – Metal Countertops

12 36 16.10 Stainless Steel Countertops	Crew	Daily Output	Labor-Hours	Unit	Material	2009 Bare Costs Labor	Equipment	Total	Total Incl O&P
0010 **STAINLESS STEEL COUNTERTOPS**									
3200 Stainless steel, custom	1 Carp	24	.333	S.F.	136	9.30		145.30	166

12 36 19 – Wood Countertops

12 36 19.10 Maple Countertops

	Crew	Daily Output	Labor-Hours	Unit	Material	2009 Bare Costs Labor	Equipment	Total	Total Incl O&P
0010 **MAPLE COUNTERTOPS**									
2900 Solid, laminated, 1-1/2" thick, no splash	1 Carp	28	.286	L.F.	59.50	8		67.50	79
3000 With square splash		28	.286	"	70.50	8		78.50	91
3400 Recessed cutting block with trim, 16" x 20" x 1"		8	1	Ea.	69	28		97	124
3411 Replace cutting block only		16	.500	"	34	14		48	61

12 36 23 – Plastic Countertops

12 36 23.13 Plastic-Laminate-Clad Countertops

	Crew	Daily Output	Labor-Hours	Unit	Material	2009 Bare Costs Labor	Equipment	Total	Total Incl O&P
0010 **PLASTIC-LAMINATE-CLAD COUNTERTOPS**									
0020 Stock, 24" wide w/ backsplash, minimum	1 Carp	30	.267	L.F.	8.65	7.45		16.10	22
0100 Maximum		25	.320		18	8.95		26.95	35
0300 Custom plastic, 7/8" thick, aluminum molding, no splash		30	.267		19.85	7.45		27.30	34.50
0400 Cove splash		30	.267		26	7.45		33.45	41
0600 1-1/4" thick, no splash		28	.286		29.50	8		37.50	46
0700 Square splash		28	.286		28.50	8		36.50	44.50
0900 Square edge, plastic face, 7/8" thick, no splash		30	.267		25	7.45		32.45	40
1000 With splash		30	.267		32	7.45		39.45	47.50
1200 For stainless channel edge, 7/8" thick, add					2.62			2.62	2.88
1300 1-1/4" thick, add					3.07			3.07	3.38
1500 For solid color suede finish, add					2.55			2.55	2.81
1700 For end splash, add				Ea.	17			17	18.70
1901 For cut outs, standard, add, minimum	1 Carp	32	.250			7		7	11.80
2000 Maximum		8	1		5.65	28		33.65	54
2010 Cut out in backsplash for elec. wall outlet		38	.211			5.90		5.90	9.95
2020 Cut out for sink		20	.400			11.20		11.20	18.90
2030 Cut out for stove top		18	.444			12.40		12.40	21
2100 Postformed, including backsplash and front edge		30	.267	L.F.	10.20	7.45		17.65	24
2110 Mitred, add		12	.667	Ea.		18.65		18.65	31.50
2200 Built-in place, 25" wide, plastic laminate		25	.320	L.F.	13.60	8.95		22.55	30

12 36 33 – Tile Countertops

12 36 33.10 Ceramic Tile Countertops

	Crew	Daily Output	Labor-Hours	Unit	Material	2009 Bare Costs Labor	Equipment	Total	Total Incl O&P
0010 **CERAMIC TILE COUNTERTOPS**									
2300 Ceramic tile mosaic	1 Carp	25	.320	L.F.	29.50	8.95		38.45	47.50

12 36 40 – Stone Countertops

12 36 40.10 Natural Stone Countertops

	Crew	Daily Output	Labor-Hours	Unit	Material	2009 Bare Costs Labor	Equipment	Total	Total Incl O&P
0010 **NATURAL STONE COUNTERTOPS**									
2500 Marble, stock, with splash, 1/2" thick, minimum	1 Bric	17	.471	L.F.	36.50	13.35		49.85	62
2700 3/4" thick, maximum		13	.615		91	17.45		108.45	129
2800 Granite, average, 1-1/4" thick, 24" wide, no splash		13.01	.615		125	17.45		142.45	166

12 36 61 – Simulated Stone Countertops

12 36 61.16 Solid Surface Countertops

	Crew	Daily Output	Labor-Hours	Unit	Material	2009 Bare Costs Labor	Equipment	Total	Total Incl O&P
0010 **SOLID SURFACE COUNTERTOPS**, Acrylic polymer									
2000 Pricing for order of 1 – 50 L.F.									
2100 25" wide, solid colors	2 Carp	20	.800	L.F.	60	22.50		82.50	104
2200 Patterned colors		20	.800		76.50	22.50		99	122
2300 Premium patterned colors		20	.800		95.50	22.50		118	143
2400 With silicone attached 4" backsplash, solid colors		19	.842		66	23.50		89.50	113

12 36 Countertops

12 36 61 – Simulated Stone Countertops

12 36 61.16 Solid Surface Countertops

		Crew	Daily Output	Labor-Hours	Unit	Material	2009 Bare Costs Labor	2009 Bare Costs Equipment	Total	Total Incl O&P
2500	Patterned colors	2 Carp	19	.842	L.F.	83.50	23.50		107	132
2600	Premium patterned colors		19	.842		104	23.50		127.50	155
2700	With hard seam attached 4" backsplash, solid colors		15	1.067		66	30		96	123
2800	Patterned colors		15	1.067		83.50	30		113.50	143
2900	Premium patterned colors		15	1.067		104	30		134	166
3800	Sinks, pricing for order of 1 – 50 units									
3900	Single bowl, hard seamed, solid colors, 13" x 17"	1 Carp	2	4	Ea.	405	112		517	635
4000	10" x 15"		4.55	1.758		188	49		237	290
4100	Cutouts for sinks		5.25	1.524			42.50		42.50	72

12 36 61.17 Solid Surface Vanity Tops

		Crew	Daily Output	Labor-Hours	Unit	Material	2009 Bare Costs Labor	2009 Bare Costs Equipment	Total	Total Incl O&P
0010	**SOLID SURFACE VANITY TOPS**									
0015	Solid surface, center bowl, 17" x 19"	1 Carp	12	.667	Ea.	189	18.65		207.65	240
0020	19" x 25"		12	.667		229	18.65		247.65	284
0030	19" x 31"		12	.667		278	18.65		296.65	335
0040	19" x 37"		12	.667		325	18.65		343.65	385
0050	22" x 25"		10	.800		202	22.50		224.50	260
0060	22" x 31"		10	.800		236	22.50		258.50	297
0070	22" x 37"		10	.800		274	22.50		296.50	340
0080	22" x 43"		10	.800		315	22.50		337.50	385
0090	22" x 49"		10	.800		345	22.50		367.50	420
0110	22" x 55"		8	1		395	28		423	485
0120	22" x 61"		8	1		450	28		478	545
0220	Double bowl, 22" x 61"		8	1		505	28		533	610
0230	Double bowl, 22" x 73"		8	1		705	28		733	825
0240	For aggregate colors, add					35%				
0250	For faucets and fittings, see Div. 22 41 39.10									

12 36 61.19 Engineered Stone Countertops

		Crew	Daily Output	Labor-Hours	Unit	Material	2009 Bare Costs Labor	2009 Bare Costs Equipment	Total	Total Incl O&P
0010	**ENGINEERED STONE COUNTERTOPS**									
0100	25" wide, 4" backsplash, color group A, minimum	2 Carp	15	1.067	L.F.	17	30		47	69
0110	Maximum		15	1.067		42	30		72	96.50
0120	Color group B, minimum		15	1.067		22	30		52	74.50
0130	Maximum		15	1.067		49	30		79	105
0140	Color group C, minimum		15	1.067		29.50	30		59.50	83
0150	Maximum		15	1.067		60	30		90	117
0160	Color group D, minimum		15	1.067		37	30		67	91
0170	Maximum		15	1.067		70	30		100	128

12 93 Site Furnishings

12 93 33 – Manufactured Planters

12 93 33.10 Planters

		Crew	Daily Output	Labor-Hours	Unit	Material	2009 Bare Costs Labor	2009 Bare Costs Equipment	Total	Total Incl O&P
0010	**PLANTERS**									
0012	Concrete, sandblasted, precast, 48" diameter, 24" high	2 Clab	15	1.067	Ea.	635	22		657	735
0300	Fiberglass, circular, 36" diameter, 24" high		15	1.067		470	22		492	555
1200	Wood, square, 48" side, 24" high		15	1.067		1,000	22		1,022	1,125
1300	Circular, 48" diameter, 30" high		10	1.600		820	33		853	960
1600	Planter/bench, 72"		5	3.200		3,225	66		3,291	3,650

12 93 43 – Site Seating and Tables

12 93 43.13 Site Seating

		Crew	Daily Output	Labor-Hours	Unit	Material	2009 Bare Costs Labor	2009 Bare Costs Equipment	Total	Total Incl O&P
0010	**SITE SEATING**									
0012	Seating, benches, park, precast conc, w/backs, wood rails, 4' long	2 Clab	5	3.200	Ea.	430	66		496	585

12 93 Site Furnishings

12 93 43 – Site Seating and Tables

12 93 43.13 Site Seating		Crew	Daily Output	Labor-Hours	Unit	Material	2009 Bare Costs Labor	Equipment	Total	Total Incl O&P
0100	8' long	2 Clab	4	4	Ea.	860	82		942	1,075
0500	Steel barstock pedestals w/backs, 2" x 3" wood rails, 4' long		10	1.600		1,050	33		1,083	1,200
0510	8' long		7	2.286		1,250	47		1,297	1,450
0800	Cast iron pedestals, back & arms, wood slats, 4' long		8	2		370	41		411	480
0820	8' long		5	3.200		1,075	66		1,141	1,300
1700	Steel frame, fir seat, 10' long		10	1.600		239	33		272	320

Division 13
Special Construction

13 11 Swimming Pools

13 11 13 – Below-Grade Swimming Pools

13 11 13.50 Swimming Pools

		Crew	Daily Output	Labor-Hours	Unit	Material	2009 Bare Costs Labor	Equipment	Total	Total Incl O&P
0010	**SWIMMING POOLS** Residential in-ground, vinyl lined, concrete									
0020	Sides including equipment, sand bottom	B-52	300	.187	SF Surf	12.60	4.32	1.65	18.57	23
0100	Metal or polystyrene sides R131113-20	B-14	410	.117	↓	10.55	2.57	.72	13.84	16.70
0200	Add for vermiculite bottom					.81			.81	.89
0500	Gunite bottom and sides, white plaster finish									
0600	12' x 30' pool	B-52	145	.386	SF Surf	23.50	8.95	3.41	35.86	45
0720	16' x 32' pool		155	.361		21	8.35	3.19	32.54	41
0750	20' x 40' pool	↓	250	.224	↓	18.90	5.20	1.98	26.08	32
0810	Concrete bottom and sides, tile finish									
0820	12' x 30' pool	B-52	80	.700	SF Surf	23.50	16.20	6.20	45.90	60.50
0830	16' x 32' pool		95	.589		19.60	13.65	5.20	38.45	50.50
0840	20' x 40' pool	↓	130	.431	↓	15.60	9.95	3.81	29.36	38
1600	For water heating system, see Div. 23 52 28.10									
1700	Filtration and deck equipment only, as % of total				Total				20%	20%
1800	Deck equipment, rule of thumb, 20' x 40' pool				SF Pool					1.30
3000	Painting pools, preparation + 3 coats, 20' x 40' pool, epoxy	2 Pord	.33	48.485	Total	1,550	1,200		2,750	3,675
3100	Rubber base paint, 18 gallons	"	.33	48.485	"	1,000	1,200		2,200	3,075

13 11 46 – Swimming Pool Accessories

13 11 46.50 Swimming Pool Equipment

		Crew	Daily Output	Labor-Hours	Unit	Material	2009 Bare Costs Labor	Equipment	Total	Total Incl O&P
0010	**SWIMMING POOL EQUIPMENT**									
0020	Diving stand, stainless steel, 3 meter	2 Carp	.40	40	Ea.	9,375	1,125		10,500	12,200
0600	Diving boards, 16' long, aluminum		2.70	5.926		3,075	166		3,241	3,650
0700	Fiberglass		2.70	5.926		2,425	166		2,591	2,950
1200	Ladders, heavy duty, stainless steel, 2 tread		7	2.286		490	64		554	650
1500	4 tread	↓	6	2.667		655	74.50		729.50	845
2100	Lights, underwater, 12 volt, with transformer, 300 watt	1 Elec	1	8		243	252		495	680
2200	110 volt, 500 watt, standard	"	1	8	↓	201	252		453	630
3000	Pool covers, reinforced vinyl	3 Clab	1800	.013	S.F.	.36	.27		.63	.86
3100	Vinyl water tube		3200	.008		.46	.15		.61	.77
3200	Maximum	↓	3000	.008	↓	.58	.16		.74	.92
3300	Slides, tubular, fiberglass, aluminum handrails & ladder, 5'-0", straight	2 Carp	1.60	10	Ea.	3,625	280		3,905	4,475
3320	8'-0", curved	"	3	5.333	"	14,600	149		14,749	16,400

13 12 Fountains

13 12 13 – Exterior Fountains

13 12 13.10 Yard Fountains

		Crew	Daily Output	Labor-Hours	Unit	Material	2009 Bare Costs Labor	Equipment	Total	Total Incl O&P
0010	**YARD FOUNTAINS**									
0100	Outdoor fountain, 48" high with bowl and figures	2 Clab	2	8	Ea.	250	164		414	555

13 17 Tubs and Pools

13 17 13 – Redwood Hot Tub System

13 17 13.10 Redwood Hot Tub System	Crew	Daily Output	Labor-Hours	Unit	Material	2009 Bare Costs Labor	2009 Bare Costs Equipment	Total	Total Incl O&P
0010 **REDWOOD HOT TUB SYSTEM**									
7050 4' diameter x 4' deep	Q-1	1	16	Ea.	2,025	465		2,490	2,975
7150 6' diameter x 4' deep		.80	20		3,125	580		3,705	4,400
7200 8' diameter x 4' deep		.80	20		4,350	580		4,930	5,725

13 17 33 – Whirlpool Tubs

13 17 33.10 Whirlpool Bath

	Crew	Daily Output	Labor-Hours	Unit	Material	2009 Bare Costs Labor	2009 Bare Costs Equipment	Total	Total Incl O&P
0010 **WHIRLPOOL BATH**									
6000 Whirlpool, bath with vented overflow, molded fiberglass									
6100 66" x 48" x 24"	Q-1	1	16	Ea.	3,050	465		3,515	4,125
6400 72" x 36" x 24"		1	16		3,000	465		3,465	4,050
6500 60" x 30" x 21"		1	16		2,575	465		3,040	3,575
6600 72" x 42" x 22"		1	16		4,025	465		4,490	5,200
6700 83" x 65"		.30	53.333		5,250	1,550		6,800	8,300

13 24 Special Activity Rooms

13 24 16 – Saunas

13 24 16.50 Saunas and Heaters

	Crew	Daily Output	Labor-Hours	Unit	Material	2009 Bare Costs Labor	2009 Bare Costs Equipment	Total	Total Incl O&P
0010 **SAUNAS AND HEATERS**									
0020 Prefabricated, incl. heater & controls, 7' high, 6' x 4', C/C	L-7	2.20	11.818	Ea.	4,625	282		4,907	5,575
0050 6' x 4', C/P		2	13		4,275	310		4,585	5,225
0400 6' x 5', C/C		2	13		5,175	310		5,485	6,200
0450 6' x 5', C/P		2	13		4,800	310		5,110	5,825
0600 6' x 6', C/C		1.80	14.444		5,475	345		5,820	6,600
0650 6' x 6', C/P		1.80	14.444		5,125	345		5,470	6,200
0800 6' x 9', C/C		1.60	16.250		6,800	390		7,190	8,125
0850 6' x 9', C/P		1.60	16.250		6,300	390		6,690	7,600
1000 8' x 12', C/C		1.10	23.636		10,600	565		11,165	12,700
1050 8' x 12', C/P		1.10	23.636		9,725	565		10,290	11,700
1400 8' x 10', C/C		1.20	21.667		8,975	515		9,490	10,700
1450 8' x 10', C/P		1.20	21.667		8,350	515		8,865	10,000
1600 10' x 12', C/C		1	26		11,200	620		11,820	13,400
1650 10' x 12', C/P		1	26		10,100	620		10,720	12,200
1700 Door only, cedar, 2'x6', with tempered insulated glass window	2 Carp	3.40	4.706		610	132		742	890
1800 Prehung, incl. jambs, pulls & hardware	"	12	1.333		615	37.50		652.50	740
2500 Heaters only (incl. above), wall mounted, to 200 C.F.					605			605	665
2750 To 300 C.F.					770			770	845
3000 Floor standing, to 720 C.F., 10,000 watts, w/controls	1 Elec	3	2.667		2,025	84		2,109	2,350
3250 To 1,000 C.F., 16,000 watts	"	3	2.667		2,525	84		2,609	2,900

13 24 26 – Steam Baths

13 24 26.50 Steam Baths and Components

	Crew	Daily Output	Labor-Hours	Unit	Material	2009 Bare Costs Labor	2009 Bare Costs Equipment	Total	Total Incl O&P
0010 **STEAM BATHS AND COMPONENTS**									
0020 Heater, timer & head, single, to 140 C.F.	1 Plum	1.20	6.667	Ea.	1,400	215		1,615	1,900
0500 To 300 C.F.	"	1.10	7.273		1,625	234		1,859	2,150
2700 Conversion unit for residential tub, including door					4,100			4,100	4,500

13 34 Fabricated Engineered Structures

13 34 13 – Glazed Structures

13 34 13.13 Greenhouses

		Crew	Daily Output	Labor-Hours	Unit	Material	2009 Bare Costs Labor	Equipment	Total	Total Incl O&P
0010	**GREENHOUSES**, Shell only, stock units, not incl. 2' stub walls,									
0020	foundation, floors, heat or compartments									
0300	Residential type, free standing, 8'-6" long x 7'-6" wide	2 Carp	59	.271	SF Flr.	47.50	7.60		55.10	65.50
0400	10'-6" wide		85	.188		37	5.25		42.25	50
0600	13'-6" wide		108	.148		35	4.14		39.14	45
0700	17'-0" wide		160	.100		35	2.80		37.80	43
0900	Lean-to type, 3'-10" wide		34	.471		38.50	13.15		51.65	64.50
1000	6'-10" wide		58	.276		37.50	7.70		45.20	54
1100	Wall mounted, to existing window, 3' x 3'	1 Carp	4	2	Ea.	445	56		501	585
1120	4' x 5'	"	3	2.667	"	665	74.50		739.50	860
1200	Deluxe quality, free standing, 7'-6" wide	2 Carp	55	.291	SF Flr.	82	8.15		90.15	104
1220	10'-6" wide		81	.198		78.50	5.50		84	95.50
1240	13'-6" wide		104	.154		71.50	4.30		75.80	86
1260	17'-0" wide		150	.107		61	2.98		63.98	72
1400	Lean-to type, 3'-10" wide		31	.516		104	14.45		118.45	140
1420	6'-10" wide		55	.291		99	8.15		107.15	123
1440	8'-0" wide		97	.165		86	4.61		90.61	102

13 34 13.19 Swimming Pool Enclosures

		Crew	Daily Output	Labor-Hours	Unit	Material	2009 Bare Costs Labor	Equipment	Total	Total Incl O&P
0010	**SWIMMING POOL ENCLOSURES** Translucent, free standing									
0020	not including foundations, heat or light									
0200	Economy, minimum	2 Carp	200	.080	SF Hor.	14.80	2.24		17.04	20
0300	Maximum		100	.160		41.50	4.47		45.97	53
0400	Deluxe, minimum		100	.160		48.50	4.47		52.97	61
0600	Maximum		70	.229		225	6.40		231.40	258

13 34 63 – Natural Fiber Construction

13 34 63.50 Straw Bale Construction

		Crew	Daily Output	Labor-Hours	Unit	Material	2009 Bare Costs Labor	Equipment	Total	Total Incl O&P
0010	**STRAW BALE CONSTRUCTION**									
2000	Straw bales wall, incl labor & material complete, minimum				SF Flr.					130
2010	Maximum				"					160
2020	Straw bales in walls w/ modified post and beam frame	[G] 2 Carp	320	.050	S.F.	4.28	1.40		5.68	7.05

Division 14
Conveying Equipment

14 21 Electric Traction Elevators

14 21 33 – Electric Traction Residential Elevators

14 21 33.20 Residential Elevators	Crew	Daily Output	Labor-Hours	Unit	Material	2009 Bare Costs Labor	Equipment	Total	Total Incl O&P
0010 **RESIDENTIAL ELEVATORS**									
7000 Residential, cab type, 1 floor, 2 stop, minimum	2 Elev	.20	80	Ea.	11,000	3,025		14,025	17,100
7100 Maximum		.10	160		18,600	6,075		24,675	30,400
7200 2 floor, 3 stop, minimum		.12	133		16,400	5,050		21,450	26,200
7300 Maximum	↓	.06	266	↓	26,700	10,100		36,800	45,900

14 42 Wheelchair Lifts

14 42 13 – Inclined Wheelchair Lifts

14 42 13.10 Inclined Wheelchair Lifts and Stairclimbers	Crew	Daily Output	Labor-Hours	Unit	Material	2009 Bare Costs Labor	Equipment	Total	Total Incl O&P
0010 **INCLINED WHEELCHAIR LIFTS AND STAIRCLIMBERS**									
7700 Stair climber (chair lift), single seat, minimum	2 Elev	1	16	Ea.	5,325	605		5,930	6,875
7800 Maximum	"	.20	80	"	7,350	3,025		10,375	13,000

Division 22
Plumbing

22 05 Common Work Results for Plumbing

22 05 05 – Selective Plumbing Demolition

22 05 05.10 Plumbing Demolition

	22 05 05.10 Plumbing Demolition	Crew	Daily Output	Labor-Hours	Unit	Material	2009 Bare Costs Labor	Equipment	Total	Total Incl O&P
0010	**PLUMBING DEMOLITION**									
1020	Fixtures, including 10' piping									
1101	Bath tubs, cast iron	1 Clab	4	2	Ea.		41		41	69.50
1121	Fiberglass		6	1.333			27.50		27.50	46.50
1141	Steel	↓	5	1.600			33		33	55.50
1200	Lavatory, wall hung	1 Plum	10	.800			26		26	42.50
1221	Counter top	1 Clab	16	.500			10.30		10.30	17.40
1301	Sink, single compartment		16	.500			10.30		10.30	17.40
1321	Double	↓	10	.800			16.45		16.45	28
1400	Water closet, floor mounted	1 Plum	8	1			32		32	53
1421	Wall mounted	1 Clab	7	1.143	↓		23.50		23.50	39.50
2001	Piping, metal, to 1-1/2" diameter		200	.040	L.F.		.82		.82	1.39
2051	2" thru 3-1/2" diameter		150	.053			1.10		1.10	1.85
2101	4" thru 6" diameter		100	.080	↓		1.64		1.64	2.78
2251	Water heater, 40 gal.	↓	6	1.333	Ea.		27.50		27.50	46.50
3000	Submersible sump pump	1 Plum	24	.333			10.75		10.75	17.60
6000	Remove and reset fixtures, minimum		6	1.333			43		43	70.50
6100	Maximum		4	2	↓		64.50		64.50	106
9000	Minimum labor/equipment charge	↓	2	4	Job		129		129	211

22 05 23 – General-Duty Valves for Plumbing Piping

22 05 23.20 Valves, Bronze

	22 05 23.20 Valves, Bronze	Crew	Daily Output	Labor-Hours	Unit	Material	2009 Bare Costs Labor	Equipment	Total	Total Incl O&P
0010	**VALVES, BRONZE**									
1750	Check, swing, class 150, regrinding disc, threaded									
1860	3/4"	1 Plum	20	.400	Ea.	64	12.90		76.90	91
1870	1"	"	19	.421	"	91.50	13.55		105.05	124
2850	Gate, N.R.S., soldered, 125 psi									
2940	3/4"	1 Plum	20	.400	Ea.	35.50	12.90		48.40	60
2950	1"	"	19	.421	"	50.50	13.55		64.05	78
5600	Relief, pressure & temperature, self-closing, ASME, threaded									
5650	1"	1 Plum	24	.333	Ea.	180	10.75		190.75	217
5660	1-1/4"	"	20	.400	"	345	12.90		357.90	400
6400	Pressure, water, ASME, threaded									
6440	3/4"	1 Plum	28	.286	Ea.	70	9.20		79.20	92
6450	1"	"	24	.333	"	149	10.75		159.75	182
6900	Reducing, water pressure									
6940	1/2"	1 Plum	24	.333	Ea.	214	10.75		224.75	254
6960	1"	"	19	.421	"	330	13.55		343.55	390
8350	Tempering, water, sweat connections									
8400	1/2"	1 Plum	24	.333	Ea.	71	10.75		81.75	95.50
8440	3/4"	"	20	.400	"	87	12.90		99.90	117
8650	Threaded connections									
8700	1/2"	1 Plum	24	.333	Ea.	87	10.75		97.75	113
8740	3/4"	"	20	.400	"	370	12.90		382.90	430

22 05 48 – Vibration and Seismic Controls for Plumbing Piping and Equipment

22 05 48.10 Seismic Bracing Supports

	22 05 48.10 Seismic Bracing Supports	Crew	Daily Output	Labor-Hours	Unit	Material	2009 Bare Costs Labor	Equipment	Total	Total Incl O&P
0010	**SEISMIC BRACING SUPPORTS**									
0020	Clamps									
0030	C-clamp, for mounting on steel beam									
0040	3/8" threaded rod	1 Skwk	160	.050	Ea.	1.81	1.40		3.21	4.36
0050	1/2" threaded rod		160	.050		2.30	1.40		3.70	4.90
0060	5/8" threaded rod		160	.050		3.81	1.40		5.21	6.55
0070	3/4" threaded rod	↓	160	.050	↓	5.80	1.40		7.20	8.75

22 05 Common Work Results for Plumbing

22 05 48 – Vibration and Seismic Controls for Plumbing Piping and Equipment

22 05 48.10 Seismic Bracing Supports		Crew	Daily Output	Labor-Hours	Unit	Material	2009 Bare Costs Labor	Equipment	Total	Total Incl O&P
0100	Brackets									
0110	Beam side or wall malleable iron									
0120	3/8" threaded rod	1 Skwk	48	.167	Ea.	2.19	4.68		6.87	10.30
0130	1/2" threaded rod		48	.167		4.22	4.68		8.90	12.55
0140	5/8" threaded rod		48	.167		5.60	4.68		10.28	14.10
0150	3/4" threaded rod		48	.167		6.75	4.68		11.43	15.30
0160	7/8" threaded rod	↓	48	.167	↓	7.85	4.68		12.53	16.55
0170	For concrete installation, add						30%			
0180	Wall, welded steel									
0190	0 size 12" wide 18" deep	1 Skwk	34	.235	Ea.	177	6.60		183.60	205
0200	1 size 18" wide 24" deep		34	.235		210	6.60		216.60	242
0210	2 size 24" wide 30" deep	↓	34	.235	↓	278	6.60		284.60	315
0300	Rod, carbon steel									
0310	Continuous thread									
0320	1/4" thread	1 Skwk	144	.056	L.F.	.14	1.56		1.70	2.78
0330	3/8" thread		144	.056		.23	1.56		1.79	2.88
0340	1/2" thread		144	.056		.46	1.56		2.02	3.14
0350	5/8" thread		144	.056		.73	1.56		2.29	3.43
0360	3/4" thread		144	.056		1.10	1.56		2.66	3.84
0370	7/8" thread	↓	144	.056	↓	1.78	1.56		3.34	4.59
0380	For galvanized, add					30%				
0400	Channel, steel									
0410	3/4" x 1-1/2"	1 Skwk	80	.100	L.F.	6.75	2.81		9.56	12.20
0420	1-1/2" x 1-1/2"		70	.114		9.05	3.21		12.26	15.35
0430	1-7/8" x 1-1/2"		60	.133		22	3.74		25.74	30.50
0440	3" x 1-1/2"	↓	50	.160	↓	25.50	4.49		29.99	35.50
0450	Spring nuts									
0460	3/8"	1 Skwk	100	.080	Ea.	2.64	2.24		4.88	6.70
0470	1/2"	"	80	.100	"	3.14	2.81		5.95	8.20
0500	Welding, field									
0510	Cleaning and welding plates, bars, or rods									
0520	To existing beams, columns, or trusses									
0530	1" weld	1 Skwk	144	.056	Ea.	.18	1.56		1.74	2.83
0540	2" weld		72	.111		.33	3.12		3.45	5.60
0550	3" weld		54	.148		.50	4.16		4.66	7.55
0560	4" weld		36	.222		.71	6.25		6.96	11.30
0570	5" weld		30	.267		.88	7.50		8.38	13.60
0580	6" weld	↓	24	.333	↓	1.01	9.35		10.36	16.90
0600	Vibration absorbers									
0610	Hangers, neoprene flex									
0620	10-120 lb. capacity	1 Skwk	8	1	Ea.	16.55	28		44.55	65.50
0630	75-550 lb. capacity		8	1		21.50	28		49.50	71
0640	250-1100 lb. capacity		6	1.333		44	37.50		81.50	111
0650	1000-4000 lb. capacity	↓	6	1.333	↓	69.50	37.50		107	139

22 05 76 – Facility Drainage Piping Cleanouts

22 05 76.10 Cleanouts

		Crew	Daily Output	Labor-Hours	Unit	Material	Labor	Equipment	Total	Total Incl O&P
0010	**CLEANOUTS**									
0080	Round or square, scoriated nickel bronze top									
0100	2" pipe size	1 Plum	10	.800	Ea.	202	26		228	265
0140	4" pipe size	"	6	1.333	"	285	43		328	385

22 05 76.20 Cleanout Tees

0010	**CLEANOUT TEES**									

22 05 Common Work Results for Plumbing

22 05 76 – Facility Drainage Piping Cleanouts

22 05 76.20 Cleanout Tees		Crew	Daily Output	Labor-Hours	Unit	Material	2009 Bare Costs Labor	Equipment	Total	Total Incl O&P
0100	Cast iron, B&S, with countersunk plug									
0220	3" pipe size	1 Plum	3.60	2.222	Ea.	205	71.50		276.50	345
0240	4" pipe size	"	3.30	2.424	"	256	78		334	410
0500	For round smooth access cover, same price									
4000	Plastic, tees and adapters. Add plugs									
4010	ABS, DWV									
4020	Cleanout tee, 1-1/2" pipe size	1 Plum	15	.533	Ea.	6.60	17.15		23.75	35.50

22 07 Plumbing Insulation

22 07 16 – Plumbing Equipment Insulation

22 07 16.10 Insulation for Plumbing Equipment			Crew	Daily Output	Labor-Hours	Unit	Material	2009 Bare Costs Labor	Equipment	Total	Total Incl O&P
0010	**INSULATION FOR PLUMBING EQUIPMENT**										
2900	Domestic water heater wrap kit										
2920	1-1/2" with vinyl jacket, 20-60 gal.	G	1 Plum	8	1	Ea.	16.05	32		48.05	70.50

22 07 19 – Plumbing Piping Insulation

22 07 19.10 Piping Insulation			Crew	Daily Output	Labor-Hours	Unit	Material	2009 Bare Costs Labor	Equipment	Total	Total Incl O&P
0010	**PIPING INSULATION**										
2930	Insulated protectors, (ADA)										
2935	For exposed piping under sinks or lavatories										
2940	Vinyl coated foam, velcro tabs										
2945	P Trap, 1-1/4" or 1-1/2"		1 Plum	32	.250	Ea.	18.50	8.05		26.55	33.50
2960	Valve and supply cover										
2965	1/2", 3/8", and 7/16" pipe size		1 Plum	32	.250	Ea.	17.60	8.05		25.65	32.50
2970	Extension drain cover										
2975	1-1/4", or 1-1/2" pipe size		1 Plum	32	.250	Ea.	19.30	8.05		27.35	34.50
2985	1-1/4" pipe size		"	32	.250	"	21.50	8.05		29.55	36.50
4000	Pipe covering (price copper tube one size less than IPS)										
6600	Fiberglass, with all service jacket										
6840	1" wall, 1/2" iron pipe size	G	Q-14	240	.067	L.F.	.89	1.75		2.64	3.96
6860	3/4" iron pipe size	G		230	.070		.97	1.82		2.79	4.18
6870	1" iron pipe size	G		220	.073		1.04	1.91		2.95	4.39
6900	2" iron pipe size	G		200	.080		1.31	2.10		3.41	5
7879	Rubber tubing, flexible closed cell foam										
8100	1/2" wall, 1/4" iron pipe size	G	1 Asbe	90	.089	L.F.	.44	2.59		3.03	4.89
8130	1/2" iron pipe size	G		89	.090		.54	2.62		3.16	5.05
8140	3/4" iron pipe size	G		89	.090		.60	2.62		3.22	5.10
8150	1" iron pipe size	G		88	.091		.67	2.65		3.32	5.25
8170	1-1/2" iron pipe size	G		87	.092		.94	2.68		3.62	5.60
8180	2" iron pipe size	G		86	.093		1.22	2.71		3.93	5.95
8300	3/4" wall, 1/4" iron pipe size	G		90	.089		.68	2.59		3.27	5.15
8330	1/2" iron pipe size	G		89	.090		.90	2.62		3.52	5.45
8340	3/4" iron pipe size	G		89	.090		1.09	2.62		3.71	5.65
8350	1" iron pipe size	G		88	.091		1.25	2.65		3.90	5.90
8380	2" iron pipe size	G		86	.093		2.23	2.71		4.94	7.05
8444	1" wall, 1/2" iron pipe size	G		86	.093		1.71	2.71		4.42	6.50
8445	3/4" iron pipe size	G		84	.095		2.07	2.77		4.84	7
8446	1" iron pipe size	G		84	.095		2.41	2.77		5.18	7.40
8447	1-1/4" iron pipe size	G		82	.098		2.72	2.84		5.56	7.85
8448	1-1/2" iron pipe size	G		82	.098		3.16	2.84		6	8.30
8449	2" iron pipe size	G		80	.100		4.23	2.91		7.14	9.60

22 07 Plumbing Insulation

22 07 19 - Plumbing Piping Insulation

22 07 19.10 Piping Insulation		Crew	Daily Output	Labor-Hours	Unit	Material	2009 Bare Costs Labor	Equipment	Total	Total Incl O&P
8450	2-1/2" iron pipe size	G 1 Asbe	80	.100	L.F.	5.50	2.91		8.41	11
8456	Rubber insulation tape, 1/8" x 2" x 30'	G			Ea.	11.40			11.40	12.50

22 11 Facility Water Distribution

22 11 13 - Facility Water Distribution Piping

22 11 13.23 Pipe, Copper

		Crew	Daily Output	Labor-Hours	Unit	Material	2009 Bare Costs Labor	Equipment	Total	Total Incl O&P
0010	**PIPE, COPPER**, Solder joints									
1000	Type K tubing, couplings & clevis hanger assemblies 10' O.C.									
1180	3/4" diameter	1 Plum	74	.108	L.F.	9.15	3.48		12.63	15.80
1200	1" diameter	"	66	.121	"	12.10	3.90		16	19.70
2000	Type L tubing, couplings & clevis hanger assemblies 10' O.C.									
2140	1/2" diameter	1 Plum	81	.099	L.F.	3.83	3.18		7.01	9.40
2160	5/8" diameter		79	.101		5.45	3.26		8.71	11.35
2180	3/4" diameter		76	.105		6	3.39		9.39	12.15
2200	1" diameter		68	.118		9.05	3.79		12.84	16.15
2220	1-1/4" diameter		58	.138		12.50	4.44		16.94	21
3000	Type M tubing, couplings & clevis hanger assemblies 10' O.C.									
3140	1/2" diameter	1 Plum	84	.095	L.F.	2.79	3.07		5.86	8.10
3180	3/4" diameter		78	.103		4.41	3.30		7.71	10.25
3200	1" diameter		70	.114		6.95	3.68		10.63	13.65
3220	1-1/4" diameter		60	.133		10.25	4.29		14.54	18.30
3240	1-1/2" diameter		54	.148		14	4.77		18.77	23.50
3260	2" diameter		44	.182		22	5.85		27.85	33.50
4000	Type DWV tubing, couplings & clevis hanger assemblies 10' O.C.									
4100	1-1/4" diameter	1 Plum	60	.133	L.F.	10.90	4.29		15.19	19.05
4120	1-1/2" diameter		54	.148		13.70	4.77		18.47	23
4140	2" diameter		44	.182		18.45	5.85		24.30	30
4160	3" diameter	Q-1	58	.276		32.50	8		40.50	49
4180	4" diameter	"	40	.400		57.50	11.60		69.10	82.50

22 11 13.25 Pipe Fittings, Copper

		Crew	Daily Output	Labor-Hours	Unit	Material	2009 Bare Costs Labor	Equipment	Total	Total Incl O&P
0010	**PIPE FITTINGS, COPPER**, Wrought unless otherwise noted									
0040	Solder joints, copper x copper									
0100	1/2"	1 Plum	20	.400	Ea.	1.23	12.90		14.13	22.50
0120	3/4"		19	.421		2.76	13.55		16.31	25.50
0250	45° elbow, 1/4"		22	.364		7.30	11.70		19	27
0280	1/2"		20	.400		2.26	12.90		15.16	23.50
0290	5/8"		19	.421		11.30	13.55		24.85	35
0300	3/4"		19	.421		3.96	13.55		17.51	27
0310	1"		16	.500		9.95	16.10		26.05	37.50
0320	1-1/4"		15	.533		13.45	17.15		30.60	43
0450	Tee, 1/4"		14	.571		8.15	18.40		26.55	39
0480	1/2"		13	.615		2.11	19.80		21.91	35
0490	5/8"		12	.667		13.55	21.50		35.05	50
0500	3/4"		12	.667		5.10	21.50		26.60	40.50
0510	1"		10	.800		15.70	26		41.70	60
0520	1-1/4"		9	.889		21.50	28.50		50	70.50
0612	Tee, reducing on the outlet, 1/4"		15	.533		13.15	17.15		30.30	42.50
0613	3/8"		15	.533		12.35	17.15		29.50	41.50
0614	1/2"		14	.571		10.85	18.40		29.25	42
0615	5/8"		13	.615		19.40	19.80		39.20	54
0616	3/4"		12	.667		7	21.50		28.50	42.50

22 11 Facility Water Distribution

22 11 13 – Facility Water Distribution Piping

22 11 13.25 Pipe Fittings, Copper

		Crew	Daily Output	Labor-Hours	Unit	Material	2009 Bare Costs Labor	Equipment	Total	Total Incl O&P
0617	1"	1 Plum	11	.727	Ea.	15.75	23.50		39.25	56
0618	1-1/4"		10	.800		29.50	26		55.50	75
0619	1-1/2"		9	.889		31.50	28.50		60	82
0620	2"	▼	8	1		44.50	32		76.50	102
0621	2-1/2"	Q-1	9	1.778		119	51.50		170.50	215
0622	3"		8	2		133	58		191	242
0623	4"		6	2.667		245	77.50		322.50	395
0624	5"	▼	5	3.200		1,250	92.50		1,342.50	1,525
0625	6"	Q-2	7	3.429		1,625	95.50		1,720.50	1,950
0626	8"	"	6	4		6,625	112		6,737	7,475
0630	Tee, reducing on the run, 1/4"	1 Plum	15	.533		16.45	17.15		33.60	46
0631	3/8"		15	.533		22	17.15		39.15	52
0632	1/2"		14	.571		14.90	18.40		33.30	46.50
0633	5/8"		13	.615		22	19.80		41.80	56.50
0634	3/4"		12	.667		6	21.50		27.50	41.50
0635	1"		11	.727		18.70	23.50		42.20	59
0636	1-1/4"		10	.800		29.50	26		55.50	75
0637	1-1/2"		9	.889		49.50	28.50		78	102
0638	2"	▼	8	1		72	32		104	132
0639	2-1/2"	Q-1	9	1.778		157	51.50		208.50	257
0640	3"		8	2		231	58		289	350
0641	4"		6	2.667		510	77.50		587.50	690
0642	5"	▼	5	3.200		1,200	92.50		1,292.50	1,450
0643	6"	Q-2	7	3.429		1,825	95.50		1,920.50	2,150
0644	8"	"	6	4		6,525	112		6,637	7,350
0650	Coupling, 1/4"	1 Plum	24	.333		.90	10.75		11.65	18.60
0680	1/2"		22	.364		.94	11.70		12.64	20
0690	5/8"		21	.381		2.81	12.25		15.06	23
0700	3/4"		21	.381		1.87	12.25		14.12	22
0710	1"		18	.444		3.73	14.30		18.03	27.50
0715	1-1/4"	▼	17	.471	▼	6.65	15.15		21.80	32.50
2000	DWV, solder joints, copper x copper									
2030	90° Elbow, 1-1/4"	1 Plum	13	.615	Ea.	11.90	19.80		31.70	45.50
2050	1-1/2"		12	.667		17.85	21.50		39.35	54.50
2070	2"	▼	10	.800		23.50	26		49.50	68
2090	3"	Q-1	10	1.600		57.50	46.50		104	140
2100	4"	"	9	1.778		281	51.50		332.50	395
2250	Tee, Sanitary, 1-1/4"	1 Plum	9	.889		23.50	28.50		52	73
2270	1-1/2"		8	1		29.50	32		61.50	85
2290	2"	▼	7	1.143		34	37		71	98
2310	3"	Q-1	7	2.286		126	66		192	248
2330	4"	"	6	2.667		320	77.50		397.50	480
2400	Coupling, 1-1/4"	1 Plum	14	.571		5.55	18.40		23.95	36
2420	1-1/2"		13	.615		6.90	19.80		26.70	40
2440	2"	▼	11	.727		9.60	23.50		33.10	49
2460	3"	Q-1	11	1.455		18.65	42		60.65	89.50
2480	4"	"	10	1.600	▼	59.50	46.50		106	142

22 11 13.44 Pipe, Steel

		Crew	Daily Output	Labor-Hours	Unit	Material	2009 Bare Costs Labor	Equipment	Total	Total Incl O&P
0010	**PIPE, STEEL**									
0050	Schedule 40, threaded, with couplings, and clevis hanger									
0060	assemblies sized for covering, 10' O.C.									
0540	Black, 1/4" diameter	1 Plum	66	.121	L.F.	2.17	3.90		6.07	8.80

22 11 Facility Water Distribution

22 11 13 – Facility Water Distribution Piping

22 11 13.44 Pipe, Steel

		Crew	Daily Output	Labor-Hours	Unit	Material	2009 Bare Costs Labor	Equipment	Total	Total Incl O&P
0570	3/4" diameter	1 Plum	61	.131	L.F.	2.64	4.22		6.86	9.85
0580	1" diameter	↓	53	.151		3.87	4.86		8.73	12.25
0590	1-1/4" diameter	Q-1	89	.180		4.89	5.20		10.09	13.95
0600	1-1/2" diameter		80	.200		5.75	5.80		11.55	15.85
0610	2" diameter	↓	64	.250	↓	7.70	7.25		14.95	20.50

22 11 13.45 Pipe Fittings, Steel, Threaded

		Crew	Daily Output	Labor-Hours	Unit	Material	2009 Bare Costs Labor	Equipment	Total	Total Incl O&P
0010	**PIPE FITTINGS, STEEL, THREADED**									
5000	Malleable iron, 150 lb.									
5020	Black									
5040	90° elbow, straight									
5090	3/4"	1 Plum	14	.571	Ea.	3.37	18.40		21.77	33.50
5100	1"	"	13	.615		4.27	19.80		24.07	37
5120	1-1/2"	Q-1	20	.800		9.25	23		32.25	48
5130	2"	"	18	.889	↓	16.70	26		42.70	61
5450	Tee, straight									
5500	3/4"	1 Plum	9	.889	Ea.	3.91	28.50		32.41	51.50
5510	1"	"	8	1		6.70	32		38.70	60.50
5520	1-1/4"	Q-1	14	1.143		10.85	33		43.85	66.50
5530	1-1/2"	↓	13	1.231		13.50	35.50		49	73.50
5540	2"	↓	11	1.455	↓	23	42		65	94.50
5650	Coupling									
5700	3/4"	1 Plum	18	.444	Ea.	3.30	14.30		17.60	27
5710	1"	"	15	.533		4.93	17.15		22.08	33.50
5730	1-1/2"	Q-1	24	.667		8.65	19.30		27.95	41
5740	2"	"	21	.762	↓	12.80	22		34.80	50

22 11 13.74 Pipe, Plastic

		Crew	Daily Output	Labor-Hours	Unit	Material	2009 Bare Costs Labor	Equipment	Total	Total Incl O&P
0010	**PIPE, PLASTIC**									
1800	PVC, couplings 10' O.C., clevis hanger assy's, 3 per 10'									
1820	Schedule 40									
1860	1/2" diameter	1 Plum	54	.148	L.F.	1.43	4.77		6.20	9.40
1870	3/4" diameter		51	.157		1.67	5.05		6.72	10.15
1880	1" diameter		46	.174		1.96	5.60		7.56	11.35
1890	1-1/4" diameter		42	.190		2.49	6.15		8.64	12.80
1900	1-1/2" diameter	↓	36	.222		2.73	7.15		9.88	14.75
1910	2" diameter	Q-1	59	.271		3.33	7.85		11.18	16.55
1920	2-1/2" diameter		56	.286		4.93	8.30		13.23	19.05
1930	3" diameter		53	.302		6.60	8.75		15.35	21.50
1940	4" diameter	↓	48	.333	↓	9.15	9.65		18.80	26
4100	DWV type, schedule 40, couplings 10' O.C., clevis hanger assy's, 3 per 10'									
4120	ABS									
4140	1-1/4" diameter	1 Plum	42	.190	L.F.	1.58	6.15		7.73	11.80
4150	1-1/2" diameter	"	36	.222		1.72	7.15		8.87	13.65
4160	2" diameter	Q-1	59	.271	↓	2.21	7.85		10.06	15.35
4400	PVC									
4410	1-1/4" diameter	1 Plum	42	.190	L.F.	1.85	6.15		8	12.10
4420	1-1/2" diameter	"	36	.222		1.82	7.15		8.97	13.75
4460	2" diameter	Q-1	59	.271		2.20	7.85		10.05	15.30
4470	3" diameter		53	.302		4.21	8.75		12.96	19
4480	4" diameter	↓	48	.333	↓	5.75	9.65		15.40	22
5360	CPVC, couplings 10' O.C., clevis hanger assy's, 3 per 10'									
5380	Schedule 40									
5460	1/2" diameter	1 Plum	54	.148	L.F.	2.84	4.77		7.61	10.95

22 11 13.74 Pipe, Plastic

		Crew	Daily Output	Labor-Hours	Unit	Material	2009 Bare Costs Labor	Equipment	Total	Total Incl O&P
5470	3/4" diameter	1 Plum	51	.157	L.F.	3.37	5.05		8.42	12
5480	1" diameter		46	.174		4.13	5.60		9.73	13.75
5490	1-1/4" diameter		42	.190		5.05	6.15		11.20	15.60
5500	1-1/2" diameter		36	.222		5.70	7.15		12.85	18
5510	2" diameter	Q-1	59	.271		7.05	7.85		14.90	20.50
6500	Residential installation, plastic pipe									
6510	Couplings 10' O.C., strap hangers 3 per 10'									
6520	PVC, Schedule 40									
6530	1/2" diameter	1 Plum	138	.058	L.F.	1.23	1.87		3.10	4.41
6540	3/4" diameter		128	.063		1.36	2.01		3.37	4.80
6550	1" diameter		119	.067		1.62	2.16		3.78	5.35
6560	1-1/4" diameter		111	.072		1.98	2.32		4.30	6
6570	1-1/2" diameter		104	.077		2.39	2.48		4.87	6.70
6580	2" diameter	Q-1	197	.081		2.85	2.35		5.20	7
6590	2-1/2" diameter		162	.099		4.86	2.86		7.72	10.05
6600	4" diameter		123	.130		7.95	3.77		11.72	14.95
6700	PVC, DWV, Schedule 40									
6720	1-1/4" diameter	1 Plum	100	.080	L.F.	2.05	2.58		4.63	6.50
6730	1-1/2" diameter	"	94	.085		2.24	2.74		4.98	6.95
6740	2" diameter	Q-1	178	.090		2.60	2.61		5.21	7.15
6760	4" diameter	"	110	.145		7.10	4.22		11.32	14.70

22 11 13.76 Pipe Fittings, Plastic

		Crew	Daily Output	Labor-Hours	Unit	Material	2009 Bare Costs Labor	Equipment	Total	Total Incl O&P
0010	**PIPE FITTINGS, PLASTIC**									
2700	PVC (white), schedule 40, socket joints									
2760	90° elbow, 1/2"	1 Plum	33.30	.240	Ea.	.41	7.75		8.16	13.15
2770	3/4"		28.60	.280		.45	9		9.45	15.30
2780	1"		25	.320		.82	10.30		11.12	17.80
2790	1-1/4"		22.20	.360		1.44	11.60		13.04	20.50
2800	1-1/2"		20	.400		1.55	12.90		14.45	22.50
2810	2"	Q-1	36.40	.440		2.42	12.75		15.17	23.50
2820	2-1/2"		26.70	.599		7.40	17.35		24.75	36.50
2830	3"		22.90	.699		8.85	20.50		29.35	42.50
2840	4"		18.20	.879		15.80	25.50		41.30	59.50
3180	Tee, 1/2"	1 Plum	22.20	.360		.50	11.60		12.10	19.60
3190	3/4"		19	.421		.58	13.55		14.13	23
3200	1"		16.70	.479		1.08	15.45		16.53	26.50
3210	1-1/4"		14.80	.541		1.68	17.40		19.08	30.50
3220	1-1/2"		13.30	.602		2.06	19.35		21.41	34.50
3230	2"	Q-1	24.20	.661		2.99	19.15		22.14	35
3240	2-1/2"		17.80	.899		9.85	26		35.85	53.50
3250	3"		15.20	1.053		12.95	30.50		43.45	64.50
3260	4"		12.10	1.322		23.50	38.50		62	89
3380	Coupling, 1/2"	1 Plum	33.30	.240		.27	7.75		8.02	13
3390	3/4"		28.60	.280		.37	9		9.37	15.20
3400	1"		25	.320		.66	10.30		10.96	17.65
3410	1-1/4"		22.20	.360		.89	11.60		12.49	20
3420	1-1/2"		20	.400		.95	12.90		13.85	22
3430	2"	Q-1	36.40	.440		1.45	12.75		14.20	22.50
3440	2-1/2"		26.70	.599		3.20	17.35		20.55	32
3450	3"		22.90	.699		5	20.50		25.50	38.50
3460	4"		18.20	.879		7.25	25.50		32.75	50
4500	DWV, ABS, non pressure, socket joints									

22 11 13.76 Pipe Fittings, Plastic	Crew	Daily Output	Labor-Hours	Unit	Material	2009 Bare Costs Labor	Equipment	Total	Total Incl O&P
4540	1/4 Bend, 1-1/4"	1 Plum	20.20	.396	Ea.	2.78	12.75		15.53	24
4560	1-1/2"	"	18.20	.440		1.96	14.15		16.11	25
4570	2"	Q-1	33.10	.483		3.04	14		17.04	26.50
4800	Tee, sanitary									
4820	1-1/4"	1 Plum	13.50	.593	Ea.	3.31	19.10		22.41	35
4830	1-1/2"	"	12.10	.661		2.67	21.50		24.17	38
4840	2"	Q-1	20	.800		4.12	23		27.12	42.50
5000	DWV, PVC, schedule 40, socket joints									
5040	1/4 bend, 1-1/4"	1 Plum	20.20	.396	Ea.	5.70	12.75		18.45	27.50
5060	1-1/2"	"	18.20	.440		1.90	14.15		16.05	25
5070	2"	Q-1	33.10	.483		2.89	14		16.89	26
5080	3"		20.80	.769		7.30	22.50		29.80	44.50
5090	4"		16.50	.970		13.75	28		41.75	61
5110	1/4 bend, long sweep, 1-1/2"	1 Plum	18.20	.440		3.41	14.15		17.56	27
5112	2"	Q-1	33.10	.483		3.68	14		17.68	27
5114	3"		20.80	.769		8.45	22.50		30.95	46
5116	4"		16.50	.970		16.15	28		44.15	64
5250	Tee, sanitary 1-1/4"	1 Plum	13.50	.593		4.10	19.10		23.20	36
5254	1-1/2"	"	12.10	.661		2.57	21.50		24.07	38
5255	2"	Q-1	20	.800		3.78	23		26.78	42
5256	3"		13.90	1.151		9.70	33.50		43.20	65
5257	4"		11	1.455		16.85	42		58.85	87.50
5259	6"		6.70	2.388		73.50	69		142.50	195
5261	8"	Q-2	6.20	3.871		184	108		292	380
5264	2" x 1-1/2"	Q-1	22	.727		3.34	21		24.34	38
5266	3" x 1-1/2"		15.50	1.032		6.15	30		36.15	56
5268	4" x 3"		12.10	1.322		21.50	38.50		60	86.50
5271	6" x 4"		6.90	2.319		79	67		146	197
5314	Combination Y & 1/8 bend, 1-1/2"	1 Plum	12.10	.661		5.60	21.50		27.10	41
5315	2"	Q-1	20	.800		7.40	23		30.40	46
5317	3"		13.90	1.151		17.90	33.50		51.40	74
5318	4"		11	1.455		32	42		74	104
5324	Combination Y & 1/8 bend, reducing									
5325	2" x 2" x 1-1/2"	Q-1	22	.727	Ea.	8	21		29	43.50
5327	3" x 3" x 1-1/2"		15.50	1.032		13.85	30		43.85	64.50
5328	3" x 3" x 2"		15.30	1.046		10.65	30.50		41.15	61
5329	4" x 4" x 2"		12.20	1.311		17.75	38		55.75	82
5331	Wye, 1-1/4"	1 Plum	13.50	.593		5.25	19.10		24.35	37.50
5332	1-1/2"	"	12.10	.661		3.93	21.50		25.43	39.50
5333	2"	Q-1	20	.800		4.60	23		27.60	43
5334	3"		13.90	1.151		11.85	33.50		45.35	67.50
5335	4"		11	1.455		22	42		64	93
5336	6"		6.70	2.388		68	69		137	189
5337	8"	Q-2	6.20	3.871		185	108		293	380
5341	2" x 1-1/2"	Q-1	22	.727		5.65	21		26.65	41
5342	3" x 1-1/2"		15.50	1.032		7.95	30		37.95	57.50
5343	4" x 3"		12.10	1.322		17.50	38.50		56	82.50
5344	6" x 4"		6.90	2.319		50	67		117	165
5345	8" x 6"	Q-2	6.40	3.750		151	105		256	340
5347	Double wye, 1-1/2"	1 Plum	9.10	.879		8.90	28.50		37.40	56.50
5348	2"	Q-1	16.60	.964		9.55	28		37.55	56.50
5349	3"		10.40	1.538		24.50	44.50		69	100
5350	4"		8.25	1.939		50	56		106	147

22 11 Facility Water Distribution

22 11 13 – Facility Water Distribution Piping

22 11 13.76 Pipe Fittings, Plastic		Crew	Daily Output	Labor-Hours	Unit	Material	2009 Bare Costs Labor	Equipment	Total	Total Incl O&P
5354	2" x 1-1/2"	Q-1	16.80	.952	Ea.	8.75	27.50		36.25	55
5355	3" x 2"		10.60	1.509		18.35	43.50		61.85	92
5356	4" x 3"		8.45	1.893		39.50	55		94.50	134
5357	6" x 4"		7.25	2.207		100	64		164	215
5410	Reducer bushing, 2" x 1-1/4"		36.50	.438		2.51	12.70		15.21	24
5412	3" x 1-1/2"		27.30	.586		6.05	17		23.05	34.50
5414	4" x 2"		18.20	.879		11.25	25.50		36.75	54.50
5416	6" x 4"		11.10	1.441		30	42		72	102
5418	8" x 6"	Q-2	10.20	2.353		73	65.50		138.50	189
5500	CPVC, Schedule 80, threaded joints									
5540	90° Elbow, 1/4"	1 Plum	32	.250	Ea.	10.20	8.05		18.25	24.50
5560	1/2"		30.30	.264		5.90	8.50		14.40	20.50
5570	3/4"		26	.308		8.85	9.90		18.75	26
5580	1"		22.70	.352		12.40	11.35		23.75	32.50
5590	1-1/4"		20.20	.396		24	12.75		36.75	47.50
5600	1-1/2"		18.20	.440		26	14.15		40.15	51.50
5610	2"	Q-1	33.10	.483		34.50	14		48.50	61
6000	Coupling, 1/4"	1 Plum	32	.250		13	8.05		21.05	27.50
6020	1/2"		30.30	.264		13	8.50		21.50	28.50
6030	3/4"		26	.308		15.60	9.90		25.50	33.50
6040	1"		22.70	.352		17.25	11.35		28.60	37.50
6050	1-1/4"		20.20	.396		21	12.75		33.75	44
6060	1-1/2"		18.20	.440		22.50	14.15		36.65	47.50
6070	2"	Q-1	33.10	.483		26.50	14		40.50	52

22 11 19 – Domestic Water Piping Specialties

22 11 19.38 Water Supply Meters

		Crew	Daily Output	Labor-Hours	Unit	Material	2009 Bare Costs Labor	Equipment	Total	Total Incl O&P
0010	**WATER SUPPLY METERS**									
2000	Domestic/commercial, bronze									
2020	Threaded									
2060	5/8" diameter, to 20 GPM	1 Plum	16	.500	Ea.	42	16.10		58.10	72.50
2080	3/4" diameter, to 30 GPM		14	.571		76.50	18.40		94.90	114
2100	1" diameter, to 50 GPM		12	.667		116	21.50		137.50	163

22 11 19.42 Backflow Preventers

		Crew	Daily Output	Labor-Hours	Unit	Material	2009 Bare Costs Labor	Equipment	Total	Total Incl O&P
0010	**BACKFLOW PREVENTERS**, Includes valves									
0020	and four test cocks, corrosion resistant, automatic operation									
4100	Threaded, bronze, valves are ball									
4120	3/4" pipe size	1 Plum	16	.500	Ea.	330	16.10		346.10	390

22 11 19.50 Vacuum Breakers

		Crew	Daily Output	Labor-Hours	Unit	Material	2009 Bare Costs Labor	Equipment	Total	Total Incl O&P
0010	**VACUUM BREAKERS**									
0013	See also backflow preventers Div. 22 11 19.42									
1000	Anti-siphon continuous pressure type									
1010	Max. 150 PSI - 210° F									
1020	Bronze body									
1030	1/2" size	1 Stpi	24	.333	Ea.	143	10.85		153.85	175
1040	3/4" size		20	.400		143	13		156	179
1050	1" size		19	.421		148	13.70		161.70	186
1060	1-1/4" size		15	.533		291	17.35		308.35	350
1070	1-1/2" size		13	.615		360	20		380	430
1080	2" size		11	.727		370	23.50		393.50	445
1200	Max. 125 PSI with atmospheric vent									
1210	Brass, in-line construction									
1220	1/4" size	1 Stpi	24	.333	Ea.	57.50	10.85		68.35	81.50

22 11 Facility Water Distribution

22 11 19 – Domestic Water Piping Specialties

22 11 19.50 Vacuum Breakers

		Crew	Daily Output	Labor-Hours	Unit	Material	2009 Bare Costs Labor	Equipment	Total	Total Incl O&P
1230	3/8" size	1 Stpi	24	.333	Ea.	57.50	10.85		68.35	81.50
1260	For polished chrome finish, add	↓			↓	13%				
2000	Anti-siphon, non-continuous pressure type									
2010	Hot or cold water 125 PSI - 210° F									
2020	Bronze body									
2030	1/4" size	1 Stpi	24	.333	Ea.	36	10.85		46.85	57.50
2040	3/8" size		24	.333		36	10.85		46.85	57.50
2050	1/2" size		24	.333		41	10.85		51.85	63
2060	3/4" size		20	.400		48.50	13		61.50	75
2070	1" size		19	.421		75.50	13.70		89.20	106
2080	1-1/4" size		15	.533		133	17.35		150.35	175
2090	1-1/2" size		13	.615		137	20		157	184
2100	2" size		11	.727		243	23.50		266.50	305
2110	2-1/2" size		8	1		695	32.50		727.50	820
2120	3" size	↓	6	1.333	↓	925	43.50		968.50	1,100
2150	For polished chrome finish, add					50%				

22 11 19.54 Water Hammer Arresters/Shock Absorbers

		Crew	Daily Output	Labor-Hours	Unit	Material	2009 Bare Costs Labor	Equipment	Total	Total Incl O&P
0010	**WATER HAMMER ARRESTERS/SHOCK ABSORBERS**									
0490	Copper									
0500	3/4" male I.P.S. For 1 to 11 fixtures	1 Plum	12	.667	Ea.	17	21.50		38.50	53.50

22 13 Facility Sanitary Sewerage

22 13 16 – Sanitary Waste and Vent Piping

22 13 16.20 Pipe, Cast Iron

		Crew	Daily Output	Labor-Hours	Unit	Material	2009 Bare Costs Labor	Equipment	Total	Total Incl O&P
0010	**PIPE, CAST IRON**, Soil, on clevis hanger assemblies, 5' O.C. R221113-50									
0020	Single hub, service wt., lead & oakum joints 10' O.C.									
2120	2" diameter	Q-1	63	.254	L.F.	6.35	7.35		13.70	19.10
2140	3" diameter		60	.267		8.90	7.75		16.65	22.50
2160	4" diameter	↓	55	.291	↓	11.50	8.45		19.95	26.50
4000	No hub, couplings 10' O.C.									
4100	1-1/2" diameter	Q-1	71	.225	L.F.	6.90	6.55		13.45	18.25
4120	2" diameter		67	.239		7.10	6.90		14	19.20
4140	3" diameter		64	.250		9.80	7.25		17.05	22.50
4160	4" diameter	↓	58	.276	↓	12.70	8		20.70	27

22 13 16.30 Pipe Fittings, Cast Iron

		Crew	Daily Output	Labor-Hours	Unit	Material	2009 Bare Costs Labor	Equipment	Total	Total Incl O&P
0010	**PIPE FITTINGS, CAST IRON**, Soil									
0040	Hub and spigot, service weight, lead & oakum joints									
0080	1/4 bend, 2"	Q-1	16	1	Ea.	13.70	29		42.70	62.50
0120	3"		14	1.143		18.40	33		51.40	75
0140	4"		13	1.231		28.50	35.50		64	90
0340	1/8 bend, 2"		16	1		9.80	29		38.80	58.50
0350	3"		14	1.143		15.35	33		48.35	71.50
0360	4"		13	1.231		22.50	35.50		58	83
0500	Sanitary tee, 2"		10	1.600		19.25	46.50		65.75	97
0540	3"		9	1.778		31	51.50		82.50	119
0620	4"	↓	8	2	↓	38	58		96	137
5990	No hub									
6000	Cplg. & labor required at joints not incl. in fitting									
6010	price. Add 1 coupling per joint for installed price									
6020	1/4 Bend, 1-1/2"				Ea.	7.70			7.70	8.45

22 13 16.30 Pipe Fittings, Cast Iron		Crew	Daily Output	Labor-Hours	Unit	Material	2009 Bare Costs Labor	Equipment	Total	Total Incl O&P
6060	2"				Ea.	8.45			8.45	9.30
6080	3"					11.60			11.60	12.80
6120	4"					16.80			16.80	18.50
6184	1/4 Bend, long sweep, 1-1/2"					18.20			18.20	20
6186	2"					18.20			18.20	20
6188	3"					21.50			21.50	23.50
6189	4"					34.50			34.50	38
6190	5"					63.50			63.50	70
6191	6"					77.50			77.50	85
6192	8"					187			187	206
6193	10"					335			335	370
6200	1/8 Bend, 1-1/2"					6.45			6.45	7.10
6210	2"					6.45			6.45	7.10
6212	3"					9.75			9.75	10.70
6214	4"					12.30			12.30	13.55
6380	Sanitary Tee, tapped, 1-1/2"					14.20			14.20	15.65
6382	2" x 1-1/2"					12.80			12.80	14.10
6384	2"					14.35			14.35	15.80
6386	3" x 2"					20			20	22
6388	3"					37			37	40.50
6390	4" x 1-1/2"					17.70			17.70	19.50
6392	4" x 2"					20			20	22
6393	4"					20			20	22
6394	6" x 1-1/2"					44			44	48.50
6396	6" x 2"					45			45	49.50
6459	Sanitary Tee, 1-1/2"					10.70			10.70	11.80
6460	2"					11.60			11.60	12.80
6470	3"					14.20			14.20	15.65
6472	4"				▼	22			22	24
8000	Coupling, standard (by CISPI Mfrs.)									
8020	1-1/2"	Q-1	48	.333	Ea.	14.45	9.65		24.10	32
8040	2"		44	.364		14.45	10.55		25	33
8080	3"		38	.421		17.25	12.20		29.45	39
8120	4"	▼	33	.485	▼	20.50	14.05		34.55	45.50

22 13 16.60 Traps

		Crew	Daily Output	Labor-Hours	Unit	Material	2009 Bare Costs Labor	Equipment	Total	Total Incl O&P
0010	**TRAPS**									
0030	Cast iron, service weight									
0050	Running P trap, without vent									
1100	2"	Q-1	16	1	Ea.	99	29		128	157
1150	4"	"	13	1.231		99	35.50		134.50	168
1160	6"	Q-2	17	1.412		430	39.50		469.50	540
3000	P trap, B&S, 2" pipe size	Q-1	16	1		23.50	29		52.50	73.50
3040	3" pipe size	"	14	1.143	▼	35	33		68	93
4700	Copper, drainage, drum trap									
4840	3" x 6" swivel, 1-1/2" pipe size	1 Plum	16	.500	Ea.	155	16.10		171.10	197
5100	P trap, standard pattern									
5200	1-1/4" pipe size	1 Plum	18	.444	Ea.	72	14.30		86.30	103
5240	1-1/2" pipe size		17	.471		69.50	15.15		84.65	102
5260	2" pipe size		15	.533		107	17.15		124.15	146
5280	3" pipe size	▼	11	.727	▼	258	23.50		281.50	325
6710	ABS DWV P trap, solvent weld joint									
6720	1-1/2" pipe size	1 Plum	18	.444	Ea.	6.75	14.30		21.05	31

22 13 Facility Sanitary Sewerage

22 13 16 – Sanitary Waste and Vent Piping

22 13 16.60 Traps

		Crew	Daily Output	Labor-Hours	Unit	Material	2009 Bare Costs Labor	2009 Bare Costs Equipment	Total	Total Incl O&P
6722	2" pipe size	1 Plum	17	.471	Ea.	9.15	15.15		24.30	35
6724	3" pipe size		15	.533		35	17.15		52.15	66.50
6726	4" pipe size		14	.571		72	18.40		90.40	109
6860	PVC DWV hub x hub, basin trap, 1-1/4" pipe size		18	.444		8.95	14.30		23.25	33.50
6870	Sink P trap, 1-1/2" pipe size		18	.444		8.95	14.30		23.25	33.50
6880	Tubular S trap, 1-1/2" pipe size	↓	17	.471	↓	16.55	15.15		31.70	43
6890	PVC sch. 40 DWV, drum trap									
6900	1-1/2" pipe size	1 Plum	16	.500	Ea.	20.50	16.10		36.60	49
6910	P trap, 1-1/2" pipe size		18	.444		5.05	14.30		19.35	29
6920	2" pipe size		17	.471		6.65	15.15		21.80	32.50
6930	3" pipe size		15	.533		23	17.15		40.15	53.50
6940	4" pipe size		14	.571		53	18.40		71.40	88
6950	P trap w/clean out, 1-1/2" pipe size		18	.444		8.35	14.30		22.65	32.50
6960	2" pipe size	↓	17	.471	↓	14.20	15.15		29.35	40.50

22 13 16.80 Vent Flashing and Caps

		Crew	Daily Output	Labor-Hours	Unit	Material	2009 Bare Costs Labor	2009 Bare Costs Equipment	Total	Total Incl O&P
0010	**VENT FLASHING AND CAPS**									
0120	Vent caps									
0140	Cast iron									
0160	1-1/4" - 1-1/2" pipe	1 Plum	23	.348	Ea.	30.50	11.20		41.70	52
0170	2" - 2-1/8" pipe		22	.364		35.50	11.70		47.20	58
0180	2-1/2" - 3-5/8" pipe		21	.381		40	12.25		52.25	64
0190	4" - 4-1/8" pipe		19	.421		48	13.55		61.55	75.50
0200	5" - 6" pipe	↓	17	.471	↓	71.50	15.15		86.65	104
0300	PVC									
0320	1-1/4" - 1-1/2" pipe	1 Plum	24	.333	Ea.	8	10.75		18.75	26.50
0330	2" - 2-1/8" pipe	"	23	.348	"	8.65	11.20		19.85	28

22 13 19 – Sanitary Waste Piping Specialties

22 13 19.13 Sanitary Drains

		Crew	Daily Output	Labor-Hours	Unit	Material	2009 Bare Costs Labor	2009 Bare Costs Equipment	Total	Total Incl O&P
0010	**SANITARY DRAINS**									
2000	Floor, medium duty, C.I., deep flange, 7" dia top									
2040	2" and 3" pipe size	Q-1	12	1.333	Ea.	141	38.50		179.50	220
2080	For galvanized body, add					67			67	74
2120	For polished bronze top, add				↓	102			102	112

22 14 Facility Storm Drainage

22 14 26 – Facility Storm Drains

22 14 26.13 Roof Drains

		Crew	Daily Output	Labor-Hours	Unit	Material	2009 Bare Costs Labor	2009 Bare Costs Equipment	Total	Total Incl O&P
0010	**ROOF DRAINS**									
3860	Roof, flat metal deck, C.I. body, 12" C.I. dome									
3890	3" pipe size	Q-1	14	1.143	Ea.	277	33		310	360

22 14 29 – Sump Pumps

22 14 29.16 Submersible Sump Pumps

		Crew	Daily Output	Labor-Hours	Unit	Material	2009 Bare Costs Labor	2009 Bare Costs Equipment	Total	Total Incl O&P
0010	**SUBMERSIBLE SUMP PUMPS**									
7000	Sump pump, automatic									
7100	Plastic, 1-1/4" discharge, 1/4 HP	1 Plum	6	1.333	Ea.	136	43		179	220
7500	Cast iron, 1-1/4" discharge, 1/4 HP	"	6	1.333	"	156	43		199	242

22 31 Domestic Water Softeners

22 31 13 – Residential Domestic Water Softeners

22 31 13.10 Residential Water Softeners	Crew	Daily Output	Labor-Hours	Unit	Material	2009 Bare Costs Labor	Equipment	Total	Total Incl O&P
0010 **RESIDENTIAL WATER SOFTENERS**									
7350 Water softener, automatic, to 30 grains per gallon	2 Plum	5	3.200	Ea.	410	103		513	620
7400 To 100 grains per gallon	"	4	4	"	680	129		809	960

22 33 Electric Domestic Water Heaters

22 33 30 – Residential, Electric Domestic Water Heaters

22 33 30.13 Residential, Small-Capacity Elec. Water Heaters

	Crew	Daily Output	Labor-Hours	Unit	Material	2009 Bare Costs Labor	Equipment	Total	Total Incl O&P
0010 **RESIDENTIAL, SMALL-CAPACITY ELECTRIC DOMESTIC WATER HEATERS**									
1000 Residential, electric, glass lined tank, 5 yr, 10 gal., single element	1 Plum	2.30	3.478	Ea.	315	112		427	530
1060 30 gallon, double element		2.20	3.636		465	117		582	705
1080 40 gallon, double element		2	4		500	129		629	755
1100 52 gallon, double element		2	4		560	129		689	825
1120 66 gallon, double element		1.80	4.444		755	143		898	1,075
1140 80 gallon, double element		1.60	5		840	161		1,001	1,200

22 34 Fuel-Fired Domestic Water Heaters

22 34 30 – Residential Gas Domestic Water Heaters

22 34 30.13 Residential, Atmos, Gas Domestic Wtr Heaters

	Crew	Daily Output	Labor-Hours	Unit	Material	2009 Bare Costs Labor	Equipment	Total	Total Incl O&P
0010 **RESIDENTIAL, ATMOSPHERIC, GAS DOMESTIC WATER HEATERS**									
2000 Gas fired, foam lined tank, 10 yr, vent not incl.,									
2040 30 gallon	1 Plum	2	4	Ea.	630	129		759	900
2100 75 gallon	"	1.50	5.333	"	1,225	172		1,397	1,625

22 34 46 – Oil-Fired Domestic Water Heaters

22 34 46.10 Residential Oil-Fired Water Heaters

	Crew	Daily Output	Labor-Hours	Unit	Material	2009 Bare Costs Labor	Equipment	Total	Total Incl O&P
0010 **RESIDENTIAL OIL-FIRED WATER HEATERS**									
3000 Oil fired, glass lined tank, 5 yr, vent not included, 30 gallon	1 Plum	2	4	Ea.	665	129		794	940
3040 50 gallon	"	1.80	4.444	"	1,175	143		1,318	1,500

22 41 Residential Plumbing Fixtures

22 41 13 – Residential Water Closets, Urinals, and Bidets

22 41 13.40 Water Closets

	Crew	Daily Output	Labor-Hours	Unit	Material	2009 Bare Costs Labor	Equipment	Total	Total Incl O&P
0010 **WATER CLOSETS**									
0150 Tank type, vitreous china, incl. seat, supply pipe w/stop									
0200 Wall hung									
0400 Two piece, close coupled	Q-1	5.30	3.019	Ea.	555	87.50		642.50	755
0960 For rough-in, supply, waste, vent and carrier		2.73	5.861		630	170		800	975
1000 Floor mounted, one piece		5.30	3.019		530	87.50		617.50	730
1020 One piece, low profile		5.30	3.019		400	87.50		487.50	590
1100 Two piece, close coupled		5.30	3.019		184	87.50		271.50	345
1960 For color, add					30%				
1980 For rough-in, supply, waste and vent	Q-1	3.05	5.246	Ea.	275	152		427	550

22 41 16 – Residential Lavatories and Sinks

22 41 16.10 Lavatories

	Crew	Daily Output	Labor-Hours	Unit	Material	2009 Bare Costs Labor	Equipment	Total	Total Incl O&P
0010 **LAVATORIES**, With trim, white unless noted otherwise									
0500 Vanity top, porcelain enamel on cast iron									
0600 20" x 18"	Q-1	6.40	2.500	Ea.	231	72.50		303.50	375

22 41 Residential Plumbing Fixtures

22 41 16 – Residential Lavatories and Sinks

22 41 16.10 Lavatories

		Crew	Daily Output	Labor-Hours	Unit	Material	2009 Bare Costs Labor	Equipment	Total	Total Incl O&P
0640	33" x 19" oval	Q-1	6.40	2.500	Ea.	505	72.50		577.50	675
0720	19" round	↓	6.40	2.500	↓	218	72.50		290.50	360
0860	For color, add					25%				
1000	Cultured marble, 19" x 17", single bowl	Q-1	6.40	2.500	Ea.	169	72.50		241.50	305
1120	25" x 22", single bowl		6.40	2.500		185	72.50		257.50	325
1160	37" x 22", single bowl	↓	6.40	2.500	↓	216	72.50		288.50	355
1560										
1900	Stainless steel, self-rimming, 25" x 22", single bowl, ledge	Q-1	6.40	2.500	Ea.	330	72.50		402.50	480
1960	17" x 22", single bowl		6.40	2.500		320	72.50		392.50	470
2600	Steel, enameled, 20" x 17", single bowl		5.80	2.759		159	80		239	305
2900	Vitreous china, 20" x 16", single bowl		5.40	2.963		260	86		346	425
3200	22" x 13", single bowl		5.40	2.963		267	86		353	435
3580	Rough-in, supply, waste and vent for all above lavatories	↓	2.30	6.957	↓	224	202		426	575
4000	Wall hung									
4040	Porcelain enamel on cast iron, 16" x 14", single bowl	Q-1	8	2	Ea.	410	58		468	545
4180	20" x 18", single bowl	"	8	2	"	297	58		355	420
4580	For color, add					30%				
6000	Vitreous china, 18" x 15", single bowl with backsplash	Q-1	7	2.286	Ea.	231	66		297	365
6060	19" x 17", single bowl		7	2.286		173	66		239	299
6960	Rough-in, supply, waste and vent for above lavatories	↓	1.66	9.639	↓	365	279		644	860

22 41 16.30 Sinks

		Crew	Daily Output	Labor-Hours	Unit	Material	2009 Bare Costs Labor	Equipment	Total	Total Incl O&P
0010	**SINKS**, With faucets and drain									
2000	Kitchen, counter top style, P.E. on C.I., 24" x 21" single bowl	Q-1	5.60	2.857	Ea.	272	83		355	435
2100	31" x 22" single bowl		5.60	2.857		278	83		361	440
2200	32" x 21" double bowl		4.80	3.333		370	96.50		466.50	570
3000	Stainless steel, self rimming, 19" x 18" single bowl		5.60	2.857		510	83		593	695
3100	25" x 22" single bowl		5.60	2.857		570	83		653	760
3200	33" x 22" double bowl		4.80	3.333		830	96.50		926.50	1,075
3300	43" x 22" double bowl		4.80	3.333		965	96.50		1,061.50	1,200
4000	Steel, enameled, with ledge, 24" x 21" single bowl		5.60	2.857		145	83		228	295
4100	32" x 21" double bowl	↓	4.80	3.333		166	96.50		262.50	340
4960	For color sinks except stainless steel, add					10%				
4980	For rough-in, supply, waste and vent, counter top sinks	Q-1	2.14	7.477	↓	266	217		483	650
5000	Kitchen, raised deck, P.E. on C.I.									
5100	32" x 21", dual level, double bowl	Q-1	2.60	6.154	Ea.	289	178		467	615
5790	For rough-in, supply, waste & vent, sinks	"	1.85	8.649	"	266	251		517	705

22 41 19 – Residential Bathtubs

22 41 19.10 Baths

		Crew	Daily Output	Labor-Hours	Unit	Material	2009 Bare Costs Labor	Equipment	Total	Total Incl O&P
0010	**BATHS** R224000-40									
0100	Tubs, recessed porcelain enamel on cast iron, with trim									
0180	48" x 42"	Q-1	4	4	Ea.	1,800	116		1,916	2,175
0220	72" x 36"	"	3	5.333	"	1,850	155		2,005	2,300
0300	Mat bottom									
0380	5' long	Q-1	4.40	3.636	Ea.	810	105		915	1,075
0480	Above floor drain, 5' long		4	4		800	116		916	1,075
0560	Corner 48" x 44"		4.40	3.636		1,800	105		1,905	2,150
2000	Enameled formed steel, 4'-6" long	↓	5.80	2.759	↓	365	80		445	530
4600	Module tub & showerwall surround, molded fiberglass									
4610	5' long x 34" wide x 76" high	Q-1	4	4	Ea.	560	116		676	805
9600	Rough-in, supply, waste and vent, for all above tubs, add	"	2.07	7.729	"	300	224		524	700

22 41 Residential Plumbing Fixtures

22 41 23 – Residential Shower Receptors and Basins

22 41 23.20 Showers		Crew	Daily Output	Labor-Hours	Unit	Material	2009 Bare Costs Labor	2009 Bare Costs Equipment	Total	Total Incl O&P
0010	**SHOWERS**									
1500	Stall, with drain only. Add for valve and door/curtain									
1520	32" square	Q-1	5	3.200	Ea.	365	92.50		457.50	550
1530	36" square		4.80	3.333		870	96.50		966.50	1,125
1540	Terrazzo receptor, 32" square		5	3.200		740	92.50		832.50	965
1560	36" square		4.80	3.333		1,150	96.50		1,246.50	1,425
1580	36" corner angle		4.80	3.333		1,275	96.50		1,371.50	1,550
3000	Fiberglass, one piece, with 3 walls, 32" x 32" square		5.50	2.909		460	84.50		544.50	650
3100	36" x 36" square		5.50	2.909		525	84.50		609.50	715
4200	Rough-in, supply, waste and vent for above showers		2.05	7.805		395	226		621	805

22 41 36 – Residential Laundry Trays

22 41 36.10 Laundry Sinks

22 41 36.10 Laundry Sinks		Crew	Daily Output	Labor-Hours	Unit	Material	2009 Bare Costs Labor	2009 Bare Costs Equipment	Total	Total Incl O&P
0010	**LAUNDRY SINKS**, With trim									
0020	Porcelain enamel on cast iron, black iron frame									
0050	24" x 21", single compartment	Q-1	6	2.667	Ea.	415	77.50		492.50	580
0100	26" x 21", single compartment	"	6	2.667	"	405	77.50		482.50	570
3000	Plastic, on wall hanger or legs									
3020	18" x 23", single compartment	Q-1	6.50	2.462	Ea.	99.50	71.50		171	227
3100	20" x 24", single compartment		6.50	2.462		133	71.50		204.50	263
3200	36" x 23", double compartment		5.50	2.909		158	84.50		242.50	310
3300	40" x 24", double compartment		5.50	2.909		236	84.50		320.50	395
5000	Stainless steel, counter top, 22" x 17" single compartment		6	2.667		58	77.50		135.50	191
5200	33" x 22", double compartment		5	3.200		70	92.50		162.50	229
9600	Rough-in, supply, waste and vent, for all laundry sinks		2.14	7.477		266	217		483	650

22 41 39 – Residential Faucets, Supplies and Trim

22 41 39.10 Faucets and Fittings

22 41 39.10 Faucets and Fittings		Crew	Daily Output	Labor-Hours	Unit	Material	2009 Bare Costs Labor	2009 Bare Costs Equipment	Total	Total Incl O&P
0010	**FAUCETS AND FITTINGS**									
0150	Bath, faucets, diverter spout combination, sweat	1 Plum	8	1	Ea.	104	32		136	167
0200	For integral stops, IPS unions, add					109			109	120
0420	Bath, press-bal mix valve w/diverter, spout, shower hd, arm/flange	1 Plum	8	1		137	32		169	204
0500	Drain, central lift, 1-1/2" IPS male		20	.400		40.50	12.90		53.40	65.50
0600	Trip lever, 1-1/2" IPS male		20	.400		41	12.90		53.90	66
1000	Kitchen sink faucets, top mount, cast spout		10	.800		55	26		81	103
1100	For spray, add		24	.333		16.15	10.75		26.90	35.50
1300	Single control lever handle									
1310	With pull out spray									
1320	Polished chrome	1 Plum	10	.800	Ea.	192	26		218	254
1330	Polished brass		10	.800		230	26		256	295
1340	White		10	.800		211	26		237	275
1348	With spray thru escutcheon									
1350	Polished chrome	1 Plum	10	.800	Ea.	152	26		178	210
1360	Polished brass		10	.800		183	26		209	244
1370	White		10	.800		164	26		190	223
2000	Laundry faucets, shelf type, IPS or copper unions		12	.667		45.50	21.50		67	85.50
2020										
2100	Lavatory faucet, centerset, without drain	1 Plum	10	.800	Ea.	44.50	26		70.50	91.50
2120	With pop-up drain	"	6.66	1.201	"	61.50	38.50		100	132
2210	Porcelain cross handles and pop-up drain									
2220	Polished chrome	1 Plum	6.66	1.201	Ea.	126	38.50		164.50	203
2230	Polished brass	"	6.66	1.201	"	189	38.50		227.50	272
2260	Single lever handle and pop-up drain									

248

22 41 Residential Plumbing Fixtures

22 41 39 – Residential Faucets, Supplies and Trim

22 41 39.10 Faucets and Fittings

22 41 39.10 Faucets and Fittings	Crew	Daily Output	Labor-Hours	Unit	Material	2009 Bare Costs Labor	Equipment	Total	Total Incl O&P
2270	Black nickel	1 Plum	6.66	1.201	Ea.	220	38.50		258.50	305
2280	Polished brass		6.66	1.201		220	38.50		258.50	305
2290	Polished chrome		6.66	1.201		169	38.50		207.50	250
2800	Self-closing, center set		10	.800		131	26		157	187
4000	Shower by-pass valve with union		18	.444		68.50	14.30		82.80	98.50
4200	Shower thermostatic mixing valve, concealed		8	1		320	32		352	405
4204	For inlet strainer, check, and stops, add					45			45	49.50
4220	Shower pressure balancing mixing valve,									
4230	With shower head, arm, flange and diverter tub spout									
4240	Chrome	1 Plum	6.14	1.303	Ea.	170	42		212	256
4250	Polished brass		6.14	1.303		238	42		280	330
4260	Satin		6.14	1.303		238	42		280	330
4270	Polished chrome/brass		6.14	1.303		196	42		238	284
5000	Sillcock, compact, brass, IPS or copper to hose		24	.333		8.05	10.75		18.80	26.50

22 42 Commercial Plumbing Fixtures

22 42 13 – Commercial Water Closets, Urinals, and Bidets

22 42 13.40 Water Closets

		Crew	Daily Output	Labor-Hours	Unit	Material	2009 Bare Costs Labor	Equipment	Total	Total Incl O&P
0010	**WATER CLOSETS**									
3000	Bowl only, with flush valve, seat									
3100	Wall hung	Q-1	5.80	2.759	Ea.	284	80		364	445
3200	For rough-in, supply, waste and vent, single WC		2.56	6.250		680	181		861	1,050
3300	Floor mounted		5.80	2.759		305	80		385	465
3400	For rough-in, supply, waste and vent, single WC		2.84	5.634		320	163		483	625

22 42 16 – Commercial Lavatories and Sinks

22 42 16.14 Lavatories

0010	**LAVATORIES**, With trim, white unless noted otherwise									
0020	Commercial lavatories same as residential. See Div. 22 41 16.10									

22 42 16.40 Service Sinks

		Crew	Daily Output	Labor-Hours	Unit	Material	2009 Bare Costs Labor	Equipment	Total	Total Incl O&P
0010	**SERVICE SINKS**									
6650	Service, floor, corner, P.E. on C.I., 28" x 28"	Q-1	4.40	3.636	Ea.	625	105		730	865
6750	Vinyl coated rim guard, add					67.50			67.50	74
6760	Mop sink, molded stone, 24" x 36"	1 Plum	3.33	2.402		248	77.50		325.50	400
6770	Mop sink, molded stone, 24" x 36", w/rim 3 sides	"	3.33	2.402		233	77.50		310.50	385
6790	For rough-in, supply, waste & vent, floor service sinks	Q-1	1.64	9.756		705	283		988	1,250

22 42 39 – Commercial Faucets, Supplies, and Trim

22 42 39.10 Faucets and Fittings

		Crew	Daily Output	Labor-Hours	Unit	Material	2009 Bare Costs Labor	Equipment	Total	Total Incl O&P
0010	**FAUCETS AND FITTINGS**									
3000	Service sink faucet, cast spout, pail hook, hose end	1 Plum	14	.571	Ea.	80	18.40		98.40	118

22 42 39.30 Carriers and Supports

		Crew	Daily Output	Labor-Hours	Unit	Material	2009 Bare Costs Labor	Equipment	Total	Total Incl O&P
0010	**CARRIERS AND SUPPORTS**, For plumbing fixtures									
0600	Plate type with studs, top back plate	1 Plum	7	1.143	Ea.	63	37		100	130
3000	Lavatory, concealed arm									
3050	Floor mounted, single									
3100	High back fixture	1 Plum	6	1.333	Ea.	350	43		393	455
3200	Flat slab fixture	"	6	1.333	"	296	43		339	395
8200	Water closet, residential									
8220	Vertical centerline, floor mount									
8240	Single, 3" caulk, 2" or 3" vent	1 Plum	6	1.333	Ea.	405	43		448	515

22 42 Commercial Plumbing Fixtures

22 42 39 – Commercial Faucets, Supplies, and Trim

22 42 39.30 Carriers and Supports	Crew	Daily Output	Labor-Hours	Unit	Material	2009 Bare Costs Labor	Equipment	Total	Total Incl O&P
8260	4" caulk, 2" or 4" vent	1 Plum	6	1.333	Ea.	520	43		563	645

22 51 Swimming Pool Plumbing Systems

22 51 19 – Swimming Pool Water Treatment Equipment

22 51 19.50 Swimming Pool Filtration Equipment

		Crew	Daily Output	Labor-Hours	Unit	Material	2009 Bare Costs Labor	Equipment	Total	Total Incl O&P
0010	**SWIMMING POOL FILTRATION EQUIPMENT**									
0900	Filter system, sand or diatomite type, incl. pump, 6,000 gal./hr.	2 Plum	1.80	8.889	Total	1,475	286		1,761	2,075
1020	Add for chlorination system, 800 S.F. pool	"	3	5.333	Ea.	78.50	172		250.50	370

Division 23
Heating, Ventilating, and Air Conditioning

23 05 Common Work Results for HVAC

23 05 05 – Selective HVAC Demolition

23 05 05.10 HVAC Demolition

	23 05 05.10 HVAC Demolition	Crew	Daily Output	Labor-Hours	Unit	Material	2009 Bare Costs Labor	Equipment	Total	Total Incl O&P
0010	**HVAC DEMOLITION**									
0100	Air conditioner, split unit, 3 ton	Q-5	2	8	Ea.		234		234	385
0150	Package unit, 3 ton	Q-6	3	8	"		226		226	370
0260	Baseboard, hydronic fin tube, 1/2"	Q-5	117	.137	L.F.		4.01		4.01	6.55
0298	Boilers									
0300	Electric, up thru 148 kW	Q-19	2	12	Ea.		360		360	590
0310	150 thru 518 kW	"	1	24			720		720	1,175
0320	550 thru 2000 kW	Q-21	.40	80			2,450		2,450	4,025
0330	2070 kW and up	"	.30	106			3,275		3,275	5,350
0340	Gas and/or oil, up thru 150 MBH	Q-7	2.20	14.545			425		425	700
0350	160 thru 2000 MBH		.80	40			1,175		1,175	1,925
0360	2100 thru 4500 MBH		.50	64			1,875		1,875	3,075
0370	4600 thru 7000 MBH		.30	106			3,125		3,125	5,125
0390	12,200 thru 25,000 MBH		.12	266			7,825		7,825	12,800
1000	Ductwork, 4" high, 8" wide	1 Clab	200	.040	L.F.		.82		.82	1.39
1100	6" high, 8" wide		165	.048			1		1	1.68
1200	10" high, 12" wide		125	.064			1.32		1.32	2.22
1300	12"-14" high, 16"-18" wide		85	.094			1.93		1.93	3.27
1500	30" high, 36" wide		56	.143			2.94		2.94	4.96
2200	Furnace, electric	Q-20	2	10	Ea.		287		287	480
2300	Gas or oil, under 120 MBH	Q-9	4	4			112		112	188
2340	Over 120 MBH	"	3	5.333			149		149	251
2800	Heat pump, package unit, 3 ton	Q-5	2.40	6.667			195		195	320
2840	Split unit, 3 ton		2	8			234		234	385
2950	Tank, steel, oil, 275 gal., above ground		10	1.600			47		47	77
2960	Remove and reset		3	5.333			156		156	256
9000	Minimum labor/equipment charge	Q-6	3	8	Job		226		226	370

23 07 HVAC Insulation

23 07 13 – Duct Insulation

23 07 13.10 Duct Thermal Insulation

	23 07 13.10 Duct Thermal Insulation		Crew	Daily Output	Labor-Hours	Unit	Material	2009 Bare Costs Labor	Equipment	Total	Total Incl O&P
0010	**DUCT THERMAL INSULATION**										
3000	Ductwork										
3020	Blanket type, fiberglass, flexible										
3140	FSK vapor barrier wrap, .75 lb. density										
3160	1" thick	G	Q-14	350	.046	S.F.	.17	1.20		1.37	2.23
3170	1-1/2" thick	G		320	.050		.20	1.31		1.51	2.46
3212	Vinyl Jacket, .75 lb. density, 1-1/2" thick	G		320	.050		.20	1.31		1.51	2.46
9600	Minimum labor/equipment charge		1 Stpi	4	2	Job		65		65	107

23 07 16 – HVAC Equipment Insulation

23 07 16.10 HVAC Equipment Thermal Insulation

	23 07 16.10 HVAC Equipment Thermal Insulation	Crew	Daily Output	Labor-Hours	Unit	Material	Labor	Equipment	Total	Total Incl O&P
0010	**HVAC EQUIPMENT THERMAL INSULATION**									

23 09 Instrumentation and Control for HVAC

23 09 53 – Pneumatic and Electric Control System for HVAC

23 09 53.10 Control Components		Crew	Daily Output	Labor-Hours	Unit	Material	2009 Bare Costs Labor	Equipment	Total	Total Incl O&P
0010	**CONTROL COMPONENTS**									
5000	Thermostats									
5030	Manual	1 Shee	8	1	Ea.	27	31		58	82
5040	1 set back, electric, timed	G	8	1		95	31		126	157
5050	2 set back, electric, timed	G	8	1		209	31		240	282

23 13 Facility Fuel-Storage Tanks

23 13 13 – Facility Underground Fuel-Oil, Storage Tanks

23 13 13.09 Single-Wall Steel Fuel-Oil Tanks

		Crew	Daily Output	Labor-Hours	Unit	Material	2009 Bare Costs Labor	Equipment	Total	Total Incl O&P
0010	**SINGLE-WALL STEEL FUEL-OIL TANKS**									
5000	Steel underground, sti-P3, set in place, not incl. hold-down bars.									
5500	Excavation, pad, pumps and piping not included									
5510	Single wall, 500 gallon capacity, 7 gauge shell	Q-5	2.70	5.926	Ea.	1,975	174		2,149	2,450
5520	1,000 gallon capacity, 7 gauge shell	"	2.50	6.400		3,150	188		3,338	3,775
5530	2,000 gallon capacity, 1/4" thick shell	Q-7	4.60	6.957		5,125	204		5,329	5,950
5535	2,500 gallon capacity, 7 gauge shell	Q-5	3	5.333		5,650	156		5,806	6,475
5610	25,000 gallon capacity, 3/8" thick shell	Q-7	1.30	24.615		25,700	720		26,420	29,500
5630	40,000 gallon capacity, 3/8" thick shell		.90	35.556		39,600	1,050		40,650	45,300
5640	50,000 gallon capacity, 3/8" thick shell		.80	40		49,500	1,175		50,675	56,500

23 13 13.23 Glass-Fiber-Reinfcd-Plastic, Fuel-Oil, Storage

		Crew	Daily Output	Labor-Hours	Unit	Material	2009 Bare Costs Labor	Equipment	Total	Total Incl O&P
0010	**GLASS-FIBER-REINFCD-PLASTIC, UNDERGRND FUEL-OIL, STORAGE**									
0210	Fiberglass, underground, single wall, U.L. listed, not including									
0220	manway or hold-down strap									
0230	1,000 gallon capacity	Q-5	2.46	6.504	Ea.	3,650	191		3,841	4,350
0240	2,000 gallon capacity	Q-7	4.57	7.002		5,750	205		5,955	6,650
0500	For manway, fittings and hold-downs, add					20%	15%			
2210	Fiberglass, underground, single wall, U.L. listed, including									
2220	hold-down straps, no manways									
2230	1,000 gallon capacity	Q-5	1.88	8.511	Ea.	4,075	249		4,324	4,875
2240	2,000 gallon capacity	Q-7	3.55	9.014	"	6,150	264		6,414	7,200

23 13 23 – Facility Aboveground Fuel-Oil, Storage Tanks

23 13 23.16 Steel

		Crew	Daily Output	Labor-Hours	Unit	Material	2009 Bare Costs Labor	Equipment	Total	Total Incl O&P
3001	**STEEL**, storage, above ground, including supports, coating									
3020	fittings, not including foundation, pumps or piping									
3040	Single wall, interior, 275 gallon	Q-5	5	3.200	Ea.	385	94		479	575
3060	550 gallon	"	2.70	5.926		2,275	174		2,449	2,800
3080	1,000 gallon	Q-7	5	6.400		2,875	188		3,063	3,450
3320	Double wall, 500 gallon capacity	Q-5	2.40	6.667		2,300	195		2,495	2,850
3330	2000 gallon capacity	Q-7	4.15	7.711		8,250	226		8,476	9,450
3340	4000 gallon capacity		3.60	8.889		14,700	260		14,960	16,600
3350	6000 gallon capacity		2.40	13.333		17,400	390		17,790	19,700
3360	8000 gallon capacity		2	16		22,300	470		22,770	25,300
3370	10000 gallon capacity		1.80	17.778		25,000	520		25,520	28,400
3380	15000 gallon capacity		1.50	21.333		37,900	625		38,525	42,700
3390	20000 gallon capacity		1.30	24.615		43,200	720		43,920	48,700
3400	25000 gallon capacity		1.15	27.826		52,500	815		53,315	59,000
3410	30000 gallon capacity		1	32		57,500	940		58,440	65,000

23 21 Hydronic Piping and Pumps

23 21 20 – Hydronic HVAC Piping Specialties

23 21 20.46 Expansion Tanks

		Crew	Daily Output	Labor-Hours	Unit	Material	2009 Bare Costs Labor	Equipment	Total	Total Incl O&P
0010	**EXPANSION TANKS**									
1507	Fiberglass and steel single / double wall storage, see Div. 23 13 13									
2000	Steel, liquid expansion, ASME, painted, 15 gallon capacity	Q-5	17	.941	Ea.	440	27.50		467.50	530
2040	30 gallon capacity		12	1.333		495	39		534	610
3000	Steel ASME expansion, rubber diaphragm, 19 gal. cap. accept.		12	1.333		1,925	39		1,964	2,200
3020	31 gallon capacity	↓	8	2	↓	2,150	58.50		2,208.50	2,475

23 21 23 – Hydronic Pumps

23 21 23.13 In-Line Centrifugal Hydronic Pumps

		Crew	Daily Output	Labor-Hours	Unit	Material	2009 Bare Costs Labor	Equipment	Total	Total Incl O&P
0010	**IN-LINE CENTRIFUGAL HYDRONIC PUMPS**									
0600	Bronze, sweat connections, 1/40 HP, in line									
0640	3/4" size	Q-1	16	1	Ea.	172	29		201	237
1000	Flange connection, 3/4" to 1-1/2" size									
1040	1/12 HP	Q-1	6	2.667	Ea.	445	77.50		522.50	615
1060	1/8 HP		6	2.667		750	77.50		827.50	950
2101	Pumps, circulating, 3/4" to 1-1/2" size, 1/3 HP	↓	6	2.667	↓	620	77.50		697.50	810

23 31 HVAC Ducts and Casings

23 31 13 – Metal Ducts

23 31 13.13 Rectangular Metal Ducts

		Crew	Daily Output	Labor-Hours	Unit	Material	2009 Bare Costs Labor	Equipment	Total	Total Incl O&P
0010	**RECTANGULAR METAL DUCTS**									
0020	Fabricated rectangular, includes fittings, joints, supports,									
0030	allowance for flexible connections, no insulation									
0031	NOTE: Fabrication and installation are combined									
0040	as LABOR cost. Approx. 25% fittings assumed.									
0100	Aluminum, alloy 3003-H14, under 100 lb.	Q-10	75	.320	Lb.	3.72	9.30		13.02	19.70
0110	100 to 500 lb.		80	.300		2.48	8.70		11.18	17.40
0120	500 to 1,000 lb.		95	.253		2.31	7.35		9.66	14.90
0140	1,000 to 2,000 lb.		120	.200		2.15	5.80		7.95	12.10
0500	Galvanized steel, under 200 lb.		235	.102		1.12	2.97		4.09	6.20
0520	200 to 500 lb.		245	.098		.99	2.85		3.84	5.85
0540	500 to 1,000 lb.	↓	255	.094	↓	.97	2.74		3.71	5.65

23 31 13.19 Metal Duct Fittings

		Crew	Daily Output	Labor-Hours	Unit	Material	2009 Bare Costs Labor	Equipment	Total	Total Incl O&P
0010	**METAL DUCT FITTINGS**									
0050	Air extractors, 12" x 4"	1 Shee	24	.333	Ea.	15.65	10.40		26.05	34.50
0100	8" x 6"	"	22	.364	"	15.65	11.35		27	36

23 33 Air Duct Accessories

23 33 13 – Dampers

23 33 13.13 Volume-Control Dampers

		Crew	Daily Output	Labor-Hours	Unit	Material	2009 Bare Costs Labor	Equipment	Total	Total Incl O&P
0010	**VOLUME-CONTROL DAMPERS**									
6000	12" x 12"	1 Shee	21	.381	Ea.	27.50	11.85		39.35	50
8000	Multi-blade dampers, parallel blade									
8100	8" x 8"	1 Shee	24	.333	Ea.	71	10.40		81.40	95.50

23 33 13.16 Fire Dampers

		Crew	Daily Output	Labor-Hours	Unit	Material	2009 Bare Costs Labor	Equipment	Total	Total Incl O&P
0010	**FIRE DAMPERS**									
3000	Fire damper, curtain type, 1-1/2 hr rated, vertical, 6" x 6"	1 Shee	24	.333	Ea.	21	10.40		31.40	40.50
3020	8" x 6"	"	22	.364	"	21	11.35		32.35	42

23 33 Air Duct Accessories

23 33 46 – Flexible Ducts

23 33 46.10 Flexible Air Ducts

23 33 46.10 Flexible Air Ducts		Crew	Daily Output	Labor-Hours	Unit	Material	2009 Bare Costs Labor	Equipment	Total	Total Incl O&P
0010	**FLEXIBLE AIR DUCTS**									
1300	Flexible, coated fiberglass fabric on corr. resist. metal helix									
1400	pressure to 12" (WG) UL-181									
1500	Non-insulated, 3" diameter	Q-9	400	.040	L.F.	.99	1.12		2.11	2.97
1540	5" diameter		320	.050		1.21	1.40		2.61	3.68
1561	Ductwork, flexible, non-insulated, 6" diameter		280	.057		1.49	1.60		3.09	4.33
1580	7" diameter		240	.067		1.76	1.87		3.63	5.10
1900	Insulated, 1" thick, PE jacket, 3" diameter [G]		380	.042		2.09	1.18		3.27	4.28
1910	4" diameter [G]		340	.047		2.09	1.32		3.41	4.51
1920	5" diameter [G]		300	.053		2.09	1.49		3.58	4.81
1940	6" diameter [G]		260	.062		2.31	1.73		4.04	5.45
1960	7" diameter [G]		220	.073		2.70	2.04		4.74	6.40
1980	8" diameter [G]		180	.089		2.92	2.49		5.41	7.40
2040	12" diameter [G]		100	.160		4.24	4.48		8.72	12.20

23 33 53 – Duct Liners

23 33 53.10 Duct Liner Board

23 33 53.10 Duct Liner Board		Crew	Daily Output	Labor-Hours	Unit	Material	2009 Bare Costs Labor	Equipment	Total	Total Incl O&P
0010	**DUCT LINER BOARD**									
3490	Board type, fiberglass liner, 3 lb. density									
3500	Fire resistant, black pigmented, 1 side									
3520	1" thick [G]	Q-14	150	.107	S.F.	.73	2.79		3.52	5.55
3540	1-1/2" thick [G]	"	130	.123	"	.92	3.22		4.14	6.50
3940	Board type, non-fibrous foam									
3950	Temperature, bacteria and fungi resistant									
3960	1" thick [G]	Q-14	150	.107	S.F.	2.25	2.79		5.04	7.25
3970	1-1/2" thick [G]		130	.123		3	3.22		6.22	8.80
3980	2" thick [G]		120	.133		3.60	3.49		7.09	9.90

23 34 HVAC Fans

23 34 23 – HVAC Power Ventilators

23 34 23.10 HVAC Power Circulators and Ventilators

23 34 23.10 HVAC Power Circulators and Ventilators		Crew	Daily Output	Labor-Hours	Unit	Material	2009 Bare Costs Labor	Equipment	Total	Total Incl O&P
0010	**HVAC POWER CIRCULATORS AND VENTILATORS**									
8020	Attic, roof type									
8030	Aluminum dome, damper & curb									
8040	6" diameter, 300 CFM	1 Elec	16	.500	Ea.	380	15.75		395.75	445
8050	7" diameter, 450 CFM		15	.533		415	16.80		431.80	485
8060	9" diameter, 900 CFM		14	.571		455	18		473	530
8080	12" diameter, 1000 CFM (gravity)		10	.800		470	25		495	560
8090	16" diameter, 1500 CFM (gravity)		9	.889		570	28		598	670
8100	20" diameter, 2500 CFM (gravity)		8	1		695	31.50		726.50	815
8160	Plastic, ABS dome									
8180	1050 CFM	1 Elec	14	.571	Ea.	139	18		157	182
8200	1600 CFM	"	12	.667	"	208	21		229	262
8240	Attic, wall type, with shutter, one speed									
8250	12" diameter, 1000 CFM	1 Elec	14	.571	Ea.	300	18		318	360
8260	14" diameter, 1500 CFM		12	.667		325	21		346	390
8270	16" diameter, 2000 CFM		9	.889		365	28		393	450
8290	Whole house, wall type, with shutter, one speed									
8300	30" diameter, 4800 CFM	1 Elec	7	1.143	Ea.	785	36		821	925
8310	36" diameter, 7000 CFM		6	1.333		855	42		897	1,000
8320	42" diameter, 10,000 CFM		5	1.600		960	50.50		1,010.50	1,125

23 34 HVAC Fans

23 34 23 - HVAC Power Ventilators

23 34 23.10 HVAC Power Circulators and Ventilators	Crew	Daily Output	Labor-Hours	Unit	Material	2009 Bare Costs Labor	Equipment	Total	Total Incl O&P
8330	48" diameter, 16,000 CFM	1 Elec	4	2	Ea.	1,200	63		1,263	1,400
8340	For two speed, add				↓	71.50			71.50	79
8350	Whole house, lay-down type, with shutter, one speed									
8360	30" diameter, 4500 CFM	1 Elec	8	1	Ea.	835	31.50		866.50	970
8370	36" diameter, 6500 CFM		7	1.143		900	36		936	1,050
8380	42" diameter, 9000 CFM		6	1.333		990	42		1,032	1,150
8390	48" diameter, 12,000 CFM	↓	5	1.600		1,125	50.50		1,175.50	1,300
8440	For two speed, add					54			54	59.50
8450	For 12 hour timer switch, add	1 Elec	32	.250	↓	54	7.90		61.90	72.50

23 37 Air Outlets and Inlets

23 37 13 - Diffusers, Registers, and Grilles

23 37 13.10 Diffusers

		Crew	Daily Output	Labor-Hours	Unit	Material	2009 Bare Costs Labor	Equipment	Total	Total Incl O&P
0010	**DIFFUSERS**, Aluminum, opposed blade damper unless noted									
0100	Ceiling, linear, also for sidewall									
0120	2" wide	1 Shee	32	.250	L.F.	39.50	7.80		47.30	56.50
0160	4" wide		26	.308	"	51.50	9.60		61.10	73
0500	Perforated, 24" x 24" lay-in panel size, 6" x 6"		16	.500	Ea.	56	15.60		71.60	87.50
0520	8" x 8"		15	.533		57	16.60		73.60	91
0530	9" x 9"		14	.571		61	17.80		78.80	97.50
0590	16" x 16"		11	.727		106	22.50		128.50	155
1000	Rectangular, 1 to 4 way blow, 6" x 6"		16	.500		45	15.60		60.60	75.50
1010	8" x 8"		15	.533		54	16.60		70.60	87.50
1014	9" x 9"		15	.533		59.50	16.60		76.10	93.50
1016	10" x 10"		15	.533		64	16.60		80.60	98.50
1020	12" x 6"		15	.533		65.50	16.60		82.10	100
1040	12" x 9"		14	.571		71.50	17.80		89.30	109
1060	12" x 12"		12	.667		72	21		93	114
1070	14" x 6"		13	.615		71	19.15		90.15	110
1074	14" x 14"		12	.667		101	21		122	146
1150	18" x 18"		9	.889		126	27.50		153.50	186
1170	24" x 12"		10	.800		134	25		159	189
1180	24" x 24"		7	1.143		256	35.50		291.50	340
1500	Round, butterfly damper, steel, diffuser size, 6" diameter		18	.444		22	13.85		35.85	47
1520	8" diameter		16	.500		23.50	15.60		39.10	52
2000	T bar mounting, 24" x 24" lay-in frame, 6" x 6"		16	.500		42.50	15.60		58.10	73
2020	8" x 8"		14	.571		44.50	17.80		62.30	78.50
2040	12" x 12"		12	.667		52	21		73	92.50
2060	16" x 16"		11	.727		70	22.50		92.50	115
2080	18" x 18"	↓	10	.800		78	25		103	128
6000	For steel diffusers instead of aluminum, deduct				↓	10%				

23 37 13.30 Grilles

		Crew	Daily Output	Labor-Hours	Unit	Material	2009 Bare Costs Labor	Equipment	Total	Total Incl O&P
0010	**GRILLES**									
0020	Aluminum									
1000	Air return, 6" x 6"	1 Shee	26	.308	Ea.	17.05	9.60		26.65	35
1020	10" x 6"		24	.333		17.05	10.40		27.45	36
1080	16" x 8"		22	.364		21.50	11.35		32.85	42.50
1100	12" x 12"		22	.364		23	11.35		34.35	44.50
1120	24" x 12"		18	.444		32	13.85		45.85	58
1180	16" x 16"	↓	22	.364		31	11.35		42.35	53

23 37 Air Outlets and Inlets

23 37 13 – Diffusers, Registers, and Grilles

23 37 13.60 Registers

		Crew	Daily Output	Labor-Hours	Unit	Material	2009 Bare Costs Labor	2009 Bare Costs Equipment	Total	Total Incl O&P
0010	**REGISTERS**									
0980	Air supply									
3000	Baseboard, hand adj. damper, enameled steel									
3012	8" x 6"	1 Shee	26	.308	Ea.	12.85	9.60		22.45	30.50
3020	10" x 6"		24	.333		14	10.40		24.40	33
3040	12" x 5"		23	.348		15.20	10.85		26.05	35
3060	12" x 6"	↓	23	.348	↓	15.20	10.85		26.05	35
4000	Floor, toe operated damper, enameled steel									
4020	4" x 8"	1 Shee	32	.250	Ea.	24	7.80		31.80	39
4040	4" x 12"	"	26	.308	"	28	9.60		37.60	46.50
4300	Spiral pipe supply register									
4310	Aluminum, double deflection, w/damper extractor									
4320	4" x 12", for 6" thru 10" diameter duct	1 Shee	25	.320	Ea.	57	9.95		66.95	80
4330	4" x 18", for 6" thru 10" diameter duct		18	.444		71	13.85		84.85	101
4340	6" x 12", for 8" thru 12" diameter duct		19	.421		62.50	13.10		75.60	90.50
4350	6" x 18", for 8" thru 12" diameter duct		18	.444		81	13.85		94.85	112
4360	6" x 24", for 8" thru 12" diameter duct		16	.500		100	15.60		115.60	136
4370	6" x 30", for 8" thru 12" diameter duct		15	.533		127	16.60		143.60	168
4380	8" x 18", for 10" thru 14" diameter duct		18	.444		86.50	13.85		100.35	118
4390	8" x 24", for 10" thru 14" diameter duct		15	.533		108	16.60		124.60	147
4400	8" x 30", for 10" thru 14" diameter duct		14	.571		144	17.80		161.80	188
4410	10" x 24", for 12" thru 18" diameter duct		13	.615		119	19.15		138.15	163
4420	10" x 30", for 12" thru 18" diameter duct		12	.667		157	21		178	207
4430	10" x 36", for 12" thru 18" diameter duct	↓	11	.727	↓	195	22.50		217.50	252

23 41 Particulate Air Filtration

23 41 13 – Panel Air Filters

23 41 13.10 Panel Type Air Filters

					Unit	Material	Labor	Equipment	Total	Total Incl O&P
0010	**PANEL TYPE AIR FILTERS**									
2950	Mechanical media filtration units									
3000	High efficiency type, with frame, non-supported	[G]			MCFM	45			45	49.50
3100	Supported type	[G]			"	60			60	66
5500	Throwaway glass or paper media type				Ea.	3.35			3.35	3.69

23 41 16 – Renewable-Media Air Filters

23 41 16.10 Disposable Media Air Filters

					Unit	Material	Labor	Equipment	Total	Total Incl O&P
0010	**DISPOSABLE MEDIA AIR FILTERS**									
5000	Renewable disposable roll				MCFM	250			250	275

23 41 19 – Washable Air Filters

23 41 19.10 Permanent Air Filters

					Unit	Material	Labor	Equipment	Total	Total Incl O&P
0010	**PERMANENT AIR FILTERS**									
4500	Permanent washable	[G]			MCFM	20			20	22

23 41 23 – Extended Surface Filters

23 41 23.10 Expanded Surface Filters

					Unit	Material	Labor	Equipment	Total	Total Incl O&P
0010	**EXPANDED SURFACE FILTERS**									
4000	Medium efficiency, extended surface	[G]			MCFM	5.50			5.50	6.05

23 42 Gas-Phase Air Filtration

23 42 13 – Activated-Carbon Air Filtration

23 42 13.10 Charcoal Type Air Filtration	Crew	Daily Output	Labor-Hours	Unit	Material	2009 Bare Costs Labor	Equipment	Total	Total Incl O&P
0010 **CHARCOAL TYPE AIR FILTRATION**									
0050 Activated charcoal type, full flow				MCFM	600			600	660
0060 Full flow, impregnated media 12" deep					225			225	248
0070 HEPA filter & frame for field erection					300			300	330
0080 HEPA filter-diffuser, ceiling install.					275			275	305

23 43 Electronic Air Cleaners

23 43 13 – Washable Electronic Air Cleaners

23 43 13.10 Electronic Air Cleaners

	Crew	Daily Output	Labor-Hours	Unit	Material	2009 Bare Costs Labor	Equipment	Total	Total Incl O&P
0010 **ELECTRONIC AIR CLEANERS**									
2000 Electronic air cleaner, duct mounted									
2150 400 – 1000 CFM	1 Shee	2.30	3.478	Ea.	1,000	108		1,108	1,275
2200 1000 – 1400 CFM		2.20	3.636		1,225	113		1,338	1,550
2250 1400 – 2000 CFM		2.10	3.810		1,400	119		1,519	1,750

23 51 Breechings, Chimneys, and Stacks

23 51 23 – Gas Vents

23 51 23.10 Gas Chimney Vents

	Crew	Daily Output	Labor-Hours	Unit	Material	2009 Bare Costs Labor	Equipment	Total	Total Incl O&P
0010 **GAS CHIMNEY VENTS**, Prefab metal, U.L. listed									
0020 Gas, double wall, galvanized steel									
0080 3" diameter	Q-9	72	.222	V.L.F.	5.35	6.25		11.60	16.35
0100 4" diameter	"	68	.235	"	6.70	6.60		13.30	18.40

23 52 Heating Boilers

23 52 13 – Electric Boilers

23 52 13.10 Electric Boilers, ASME

	Crew	Daily Output	Labor-Hours	Unit	Material	2009 Bare Costs Labor	Equipment	Total	Total Incl O&P
0010 **ELECTRIC BOILERS, ASME**, Standard controls and trim.									
1000 Steam, 6 KW, 20.5 MBH	Q-19	1.20	20	Ea.	3,375	600		3,975	4,700
1160 60 KW, 205 MBH		1	24		5,800	720		6,520	7,550
2000 Hot water, 7.5 KW, 25.6 MBH		1.30	18.462		3,575	555		4,130	4,850
2040 30 KW, 102 MBH		1.20	20		3,825	600		4,425	5,175
2060 45 KW, 164 MBH		1.20	20		4,275	600		4,875	5,675

23 52 23 – Cast-Iron Boilers

23 52 23.20 Gas-Fired Boilers

	Crew	Daily Output	Labor-Hours	Unit	Material	2009 Bare Costs Labor	Equipment	Total	Total Incl O&P
0010 **GAS-FIRED BOILERS,** Natural or propane, standard controls, packaged.									
1000 Cast iron, with insulated jacket									
3000 Hot water, gross output, 80 MBH	Q-7	1.46	21.918	Ea.	1,700	640		2,340	2,925
3020 100 MBH	"	1.35	23.704		1,950	695		2,645	3,300
7000 For tankless water heater, add					10%				
7050 For additional zone valves up to 312 MBH add					145			145	159

23 52 23.30 Gas/Oil Fired Boilers

	Crew	Daily Output	Labor-Hours	Unit	Material	2009 Bare Costs Labor	Equipment	Total	Total Incl O&P
0010 **GAS/OIL FIRED BOILERS,** Combination with burners and controls, packaged.									
1000 Cast iron with insulated jacket									
2000 Steam, gross output, 720 MBH	Q-7	.43	74.074	Ea.	13,400	2,175		15,575	18,400
2900 Hot water, gross output									
2910 200 MBH	Q-6	.62	39.024	Ea.	7,750	1,100		8,850	10,300
2920 300 MBH		.49	49.080		7,750	1,375		9,125	10,800

23 52 Heating Boilers

23 52 23 – Cast-Iron Boilers

23 52 23.30 Gas/Oil Fired Boilers

		Crew	Daily Output	Labor-Hours	Unit	Material	2009 Bare Costs Labor	Equipment	Total	Total Incl O&P
2930	400 MBH	Q-6	.41	57.971	Ea.	9,075	1,625		10,700	12,700
2940	500 MBH	↓	.36	67.039		9,775	1,900		11,675	13,900
3000	584 MBH	Q-7	.44	72.072	↓	13,200	2,100		15,300	18,000

23 52 23.40 Oil-Fired Boilers

		Crew	Daily Output	Labor-Hours	Unit	Material	2009 Bare Costs Labor	Equipment	Total	Total Incl O&P
0010	**OIL-FIRED BOILERS,** Standard controls, flame retention burner, packaged									
1000	Cast iron, with insulated flush jacket									
2000	Steam, gross output, 109 MBH	Q-7	1.20	26.667	Ea.	1,875	780		2,655	3,350
2060	207 MBH	"	.90	35.556	"	2,575	1,050		3,625	4,525
3000	Hot water, same price as steam									

23 52 26 – Steel Boilers

23 52 26.40 Oil-Fired Boilers

		Crew	Daily Output	Labor-Hours	Unit	Material	2009 Bare Costs Labor	Equipment	Total	Total Incl O&P
0010	**OIL-FIRED BOILERS,** Standard controls, flame retention burner									
5000	Steel, with insulated flush jacket									
7000	Hot water, gross output, 103 MBH	Q-6	1.60	15	Ea.	1,675	425		2,100	2,550
7020	122 MBH		1.45	16.506		1,775	465		2,240	2,725
7060	168 MBH		1.30	18.405		2,000	520		2,520	3,025
7080	225 MBH	↓	1.22	19.704	↓	2,625	555		3,180	3,800

23 52 28 – Swimming Pool Boilers

23 52 28.10 Swimming Pool Heaters

		Crew	Daily Output	Labor-Hours	Unit	Material	2009 Bare Costs Labor	Equipment	Total	Total Incl O&P
0010	**SWIMMING POOL HEATERS,** Not including wiring, external									
0020	piping, base or pad,									
0160	Gas fired, input, 155 MBH	Q-6	1.50	16	Ea.	1,575	450		2,025	2,500
0200	199 MBH		1	24		1,700	675		2,375	2,975
0280	500 MBH	↓	.40	60		7,225	1,700		8,925	10,700
2000	Electric, 12 KW, 4,800 gallon pool	Q-19	3	8		1,950	240		2,190	2,550
2020	15 KW, 7,200 gallon pool		2.80	8.571		2,000	257		2,257	2,625
2040	24 KW, 9,600 gallon pool		2.40	10		2,275	300		2,575	3,000
2100	57 KW, 24,000 gallon pool	↓	1.20	20	↓	3,325	600		3,925	4,625

23 52 88 – Burners

23 52 88.10 Replacement Type Burners

		Crew	Daily Output	Labor-Hours	Unit	Material	2009 Bare Costs Labor	Equipment	Total	Total Incl O&P
0010	**REPLACEMENT TYPE BURNERS**									
0990	Residential, conversion, gas fired, LP or natural									
1000	Gun type, atmospheric input 50 to 225 MBH	Q-1	2.50	6.400	Ea.	645	185		830	1,025
1020	100 to 400 MBH		2	8		1,075	232		1,307	1,550
1040	300 to 1000 MBH	↓	1.70	9.412	↓	3,300	273		3,573	4,075

23 54 Furnaces

23 54 13 – Electric-Resistance Furnaces

23 54 13.10 Electric Furnaces

		Crew	Daily Output	Labor-Hours	Unit	Material	2009 Bare Costs Labor	Equipment	Total	Total Incl O&P
0010	**ELECTRIC FURNACES,** Hot air, blowers, std. controls									
0011	not including gas, oil or flue piping									
1000	Electric, UL listed									
1020	10.2 MBH	Q-20	5	4	Ea.	360	115		475	585
1100	34.1 MBH	"	4.40	4.545	"	465	131		596	730

23 54 16 – Fuel-Fired Furnaces

23 54 16.13 Gas-Fired Furnaces

		Crew	Daily Output	Labor-Hours	Unit	Material	2009 Bare Costs Labor	Equipment	Total	Total Incl O&P
0010	**GAS-FIRED FURNACES**									
3000	Gas, AGA certified, upflow, direct drive models									

23 54 Furnaces

23 54 16 – Fuel-Fired Furnaces

23 54 16.13 Gas-Fired Furnaces

		Crew	Daily Output	Labor-Hours	Unit	Material	2009 Bare Costs Labor	2009 Bare Costs Equipment	Total	Total Incl O&P
3020	45 MBH input	Q-9	4	4	Ea.	575	112		687	825
3040	60 MBH input		3.80	4.211		615	118		733	875
3060	75 MBH input		3.60	4.444		685	125		810	965
3100	100 MBH input		3.20	5		695	140		835	995
3120	125 MBH input		3	5.333		700	149		849	1,025
3130	150 MBH input		2.80	5.714		705	160		865	1,050
3140	200 MBH input		2.60	6.154		2,425	172		2,597	2,950
4000	For starter plenum, add		16	1		80	28		108	135

23 54 16.16 Oil-Fired Furnaces

		Crew	Daily Output	Labor-Hours	Unit	Material	2009 Bare Costs Labor	2009 Bare Costs Equipment	Total	Total Incl O&P
0010	**OIL-FIRED FURNACES**									
6000	Oil, UL listed, atomizing gun type burner									
6020	56 MBH output	Q-9	3.60	4.444	Ea.	1,850	125		1,975	2,225
6030	84 MBH output		3.50	4.571		1,850	128		1,978	2,275
6040	95 MBH output		3.40	4.706		1,875	132		2,007	2,300
6060	134 MBH output		3.20	5		1,950	140		2,090	2,375
6080	151 MBH output		3	5.333		2,275	149		2,424	2,750
6100	200 MBH input		2.60	6.154		2,325	172		2,497	2,850

23 54 16.21 Solid Fuel-Fired Furnaces

			Crew	Daily Output	Labor-Hours	Unit	Material	2009 Bare Costs Labor	2009 Bare Costs Equipment	Total	Total Incl O&P
0010	**SOLID FUEL-FIRED FURNACES**										
6020	Wood fired furnaces										
6030	Includes hot water coil, thermostat, and auto draft control										
6040	24" long firebox	G	Q-9	4	4	Ea.	3,600	112		3,712	4,175
6050	30" long firebox	G		3.60	4.444		4,200	125		4,325	4,800
6060	With fireplace glass doors	G		3.20	5		5,400	140		5,540	6,175
6200	Wood / oil fired furnaces, includes two thermostats										
6210	Includes hot water coil and auto draft control										
6240	24" long firebox	G	Q-9	3.40	4.706	Ea.	4,400	132		4,532	5,050
6250	30" long firebox	G		3	5.333		4,875	149		5,024	5,600
6260	With fireplace glass doors	G		2.80	5.714		5,400	160		5,560	6,225
6400	Wood / gas fired furnaces, includes two thermostats										
6410	Includes hot water coil and auto draft control										
6440	24" long firebox	G	Q-9	2.80	5.714	Ea.	4,925	160		5,085	5,700
6450	30" long firebox	G		2.40	6.667		5,525	187		5,712	6,400
6460	With fireplace glass doors	G		2	8		6,000	224		6,224	6,975
6600	Wood / oil /gas fired furnaces, optional accessories										
6610	Hot air plenum		Q-9	16	1	Ea.	81	28		109	136
6620	Safety heat dump			24	.667		72	18.70		90.70	111
6630	Auto air intake			18	.889		119	25		144	173
6640	Cold air return package			14	1.143		125	32		157	192
6650	Wood fork						39			39	43
6700	Wood fired outdoor furnace										
6740	24" long firebox	G	Q-9	3.80	4.211	Ea.	3,775	118		3,893	4,350
6760	Wood fired outdoor furnace, optional accessories										
6770	Chimney section, stainless steel, 6" ID x 3' lg.	G	Q-9	36	.444	Ea.	95	12.45		107.45	126
6780	Chimney cap, stainless steel	G	"	40	.400	"	51	11.20		62.20	75
6800	Wood fired hot water furnace										
6820	Includes 200 gal. hot water storage, thermostat, and auto draft control										
6840	30" long firebox	G	Q-9	2.10	7.619	Ea.	7,450	214		7,664	8,525
6850	Water to air heat exchanger										
6870	Includes mounting kit and blower relay										
6880	140 MBH, 18.75" W x 18.75" L		Q-9	7.50	2.133	Ea.	370	60		430	505
6890	200 MBH, 24" W x 24" L		"	7	2.286	"	525	64		589	690

23 54 Furnaces

23 54 16 – Fuel-Fired Furnaces

23 54 16.21 Solid Fuel-Fired Furnaces

		Crew	Daily Output	Labor-Hours	Unit	Material	2009 Bare Costs Labor	Equipment	Total	Total Incl O&P
6900	Water to water heat exchanger									
6940	100 MBH	Q-9	6.50	2.462	Ea.	370	69		439	525
6960	290 MBH	"	6	2.667	"	535	75		610	715
7000	Optional accessories									
7010	Large volume circulation pump, (2 Included)	Q-9	14	1.143	Ea.	242	32		274	320
7020	Air bleed fittings, (package)		24	.667		49	18.70		67.70	85.50
7030	Domestic water preheater		6	2.667		219	75		294	365
7040	Smoke pipe kit		4	4		62	112		174	256

23 54 24 – Furnace Components for Cooling

23 54 24.10 Furnace Components and Combinations

		Crew	Daily Output	Labor-Hours	Unit	Material	2009 Bare Costs Labor	Equipment	Total	Total Incl O&P
0010	**FURNACE COMPONENTS AND COMBINATIONS**									
0080	Coils, A/C evaporator, for gas or oil furnaces									
0090	Add-on, with holding charge									
0100	Upflow									
0120	1-1/2 ton cooling	Q-5	4	4	Ea.	158	117		275	365
0130	2 ton cooling		3.70	4.324		188	127		315	415
0140	3 ton cooling		3.30	4.848		238	142		380	495
0150	4 ton cooling		3	5.333		325	156		481	615
0160	5 ton cooling		2.70	5.926		385	174		559	710
0300	Downflow									
0330	2-1/2 ton cooling	Q-5	3	5.333	Ea.	232	156		388	510
0340	3-1/2 ton cooling		2.60	6.154		276	180		456	600
0350	5 ton cooling		2.20	7.273		385	213		598	775
0600	Horizontal									
0630	2 ton cooling	Q-5	3.90	4.103	Ea.	156	120		276	370
0640	3 ton cooling		3.50	4.571		184	134		318	420
0650	4 ton cooling		3.20	5		224	147		371	485
0660	5 ton cooling		2.90	5.517		224	162		386	510
2000	Cased evaporator coils for air handlers									
2100	1-1/2 ton cooling	Q-5	4.40	3.636	Ea.	244	107		351	445
2110	2 ton cooling		4.10	3.902		248	114		362	460
2120	2-1/2 ton cooling		3.90	4.103		259	120		379	480
2130	3 ton cooling		3.70	4.324		286	127		413	525
2140	3-1/2 ton cooling		3.50	4.571		299	134		433	550
2150	4 ton cooling		3.20	5		360	147		507	635
2160	5 ton cooling		2.90	5.517		415	162		577	720
3010	Air handler, modular									
3100	With cased evaporator cooling coil									
3120	1-1/2 ton cooling	Q-5	3.80	4.211	Ea.	775	123		898	1,050
3130	2 ton cooling		3.50	4.571		865	134		999	1,175
3140	2-1/2 ton cooling		3.30	4.848		885	142		1,027	1,200
3150	3 ton cooling		3.10	5.161		945	151		1,096	1,275
3160	3-1/2 ton cooling		2.90	5.517		1,000	162		1,162	1,375
3170	4 ton cooling		2.50	6.400		1,125	188		1,313	1,550
3180	5 ton cooling		2.10	7.619		1,275	223		1,498	1,775
3500	With no cooling coil									
3520	1-1/2 ton coil size	Q-5	12	1.333	Ea.	575	39		614	695
3530	2 ton coil size		10	1.600		605	47		652	740
3540	2-1/2 ton coil size		10	1.600		605	47		652	740
3554	3 ton coil size		9	1.778		685	52		737	840
3560	3-1/2 ton coil size		9	1.778		745	52		797	905
3570	4 ton coil size		8.50	1.882		860	55		915	1,025

23 54 Furnaces

23 54 24 – Furnace Components for Cooling

23 54 24.10 Furnace Components and Combinations	Crew	Daily Output	Labor-Hours	Unit	Material	2009 Bare Costs Labor	Equipment	Total	Total Incl O&P
3580	5 ton coil size	Q-5	8	2	Ea.	1,000	58.50		1,058.50	1,200
4000	With heater									
4120	5 kW, 17.1 MBH	Q-5	16	1	Ea.	585	29.50		614.50	695
4130	7.5 kW, 25.6 MBH		15.60	1.026		590	30		620	700
4140	10 kW, 34.2 MBH		15.20	1.053		825	31		856	955
4150	12.5 KW, 42.7 MBH		14.80	1.081		910	31.50		941.50	1,050
4160	15 KW, 51.2 MBH		14.40	1.111		935	32.50		967.50	1,075
4170	25 KW, 85.4 MBH		14	1.143		1,050	33.50		1,083.50	1,200
4180	30 KW, 102 MBH		13	1.231		1,225	36		1,261	1,400

23 62 Packaged Compressor and Condenser Units

23 62 13 – Packaged Air-Cooled Refrigerant Compressor and Condenser Units

23 62 13.10 Packaged Air-Cooled Refrig. Condensing Units

		Crew	Daily Output	Labor-Hours	Unit	Material	2009 Bare Costs Labor	Equipment	Total	Total Incl O&P
0010	**PACKAGED AIR-COOLED REFRIGERANT CONDENSING UNITS**									
0030	Air cooled, compressor, standard controls									
0050	1.5 ton	Q-5	2.50	6.400	Ea.	1,125	188		1,313	1,550
0100	2 ton		2.10	7.619		1,150	223		1,373	1,650
0200	2.5 ton		1.70	9.412		1,250	276		1,526	1,825
0300	3 ton		1.30	12.308		1,300	360		1,660	2,025
0350	3.5 ton		1.10	14.545		1,500	425		1,925	2,350
0400	4 ton		.90	17.778		1,700	520		2,220	2,700
0500	5 ton		.60	26.667		2,050	780		2,830	3,525

23 74 Packaged Outdoor HVAC Equipment

23 74 33 – Packaged, Outdoor, Heating and Cooling Makeup Air-Conditioners

23 74 33.10 Roof Top Air Conditioners

		Crew	Daily Output	Labor-Hours	Unit	Material	2009 Bare Costs Labor	Equipment	Total	Total Incl O&P
0010	**ROOF TOP AIR CONDITIONERS**, Standard controls, curb, economizer.									
1000	Single zone, electric cool, gas heat									
1140	5 ton cooling, 112 MBH heating	Q-5	.56	28.521	Ea.	4,150	835		4,985	5,950
1160	10 ton cooling, 200 MBH heating	Q-6	.67	35.982	"	8,300	1,025		9,325	10,800

23 81 Decentralized Unitary HVAC Equipment

23 81 13 – Packaged Terminal Air-Conditioners

23 81 13.10 Packaged Cabinet Type Air-Conditioners

		Crew	Daily Output	Labor-Hours	Unit	Material	2009 Bare Costs Labor	Equipment	Total	Total Incl O&P
0010	**PACKAGED CABINET TYPE AIR-CONDITIONERS**, Cabinet, wall sleeve,									
0100	louver, electric heat, thermostat, manual changeover, 208 V									
0200	6,000 BTUH cooling, 8800 BTU heat	Q-5	6	2.667	Ea.	625	78		703	820
0220	9,000 BTUH cooling, 13,900 BTU heat		5	3.200		720	94		814	945
0240	12,000 BTUH cooling, 13,900 BTU heat		4	4		780	117		897	1,050
0260	15,000 BTUH cooling, 13,900 BTU heat		3	5.333		865	156		1,021	1,200

23 81 19 – Self-Contained Air-Conditioners

23 81 19.10 Window Unit Air Conditioners

		Crew	Daily Output	Labor-Hours	Unit	Material	2009 Bare Costs Labor	Equipment	Total	Total Incl O&P
0010	**WINDOW UNIT AIR CONDITIONERS**									
4000	Portable/window, 15 amp 125V grounded receptacle required									
4060	5000 BTUH	1 Carp	8	1	Ea.	163	28		191	227
4340	6000 BTUH		8	1		265	28		293	340
4480	8000 BTUH		6	1.333		310	37.50		347.50	410

23 81 Decentralized Unitary HVAC Equipment

23 81 19 – Self-Contained Air-Conditioners

23 81 19.10 Window Unit Air Conditioners

		Crew	Daily Output	Labor-Hours	Unit	Material	2009 Bare Costs Labor	2009 Bare Costs Equipment	Total	Total Incl O&P
4500	10,000 BTUH	1 Carp	6	1.333	Ea.	395	37.50		432.50	500
4520	12,000 BTUH	L-2	8	2	↓	510	49		559	645
4600	Window/thru-the-wall, 15 amp 230V grounded receptacle required									
4780	17,000 BTUH	L-2	6	2.667	Ea.	695	65		760	875
4940	25,000 BTUH	↓	4	4		880	97.50		977.50	1,125
4960	29,000 BTUH	↓	4	4	↓	950	97.50		1,047.50	1,225

23 81 19.20 Self-Contained Single Package

		Crew	Daily Output	Labor-Hours	Unit	Material	2009 Bare Costs Labor	2009 Bare Costs Equipment	Total	Total Incl O&P
0010	**SELF-CONTAINED SINGLE PACKAGE**									
0100	Air cooled, for free blow or duct, not incl. remote condenser									
0200	3 ton cooling	Q-5	1	16	Ea.	3,200	470		3,670	4,300
0210	4 ton cooling	"	.80	20	"	3,500	585		4,085	4,800
1000	Water cooled for free blow or duct, not including tower									
1100	3 ton cooling	Q-6	1	24	Ea.	3,075	675		3,750	4,500

23 81 26 – Split-System Air-Conditioners

23 81 26.10 Split Ductless Systems

		Crew	Daily Output	Labor-Hours	Unit	Material	2009 Bare Costs Labor	2009 Bare Costs Equipment	Total	Total Incl O&P
0010	**SPLIT DUCTLESS SYSTEMS**									
0100	Cooling only, single zone									
0110	Wall mount									
0120	3/4 ton cooling	Q-5	2	8	Ea.	1,175	234		1,409	1,650
0130	1 ton cooling	↓	1.80	8.889		1,300	260		1,560	1,850
0140	1-1/2 ton cooling		1.60	10		1,800	293		2,093	2,450
0150	2 ton cooling	↓	1.40	11.429	↓	2,125	335		2,460	2,900
1000	Ceiling mount									
1020	2 ton cooling	Q-5	1.40	11.429	Ea.	1,450	335		1,785	2,125
1030	3 ton cooling	"	1.20	13.333	"	3,900	390		4,290	4,950
2000	T-Bar mount									
2010	2 ton cooling	Q-5	1.40	11.429	Ea.	3,400	335		3,735	4,300
2020	3 ton cooling	↓	1.20	13.333		3,800	390		4,190	4,825
2030	3-1/2 ton cooling	↓	1.10	14.545	↓	4,200	425		4,625	5,325
3000	Multizone									
3010	Wall mount									
3020	2 @ 3/4 ton cooling	Q-5	1.80	8.889	Ea.	1,475	260		1,735	2,050
5000	Cooling / Heating									
5110	1 ton cooling	Q-5	1.70	9.412	Ea.	1,100	276		1,376	1,650
5120	1-1/2 ton cooling	"	1.50	10.667	"	1,750	315		2,065	2,450
5300	Ceiling mount									
5310	3 ton cooling	Q-5	1	16	Ea.	4,125	470		4,595	5,325
7000	Accessories for all split ductless systems									
7010	Add for ambient frost control	Q-5	8	2	Ea.	130	58.50		188.50	239
7020	Add for tube / wiring kit									
7030	15' kit	Q-5	32	.500	Ea.	38.50	14.65		53.15	66.50
7040	35' kit	"	24	.667	"	125	19.55		144.55	169

23 81 43 – Air-Source Unitary Heat Pumps

23 81 43.10 Air-Source Heat Pumps

		Crew	Daily Output	Labor-Hours	Unit	Material	2009 Bare Costs Labor	2009 Bare Costs Equipment	Total	Total Incl O&P
0010	**AIR-SOURCE HEAT PUMPS**, Not including interconnecting tubing.									
1000	Air to air, split system, not including curbs, pads, fan coil and ductwork									
1012	Outside condensing unit only, for fan coil see Div. 23 82 19.10									
1020	2 ton cooling, 8.5 MBH heat @ 0° F	Q-5	2	8	Ea.	1,650	234		1,884	2,200
1054	4 ton cooling, 24 MBH heat @ 0° F	"	.80	20	"	2,450	585		3,035	3,650
1500	Single package, not including curbs, pads, or plenums									
1520	2 ton cooling, 6.5 MBH heat @ 0° F	Q-5	1.50	10.667	Ea.	2,575	315		2,890	3,350

263

23 81 Decentralized Unitary HVAC Equipment

23 81 43 – Air-Source Unitary Heat Pumps

23 81 43.10 Air-Source Heat Pumps	Crew	Daily Output	Labor-Hours	Unit	Material	2009 Bare Costs Labor	Equipment	Total	Total Incl O&P
1580	4 ton cooling, 13 MBH heat @ 0° F	Q-5	.96	16.667	Ea.	3,650	490		4,140	4,825

23 81 46 – Water-Source Unitary Heat Pumps

23 81 46.10 Water Source Heat Pumps	Crew	Daily Output	Labor-Hours	Unit	Material	2009 Bare Costs Labor	Equipment	Total	Total Incl O&P
0010	**WATER SOURCE HEAT PUMPS**, Not incl. connecting tubing or water source									
2000	Water source to air, single package									
2100	1 ton cooling, 13 MBH heat @ 75° F	Q-5	2	8	Ea.	1,300	234		1,534	1,800
2200	4 ton cooling, 31 MBH heat @ 75° F	"	1.20	13.333	"	2,100	390		2,490	2,950

23 82 Convection Heating and Cooling Units

23 82 19 – Fan Coil Units

23 82 19.10 Fan Coil Air Conditioning

		Crew	Daily Output	Labor-Hours	Unit	Material	2009 Bare Costs Labor	Equipment	Total	Total Incl O&P
0010	**FAN COIL AIR CONDITIONING** Cabinet mounted, filters, controls									
0100	Chilled water, 1/2 ton cooling	Q-5	8	2	Ea.	675	58.50		733.50	840
0110	3/4 ton cooling		7	2.286		700	67		767	880
0120	1 ton cooling		6	2.667		775	78		853	985

23 82 29 – Radiators

23 82 29.10 Hydronic Heating

		Crew	Daily Output	Labor-Hours	Unit	Material	2009 Bare Costs Labor	Equipment	Total	Total Incl O&P
0010	**HYDRONIC HEATING**, Terminal units, not incl. main supply pipe									
1000	Radiation									
3000	Radiators, cast iron									
3100	Free standing or wall hung, 6 tube, 25" high	Q-5	96	.167	Section	30	4.88		34.88	41

23 82 36 – Finned-Tube Radiation Heaters

23 82 36.10 Finned Tube Radiation

		Crew	Daily Output	Labor-Hours	Unit	Material	2009 Bare Costs Labor	Equipment	Total	Total Incl O&P
0010	**FINNED TUBE RADIATION**, Terminal units, not incl. main supply pipe									
1310	Baseboard, pkgd, 1/2" copper tube, alum. fin, 7" high	Q-5	60	.267	L.F.	7.20	7.80		15	20.50
1320	3/4" copper tube, alum. fin, 7" high		58	.276		7.65	8.10		15.75	21.50
1340	1" copper tube, alum. fin, 8-7/8" high		56	.286		24.50	8.35		32.85	40.50
1360	1-1/4" copper tube, alum. fin, 8-7/8" high		54	.296		25	8.70		33.70	42

23 83 Radiant Heating Units

23 83 16 – Radiant-Heating Hydronic Piping

23 83 16.10 Radiant Floor Heating

		Crew	Daily Output	Labor-Hours	Unit	Material	2009 Bare Costs Labor	Equipment	Total	Total Incl O&P
0010	**RADIANT FLOOR HEATING**									
0100	Tubing, PEX (cross-linked polyethylene)									
0110	Oxygen barrier type for systems with ferrous materials									
0120	1/2"	Q-5	800	.020	L.F.	.94	.59		1.53	1.99
0130	3/4"		535	.030		1.15	.88		2.03	2.71
0140	1"		400	.040		2.07	1.17		3.24	4.20
0200	Non barrier type for ferrous free systems									
0210	1/2"	Q-5	800	.020	L.F.	.51	.59		1.10	1.52
0220	3/4"		535	.030		.92	.88		1.80	2.45
0230	1"		400	.040		1.60	1.17		2.77	3.68
1000	Manifolds									
1110	Brass									
1120	Valved									
1130	1", 2 circuit	Q-5	13	1.231	Ea.	47	36		83	111
1140	1", 3 circuit		11	1.455		63.50	42.50		106	140
1150	1", 4 circuit		9	1.778		87.50	52		139.50	182

23 83 16 – Radiant-Heating Hydronic Piping

23 83 16.10 Radiant Floor Heating	Crew	Daily Output	Labor-Hours	Unit	Material	2009 Bare Costs Labor	Equipment	Total	Total Incl O&P
1160	1-1/4", 3 circuit	Q-5	10	1.600	Ea.	73	47		120	157
1170	1-1/4", 4 circuit	↓	8	2	↓	88.50	58.50		147	194
1300	Valveless									
1310	1", 2 circuit	Q-5	14	1.143	Ea.	24.50	33.50		58	81.50
1320	1", 3 circuit		12	1.333		32	39		71	99.50
1330	1", 4 circuit		10	1.600		41.50	47		88.50	123
1340	1-1/4", 3 circuit		11	1.455		43.50	42.50		86	118
1350	1-1/4", 4 circuit	↓	9	1.778	↓	45	52		97	135
1400	Manifold connector kit									
1410	Mounting brackets, couplings, end cap, air vent, valve, plug									
1420	1"	Q-5	8	2	Ea.	81	58.50		139.50	185
1430	1-1/4"	"	7	2.286	"	95.50	67		162.50	215
1610	Copper, (cut to size)									
1620	1" x 1/2" x 72" Lg, 2" sweat drops, 24 circuit	Q-5	3.33	4.805	Ea.	84.50	141		225.50	325
1630	1-1/4" x 1/2" x 72" Lg, 2" sweat drops, 24 circuit		3.20	5		98	147		245	350
1640	1-1/2" x 1/2" x 72" Lg, 2" sweat drops, 24 circuit		3	5.333		120	156		276	390
1650	1-1/4" x 3/4" x 72" Lg, 2" sweat drops, 24 circuit		3.10	5.161		105	151		256	365
1660	1-1/2" x 3/4" x 72" Lg, 2" sweat drops, 24 circuit	↓	2.90	5.517	↓	127	162		289	405
1710	Copper, (modular)									
1720	1" x 1/2", 2" sweat drops, 2 circuit	Q-5	17.50	.914	Ea.	8.60	27		35.60	53.50
1730	1" x 1/2", 2" sweat drops, 3 circuit		14.50	1.103		12.05	32.50		44.55	66.50
1740	1" x 1/2", 2" sweat drops, 4 circuit		12.30	1.301		13.75	38		51.75	77.50
1750	1-1/4" x 1/2", 2" sweat drops, 2 circuit		17	.941		9.45	27.50		36.95	56
1760	1-1/4" x 1/2", 2" sweat drops, 3 circuit		14	1.143		13.75	33.50		47.25	70
1770	1-1/4" x 1/2", 2" sweat drops, 4 circuit	↓	11.80	1.356	↓	18.05	39.50		57.55	85
3000	Valves									
3110	Motorized zone valve operator, 24V	Q-5	30	.533	Ea.	49	15.65		64.65	79.50
3120	Motorized zone valve with operator complete, 24V									
3130	3/4"	Q-5	35	.457	Ea.	85	13.40		98.40	116
3140	1"		32	.500		96.50	14.65		111.15	130
3150	1-1/4"		29.60	.541		119	15.85		134.85	157
3200	Flow control/isolation valve, 1/2"	↓	32	.500	↓	27.50	14.65		42.15	54.50
3300	Thermostatic mixing valves									
3310	1/2"	Q-5	26.60	.602	Ea.	88.50	17.60		106.10	127
3320	3/4"		25	.640		107	18.75		125.75	149
3330	1"	↓	21.40	.748	↓	115	22		137	163
3400	3 Way mixing / diverting valve, manual, brass									
3410	1/2"	Q-5	21.30	.751	Ea.	117	22		139	165
3420	3/4"		20	.800		120	23.50		143.50	171
3430	1"	↓	17.80	.899	↓	139	26.50		165.50	196
3500	4 Way mixing / diverting valve, manual, brass									
3510	1/2"	Q-5	16	1	Ea.	122	29.50		151.50	182
3520	3/4"		15	1.067		136	31.50		167.50	201
3530	1"		13.30	1.203		152	35.50		187.50	225
3540	1-1/4"		11.40	1.404		171	41		212	256
3550	1-1/2"		10.60	1.509		224	44		268	320
3560	2"		10	1.600		242	47		289	345
3570	2-1/2"		8	2		665	58.50		723.50	830
3800	Motor control, 3 or 4 way, for valves up to 1-1/2"	↓	30	.533	↓	169	15.65		184.65	211
5000	Radiant floor heating, zone control panel									
5110	6 Zone control									
5120	For zone valves	Q-5	16	1	Ea.	158	29.50		187.50	222
5130	For circulators	"	16	1	"	225	29.50		254.50	296

23 83 Radiant Heating Units

23 83 16 – Radiant-Heating Hydronic Piping

23 83 16.10 Radiant Floor Heating

		Crew	Daily Output	Labor-Hours	Unit	Material	2009 Bare Costs Labor	Equipment	Total	Total Incl O&P
7000	PEX tubing fittings, compression type									
7110	PEX tubing to brass valve conn., 1/2"	Q-5	54	.296	Ea.	2.72	8.70		11.42	17.25
7200	PEX x male NPT									
7210	1/2" x 1/2"	Q-5	54	.296	Ea.	5.30	8.70		14	20
7220	3/4" x 3/4"		46	.348		9.05	10.20		19.25	26.50
7230	1" x 1"	↓	40	.400	↓	21.50	11.70		33.20	43
7300	PEX coupling									
7310	1/2" x 1/2"	Q-5	50	.320	Ea.	6.45	9.40		15.85	22.50
7320	3/4" x 3/4"		42	.381		18.50	11.15		29.65	39
7330	1" x 1"	↓	36	.444	↓	25	13		38	49
7400	PEX x copper sweat									
7410	1/2" x 1/2"	Q-5	38	.421	Ea.	4.73	12.35		17.08	25
7420	3/4" x 3/4"		34	.471		9.90	13.80		23.70	33.50
7430	1" x 1"	↓	29	.552	↓	21.50	16.15		37.65	50

23 83 33 – Electric Radiant Heaters

23 83 33.10 Electric Heating

		Crew	Daily Output	Labor-Hours	Unit	Material	2009 Bare Costs Labor	Equipment	Total	Total Incl O&P
0010	**ELECTRIC HEATING**, not incl. conduit or feed wiring.									
1100	Rule of thumb: Baseboard units, including control	1 Elec	4.40	1.818	kW	92.50	57.50		150	196
1300	Baseboard heaters, 2' long, 375 watt		8	1	Ea.	40	31.50		71.50	95.50
1400	3' long, 500 watt		8	1		46	31.50		77.50	102
1600	4' long, 750 watt		6.70	1.194		54	37.50		91.50	121
1800	5' long, 935 watt		5.70	1.404		68	44		112	147
2000	6' long, 1125 watt		5	1.600		74	50.50		124.50	164
2400	8' long, 1500 watt		4	2		92	63		155	204
2800	10' long, 1875 watt	↓	3.30	2.424	↓	193	76.50		269.50	335
2950	Wall heaters with fan, 120 to 277 volt									
3600	Thermostats, integral	1 Elec	16	.500	Ea.	20.50	15.75		36.25	48
3800	Line voltage, 1 pole		8	1		31	31.50		62.50	85.50
5000	Radiant heating ceiling panels, 2' x 4', 500 watt		16	.500		253	15.75		268.75	305
5050	750 watt		16	.500		279	15.75		294.75	330
5300	Infrared quartz heaters, 120 volts, 1000 watts		6.70	1.194		164	37.50		201.50	242
5350	1500 watt		5	1.600		164	50.50		214.50	262
5400	240 volts, 1500 watt		5	1.600		164	50.50		214.50	262
5450	2000 watt		4	2		164	63		227	283
5500	3000 watt	↓	3	2.667	↓	189	84		273	345

Division 26
Electrical

26 05 05.10 Electrical Demolition	Crew	Daily Output	Labor-Hours	Unit	Material	2009 Bare Costs Labor	Equipment	Total	Total Incl O&P
0010 **ELECTRICAL DEMOLITION**									
0020 Conduit to 15' high, including fittings & hangers									
0100 Rigid galvanized steel, 1/2" to 1" diameter	1 Elec	242	.033	L.F.		1.04		1.04	1.70
0120 1-1/4" to 2"	"	200	.040	"		1.26		1.26	2.05
0270 Armored cable, (BX) avg. 50' runs									
0290 #14, 3 wire	1 Elec	571	.014	L.F.		.44		.44	.72
0300 #12, 2 wire		605	.013			.42		.42	.68
0310 #12, 3 wire		514	.016			.49		.49	.80
0320 #10, 2 wire		514	.016			.49		.49	.80
0330 #10, 3 wire		425	.019			.59		.59	.97
0340 #8, 3 wire	↓	342	.023	↓		.74		.74	1.20
0350 Non metallic sheathed cable (Romex)									
0360 #14, 2 wire	1 Elec	720	.011	L.F.		.35		.35	.57
0370 #14, 3 wire		657	.012			.38		.38	.62
0380 #12, 2 wire		629	.013			.40		.40	.65
0390 #10, 3 wire	↓	450	.018	↓		.56		.56	.91
0400 Wiremold raceway, including fittings & hangers									
0420 No. 3000	1 Elec	250	.032	L.F.		1.01		1.01	1.64
0440 No. 4000		217	.037			1.16		1.16	1.89
0460 No. 6000		166	.048	↓		1.52		1.52	2.47
0465 Telephone/power pole		12	.667	Ea.		21		21	34
0470 Non-metallic, straight section	↓	480	.017	L.F.		.53		.53	.86
0500 Channels, steel, including fittings & hangers									
0520 3/4" x 1-1/2"	1 Elec	308	.026	L.F.		.82		.82	1.33
0540 1-1/2" x 1-1/2"		269	.030			.94		.94	1.53
0560 1-1/2" x 1-7/8"	↓	229	.035	↓		1.10		1.10	1.79
1180 400 amp	2 Elec	6.80	2.353	Ea.		74		74	121
1210 Panel boards, incl. removal of all breakers,									
1220 conduit terminations & wire connections									
1230 3 wire, 120/240V, 100A, to 20 circuits	1 Elec	2.60	3.077	Ea.		97		97	158
1240 200 amps, to 42 circuits	2 Elec	2.60	6.154			194		194	315
1260 4 wire, 120/208V, 125A, to 20 circuits	1 Elec	2.40	3.333			105		105	171
1270 200 amps, to 42 circuits	2 Elec	2.40	6.667			210		210	340
1720 Junction boxes, 4" sq. & oct.	1 Elec	80	.100			3.15		3.15	5.15
1760 Switch box		107	.075			2.36		2.36	3.84
1780 Receptacle & switch plates	↓	257	.031	↓		.98		.98	1.60
1800 Wire, THW-THWN-THHN, removed from									
1810 in place conduit, to 15' high									
1830 #14	1 Elec	65	.123	C.L.F.		3.88		3.88	6.30
1840 #12		55	.145			4.58		4.58	7.45
1850 #10	↓	45.50	.176	↓		5.55		5.55	9
2000 Interior fluorescent fixtures, incl. supports									
2010 & whips, to 15' high									
2100 Recessed drop-in 2' x 2', 2 lamp	2 Elec	35	.457	Ea.		14.40		14.40	23.50
2140 2' x 4', 4 lamp	"	30	.533	"		16.80		16.80	27.50
2180 Surface mount, acrylic lens & hinged frame									
2220 2' x 2', 2 lamp	2 Elec	44	.364	Ea.		11.45		11.45	18.65
2260 2' x 4', 4 lamp	"	33	.485	"		15.25		15.25	25
2300 Strip fixtures, surface mount									
2320 4' long, 1 lamp	2 Elec	53	.302	Ea.		9.50		9.50	15.50
2380 8' long, 2 lamp	"	40	.400	"		12.60		12.60	20.50
2460 Interior incandescent, surface, ceiling									
2470 or wall mount, to 12' high									

26 05 Common Work Results for Electrical

26 05 05 – Selective Electrical Demolition

26 05 05.10 Electrical Demolition

		Crew	Daily Output	Labor-Hours	Unit	Material	2009 Bare Costs Labor	2009 Bare Costs Equipment	Total	Total Incl O&P
2480	Metal cylinder type, 75 Watt	2 Elec	62	.258	Ea.		8.15		8.15	13.25
2600	Exterior fixtures, incandescent, wall mount									
2620	100 Watt	2 Elec	50	.320	Ea.		10.10		10.10	16.40
3000	Ceiling fan, tear out and remove	1 Elec	18	.444	"		14		14	23
9000	Minimum labor/equipment charge	"	4	2	Job		63		63	103

26 05 19 – Low-Voltage Electrical Power Conductors and Cables

26 05 19.20 Armored Cable

		Crew	Daily Output	Labor-Hours	Unit	Material	2009 Bare Costs Labor	2009 Bare Costs Equipment	Total	Total Incl O&P
0010	**ARMORED CABLE**									
0051	600 volt, copper (BX), #14, 2 conductor, solid	1 Elec	240	.033	L.F.	.77	1.05		1.82	2.56
0101	3 conductor, solid		200	.040		1.21	1.26		2.47	3.38
0151	#12, 2 conductor, solid		210	.038		.79	1.20		1.99	2.82
0201	3 conductor, solid		180	.044		1.28	1.40		2.68	3.69
0251	#10, 2 conductor, solid		180	.044		1.50	1.40		2.90	3.93
0301	3 conductor, solid		150	.053		2.21	1.68		3.89	5.15
0351	#8, 3 conductor, solid		120	.067		4.51	2.10		6.61	8.40

26 05 19.55 Non-Metallic Sheathed Cable

		Crew	Daily Output	Labor-Hours	Unit	Material	2009 Bare Costs Labor	2009 Bare Costs Equipment	Total	Total Incl O&P
0010	**NON-METALLIC SHEATHED CABLE** 600 volt									
0100	Copper with ground wire, (Romex)									
0151	#14, 2 wire	1 Elec	250	.032	L.F.	.32	1.01		1.33	2
0201	3 wire		230	.035		.50	1.10		1.60	2.33
0251	#12, 2 wire		220	.036		.50	1.15		1.65	2.42
0301	3 wire		200	.040		.70	1.26		1.96	2.82
0351	#10, 2 wire		200	.040		.78	1.26		2.04	2.91
0401	3 wire		140	.057		1.11	1.80		2.91	4.15
0451	#8, 3 conductor		130	.062		1.85	1.94		3.79	5.20
0501	#6, 3 wire		120	.067		3	2.10		5.10	6.70
0550	SE type SER aluminum cable, 3 RHW and									
0601	1 bare neutral, 3 #8 & 1 #8	1 Elec	150	.053	L.F.	1.61	1.68		3.29	4.51
0651	3 #6 & 1 #6	"	130	.062		1.82	1.94		3.76	5.15
0701	3 #4 & 1 #6	2 Elec	220	.073		2.04	2.29		4.33	5.95
0751	3 #2 & 1 #4		200	.080		3	2.52		5.52	7.40
0801	3 #1/0 & 1 #2		180	.089		4.55	2.80		7.35	9.55
0851	3 #2/0 & 1 #1		160	.100		5.35	3.15		8.50	11.05
0901	3 #4/0 & 1 #2/0		140	.114		7.65	3.60		11.25	14.25
2401	SEU service entrance cable, copper 2 conductors, #8 + #8 neutral	1 Elec	150	.053		1.52	1.68		3.20	4.41
2601	#6 + #8 neutral		130	.062		2.27	1.94		4.21	5.65
2801	#6 + #6 neutral		130	.062		2.58	1.94		4.52	6
3001	#4 + #6 neutral	2 Elec	220	.073		3.53	2.29		5.82	7.60
3201	#4 + #4 neutral		220	.073		3.99	2.29		6.28	8.10
3401	#3 + #5 neutral		210	.076		4.67	2.40		7.07	9.05
6500	Service entrance cap for copper SEU									
6600	100 amp	1 Elec	12	.667	Ea.	12.15	21		33.15	47.50
6700	150 amp		10	.800		22	25		47	65.50
6800	200 amp		8	1		33.50	31.50		65	88.50

26 05 19.90 Wire

		Crew	Daily Output	Labor-Hours	Unit	Material	2009 Bare Costs Labor	2009 Bare Costs Equipment	Total	Total Incl O&P
0010	**WIRE**									
0021	600 volt type THW, copper solid, #14	1 Elec	1300	.006	L.F.	.10	.19		.29	.43
0031	#12		1100	.007		.16	.23		.39	.55
0041	#10		1000	.008		.25	.25		.50	.69
0161	#6		650	.012		.68	.39		1.07	1.37
0181	#4	2 Elec	1060	.015		1.06	.48		1.54	1.94
0201	#3		1000	.016		1.34	.50		1.84	2.30

26 05 Common Work Results for Electrical

26 05 19 – Low-Voltage Electrical Power Conductors and Cables

26 05 19.90 Wire

		Crew	Daily Output	Labor-Hours	Unit	Material	2009 Bare Costs Labor	Equipment	Total	Total Incl O&P
0221	#2	2 Elec	900	.018	L.F.	1.68	.56		2.24	2.76
0241	#1		800	.020		2.13	.63		2.76	3.37
0261	1/0		660	.024		2.59	.76		3.35	4.09
0281	2/0		580	.028		3.25	.87		4.12	5
0301	3/0		500	.032		4.08	1.01		5.09	6.15
0351	4/0	↓	440	.036	↓	5.15	1.15		6.30	7.50

26 05 26 – Grounding and Bonding for Electrical Systems

26 05 26.80 Grounding

		Crew	Daily Output	Labor-Hours	Unit	Material	2009 Bare Costs Labor	Equipment	Total	Total Incl O&P
0010	**GROUNDING**									
0030	Rod, copper clad, 8' long, 1/2" diameter	1 Elec	5.50	1.455	Ea.	14.80	46		60.80	91
0050	3/4" diameter		5.30	1.509		27.50	47.50		75	108
0080	10' long, 1/2" diameter		4.80	1.667		17.35	52.50		69.85	105
0100	3/4" diameter		4.40	1.818	↓	33	57.50		90.50	130
0261	Wire, ground bare armored, #8-1 conductor		200	.040	L.F.	1.09	1.26		2.35	3.25
0271	#6-1 conductor		180	.044	"	1.31	1.40		2.71	3.72
0390	Bare copper wire, stranded, #8		11	.727	C.L.F.	42	23		65	83.50
0401	Bare copper, #6 wire		1000	.008	L.F.	.66	.25		.91	1.14
0601	#2 stranded	2 Elec	1000	.016	"	1.67	.50		2.17	2.65
1800	Water pipe ground clamps, heavy duty									
2000	Bronze, 1/2" to 1" diameter	1 Elec	8	1	Ea.	26	31.50		57.50	80

26 05 33 – Raceway and Boxes for Electrical Systems

26 05 33.05 Conduit

		Crew	Daily Output	Labor-Hours	Unit	Material	2009 Bare Costs Labor	Equipment	Total	Total Incl O&P
0010	**CONDUIT** To 15' high, includes 2 terminations, 2 elbows,									
0020	11 beam clamps, and 11 couplings per 100 L.F.									
1750	Rigid galvanized steel, 1/2" diameter	1 Elec	90	.089	L.F.	2.96	2.80		5.76	7.80
1770	3/4" diameter		80	.100		3.23	3.15		6.38	8.70
1800	1" diameter		65	.123		4.44	3.88		8.32	11.20
1830	1-1/4" diameter		60	.133		6.15	4.20		10.35	13.60
1850	1-1/2" diameter		55	.145		7.15	4.58		11.73	15.35
1870	2" diameter		45	.178		9.25	5.60		14.85	19.30
5000	Electric metallic tubing (EMT), 1/2" diameter		170	.047		.66	1.48		2.14	3.13
5020	3/4" diameter		130	.062		1.05	1.94		2.99	4.32
5040	1" diameter		115	.070		1.84	2.19		4.03	5.60
5060	1-1/4" diameter		100	.080		2.95	2.52		5.47	7.35
5080	1-1/2" diameter		90	.089		3.78	2.80		6.58	8.70
9100	PVC, schedule 40, 1/2" diameter		190	.042		1.19	1.33		2.52	3.47
9110	3/4" diameter		145	.055		1.46	1.74		3.20	4.44
9120	1" diameter		125	.064		2.42	2.02		4.44	5.95
9130	1-1/4" diameter		110	.073		3.21	2.29		5.50	7.25
9140	1-1/2" diameter		100	.080		3.68	2.52		6.20	8.15
9150	2" diameter	↓	90	.089	↓	4.67	2.80		7.47	9.70

26 05 33.25 Conduit Fittings for Rigid Galvanized Steel

		Crew	Daily Output	Labor-Hours	Unit	Material	2009 Bare Costs Labor	Equipment	Total	Total Incl O&P
0010	**CONDUIT FITTINGS FOR RIGID GALVANIZED STEEL**									
2280	LB, LR or LL fittings & covers, 1/2" diameter	1 Elec	16	.500	Ea.	11.95	15.75		27.70	38.50
2290	3/4" diameter		13	.615		13.95	19.40		33.35	47
2300	1" diameter		11	.727		20.50	23		43.50	60.50
2330	1-1/4" diameter		8	1		31.50	31.50		63	86
2350	1-1/2" diameter		6	1.333		39	42		81	112
2370	2" diameter		5	1.600		65	50.50		115.50	154
5280	Service entrance cap, 1/2" diameter		16	.500		9.05	15.75		24.80	35.50
5300	3/4" diameter	↓	13	.615	↓	10.50	19.40		29.90	43

26 05 Common Work Results for Electrical

26 05 33 – Raceway and Boxes for Electrical Systems

26 05 33.25 Conduit Fittings for Rigid Galvanized Steel

		Crew	Daily Output	Labor-Hours	Unit	Material	2009 Bare Costs Labor	Equipment	Total	Total Incl O&P
5320	1" diameter	1 Elec	10	.800	Ea.	12.15	25		37.15	54.50
5340	1-1/4" diameter		8	1		18.35	31.50		49.85	71.50
5360	1-1/2" diameter		6.50	1.231		33	39		72	99.50
5380	2" diameter	↓	5.50	1.455	↓	50.50	46		96.50	130

26 05 33.35 Flexible Metallic Conduit

		Crew	Daily Output	Labor-Hours	Unit	Material	2009 Bare Costs Labor	Equipment	Total	Total Incl O&P
0010	**FLEXIBLE METALLIC CONDUIT**									
0050	Steel, 3/8" diameter	1 Elec	200	.040	L.F.	.43	1.26		1.69	2.52
0100	1/2" diameter		200	.040		.49	1.26		1.75	2.59
0200	3/4" diameter		160	.050		.66	1.58		2.24	3.30
0250	1" diameter		100	.080		1.30	2.52		3.82	5.55
0300	1-1/4" diameter		70	.114		1.59	3.60		5.19	7.60
0350	1-1/2" diameter		50	.160		2.64	5.05		7.69	11.10
0370	2" diameter	↓	40	.200	↓	3.30	6.30		9.60	13.90

26 05 33.50 Outlet Boxes

		Crew	Daily Output	Labor-Hours	Unit	Material	2009 Bare Costs Labor	Equipment	Total	Total Incl O&P
0010	**OUTLET BOXES**									
0021	Pressed steel, octagon, 4"	1 Elec	18	.444	Ea.	3.09	14		17.09	26.50
0060	Covers, blank		64	.125		1.29	3.94		5.23	7.80
0100	Extension rings		40	.200		5.15	6.30		11.45	15.90
0151	Square 4"		18	.444		2.38	14		16.38	25.50
0200	Extension rings		40	.200		5.20	6.30		11.50	15.95
0250	Covers, blank		64	.125		1.40	3.94		5.34	7.95
0300	Plaster rings		64	.125		2.84	3.94		6.78	9.50
0651	Switchbox		24	.333		4.97	10.50		15.47	22.50
1100	Concrete, floor, 1 gang	↓	5.30	1.509	↓	85	47.50		132.50	171

26 05 33.60 Outlet Boxes, Plastic

		Crew	Daily Output	Labor-Hours	Unit	Material	2009 Bare Costs Labor	Equipment	Total	Total Incl O&P
0010	**OUTLET BOXES, PLASTIC**									
0051	4" diameter, round, with 2 mounting nails	1 Elec	23	.348	Ea.	2.96	10.95		13.91	21
0101	Bar hanger mounted		23	.348		5.35	10.95		16.30	24
0201	Square with 2 mounting nails		23	.348		4.44	10.95		15.39	22.50
0300	Plaster ring		64	.125		1.84	3.94		5.78	8.40
0401	Switch box with 2 mounting nails, 1 gang		27	.296		2.01	9.35		11.36	17.40
0501	2 gang		23	.348		4.12	10.95		15.07	22.50
0601	3 gang	↓	18	.444	↓	6.50	14		20.50	30

26 05 33.65 Pull Boxes

		Crew	Daily Output	Labor-Hours	Unit	Material	2009 Bare Costs Labor	Equipment	Total	Total Incl O&P
0010	**PULL BOXES**									
0100	Sheet metal, pull box, NEMA 1, type SC, 6" W x 6" H x 4" D	1 Elec	8	1	Ea.	12.55	31.50		44.05	65.50
0200	8" W x 8" H x 4" D		8	1		17.20	31.50		48.70	70.50
0300	10" W x 12" H x 6" D	↓	5.30	1.509	↓	30.50	47.50		78	111

26 05 39 – Underfloor Raceways for Electrical Systems

26 05 39.30 Conduit In Concrete Slab

		Crew	Daily Output	Labor-Hours	Unit	Material	2009 Bare Costs Labor	Equipment	Total	Total Incl O&P
0010	**CONDUIT IN CONCRETE SLAB** Including terminations,									
0020	fittings and supports									
3230	PVC, schedule 40, 1/2" diameter	1 Elec	270	.030	L.F.	.93	.93		1.86	2.54
3250	3/4" diameter		230	.035		1.14	1.10		2.24	3.03
3270	1" diameter		200	.040		1.55	1.26		2.81	3.76
3300	1-1/4" diameter		170	.047		2.18	1.48		3.66	4.81
3330	1-1/2" diameter		140	.057		2.60	1.80		4.40	5.80
3350	2" diameter		120	.067		3.35	2.10		5.45	7.10
4350	Rigid galvanized steel, 1/2" diameter		200	.040		2.73	1.26		3.99	5.05
4400	3/4" diameter		170	.047		3.01	1.48		4.49	5.70
4450	1" diameter	↓	130	.062	↓	4.21	1.94		6.15	7.80

26 05 39 – Underfloor Raceways for Electrical Systems

26 05 39.30 Conduit In Concrete Slab

		Crew	Daily Output	Labor-Hours	Unit	Material	2009 Bare Costs Labor	Equipment	Total	Total Incl O&P
4500	1-1/4" diameter	1 Elec	110	.073	L.F.	5.65	2.29		7.94	9.95
4600	1-1/2" diameter		100	.080		6.70	2.52		9.22	11.45
4800	2" diameter	↓	90	.089	↓	8.55	2.80		11.35	14

26 05 39.40 Conduit In Trench

		Crew	Daily Output	Labor-Hours	Unit	Material	2009 Bare Costs Labor	Equipment	Total	Total Incl O&P
0010	**CONDUIT IN TRENCH** Includes terminations and fittings									
0200	Rigid galvanized steel, 2" diameter	1 Elec	150	.053	L.F.	8.20	1.68		9.88	11.80
0400	2-1/2" diameter	"	100	.080		15.60	2.52		18.12	21.50
0600	3" diameter	2 Elec	160	.100		19.60	3.15		22.75	26.50
0800	3-1/2" diameter	"	140	.114	↓	25.50	3.60		29.10	34

26 05 80 – Wiring Connections

26 05 80.10 Motor Connections

		Crew	Daily Output	Labor-Hours	Unit	Material	2009 Bare Costs Labor	Equipment	Total	Total Incl O&P
0010	**MOTOR CONNECTIONS**									
0020	Flexible conduit and fittings, 115 volt, 1 phase, up to 1 HP motor	1 Elec	8	1	Ea.	9.10	31.50		40.60	61.50

26 05 90 – Residential Applications

26 05 90.10 Residential Wiring

		Crew	Daily Output	Labor-Hours	Unit	Material	2009 Bare Costs Labor	Equipment	Total	Total Incl O&P
0010	**RESIDENTIAL WIRING**									
0020	20' avg. runs and #14/2 wiring incl. unless otherwise noted									
1000	Service & panel, includes 24' SE-AL cable, service eye, meter,									
1010	Socket, panel board, main bkr., ground rod, 15 or 20 amp									
1020	1-pole circuit breakers, and misc. hardware									
1100	100 amp, with 10 branch breakers	1 Elec	1.19	6.723	Ea.	585	212		797	985
1110	With PVC conduit and wire		.92	8.696		660	274		934	1,175
1120	With RGS conduit and wire		.73	10.959		840	345		1,185	1,475
1150	150 amp, with 14 branch breakers		1.03	7.767		895	245		1,140	1,375
1170	With PVC conduit and wire		.82	9.756		1,050	305		1,355	1,650
1180	With RGS conduit and wire	↓	.67	11.940		1,400	375		1,775	2,150
1200	200 amp, with 18 branch breakers	2 Elec	1.80	8.889		1,175	280		1,455	1,750
1220	With PVC conduit and wire		1.46	10.959		1,325	345		1,670	2,025
1230	With RGS conduit and wire	↓	1.24	12.903		1,800	405		2,205	2,625
1800	Lightning surge suppressor for above services, add	1 Elec	32	.250	↓	50.50	7.90		58.40	68.50
2000	Switch devices									
2100	Single pole, 15 amp, Ivory, with a 1-gang box, cover plate,									
2110	Type NM (Romex) cable	1 Elec	17.10	.468	Ea.	11.70	14.75		26.45	37
2120	Type MC (BX) cable		14.30	.559		28	17.60		45.60	59.50
2130	EMT & wire		5.71	1.401		34.50	44		78.50	110
2150	3-way, #14/3, type NM cable		14.55	.550		17.25	17.30		34.55	47
2170	Type MC cable		12.31	.650		39	20.50		59.50	76.50
2180	EMT & wire		5	1.600		38.50	50.50		89	125
2200	4-way, #14/3, type NM cable		14.55	.550		33.50	17.30		50.80	65
2220	Type MC cable		12.31	.650		55.50	20.50		76	94.50
2230	EMT & wire		5	1.600		55	50.50		105.50	143
2250	S.P., 20 amp, #12/2, type NM cable		13.33	.600		19.70	18.90		38.60	52.50
2270	Type MC cable		11.43	.700		33	22		55	72.50
2280	EMT & wire		4.85	1.649		44.50	52		96.50	134
2290	S.P. rotary dimmer, 600W, no wiring		17	.471		19.90	14.80		34.70	46
2300	S.P. rotary dimmer, 600W, type NM cable		14.55	.550		26.50	17.30		43.80	57
2320	Type MC cable		12.31	.650		43	20.50		63.50	80.50
2330	EMT & wire		5	1.600		51.50	50.50		102	139
2350	3-way rotary dimmer, type NM cable		13.33	.600		22.50	18.90		41.40	55.50
2370	Type MC cable		11.43	.700		39	22		61	79
2380	EMT & wire	↓	4.85	1.649	↓	47.50	52		99.50	137

26 05 90 – Residential Applications

26 05 90.10 Residential Wiring	Crew	Daily Output	Labor-Hours	Unit	Material	2009 Bare Costs Labor	Equipment	Total	Total Incl O&P
2400	Interval timer wall switch, 20 amp, 1-30 min., #12/2									
2410	Type NM cable	1 Elec	14.55	.550	Ea.	53	17.30		70.30	86
2420	Type MC cable		12.31	.650		60	20.50		80.50	99.50
2430	EMT & wire		5	1.600		77.50	50.50		128	168
2500	Decorator style									
2510	S.P., 15 amp, type NM cable	1 Elec	17.10	.468	Ea.	16.15	14.75		30.90	42
2520	Type MC cable		14.30	.559		32.50	17.60		50.10	64.50
2530	EMT & wire		5.71	1.401		39	44		83	115
2550	3-way, #14/3, type NM cable		14.55	.550		21.50	17.30		38.80	52
2570	Type MC cable		12.31	.650		43.50	20.50		64	81
2580	EMT & wire		5	1.600		43	50.50		93.50	130
2600	4-way, #14/3, type NM cable		14.55	.550		38	17.30		55.30	70
2620	Type MC cable		12.31	.650		60	20.50		80.50	99
2630	EMT & wire		5	1.600		59.50	50.50		110	148
2650	S.P., 20 amp, #12/2, type NM cable		13.33	.600		24	18.90		42.90	57.50
2670	Type MC cable		11.43	.700		37.50	22		59.50	77.50
2680	EMT & wire		4.85	1.649		49	52		101	139
2700	S.P., slide dimmer, type NM cable		17.10	.468		30	14.75		44.75	57.50
2720	Type MC cable		14.30	.559		46.50	17.60		64.10	80
2730	EMT & wire		5.71	1.401		55	44		99	133
2750	S.P., touch dimmer, type NM cable		17.10	.468		27	14.75		41.75	54
2770	Type MC cable		14.30	.559		43.50	17.60		61.10	76.50
2780	EMT & wire		5.71	1.401		52	44		96	130
2800	3-way touch dimmer, type NM cable		13.33	.600		47	18.90		65.90	83
2820	Type MC cable		11.43	.700		63.50	22		85.50	106
2830	EMT & wire		4.85	1.649		72	52		124	164
3000	Combination devices									
3100	S.P. switch/15 amp recpt., Ivory, 1-gang box, plate									
3110	Type NM cable	1 Elec	11.43	.700	Ea.	22.50	22		44.50	61
3120	Type MC cable		10	.800		39	25		64	84
3130	EMT & wire		4.40	1.818		47.50	57.50		105	146
3150	S.P. switch/pilot light, type NM cable		11.43	.700		23.50	22		45.50	61.50
3170	Type MC cable		10	.800		40	25		65	85
3180	EMT & wire		4.43	1.806		48	57		105	146
3190	2-S.P. switches, 2-#14/2, no wiring		14	.571		7.80	18		25.80	38
3200	2-S.P. switches, 2-#14/2, type NM cables		10	.800		28	25		53	72
3220	Type MC cable		8.89	.900		53.50	28.50		82	105
3230	EMT & wire		4.10	1.951		52.50	61.50		114	158
3250	3-way switch/15 amp recpt., #14/3, type NM cable		10	.800		31.50	25		56.50	76
3270	Type MC cable		8.89	.900		53.50	28.50		82	105
3280	EMT & wire		4.10	1.951		53	61.50		114.50	159
3300	2-3 way switches, 2-#14/3, type NM cables		8.89	.900		44.50	28.50		73	94.50
3320	Type MC cable		8	1		80	31.50		111.50	140
3330	EMT & wire		4	2		62	63		125	171
3350	S.P. switch/20 amp recpt., #12/2, type NM cable		10	.800		36	25		61	80.50
3370	Type MC cable		8.89	.900		43	28.50		71.50	93.50
3380	EMT & wire		4.10	1.951		60.50	61.50		122	167
3400	Decorator style									
3410	S.P. switch/15 amp recpt., type NM cable	1 Elec	11.43	.700	Ea.	27	22		49	65.50
3420	Type MC cable		10	.800		43.50	25		68.50	89
3430	EMT & wire		4.40	1.818		52	57.50		109.50	151
3450	S.P. switch/pilot light, type NM cable		11.43	.700		28	22		50	66.50
3470	Type MC cable		10	.800		44.50	25		69.50	89.50

26 05 90.10 Residential Wiring		Crew	Daily Output	Labor-Hours	Unit	Material	2009 Bare Costs Labor	Equipment	Total	Total Incl O&P
3480	EMT & wire	1 Elec	4.40	1.818	Ea.	52.50	57.50		110	152
3500	2-S.P. switches, 2-#14/2, type NM cables		10	.800		32.50	25		57.50	77
3520	Type MC cable		8.89	.900		58	28.50		86.50	110
3530	EMT & wire		4.10	1.951		57	61.50		118.50	163
3550	3-way/15 amp recpt., #14/3, type NM cable		10	.800		36	25		61	80.50
3570	Type MC cable		8.89	.900		58	28.50		86.50	110
3580	EMT & wire		4.10	1.951		57.50	61.50		119	164
3650	2-3 way switches, 2-#14/3, type NM cables		8.89	.900		48.50	28.50		77	99.50
3670	Type MC cable		8	1		84.50	31.50		116	145
3680	EMT & wire		4	2		66.50	63		129.50	176
3700	S.P. switch/20 amp recpt., #12/2, type NM cable		10	.800		40.50	25		65.50	85.50
3720	Type MC cable		8.89	.900		47.50	28.50		76	98.50
3730	EMT & wire	▼	4.10	1.951	▼	65	61.50		126.50	172
4000	Receptacle devices									
4010	Duplex outlet, 15 amp recpt., Ivory, 1-gang box, plate									
4015	Type NM cable	1 Elec	14.55	.550	Ea.	10.20	17.30		27.50	39
4020	Type MC cable		12.31	.650		26.50	20.50		47	63
4030	EMT & wire		5.33	1.501		33	47.50		80.50	114
4050	With #12/2, type NM cable		12.31	.650		13.65	20.50		34.15	48.50
4070	Type MC cable		10.67	.750		27	23.50		50.50	68
4080	EMT & wire		4.71	1.699		38.50	53.50		92	130
4100	20 amp recpt., #12/2, type NM cable		12.31	.650		22.50	20.50		43	58
4120	Type MC cable		10.67	.750		35.50	23.50		59	77.50
4130	EMT & wire	▼	4.71	1.699	▼	47	53.50		100.50	139
4140	For GFI see Div. 26 05 90.10 line 4300 below									
4150	Decorator style, 15 amp recpt., type NM cable	1 Elec	14.55	.550	Ea.	14.65	17.30		31.95	44
4170	Type MC cable		12.31	.650		31	20.50		51.50	67.50
4180	EMT & wire		5.33	1.501		37.50	47.50		85	118
4200	With #12/2, type NM cable		12.31	.650		18.10	20.50		38.60	53.50
4220	Type MC cable		10.67	.750		31.50	23.50		55	73
4230	EMT & wire		4.71	1.699		43	53.50		96.50	134
4250	20 amp recpt. #12/2, type NM cable		12.31	.650		26.50	20.50		47	63
4270	Type MC cable		10.67	.750		40	23.50		63.50	82.50
4280	EMT & wire		4.71	1.699		51.50	53.50		105	144
4300	GFI, 15 amp recpt., type NM cable		12.31	.650		41.50	20.50		62	79
4320	Type MC cable		10.67	.750		58	23.50		81.50	102
4330	EMT & wire		4.71	1.699		64.50	53.50		118	158
4350	GFI with #12/2, type NM cable		10.67	.750		45	23.50		68.50	88
4370	Type MC cable		9.20	.870		58.50	27.50		86	109
4380	EMT & wire		4.21	1.900		69.50	60		129.50	174
4400	20 amp recpt., #12/2 type NM cable		10.67	.750		47	23.50		70.50	90
4420	Type MC cable		9.20	.870		60	27.50		87.50	111
4430	EMT & wire		4.21	1.900		71.50	60		131.50	177
4500	Weather-proof cover for above receptacles, add	▼	32	.250	▼	5.25	7.90		13.15	18.60
4550	Air conditioner outlet, 20 amp-240 volt recpt.									
4560	30' of #12/2, 2 pole circuit breaker									
4570	Type NM cable	1 Elec	10	.800	Ea.	63	25		88	110
4580	Type MC cable		9	.889		79	28		107	133
4590	EMT & wire		4	2		87.50	63		150.50	199
4600	Decorator style, type NM cable		10	.800		67.50	25		92.50	115
4620	Type MC cable		9	.889		84	28		112	138
4630	EMT & wire	▼	4	2	▼	92	63		155	204
4650	Dryer outlet, 30 amp-240 volt recpt., 20' of #10/3									

26 05 90.10 Residential Wiring	Crew	Daily Output	Labor-Hours	Unit	Material	2009 Bare Costs Labor	Equipment	Total	Total Incl O&P
4660	2 pole circuit breaker									
4670	Type NM cable	1 Elec	6.41	1.248	Ea.	75	39.50		114.50	147
4680	Type MC cable		5.71	1.401		83	44		127	164
4690	EMT & wire		3.48	2.299		89	72.50		161.50	216
4700	Range outlet, 50 amp-240 volt recpt., 30' of #8/3									
4710	Type NM cable	1 Elec	4.21	1.900	Ea.	110	60		170	218
4720	Type MC cable		4	2		195	63		258	315
4730	EMT & wire		2.96	2.703		121	85		206	272
4750	Central vacuum outlet, Type NM cable		6.40	1.250		69.50	39.50		109	141
4770	Type MC cable		5.71	1.401		94	44		138	175
4780	EMT & wire		3.48	2.299		96.50	72.50		169	224
4800	30 amp-110 volt locking recpt., #10/2 circ. bkr.									
4810	Type NM cable	1 Elec	6.20	1.290	Ea.	80.50	40.50		121	155
4820	Type MC cable		5.40	1.481		111	46.50		157.50	198
4830	EMT & wire		3.20	2.500		110	79		189	249
4900	Low voltage outlets									
4910	Telephone recpt., 20' of 4/C phone wire	1 Elec	26	.308	Ea.	10.30	9.70		20	27
4920	TV recpt., 20' of RG59U coax wire, F type connector	"	16	.500	"	18.35	15.75		34.10	45.50
4950	Door bell chime, transformer, 2 buttons, 60' of bellwire									
4970	Economy model	1 Elec	11.50	.696	Ea.	75	22		97	118
4980	Custom model		11.50	.696		116	22		138	164
4990	Luxury model, 3 buttons		9.50	.842		310	26.50		336.50	385
6000	Lighting outlets									
6050	Wire only (for fixture), type NM cable	1 Elec	32	.250	Ea.	8.45	7.90		16.35	22
6070	Type MC cable		24	.333		18.55	10.50		29.05	37.50
6080	EMT & wire		10	.800		23.50	25		48.50	66.50
6100	Box (4"), and wire (for fixture), type NM cable		25	.320		17.45	10.10		27.55	35.50
6120	Type MC cable		20	.400		27.50	12.60		40.10	51
6130	EMT & wire		11	.727		32.50	23		55.50	73
6200	Fixtures (use with lines 6050 or 6100 above)									
6210	Canopy style, economy grade	1 Elec	40	.200	Ea.	29	6.30		35.30	42.50
6220	Custom grade		40	.200		51	6.30		57.30	66.50
6250	Dining room chandelier, economy grade		19	.421		82.50	13.25		95.75	113
6260	Custom grade		19	.421		243	13.25		256.25	289
6270	Luxury grade		15	.533		540	16.80		556.80	625
6310	Kitchen fixture (fluorescent), economy grade		30	.267		61.50	8.40		69.90	81
6320	Custom grade		25	.320		168	10.10		178.10	201
6350	Outdoor, wall mounted, economy grade		30	.267		29.50	8.40		37.90	46
6360	Custom grade		30	.267		108	8.40		116.40	133
6370	Luxury grade		25	.320		244	10.10		254.10	284
6410	Outdoor PAR floodlights, 1 lamp, 150 watt		20	.400		32	12.60		44.60	55.50
6420	2 lamp, 150 watt each		20	.400		52	12.60		64.60	77.50
6425	Motion sensing, 2 lamp, 150 watt each		20	.400		70	12.60		82.60	97.50
6430	For infrared security sensor, add		32	.250		111	7.90		118.90	135
6450	Outdoor, quartz-halogen, 300 watt flood		20	.400		41	12.60		53.60	65.50
6600	Recessed downlight, round, pre-wired, 50 or 75 watt trim		30	.267		42	8.40		50.40	59.50
6610	With shower light trim		30	.267		52	8.40		60.40	70.50
6620	With wall washer trim		28	.286		62.50	9		71.50	83.50
6630	With eye-ball trim		28	.286		62.50	9		71.50	83.50
6640	For direct contact with insulation, add					2.10			2.10	2.31
6700	Porcelain lamp holder	1 Elec	40	.200		4.10	6.30		10.40	14.75
6710	With pull switch		40	.200		4.59	6.30		10.89	15.30
6750	Fluorescent strip, 1-20 watt tube, wrap around diffuser, 24"		24	.333		54	10.50		64.50	76.50

26 05 90.10 Residential Wiring	Crew	Daily Output	Labor-Hours	Unit	Material	2009 Bare Costs Labor	2009 Bare Costs Equipment	Total	Total Incl O&P
6760 1-40 watt tube, 48"	1 Elec	24	.333	Ea.	66	10.50		76.50	89.50
6770 2-40 watt tubes, 48"		20	.400		80	12.60		92.60	109
6780 With residential ballast		20	.400		90	12.60		102.60	120
6800 Bathroom heat lamp, 1-250 watt		28	.286		46	9		55	65
6810 2-250 watt lamps		28	.286		71.50	9		80.50	93
6820 For timer switch, see Div. 26 05 90.10 line 2400									
6900 Outdoor post lamp, incl. post, fixture, 35' of #14/2									
6910 Type NMC cable	1 Elec	3.50	2.286	Ea.	203	72		275	340
6920 Photo-eye, add		27	.296		32.50	9.35		41.85	51
6950 Clock dial time switch, 24 hr., w/enclosure, type NM cable		11.43	.700		63	22		85	105
6970 Type MC cable		11	.727		79.50	23		102.50	125
6980 EMT & wire		4.85	1.649		85.50	52		137.50	179
7000 Alarm systems									
7050 Smoke detectors, box, #14/3, type NM cable	1 Elec	14.55	.550	Ea.	37.50	17.30		54.80	69
7070 Type MC cable		12.31	.650		53	20.50		73.50	91.50
7080 EMT & wire		5	1.600		52.50	50.50		103	140
7090 For relay output to security system, add					12.90			12.90	14.20
8000 Residential equipment									
8050 Disposal hook-up, incl. switch, outlet box, 3' of flex									
8060 20 amp-1 pole circ. bkr., and 25' of #12/2									
8070 Type NM cable	1 Elec	10	.800	Ea.	33.50	25		58.50	78
8080 Type MC cable		8	1		48.50	31.50		80	105
8090 EMT & wire		5	1.600		61	50.50		111.50	150
8100 Trash compactor or dishwasher hook-up, incl. outlet box,									
8110 3' of flex, 15 amp-1 pole circ. bkr., and 25' of #14/2									
8130 Type MC cable	1 Elec	8	1	Ea.	43.50	31.50		75	99.50
8140 EMT & wire	"	5	1.600	"	52.50	50.50		103	140
8150 Hot water sink dispensor hook-up, use line 8100									
8200 Vent/exhaust fan hook-up, type NM cable	1 Elec	32	.250	Ea.	8.45	7.90		16.35	22
8220 Type MC cable		24	.333		18.55	10.50		29.05	37.50
8230 EMT & wire		10	.800		23.50	25		48.50	66.50
8250 Bathroom vent fan, 50 CFM (use with above hook-up)									
8260 Economy model	1 Elec	15	.533	Ea.	24.50	16.80		41.30	54.50
8270 Low noise model		15	.533		32	16.80		48.80	62.50
8280 Custom model		12	.667		117	21		138	163
8300 Bathroom or kitchen vent fan, 110 CFM									
8310 Economy model	1 Elec	15	.533	Ea.	66.50	16.80		83.30	101
8320 Low noise model	"	15	.533	"	86	16.80		102.80	122
8350 Paddle fan, variable speed (w/o lights)									
8360 Economy model (AC motor)	1 Elec	10	.800	Ea.	105	25		130	157
8370 Custom model (AC motor)		10	.800		182	25		207	241
8380 Luxury model (DC motor)		8	1		360	31.50		391.50	445
8390 Remote speed switch for above, add		12	.667		28.50	21		49.50	65.50
8500 Whole house exhaust fan, ceiling mount, 36", variable speed									
8510 Remote switch, incl. shutters, 20 amp-1 pole circ. bkr.									
8520 30' of #12/2, type NM cable	1 Elec	4	2	Ea.	985	63		1,048	1,175
8530 Type MC cable		3.50	2.286		1,000	72		1,072	1,225
8540 EMT & wire		3	2.667		1,025	84		1,109	1,250
8600 Whirlpool tub hook-up, incl. timer switch, outlet box									
8610 3' of flex, 20 amp-1 pole GFI circ. bkr.									
8620 30' of #12/2, type NM cable	1 Elec	5	1.600	Ea.	127	50.50		177.50	221
8630 Type MC cable		4.20	1.905		135	60		195	247
8640 EMT & wire		3.40	2.353		146	74		220	282

26 05 Common Work Results for Electrical

26 05 90 – Residential Applications

26 05 90.10 Residential Wiring

		Crew	Daily Output	Labor-Hours	Unit	Material	Labor	Equipment	Total	Total Incl O&P
8650	Hot water heater hook-up, incl. 1-2 pole circ. bkr., box;									
8660	3' of flex, 20' of #10/2, type NM cable	1 Elec	5	1.600	Ea.	36	50.50		86.50	122
8670	Type MC cable		4.20	1.905		58	60		118	162
8680	EMT & wire		3.40	2.353		55.50	74		129.50	182
9000	Heating/air conditioning									
9050	Furnace/boiler hook-up, incl. firestat, local on-off switch									
9060	Emergency switch, and 40' of type NM cable	1 Elec	4	2	Ea.	57	63		120	166
9070	Type MC cable		3.50	2.286		82.50	72		154.50	208
9080	EMT & wire		1.50	5.333		94	168		262	375
9100	Air conditioner hook-up, incl. local 60 amp disc. switch									
9110	3' sealtite, 40 amp, 2 pole circuit breaker									
9130	40' of #8/2, type NM cable	1 Elec	3.50	2.286	Ea.	229	72		301	370
9140	Type MC cable		3	2.667		360	84		444	530
9150	EMT & wire		1.30	6.154		268	194		462	610
9200	Heat pump hook-up, 1-40 & 1-100 amp 2 pole circ. bkr.									
9210	Local disconnect switch, 3' sealtite									
9220	40' of #8/2 & 30' of #3/2									
9230	Type NM cable	1 Elec	1.30	6.154	Ea.	595	194		789	970
9240	Type MC cable		1.08	7.407		865	233		1,098	1,325
9250	EMT & wire		.94	8.511		675	268		943	1,175
9500	Thermostat hook-up, using low voltage wire									
9520	Heating only, 25' of #18-3	1 Elec	24	.333	Ea.	7.70	10.50		18.20	25.50
9530	Heating/cooling, 25' of #18-4	"	20	.400	"	9.15	12.60		21.75	30.50

26 24 Switchboards and Panelboards

26 24 16 – Panelboards

26 24 16.10 Load Centers

		Crew	Daily Output	Labor-Hours	Unit	Material	Labor	Equipment	Total	Total Incl O&P
0010	**LOAD CENTERS** (residential type)									
0100	3 wire, 120/240V, 1 phase, including 1 pole plug-in breakers									
0200	100 amp main lugs, indoor, 8 circuits	1 Elec	1.40	5.714	Ea.	176	180		356	485
0300	12 circuits		1.20	6.667		246	210		456	610
0400	Rainproof, 8 circuits		1.40	5.714		211	180		391	525
0500	12 circuits		1.20	6.667		298	210		508	665
0600	200 amp main lugs, indoor, 16 circuits	R-1A	1.80	8.889		340	233		573	760
0700	20 circuits		1.50	10.667		425	279		704	930
0800	24 circuits		1.30	12.308		560	320		880	1,150
1200	Rainproof, 16 circuits		1.80	8.889		405	233		638	830
1300	20 circuits		1.50	10.667		485	279		764	995
1400	24 circuits		1.30	12.308		725	320		1,045	1,325

26 27 Low-Voltage Distribution Equipment

26 27 13 – Electricity Metering

26 27 13.10 Meter Centers and Sockets		Crew	Daily Output	Labor-Hours	Unit	Material	2009 Bare Costs		Total	Total Incl O&P
							Labor	Equipment		
0010	**METER CENTERS AND SOCKETS**									
0100	Sockets, single position, 4 terminal, 100 amp	1 Elec	3.20	2.500	Ea.	43	79		122	176
0200	150 amp		2.30	3.478		50.50	110		160.50	234
0300	200 amp		1.90	4.211		72.50	133		205.50	296
0500	Double position, 4 terminal, 100 amp		2.80	2.857		177	90		267	340
0600	150 amp		2.10	3.810		208	120		328	425
0700	200 amp		1.70	4.706		430	148		578	710
2590	Basic meter device									
2600	1P 3W 120/240V 4 jaw 125A sockets, 3 meter	2 Elec	1	16	Ea.	685	505		1,190	1,575
2620	5 meter		.80	20		1,025	630		1,655	2,150
2640	7 meter		.56	28.571		1,500	900		2,400	3,125
2660	10 meter		.48	33.333		2,050	1,050		3,100	3,950
2680	Rainproof 1P 3W 120/240V 4 jaw 125A sockets									
2690	3 meter	2 Elec	1	16	Ea.	685	505		1,190	1,575
2710	6 meter		.60	26.667		1,175	840		2,015	2,675
2730	8 meter		.52	30.769		1,650	970		2,620	3,375
2750	1P 3W 120/240V 4 jaw sockets									
2760	with 125A circuit breaker, 3 meter	2 Elec	1	16	Ea.	1,275	505		1,780	2,225
2780	5 meter		.80	20		2,025	630		2,655	3,250
2800	7 meter		.56	28.571		2,900	900		3,800	4,650
2820	10 meter		.48	33.333		4,050	1,050		5,100	6,150
2830	Rainproof 1P 3W 120/240V 4 jaw sockets									
2840	with 125A circuit breaker, 3 meter	2 Elec	1	16	Ea.	1,275	505		1,780	2,225
2870	6 meter		.60	26.667		2,375	840		3,215	3,975
2890	8 meter		.52	30.769		3,225	970		4,195	5,125
3250	1P 3W 120/240V 4 jaw sockets									
3260	with 200A circuit breaker, 3 meter	2 Elec	1	16	Ea.	1,925	505		2,430	2,925
3290	6 meter		.60	26.667		3,850	840		4,690	5,600
3310	8 meter		.56	28.571		5,200	900		6,100	7,175
3330	Rainproof 1P 3W 120/240V 4 jaw sockets									
3350	with 200A circuit breaker, 3 meter	2 Elec	1	16	Ea.	1,925	505		2,430	2,925
3380	6 meter		.60	26.667		3,850	840		4,690	5,600
3400	8 meter		.52	30.769		5,200	970		6,170	7,275

26 27 23 – Indoor Service Poles

26 27 23.40 Surface Raceway

		Crew	Daily Output	Labor-Hours	Unit	Material	Labor	Equipment	Total	Total Incl O&P
0010	**SURFACE RACEWAY**									
0100	No. 500	1 Elec	100	.080	L.F.	.98	2.52		3.50	5.20
0110	No. 700		100	.080		1.10	2.52		3.62	5.30
0200	No. 1000		90	.089		1.87	2.80		4.67	6.60
0400	No. 1500, small pancake		90	.089		1.99	2.80		4.79	6.75
0600	No. 2000, base & cover, blank		90	.089		2.03	2.80		4.83	6.80
0800	No. 3000, base & cover, blank		75	.107		3.91	3.36		7.27	9.75
2400	Fittings, elbows, No. 500		40	.200	Ea.	1.76	6.30		8.06	12.20
2800	Elbow cover, No. 2000		40	.200		3.35	6.30		9.65	13.95
2880	Tee, No. 500		42	.190		3.40	6		9.40	13.50
2900	No. 2000		27	.296		11.15	9.35		20.50	27.50
3000	Switch box, No. 500		16	.500		11.55	15.75		27.30	38
3400	Telephone outlet, No. 1500		16	.500		13.80	15.75		29.55	40.50
3600	Junction box, No. 1500		16	.500		9.40	15.75		25.15	36
3800	Plugmold wired sections, No. 2000									
4000	1 circuit, 6 outlets, 3 ft. long	1 Elec	8	1	Ea.	35.50	31.50		67	91
4100	2 circuits, 8 outlets, 6 ft. long	"	5.30	1.509	"	57	47.50		104.50	140

26 27 Low-Voltage Distribution Equipment

26 27 26 – Wiring Devices

26 27 26.10 Low Voltage Switching

		Crew	Daily Output	Labor-Hours	Unit	Material	2009 Bare Costs Labor	Equipment	Total	Total Incl O&P
0010	**LOW VOLTAGE SWITCHING**									
3600	Relays, 120 V or 277 V standard	1 Elec	12	.667	Ea.	31	21		52	68
3800	Flush switch, standard		40	.200		10.75	6.30		17.05	22
4000	Interchangeable		40	.200		14.05	6.30		20.35	25.50
4100	Surface switch, standard		40	.200		7.90	6.30		14.20	18.95
4200	Transformer 115 V to 25 V		12	.667		110	21		131	155
4400	Master control, 12 circuit, manual		4	2		112	63		175	227
4500	25 circuit, motorized		4	2		122	63		185	237
4600	Rectifier, silicon		12	.667		36.50	21		57.50	74
4800	Switchplates, 1 gang, 1, 2 or 3 switch, plastic		80	.100		3.59	3.15		6.74	9.10
5000	Stainless steel		80	.100		9.65	3.15		12.80	15.80
5400	2 gang, 3 switch, stainless steel		53	.151		18.60	4.75		23.35	28.50
5500	4 switch, plastic		53	.151		8	4.75		12.75	16.55
5600	2 gang, 4 switch, stainless steel		53	.151		19.45	4.75		24.20	29.50
5700	6 switch, stainless steel		53	.151		43	4.75		47.75	55
5800	3 gang, 9 switch, stainless steel		32	.250		59.50	7.90		67.40	78.50

26 27 26.20 Wiring Devices Elements

		Crew	Daily Output	Labor-Hours	Unit	Material	2009 Bare Costs Labor	Equipment	Total	Total Incl O&P
0010	**WIRING DEVICES ELEMENTS**									
0200	Toggle switch, quiet type, single pole, 15 amp	1 Elec	40	.200	Ea.	5.35	6.30		11.65	16.15
0600	3 way, 15 amp		23	.348		7.65	10.95		18.60	26.50
0900	4 way, 15 amp		15	.533		25	16.80		41.80	55
1650	Dimmer switch, 120 volt, incandescent, 600 watt, 1 pole [G]		16	.500		13.10	15.75		28.85	40
2460	Receptacle, duplex, 120 volt, grounded, 15 amp		40	.200		1.35	6.30		7.65	11.75
2470	20 amp		27	.296		9.95	9.35		19.30	26
2490	Dryer, 30 amp		15	.533		6.05	16.80		22.85	34
2500	Range, 50 amp		11	.727		13.10	23		36.10	52
2600	Wall plates, stainless steel, 1 gang		80	.100		2.30	3.15		5.45	7.70
2800	2 gang		53	.151		4.20	4.75		8.95	12.35
3200	Lampholder, keyless		26	.308		12.30	9.70		22	29.50
3400	Pullchain with receptacle		22	.364		13.35	11.45		24.80	33.50

26 27 73 – Door Chimes

26 27 73.10 Doorbell System

		Crew	Daily Output	Labor-Hours	Unit	Material	2009 Bare Costs Labor	Equipment	Total	Total Incl O&P
0010	**DOORBELL SYSTEM**, incl. transformer, button & signal									
1000	Door chimes, 2 notes, minimum	1 Elec	16	.500	Ea.	29.50	15.75		45.25	57.50
1020	Maximum		12	.667		150	21		171	199
1100	Tube type, 3 tube system		12	.667		222	21		243	279
1180	4 tube system		10	.800		355	25		380	430
1900	For transformer & button, minimum add		5	1.600		16.15	50.50		66.65	100
1960	Maximum, add		4.50	1.778		48.50	56		104.50	145
3000	For push button only, minimum		24	.333		3.27	10.50		13.77	20.50
3100	Maximum		20	.400		25.50	12.60		38.10	48.50

26 28 Low-Voltage Circuit Protective Devices

26 28 16 – Enclosed Switches and Circuit Breakers

26 28 16.10 Circuit Breakers	Crew	Daily Output	Labor-Hours	Unit	Material	2009 Bare Costs Labor	Equipment	Total	Total Incl O&P
0010 **CIRCUIT BREAKERS** (in enclosure)									
0100 Enclosed (NEMA 1), 600 volt, 3 pole, 30 amp	1 Elec	3.20	2.500	Ea.	610	79		689	800
0200 60 amp		2.80	2.857		610	90		700	815
0400 100 amp		2.30	3.478		700	110		810	950

26 28 16.20 Safety Switches

	Crew	Daily Output	Labor-Hours	Unit	Material	Labor	Equipment	Total	Total Incl O&P
0010 **SAFETY SWITCHES**									
0100 General duty 240 volt, 3 pole NEMA 1, fusible, 30 amp	1 Elec	3.20	2.500	Ea.	113	79		192	253
0200 60 amp		2.30	3.478		192	110		302	390
0300 100 amp		1.90	4.211		330	133		463	575
0400 200 amp		1.30	6.154		710	194		904	1,100
0500 400 amp	2 Elec	1.80	8.889		1,775	280		2,055	2,425
9010 Disc. switch, 600V 3 pole fusible, 30 amp, to 10 HP motor	1 Elec	3.20	2.500		330	79		409	495
9050 60 amp, to 30 HP motor		2.30	3.478		760	110		870	1,025
9070 100 amp, to 60 HP motor		1.90	4.211		760	133		893	1,050

26 28 16.40 Time Switches

	Crew	Daily Output	Labor-Hours	Unit	Material	Labor	Equipment	Total	Total Incl O&P
0010 **TIME SWITCHES**									
0100 Single pole, single throw, 24 hour dial	1 Elec	4	2	Ea.	107	63		170	221
0200 24 hour dial with reserve power		3.60	2.222		460	70		530	620
0300 Astronomic dial		3.60	2.222		184	70		254	315
0400 Astronomic dial with reserve power		3.30	2.424		595	76.50		671.50	780
0500 7 day calendar dial		3.30	2.424		146	76.50		222.50	285
0600 7 day calendar dial with reserve power		3.20	2.500		655	79		734	850
0700 Photo cell 2000 watt		8	1		19	31.50		50.50	72.50

26 32 Packaged Generator Assemblies

26 32 13 – Engine Generators

26 32 13.16 Gas-Engine-Driven Generator Sets

	Crew	Daily Output	Labor-Hours	Unit	Material	Labor	Equipment	Total	Total Incl O&P
0010 **GAS-ENGINE-DRIVEN GENERATOR SETS**									
0020 Gas or gasoline operated, includes battery,									
0050 charger, muffler & transfer switch									
0200 3 phase 4 wire, 277/480 volt, 7.5 kW	R-3	.83	24.096	Ea.	7,700	755	156	8,611	9,875
0300 11.5 kW		.71	28.169		10,900	880	182	11,962	13,600
0400 20 kW		.63	31.746		12,900	995	205	14,100	16,000
0500 35 kW		.55	36.364		15,300	1,125	235	16,660	18,900

26 36 Transfer Switches

26 36 23 – Automatic Transfer Switches

26 36 23.10 Automatic Transfer Switch Devices

	Crew	Daily Output	Labor-Hours	Unit	Material	Labor	Equipment	Total	Total Incl O&P
0010 **AUTOMATIC TRANSFER SWITCH DEVICES**									
0100 Switches, enclosed 480 volt, 3 pole, 30 amp	1 Elec	2.30	3.478	Ea.	3,275	110		3,385	3,775
0200 60 amp	"	1.90	4.211	"	3,275	133		3,408	3,825

26 41 Facility Lightning Protection

26 41 13 – Lightning Protection for Structures

26 41 13.13 Lightning Protection for Buildings

		Crew	Daily Output	Labor-Hours	Unit	Material	2009 Bare Costs Labor	Equipment	Total	Total Incl O&P
0010	**LIGHTNING PROTECTION FOR BUILDINGS**									
0200	Air terminals & base, copper									
0400	3/8" diameter x 10" (to 75' high)	1 Elec	8	1	Ea.	23	31.50		54.50	77
1000	Aluminum, 1/2" diameter x 12" (to 75' high)		8	1	"	3.53	31.50		35.03	55.50
2000	Cable, copper, 220 lb. per thousand ft. (to 75' high)		320	.025	L.F.	2.73	.79		3.52	4.28
2500	Aluminum, 101 lb. per thousand ft. (to 75' high)		280	.029	"	.82	.90		1.72	2.37
3000	Arrester, 175 volt AC to ground		8	1	Ea.	59	31.50		90.50	117

26 51 Interior Lighting

26 51 13 – Interior Lighting Fixtures, Lamps, and Ballasts

26 51 13.50 Interior Lighting Fixtures

		Crew	Daily Output	Labor-Hours	Unit	Material	2009 Bare Costs Labor	Equipment	Total	Total Incl O&P
0010	**INTERIOR LIGHTING FIXTURES** Including lamps, mounting									
0030	hardware and connections									
0100	Fluorescent, C.W. lamps, troffer, recess mounted in grid, RS									
0130	grid ceiling mount									
0200	Acrylic lens, 1'W x 4'L, two 40 watt	1 Elec	5.70	1.404	Ea.	53	44		97	130
0300	2'W x 2'L, two U40 watt		5.70	1.404		56	44		100	134
0600	2'W x 4'L, four 40 watt		4.70	1.702		64	53.50		117.50	158
1000	Surface mounted, RS									
1030	Acrylic lens with hinged & latched door frame									
1100	1'W x 4'L, two 40 watt	1 Elec	7	1.143	Ea.	74	36		110	140
1200	2'W x 2'L, two U40 watt		7	1.143		79.50	36		115.50	146
1500	2'W x 4'L, four 40 watt		5.30	1.509		94	47.50		141.50	181
2100	Strip fixture									
2200	4' long, one 40 watt, RS	1 Elec	8.50	.941	Ea.	31	29.50		60.50	82.50
2300	4' long, two 40 watt, RS	"	8	1		33.50	31.50		65	88.50
2600	8' long, one 75 watt, SL	2 Elec	13.40	1.194		46.50	37.50		84	113
2700	8' long, two 75 watt, SL	"	12.40	1.290		56	40.50		96.50	128
4450	Incandescent, high hat can, round alzak reflector, prewired									
4470	100 watt	1 Elec	8	1	Ea.	69	31.50		100.50	128
4480	150 watt	"	8	1	"	102	31.50		133.50	164
5200	Ceiling, surface mounted, opal glass drum									
5300	8", one 60 watt lamp	1 Elec	10	.800	Ea.	43.50	25		68.50	89
5400	10", two 60 watt lamps		8	1		49	31.50		80.50	106
5500	12", four 60 watt lamps		6.70	1.194		67	37.50		104.50	135
6900	Mirror light, fluorescent, RS, acrylic enclosure, two 40 watt		8	1		116	31.50		147.50	180
6910	One 40 watt		8	1		94	31.50		125.50	155
6920	One 20 watt		12	.667		74	21		95	116

26 51 13.70 Residential Fixtures

		Crew	Daily Output	Labor-Hours	Unit	Material	2009 Bare Costs Labor	Equipment	Total	Total Incl O&P
0010	**RESIDENTIAL FIXTURES**									
0400	Fluorescent, interior, surface, circline, 32 watt & 40 watt	1 Elec	20	.400	Ea.	91	12.60		103.60	121
0500	2' x 2', two U 40 watt		8	1		110	31.50		141.50	173
0700	Shallow under cabinet, two 20 watt		16	.500		47.50	15.75		63.25	78
0900	Wall mounted, 4'L, one 40 watt, with baffle		10	.800		135	25		160	190
2000	Incandescent, exterior lantern, wall mounted, 60 watt		16	.500		37	15.75		52.75	66
2100	Post light, 150W, with 7' post		4	2		117	63		180	232
2500	Lamp holder, weatherproof with 150W PAR		16	.500		23	15.75		38.75	50.50
2550	With reflector and guard		12	.667		58.50	21		79.50	98.50
2600	Interior pendent, globe with shade, 150 watt		20	.400		152	12.60		164.60	188

26 55 Special Purpose Lighting

26 55 59 – Display Lighting

26 55 59.10 Track Lighting		Crew	Daily Output	Labor-Hours	Unit	Material	2009 Bare Costs Labor	2009 Bare Costs Equipment	Total	Total Incl O&P
0010	**TRACK LIGHTING**									
0100	8' section	1 Elec	5.30	1.509	Ea.	81	47.50		128.50	167
0300	3 circuits, 4' section		6.70	1.194		68	37.50		105.50	137
0400	8' section		5.30	1.509		106	47.50		153.50	195
0500	12' section		4.40	1.818		176	57.50		233.50	288
1000	Feed kit, surface mounting		16	.500		12.20	15.75		27.95	39
1100	End cover		24	.333		4.75	10.50		15.25	22.50
1200	Feed kit, stem mounting, 1 circuit		16	.500		36.50	15.75		52.25	66
1300	3 circuit		16	.500		36.50	15.75		52.25	66
2000	Electrical joiner, for continuous runs, 1 circuit		32	.250		17.60	7.90		25.50	32
2100	3 circuit		32	.250		42.50	7.90		50.40	59.50
2200	Fixtures, spotlight, 75W PAR halogen		16	.500		105	15.75		120.75	142
2210	50W MR16 halogen		16	.500		128	15.75		143.75	167
3000	Wall washer, 250 watt tungsten halogen		16	.500		122	15.75		137.75	160
3100	Low voltage, 25/50 watt, 1 circuit		16	.500		123	15.75		138.75	161
3120	3 circuit		16	.500		127	15.75		142.75	166

26 56 Exterior Lighting

26 56 13 – Lighting Poles and Standards

26 56 13.10 Lighting Poles

		Crew	Daily Output	Labor-Hours	Unit	Material	Labor	Equipment	Total	Total Incl O&P
0010	**LIGHTING POLES**									
6420	Wood pole, 4-1/2" x 5-1/8", 8' high	1 Elec	6	1.333	Ea.	305	42		347	405
6440	12' high		5.70	1.404		435	44		479	550
6460	20' high		4	2		615	63		678	780

26 56 23 – Area Lighting

26 56 23.10 Exterior Fixtures

		Crew	Daily Output	Labor-Hours	Unit	Material	Labor	Equipment	Total	Total Incl O&P
0010	**EXTERIOR FIXTURES** With lamps									
0400	Quartz, 500 watt	1 Elec	5.30	1.509	Ea.	54	47.50		101.50	137
1100	Wall pack, low pressure sodium, 35 watt		4	2		234	63		297	360
1150	55 watt		4	2		278	63		341	410

26 56 26 – Landscape Lighting

26 56 26.20 Landscape Fixtures

		Crew	Daily Output	Labor-Hours	Unit	Material	Labor	Equipment	Total	Total Incl O&P
0010	**LANDSCAPE FIXTURES**									
7380	Landscape recessed uplight, incl. housing, ballast, transformer									
7390	& reflector, not incl. conduit, wire, trench									
7420	Incandescent, 250 watt	1 Elec	5	1.600	Ea.	550	50.50		600.50	685
7440	Quartz, 250 watt	"	5	1.600	"	525	50.50		575.50	655

26 56 33 – Walkway Lighting

26 56 33.10 Walkway Luminaire

		Crew	Daily Output	Labor-Hours	Unit	Material	Labor	Equipment	Total	Total Incl O&P
0010	**WALKWAY LUMINAIRE**									
6500	Bollard light, lamp & ballast, 42" high with polycarbonate lens									
7200	Incandescent, 150 watt	1 Elec	3	2.667	Ea.	540	84		624	730

26 61 Lighting Systems and Accessories

26 61 23 – Lamps Applications

26 61 23.10 Lamps		Crew	Daily Output	Labor-Hours	Unit	Material	2009 Bare Costs Labor	Equipment	Total	Total Incl O&P
0010	**LAMPS**									
0081	Fluorescent, rapid start, cool white, 2' long, 20 watt	1 Elec	100	.080	Ea.	3.34	2.52		5.86	7.75
0101	4' long, 40 watt		90	.089		3.64	2.80		6.44	8.55
1351	High pressure sodium, 70 watt		30	.267		45.50	8.40		53.90	63.50
1371	150 watt		30	.267		48.50	8.40		56.90	67

Division 27
Communications

27 41 Audio-Video Systems

27 41 19 – Portable Audio-Video Equipment

27 41 19.10 T.V. Systems	Crew	Daily Output	Labor-Hours	Unit	Material	2009 Bare Costs Labor	Equipment	Total	Total Incl O&P
0010 **T.V. SYSTEMS**, not including rough-in wires, cables & conduits									
0100 Master TV antenna system									
0200 VHF reception & distribution, 12 outlets	1 Elec	6	1.333	Outlet	203	42		245	292
0800 VHF & UHF reception & distribution, 12 outlets		6	1.333	"	202	42		244	291
5000 T.V. Antenna only, minimum		6	1.333	Ea.	44.50	42		86.50	118
5100 Maximum		4	2	"	187	63		250	310

Division 28
Electronic Safety and Security

28 16 Intrusion Detection

28 16 16 – Intrusion Detection Systems Infrastructure

28 16 16.50 Intrusion Detection		Crew	Daily Output	Labor-Hours	Unit	Material	2009 Bare Costs		Total	Total Incl O&P
							Labor	Equipment		
0010	**INTRUSION DETECTION**, not including wires & conduits									
0100	Burglar alarm, battery operated, mechanical trigger	1 Elec	4	2	Ea.	272	63		335	400
0200	Electrical trigger		4	2		325	63		388	460
0400	For outside key control, add		8	1		77	31.50		108.50	136
0600	For remote signaling circuitry, add		8	1		122	31.50		153.50	187
0800	Card reader, flush type, standard		2.70	2.963		910	93.50		1,003.50	1,150
1000	Multi-code		2.70	2.963		1,175	93.50		1,268.50	1,450
1200	Door switches, hinge switch		5.30	1.509		57.50	47.50		105	141
1400	Magnetic switch		5.30	1.509		67.50	47.50		115	152
2800	Ultrasonic motion detector, 12 volt		2.30	3.478		225	110		335	425
3000	Infrared photoelectric detector		2.30	3.478		186	110		296	380
3200	Passive infrared detector		2.30	3.478		278	110		388	485
3420	Switchmats, 30" x 5'		5.30	1.509		83	47.50		130.50	169
3440	30" x 25'		4	2		199	63		262	320
3460	Police connect panel		4	2		239	63		302	365
3480	Telephone dialer		5.30	1.509		375	47.50		422.50	495
3500	Alarm bell		4	2		76	63		139	187
3520	Siren		4	2		143	63		206	260

28 31 Fire Detection and Alarm

28 31 23 – Fire Detection and Alarm Annunciation Panels and Fire Stations

28 31 23.50 Alarm Panels and Devices

		Crew	Daily Output	Labor-Hours	Unit	Material	Labor	Equipment	Total	Total Incl O&P
0010	**ALARM PANELS AND DEVICES**, not including wires & conduits									
5600	Strobe and horn	1 Elec	5.30	1.509	Ea.	105	47.50		152.50	194
5800	Fire alarm horn		6.70	1.194		40.50	37.50		78	106
6600	Drill switch		8	1		96	31.50		127.50	157
6800	Master box		2.70	2.963		3,425	93.50		3,518.50	3,925
7800	Remote annunciator, 8 zone lamp		1.80	4.444		201	140		341	450
8000	12 zone lamp	2 Elec	2.60	6.154		345	194		539	695
8200	16 zone lamp	"	2.20	7.273		345	229		574	755
8400	Standpipe or sprinkler alarm, alarm device	1 Elec	8	1		139	31.50		170.50	204
8600	Actuating device	"	8	1		320	31.50		351.50	405

28 31 46 – Smoke Detection Sensors

28 31 46.50 Smoke Detectors

		Crew	Daily Output	Labor-Hours	Unit	Material	Labor	Equipment	Total	Total Incl O&P
0010	**SMOKE DETECTORS**									
5200	Smoke detector, ceiling type	1 Elec	6.20	1.290	Ea.	87.50	40.50		128	162

Division 31
Earthwork

31 05 Common Work Results for Earthwork

31 05 13 – Soils for Earthwork

31 05 13.10 Borrow	Crew	Daily Output	Labor-Hours	Unit	Material	2009 Bare Costs Labor	2009 Bare Costs Equipment	Total	Total Incl O&P
0010 **BORROW**									
0020 Spread, 200 H.P. dozer, no compaction, 2 mi. RT haul									
0200 Common borrow	B-15	600	.047	C.Y.	7.35	1.12	3.58	12.05	13.90

31 05 16 – Aggregates for Earthwork

31 05 16.10 Borrow	Crew	Daily Output	Labor-Hours	Unit	Material	2009 Bare Costs Labor	2009 Bare Costs Equipment	Total	Total Incl O&P
0010 **BORROW**									
0020 Spread, with 200 H.P. dozer, no compaction, 2 mi. RT haul									
0100 Bank run gravel	B-15	600	.047	C.Y.	27.50	1.12	3.58	32.20	36
0300 Crushed stone (1.40 tons per CY) , 1-1/2"		600	.047		43.50	1.12	3.58	48.20	54
0320 3/4"		600	.047		43.50	1.12	3.58	48.20	54
0340 1/2"		600	.047		37.50	1.12	3.58	42.20	47.50
0360 3/8"		600	.047		37	1.12	3.58	41.70	46.50
0400 Sand, washed, concrete		600	.047		34.50	1.12	3.58	39.20	44
0500 Dead or bank sand		600	.047		10.50	1.12	3.58	15.20	17.35

31 11 Clearing and Grubbing Land

31 11 10 – Clearing and Grubbing Land

31 11 10.10 Clear and Grub Site	Crew	Daily Output	Labor-Hours	Unit	Material	2009 Bare Costs Labor	2009 Bare Costs Equipment	Total	Total Incl O&P
0010 **CLEAR AND GRUB SITE**									
0020 Cut & chip light trees to 6" diam.	B-7	1	48	Acre		1,075	1,300	2,375	3,225
0150 Grub stumps and remove	B-30	2	12			296	965	1,261	1,550
0200 Cut & chip medium, trees to 12" diam.	B-7	.70	68.571			1,525	1,850	3,375	4,575
0250 Grub stumps and remove	B-30	1	24			590	1,925	2,515	3,100
0300 Cut & chip heavy, trees to 24" diam.	B-7	.30	160			3,550	4,300	7,850	10,700
0350 Grub stumps and remove	B-30	.50	48			1,175	3,875	5,050	6,225
0400 If burning is allowed, reduce cut & chip									40%

31 13 Selective Tree and Shrub Removal and Trimming

31 13 13 – Selective Tree and Shrub Removal

31 13 13.10 Selective Clearing	Crew	Daily Output	Labor-Hours	Unit	Material	2009 Bare Costs Labor	2009 Bare Costs Equipment	Total	Total Incl O&P
0010 **SELECTIVE CLEARING**									
0020 Clearing brush with brush saw	A-1C	.25	32	Acre		660	103	763	1,225
0100 By hand	1 Clab	.12	66.667			1,375		1,375	2,325
0300 With dozer, ball and chain, light clearing	B-11A	2	8			196	540	736	920
0400 Medium clearing	"	1.50	10.667			262	720	982	1,225

31 14 Earth Stripping and Stockpiling

31 14 13 – Soil Stripping and Stockpiling

31 14 13.23 Topsoil Stripping and Stockpiling	Crew	Daily Output	Labor-Hours	Unit	Material	2009 Bare Costs Labor	2009 Bare Costs Equipment	Total	Total Incl O&P
0010 **TOPSOIL STRIPPING AND STOCKPILING**									
1400 Loam or topsoil, remove and stockpile on site									
1420 6" deep, 200' haul	B-10B	865	.009	C.Y.		.26	1.25	1.51	1.81
1430 300' haul		520	.015			.44	2.08	2.52	3.01
1440 500' haul		225	.036			1.02	4.81	5.83	6.95
1450 Alternate method: 6" deep, 200' haul		5090	.002	S.Y.		.04	.21	.25	.30
1460 500' haul		1325	.006	"		.17	.82	.99	1.18

31 22 Grading

31 22 16 – Fine Grading

31 22 16.10 Finish Grading

		Crew	Daily Output	Labor-Hours	Unit	Material	2009 Bare Costs Labor	Equipment	Total	Total Incl O&P
0010	**FINISH GRADING**									
0012	Finish grading area to be paved with grader, small area	B-11L	400	.040	S.Y.		.98	1.38	2.36	3.14
0100	Large area	↓	2000	.008			.20	.28	.48	.63
0200	Grade subgrade for base course, roadways		3500	.005			.11	.16	.27	.36
1020	For large parking lots	B-32C	5000	.010			.24	.37	.61	.81
1050	For small irregular areas	"	2000	.024			.60	.93	1.53	2.02
1100	Fine grade for slab on grade, machine	B-11L	1040	.015			.38	.53	.91	1.21
1150	Hand grading	B-18	700	.034			.73	.06	.79	1.29
1200	Fine grade granular base for sidewalks and bikeways	B-62	1200	.020	↓		.45	.12	.57	.90
2550	Hand grade select gravel	2 Clab	60	.267	C.S.F.		5.50		5.50	9.25
3000	Hand grade select gravel, including compaction, 4" deep	B-18	555	.043	S.Y.		.92	.07	.99	1.63
3100	6" deep		400	.060			1.27	.10	1.37	2.26
3120	8" deep	↓	300	.080			1.70	.13	1.83	3.01
3300	Finishing grading slopes, gentle	B-11L	8900	.002			.04	.06	.10	.14
3310	Steep slopes	"	7100	.002	↓		.06	.08	.14	.18

31 23 Excavation and Fill

31 23 16 – Excavation

31 23 16.13 Excavating, Trench

		Crew	Daily Output	Labor-Hours	Unit	Material	2009 Bare Costs Labor	Equipment	Total	Total Incl O&P
0010	**EXCAVATING, TRENCH**									
0011	Or continuous footing									
0050	1' to 4' deep, 3/8 C.Y. excavator	B-11C	150	.107	B.C.Y.		2.62	1.96	4.58	6.50
0060	1/2 C.Y. excavator	B-11M	200	.080			1.96	1.72	3.68	5.15
0090	4' to 6' deep, 1/2 C.Y. excavator	"	200	.080			1.96	1.72	3.68	5.15
0100	5/8 C.Y. excavator	B-12Q	250	.064			1.60	2.07	3.67	4.93
0300	1/2 C.Y. excavator, truck mounted	B-12J	200	.080			2	4.79	6.79	8.55
1352	4' to 6' deep, 1/2 C.Y. excavator w/ trench box	B-13H	188	.085			2.12	5.70	7.82	9.80
1354	5/8 C.Y. excavator	"	235	.068			1.70	4.56	6.26	7.80
1400	By hand with pick and shovel 2' to 6' deep, light soil	1 Clab	8	1			20.50		20.50	35
1500	Heavy soil	"	4	2	↓		41		41	69.50
5020	Loam & Sandy clay with no sheeting or dewatering included									
5050	1' to 4' deep, 3/8 C.Y. tractor loader/backhoe	B-11C	162	.099	B.C.Y.		2.42	1.81	4.23	6.05
5060	1/2 C.Y. excavator	B-11M	216	.074			1.82	1.60	3.42	4.78
5080	4' to 6' deep, 1/2 C.Y. excavator	"	216	.074			1.82	1.60	3.42	4.78
5090	5/8 C.Y. excavator	B-12Q	276	.058			1.45	1.87	3.32	4.46
5130	1/2 C.Y. excavator, truck mounted	B-12J	216	.074			1.85	4.44	6.29	7.95
5352	4' to 6' deep, 1/2 C.Y. excavator w/trench box	B-13H	205	.078			1.95	5.25	7.20	9
5354	5/8 C.Y. excavator w/trench box	"	257	.062	↓		1.55	4.17	5.72	7.15
6020	Sand & gravel with no sheeting or dewatering included									
6050	1' to 4' deep, 3/8 C.Y. excavator	B-11C	165	.097	B.C.Y.		2.38	1.78	4.16	5.90
6060	1/2 C.Y. excavator	B-11M	220	.073			1.79	1.57	3.36	4.70
6080	4' to 6' deep, 1/2 C.Y. excavator	"	220	.073			1.79	1.57	3.36	4.70
6090	5/8 C.Y. excavator	B-12Q	275	.058			1.45	1.88	3.33	4.48
6130	1/2 C.Y. excavator, truck mounted	B-12J	220	.073			1.81	4.36	6.17	7.80
6352	4' to 6' deep, 1/2 C.Y. excavator w/ trench box	B-13H	209	.077			1.91	5.15	7.06	8.85
6354	5/8 C.Y. excavator w/ trench box	"	261	.061	↓		1.53	4.11	5.64	7.05
7020	Dense hard clay with no sheeting or dewatering included									
7050	1' to 4' deep, 3/8 C.Y. excavator	B-11C	132	.121	B.C.Y.		2.98	2.23	5.21	7.40
7060	1/2 C.Y. excavator	B-11M	176	.091			2.23	1.96	4.19	5.85
7080	4' to 6' deep, 1/2 C.Y. excavator	"	176	.091			2.23	1.96	4.19	5.85
7090	5/8 C.Y. excavator	B-12Q	220	.073	↓		1.81	2.35	4.16	5.60

31 23 Excavation and Fill

31 23 16 – Excavation

31 23 16.13 Excavating, Trench	Crew	Daily Output	Labor-Hours	Unit	Material	2009 Bare Costs Labor	Equipment	Total	Total Incl O&P
7130 1/2 C.Y. excavator, truck mounted	B-12J	176	.091	B.C.Y.		2.27	5.45	7.72	9.75

31 23 16.14 Excavating, Utility Trench

	Crew	Daily Output	Labor-Hours	Unit	Material	Labor	Equipment	Total	Total Incl O&P
0010 **EXCAVATING, UTILITY TRENCH**									
0011 Common earth									
0050 Trenching with chain trencher, 12 H.P., operator walking									
0100 4" wide trench, 12" deep	B-53	800	.010	L.F.		.21	.07	.28	.43
1000 Backfill by hand including compaction, add									
1050 4" wide trench, 12" deep	A-1G	800	.010	L.F.		.21	.06	.27	.42

31 23 16.16 Structural Excavation for Minor Structures

	Crew	Daily Output	Labor-Hours	Unit	Material	Labor	Equipment	Total	Total Incl O&P
0010 **STRUCTURAL EXCAVATION FOR MINOR STRUCTURES**									
0015 Hand, pits to 6' deep, sandy soil	1 Clab	8	1	B.C.Y.		20.50		20.50	35
0100 Heavy soil or clay		4	2			41		41	69.50
1100 Hand loading trucks from stock pile, sandy soil		12	.667			13.70		13.70	23
1300 Heavy soil or clay	↓	8	1	↓		20.50		20.50	35
1500 For wet or muck hand excavation, add to above								50%	50%

31 23 16.42 Excavating, Bulk Bank Measure

	Crew	Daily Output	Labor-Hours	Unit	Material	Labor	Equipment	Total	Total Incl O&P
0010 **EXCAVATING, BULK BANK MEASURE**									
0011 Common earth piled									
0020 For loading onto trucks, add								15%	15%
0200 Excavator, hydraulic, crawler mtd., 1 C.Y. cap. = 100 C.Y./hr.	B-12A	800	.020	B.C.Y.		.50	.84	1.34	1.75
0310 Wheel mounted, 1/2 C.Y. cap. = 40 C.Y./hr.	B-12E	320	.050			1.25	1.23	2.48	3.43
1200 Front end loader, track mtd., 1-1/2 C.Y. cap. = 70 C.Y./hr.	B-10N	560	.014			.41	.66	1.07	1.39
1500 Wheel mounted, 3/4 C.Y. cap. = 45 C.Y./hr.	B-10R	360	.022	↓		.63	.66	1.29	1.77
5000 Excavating, bulk bank measure, sandy clay & loam piled									
5020 For loading onto trucks, add								15%	15%
5100 Excavator, hydraulic, crawler mtd., 1 C.Y. cap. = 120 C.Y./hr.	B-12A	960	.017	B.C.Y.		.42	.70	1.12	1.46
5610 Wheel mounted, 1/2 C.Y. cap. = 44 C.Y./hr.	B-12E	352	.045	"		1.13	1.12	2.25	3.12
8000 For hauling excavated material, see Div. 31 23 23.20									

31 23 23 – Fill

31 23 23.13 Backfill

	Crew	Daily Output	Labor-Hours	Unit	Material	Labor	Equipment	Total	Total Incl O&P
0010 **BACKFILL**									
0015 By hand, no compaction, light soil	1 Clab	14	.571	L.C.Y.		11.75		11.75	19.85
0100 Heavy soil		11	.727	"		14.95		14.95	25.50
0300 Compaction in 6" layers, hand tamp, add to above	↓	20.60	.388	E.C.Y.		8		8	13.50
0500 Air tamp, add	B-9D	190	.211			4.41	1.17	5.58	8.75
0600 Vibrating plate, add	A-1D	60	.133			2.74	.53	3.27	5.20
0800 Compaction in 12" layers, hand tamp, add to above	1 Clab	34	.235	↓		4.84		4.84	8.20
1300 Dozer backfilling, bulk, up to 300' haul, no compaction	B-10B	1200	.007	L.C.Y.		.19	.90	1.09	1.30
1400 Air tamped, add	B-11B	80	.200	E.C.Y.		4.75	3.10	7.85	11.30

31 23 23.16 Fill By Borrow and Utility Bedding

	Crew	Daily Output	Labor-Hours	Unit	Material	Labor	Equipment	Total	Total Incl O&P
0010 **FILL BY BORROW AND UTILITY BEDDING**									
0049 Utility bedding, for pipe & conduit, not incl. compaction									
0050 Crushed or screened bank run gravel	B-6	150	.160	L.C.Y.	31	3.63	1.96	36.59	42
0100 Crushed stone 3/4" to 1/2"		150	.160		43.50	3.63	1.96	49.09	56
0200 Sand, dead or bank	↓	150	.160	↓	10.50	3.63	1.96	16.09	19.75
0500 Compacting bedding in trench	A-1D	90	.089	E.C.Y.		1.83	.35	2.18	3.48
0600 If material source exceeds 2 miles, add for extra mileage.									
0610 See Div. 31 23 23 .20 for hauling mileage add.									

31 23 23.17 General Fill

	Crew	Daily Output	Labor-Hours	Unit	Material	Labor	Equipment	Total	Total Incl O&P
0010 **GENERAL FILL**									
0011 Spread dumped material, no compaction									

31 23 Excavation and Fill

31 23 23 – Fill

31 23 23.17 General Fill

		Crew	Daily Output	Labor-Hours	Unit	Material	2009 Bare Costs Labor	2009 Bare Costs Equipment	Total	Total Incl O&P
0020	By dozer, no compaction	B-10B	1000	.008	L.C.Y.		.23	1.08	1.31	1.57
0100	By hand	1 Clab	12	.667	"		13.70		13.70	23
0500	Gravel fill, compacted, under floor slabs, 4" deep	B-37	10000	.005	S.F.	.25	.11	.01	.37	.48
0600	6" deep		8600	.006		.38	.12	.02	.52	.65
0700	9" deep		7200	.007		.63	.15	.02	.80	.96
0800	12" deep		6000	.008		.88	.18	.02	1.08	1.30
1000	Alternate pricing method, 4" deep		120	.400	E.C.Y.	18.90	8.80	1.15	28.85	37
1100	6" deep		160	.300		18.90	6.60	.86	26.36	33
1200	9" deep		200	.240		18.90	5.25	.69	24.84	30.50
1300	12" deep		220	.218		18.90	4.79	.63	24.32	29.50

31 23 23.20 Hauling

		Crew	Daily Output	Labor-Hours	Unit	Material	2009 Bare Costs Labor	2009 Bare Costs Equipment	Total	Total Incl O&P
0010	**HAULING**									
0011	Excavated or borrow, loose cubic yards									
0012	no loading equipment, including hauling, waiting, loading/dumping									
0013	time per cycle (wait, load, travel, unload or dump & return)									
0014	8 CY truck, 15 MPH ave, cycle 0.5 miles, 10 min. wait/Ld./Uld.	B-34A	320	.025	L.C.Y.		.57	1.02	1.59	2.07
0016	cycle 1 mile		272	.029			.67	1.20	1.87	2.44
0018	cycle 2 miles		208	.038			.87	1.57	2.44	3.20
0020	cycle 4 miles		144	.056			1.26	2.27	3.53	4.61
0022	cycle 6 miles		112	.071			1.62	2.91	4.53	5.95
0024	cycle 8 miles		88	.091			2.06	3.71	5.77	7.55
0026	20 MPH ave, cycle 0.5 mile		336	.024			.54	.97	1.51	1.98
0028	cycle 1 mile		296	.027			.61	1.10	1.71	2.24
0030	cycle 2 miles		240	.033			.76	1.36	2.12	2.77
0032	cycle 4 miles		176	.045			1.03	1.85	2.88	3.77
0034	cycle 6 miles		136	.059			1.34	2.40	3.74	4.88
0036	cycle 8 miles		112	.071			1.62	2.91	4.53	5.95
0044	25 MPH ave, cycle 4 miles		192	.042			.95	1.70	2.65	3.46
0046	cycle 6 miles		160	.050			1.14	2.04	3.18	4.15
0048	cycle 8 miles		128	.063			1.42	2.55	3.97	5.20
0050	30 MPH ave, cycle 4 miles		216	.037			.84	1.51	2.35	3.07
0052	cycle 6 miles		176	.045			1.03	1.85	2.88	3.77
0054	cycle 8 miles		144	.056			1.26	2.27	3.53	4.61
0114	15 MPH ave, cycle 0.5 mile, 15 min. wait/Ld./Uld.		224	.036			.81	1.46	2.27	2.96
0116	cycle 1 mile		200	.040			.91	1.63	2.54	3.32
0118	cycle 2 miles		168	.048			1.08	1.94	3.02	3.95
0120	cycle 4 miles		120	.067			1.51	2.72	4.23	5.55
0122	cycle 6 miles		96	.083			1.89	3.40	5.29	6.90
0124	cycle 8 miles		80	.100			2.27	4.08	6.35	8.30
0126	20 MPH ave, cycle 0.5 mile		232	.034			.78	1.41	2.19	2.86
0128	cycle 1 mile		208	.038			.87	1.57	2.44	3.20
0130	cycle 2 miles		184	.043			.99	1.77	2.76	3.61
0132	cycle 4 miles		144	.056			1.26	2.27	3.53	4.61
0134	cycle 6 miles		112	.071			1.62	2.91	4.53	5.95
0136	cycle 8 miles		96	.083			1.89	3.40	5.29	6.90
0144	25 MPH ave, cycle 4 miles		152	.053			1.19	2.15	3.34	4.37
0146	cycle 6 miles		128	.063			1.42	2.55	3.97	5.20
0148	cycle 8 miles		112	.071			1.62	2.91	4.53	5.95
0150	30 MPH ave, cycle 4 miles		168	.048			1.08	1.94	3.02	3.95
0152	cycle 6 miles		144	.056			1.26	2.27	3.53	4.61
0154	cycle 8 miles		120	.067			1.51	2.72	4.23	5.55
0214	15 MPH ave, cycle 0.5 mile, 20 min wait/Ld./Uld.		176	.045			1.03	1.85	2.88	3.77

31 23 23 – Fill

31 23 23.20 Hauling		Crew	Daily Output	Labor-Hours	Unit	Material	2009 Bare Costs		Total	Total Incl O&P
							Labor	Equipment		
0216	cycle 1 mile	B-34A	160	.050	L.C.Y.		1.14	2.04	3.18	4.15
0218	cycle 2 miles		136	.059			1.34	2.40	3.74	4.88
0220	cycle 4 miles		104	.077			1.75	3.14	4.89	6.40
0222	cycle 6 miles		88	.091			2.06	3.71	5.77	7.55
0224	cycle 8 miles		72	.111			2.52	4.53	7.05	9.20
0226	20 MPH ave, cycle 0.5 mile		176	.045			1.03	1.85	2.88	3.77
0228	cycle 1 mile		168	.048			1.08	1.94	3.02	3.95
0230	cycle 2 miles		144	.056			1.26	2.27	3.53	4.61
0232	cycle 4 miles		120	.067			1.51	2.72	4.23	5.55
0234	cycle 6 miles		96	.083			1.89	3.40	5.29	6.90
0236	cycle 8 miles		88	.091			2.06	3.71	5.77	7.55
0244	25 MPH ave, cycle 4 miles		128	.063			1.42	2.55	3.97	5.20
0246	cycle 6 miles		112	.071			1.62	2.91	4.53	5.95
0248	cycle 8 miles		96	.083			1.89	3.40	5.29	6.90
0250	30 MPH ave, cycle 4 miles		136	.059			1.34	2.40	3.74	4.88
0252	cycle 6 miles		120	.067			1.51	2.72	4.23	5.55
0254	cycle 8 miles		104	.077			1.75	3.14	4.89	6.40
0314	15 MPH ave, cycle 0.5 mile, 25 min wait/Ld./Uld.		144	.056			1.26	2.27	3.53	4.61
0316	cycle 1 mile		128	.063			1.42	2.55	3.97	5.20
0318	cycle 2 miles		112	.071			1.62	2.91	4.53	5.95
0320	cycle 4 miles		96	.083			1.89	3.40	5.29	6.90
0322	cycle 6 miles		80	.100			2.27	4.08	6.35	8.30
0324	cycle 8 miles		64	.125			2.84	5.10	7.94	10.35
0326	20 MPH ave, cycle 0.5 mile		144	.056			1.26	2.27	3.53	4.61
0328	cycle 1 mile		136	.059			1.34	2.40	3.74	4.88
0330	cycle 2 miles		120	.067			1.51	2.72	4.23	5.55
0332	cycle 4 miles		104	.077			1.75	3.14	4.89	6.40
0334	cycle 6 miles		88	.091			2.06	3.71	5.77	7.55
0336	cycle 8 miles		80	.100			2.27	4.08	6.35	8.30
0344	25 MPH ave, cycle 4 miles		112	.071			1.62	2.91	4.53	5.95
0346	cycle 6 miles		96	.083			1.89	3.40	5.29	6.90
0348	cycle 8 miles		88	.091			2.06	3.71	5.77	7.55
0350	30 MPH ave, cycle 4 miles		112	.071			1.62	2.91	4.53	5.95
0352	cycle 6 miles		104	.077			1.75	3.14	4.89	6.40
0354	cycle 8 miles		96	.083			1.89	3.40	5.29	6.90
0414	15 MPH ave, cycle 0.5 mile, 30 min wait/Ld./Uld.		120	.067			1.51	2.72	4.23	5.55
0416	cycle 1 mile		112	.071			1.62	2.91	4.53	5.95
0418	cycle 2 miles		96	.083			1.89	3.40	5.29	6.90
0420	cycle 4 miles		80	.100			2.27	4.08	6.35	8.30
0422	cycle 6 miles		72	.111			2.52	4.53	7.05	9.20
0424	cycle 8 miles		64	.125			2.84	5.10	7.94	10.35
0426	20 MPH ave, cycle 0.5 mile		120	.067			1.51	2.72	4.23	5.55
0428	cycle 1 mile		112	.071			1.62	2.91	4.53	5.95
0430	cycle 2 miles		104	.077			1.75	3.14	4.89	6.40
0432	cycle 4 miles		88	.091			2.06	3.71	5.77	7.55
0434	cycle 6 miles		80	.100			2.27	4.08	6.35	8.30
0436	cycle 8 miles		72	.111			2.52	4.53	7.05	9.20
0444	25 MPH ave, cycle 4 miles		96	.083			1.89	3.40	5.29	6.90
0446	cycle 6 miles		88	.091			2.06	3.71	5.77	7.55
0448	cycle 8 miles		80	.100			2.27	4.08	6.35	8.30
0450	30 MPH ave, cycle 4 miles		96	.083			1.89	3.40	5.29	6.90
0452	cycle 6 miles		88	.091			2.06	3.71	5.77	7.55
0454	cycle 8 miles		80	.100			2.27	4.08	6.35	8.30

31 23 23.20 Hauling	Crew	Daily Output	Labor-Hours	Unit	Material	2009 Bare Costs Labor	2009 Bare Costs Equipment	Total	Total Incl O&P
0514	15 MPH ave, cycle 0.5 mile, 35 min wait/Ld./Uld.	B-34A	104	.077	L.C.Y.		1.75	3.14	4.89	6.40
0516	cycle 1 mile		96	.083			1.89	3.40	5.29	6.90
0518	cycle 2 miles		88	.091			2.06	3.71	5.77	7.55
0520	cycle 4 miles		72	.111			2.52	4.53	7.05	9.20
0522	cycle 6 miles		64	.125			2.84	5.10	7.94	10.35
0524	cycle 8 miles		56	.143			3.24	5.85	9.09	11.85
0526	20 MPH ave, cycle 0.5 mile		104	.077			1.75	3.14	4.89	6.40
0528	cycle 1 mile		96	.083			1.89	3.40	5.29	6.90
0530	cycle 2 miles		96	.083			1.89	3.40	5.29	6.90
0532	cycle 4 miles		80	.100			2.27	4.08	6.35	8.30
0534	cycle 6 miles		72	.111			2.52	4.53	7.05	9.20
0536	cycle 8 miles		64	.125			2.84	5.10	7.94	10.35
0544	25 MPH ave, cycle 4 miles		88	.091			2.06	3.71	5.77	7.55
0546	cycle 6 miles		80	.100			2.27	4.08	6.35	8.30
0548	cycle 8 miles		72	.111			2.52	4.53	7.05	9.20
0550	30 MPH ave, cycle 4 miles		88	.091			2.06	3.71	5.77	7.55
0552	cycle 6 miles		80	.100			2.27	4.08	6.35	8.30
0554	cycle 8 miles		72	.111			2.52	4.53	7.05	9.20
1014	12 CY truck, cycle 0.5 mile, 15 MPH ave, 15 min. wait/Ld./Uld.	B-34B	336	.024			.54	1.59	2.13	2.66
1016	cycle 1 mile		300	.027			.61	1.78	2.39	2.97
1018	cycle 2 miles		252	.032			.72	2.12	2.84	3.54
1020	cycle 4 miles		180	.044			1.01	2.96	3.97	4.95
1022	cycle 6 miles		144	.056			1.26	3.70	4.96	6.20
1024	cycle 8 miles		120	.067			1.51	4.44	5.95	7.45
1025	cycle 10 miles		96	.083			1.89	5.55	7.44	9.25
1026	20 MPH ave, cycle 0.5 mile		348	.023			.52	1.53	2.05	2.56
1028	cycle 1 mile		312	.026			.58	1.71	2.29	2.86
1030	cycle 2 miles		276	.029			.66	1.93	2.59	3.22
1032	cycle 4 miles		216	.037			.84	2.47	3.31	4.12
1034	cycle 6 miles		168	.048			1.08	3.17	4.25	5.30
1036	cycle 8 miles		144	.056			1.26	3.70	4.96	6.20
1038	cycle 10 miles		120	.067			1.51	4.44	5.95	7.45
1040	25 MPH ave, cycle 4 miles		228	.035			.80	2.34	3.14	3.91
1042	cycle 6 miles		192	.042			.95	2.78	3.73	4.64
1044	cycle 8 miles		168	.048			1.08	3.17	4.25	5.30
1046	cycle 10 miles		144	.056			1.26	3.70	4.96	6.20
1050	30 MPH ave, cycle 4 miles		252	.032			.72	2.12	2.84	3.54
1052	cycle 6 miles		216	.037			.84	2.47	3.31	4.12
1054	cycle 8 miles		180	.044			1.01	2.96	3.97	4.95
1056	cycle 10 miles		156	.051			1.16	3.42	4.58	5.70
1060	35 MPH ave, cycle 4 miles		264	.030			.69	2.02	2.71	3.37
1062	cycle 6 miles		228	.035			.80	2.34	3.14	3.91
1064	cycle 8 miles		204	.039			.89	2.61	3.50	4.36
1066	cycle 10 miles		180	.044			1.01	2.96	3.97	4.95
1068	cycle 20 miles		120	.067			1.51	4.44	5.95	7.45
1069	cycle 30 miles		84	.095			2.16	6.35	8.51	10.65
1070	cycle 40 miles		72	.111			2.52	7.40	9.92	12.40
1072	40 MPH ave, cycle 6 miles		240	.033			.76	2.22	2.98	3.71
1074	cycle 8 miles		216	.037			.84	2.47	3.31	4.12
1076	cycle 10 miles		192	.042			.95	2.78	3.73	4.64
1078	cycle 20 miles		120	.067			1.51	4.44	5.95	7.45
1080	cycle 30 miles		96	.083			1.89	5.55	7.44	9.25
1082	cycle 40 miles		72	.111			2.52	7.40	9.92	12.40

31 23 23.20 **Hauling**	Crew	Daily Output	Labor-Hours	Unit	Material	Labor	Equipment	Total	Total Incl O&P
						2009 Bare Costs			
1084	cycle 50 miles	B-34B	60	.133	L.C.Y.		3.03	8.90	11.93	14.85
1094	45 MPH ave, cycle 8 miles		216	.037			.84	2.47	3.31	4.12
1096	cycle 10 miles		204	.039			.89	2.61	3.50	4.36
1098	cycle 20 miles		132	.061			1.38	4.04	5.42	6.75
1100	cycle 30 miles		108	.074			1.68	4.94	6.62	8.25
1102	cycle 40 miles		84	.095			2.16	6.35	8.51	10.65
1104	cycle 50 miles		72	.111			2.52	7.40	9.92	12.40
1106	50 MPH ave, cycle 10 miles		216	.037			.84	2.47	3.31	4.12
1108	cycle 20 miles		144	.056			1.26	3.70	4.96	6.20
1110	cycle 30 miles		108	.074			1.68	4.94	6.62	8.25
1112	cycle 40 miles		84	.095			2.16	6.35	8.51	10.65
1114	cycle 50 miles		72	.111			2.52	7.40	9.92	12.40
1214	15 MPH ave, cycle 0.5 mile, 20 min. wait/Ld./Uld.		264	.030			.69	2.02	2.71	3.37
1216	cycle 1 mile		240	.033			.76	2.22	2.98	3.71
1218	cycle 2 miles		204	.039			.89	2.61	3.50	4.36
1220	cycle 4 miles		156	.051			1.16	3.42	4.58	5.70
1222	cycle 6 miles		132	.061			1.38	4.04	5.42	6.75
1224	cycle 8 miles		108	.074			1.68	4.94	6.62	8.25
1225	cycle 10 miles		96	.083			1.89	5.55	7.44	9.25
1226	20 MPH ave, cycle 0.5 mile		264	.030			.69	2.02	2.71	3.37
1228	cycle 1 mile		252	.032			.72	2.12	2.84	3.54
1230	cycle 2 miles		216	.037			.84	2.47	3.31	4.12
1232	cycle 4 miles		180	.044			1.01	2.96	3.97	4.95
1234	cycle 6 miles		144	.056			1.26	3.70	4.96	6.20
1236	cycle 8 miles		132	.061			1.38	4.04	5.42	6.75
1238	cycle 10 miles		108	.074			1.68	4.94	6.62	8.25
1240	25 MPH ave, cycle 4 miles		192	.042			.95	2.78	3.73	4.64
1242	cycle 6 miles		168	.048			1.08	3.17	4.25	5.30
1244	cycle 8 miles		144	.056			1.26	3.70	4.96	6.20
1246	cycle 10 miles		132	.061			1.38	4.04	5.42	6.75
1250	30 MPH ave, cycle 4 miles		204	.039			.89	2.61	3.50	4.36
1252	cycle 6 miles		180	.044			1.01	2.96	3.97	4.95
1254	cycle 8 miles		156	.051			1.16	3.42	4.58	5.70
1256	cycle 10 miles		144	.056			1.26	3.70	4.96	6.20
1260	35 MPH ave, cycle 4 miles		216	.037			.84	2.47	3.31	4.12
1262	cycle 6 miles		192	.042			.95	2.78	3.73	4.64
1264	cycle 8 miles		168	.048			1.08	3.17	4.25	5.30
1266	cycle 10 miles		156	.051			1.16	3.42	4.58	5.70
1268	cycle 20 miles		108	.074			1.68	4.94	6.62	8.25
1269	cycle 30 miles		72	.111			2.52	7.40	9.92	12.40
1270	cycle 40 miles		60	.133			3.03	8.90	11.93	14.85
1272	40 MPH ave, cycle 6 miles		192	.042			.95	2.78	3.73	4.64
1274	cycle 8 miles		180	.044			1.01	2.96	3.97	4.95
1276	cycle 10 miles		156	.051			1.16	3.42	4.58	5.70
1278	cycle 20 miles		108	.074			1.68	4.94	6.62	8.25
1280	cycle 30 miles		84	.095			2.16	6.35	8.51	10.65
1282	cycle 40 miles		72	.111			2.52	7.40	9.92	12.40
1284	cycle 50 miles		60	.133			3.03	8.90	11.93	14.85
1294	45 MPH ave, cycle 8 miles		180	.044			1.01	2.96	3.97	4.95
1296	cycle 10 miles		168	.048			1.08	3.17	4.25	5.30
1298	cycle 20 miles		120	.067			1.51	4.44	5.95	7.45
1300	cycle 30 miles		96	.083			1.89	5.55	7.44	9.25
1302	cycle 40 miles		72	.111			2.52	7.40	9.92	12.40

31 23 Excavation and Fill

31 23 23 – Fill

31 23 23.20 Hauling

		Crew	Daily Output	Labor-Hours	Unit	Material	2009 Bare Costs Labor	2009 Bare Costs Equipment	Total	Total Incl O&P
1304	cycle 50 miles	B-34B	60	.133	L.C.Y.		3.03	8.90	11.93	14.85
1306	50 MPH ave, cycle 10 miles		180	.044			1.01	2.96	3.97	4.95
1308	cycle 20 miles		132	.061			1.38	4.04	5.42	6.75
1310	cycle 30 miles		96	.083			1.89	5.55	7.44	9.25
1312	cycle 40 miles		84	.095			2.16	6.35	8.51	10.65
1314	cycle 50 miles		72	.111			2.52	7.40	9.92	12.40
1414	15 MPH ave, cycle 0.5 mile, 25 min. wait/Ld./Uld.		204	.039			.89	2.61	3.50	4.36
1416	cycle 1 mile		192	.042			.95	2.78	3.73	4.64
1418	cycle 2 miles		168	.048			1.08	3.17	4.25	5.30
1420	cycle 4 miles		132	.061			1.38	4.04	5.42	6.75
1422	cycle 6 miles		120	.067			1.51	4.44	5.95	7.45
1424	cycle 8 miles		96	.083			1.89	5.55	7.44	9.25
1425	cycle 10 miles		84	.095			2.16	6.35	8.51	10.65
1426	20 MPH ave, cycle 0.5 mile		216	.037			.84	2.47	3.31	4.12
1428	cycle 1 mile		204	.039			.89	2.61	3.50	4.36
1430	cycle 2 miles		180	.044			1.01	2.96	3.97	4.95
1432	cycle 4 miles		156	.051			1.16	3.42	4.58	5.70
1434	cycle 6 miles		132	.061			1.38	4.04	5.42	6.75
1436	cycle 8 miles		120	.067			1.51	4.44	5.95	7.45
1438	cycle 10 miles		96	.083			1.89	5.55	7.44	9.25
1440	25 MPH ave, cycle 4 miles		168	.048			1.08	3.17	4.25	5.30
1442	cycle 6 miles		144	.056			1.26	3.70	4.96	6.20
1444	cycle 8 miles		132	.061			1.38	4.04	5.42	6.75
1446	cycle 10 miles		108	.074			1.68	4.94	6.62	8.25
1450	30 MPH ave, cycle 4 miles		168	.048			1.08	3.17	4.25	5.30
1452	cycle 6 miles		156	.051			1.16	3.42	4.58	5.70
1454	cycle 8 miles		132	.061			1.38	4.04	5.42	6.75
1456	cycle 10 miles		120	.067			1.51	4.44	5.95	7.45
1460	35 MPH ave, cycle 4 miles		180	.044			1.01	2.96	3.97	4.95
1462	cycle 6 miles		156	.051			1.16	3.42	4.58	5.70
1464	cycle 8 miles		144	.056			1.26	3.70	4.96	6.20
1466	cycle 10 miles		132	.061			1.38	4.04	5.42	6.75
1468	cycle 20 miles		96	.083			1.89	5.55	7.44	9.25
1469	cycle 30 miles		72	.111			2.52	7.40	9.92	12.40
1470	cycle 40 miles		60	.133			3.03	8.90	11.93	14.85
1472	40 MPH ave, cycle 6 miles		168	.048			1.08	3.17	4.25	5.30
1474	cycle 8 miles		156	.051			1.16	3.42	4.58	5.70
1476	cycle 10 miles		144	.056			1.26	3.70	4.96	6.20
1478	cycle 20 miles		96	.083			1.89	5.55	7.44	9.25
1480	cycle 30 miles		84	.095			2.16	6.35	8.51	10.65
1482	cycle 40 miles		60	.133			3.03	8.90	11.93	14.85
1484	cycle 50 miles		60	.133			3.03	8.90	11.93	14.85
1494	45 MPH ave, cycle 8 miles		156	.051			1.16	3.42	4.58	5.70
1496	cycle 10 miles		144	.056			1.26	3.70	4.96	6.20
1498	cycle 20 miles		108	.074			1.68	4.94	6.62	8.25
1500	cycle 30 miles		84	.095			2.16	6.35	8.51	10.65
1502	cycle 40 miles		72	.111			2.52	7.40	9.92	12.40
1504	cycle 50 miles		60	.133			3.03	8.90	11.93	14.85
1506	50 MPH ave, cycle 10 miles		156	.051			1.16	3.42	4.58	5.70
1508	cycle 20 miles		120	.067			1.51	4.44	5.95	7.45
1510	cycle 30 miles		96	.083			1.89	5.55	7.44	9.25
1512	cycle 40 miles		72	.111			2.52	7.40	9.92	12.40
1514	cycle 50 miles		60	.133			3.03	8.90	11.93	14.85

31 23 23.20 Hauling		Crew	Daily Output	Labor-Hours	Unit	Material	2009 Bare Costs			Total Incl O&P
							Labor	Equipment	Total	
1614	15 MPH, cycle 0.5 mile, 30 min. wait/Ld./Uld.	B-34B	180	.044	L.C.Y.		1.01	2.96	3.97	4.95
1616	cycle 1 mile		168	.048			1.08	3.17	4.25	5.30
1618	cycle 2 miles		144	.056			1.26	3.70	4.96	6.20
1620	cycle 4 miles		120	.067			1.51	4.44	5.95	7.45
1622	cycle 6 miles		108	.074			1.68	4.94	6.62	8.25
1624	cycle 8 miles		84	.095			2.16	6.35	8.51	10.65
1625	cycle 10 miles		84	.095			2.16	6.35	8.51	10.65
1626	20 MPH ave, cycle 0.5 mile		180	.044			1.01	2.96	3.97	4.95
1628	cycle 1 mile		168	.048			1.08	3.17	4.25	5.30
1630	cycle 2 miles		156	.051			1.16	3.42	4.58	5.70
1632	cycle 4 miles		132	.061			1.38	4.04	5.42	6.75
1634	cycle 6 miles		120	.067			1.51	4.44	5.95	7.45
1636	cycle 8 miles		108	.074			1.68	4.94	6.62	8.25
1638	cycle 10 miles		96	.083			1.89	5.55	7.44	9.25
1640	25 MPH ave, cycle 4 miles		144	.056			1.26	3.70	4.96	6.20
1642	cycle 6 miles		132	.061			1.38	4.04	5.42	6.75
1644	cycle 8 miles		108	.074			1.68	4.94	6.62	8.25
1646	cycle 10 miles		108	.074			1.68	4.94	6.62	8.25
1650	30 MPH ave, cycle 4 miles		144	.056			1.26	3.70	4.96	6.20
1652	cycle 6 miles		132	.061			1.38	4.04	5.42	6.75
1654	cycle 8 miles		120	.067			1.51	4.44	5.95	7.45
1656	cycle 10 miles		108	.074			1.68	4.94	6.62	8.25
1660	35 MPH ave, cycle 4 miles		156	.051			1.16	3.42	4.58	5.70
1662	cycle 6 miles		144	.056			1.26	3.70	4.96	6.20
1664	cycle 8 miles		132	.061			1.38	4.04	5.42	6.75
1666	cycle 10 miles		120	.067			1.51	4.44	5.95	7.45
1668	cycle 20 miles		84	.095			2.16	6.35	8.51	10.65
1669	cycle 30 miles		72	.111			2.52	7.40	9.92	12.40
1670	cycle 40 miles		60	.133			3.03	8.90	11.93	14.85
1672	40 MPH, cycle 6 miles		144	.056			1.26	3.70	4.96	6.20
1674	cycle 8 miles		132	.061			1.38	4.04	5.42	6.75
1676	cycle 10 miles		120	.067			1.51	4.44	5.95	7.45
1678	cycle 20 miles		96	.083			1.89	5.55	7.44	9.25
1680	cycle 30 miles		72	.111			2.52	7.40	9.92	12.40
1682	cycle 40 miles		60	.133			3.03	8.90	11.93	14.85
1684	cycle 50 miles		48	.167			3.78	11.10	14.88	18.55
1694	45 MPH ave, cycle 8 miles		144	.056			1.26	3.70	4.96	6.20
1696	cycle 10 miles		132	.061			1.38	4.04	5.42	6.75
1698	cycle 20 miles		96	.083			1.89	5.55	7.44	9.25
1700	cycle 30 miles		84	.095			2.16	6.35	8.51	10.65
1702	cycle 40 miles		60	.133			3.03	8.90	11.93	14.85
1704	cycle 50 miles		60	.133			3.03	8.90	11.93	14.85
1706	50 MPH ave, cycle 10 miles		132	.061			1.38	4.04	5.42	6.75
1708	cycle 20 miles		108	.074			1.68	4.94	6.62	8.25
1710	cycle 30 miles		84	.095			2.16	6.35	8.51	10.65
1712	cycle 40 miles		72	.111			2.52	7.40	9.92	12.40
1714	cycle 50 miles		60	.133			3.03	8.90	11.93	14.85
2000	Hauling, 8 CY truck, small project cost per hour	B-34A	8	1	Hr.		22.50	41	63.50	83
2100	12 CY Truck	B-34B	8	1			22.50	66.50	89	112
2150	16.5 CY Truck	B-34C	8	1			22.50	74.50	97	120
2200	20 CY Truck	B-34D	8	1			22.50	76	98.50	122
3014	16.5 CY truck, 15 min. wait/Ld./Uld., 15 MPH, cycle 0.5 mile	B-34C	462	.017	L.C.Y.		.39	1.29	1.68	2.08
3016	cycle 1 mile		413	.019			.44	1.44	1.88	2.32

31 23 23 – Fill

31 23 23.20 Hauling		Crew	Daily Output	Labor-Hours	Unit	Material	2009 Bare Costs Labor	2009 Bare Costs Equipment	Total	Total Incl O&P
3018	cycle 2 miles	B-34C	347	.023	L.C.Y.		.52	1.71	2.23	2.76
3020	cycle 4 miles		248	.032			.73	2.40	3.13	3.87
3022	cycle 6 miles		198	.040			.92	3	3.92	4.84
3024	cycle 8 miles		165	.048			1.10	3.60	4.70	5.80
3025	cycle 10 miles		132	.061			1.38	4.50	5.88	7.25
3026	20 MPH ave, cycle 0.5 mile		479	.017			.38	1.24	1.62	2
3028	cycle 1 mile		429	.019			.42	1.39	1.81	2.23
3030	cycle 2 miles		380	.021			.48	1.56	2.04	2.52
3032	cycle 4 miles		281	.028			.65	2.12	2.77	3.41
3034	cycle 6 miles		231	.035			.79	2.57	3.36	4.15
3036	cycle 8 miles		198	.040			.92	3	3.92	4.84
3038	cycle 10 miles		165	.048			1.10	3.60	4.70	5.80
3040	25 MPH ave, cycle 4 miles		314	.025			.58	1.89	2.47	3.05
3042	cycle 6 miles		264	.030			.69	2.25	2.94	3.63
3044	cycle 8 miles		231	.035			.79	2.57	3.36	4.15
3046	cycle 10 miles		198	.040			.92	3	3.92	4.84
3050	30 MPH ave, cycle 4 miles		347	.023			.52	1.71	2.23	2.76
3052	cycle 6 miles		281	.028			.65	2.12	2.77	3.41
3054	cycle 8 miles		248	.032			.73	2.40	3.13	3.87
3056	cycle 10 miles		215	.037			.84	2.76	3.60	4.46
3060	35 MPH ave, cycle 4 miles		363	.022			.50	1.64	2.14	2.64
3062	cycle 6 miles		314	.025			.58	1.89	2.47	3.05
3064	cycle 8 miles		264	.030			.69	2.25	2.94	3.63
3066	cycle 10 miles		248	.032			.73	2.40	3.13	3.87
3068	cycle 20 miles		149	.054			1.22	3.99	5.21	6.45
3070	cycle 30 miles		116	.069			1.57	5.10	6.67	8.30
3072	cycle 40 miles		83	.096			2.19	7.15	9.34	11.55
3074	40 MPH ave, cycle 6 miles		330	.024			.55	1.80	2.35	2.90
3076	cycle 8 miles		281	.028			.65	2.12	2.77	3.41
3078	cycle 10 miles		264	.030			.69	2.25	2.94	3.63
3080	cycle 20 miles		165	.048			1.10	3.60	4.70	5.80
3082	cycle 30 miles		132	.061			1.38	4.50	5.88	7.25
3084	cycle 40 miles		99	.081			1.83	6	7.83	9.70
3086	cycle 50 miles		83	.096			2.19	7.15	9.34	11.55
3094	45 MPH ave, cycle 8 miles		297	.027			.61	2	2.61	3.23
3096	cycle 10 miles		281	.028			.65	2.12	2.77	3.41
3098	cycle 20 miles		182	.044			1	3.27	4.27	5.25
3100	cycle 30 miles		132	.061			1.38	4.50	5.88	7.25
3102	cycle 40 miles		116	.069			1.57	5.10	6.67	8.30
3104	cycle 50 miles		99	.081			1.83	6	7.83	9.70
3106	50 MPH ave, cycle 10 miles		281	.028			.65	2.12	2.77	3.41
3108	cycle 20 miles		198	.040			.92	3	3.92	4.84
3110	cycle 30 miles		149	.054			1.22	3.99	5.21	6.45
3112	cycle 40 miles		116	.069			1.57	5.10	6.67	8.30
3114	cycle 50 miles		99	.081			1.83	6	7.83	9.70
3214	20 min. wait/Ld./Uld., 15 MPH, cycle 0.5 mile		363	.022			.50	1.64	2.14	2.64
3216	cycle 1 mile		330	.024			.55	1.80	2.35	2.90
3218	cycle 2 miles		281	.028			.65	2.12	2.77	3.41
3220	cycle 4 miles		215	.037			.84	2.76	3.60	4.46
3222	cycle 6 miles		182	.044			1	3.27	4.27	5.25
3224	cycle 8 miles		149	.054			1.22	3.99	5.21	6.45
3225	cycle 10 miles		132	.061			1.38	4.50	5.88	7.25
3226	20 MPH ave, cycle 0.5 mile		363	.022			.50	1.64	2.14	2.64

31 23 23 – Fill

31 23 23.20 Hauling		Crew	Daily Output	Labor-Hours	Unit	Material	2009 Bare Costs		Total	Total Incl O&P
							Labor	Equipment		
3228	cycle 1 mile	B-34C	347	.023	L.C.Y.		.52	1.71	2.23	2.76
3230	cycle 2 miles		297	.027			.61	2	2.61	3.23
3232	cycle 4 miles		248	.032			.73	2.40	3.13	3.87
3234	cycle 6 miles		198	.040			.92	3	3.92	4.84
3236	cycle 8 miles		182	.044			1	3.27	4.27	5.25
3238	cycle 10 miles		149	.054			1.22	3.99	5.21	6.45
3240	25 MPH ave, cycle 4 miles		264	.030			.69	2.25	2.94	3.63
3242	cycle 6 miles		231	.035			.79	2.57	3.36	4.15
3244	cycle 8 miles		198	.040			.92	3	3.92	4.84
3246	cycle 10 miles		182	.044			1	3.27	4.27	5.25
3250	30 MPH ave, cycle 4 miles		281	.028			.65	2.12	2.77	3.41
3252	cycle 6 miles		248	.032			.73	2.40	3.13	3.87
3254	cycle 8 miles		215	.037			.84	2.76	3.60	4.46
3256	cycle 10 miles		198	.040			.92	3	3.92	4.84
3260	35 MPH ave, cycle 4 miles		297	.027			.61	2	2.61	3.23
3262	cycle 6 miles		264	.030			.69	2.25	2.94	3.63
3264	cycle 8 miles		231	.035			.79	2.57	3.36	4.15
3266	cycle 10 miles		215	.037			.84	2.76	3.60	4.46
3268	cycle 20 miles		149	.054			1.22	3.99	5.21	6.45
3270	cycle 30 miles		99	.081			1.83	6	7.83	9.70
3272	cycle 40 miles		83	.096			2.19	7.15	9.34	11.55
3274	40 MPH ave, cycle 6 miles		264	.030			.69	2.25	2.94	3.63
3276	cycle 8 miles		248	.032			.73	2.40	3.13	3.87
3278	cycle 10 miles		215	.037			.84	2.76	3.60	4.46
3280	cycle 20 miles		149	.054			1.22	3.99	5.21	6.45
3282	cycle 30 miles		116	.069			1.57	5.10	6.67	8.30
3284	cycle 40 miles		99	.081			1.83	6	7.83	9.70
3286	cycle 50 miles		83	.096			2.19	7.15	9.34	11.55
3294	45 MPH ave, cycle 8 miles		248	.032			.73	2.40	3.13	3.87
3296	cycle 10 miles		231	.035			.79	2.57	3.36	4.15
3298	cycle 20 miles		165	.048			1.10	3.60	4.70	5.80
3300	cycle 30 miles		132	.061			1.38	4.50	5.88	7.25
3302	cycle 40 miles		99	.081			1.83	6	7.83	9.70
3304	cycle 50 miles		83	.096			2.19	7.15	9.34	11.55
3306	50 MPH ave, cycle 10 miles		248	.032			.73	2.40	3.13	3.87
3308	cycle 20 miles		182	.044			1	3.27	4.27	5.25
3310	cycle 30 miles		132	.061			1.38	4.50	5.88	7.25
3312	cycle 40 miles		116	.069			1.57	5.10	6.67	8.30
3314	cycle 50 miles		99	.081			1.83	6	7.83	9.70
3414	25 min. wait/Ld./Uld., 15 MPH, cycle 0.5 mile		281	.028			.65	2.12	2.77	3.41
3416	cycle 1 mile		264	.030			.69	2.25	2.94	3.63
3418	cycle 2 miles		231	.035			.79	2.57	3.36	4.15
3420	cycle 4 miles		182	.044			1	3.27	4.27	5.25
3422	cycle 6 miles		165	.048			1.10	3.60	4.70	5.80
3424	cycle 8 miles		132	.061			1.38	4.50	5.88	7.25
3425	cycle 10 miles		116	.069			1.57	5.10	6.67	8.30
3426	20 MPH ave, cycle 0.5 mile		297	.027			.61	2	2.61	3.23
3428	cycle 1 mile		281	.028			.65	2.12	2.77	3.41
3430	cycle 2 miles		248	.032			.73	2.40	3.13	3.87
3432	cycle 4 miles		215	.037			.84	2.76	3.60	4.46
3434	cycle 6 miles		182	.044			1	3.27	4.27	5.25
3436	cycle 8 miles		165	.048			1.10	3.60	4.70	5.80
3438	cycle 10 miles		132	.061			1.38	4.50	5.88	7.25

31 23 23 – Fill

31 23 23.20 Hauling	Crew	Daily Output	Labor-Hours	Unit	Material	2009 Bare Costs Labor	2009 Bare Costs Equipment	Total	Total Incl O&P
3440	25 MPH ave, cycle 4 miles	B-34C	231	.035	L.C.Y.		.79	2.57	3.36	4.15
3442	cycle 6 miles		198	.040			.92	3	3.92	4.84
3444	cycle 8 miles		182	.044			1	3.27	4.27	5.25
3446	cycle 10 miles		165	.048			1.10	3.60	4.70	5.80
3450	30 MPH ave, cycle 4 miles		231	.035			.79	2.57	3.36	4.15
3452	cycle 6 miles		215	.037			.84	2.76	3.60	4.46
3454	cycle 8 miles		182	.044			1	3.27	4.27	5.25
3456	cycle 10 miles		165	.048			1.10	3.60	4.70	5.80
3460	35 MPH ave, cycle 4 miles		248	.032			.73	2.40	3.13	3.87
3462	cycle 6 miles		215	.037			.84	2.76	3.60	4.46
3464	cycle 8 miles		198	.040			.92	3	3.92	4.84
3466	cycle 10 miles		182	.044			1	3.27	4.27	5.25
3468	cycle 20 miles		132	.061			1.38	4.50	5.88	7.25
3470	cycle 30 miles		99	.081			1.83	6	7.83	9.70
3472	cycle 40 miles		83	.096			2.19	7.15	9.34	11.55
3474	40 MPH, cycle 6 miles		231	.035			.79	2.57	3.36	4.15
3476	cycle 8 miles		215	.037			.84	2.76	3.60	4.46
3478	cycle 10 miles		198	.040			.92	3	3.92	4.84
3480	cycle 20 miles		132	.061			1.38	4.50	5.88	7.25
3482	cycle 30 miles		116	.069			1.57	5.10	6.67	8.30
3484	cycle 40 miles		83	.096			2.19	7.15	9.34	11.55
3486	cycle 50 miles		83	.096			2.19	7.15	9.34	11.55
3494	45 MPH ave, cycle 8 miles		215	.037			.84	2.76	3.60	4.46
3496	cycle 10 miles		198	.040			.92	3	3.92	4.84
3498	cycle 20 miles		149	.054			1.22	3.99	5.21	6.45
3500	cycle 30 miles		116	.069			1.57	5.10	6.67	8.30
3502	cycle 40 miles		99	.081			1.83	6	7.83	9.70
3504	cycle 50 miles		83	.096			2.19	7.15	9.34	11.55
3506	50 MPH ave, cycle 10 miles		215	.037			.84	2.76	3.60	4.46
3508	cycle 20 miles		165	.048			1.10	3.60	4.70	5.80
3510	cycle 30 miles		132	.061			1.38	4.50	5.88	7.25
3512	cycle 40 miles		99	.081			1.83	6	7.83	9.70
3514	cycle 50 miles		83	.096			2.19	7.15	9.34	11.55
3614	30 min. wait/Ld./Uld., 15 MPH, cycle 0.5 mile		248	.032			.73	2.40	3.13	3.87
3616	cycle 1 mile		231	.035			.79	2.57	3.36	4.15
3618	cycle 2 miles		198	.040			.92	3	3.92	4.84
3620	cycle 4 miles		165	.048			1.10	3.60	4.70	5.80
3622	cycle 6 miles		149	.054			1.22	3.99	5.21	6.45
3624	cycle 8 miles		116	.069			1.57	5.10	6.67	8.30
3625	cycle 10 miles		116	.069			1.57	5.10	6.67	8.30
3626	20 MPH ave, cycle 0.5 mile		248	.032			.73	2.40	3.13	3.87
3628	cycle 1 mile		231	.035			.79	2.57	3.36	4.15
3630	cycle 2 miles		215	.037			.84	2.76	3.60	4.46
3632	cycle 4 miles		182	.044			1	3.27	4.27	5.25
3634	cycle 6 miles		165	.048			1.10	3.60	4.70	5.80
3636	cycle 8 miles		149	.054			1.22	3.99	5.21	6.45
3638	cycle 10 miles		132	.061			1.38	4.50	5.88	7.25
3640	25 MPH ave, cycle 4 miles		198	.040			.92	3	3.92	4.84
3642	cycle 6 miles		182	.044			1	3.27	4.27	5.25
3644	cycle 8 miles		149	.054			1.22	3.99	5.21	6.45
3646	cycle 10 miles		149	.054			1.22	3.99	5.21	6.45
3650	30 MPH ave, cycle 4 miles		198	.040			.92	3	3.92	4.84
3652	cycle 6 miles		182	.044			1	3.27	4.27	5.25

31 23 23 – Fill

31 23 23.20 Hauling		Crew	Daily Output	Labor-Hours	Unit	Material	2009 Bare Costs Labor	2009 Bare Costs Equipment	Total	Total Incl O&P
3654	cycle 8 miles	B-34C	165	.048	L.C.Y.		1.10	3.60	4.70	5.80
3656	cycle 10 miles		149	.054			1.22	3.99	5.21	6.45
3660	35 MPH ave, cycle 4 miles		215	.037			.84	2.76	3.60	4.46
3662	cycle 6 miles		198	.040			.92	3	3.92	4.84
3664	cycle 8 miles		182	.044			1	3.27	4.27	5.25
3666	cycle 10 miles		165	.048			1.10	3.60	4.70	5.80
3668	cycle 20 miles		116	.069			1.57	5.10	6.67	8.30
3670	cycle 30 miles		99	.081			1.83	6	7.83	9.70
3672	cycle 40 miles		83	.096			2.19	7.15	9.34	11.55
3674	40 MPH, cycle 6 miles		198	.040			.92	3	3.92	4.84
3676	cycle 8 miles		182	.044			1	3.27	4.27	5.25
3678	cycle 10 miles		165	.048			1.10	3.60	4.70	5.80
3680	cycle 20 miles		132	.061			1.38	4.50	5.88	7.25
3682	cycle 30 miles		99	.081			1.83	6	7.83	9.70
3684	cycle 40 miles		83	.096			2.19	7.15	9.34	11.55
3686	cycle 50 miles		66	.121			2.75	9	11.75	14.50
3694	45 MPH ave, cycle 8 miles		198	.040			.92	3	3.92	4.84
3696	cycle 10 miles		182	.044			1	3.27	4.27	5.25
3698	cycle 20 miles		132	.061			1.38	4.50	5.88	7.25
3700	cycle 30 miles		116	.069			1.57	5.10	6.67	8.30
3702	cycle 40 miles		83	.096			2.19	7.15	9.34	11.55
3704	cycle 50 miles		83	.096			2.19	7.15	9.34	11.55
3706	50 MPH ave, cycle 10 miles		182	.044			1	3.27	4.27	5.25
3708	cycle 20 miles		149	.054			1.22	3.99	5.21	6.45
3710	cycle 30 miles		116	.069			1.57	5.10	6.67	8.30
3712	cycle 40 miles		99	.081			1.83	6	7.83	9.70
3714	cycle 50 miles		83	.096			2.19	7.15	9.34	11.55
4014	20 CY truck, 15 min. wait/Ld./Uld., 15 MPH, cycle 0.5 mile	B-34D	560	.014			.32	1.09	1.41	1.74
4016	cycle 1 mile		500	.016			.36	1.22	1.58	1.95
4018	cycle 2 miles		420	.019			.43	1.45	1.88	2.33
4020	cycle 4 miles		300	.027			.61	2.03	2.64	3.25
4022	cycle 6 miles		240	.033			.76	2.54	3.30	4.06
4024	cycle 8 miles		200	.040			.91	3.05	3.96	4.87
4025	cycle 10 miles		160	.050			1.14	3.81	4.95	6.10
4026	20 MPH ave, cycle 0.5 mile		580	.014			.31	1.05	1.36	1.68
4028	cycle 1 mile		520	.015			.35	1.17	1.52	1.88
4030	cycle 2 miles		460	.017			.39	1.32	1.71	2.12
4032	cycle 4 miles		340	.024			.53	1.79	2.32	2.87
4034	cycle 6 miles		280	.029			.65	2.18	2.83	3.48
4036	cycle 8 miles		240	.033			.76	2.54	3.30	4.06
4038	cycle 10 miles		200	.040			.91	3.05	3.96	4.87
4040	25 MPH ave, cycle 4 miles		380	.021			.48	1.60	2.08	2.56
4042	cycle 6 miles		320	.025			.57	1.90	2.47	3.04
4044	cycle 8 miles		280	.029			.65	2.18	2.83	3.48
4046	cycle 10 miles		240	.033			.76	2.54	3.30	4.06
4050	30 MPH ave, cycle 4 miles		420	.019			.43	1.45	1.88	2.33
4052	cycle 6 miles		340	.024			.53	1.79	2.32	2.87
4054	cycle 8 miles		300	.027			.61	2.03	2.64	3.25
4056	cycle 10 miles		260	.031			.70	2.34	3.04	3.75
4060	35 MPH ave, cycle 4 miles		440	.018			.41	1.38	1.79	2.21
4062	cycle 6 miles		380	.021			.48	1.60	2.08	2.56
4064	cycle 8 miles		320	.025			.57	1.90	2.47	3.04
4066	cycle 10 miles		300	.027			.61	2.03	2.64	3.25

31 23 23.20 Hauling		Crew	Daily Output	Labor-Hours	Unit	Material	2009 Bare Costs Labor	2009 Bare Costs Equipment	Total	Total Incl O&P
4068	cycle 20 miles	B-34D	180	.044	L.C.Y.		1.01	3.38	4.39	5.40
4070	cycle 30 miles		140	.057			1.30	4.35	5.65	6.95
4072	cycle 40 miles		100	.080			1.82	6.10	7.92	9.75
4074	40 MPH ave, cycle 6 miles		400	.020			.45	1.52	1.97	2.43
4076	cycle 8 miles		340	.024			.53	1.79	2.32	2.87
4078	cycle 10 miles		320	.025			.57	1.90	2.47	3.04
4080	cycle 20 miles		200	.040			.91	3.05	3.96	4.87
4082	cycle 30 miles		160	.050			1.14	3.81	4.95	6.10
4084	cycle 40 miles		120	.067			1.51	5.10	6.61	8.15
4086	cycle 50 miles		100	.080			1.82	6.10	7.92	9.75
4094	45 MPH ave, cycle 8 miles		360	.022			.50	1.69	2.19	2.71
4096	cycle 10 miles		340	.024			.53	1.79	2.32	2.87
4098	cycle 20 miles		220	.036			.83	2.77	3.60	4.43
4100	cycle 30 miles		160	.050			1.14	3.81	4.95	6.10
4102	cycle 40 miles		140	.057			1.30	4.35	5.65	6.95
4104	cycle 50 miles		120	.067			1.51	5.10	6.61	8.15
4106	50 MPH ave, cycle 10 miles		340	.024			.53	1.79	2.32	2.87
4108	cycle 20 miles		240	.033			.76	2.54	3.30	4.06
4110	cycle 30 miles		180	.044			1.01	3.38	4.39	5.40
4112	cycle 40 miles		140	.057			1.30	4.35	5.65	6.95
4114	cycle 50 miles		120	.067			1.51	5.10	6.61	8.15
4214	20 min. wait/Ld./Uld., 15 MPH, cycle 0.5 mile		440	.018			.41	1.38	1.79	2.21
4216	cycle 1 mile		400	.020			.45	1.52	1.97	2.43
4218	cycle 2 miles		340	.024			.53	1.79	2.32	2.87
4220	cycle 4 miles		260	.031			.70	2.34	3.04	3.75
4222	cycle 6 miles		220	.036			.83	2.77	3.60	4.43
4224	cycle 8 miles		180	.044			1.01	3.38	4.39	5.40
4225	cycle 10 miles		160	.050			1.14	3.81	4.95	6.10
4226	20 MPH ave, cycle 0.5 mile		440	.018			.41	1.38	1.79	2.21
4228	cycle 1 mile		420	.019			.43	1.45	1.88	2.33
4230	cycle 2 miles		360	.022			.50	1.69	2.19	2.71
4232	cycle 4 miles		300	.027			.61	2.03	2.64	3.25
4234	cycle 6 miles		240	.033			.76	2.54	3.30	4.06
4236	cycle 8 miles		220	.036			.83	2.77	3.60	4.43
4238	cycle 10 miles		180	.044			1.01	3.38	4.39	5.40
4240	25 MPH ave, cycle 4 miles		320	.025			.57	1.90	2.47	3.04
4242	cycle 6 miles		280	.029			.65	2.18	2.83	3.48
4244	cycle 8 miles		240	.033			.76	2.54	3.30	4.06
4246	cycle 10 miles		220	.036			.83	2.77	3.60	4.43
4250	30 MPH ave, cycle 4 miles		340	.024			.53	1.79	2.32	2.87
4252	cycle 6 miles		300	.027			.61	2.03	2.64	3.25
4254	cycle 8 miles		260	.031			.70	2.34	3.04	3.75
4256	cycle 10 miles		240	.033			.76	2.54	3.30	4.06
4260	35 MPH ave, cycle 4 miles		360	.022			.50	1.69	2.19	2.71
4262	cycle 6 miles		320	.025			.57	1.90	2.47	3.04
4264	cycle 8 miles		280	.029			.65	2.18	2.83	3.48
4266	cycle 10 miles		260	.031			.70	2.34	3.04	3.75
4268	cycle 20 miles		180	.044			1.01	3.38	4.39	5.40
4270	cycle 30 miles		120	.067			1.51	5.10	6.61	8.15
4272	cycle 40 miles		100	.080			1.82	6.10	7.92	9.75
4274	40 MPH ave, cycle 6 miles		320	.025			.57	1.90	2.47	3.04
4276	cycle 8 miles		300	.027			.61	2.03	2.64	3.25
4278	cycle 10 miles		260	.031			.70	2.34	3.04	3.75

31 23 23.20 Hauling		Crew	Daily Output	Labor-Hours	Unit	Material	Labor	Equipment	Total	Total Incl O&P
								2009 Bare Costs		
4280	cycle 20 miles	B-34D	180	.044	L.C.Y.		1.01	3.38	4.39	5.40
4282	cycle 30 miles		140	.057			1.30	4.35	5.65	6.95
4284	cycle 40 miles		120	.067			1.51	5.10	6.61	8.15
4286	cycle 50 miles		100	.080			1.82	6.10	7.92	9.75
4294	45 MPH ave, cycle 8 miles		300	.027			.61	2.03	2.64	3.25
4296	cycle 10 miles		280	.029			.65	2.18	2.83	3.48
4298	cycle 20 miles		200	.040			.91	3.05	3.96	4.87
4300	cycle 30 miles		160	.050			1.14	3.81	4.95	6.10
4302	cycle 40 miles		120	.067			1.51	5.10	6.61	8.15
4304	cycle 50 miles		100	.080			1.82	6.10	7.92	9.75
4306	50 MPH ave, cycle 10 miles		300	.027			.61	2.03	2.64	3.25
4308	cycle 20 miles		220	.036			.83	2.77	3.60	4.43
4310	cycle 30 miles		180	.044			1.01	3.38	4.39	5.40
4312	cycle 40 miles		140	.057			1.30	4.35	5.65	6.95
4314	cycle 50 miles		120	.067			1.51	5.10	6.61	8.15
4414	25 min. wait/Ld./Uld., 15 MPH, cycle 0.5 mile		340	.024			.53	1.79	2.32	2.87
4416	cycle 1 mile		320	.025			.57	1.90	2.47	3.04
4418	cycle 2 miles		280	.029			.65	2.18	2.83	3.48
4420	cycle 4 miles		220	.036			.83	2.77	3.60	4.43
4422	cycle 6 miles		200	.040			.91	3.05	3.96	4.87
4424	cycle 8 miles		160	.050			1.14	3.81	4.95	6.10
4425	cycle 10 miles		140	.057			1.30	4.35	5.65	6.95
4426	20 MPH ave, cycle 0.5 mile		360	.022			.50	1.69	2.19	2.71
4428	cycle 1 mile		340	.024			.53	1.79	2.32	2.87
4430	cycle 2 miles		300	.027			.61	2.03	2.64	3.25
4432	cycle 4 miles		260	.031			.70	2.34	3.04	3.75
4434	cycle 6 miles		220	.036			.83	2.77	3.60	4.43
4436	cycle 8 miles		200	.040			.91	3.05	3.96	4.87
4438	cycle 10 miles		160	.050			1.14	3.81	4.95	6.10
4440	25 MPH ave, cycle 4 miles		280	.029			.65	2.18	2.83	3.48
4442	cycle 6 miles		240	.033			.76	2.54	3.30	4.06
4444	cycle 8 miles		220	.036			.83	2.77	3.60	4.43
4446	cycle 10 miles		200	.040			.91	3.05	3.96	4.87
4450	30 MPH ave, cycle 4 miles		280	.029			.65	2.18	2.83	3.48
4452	cycle 6 miles		260	.031			.70	2.34	3.04	3.75
4454	cycle 8 miles		220	.036			.83	2.77	3.60	4.43
4456	cycle 10 miles		200	.040			.91	3.05	3.96	4.87
4460	35 MPH ave, cycle 4 miles		300	.027			.61	2.03	2.64	3.25
4462	cycle 6 miles		260	.031			.70	2.34	3.04	3.75
4464	cycle 8 miles		240	.033			.76	2.54	3.30	4.06
4466	cycle 10 miles		220	.036			.83	2.77	3.60	4.43
4468	cycle 20 miles		160	.050			1.14	3.81	4.95	6.10
4470	cycle 30 miles		120	.067			1.51	5.10	6.61	8.15
4472	cycle 40 miles		100	.080			1.82	6.10	7.92	9.75
4474	40 MPH, cycle 6 miles		280	.029			.65	2.18	2.83	3.48
4476	cycle 8 miles		260	.031			.70	2.34	3.04	3.75
4478	cycle 10 miles		240	.033			.76	2.54	3.30	4.06
4480	cycle 20 miles		160	.050			1.14	3.81	4.95	6.10
4482	cycle 30 miles		140	.057			1.30	4.35	5.65	6.95
4484	cycle 40 miles		100	.080			1.82	6.10	7.92	9.75
4486	cycle 50 miles		100	.080			1.82	6.10	7.92	9.75
4494	45 MPH ave, cycle 8 miles		260	.031			.70	2.34	3.04	3.75
4496	cycle 10 miles		240	.033			.76	2.54	3.30	4.06

31 23 Excavation and Fill

31 23 23 – Fill

31 23 23.20 Hauling		Crew	Daily Output	Labor-Hours	Unit	Material	2009 Bare Costs Labor	Equipment	Total	Total Incl O&P
4498	cycle 20 miles	B-34D	180	.044	L.C.Y.		1.01	3.38	4.39	5.40
4500	cycle 30 miles		140	.057			1.30	4.35	5.65	6.95
4502	cycle 40 miles		120	.067			1.51	5.10	6.61	8.15
4504	cycle 50 miles		100	.080			1.82	6.10	7.92	9.75
4506	50 MPH ave, cycle 10 miles		260	.031			.70	2.34	3.04	3.75
4508	cycle 20 miles		200	.040			.91	3.05	3.96	4.87
4510	cycle 30 miles		160	.050			1.14	3.81	4.95	6.10
4512	cycle 40 miles		120	.067			1.51	5.10	6.61	8.15
4514	cycle 50 miles		100	.080			1.82	6.10	7.92	9.75
4614	30 min. wait/Ld./Uld., 15 MPH, cycle 0.5 mile		300	.027			.61	2.03	2.64	3.25
4616	cycle 1 mile		280	.029			.65	2.18	2.83	3.48
4618	cycle 2 miles		240	.033			.76	2.54	3.30	4.06
4620	cycle 4 miles		200	.040			.91	3.05	3.96	4.87
4622	cycle 6 miles		180	.044			1.01	3.38	4.39	5.40
4624	cycle 8 miles		140	.057			1.30	4.35	5.65	6.95
4625	cycle 10 miles		140	.057			1.30	4.35	5.65	6.95
4626	20 MPH ave, cycle 0.5 mile		300	.027			.61	2.03	2.64	3.25
4628	cycle 1 mile		280	.029			.65	2.18	2.83	3.48
4630	cycle 2 miles		260	.031			.70	2.34	3.04	3.75
4632	cycle 4 miles		220	.036			.83	2.77	3.60	4.43
4634	cycle 6 miles		200	.040			.91	3.05	3.96	4.87
4636	cycle 8 miles		180	.044			1.01	3.38	4.39	5.40
4638	cycle 10 miles		160	.050			1.14	3.81	4.95	6.10
4640	25 MPH ave, cycle 4 miles		240	.033			.76	2.54	3.30	4.06
4642	cycle 6 miles		220	.036			.83	2.77	3.60	4.43
4644	cycle 8 miles		180	.044			1.01	3.38	4.39	5.40
4646	cycle 10 miles		180	.044			1.01	3.38	4.39	5.40
4650	30 MPH ave, cycle 4 miles		240	.033			.76	2.54	3.30	4.06
4652	cycle 6 miles		220	.036			.83	2.77	3.60	4.43
4654	cycle 8 miles		200	.040			.91	3.05	3.96	4.87
4656	cycle 10 miles		180	.044			1.01	3.38	4.39	5.40
4660	35 MPH ave, cycle 4 miles		260	.031			.70	2.34	3.04	3.75
4662	cycle 6 miles		240	.033			.76	2.54	3.30	4.06
4664	cycle 8 miles		220	.036			.83	2.77	3.60	4.43
4666	cycle 10 miles		200	.040			.91	3.05	3.96	4.87
4668	cycle 20 miles		140	.057			1.30	4.35	5.65	6.95
4670	cycle 30 miles		120	.067			1.51	5.10	6.61	8.15
4672	cycle 40 miles		100	.080			1.82	6.10	7.92	9.75
4674	40 MPH, cycle 6 miles		240	.033			.76	2.54	3.30	4.06
4676	cycle 8 miles		220	.036			.83	2.77	3.60	4.43
4678	cycle 10 miles		200	.040			.91	3.05	3.96	4.87
4680	cycle 20 miles		160	.050			1.14	3.81	4.95	6.10
4682	cycle 30 miles		120	.067			1.51	5.10	6.61	8.15
4684	cycle 40 miles		100	.080			1.82	6.10	7.92	9.75
4686	cycle 50 miles		80	.100			2.27	7.60	9.87	12.15
4694	45 MPH ave, cycle 8 miles		220	.036			.83	2.77	3.60	4.43
4696	cycle 10 miles		220	.036			.83	2.77	3.60	4.43
4698	cycle 20 miles		160	.050			1.14	3.81	4.95	6.10
4700	cycle 30 miles		140	.057			1.30	4.35	5.65	6.95
4702	cycle 40 miles		100	.080			1.82	6.10	7.92	9.75
4704	cycle 50 miles		100	.080			1.82	6.10	7.92	9.75
4706	50 MPH ave, cycle 10 miles		220	.036			.83	2.77	3.60	4.43
4708	cycle 20 miles		180	.044			1.01	3.38	4.39	5.40

31 23 23 – Fill

31 23 23.20 Hauling		Crew	Daily Output	Labor-Hours	Unit	Material	2009 Bare Costs Labor	Equipment	Total	Total Incl O&P
4710	cycle 30 miles	B-34D	140	.057	L.C.Y.		1.30	4.35	5.65	6.95
4712	cycle 40 miles		120	.067			1.51	5.10	6.61	8.15
4714	cycle 50 miles	↓	100	.080			1.82	6.10	7.92	9.75
5000	22 CY off-road, 15 min. wait/Ld./Uld., 5 MPH, cycle 2000 ft	B-34F	528	.015			.34	2.17	2.51	2.96
5010	cycle 3000 ft		484	.017			.38	2.36	2.74	3.23
5020	cycle 4000 ft		440	.018			.41	2.60	3.01	3.55
5030	cycle 0.5 mile		506	.016			.36	2.26	2.62	3.09
5040	cycle 1 mile		374	.021			.49	3.06	3.55	4.18
5050	cycle 2 miles		264	.030			.69	4.33	5.02	5.90
5060	10 MPH, cycle 2000 ft		594	.013			.31	1.93	2.24	2.63
5070	cycle 3000 ft		572	.014			.32	2	2.32	2.73
5080	cycle 4000 ft		528	.015			.34	2.17	2.51	2.96
5090	cycle 0.5 mile		572	.014			.32	2	2.32	2.73
5100	cycle 1 mile		506	.016			.36	2.26	2.62	3.09
5110	cycle 2 miles		374	.021			.49	3.06	3.55	4.18
5120	cycle 4 miles		264	.030			.69	4.33	5.02	5.90
5130	15 MPH, cycle 2000 ft		638	.013			.28	1.79	2.07	2.45
5140	cycle 3000 ft		594	.013			.31	1.93	2.24	2.63
5150	cycle 4000 ft		572	.014			.32	2	2.32	2.73
5160	cycle 0.5 mile		616	.013			.29	1.86	2.15	2.53
5170	cycle 1 mile		550	.015			.33	2.08	2.41	2.84
5180	cycle 2 miles		462	.017			.39	2.48	2.87	3.38
5190	cycle 4 miles		330	.024			.55	3.47	4.02	4.73
5200	20 MPH, cycle 2 miles		506	.016			.36	2.26	2.62	3.09
5210	cycle 4 miles		374	.021			.49	3.06	3.55	4.18
5220	25 MPH, cycle 2 miles		528	.015			.34	2.17	2.51	2.96
5230	cycle 4 miles		418	.019			.43	2.74	3.17	3.74
5300	20 min. wait/Ld./Uld., 5 MPH, cycle 2000 ft		418	.019			.43	2.74	3.17	3.74
5310	cycle 3000 ft		396	.020			.46	2.89	3.35	3.95
5320	cycle 4000 ft		352	.023			.52	3.25	3.77	4.45
5330	cycle 0.5 mile		396	.020			.46	2.89	3.35	3.95
5340	cycle 1 mile		330	.024			.55	3.47	4.02	4.73
5350	cycle 2 miles		242	.033			.75	4.73	5.48	6.45
5360	10 MPH, cycle 2000 ft		462	.017			.39	2.48	2.87	3.38
5370	cycle 3000 ft		440	.018			.41	2.60	3.01	3.55
5380	cycle 4000 ft		418	.019			.43	2.74	3.17	3.74
5390	cycle 0.5 mile		462	.017			.39	2.48	2.87	3.38
5400	cycle 1 mile		396	.020			.46	2.89	3.35	3.95
5410	cycle 2 miles		330	.024			.55	3.47	4.02	4.73
5420	cycle 4 miles		242	.033			.75	4.73	5.48	6.45
5430	15 MPH, cycle 2000 ft		484	.017			.38	2.36	2.74	3.23
5440	cycle 3000 ft		462	.017			.39	2.48	2.87	3.38
5450	cycle 4000 ft		462	.017			.39	2.48	2.87	3.38
5460	cycle 0.5 mile		484	.017			.38	2.36	2.74	3.23
5470	cycle 1 mile		440	.018			.41	2.60	3.01	3.55
5480	cycle 2 miles		374	.021			.49	3.06	3.55	4.18
5490	cycle 4 miles		286	.028			.63	4	4.63	5.45
5500	20 MPH, cycle 2 miles		396	.020			.46	2.89	3.35	3.95
5510	cycle 4 miles		330	.024			.55	3.47	4.02	4.73
5520	25 MPH, cycle 2 miles		418	.019			.43	2.74	3.17	3.74
5530	cycle 4 miles		352	.023			.52	3.25	3.77	4.45
5600	25 min. wait/Ld./Uld., 5 MPH, cycle 2000 ft		352	.023			.52	3.25	3.77	4.45
5610	cycle 3000 ft	↓	330	.024	↓		.55	3.47	4.02	4.73

31 23 23 – Fill

31 23 23.20 Hauling		Crew	Daily Output	Labor-Hours	Unit	Material	2009 Bare Costs Labor	2009 Bare Costs Equipment	Total	Total Incl O&P
5620	cycle 4000 ft	B-34F	308	.026	L.C.Y.		.59	3.71	4.30	5.10
5630	cycle 0.5 mile		330	.024			.55	3.47	4.02	4.73
5640	cycle 1 mile		286	.028			.63	4	4.63	5.45
5650	cycle 2 miles		220	.036			.83	5.20	6.03	7.10
5660	10 MPH, cycle 2000 ft		374	.021			.49	3.06	3.55	4.18
5670	cycle 3000 ft		374	.021			.49	3.06	3.55	4.18
5680	cycle 4000 ft		352	.023			.52	3.25	3.77	4.45
5690	cycle 0.5 mile		374	.021			.49	3.06	3.55	4.18
5700	cycle 1 mile		330	.024			.55	3.47	4.02	4.73
5710	cycle 2 miles		286	.028			.63	4	4.63	5.45
5720	cycle 4 miles		220	.036			.83	5.20	6.03	7.10
5730	15 MPH, cycle 2000 ft		396	.020			.46	2.89	3.35	3.95
5740	cycle 3000 ft		374	.021			.49	3.06	3.55	4.18
5750	cycle 4000 ft		374	.021			.49	3.06	3.55	4.18
5760	cycle 0.5 mile		374	.021			.49	3.06	3.55	4.18
5770	cycle 1 mile		352	.023			.52	3.25	3.77	4.45
5780	cycle 2 miles		308	.026			.59	3.71	4.30	5.10
5790	cycle 4 miles		242	.033			.75	4.73	5.48	6.45
5800	20 MPH, cycle 2 miles		330	.024			.55	3.47	4.02	4.73
5810	cycle 4 miles		286	.028			.63	4	4.63	5.45
5820	25 MPH, cycle 2 miles		352	.023			.52	3.25	3.77	4.45
5830	cycle 4 miles		308	.026			.59	3.71	4.30	5.10
6000	34 CY off-road, 15 min. wait/Ld./Uld., 5 MPH, cycle 2000 ft	B-34G	816	.010			.22	1.94	2.16	2.50
6010	cycle 3000 ft		748	.011			.24	2.12	2.36	2.74
6020	cycle 4000 ft		680	.012			.27	2.33	2.60	3.01
6030	cycle 0.5 mile		782	.010			.23	2.02	2.25	2.62
6040	cycle 1 mile		578	.014			.31	2.74	3.05	3.54
6050	cycle 2 miles		408	.020			.45	3.88	4.33	5
6060	10 MPH, cycle 2000 ft		918	.009			.20	1.72	1.92	2.22
6070	cycle 3000 ft		884	.009			.21	1.79	2	2.31
6080	cycle 4000 ft		816	.010			.22	1.94	2.16	2.50
6090	cycle 0.5 mile		884	.009			.21	1.79	2	2.31
6100	cycle 1 mile		782	.010			.23	2.02	2.25	2.62
6110	cycle 2 miles		578	.014			.31	2.74	3.05	3.54
6120	cycle 4 miles		408	.020			.45	3.88	4.33	5
6130	15 MPH, cycle 2000 ft		986	.008			.18	1.60	1.78	2.07
6140	cycle 3000 ft		918	.009			.20	1.72	1.92	2.22
6150	cycle 4000 ft		884	.009			.21	1.79	2	2.31
6160	cycle 0.5 mile		952	.008			.19	1.66	1.85	2.15
6170	cycle 1 mile		850	.009			.21	1.86	2.07	2.41
6180	cycle 2 miles		714	.011			.25	2.21	2.46	2.87
6190	cycle 4 miles		510	.016			.36	3.10	3.46	4.01
6200	20 MPH, cycle 2 miles		782	.010			.23	2.02	2.25	2.62
6210	cycle 4 miles		578	.014			.31	2.74	3.05	3.54
6220	25 MPH, cycle 2 miles		816	.010			.22	1.94	2.16	2.50
6230	cycle 4 miles		646	.012			.28	2.45	2.73	3.16
6300	20 min. wait/Ld./Uld., 5 MPH, cycle 2000 ft		646	.012			.28	2.45	2.73	3.16
6310	cycle 3000 ft		612	.013			.30	2.58	2.88	3.34
6320	cycle 4000 ft		544	.015			.33	2.91	3.24	3.76
6330	cycle 0.5 mile		612	.013			.30	2.58	2.88	3.34
6340	cycle 1 mile		510	.016			.36	3.10	3.46	4.01
6350	cycle 2 miles		374	.021			.49	4.23	4.72	5.45
6360	10 MPH, cycle 2000 ft		714	.011			.25	2.21	2.46	2.87

31 23 23.20 Hauling		Crew	Daily Output	Labor-Hours	Unit	Material	2009 Bare Costs Labor	2009 Bare Costs Equipment	Total	Total Incl O&P
6370	cycle 3000 ft	B-34G	680	.012	L.C.Y.		.27	2.33	2.60	3.01
6380	cycle 4000 ft		646	.012			.28	2.45	2.73	3.16
6390	cycle 0.5 mile		714	.011			.25	2.21	2.46	2.87
6400	cycle 1 mile		612	.013			.30	2.58	2.88	3.34
6410	cycle 2 miles		510	.016			.36	3.10	3.46	4.01
6420	cycle 4 miles		374	.021			.49	4.23	4.72	5.45
6430	15 MPH, cycle 2000 ft		748	.011			.24	2.12	2.36	2.74
6440	cycle 3000 ft		714	.011			.25	2.21	2.46	2.87
6450	cycle 4000 ft		714	.011			.25	2.21	2.46	2.87
6460	cycle 0.5 mile		748	.011			.24	2.12	2.36	2.74
6470	cycle 1 mile		680	.012			.27	2.33	2.60	3.01
6480	cycle 2 miles		578	.014			.31	2.74	3.05	3.54
6490	cycle 4 miles		442	.018			.41	3.58	3.99	4.63
6500	20 MPH, cycle 2 miles		612	.013			.30	2.58	2.88	3.34
6510	cycle 4 miles		510	.016			.36	3.10	3.46	4.01
6520	25 MPH, cycle 2 miles		646	.012			.28	2.45	2.73	3.16
6530	cycle 4 miles		544	.015			.33	2.91	3.24	3.76
6600	25 min. wait/Ld./Uld., 5 MPH, cycle 2000 ft		544	.015			.33	2.91	3.24	3.76
6610	cycle 3000 ft		510	.016			.36	3.10	3.46	4.01
6620	cycle 4000 ft		476	.017			.38	3.32	3.70	4.30
6630	cycle 0.5 mile		510	.016			.36	3.10	3.46	4.01
6640	cycle 1 mile		442	.018			.41	3.58	3.99	4.63
6650	cycle 2 miles		340	.024			.53	4.65	5.18	6
6660	10 MPH, cycle 2000 ft		578	.014			.31	2.74	3.05	3.54
6670	cycle 3000 ft		578	.014			.31	2.74	3.05	3.54
6680	cycle 4000 ft		544	.015			.33	2.91	3.24	3.76
6690	cycle 0.5 mile		578	.014			.31	2.74	3.05	3.54
6700	cycle 1 mile		510	.016			.36	3.10	3.46	4.01
6710	cycle 2 miles		442	.018			.41	3.58	3.99	4.63
6720	cycle 4 miles		340	.024			.53	4.65	5.18	6
6730	15 MPH, cycle 2000 ft		612	.013			.30	2.58	2.88	3.34
6740	cycle 3000 ft		578	.014			.31	2.74	3.05	3.54
6750	cycle 4000 ft		578	.014			.31	2.74	3.05	3.54
6760	cycle 0.5 mile		612	.013			.30	2.58	2.88	3.34
6770	cycle 1 mile		544	.015			.33	2.91	3.24	3.76
6780	cycle 2 miles		476	.017			.38	3.32	3.70	4.30
6790	cycle 4 miles		374	.021			.49	4.23	4.72	5.45
6800	20 MPH, cycle 2 miles		510	.016			.36	3.10	3.46	4.01
6810	cycle 4 miles		442	.018			.41	3.58	3.99	4.63
6820	25 MPH, cycle 2 miles		544	.015			.33	2.91	3.24	3.76
6830	cycle 4 miles		476	.017			.38	3.32	3.70	4.30
7000	42 CY off-road, 20 min. wait/Ld./Uld., 5 MPH, cycle 2000 ft	B-34H	798	.010			.23	2	2.23	2.58
7010	cycle 3000 ft		756	.011			.24	2.11	2.35	2.72
7020	cycle 4000 ft		672	.012			.27	2.37	2.64	3.06
7030	cycle 0.5 mile		756	.011			.24	2.11	2.35	2.72
7040	cycle 1 mile		630	.013			.29	2.53	2.82	3.26
7050	cycle 2 miles		462	.017			.39	3.45	3.84	4.46
7060	10 MPH, cycle 2000 ft		882	.009			.21	1.81	2.02	2.34
7070	cycle 3000 ft		840	.010			.22	1.90	2.12	2.45
7080	cycle 4000 ft		798	.010			.23	2	2.23	2.58
7090	cycle 0.5 mile		882	.009			.21	1.81	2.02	2.34
7100	cycle 1 mile		798	.010			.23	2	2.23	2.58
7110	cycle 2 miles		630	.013			.29	2.53	2.82	3.26

31 23 23 – Fill

31 23 23.20 Hauling	Crew	Daily Output	Labor-Hours	Unit	Material	2009 Bare Costs Labor	2009 Bare Costs Equipment	Total	Total Incl O&P
7120	cycle 4 miles	B-34H	462	.017	L.C.Y.		.39	3.45	3.84	4.46
7130	15 MPH, cycle 2000 ft		924	.009			.20	1.73	1.93	2.23
7140	cycle 3000 ft		882	.009			.21	1.81	2.02	2.34
7150	cycle 4000 ft		882	.009			.21	1.81	2.02	2.34
7160	cycle 0.5 mile		882	.009			.21	1.81	2.02	2.34
7170	cycle 1 mile		840	.010			.22	1.90	2.12	2.45
7180	cycle 2 miles		714	.011			.25	2.23	2.48	2.88
7190	cycle 4 miles		546	.015			.33	2.92	3.25	3.77
7200	20 MPH, cycle 2 miles		756	.011			.24	2.11	2.35	2.72
7210	cycle 4 miles		630	.013			.29	2.53	2.82	3.26
7220	25 MPH, cycle 2 miles		798	.010			.23	2	2.23	2.58
7230	cycle 4 miles		672	.012			.27	2.37	2.64	3.06
7300	25 min. wait/Ld./Uld., 5 MPH, cycle 2000 ft		672	.012			.27	2.37	2.64	3.06
7310	cycle 3000 ft		630	.013			.29	2.53	2.82	3.26
7320	cycle 4000 ft		588	.014			.31	2.71	3.02	3.50
7330	cycle 0.5 mile		630	.013			.29	2.53	2.82	3.26
7340	cycle 1 mile		546	.015			.33	2.92	3.25	3.77
7350	cycle 2 miles		378	.021			.48	4.22	4.70	5.45
7360	10 MPH, cycle 2000 ft		714	.011			.25	2.23	2.48	2.88
7370	cycle 3000 ft		714	.011			.25	2.23	2.48	2.88
7380	cycle 4000 ft		672	.012			.27	2.37	2.64	3.06
7390	cycle 0.5 mile		714	.011			.25	2.23	2.48	2.88
7400	cycle 1 mile		630	.013			.29	2.53	2.82	3.26
7410	cycle 2 miles		546	.015			.33	2.92	3.25	3.77
7420	cycle 4 miles		378	.021			.48	4.22	4.70	5.45
7430	15 MPH, cycle 2000 ft		756	.011			.24	2.11	2.35	2.72
7440	cycle 3000 ft		714	.011			.25	2.23	2.48	2.88
7450	cycle 4000 ft		714	.011			.25	2.23	2.48	2.88
7460	cycle 0.5 mile		714	.011			.25	2.23	2.48	2.88
7470	cycle 1 mile		672	.012			.27	2.37	2.64	3.06
7480	cycle 2 miles		588	.014			.31	2.71	3.02	3.50
7490	cycle 4 miles		462	.017			.39	3.45	3.84	4.46
7500	20 MPH, cycle 2 miles		630	.013			.29	2.53	2.82	3.26
7510	cycle 4 miles		546	.015			.33	2.92	3.25	3.77
7520	25 MPH, cycle 2 miles		672	.012			.27	2.37	2.64	3.06
7530	cycle 4 miles		588	.014			.31	2.71	3.02	3.50
8000	60 CY off-road, 20 min. wait/Ld./Uld., 5 MPH, cycle 2000 ft	B-34J	1140	.007			.16	1.80	1.96	2.25
8010	cycle 3000 ft		1080	.007			.17	1.90	2.07	2.37
8020	cycle 4000 ft		960	.008			.19	2.13	2.32	2.66
8030	cycle 0.5 mile		1080	.007			.17	1.90	2.07	2.37
8040	cycle 1 mile		900	.009			.20	2.27	2.47	2.84
8050	cycle 2 miles		660	.012			.28	3.10	3.38	3.87
8060	10 MPH, cycle 2000 ft		1260	.006			.14	1.62	1.76	2.03
8070	cycle 3000 ft		1200	.007			.15	1.71	1.86	2.13
8080	cycle 4000 ft		1140	.007			.16	1.80	1.96	2.25
8090	cycle 0.5 mile		1260	.006			.14	1.62	1.76	2.03
8100	cycle 1 mile		1080	.007			.17	1.90	2.07	2.37
8110	cycle 2 miles		900	.009			.20	2.27	2.47	2.84
8120	cycle 4 miles		660	.012			.28	3.10	3.38	3.87
8130	15 MPH, cycle 2000 ft		1320	.006			.14	1.55	1.69	1.94
8140	cycle 3000 ft		1260	.006			.14	1.62	1.76	2.03
8150	cycle 4000 ft		1260	.006			.14	1.62	1.76	2.03
8160	cycle 0.5 mile		1320	.006			.14	1.55	1.69	1.94

31 23 Excavation and Fill

31 23 23 – Fill

31 23 23.20 Hauling

		Crew	Daily Output	Labor-Hours	Unit	Material	2009 Bare Costs Labor	2009 Bare Costs Equipment	Total	Total Incl O&P
8170	cycle 1 mile	B-34J	1200	.007	L.C.Y.		.15	1.71	1.86	2.13
8180	cycle 2 miles		1020	.008			.18	2.01	2.19	2.51
8190	cycle 4 miles		780	.010			.23	2.63	2.86	3.28
8200	20 MPH, cycle 2 miles		1080	.007			.17	1.90	2.07	2.37
8210	cycle 4 miles		900	.009			.20	2.27	2.47	2.84
8220	25 MPH, cycle 2 miles		1140	.007			.16	1.80	1.96	2.25
8230	cycle 4 miles		960	.008			.19	2.13	2.32	2.66
8300	25 min. wait/Ld./Uld., 5 MPH, cycle 2000 ft		960	.008			.19	2.13	2.32	2.66
8310	cycle 3000 ft		900	.009			.20	2.27	2.47	2.84
8320	cycle 4000 ft		840	.010			.22	2.44	2.66	3.04
8330	cycle 0.5 mile		900	.009			.20	2.27	2.47	2.84
8340	cycle 1 mile		780	.010			.23	2.63	2.86	3.28
8350	cycle 2 miles		600	.013			.30	3.41	3.71	4.26
8360	10 MPH, cycle 2000 ft		1020	.008			.18	2.01	2.19	2.51
8370	cycle 3000 ft		1020	.008			.18	2.01	2.19	2.51
8380	cycle 4000 ft		960	.008			.19	2.13	2.32	2.66
8390	cycle 0.5 mile		1020	.008			.18	2.01	2.19	2.51
8400	cycle 1 mile		900	.009			.20	2.27	2.47	2.84
8410	cycle 2 miles		780	.010			.23	2.63	2.86	3.28
8420	cycle 4 miles		600	.013			.30	3.41	3.71	4.26
8430	15 MPH, cycle 2000 ft		1080	.007			.17	1.90	2.07	2.37
8440	cycle 3000 ft		1020	.008			.18	2.01	2.19	2.51
8450	cycle 4000 ft		1020	.008			.18	2.01	2.19	2.51
8460	cycle 0.5 mile		1080	.007			.17	1.90	2.07	2.37
8470	cycle 1 mile		960	.008			.19	2.13	2.32	2.66
8480	cycle 2 miles		840	.010			.22	2.44	2.66	3.04
8490	cycle 4 miles		660	.012			.28	3.10	3.38	3.87
8500	20 MPH, cycle 2 miles		900	.009			.20	2.27	2.47	2.84
8510	cycle 4 miles		780	.010			.23	2.63	2.86	3.28
8520	25 MPH, cycle 2 miles		960	.008			.19	2.13	2.32	2.66
8530	cycle 4 miles	▼	840	.010	▼		.22	2.44	2.66	3.04

31 23 23.24 Compaction, Structural

		Crew	Daily Output	Labor-Hours	Unit	Material	2009 Bare Costs Labor	2009 Bare Costs Equipment	Total	Total Incl O&P
0010	**COMPACTION, STRUCTURAL**									
0020	Steel wheel tandem roller, 5 tons	B-10E	8	1	Hr.		28.50	17.25	45.75	66
0050	Air tamp, 6" to 8" lifts, common fill	B-9	250	.160	E.C.Y.		3.35	.77	4.12	6.50
0060	Select fill	"	300	.133			2.79	.64	3.43	5.40
0600	Vibratory plate, 8" lifts, common fill	A-1D	200	.040			.82	.16	.98	1.56
0700	Select fill	"	216	.037	▼		.76	.15	.91	1.45

31 25 Erosion and Sedimentation Controls

31 25 13 – Erosion Controls

31 25 13.10 Synthetic Erosion Control

			Crew	Daily Output	Labor-Hours	Unit	Material	2009 Bare Costs Labor	2009 Bare Costs Equipment	Total	Total Incl O&P
0010	**SYNTHETIC EROSION CONTROL**										
0020	Jute mesh, 100 SY per roll, 4' wide, stapled	G	B-80A	2400	.010	S.Y.	1.06	.21	.10	1.37	1.63
0100	Plastic netting, stapled, 2" x 1" mesh, 20 mil	G	B-1	2500	.010		.81	.20		1.01	1.23
0200	Polypropylene mesh, stapled, 6.5 oz./S.Y.	G		2500	.010		1.67	.20		1.87	2.18
0300	Tobacco netting, or jute mesh #2, stapled	G	▼	2500	.010	▼	.09	.20		.29	.44
1000	Silt fence, polypropylene, 3' high, ideal conditions	G	2 Clab	1600	.010	L.F.	.39	.21		.60	.78
1100	Adverse conditions	G	"	950	.017	"	.39	.35		.74	1.02

31 31 Soil Treatment

31 31 16 – Termite Control

31 31 16.13 Chemical Termite Control	Crew	Daily Output	Labor-Hours	Unit	Material	2009 Bare Costs Labor	Equipment	Total	Total Incl O&P
0010 **CHEMICAL TERMITE CONTROL**									
0020 Slab and walls, residential	1 Skwk	1200	.007	SF Flr.	.32	.19		.51	.67
0400 Insecticides for termite control, minimum		14.20	.563	Gal.	13.05	15.80		28.85	41
0500 Maximum		11	.727	"	22.50	20.50		43	59

Division 32
Exterior Improvements

32 01 Operation and Maintenance of Exterior Improvements

32 01 13 – Flexible Paving Surface Treatment

32 01 13.66 Fog Seal

		Crew	Daily Output	Labor-Hours	Unit	Material	2009 Bare Costs Labor	Equipment	Total	Total Incl O&P
0010	**FOG SEAL**									
0012	Sealcoating, 2 coat coal tar pitch emulsion over 10,000 SY	B-45	5000	.003	S.Y.	1	.07	.06	1.13	1.29
0030	1000 to 10,000 S.Y.	"	3000	.005		1	.12	.10	1.22	1.40
0100	Under 1000 S.Y.	B-1	1050	.023		1	.49		1.49	1.92
0300	Petroleum resistant, over 10,000 S.Y.	B-45	5000	.003		1.24	.07	.06	1.37	1.55
0320	1000 to 10,000 S.Y.	"	3000	.005		1.24	.12	.10	1.46	1.66
0400	Under 1000 S.Y.	B-1	1050	.023		1.24	.49		1.73	2.18

32 06 Schedules for Exterior Improvements

32 06 10 – Schedules for Bases, Ballasts, and Paving

32 06 10.10 Sidewalks, Driveways and Patios

		Crew	Daily Output	Labor-Hours	Unit	Material	2009 Bare Costs Labor	Equipment	Total	Total Incl O&P
0010	**SIDEWALKS, DRIVEWAYS AND PATIOS** No base									
0021	Asphaltic concrete, 2" thick	B-37	6480	.007	S.F.	.67	.16	.02	.85	1.03
0101	2-1/2" thick	"	5950	.008	"	.85	.18	.02	1.05	1.27
0300	Concrete, 3000 psi, CIP, 6 x 6 - W1.4 x W1.4 mesh,									
0310	broomed finish, no base, 4" thick	B-24	600	.040	S.F.	1.68	1		2.68	3.52
0350	5" thick		545	.044		2.24	1.11		3.35	4.31
0400	6" thick		510	.047		2.62	1.18		3.80	4.84
0450	For bank run gravel base, 4" thick, add	B-18	2500	.010		.64	.20	.02	.86	1.06
0520	8" thick, add	"	1600	.015		1.28	.32	.02	1.62	1.98
1000	Crushed stone, 1" thick, white marble	2 Clab	1700	.009		.22	.19		.41	.57
1050	Bluestone	"	1700	.009		.24	.19		.43	.60
1700	Redwood, prefabricated, 4' x 4' sections	2 Carp	316	.051		8.25	1.42		9.67	11.50
1750	Redwood planks, 1" thick, on sleepers	"	240	.067		5.80	1.86		7.66	9.50
2250	Stone dust, 4" thick	B-62	900	.027	S.Y.	3.33	.60	.17	4.10	4.86

32 06 10.20 Steps

		Crew	Daily Output	Labor-Hours	Unit	Material	2009 Bare Costs Labor	Equipment	Total	Total Incl O&P
0010	**STEPS**									
0011	Incl. excav., borrow & concrete base as required									
0100	Brick steps	B-24	35	.686	LF Riser	11.40	17.20		28.60	41
0200	Railroad ties	2 Clab	25	.640		3.30	13.15		16.45	25.50
0300	Bluestone treads, 12" x 2" or 12" x 1-1/2"	B-24	30	.800		27	20		47	63
0600	Precast concrete, see Div. 03 41 23.50									
4025	Steel edge strips, incl. stakes, 1/4" x 5"	B-1	390	.062	L.F.	4.25	1.31		5.56	6.90
4050	Edging, landscape timber or railroad ties, 6" x 8"	2 Carp	170	.094	"	3.10	2.63		5.73	7.85

32 11 Base Courses

32 11 23 – Aggregate Base Courses

32 11 23.23 Base Course Drainage Layers

		Crew	Daily Output	Labor-Hours	Unit	Material	2009 Bare Costs Labor	Equipment	Total	Total Incl O&P
0010	**BASE COURSE DRAINAGE LAYERS**									
0011	For roadways and large areas									
0051	3/4" stone compacted to 3" deep	B-36	36000	.001	S.F.	.47	.03	.04	.54	.60
0101	6" deep		35100	.001		.94	.03	.04	1.01	1.13
0201	9" deep		25875	.002		1.37	.04	.05	1.46	1.63
0305	12" deep		21150	.002		1.94	.05	.06	2.05	2.29
0306	Crushed 1-1/2" stone base, compacted to 4" deep		47000	.001		.07	.02	.03	.12	.14
0307	6" deep		35100	.001		.97	.03	.04	1.04	1.16
0308	8" deep		27000	.001		1.30	.04	.05	1.39	1.54
0309	12" deep		16200	.002		1.94	.06	.08	2.08	2.33
0350	Bank run gravel, spread and compacted									

32 11 Base Courses

32 11 23 – Aggregate Base Courses

32 11 23.23 Base Course Drainage Layers

		Crew	Daily Output	Labor-Hours	Unit	Material	2009 Bare Costs Labor	2009 Bare Costs Equipment	Total	Total Incl O&P
0371	6" deep	B-32	54000	.001	S.F.	.59	.02	.03	.64	.72
0391	9" deep		39600	.001		.86	.02	.05	.93	1.04
0401	12" deep		32400	.001		1.18	.03	.06	1.27	1.40
6900	For small and irregular areas, add						50%	50%		

32 11 26 – Asphaltic Base Courses

32 11 26.19 Bituminous-Stabilized Base Courses

		Crew	Daily Output	Labor-Hours	Unit	Material	2009 Bare Costs Labor	2009 Bare Costs Equipment	Total	Total Incl O&P
0010	**BITUMINOUS-STABILIZED BASE COURSES**									
0020	And large paved areas									
0700	Liquid application to gravel base, asphalt emulsion	B-45	6000	.003	Gal.	5.85	.06	.05	5.96	6.55
0800	Prime and seal, cut back asphalt		6000	.003	"	6.90	.06	.05	7.01	7.75
1000	Macadam penetration crushed stone, 2 gal. per S.Y., 4" thick		6000	.003	S.Y.	11.70	.06	.05	11.81	13
1100	6" thick, 3 gal. per S.Y.		4000	.004		17.50	.09	.08	17.67	19.50
1200	8" thick, 4 gal. per S.Y.		3000	.005		23.50	.12	.10	23.72	26
8900	For small and irregular areas, add						50%	50%		

32 12 Flexible Paving

32 12 16 – Asphalt Paving

32 12 16.14 Paving Asphaltic Concrete

		Crew	Daily Output	Labor-Hours	Unit	Material	2009 Bare Costs Labor	2009 Bare Costs Equipment	Total	Total Incl O&P
0010	**PAVING ASPHALTIC CONCRETE**									
0020	6" stone base, 2" binder course, 1" topping	B-25C	9000	.005	S.F.	2.07	.13	.24	2.44	2.75
0300	Binder course, 1-1/2" thick		35000	.001		.52	.03	.06	.61	.69
0400	2" thick		25000	.002		.67	.05	.09	.81	.92
0500	3" thick		15000	.003		1.04	.08	.14	1.26	1.43
0600	4" thick		10800	.004		1.36	.10	.20	1.66	1.88
0800	Sand finish course, 3/4" thick		41000	.001		.30	.03	.05	.38	.43
0900	1" thick		34000	.001		.36	.03	.06	.45	.53
1000	Fill pot holes, hot mix, 2" thick	B-16	4200	.008		.77	.16	.13	1.06	1.26
1100	4" thick		3500	.009		1.12	.20	.15	1.47	1.73
1120	6" thick		3100	.010		1.50	.22	.17	1.89	2.22
1140	Cold patch, 2" thick	B-51	3000	.016		.78	.34	.06	1.18	1.49
1160	4" thick		2700	.018		1.48	.38	.07	1.93	2.33
1180	6" thick		1900	.025		2.30	.53	.10	2.93	3.54

32 13 Rigid Paving

32 13 13 – Concrete Paving

32 13 13.23 Concrete Paving Surface Treatment

		Crew	Daily Output	Labor-Hours	Unit	Material	2009 Bare Costs Labor	2009 Bare Costs Equipment	Total	Total Incl O&P
0010	**CONCRETE PAVING SURFACE TREATMENT**									
0015	Including joints, finishing and curing									
0021	Fixed form, 12' pass, unreinforced, 6" thick	B-26	18000	.005	S.F.	3.15	.12	.17	3.44	3.84
0101	8" thick	"	13500	.007		4.23	.15	.23	4.61	5.15
0701	Finishing, broom finish small areas	2 Cefi	1215	.013			.35		.35	.57

32 14 Unit Paving

32 14 16 – Brick Unit Paving

32 14 16.10 Brick Paving

		Crew	Daily Output	Labor-Hours	Unit	Material	2009 Bare Costs Labor	Equipment	Total	Total Incl O&P
0010	**BRICK PAVING**									
0012	4" x 8" x 1-1/2", without joints (4.5 brick/S.F.)	D-1	110	.145	S.F.	2.78	3.70		6.48	9.20
0100	Grouted, 3/8" joint (3.9 brick/S.F.)		90	.178		3.38	4.52		7.90	11.20
0200	4" x 8" x 2-1/4", without joints (4.5 bricks/S.F.)		110	.145		3.74	3.70		7.44	10.25
0300	Grouted, 3/8" joint (3.9 brick/S.F.)		90	.178		3.45	4.52		7.97	11.30
0500	Bedding, asphalt, 3/4" thick	B-25	5130	.017		.57	.39	.48	1.44	1.82
0540	Course washed sand bed, 1" thick	B-18	5000	.005		.28	.10	.01	.39	.48
0580	Mortar, 1" thick	D-1	300	.053		.64	1.36		2	2.95
0620	2" thick		200	.080		1.28	2.03		3.31	4.78
1500	Brick on 1" thick sand bed laid flat, 4.5 per S.F.		100	.160		2.99	4.07		7.06	10.05
2000	Brick pavers, laid on edge, 7.2 per S.F.		70	.229		2.74	5.80		8.54	12.65

32 14 23 – Asphalt Unit Paving

32 14 23.10 Asphalt Blocks

		Crew	Daily Output	Labor-Hours	Unit	Material	2009 Bare Costs Labor	Equipment	Total	Total Incl O&P
0010	**ASPHALT BLOCKS**									
0020	Rectangular, 6" x 12" x 1-1/4", w/bed & neopr. adhesive	D-1	135	.119	S.F.	7.65	3.01		10.66	13.45
0100	3" thick		130	.123		10.75	3.13		13.88	17
0300	Hexagonal tile, 8" wide, 1-1/4" thick		135	.119		7.65	3.01		10.66	13.45
0400	2" thick		130	.123		10.75	3.13		13.88	17
0500	Square, 8" x 8", 1-1/4" thick		135	.119		7.65	3.01		10.66	13.45
0600	2" thick		130	.123		10.75	3.13		13.88	17

32 14 40 – Stone Paving

32 14 40.10 Stone Pavers

		Crew	Daily Output	Labor-Hours	Unit	Material	2009 Bare Costs Labor	Equipment	Total	Total Incl O&P
0010	**STONE PAVERS**									
1100	Flagging, bluestone, irregular, 1" thick,	D-1	81	.198	S.F.	5.95	5		10.95	14.90
1150	Snapped random rectangular, 1" thick		92	.174		9	4.42		13.42	17.25
1200	1-1/2" thick		85	.188		10.80	4.79		15.59	19.85
1250	2" thick		83	.193		12.60	4.90		17.50	22
1300	Slate, natural cleft, irregular, 3/4" thick		92	.174		6.95	4.42		11.37	15
1310	1" thick		85	.188		8.10	4.79		12.89	16.85
1351	Random rectangular, gauged, 1/2" thick		105	.152		15.05	3.88		18.93	23
1400	Random rectangular, butt joint, gauged, 1/4" thick		150	.107		16.20	2.71		18.91	22.50
1450	For sand rubbed finish, add					7.55			7.55	8.30
1500	For interior setting, add								25%	25%
1550	Granite blocks, 3-1/2" x 3-1/2" x 3-1/2"	D-1	92	.174	S.F.	10	4.42		14.42	18.35

32 16 Curbs and Gutters

32 16 13 – Concrete Curbs and Gutters

32 16 13.13 Cast-in-Place Concrete Curbs and Gutters

		Crew	Daily Output	Labor-Hours	Unit	Material	2009 Bare Costs Labor	Equipment	Total	Total Incl O&P
0010	**CAST-IN-PLACE CONCRETE CURBS AND GUTTERS**									
0300	Concrete, wood forms, 6" x 18", straight	C-2A	500	.096	L.F.	4.84	2.58		7.42	9.60
0400	6" x 18", radius	"	200	.240	"	4.92	6.45		11.37	16.20

32 16 13.26 Precast Concrete Curbs

		Crew	Daily Output	Labor-Hours	Unit	Material	2009 Bare Costs Labor	Equipment	Total	Total Incl O&P
0010	**PRECAST CONCRETE CURBS**									
0550	Precast, 6" x 18", straight	B-29	700	.069	L.F.	10.50	1.53	1.37	13.40	15.65
0600	6" x 18", radius	"	325	.148	"	12.10	3.30	2.95	18.35	22

32 16 19 – Asphalt Curbs

32 16 19.10 Bituminous Concrete Curbs

		Crew	Daily Output	Labor-Hours	Unit	Material	2009 Bare Costs Labor	Equipment	Total	Total Incl O&P
0010	**BITUMINOUS CONCRETE CURBS**									
0012	Curbs, asphaltic, machine formed, 8" wide, 6" high, 40 L.F./ton	B-27	1000	.032	L.F.	1.24	.67	.27	2.18	2.80

32 16 Curbs and Gutters

32 16 19 – Asphalt Curbs

32 16 19.10 Bituminous Concrete Curbs

		Crew	Daily Output	Labor-Hours	Unit	Material	2009 Bare Costs Labor	Equipment	Total	Total Incl O&P
0100	8" wide, 8" high, 30 L.F. per ton	B-27	900	.036	L.F.	1.43	.75	.30	2.48	3.17
0150	Asphaltic berm, 12" W, 3"-6" H, 35 L.F./ton, before pavement	↓	700	.046		1.59	.96	.39	2.94	3.81
0200	12" W, 1-1/2" to 4" H, 60 L.F. per ton, laid with pavement	B-2	1050	.038	↓	.97	.80		1.77	2.41

32 16 40 – Stone Curbs

32 16 40.13 Cut Stone Curbs

		Crew	Daily Output	Labor-Hours	Unit	Material	2009 Bare Costs Labor	Equipment	Total	Total Incl O&P
0010	**CUT STONE CURBS**									
1000	Granite, split face, straight, 5" x 16"	D-13	500	.096	L.F.	12.05	2.58	1.20	15.83	18.85
1100	6" x 18"	"	450	.107	↓	15.85	2.87	1.34	20.06	23.50
1300	Radius curbing, 6" x 18", over 10' radius	B-29	260	.185	↓	19.40	4.13	3.69	27.22	32.50
1400	Corners, 2' radius		80	.600	Ea.	65	13.40	12	90.40	107
1600	Edging, 4-1/2" x 12", straight		300	.160	L.F.	6.05	3.58	3.20	12.83	16.15
1800	Curb inlets, (guttermouth) straight	↓	41	1.171	Ea.	145	26	23.50	194.50	229
2000	Indian granite (belgian block)									
2100	Jumbo, 10-1/2" x 7-1/2" x 4", grey	D-1	150	.107	L.F.	2.21	2.71		4.92	6.95
2150	Pink		150	.107		2.88	2.71		5.59	7.65
2200	Regular, 9" x 4-1/2" x 4-1/2", grey		160	.100		2	2.54		4.54	6.40
2250	Pink		160	.100		2.76	2.54		5.30	7.25
2300	Cubes, 4" x 4" x 4", grey		175	.091		1.91	2.33		4.24	5.95
2350	Pink		175	.091		2	2.33		4.33	6.05
2400	6" x 6" x 6", pink	↓	155	.103	↓	4.96	2.63		7.59	9.80
2500	Alternate pricing method for indian granite									
2550	Jumbo, 10-1/2" x 7-1/2" x 4" (30 lb), grey				Ton	126			126	139
2600	Pink					167			167	184
2650	Regular, 9" x 4-1/2" x 4-1/2" (20 lb), grey					141			141	156
2700	Pink					193			193	212
2750	Cubes, 4" x 4" x 4" (5 lb), grey					231			231	255
2800	Pink					257			257	283
2850	6" x 6" x 6" (25 lb), pink					193			193	212
2900	For pallets, add				↓	16.50			16.50	18.15

32 31 Fences and Gates

32 31 13 – Chain Link Fences and Gates

32 31 13.15 Chain Link Fence

		Crew	Daily Output	Labor-Hours	Unit	Material	2009 Bare Costs Labor	Equipment	Total	Total Incl O&P
0010	**CHAIN LINK FENCE**									
0020	1-5/8" post 10'O.C.,1-3/8" top rail,2" corner post galv. stl., 3' high	B-1	185	.130	L.F.	8.85	2.75		11.60	14.40
0050	4' high		170	.141		10.60	3		13.60	16.75
0100	6' high		115	.209	↓	12.05	4.43		16.48	21
0150	Add for gate 3' wide, 1-3/8" frame 3' high		12	2	Ea.	64.50	42.50		107	143
0170	4' high		10	2.400		62.50	51		113.50	155
0190	6' high		10	2.400		68	51		119	161
0200	Add for gate 4' wide, 1-3/8" frame 3' high		9	2.667		66	56.50		122.50	168
0220	4' high		9	2.667		71	56.50		127.50	174
0240	6' high		8	3	↓	178	63.50		241.50	305
0350	Aluminized steel, 9 ga. wire, 3' high		185	.130	L.F.	10.50	2.75		13.25	16.20
0380	4' high		170	.141		11.95	3		14.95	18.15
0400	6' high		115	.209	↓	15.30	4.43		19.73	24.50
0450	Add for gate 3' wide, 1-3/8" frame 3' high		12	2	Ea.	102	42.50		144.50	184
0470	4' high		10	2.400		139	51		190	239
0490	6' high		10	2.400		209	51		260	315
0500	Add for gate 4' wide, 1-3/8" frame 3' high	↓	10	2.400		139	51		190	239

32 31 13 – Chain Link Fences and Gates

32 31 13.15 Chain Link Fence	Crew	Daily Output	Labor-Hours	Unit	Material	2009 Bare Costs Labor	Equipment	Total	Total Incl O&P
0520	4' high	B-1	9	2.667	Ea.	186	56.50		242.50	300
0540	6' high		8	3		290	63.50		353.50	430
0620	Vinyl covered 9 ga. wire, 3' high		185	.130	L.F.	9.30	2.75		12.05	14.85
0640	4' high		170	.141		13.95	3		16.95	20.50
0660	6' high		115	.209		15.95	4.43		20.38	25
0720	Add for gate 3' wide, 1-3/8" frame 3' high		12	2	Ea.	99.50	42.50		142	182
0740	4' high		10	2.400		130	51		181	229
0760	6' high		10	2.400		199	51		250	305
0780	Add for gate 4' wide, 1-3/8" frame 3' high		10	2.400		136	51		187	235
0800	4' high		9	2.667		179	56.50		235.50	293
0820	6' high		8	3		259	63.50		322.50	395
0860	Tennis courts, 11 ga. wire, 2 1/2" post 10' O.C., 1-5/8" top rail									
0900	2-1/2" corner post, 10' high	B-1	95	.253	L.F.	19.90	5.35		25.25	31
0920	12' high		80	.300	"	24	6.35		30.35	37.50
1000	Add for gate 3' wide, 1-5/8" frame 10' high		10	2.400	Ea.	249	51		300	360
1040	Aluminized, 11 ga. wire 10' high		95	.253	L.F.	32.50	5.35		37.85	45
1100	12' high		80	.300	"	33.50	6.35		39.85	48
1140	Add for gate 3' wide, 1-5/8" frame, 10' high		10	2.400	Ea.	370	51		421	495
1250	Vinyl covered 11 ga. wire, 10' high		95	.253	L.F.	25	5.35		30.35	36.50
1300	12' high		80	.300	"	29.50	6.35		35.85	43
1400	Add for gate 3' wide, 1-3/8" frame, 10' high		10	2.400	Ea.	375	51		426	500

32 31 23 – Plastic Fences and Gates

32 31 23.10 Fence, Vinyl	Crew	Daily Output	Labor-Hours	Unit	Material	2009 Bare Costs Labor	Equipment	Total	Total Incl O&P
0010	**FENCE, VINYL**									
0011	White, steel reinforced, stainless steel fasteners									
0020	Picket, 4" x 4" posts @ 6' - 0" OC, 3' high	B-1	140	.171	L.F.	19.85	3.64		23.49	28
0030	4' high		130	.185		23	3.92		26.92	31.50
0040	5' high		120	.200		26	4.24		30.24	36
0100	Board (semi-privacy), 5" x 5" posts @ 7' - 6" OC, 5' high		130	.185		27.50	3.92		31.42	36.50
0120	6' high		125	.192		31	4.07		35.07	41
0200	Basketweave, 5" x 5" posts @ 7' - 6" OC, 5' high		160	.150		24	3.18		27.18	32
0220	6' high		150	.160		28.50	3.40		31.90	37.50
0300	Privacy, 5" x 5" posts @ 7' - 6" OC, 5' high		130	.185		29.50	3.92		33.42	39
0320	6' high		150	.160		33.50	3.40		36.90	43
0350	Gate, 5' high		9	2.667	Ea.	415	56.50		471.50	550
0360	6' high		9	2.667		425	56.50		481.50	560
0400	For posts set in concrete, add		25	.960		8.45	20.50		28.95	44

32 31 26 – Wire Fences and Gates

32 31 26.10 Fences, Misc. Metal	Crew	Daily Output	Labor-Hours	Unit	Material	2009 Bare Costs Labor	Equipment	Total	Total Incl O&P
0010	**FENCES, MISC. METAL**									
0012	Chicken wire, posts @ 4', 1" mesh, 4' high	B-80C	410	.059	L.F.	2.16	1.23	.49	3.88	5
0100	2" mesh, 6' high		350	.069		1.95	1.44	.58	3.97	5.20
0200	Galv. steel, 12 ga., 2" x 4" mesh, posts 5' O.C., 3' high		300	.080		3.12	1.68	.67	5.47	7
0300	5' high		300	.080		4.16	1.68	.67	6.51	8.15
0400	14 ga., 1" x 2" mesh, 3' high		300	.080		3.31	1.68	.67	5.66	7.20
0500	5' high		300	.080		4.58	1.68	.67	6.93	8.65
1000	Kennel fencing, 1-1/2" mesh, 6' long, 3'-6" wide, 6'-2" high	2 Clab	4	4	Ea.	520	82		602	710
1050	12' long		4	4		625	82		707	825
1200	Top covers, 1-1/2" mesh, 6' long		15	1.067		106	22		128	153
1250	12' long		12	1.333		169	27.50		196.50	233

32 31 Fences and Gates

32 31 29 – Wood Fences and Gates

32 31 29.10 Fence, Wood

		Crew	Daily Output	Labor-Hours	Unit	Material	2009 Bare Costs Labor	2009 Bare Costs Equipment	Total	Total Incl O&P
0010	**FENCE, WOOD**									
0011	Basket weave, 3/8" x 4" boards, 2" x 4"									
0020	stringers on spreaders, 4" x 4" posts									
0050	No. 1 cedar, 6' high	B-80C	160	.150	L.F.	9.10	3.15	1.26	13.51	16.70
0070	Treated pine, 6' high		150	.160		10.95	3.36	1.35	15.66	19.20
0090	Vertical weave 6' high		145	.166		13	3.48	1.39	17.87	21.50
0200	Board fence, 1" x 4" boards, 2" x 4" rails, 4" x 4" post									
0220	Preservative treated, 2 rail, 3' high	B-80C	145	.166	L.F.	6.75	3.48	1.39	11.62	14.85
0240	4' high		135	.178		7.40	3.74	1.50	12.64	16.10
0260	3 rail, 5' high		130	.185		8.35	3.88	1.55	13.78	17.45
0300	6' high		125	.192		9.70	4.04	1.61	15.35	19.30
0320	No. 2 grade western cedar, 2 rail, 3' high		145	.166		7.35	3.48	1.39	12.22	15.50
0340	4' high		135	.178		8.80	3.74	1.50	14.04	17.65
0360	3 rail, 5' high		130	.185		10.15	3.88	1.55	15.58	19.45
0400	6' high		125	.192		11.10	4.04	1.61	16.75	21
0420	No. 1 grade cedar, 2 rail, 3' high		145	.166		11.15	3.48	1.39	16.02	19.70
0440	4' high		135	.178		12.60	3.74	1.50	17.84	22
0460	3 rail, 5' high		130	.185		14.85	3.88	1.55	20.28	24.50
0500	6' high		125	.192		16.60	4.04	1.61	22.25	27
0540	Shadow box, 1" x 6" board, 2" x 4" rail, 4" x 4"post									
0560	Pine, pressure treated, 3 rail, 6' high	B-80C	150	.160	L.F.	12.45	3.36	1.35	17.16	21
0600	Gate, 3'-6" wide		8	3	Ea.	75.50	63	25	163.50	217
0620	No. 1 cedar, 3 rail, 4' high		130	.185	L.F.	15.25	3.88	1.55	20.68	25
0640	6' high		125	.192		18.85	4.04	1.61	24.50	29.50
0860	Open rail fence, split rails, 2 rail 3' high, no. 1 cedar		160	.150		6.15	3.15	1.26	10.56	13.50
0870	No. 2 cedar		160	.150		4.86	3.15	1.26	9.27	12.05
0880	3 rail, 4' high, no. 1 cedar		150	.160		8.25	3.36	1.35	12.96	16.25
0890	No. 2 cedar		150	.160		5.80	3.36	1.35	10.51	13.55
0920	Rustic rails, 2 rail 3' high, no. 1 cedar		160	.150		3.98	3.15	1.26	8.39	11.05
0930	No. 2 cedar		160	.150		3.87	3.15	1.26	8.28	10.95
0940	3 rail, 4' high		150	.160		5.40	3.36	1.35	10.11	13.05
0950	No. 2 cedar		150	.160		4.10	3.36	1.35	8.81	11.65
0960	Picket fence, gothic, pressure treated pine									
1000	2 rail, 3' high	B-80C	140	.171	L.F.	4.60	3.60	1.44	9.64	12.75
1020	3 rail, 4' high		130	.185	"	5.45	3.88	1.55	10.88	14.25
1040	Gate, 3'-6" wide		9	2.667	Ea.	48.50	56	22.50	127	172
1060	No. 2 cedar, 2 rail, 3' high		140	.171	L.F.	5.75	3.60	1.44	10.79	14.05
1100	3 rail, 4' high		130	.185	"	5.90	3.88	1.55	11.33	14.75
1120	Gate, 3'-6" wide		9	2.667	Ea.	57.50	56	22.50	136	182
1140	No. 1 cedar, 2 rail 3' high		140	.171	L.F.	11.50	3.60	1.44	16.54	20.50
1160	3 rail, 4' high		130	.185		13.40	3.88	1.55	18.83	23
1200	Rustic picket, molded pine, 2 rail, 3' high		140	.171		5.50	3.60	1.44	10.54	13.75
1220	No. 1 cedar, 2 rail, 3' high		140	.171		7.35	3.60	1.44	12.39	15.80
1240	Stockade fence, no. 1 cedar, 3-1/4" rails, 6' high		160	.150		10.85	3.15	1.26	15.26	18.65
1260	8' high		155	.155		14	3.25	1.30	18.55	22.50
1300	No. 2 cedar, treated wood rails, 6' high		160	.150		10.85	3.15	1.26	15.26	18.65
1320	Gate, 3'-6" wide		8	3	Ea.	63.50	63	25	151.50	204
1360	Treated pine, treated rails, 6' high		160	.150	L.F.	10.60	3.15	1.26	15.01	18.35
1400	8' high		150	.160	"	15.90	3.36	1.35	20.61	24.50

32 31 29.20 Fence, Wood Rail

		Crew	Daily Output	Labor-Hours	Unit	Material	2009 Bare Costs Labor	2009 Bare Costs Equipment	Total	Total Incl O&P
0010	**FENCE, WOOD RAIL**									
0012	Picket, No. 2 cedar, Gothic, 2 rail, 3' high	B-1	160	.150	L.F.	5.85	3.18		9.03	11.85

32 31 Fences and Gates

32 31 29 – Wood Fences and Gates

32 31 29.20 Fence, Wood Rail

		Crew	Daily Output	Labor-Hours	Unit	Material	2009 Bare Costs Labor	Equipment	Total	Total Incl O&P
0050	Gate, 3'-6" wide	B-80C	9	2.667	Ea.	50.50	56	22.50	129	175
0400	3 rail, 4' high		150	.160	L.F.	6.75	3.36	1.35	11.46	14.55
0500	Gate, 3'-6" wide		9	2.667	Ea.	60.50	56	22.50	139	186
1200	Stockade, No. 2 cedar, treated wood rails, 6' high		160	.150	L.F.	7.30	3.15	1.26	11.71	14.75
1250	Gate, 3' wide		9	2.667	Ea.	60	56	22.50	138.50	185
1300	No. 1 cedar, 3-1/4" cedar rails, 6' high		160	.150	L.F.	18.05	3.15	1.26	22.46	26.50
1500	Gate, 3' wide		9	2.667	Ea.	152	56	22.50	230.50	286
2700	Prefabricated redwood or cedar, 4' high		160	.150	L.F.	13.50	3.15	1.26	17.91	21.50
2800	6' high		150	.160		17.90	3.36	1.35	22.61	27
3300	Board, shadow box, 1" x 6", treated pine, 6' high		160	.150		10.50	3.15	1.26	14.91	18.25
3400	No. 1 cedar, 6' high		150	.160		20.50	3.36	1.35	25.21	30
3900	Basket weave, No. 1 cedar, 6' high	↓	160	.150	↓	20.50	3.15	1.26	24.91	29
4200	Gate, 3'-6" wide	B-1	9	2.667	Ea.	65.50	56.50		122	168
5000	Fence rail, redwood, 2" x 4", merch grade 8'	"	2400	.010	L.F.	1.15	.21		1.36	1.63

32 32 Retaining Walls

32 32 13 – Cast-in-Place Concrete Retaining Walls

32 32 13.10 Retaining Walls, Cast Concrete

		Crew	Daily Output	Labor-Hours	Unit	Material	2009 Bare Costs Labor	Equipment	Total	Total Incl O&P
0010	**RETAINING WALLS, CAST CONCRETE**									
1800	Concrete gravity wall with vertical face including excavation & backfill									
1850	No reinforcing									
1900	6' high, level embankment	C-17C	36	2.306	L.F.	74	65.50	13.50	153	207
2000	33° slope embankment	"	32	2.594	"	67.50	74	15.20	156.70	216
2800	Reinforced concrete cantilever, incl. excavation, backfill & reinf.									
2900	6' high, 33° slope embankment	C-17C	35	2.371	L.F.	67.50	67.50	13.90	148.90	203

32 32 23 – Segmental Retaining Walls

32 32 23.13 Segmental Conc. Unit Masonry Retaining Walls

		Crew	Daily Output	Labor-Hours	Unit	Material	2009 Bare Costs Labor	Equipment	Total	Total Incl O&P
0010	**SEGMENTAL CONC. UNIT MASONRY RETAINING WALLS**									
7100	Segmental Retaining Wall system, incl pins, and void fill									
7120	base not included									
7140	Large unit, 8" high x 18" wide x 20" deep, 3 plane split	B-62	300	.080	S.F.	10.30	1.81	.50	12.61	14.90
7150	Straight split		300	.080		10.30	1.81	.50	12.61	14.90
7160	Medium, ltwt, 8" high x 18" wide x 12" deep, 3 plane split		400	.060		11.50	1.36	.37	13.23	15.35
7170	Straight split		400	.060		8.65	1.36	.37	10.38	12.20
7180	Small unit, 4" x 18" x 10" deep, 3 plane split		400	.060		12.25	1.36	.37	13.98	16.20
7190	Straight split		400	.060		12.95	1.36	.37	14.68	16.95
7200	Cap unit, 3 plane split		300	.080		14.50	1.81	.50	16.81	19.55
7210	Cap unit, straight split	↓	300	.080	↓	14.50	1.81	.50	16.81	19.55
7260	For reinforcing, add								2.20	2.75
8000	For higher walls, add components as necessary									

32 32 26 – Metal Crib Retaining Walls

32 32 26.10 Metal Bin Retaining Walls

		Crew	Daily Output	Labor-Hours	Unit	Material	2009 Bare Costs Labor	Equipment	Total	Total Incl O&P
0010	**METAL BIN RETAINING WALLS**									
0011	Aluminized steel bin, excavation									
0020	and backfill not included, 10' wide									
0100	4' high, 5.5' deep	B-13	650	.074	S.F.	23.50	1.65	1.21	26.36	30
0200	8' high, 5.5' deep		615	.078		27	1.74	1.28	30.02	34
0300	10' high, 7.7' deep		580	.083		28.50	1.85	1.36	31.71	35.50
0400	12' high, 7.7' deep		530	.091		30.50	2.02	1.49	34.01	38.50
0500	16' high, 7.7' deep	↓	515	.093	↓	32.50	2.08	1.53	36.11	40.50

32 32 Retaining Walls

32 32 60 – Stone Retaining Walls

32 32 60.10 Retaining Walls, Stone

32 32 60.10 Retaining Walls, Stone	Crew	Daily Output	Labor-Hours	Unit	Material	2009 Bare Costs Labor	Equipment	Total	Total Incl O&P
0010 **RETAINING WALLS, STONE**									
0015 Including excavation, concrete footing and									
0020 stone 3' below grade. Price is exposed face area.									
0200 Decorative random stone, to 6' high, 1'-6" thick, dry set	D-1	35	.457	S.F.	47	11.65		58.65	71.50
0300 Mortar set		40	.400		47	10.15		57.15	69
0500 Cut stone, to 6' high, 1'-6" thick, dry set		35	.457		47	11.65		58.65	71.50
0600 Mortar set		40	.400		47	10.15		57.15	69
0800 Random stone, 6' to 10' high, 2' thick, dry set		45	.356		47	9.05		56.05	67
0900 Mortar set		50	.320		47	8.15		55.15	65.50
1100 Cut stone, 6' to 10' high, 2' thick, dry set		45	.356		47	9.05		56.05	67
1200 Mortar set		50	.320		47	8.15		55.15	65.50

32 84 Planting Irrigation

32 84 23 – Underground Sprinklers

32 84 23.10 Sprinkler Irrigation System

32 84 23.10 Sprinkler Irrigation System	Crew	Daily Output	Labor-Hours	Unit	Material	2009 Bare Costs Labor	Equipment	Total	Total Incl O&P
0010 **SPRINKLER IRRIGATION SYSTEM**									
0011 For lawns									
0800 Residential system, custom, 1" supply	B-20	2000	.012	S.F.	.32	.25		.57	.78
0900 1-1/2" supply	"	1800	.013	"	.39	.28		.67	.91

32 91 Planting Preparation

32 91 13 – Soil Preparation

32 91 13.16 Mulching

32 91 13.16 Mulching	Crew	Daily Output	Labor-Hours	Unit	Material	2009 Bare Costs Labor	Equipment	Total	Total Incl O&P
0010 **MULCHING**									
0100 Aged barks, 3" deep, hand spread	1 Clab	100	.080	S.Y.	2.88	1.64		4.52	5.95
0150 Skid steer loader	B-63	13.50	2.963	M.S.F.	320	61	11.05	392.05	465
0200 Hay, 1" deep, hand spread	1 Clab	475	.017	S.Y.	.47	.35		.82	1.11
0250 Power mulcher, small	B-64	180	.089	M.S.F.	52	1.89	1.84	55.73	62.50
0350 Large	B-65	530	.030	"	52	.64	.88	53.52	59.50
0400 Humus peat, 1" deep, hand spread	1 Clab	700	.011	S.Y.	4.50	.23		4.73	5.35
0450 Push spreader	"	2500	.003	"	4.50	.07		4.57	5.05
0550 Tractor spreader	B-66	700	.011	M.S.F.	500	.31	.33	500.64	550
0600 Oat straw, 1" deep, hand spread	1 Clab	475	.017	S.Y.	.46	.35		.81	1.10
0650 Power mulcher, small	B-64	180	.089	M.S.F.	51	1.89	1.84	54.73	61
0700 Large	B-65	530	.030	"	51	.64	.88	52.52	58
0750 Add for asphaltic emulsion	B-45	1770	.009	Gal.	2.24	.20	.17	2.61	2.98
0800 Peat moss, 1" deep, hand spread	1 Clab	900	.009	S.Y.	1.85	.18		2.03	2.35
0850 Push spreader	"	2500	.003	"	1.85	.07		1.92	2.15
0950 Tractor spreader	B-66	700	.011	M.S.F.	206	.31	.33	206.64	227
1000 Polyethylene film, 6 mil.	2 Clab	2000	.008	S.Y.	.19	.16		.35	.49
1100 Redwood nuggets, 3" deep, hand spread	1 Clab	150	.053	"	4.65	1.10		5.75	6.95
1150 Skid steer loader	B-63	13.50	2.963	M.S.F.	515	61	11.05	587.05	685
1200 Stone mulch, hand spread, ceramic chips, economy	1 Clab	125	.064	S.Y.	6.90	1.32		8.22	9.80
1250 Deluxe	"	95	.084	"	10.55	1.73		12.28	14.60
1300 Granite chips	B-1	10	2.400	C.Y.	34.50	51		85.50	124
1400 Marble chips		10	2.400		129	51		180	228
1600 Pea gravel		28	.857		68.50	18.20		86.70	106
1700 Quartz		10	2.400		166	51		217	268
1800 Tar paper, 15 Lb. felt	1 Clab	800	.010	S.Y.	.43	.21		.64	.82

32 91 Planting Preparation

32 91 13 – Soil Preparation

32 91 13.16 Mulching

		Crew	Daily Output	Labor-Hours	Unit	Material	2009 Bare Costs Labor	2009 Bare Costs Equipment	Total	Total Incl O&P
1900	Wood chips, 2" deep, hand spread	1 Clab	220	.036	S.Y.	1.88	.75		2.63	3.33
1950	Skid steer loader	B-63	20.30	1.970	M.S.F.	209	40.50	7.35	256.85	305

32 91 13.26 Planting Beds

		Crew	Daily Output	Labor-Hours	Unit	Material	2009 Bare Costs Labor	2009 Bare Costs Equipment	Total	Total Incl O&P
0010	**PLANTING BEDS**									
0100	Backfill planting pit, by hand, on site topsoil	2 Clab	18	.889	C.Y.		18.25		18.25	31
0200	Prepared planting mix, by hand	"	24	.667			13.70		13.70	23
0300	Skid steer loader, on site topsoil	B-62	340	.071			1.60	.44	2.04	3.16
0400	Prepared planting mix	"	410	.059			1.33	.36	1.69	2.62
1000	Excavate planting pit, by hand, sandy soil	2 Clab	16	1			20.50		20.50	35
1100	Heavy soil or clay	"	8	2			41		41	69.50
1200	1/2 C.Y. backhoe, sandy soil	B-11C	150	.107			2.62	1.96	4.58	6.50
1300	Heavy soil or clay	"	115	.139			3.42	2.55	5.97	8.50
2000	Mix planting soil, incl. loam, manure, peat, by hand	2 Clab	60	.267		39.50	5.50		45	53
2100	Skid steer loader	B-62	150	.160		39.50	3.63	1	44.13	50.50
3000	Pile sod, skid steer loader	"	2800	.009	S.Y.		.19	.05	.24	.39
3100	By hand	2 Clab	400	.040			.82		.82	1.39
4000	Remove sod, F.E. loader	B-10S	2000	.004			.11	.16	.27	.37
4100	Sod cutter	B-12K	3200	.005			.12	.35	.47	.60
4200	By hand	2 Clab	240	.067			1.37		1.37	2.32

32 91 19 – Landscape Grading

32 91 19.13 Topsoil Placement and Grading

		Crew	Daily Output	Labor-Hours	Unit	Material	2009 Bare Costs Labor	2009 Bare Costs Equipment	Total	Total Incl O&P
0010	**TOPSOIL PLACEMENT AND GRADING**									
0300	Fine grade, base course for paving, see Div. 32 11 23.23									
0701	Furnish and place, truck dumped, unscreened, 4" deep	B-10S	12000	.001	S.F.	.38	.02	.03	.43	.47
0801	6" deep	"	7400	.001	"	.37	.03	.04	.44	.51
0900	Fine grading and seeding, incl. lime, fertilizer & seed,									
1001	With equipment	B-14	9000	.005	S.F.	.06	.12	.03	.21	.30

32 92 Turf and Grasses

32 92 19 – Seeding

32 92 19.13 Mechanical Seeding

		Crew	Daily Output	Labor-Hours	Unit	Material	2009 Bare Costs Labor	2009 Bare Costs Equipment	Total	Total Incl O&P
0010	**MECHANICAL SEEDING** R329219-50									
0020	Mechanical seeding, 215 lb./acre	B-66	1.50	5.333	Acre	565	144	154	863	1,025
0101	$2.00/lb., 44 lb./M.S.Y.	1 Clab	13950	.001	S.F.	.02	.01		.03	.04
0300	Fine grading and seeding incl. lime, fertilizer & seed,									
0310	with equipment	B-14	1000	.048	S.Y.	.19	1.05	.29	1.53	2.30
0600	Limestone hand push spreader, 50 lbs. per M.S.F.	1 Clab	180	.044	M.S.F.	3.97	.91		4.88	5.90
0800	Grass seed hand push spreader, 4.5 lbs. per M.S.F.	"	180	.044	"	18.90	.91		19.81	22.50

32 92 23 – Sodding

32 92 23.10 Sodding Systems

		Crew	Daily Output	Labor-Hours	Unit	Material	2009 Bare Costs Labor	2009 Bare Costs Equipment	Total	Total Incl O&P
0010	**SODDING SYSTEMS**									
0020	Sodding, 1" deep, bluegrass sod, on level ground, over 8 MSF	B-63	22	1.818	M.S.F.	245	37.50	6.80	289.30	340
0200	4 M.S.F.		17	2.353		270	48.50	8.80	327.30	390
0300	1000 S.F.		13.50	2.963		295	61	11.05	367.05	440
0500	Sloped ground, over 8 M.S.F.		6	6.667		245	137	25	407	530
0600	4 M.S.F.		5	8		270	164	30	464	610
0700	1000 S.F.		4	10		295	206	37.50	538.50	715
1000	Bent grass sod, on level ground, over 6 M.S.F.		20	2		545	41	7.45	593.45	675
1100	3 M.S.F.		18	2.222		605	45.50	8.30	658.80	750
1200	Sodding 1000 S.F. or less		14	2.857		695	58.50	10.65	764.15	870

32 92 Turf and Grasses

32 92 23 – Sodding

	32 92 23.10 Sodding Systems	Crew	Daily Output	Labor-Hours	Unit	Material	2009 Bare Costs Labor	Equipment	Total	Total Incl O&P
1500	Sloped ground, over 6 M.S.F.	B-63	15	2.667	M.S.F.	545	55	9.95	609.95	700
1600	3 M.S.F.		13.50	2.963		605	61	11.05	677.05	780
1700	1000 S.F.	↓	12	3.333	↓	695	68.50	12.45	775.95	890

32 93 Plants

32 93 13 – Ground Covers

32 93 13.10 Ground Cover Plants

	32 93 13.10 Ground Cover Plants	Crew	Daily Output	Labor-Hours	Unit	Material	Labor	Equipment	Total	Total Incl O&P
0010	**GROUND COVER PLANTS**									
0012	Plants, pachysandra, in prepared beds	B-1	15	1.600	C	27	34		61	87
0200	Vinca minor, 1 yr, bare root, in prepared beds		12	2	"	40	42.50		82.50	116
0600	Stone chips, in 50 lb. bags, Georgia marble		520	.046	Bag	5.20	.98		6.18	7.35
0700	Onyx gemstone		260	.092	↓	22.50	1.96		24.46	28.50
0800	Quartz		260	.092	↓	8.45	1.96		10.41	12.60
0900	Pea gravel, truckload lots	↓	28	.857	Ton	34	18.20		52.20	68.50

32 93 33 – Shrubs

32 93 33.10 Shrubs and Trees

	32 93 33.10 Shrubs and Trees	Crew	Daily Output	Labor-Hours	Unit	Material	Labor	Equipment	Total	Total Incl O&P
0010	**SHRUBS AND TREES**									
0011	Evergreen, in prepared beds, B & B									
0100	Arborvitae pyramidal, 4'-5'	B-17	30	1.067	Ea.	47	24	20.50	91.50	115
0150	Globe, 12"-15"	B-1	96	.250		21	5.30		26.30	32
0300	Cedar, blue, 8'-10'	B-17	18	1.778		233	40.50	34.50	308	360
0500	Hemlock, canadian, 2-1/2'-3'	B-1	36	.667		33.50	14.15		47.65	61
0550	Holly, Savannah, 8' - 10' H		9.68	2.479		475	52.50		527.50	615
0600	Juniper, andorra, 18"-24"		80	.300		16.75	6.35		23.10	29
0620	Wiltoni, 15"-18"	↓	80	.300		18.10	6.35		24.45	30.50
0640	Skyrocket, 4-1/2'-5'	B-17	55	.582		55	13.20	11.30	79.50	95
0660	Blue pfitzer, 2'-2-1/2'	B-1	44	.545		25.50	11.55		37.05	47.50
0680	Ketleerie, 2-1/2'-3'		50	.480		33.50	10.20		43.70	54
0700	Pine, black, 2-1/2'-3'		50	.480		45.50	10.20		55.70	67
0720	Mugo, 18"-24"	↓	60	.400		36	8.50		44.50	54
0740	White, 4'-5'	B-17	75	.427		56.50	9.70	8.25	74.45	88
0800	Spruce, blue, 18"-24"	B-1	60	.400		38	8.50		46.50	56.50
0840	Norway, 4'-5'	B-17	75	.427		65	9.70	8.25	82.95	97
0900	Yew, denisforma, 12"-15"	B-1	60	.400		25.50	8.50		34	43
1000	Capitata, 18"-24"		30	.800		24	17		41	54.50
1100	Hicksi, 2'-2-1/2'	↓	30	.800	↓	32	17		49	63.50

32 93 33.20 Shrubs

	32 93 33.20 Shrubs	Crew	Daily Output	Labor-Hours	Unit	Material	Labor	Equipment	Total	Total Incl O&P
0010	**SHRUBS**									
0011	Broadleaf Evergreen, planted in prepared beds									
0100	Andromeda, 15"-18", container	B-1	96	.250	Ea.	26.50	5.30		31.80	38
0200	Azalea, 15" - 18", container		96	.250		32	5.30		37.30	44
0300	Barberry, 9"-12", container		130	.185		11.75	3.92		15.67	19.55
0400	Boxwood, 15"-18", B&B		96	.250		29	5.30		34.30	40.50
0500	Euonymus, emerald gaiety, 12" to 15", container		115	.209		21	4.43		25.43	30.50
0600	Holly, 15"-18", B & B		96	.250		18.60	5.30		23.90	29.50
0900	Mount laurel, 18" - 24", B & B		80	.300		56	6.35		62.35	72.50
1000	Paxistema, 9 – 12" high		130	.185		18.25	3.92		22.17	26.50
1100	Rhododendron, 18"-24", container		48	.500		31.50	10.60		42.10	52.50
1200	Rosemary, 1 gal container		600	.040		62.50	.85		63.35	70.50
2000	Deciduous, planted in prepared beds, amelanchier, 2'-3', B & B	↓	57	.421	↓	90	8.95		98.95	114

32 93 33 – Shrubs

32 93 33.20 Shrubs

		Crew	Daily Output	Labor-Hours	Unit	Material	2009 Bare Costs Labor	Equipment	Total	Total Incl O&P
2100	Azalea, 15"-18", B & B	B-1	96	.250	Ea.	24.50	5.30		29.80	36
2300	Bayberry, 2'-3', B & B		57	.421		28	8.95		36.95	45.50
2600	Cotoneaster, 15"-18", B & B		80	.300		16.50	6.35		22.85	29
2800	Dogwood, 3'-4', B & B	B-17	40	.800		26	18.15	15.50	59.65	76
2900	Euonymus, alatus compacta, 15" to 18", container	B-1	80	.300		22	6.35		28.35	35
3200	Forsythia, 2'-3', container	"	60	.400		20	8.50		28.50	36.50
3300	Hibiscus, 3'-4', B & B	B-17	75	.427		15	9.70	8.25	32.95	42
3400	Honeysuckle, 3'-4', B & B	B-1	60	.400		23	8.50		31.50	40
3500	Hydrangea, 2'-3', B & B	"	57	.421		28.50	8.95		37.45	46.50
3600	Lilac, 3'-4', B & B	B-17	40	.800		27	18.15	15.50	60.65	77
3900	Privet, bare root, 18"-24"	B-1	80	.300		13.15	6.35		19.50	25
4100	Quince, 2'-3', B & B	"	57	.421		24	8.95		32.95	41.50
4200	Russian olive, 3'-4', B & B	B-17	75	.427		24	9.70	8.25	41.95	51.50
4400	Spirea, 3'-4', B & B	B-1	70	.343		17.15	7.30		24.45	31
4500	Viburnum, 3'-4', B & B	B-17	40	.800		29.50	18.15	15.50	63.15	80

32 93 43 – Trees

32 93 43.20 Trees

			Crew	Daily Output	Labor-Hours	Unit	Material	2009 Bare Costs Labor	Equipment	Total	Total Incl O&P
0010	**TREES**										
0011	Deciduous, in prep. beds, balled & burlapped (B&B)										
0100	Ash, 2" caliper	G	B-17	8	4	Ea.	115	91	77.50	283.50	365
0200	Beech, 5'-6'	G		50	.640		223	14.50	12.40	249.90	283
0300	Birch, 6'-8', 3 stems	G		20	1.600		123	36.50	31	190.50	230
0500	Crabapple, 6'-8'	G		20	1.600		160	36.50	31	227.50	271
0600	Dogwood, 4'-5'	G		40	.800		68.50	18.15	15.50	102.15	123
0700	Eastern redbud 4'-5'	G		40	.800		133	18.15	15.50	166.65	195
0800	Elm, 8'-10'	G		20	1.600		119	36.50	31	186.50	226
0900	Ginkgo, 6'-7'	G		24	1.333		163	30.50	26	219.50	258
1000	Hawthorn, 8'-10', 1" caliper	G		20	1.600		136	36.50	31	203.50	245
1100	Honeylocust, 10'-12', 1-1/2" caliper	G		10	3.200		149	72.50	62	283.50	355
1300	Larch, 8'	G		32	1		95	22.50	19.40	136.90	165
1400	Linden, 8'-10', 1" caliper	G		20	1.600		107	36.50	31	174.50	213
1500	Magnolia, 4'-5'	G		20	1.600		79.50	36.50	31	147	183
1600	Maple, red, 8'-10', 1-1/2" caliper	G		10	3.200		174	72.50	62	308.50	380
1700	Mountain ash, 8'-10', 1" caliper	G		16	2		181	45.50	39	265.50	320
1800	Oak, 2-1/2"-3" caliper	G		6	5.333		258	121	103	482	600
2100	Planetree, 9'-11', 1-1/4" caliper	G		10	3.200		105	72.50	62	239.50	305
2200	Plum, 6'-8', 1" caliper	G		20	1.600		91	36.50	31	158.50	195
2300	Poplar, 9'-11', 1-1/4" caliper	G		10	3.200		50	72.50	62	184.50	245
2500	Sumac, 2'-3'	G		75	.427		23	9.70	8.25	40.95	50.50
2700	Tulip, 5'-6'	G		40	.800		49	18.15	15.50	82.65	101
2800	Willow, 6'-8', 1" caliper	G		20	1.600		64	36.50	31	131.50	165

32 94 Planting Accessories

32 94 13 – Landscape Edging

32 94 13.20 Edging

		Crew	Daily Output	Labor-Hours	Unit	Material	2009 Bare Costs Labor	Equipment	Total	Total Incl O&P
0010	**EDGING**									
0050	Aluminum alloy, including stakes, 1/8" x 4", mill finish	B-1	390	.062	L.F.	3.10	1.31		4.41	5.60
0051	Black paint		390	.062		3.60	1.31		4.91	6.15
0052	Black anodized	↓	390	.062		4.15	1.31		5.46	6.80
0100	Brick, set horizontally, 1-1/2 bricks per L.F.	D-1	370	.043		1.67	1.10		2.77	3.66
0150	Set vertically, 3 bricks per L.F.	"	135	.119		4.20	3.01		7.21	9.60
0200	Corrugated aluminum, roll, 4" wide	1 Carp	650	.012		.70	.34		1.04	1.35
0250	6" wide	"	550	.015		.88	.41		1.29	1.65
0600	Railroad ties, 6" x 8"	2 Carp	170	.094		3.10	2.63		5.73	7.85
0650	7" x 9"	↓	136	.118		3.44	3.29		6.73	9.35
0750	Redwood 2" x 4"	↓	330	.048		2.27	1.36		3.63	4.78
0800	Steel edge strips, incl. stakes, 1/4" x 5"	B-1	390	.062		4.25	1.31		5.56	6.90
0850	3/16" x 4"	"	390	.062	↓	3.36	1.31		4.67	5.90

32 94 50 – Tree Guying

32 94 50.10 Tree Guying Systems

		Crew	Daily Output	Labor-Hours	Unit	Material	2009 Bare Costs Labor	Equipment	Total	Total Incl O&P
0010	**TREE GUYING SYSTEMS**									
0015	Tree guying Including stakes, guy wire and wrap									
0100	Less than 3" caliper, 2 stakes	2 Clab	35	.457	Ea.	17.50	9.40		26.90	35
0200	3" to 4" caliper, 3 stakes	"	21	.762	"	20	15.65		35.65	48.50
1000	Including arrowhead anchor, cable, turnbuckles and wrap									
1100	Less than 3" caliper, 3" anchors	2 Clab	20	.800	Ea.	51	16.45		67.45	84
1200	3" to 6" caliper, 4" anchors		15	1.067		51	22		73	93
1300	6" caliper, 6" anchors		12	1.333		95	27.50		122.50	152
1400	8" caliper, 8" anchors	↓	9	1.778	↓	107	36.50		143.50	179

32 96 Transplanting

32 96 23 – Plant and Bulb Transplanting

32 96 23.23 Planting

		Crew	Daily Output	Labor-Hours	Unit	Material	2009 Bare Costs Labor	Equipment	Total	Total Incl O&P
0010	**PLANTING**									
0012	Moving shrubs on site, 12" ball	B-62	28	.857	Ea.		19.45	5.35	24.80	38.50
0100	24" ball	"	22	1.091	"		24.50	6.80	31.30	49

32 96 23.43 Moving Trees

		Crew	Daily Output	Labor-Hours	Unit	Material	2009 Bare Costs Labor	Equipment	Total	Total Incl O&P
0290	**MOVING TREES**, On site									
0300	Moving trees on site, 36" ball	B-6	3.75	6.400	Ea.		145	78.50	223.50	330
0400	60" ball	"	1	24	"		545	294	839	1,225

Division 33
Utilities

33 05 Common Work Results for Utilities

33 05 16 – Utility Structures

33 05 16.13 Utility Boxes

		Crew	Daily Output	Labor-Hours	Unit	Material	2009 Bare Costs Labor	Equipment	Total	Total Incl O&P
0010	**UTILITY BOXES** Precast concrete, 6" thick									
0050	5' x 10' x 6' high, I.D.	B-13	2	24	Ea.	2,225	535	395	3,155	3,775
0350	Hand hole, precast concrete, 1-1/2" thick									
0400	1'-0" x 2'-0" x 1'-9", I.D., light duty	B-1	4	6	Ea.	380	127		507	635
0450	4'-6" x 3'-2" x 2'-0", O.D., heavy duty	B-6	3	8	"	1,175	181	98	1,454	1,700

33 05 23 – Trenchless Utility Installation

33 05 23.19 Microtunneling

		Crew	Daily Output	Labor-Hours	Unit	Material	2009 Bare Costs Labor	Equipment	Total	Total Incl O&P
0010	**MICROTUNNELING**									
0011	Not including excavation, backfill, shoring,									
0020	or dewatering, average 50'/day, slurry method									
0100	24" to 48" outside diameter, minimum				L.F.					640
0110	Adverse conditions, add				%					50%
1000	Rent microtunneling machine, average monthly lease				Month					85,500
1010	Operating technician				Day					640
1100	Mobilization and demobilization, minimum				Job					42,800
1110	Maximum				"					430,000

33 11 Water Utility Distribution Piping

33 11 13 – Public Water Utility Distribution Piping

33 11 13.15 Water Supply, Ductile Iron Pipe

		Crew	Daily Output	Labor-Hours	Unit	Material	2009 Bare Costs Labor	Equipment	Total	Total Incl O&P
0010	**WATER SUPPLY, DUCTILE IRON PIPE**									
0020	Not including excavation or backfill									
2000	Pipe, class 50 water piping, 18' lengths									
2020	Mechanical joint, 4" diameter	B-21A	200	.200	L.F.	15.35	5.20	3.01	23.56	29
2040	6" diameter		160	.250		17.80	6.50	3.76	28.06	34.50
3000	Tyton, push-on joint, 4" diameter		400	.100		11.50	2.61	1.51	15.62	18.65
3020	6" diameter		333.33	.120		14.70	3.13	1.81	19.64	23.50
8000	Fittings, mechanical joint									
8006	90° bend, 4" diameter	B-20A	16	2	Ea.	213	50.50		263.50	320
8020	6" diameter		12.80	2.500		292	63		355	425
8200	Wye or tee, 4" diameter		10.67	2.999		325	76		401	480
8220	6" diameter		8.53	3.751		435	95		530	635
8398	45° bends, 4" diameter		16	2		167	50.50		217.50	268
8400	6" diameter		12.80	2.500		231	63		294	360
8450	Decreaser, 6" x 4" diameter		14.22	2.250		179	57		236	292
8460	8" x 6" diameter		11.64	2.749		229	69.50		298.50	365
8550	Piping, butterfly valves, cast iron									
8560	4" diameter	B-20	6	4	Ea.	860	85		945	1,100
8700	Joint restraint, ductile iron mechanical joints									
8710	4" diameter	B-20A	32	1	Ea.	33	25.50		58.50	78
8720	6" diameter		25.60	1.250		39.50	31.50		71	96
8730	8" diameter		21.33	1.500		57	38		95	126
8740	10" diameter		18.28	1.751		83	44		127	165
8750	12" diameter		16.84	1.900		116	48		164	208
8760	14" diameter		16	2		159	50.50		209.50	259
8770	16" diameter		11.64	2.749		210	69.50		279.50	345
8780	18" diameter		11.03	2.901		292	73.50		365.50	440
8785	20" diameter		9.14	3.501		360	88.50		448.50	540
8790	24" diameter		7.53	4.250		435	107		542	660
9600	Steel sleeve and tap, 4" diameter	B-20	3	8		610	170		780	955

33 11 Water Utility Distribution Piping

33 11 13 – Public Water Utility Distribution Piping

33 11 13.15 Water Supply, Ductile Iron Pipe

		Crew	Daily Output	Labor-Hours	Unit	Material	2009 Bare Costs Labor	Equipment	Total	Total Incl O&P
9620	6" diameter	B-20	2	12	Ea.	650	255		905	1,150

33 11 13.25 Water Supply, Polyvinyl Chloride Pipe

		Crew	Daily Output	Labor-Hours	Unit	Material	2009 Bare Costs Labor	Equipment	Total	Total Incl O&P
0010	**WATER SUPPLY, POLYVINYL CHLORIDE PIPE**									
2100	AWWA Class 150, S.D.R. 18, 1-1/2" diameter	Q-1A	750	.013	L.F.	.34	.43		.77	1.08
2120	2" diameter		686	.015		.51	.48		.99	1.34
2140	2-1/2" diameter		500	.020		.87	.65		1.52	2.03
2160	3" diameter	B-20	430	.056		1.09	1.18		2.27	3.20
2180	4" diameter	"	375	.064		1.65	1.36		3.01	4.12
8700	PVC pipe, joint restraint									
8710	4" diameter	B-20A	32	1	Ea.	34.50	25.50		60	80
8720	6" diameter		25.60	1.250		42	31.50		73.50	98.50
8730	8" diameter		21.33	1.500		62	38		100	131
8740	10" diameter		18.28	1.751		117	44		161	203
8750	12" diameter		16.84	1.900		123	48		171	215
8760	14" diameter		16	2		196	50.50		246.50	300
8770	16" diameter		11.64	2.749		264	69.50		333.50	405
8780	18" diameter		11.03	2.901		325	73.50		398.50	480
8785	20" diameter		9.14	3.501		405	88.50		493.50	590
8790	24" diameter		7.53	4.250		470	107		577	695

33 12 Water Utility Distribution Equipment

33 12 13 – Water Service Connections

33 12 13.15 Tapping, Crosses and Sleeves

		Crew	Daily Output	Labor-Hours	Unit	Material	2009 Bare Costs Labor	Equipment	Total	Total Incl O&P
0010	**TAPPING, CROSSES AND SLEEVES**									
4000	Drill and tap pressurized main (labor only)									
4100	6" main, 1" to 2" service	Q-1	3	5.333	Ea.		155		155	254
4150	8" main, 1" to 2" service	"	2.75	5.818	"		169		169	277
4500	Tap and insert gate valve									
4600	8" main, 4" branch	B-21	3.20	8.750	Ea.		196	40.50	236.50	375
4650	6" branch		2.70	10.370			232	48	280	445
4700	10" Main, 4" branch		2.70	10.370			232	48	280	445
4750	6" branch		2.35	11.915			267	55	322	510
4800	12" main, 6" branch		2.35	11.915			267	55	322	510

33 21 Water Supply Wells

33 21 13 – Public Water Supply Wells

33 21 13.10 Wells and Accessories

		Crew	Daily Output	Labor-Hours	Unit	Material	2009 Bare Costs Labor	Equipment	Total	Total Incl O&P
0010	**WELLS & ACCESSORIES**									
0011	Domestic									
0100	Drilled, 4" to 6" diameter	B-23	120	.333	L.F.		7	23	30	37.50
1500	Pumps, installed in wells to 100' deep, 4" submersible									
1520	3/4 H.P.	Q-1	2.66	6.015	Ea.	565	174		739	905
1600	1 H.P.	"	2.29	6.987	"	615	202		817	1,000

33 31 Sanitary Utility Sewerage Piping

33 31 13 – Public Sanitary Utility Sewerage Piping

33 31 13.15 Sewage Collection, Concrete Pipe	Crew	Daily Output	Labor-Hours	Unit	Material	2009 Bare Costs Labor	Equipment	Total	Total Incl O&P
0010 **SEWAGE COLLECTION, CONCRETE PIPE**									
0020 See Div. 33 41 13.60 for sewage/drainage collection, concrete pipe									

33 31 13.25 Sewage Collection, Polyvinyl Chloride Pipe	Crew	Daily Output	Labor-Hours	Unit	Material	Labor	Equipment	Total	Total Incl O&P
0010 **SEWAGE COLLECTION, POLYVINYL CHLORIDE PIPE**									
0020 Not including excavation or backfill									
2000 20' lengths, S.D.R. 35, B&S, 4" diameter	B-20	375	.064	L.F.	1.66	1.36		3.02	4.13
2040 6" diameter		350	.069		3.40	1.46		4.86	6.20
2080 13' lengths , S.D.R. 35, B&S, 8" diameter	↓	335	.072		7.05	1.52		8.57	10.30
2120 10" diameter	B-21	330	.085		11.15	1.90	.39	13.44	15.85
4000 Piping, DWV PVC, no exc/bkfill, 10' L, Sch 40, 4" diameter	B-20	375	.064		3.84	1.36		5.20	6.50
4010 6" diameter		350	.069		7.65	1.46		9.11	10.85
4020 8" diameter	↓	335	.072	↓	13.30	1.52		14.82	17.20

33 36 Utility Septic Tanks

33 36 13 – Utility Septic Tank and Effluent Wet Wells

33 36 13.10 Septic Tanks

33 36 13.10 Septic Tanks	Crew	Daily Output	Labor-Hours	Unit	Material	Labor	Equipment	Total	Total Incl O&P
0010 **SEPTIC TANKS**									
0011 Not including excavation or piping									
0015 Septic tanks, not incl exc or piping, precast, 1,000 gal	B-21	8	3.500	Ea.	730	78.50	16.15	824.65	950
0100 2,000 gallon		5	5.600		1,625	125	26	1,776	2,050
0600 High density polyethylene, 1,000 gallon		6	4.667		1,125	104	21.50	1,250.50	1,425
0700 1,500 gallon	↓	4	7		1,450	157	32.50	1,639.50	1,900
1000 Distribution boxes, concrete, 7 outlets	2 Clab	16	1		75	20.50		95.50	118
1100 9 outlets	"	8	2		440	41		481	555
1150 Leaching field chambers, 13' x 3'-7" x 1'-4", standard	B-13	16	3		510	67	49.50	626.50	730
1420 Leaching pit, 6', dia, 3' deep complete				↓	1,025			1,025	1,125
2200 Excavation for septic tank, 3/4 C.Y. backhoe	B-12F	145	.110	C.Y.		2.75	4.13	6.88	9.10
2400 4' trench for disposal field, 3/4 C.Y. backhoe	"	335	.048	L.F.		1.19	1.79	2.98	3.95
2600 Gravel fill, run of bank	B-6	150	.160	C.Y.	25	3.63	1.96	30.59	35.50
2800 Crushed stone, 3/4"	"	150	.160	"	29	3.63	1.96	34.59	39.50

33 41 Storm Utility Drainage Piping

33 41 13 – Public Storm Utility Drainage Piping

33 41 13.60 Sewage/Drainage Collection, Concrete Pipe

33 41 13.60 Sewage/Drainage Collection, Concrete Pipe	Crew	Daily Output	Labor-Hours	Unit	Material	Labor	Equipment	Total	Total Incl O&P
0010 **SEWAGE/DRAINAGE COLLECTION, CONCRETE PIPE**									
0020 Not including excavation or backfill									
1020 8" diameter	B-14	224	.214	L.F.	6.70	4.70	1.31	12.71	16.70
1030 10" diameter	"	216	.222	"	7.40	4.88	1.36	13.64	17.85
3780 Concrete slotted pipe, class 4 mortar joint									
3800 12" diameter	B-21	168	.167	L.F.	18.80	3.73	.77	23.30	27.50
3840 18" diameter	"	152	.184	"	29	4.12	.85	33.97	40
3900 Class 4 O-ring									
3940 12" diameter	B-21	168	.167	L.F.	19.70	3.73	.77	24.20	28.50
3960 18" diameter	"	152	.184	"	26.50	4.12	.85	31.47	37

33 44 Storm Utility Water Drains

33 44 13 – Utility Area Drains

33 44 13.13 Catchbasins

33 44 13.13 Catchbasins	Crew	Daily Output	Labor-Hours	Unit	Material	2009 Bare Costs Labor	2009 Bare Costs Equipment	Total	Total Incl O&P
0010 **CATCHBASINS**									
0011 Not including footing & excavation									
1600 Frames & covers, C.I., 24" square, 500 lb.	B-6	7.80	3.077	Ea.	298	70	37.50	405.50	490

33 46 Subdrainage

33 46 16 – Subdrainage Piping

33 46 16.20 Piping, Subdrainage, Concrete

	Crew	Daily Output	Labor-Hours	Unit	Material	Labor	Equipment	Total	Total Incl O&P
0010 **PIPING, SUBDRAINAGE, CONCRETE**									
0021 Not including excavation and backfill									
3000 Porous wall concrete underdrain, std. strength, 4" diameter	B-20	335	.072	L.F.	3.30	1.52		4.82	6.20
3020 6" diameter	"	315	.076		4.29	1.62		5.91	7.45
3040 8" diameter	B-21	310	.090		5.30	2.02	.42	7.74	9.70

33 46 16.25 Piping, Subdrainage, Corrugated Metal

	Crew	Daily Output	Labor-Hours	Unit	Material	Labor	Equipment	Total	Total Incl O&P
0010 **PIPING, SUBDRAINAGE, CORRUGATED METAL**									
0021 Not including excavation and backfill									
2010 Aluminum, perforated									
2020 6" diameter, 18 ga.	B-14	380	.126	L.F.	4.70	2.77	.77	8.24	10.65
2200 8" diameter, 16 ga.		370	.130		7.15	2.85	.79	10.79	13.50
2220 10" diameter, 16 ga.		360	.133		8.90	2.93	.82	12.65	15.60
3000 Uncoated galvanized, perforated									
3020 6" diameter, 18 ga.	B-20	380	.063	L.F.	8.25	1.34		9.59	11.30
3200 8" diameter, 16 ga.	"	370	.065		11.35	1.38		12.73	14.80
3220 10" diameter, 16 ga.	B-21	360	.078		17	1.74	.36	19.10	22
3240 12" diameter, 16 ga.	"	285	.098		17.80	2.20	.45	20.45	24
4000 Steel, perforated, asphalt coated									
4020 6" diameter 18 ga.	B-20	380	.063	L.F.	6.60	1.34		7.94	9.50
4030 8" diameter 18 ga.	"	370	.065		10.30	1.38		11.68	13.70
4040 10" diameter 16 ga.	B-21	360	.078		11.85	1.74	.36	13.95	16.30
4050 12" diameter 16 ga.		285	.098		13.60	2.20	.45	16.25	19.15
4060 18" diameter 16 ga.		205	.137		18.50	3.06	.63	22.19	26.50

33 49 Storm Drainage Structures

33 49 13 – Storm Drainage Manholes, Frames, and Covers

33 49 13.10 Storm Drainage Manholes, Frames and Covers

	Crew	Daily Output	Labor-Hours	Unit	Material	Labor	Equipment	Total	Total Incl O&P
0010 **STORM DRAINAGE MANHOLES, FRAMES & COVERS**									
0020 Excludes footing, excavation, backfill (See line items for frame & cover)									
0050 Brick, 4' inside diameter, 4' deep	D-1	1	16	Ea.	400	405		805	1,125
1110 Precast, 4' I.D., 4' deep	B-22	4.10	7.317	"	880	167	47.50	1,094.50	1,300

33 51 Natural-Gas Distribution

33 51 13 – Natural-Gas Piping

33 51 13.10 Piping, Gas Service and Distribution, P.E.

	33 51 13.10 Piping, Gas Service and Distribution, P.E.	Crew	Daily Output	Labor-Hours	Unit	Material	2009 Bare Costs Labor	Equipment	Total	Total Incl O&P
0010	**PIPING, GAS SERVICE AND DISTRIBUTION, POLYETHYLENE**									
0020	Not including excavation or backfill									
1000	60 psi coils, compression coupling @ 100', 1/2" diameter, SDR 11	B-20A	608	.053	L.F.	1.61	1.33		2.94	3.98
1040	1-1/4" diameter, SDR 11		544	.059		2.49	1.49		3.98	5.20
1100	2" diameter, SDR 11		488	.066		3.09	1.66		4.75	6.15
1160	3" diameter, SDR 11		408	.078		6.45	1.98		8.43	10.40
1500	60 PSI 40' joints with coupling, 3" diameter, SDR 11	B-21A	408	.098		6.45	2.56	1.48	10.49	12.95
1540	4" diameter, SDR 11		352	.114		14.75	2.96	1.71	19.42	23
1600	6" diameter, SDR 11		328	.122		46	3.18	1.84	51.02	58
1640	8" diameter, SDR 11		272	.147		62.50	3.84	2.21	68.55	78

33 51 13.20 Piping, Gas Service and Distribution, Steel

	33 51 13.20 Piping, Gas Service and Distribution, Steel	Crew	Daily Output	Labor-Hours	Unit	Material	2009 Bare Costs Labor	Equipment	Total	Total Incl O&P
0010	**PIPING, GAS SERVICE & DISTRIBUTION, STEEL**									
0020	Not including excavation or backfill, tar coated and wrapped									
4000	Pipe schedule 40, plain end									
4040	1" diameter	Q-4	300	.107	L.F.	7.55	3.26	.19	11	13.85
4080	2" diameter	Q-4	280	.114	L.F.	11.85	3.50	.20	15.55	18.95

Reference Section

All the reference information is in one section, making it easy to find what you need to know . . . and easy to use the book on a daily basis. This section is visually identified by a vertical gray bar on the page edges.

In this Reference Section, we've included Equipment Rental Costs, a listing of rental and operating costs; Crew Listings, a full listing of all crews, equipment, and their costs; Location Factors for adjusting costs to the region you are in; Reference Tables, where you will find explanations, estimating information and procedures, or technical data; and an explanation of all the Abbreviations in the book.

Table of Contents

333

Equipment Rental Costs

01 54 33 | Equipment Rental

			UNIT	HOURLY OPER. COST	RENT PER DAY	RENT PER WEEK	RENT PER MONTH	EQUIPMENT COST/DAY	
10	0010	**CONCRETE EQUIPMENT RENTAL** without operators							**10**
	0200	Bucket, concrete lightweight, 1/2 C.Y.	Ea.	.65	20	60	180	17.20	
	0300	1 C.Y.		.70	24.50	73	219	20.20	
	0400	1-1/2 C.Y.		.90	33	99	297	27	
	0500	2 C.Y.		1.00	40	120	360	32	
	0580	8 C.Y.		5.50	262	785	2,350	201	
	0600	Cart, concrete, self propelled, operator walking, 10 C.F.		2.65	56.50	170	510	55.20	
	0700	Operator riding, 18 C.F.		4.30	86.50	260	780	86.40	
	0800	Conveyer for concrete, portable, gas, 16" wide, 26' long		10.00	123	370	1,100	154	
	0900	46' long		10.40	150	450	1,350	173.20	
	1000	56' long		10.50	158	475	1,425	179	
	1100	Core drill, electric, 2-1/2 H.P., 1" to 8" bit diameter		1.72	65.50	196	590	52.95	
	1150	11 HP, 8" to 18" cores		5.15	115	345	1,025	110.20	
	1200	Finisher, concrete floor, gas, riding trowel, 96" wide		9.95	145	435	1,300	166.60	
	1300	Gas, walk-behind, 3 blade, 36" trowel		1.65	19.65	59	177	25	
	1400	4 blade, 48" trowel		3.20	28	84	252	42.40	
	1500	Float, hand-operated (Bull float) 48" wide		.08	13.65	41	123	8.85	
	1570	Curb builder, 14 H.P., gas, single screw		11.85	243	730	2,200	240.80	
	1590	Double screw		12.55	285	855	2,575	271.40	
	1600	Grinder, concrete and terrazzo, electric, floor		2.45	103	310	930	81.60	
	1700	Wall grinder		1.23	51.50	155	465	40.85	
	1800	Mixer, powered, mortar and concrete, gas, 6 C.F., 18 H.P.		6.95	118	355	1,075	126.60	
	1900	10 C.F., 25 H.P.		8.65	143	430	1,300	155.20	
	2000	16 C.F.		8.95	167	500	1,500	171.60	
	2100	Concrete, stationary, tilt drum, 2 C.Y.		6.10	232	695	2,075	187.80	
	2120	Pump, concrete, truck mounted 4" line 80' boom		22.85	925	2,775	8,325	737.80	
	2140	5" line, 110' boom		29.75	1,225	3,695	11,100	977	
	2160	Mud jack, 50 C.F. per hr.		6.10	130	390	1,175	126.80	
	2180	225 C.F. per hr.		7.95	147	440	1,325	151.60	
	2190	Shotcrete pump rig, 12 CY/hr		12.60	228	685	2,050	237.80	
	2600	Saw, concrete, manual, gas, 18 H.P.		5.15	36.50	110	330	63.20	
	2650	Self-propelled, gas, 30 H.P.		10.25	98.50	295	885	141	
	2700	Vibrators, concrete, electric, 60 cycle, 2 H.P.		.46	8.35	25	75	8.70	
	2800	3 H.P.		.64	10.35	31	93	11.30	
	2900	Gas engine, 5 H.P.		1.55	14.35	43	129	21	
	3000	8 H.P.		2.10	15.35	46	138	26	
	3050	Vibrating screed, gas engine, 8 H.P.		3.32	73	219	655	70.35	
	3120	Concrete transit mixer, 6 x 4, 250 H.P., 8 C.Y., rear discharge		47.15	575	1,720	5,150	721.20	
	3200	Front discharge		55.50	710	2,125	6,375	869	
	3300	6 x 6, 285 H.P., 12 C.Y., rear discharge		54.60	665	2,000	6,000	836.80	
	3400	Front discharge		57.05	715	2,150	6,450	886.40	
20	0010	**EARTHWORK EQUIPMENT RENTAL** without operators							**20**
	0040	Aggregate spreader, push type 8' to 12' wide	Ea.	2.45	25.50	77	231	35	
	0045	Tailgate type, 8' wide		2.35	33.50	100	300	38.80	
	0055	Earth auger, truck-mounted, for fence & sign posts, utility poles		13.50	505	1,520	4,550	412	
	0060	For borings and monitoring wells		38.20	670	2,005	6,025	706.60	
	0070	Portable, trailer mounted		2.65	26	78	234	36.80	
	0075	Truck-mounted, for caissons, water wells		85.35	3,075	9,250	27,800	2,533	
	0080	Horizontal boring machine, 12" to 36" diameter, 45 H.P.		20.30	202	605	1,825	283.40	
	0090	12" to 48" diameter, 65 H.P.		28.65	360	1,075	3,225	444.20	
	0095	Auger, for fence posts, gas engine, hand held		.50	5.65	17	51	7.40	
	0100	Excavator, diesel hydraulic, crawler mounted, 1/2 C.Y. cap.		20.95	380	1,135	3,400	394.60	
	0120	5/8 C.Y. capacity		25.40	525	1,570	4,700	517.20	
	0140	3/4 C.Y. capacity		29.70	600	1,805	5,425	598.60	
	0150	1 C.Y. capacity		36.90	625	1,870	5,600	669.20	
	0200	1-1/2 C.Y. capacity		45.10	840	2,525	7,575	865.80	
	0300	2 C.Y. capacity		61.75	1,150	3,440	10,300	1,182	

R015433 -10

01 54 33 | Equipment Rental

		UNIT	HOURLY OPER. COST	RENT PER DAY	RENT PER WEEK	RENT PER MONTH	EQUIPMENT COST/DAY	
20	0320	2-1/2 C.Y. capacity	Ea.	83.30	1,600	4,765	14,300	1,619	20
	0325	3-1/2 C.Y. capacity		111.30	2,125	6,360	19,100	2,162	
	0330	4-1/2 C.Y. capacity		134.65	2,600	7,790	23,400	2,635	
	0335	6 C.Y. capacity		168.45	2,875	8,660	26,000	3,080	
	0340	7 C.Y. capacity		173.20	3,125	9,350	28,100	3,256	
	0342	Excavator attachments, bucket thumbs		2.70	215	645	1,925	150.60	
	0345	Grapples		2.45	188	565	1,700	132.60	
	0347	Hydraulic hammer for boom mounting, 5000 ft-lb		12.10	405	1,220	3,650	340.80	
	0349	11,000 ft-lb		20.40	730	2,190	6,575	601.20	
	0350	Gradall type, truck mounted, 3 ton @ 15' radius, 5/8 C.Y.		49.60	935	2,810	8,425	958.80	
	0370	1 C.Y. capacity		55.10	1,150	3,445	10,300	1,130	
	0400	Backhoe-loader, 40 to 45 H.P., 5/8 C.Y. capacity		11.90	227	680	2,050	231.20	
	0450	45 H.P. to 60 H.P., 3/4 C.Y. capacity		15.85	278	835	2,500	293.80	
	0460	80 H.P., 1-1/4 C.Y. capacity		19.00	320	965	2,900	345	
	0470	112 H.P., 1-1/2 C.Y. capacity		25.45	465	1,400	4,200	483.60	
	0482	Backhoe-loader attachment, compactor, 20,000 lb		4.70	127	380	1,150	113.60	
	0485	Hydraulic hammer, 750 ft-lbs		2.80	90	270	810	76.40	
	0486	Hydraulic hammer, 1200 ft-lbs		5.25	185	555	1,675	153	
	0500	Brush chipper, gas engine, 6" cutter head, 35 H.P.		8.95	108	325	975	136.60	
	0550	12" cutter head, 130 H.P.		14.95	177	530	1,600	225.60	
	0600	15" cutter head, 165 H.P.		22.45	217	650	1,950	309.60	
	0750	Bucket, clamshell, general purpose, 3/8 C.Y.		1.10	36.50	110	330	30.80	
	0800	1/2 C.Y.		1.20	45	135	405	36.60	
	0850	3/4 C.Y.		1.35	55	165	495	43.80	
	0900	1 C.Y.		1.40	58.50	175	525	46.20	
	0950	1-1/2 C.Y.		2.25	80	240	720	66	
	1000	2 C.Y.		2.40	88.50	265	795	72.20	
	1010	Bucket, dragline, medium duty, 1/2 C.Y.		.65	23.50	70	210	19.20	
	1020	3/4 C.Y.		.65	24.50	74	222	20	
	1030	1 C.Y.		.70	26.50	80	240	21.60	
	1040	1-1/2 C.Y.		1.10	40	120	360	32.80	
	1050	2 C.Y.		1.15	45	135	405	36.20	
	1070	3 C.Y.		1.65	55	165	495	46.20	
	1200	Compactor, manually guided 2-drum vibratory smooth roller, 7.5 H.P.		5.65	167	500	1,500	145.20	
	1250	Rammer/tamper, gas, 8"		2.30	41.50	125	375	43.40	
	1260	15"		2.55	48.50	145	435	49.40	
	1300	Vibratory plate, gas, 18" plate, 3000 lb. blow		2.25	23	69	207	31.80	
	1350	21" plate, 5000 lb. blow		2.80	28	84	252	39.20	
	1370	Curb builder/extruder, 14 H.P., gas, single screw		11.85	243	730	2,200	240.80	
	1390	Double screw		12.55	285	855	2,575	271.40	
	1500	Disc harrow attachment, for tractor		.39	65.50	197	590	42.50	
	1810	Feller buncher, shearing & accumulating trees, 100 H.P.		28.55	555	1,665	5,000	561.40	
	1860	Grader, self-propelled, 25,000 lb.		24.85	450	1,345	4,025	467.80	
	1910	30,000 lb.		29.65	520	1,565	4,700	550.20	
	1920	40,000 lb.		45.50	995	2,985	8,950	961	
	1930	55,000 lb.		59.10	1,400	4,215	12,600	1,316	
	1950	Hammer, pavement breaker, self-propelled, diesel, 1000 to 1250 lb		23.55	340	1,025	3,075	393.40	
	2000	1300 to 1500 lb.		35.33	685	2,050	6,150	692.65	
	2050	Pile driving hammer, steam or air, 4150 ft.-lb. @ 225 BPM		9.35	470	1,405	4,225	355.80	
	2100	8750 ft.-lb. @ 145 BPM		11.35	650	1,955	5,875	481.80	
	2150	15,000 ft.-lb. @ 60 BPM		12.95	790	2,375	7,125	578.60	
	2200	24,450 ft.-lb. @ 111 BPM		13.95	885	2,655	7,975	642.60	
	2250	Leads, 60' high for pile driving hammers up to 20,000 ft.-lb.		2.85	78	234	700	69.60	
	2300	90' high for hammers over 20,000 ft.-lb.		4.35	140	420	1,250	118.80	
	2350	Diesel type hammer, 22,400 ft.-lb.		25.75	640	1,920	5,750	590	
	2400	41,300 ft.-lb.		32.70	695	2,080	6,250	677.60	
	2450	141,000 ft.-lb.		49.25	1,200	3,570	10,700	1,108	
	2500	Vib. elec. hammer/extractor, 200 KW diesel generator, 34 H.P.		41.05	665	2,000	6,000	728.40	

		UNIT	HOURLY OPER. COST	RENT PER DAY	RENT PER WEEK	RENT PER MONTH	EQUIPMENT COST/DAY	
20	2550	80 H.P.	Ea.	73.75	980	2,940	8,825	1,178	**20**
	2600	150 H.P.		139.15	1,900	5,720	17,200	2,257	
	2700	Extractor, steam or air, 700 ft.-lb.		17.15	510	1,530	4,600	443.20	
	2750	1000 ft.-lb.		19.55	625	1,875	5,625	531.40	
	2800	Log chipper, up to 22" diam, 600 H.P.		40.05	445	1,340	4,025	588.40	
	2850	Logger, for skidding & stacking logs, 150 H.P.		44.95	895	2,690	8,075	897.60	
	2860	Mulcher, diesel powered, trailer mounted		17.60	218	655	1,975	271.80	
	2900	Rake, spring tooth, with tractor		11.34	278	835	2,500	257.70	
	3000	Roller, vibratory, tandem, smooth drum, 20 H.P.		7.35	132	395	1,175	137.80	
	3050	35 H.P.		9.50	262	785	2,350	233	
	3100	Towed type vibratory compactor, smooth drum, 50 H.P.		22.75	380	1,145	3,425	411	
	3150	Sheepsfoot, 50 H.P.		25.05	440	1,325	3,975	465.40	
	3170	Landfill compactor, 220 HP		69.10	1,450	4,315	12,900	1,416	
	3200	Pneumatic tire roller, 80 H.P.		13.10	340	1,015	3,050	307.80	
	3250	120 H.P.		19.90	600	1,805	5,425	520.20	
	3300	Sheepsfoot vibratory roller, 240 H.P.		56.65	1,150	3,445	10,300	1,142	
	3320	340 H.P.		75.60	1,425	4,240	12,700	1,453	
	3350	Smooth drum vibratory roller, 75 H.P.		19.70	520	1,560	4,675	469.60	
	3400	125 H.P.		25.70	650	1,945	5,825	594.60	
	3410	Rotary mower, brush, 60", with tractor		15.30	252	755	2,275	273.40	
	3420	Rototiller, walk-behind, gas, 5 H.P.		2.43	62.50	188	565	57.05	
	3422	8 H.P.		3.89	100	300	900	91.10	
	3440	Scrapers, towed type, 7 CY capacity		4.90	105	315	945	102.20	
	3450	10 C.Y. capacity		6.20	173	520	1,550	153.60	
	3500	15 C.Y. capacity		6.85	190	570	1,700	168.80	
	3525	Self-propelled, single engine, 14 C.Y. capacity		83.90	1,400	4,190	12,600	1,509	
	3550	Dual engine, 21 C.Y. capacity		119.25	1,625	4,850	14,600	1,924	
	3600	31 C.Y. capacity		165.60	2,600	7,765	23,300	2,878	
	3640	44 C.Y. capacity		195.70	3,150	9,430	28,300	3,452	
	3650	Elevating type, single engine, 11 CY capacity		56.70	1,075	3,205	9,625	1,095	
	3700	22 C.Y. capacity		105.60	2,125	6,410	19,200	2,127	
	3710	Screening plant 110 H.P. w/ 5' x 10' screen		30.75	410	1,230	3,700	492	
	3720	5' x 16' screen		32.90	515	1,540	4,625	571.20	
	3850	Shovel, crawler-mounted, front-loading, 7 C.Y. capacity		171.90	2,850	8,575	25,700	3,090	
	3855	12 C.Y. capacity		244.40	3,650	10,965	32,900	4,148	
	3860	Shovel/backhoe bucket, 1/2 C.Y.		2.10	60	180	540	52.80	
	3870	3/4 C.Y.		2.15	68.50	205	615	58.20	
	3880	1 C.Y.		2.25	76.50	230	690	64	
	3890	1-1/2 C.Y.		2.40	91.50	275	825	74.20	
	3910	3 C.Y.		2.70	127	380	1,150	97.60	
	3950	Stump chipper, 18" deep, 30 H.P.		8.99	215	646	1,950	201.10	
	4110	Tractor, crawler, with bulldozer, torque converter, diesel 80 H.P.		21.90	375	1,120	3,350	399.20	
	4150	105 H.P.		31.70	580	1,735	5,200	600.60	
	4200	140 H.P.		36.50	650	1,950	5,850	682	
	4260	200 H.P.		55.85	1,050	3,175	9,525	1,082	
	4310	300 H.P.		71.45	1,425	4,255	12,800	1,423	
	4360	410 H.P.		95.30	1,800	5,435	16,300	1,849	
	4370	500 H.P.		127.20	2,400	7,230	21,700	2,464	
	4380	700 H.P.		204.15	4,325	12,995	39,000	4,232	
	4400	Loader, crawler, torque conv., diesel, 1-1/2 C.Y., 80 H.P.		20.00	345	1,035	3,100	367	
	4450	1-1/2 to 1-3/4 C.Y., 95 H.P.		23.75	455	1,365	4,100	463	
	4510	1-3/4 to 2-1/4 C.Y., 130 H.P.		35.05	810	2,435	7,300	767.40	
	4530	2-1/2 to 3-1/4 C.Y., 190 H.P.		47.75	1,025	3,040	9,125	990	
	4560	3-1/2 to 5 C.Y., 275 H.P.		68.30	1,500	4,520	13,600	1,450	
	4610	Front end loader, 4WD, articulated frame, 1 to 1-1/4 C.Y., 70 H.P.		14.25	208	625	1,875	239	
	4620	1-1/2 to 1-3/4 C.Y., 95 H.P.		18.90	295	885	2,650	328.20	
	4650	1-3/4 to 2 C.Y., 130 H.P.		21.60	345	1,035	3,100	379.80	
	4710	2-1/2 to 3-1/2 C.Y., 145 H.P.		22.85	355	1,065	3,200	395.80	

Reed Construction Data
The leader in construction information and BIM solutions

Reed Construction Data, Inc. is a leading provider of construction information and building information modeling (BIM) solutions. The company's portfolio of information products and services is designed specifically to help construction industry professionals advance their business with timely and accurate project, product, and cost data. Reed Construction Data is a division of Reed Business Information, a member of the Reed Elsevier PLC group of companies.

Cost Information

RSMeans, the undisputed market leader in construction costs, provides current cost and estimating information through its innovative MeansCostworks.com® web-based solution. In addition, RSMeans publishes annual cost books, estimating software, and a rich library of reference books. RSMeans also conducts a series of professional seminars and provides construction cost consulting for owners, manufacturers, designers, and contractors to sharpen personal skills and maximize the effective use of cost estimating and management tools.

Project Data

Reed Construction Data assembles one of the largest databases of public and private project data for use by contractors, distributors, and building product manufacturers in the U.S. and Canadian markets. In addition, Reed Construction Data is the North American construction community's premier resource for project leads and bid documents. Reed Bulletin and Reed CONNECT™ provide project leads and project data through all stages of construction for many of the country's largest public sector, commercial, industrial, and multi-family residential projects.

Research and Analytics

Reed Construction Data's forecasting tools cover most aspects of the construction business in the U.S. and Canada. With a vast network of resources, Reed Construction Data is uniquely qualified to give you the information you need to keep your business profitable.

SmartBIM Solutions

Reed Construction Data has emerged as the leader in the field of building information modeling (BIM) with products and services that have helped to advance the evolution of BIM. Through an in-depth SmartBIM Object Creation Program, BPMs can rely on Reed to create high-quality, real world objects embedded with superior cost data (RSMeans). In addition, Reed has made it easy for architects to manage these manufacturer-specific objects as well as generic objects in Revit with the SmartBIM Library.

SmartBuilding Index

The leading industry source for product research, product documentation, BIM objects, design ideas, and source locations. Search, select, and specify available building products with our online directories of manufacturer profiles, MANU-SPEC, SPEC-DATA, guide specs, manufacturer catalogs, building codes, historical project data, and BIM objects.

SmartBuilding Studio

A vibrant online resource library which brings together a comprehensive catalog of commercial interior finishes and products under a single standard for high-definition imagery, product data, and searchable attributes.

Associated Construction Publications (ACP)

Reed Construction Data's regional construction magazines cover the nation through a network of 14 regional magazines. Serving the construction market for more than 100 years, our magazines are a trusted source of news and information in the local and national construction communities.

For more information, please visit our website at www.reedconstructiondata.com

The leader in construction information and BIM solutions.

Reed Construction Data®

01 54 33 | Equipment Rental

		UNIT	HOURLY OPER. COST	RENT PER DAY	RENT PER WEEK	RENT PER MONTH	EQUIPMENT COST/DAY	
20	4730	3 to 4-1/2 C.Y., 185 H.P.	Ea.	29.75	505	1,515	4,550	541	**20**
	4760	5-1/4 to 5-3/4 C.Y., 270 H.P.		48.05	785	2,350	7,050	854.40	
	4810	7 to 9 C.Y., 475 H.P.		81.85	1,400	4,210	12,600	1,497	
	4870	9 - 11 C.Y., 620 H.P.		112.20	2,200	6,635	19,900	2,225	
	4880	Skid steer loader, wheeled, 10 C.F., 30 H.P. gas		8.15	140	420	1,250	149.20	
	4890	1 C.Y., 78 H.P., diesel		14.95	233	700	2,100	259.60	
	4892	Skid-steer attachment, auger		.54	89.50	268	805	57.90	
	4893	Backhoe		.67	112	336	1,000	72.55	
	4894	Broom		.70	117	352	1,050	76	
	4895	Forks		.24	39.50	118	355	25.50	
	4896	Grapple		.57	95.50	287	860	61.95	
	4897	Concrete hammer		1.06	176	528	1,575	114.10	
	4898	Tree spade		.80	133	398	1,200	86	
	4899	Trencher		.77	128	385	1,150	83.15	
	4900	Trencher, chain, boom type, gas, operator walking, 12 H.P.		3.95	46.50	140	420	59.60	
	4910	Operator riding, 40 H.P.		15.50	295	885	2,650	301	
	5000	Wheel type, diesel, 4' deep, 12" wide		65.30	810	2,435	7,300	1,009	
	5100	6' deep, 20" wide		77.15	1,900	5,680	17,000	1,753	
	5150	Chain type, diesel, 5' deep, 8" wide		28.90	580	1,745	5,225	580.20	
	5200	Diesel, 8' deep, 16" wide		74.70	1,975	5,895	17,700	1,777	
	5210	Tree spade, self-propelled		16.78	267	800	2,400	294.25	
	5250	Truck, dump, 2-axle, 12 ton, 8 CY payload, 220 H.P.		24.55	217	650	1,950	326.40	
	5300	Three axle dump, 16 ton, 12 CY payload, 400 H.P.		43.25	310	935	2,800	533	
	5350	Dump trailer only, rear dump, 16-1/2 C.Y.		4.70	130	390	1,175	115.60	
	5400	20 C.Y.		5.15	148	445	1,325	130.20	
	5450	Flatbed, single axle, 1-1/2 ton rating		19.30	66.50	200	600	194.40	
	5500	3 ton rating		23.20	95	285	855	242.60	
	5550	Off highway rear dump, 25 ton capacity		59.95	1,325	4,000	12,000	1,280	
	5600	35 ton capacity		55.80	1,175	3,490	10,500	1,144	
	5610	50 ton capacity		76.40	1,625	4,855	14,600	1,582	
	5620	65 ton capacity		78.80	1,600	4,820	14,500	1,594	
	5630	100 ton capacity		101.35	2,050	6,180	18,500	2,047	
	6000	Vibratory plow, 25 H.P., walking	▼	6.65	61.50	185	555	90.20	
40	0010	**GENERAL EQUIPMENT RENTAL** without operators R015433-10							**40**
	0150	Aerial lift, scissor type, to 15' high, 1000 lb. cap., electric	Ea.	2.70	56.50	170	510	55.60	
	0160	To 25' high, 2000 lb. capacity		3.15	80	240	720	73.20	
	0170	Telescoping boom to 40' high, 500 lb. capacity, gas		17.10	320	960	2,875	328.80	
	0180	To 45' high, 500 lb. capacity		18.10	365	1,100	3,300	364.80	
	0190	To 60' high, 600 lb. capacity		20.45	505	1,510	4,525	465.60	
	0195	Air compressor, portable, 6.5 CFM, electric		.47	12.65	38	114	11.35	
	0196	Gasoline		.96	19	57	171	19.10	
	0200	Towed type, gas engine, 60 C.F.M.		11.70	48.50	145	435	122.60	
	0300	160 C.F.M.		13.60	50	150	450	138.80	
	0400	Diesel engine, rotary screw, 250 C.F.M.		12.80	100	300	900	162.40	
	0500	365 C.F.M.		17.15	123	370	1,100	211.20	
	0550	450 C.F.M.		21.75	153	460	1,375	266	
	0600	600 C.F.M.		38.05	217	650	1,950	434.40	
	0700	750 C.F.M.	▼	38.20	223	670	2,000	439.60	
	0800	For silenced models, small sizes, add to rent		3%	5%	5%	5%		
	0900	Large sizes, add to rent		5%	7%	7%	7%		
	0930	Air tools, breaker, pavement, 60 lb.	Ea.	.40	8.35	25	75	8.20	
	0940	80 lb.		.40	8.35	25	75	8.20	
	0950	Drills, hand (jackhammer) 65 lb.		.50	16	48	144	13.60	
	0960	Track or wagon, swing boom, 4" drifter		49.20	760	2,285	6,850	850.60	
	0970	5" drifter		60.00	825	2,470	7,400	974	
	0975	Track mounted quarry drill, 6" diameter drill		82.25	1,250	3,780	11,300	1,414	
	0980	Dust control per drill	▼	.99	19	57	171	19.30	

01 54 33 | Equipment Rental

			UNIT	HOURLY OPER. COST	RENT PER DAY	RENT PER WEEK	RENT PER MONTH	EQUIPMENT COST/DAY	
40	0990	Hammer, chipping, 12 lb.	Ea.	.45	22.50	67	201	17	40
	1000	Hose, air with couplings, 50' long, 3/4" diameter		.03	4.33	13	39	2.85	
	1100	1" diameter		.04	6.35	19	57	4.10	
	1200	1-1/2" diameter		.06	9.65	29	87	6.30	
	1300	2" diameter		.13	21.50	65	195	14.05	
	1400	2-1/2" diameter		.12	19.65	59	177	12.75	
	1410	3" diameter		.15	25	75	225	16.20	
	1450	Drill, steel, 7/8" x 2'		.05	8.35	25	75	5.40	
	1460	7/8" x 6'		.06	9.35	28	84	6.10	
	1520	Moil points		.02	3.67	11	33	2.35	
	1525	Pneumatic nailer w/accessories		.45	30	90	270	21.60	
	1530	Sheeting driver for 60 lb. breaker		.04	6	18	54	3.90	
	1540	For 90 lb. breaker		.12	8	24	72	5.75	
	1550	Spade, 25 lb.		.40	6	18	54	6.80	
	1560	Tamper, single, 35 lb.		.50	33.50	100	300	24	
	1570	Triple, 140 lb.		.75	50	150	450	36	
	1580	Wrenches, impact, air powered, up to 3/4" bolt		.30	8	24	72	7.20	
	1590	Up to 1-1/4" bolt		.45	21	63	189	16.20	
	1600	Barricades, barrels, reflectorized, 1 to 50 barrels		.03	4.60	13.80	41.50	3	
	1610	100 to 200 barrels		.02	3.53	10.60	32	2.30	
	1620	Barrels with flashers, 1 to 50 barrels		.03	5.25	15.80	47.50	3.40	
	1630	100 to 200 barrels		.03	4.20	12.60	38	2.75	
	1640	Barrels with steady burn type C lights		.04	7	21	63	4.50	
	1650	Illuminated board, trailer mounted, with generator		.70	120	360	1,075	77.60	
	1670	Portable barricade, stock, with flashers, 1 to 6 units		.03	5.25	15.80	47.50	3.40	
	1680	25 to 50 units		.03	4.90	14.70	44	3.20	
	1690	Butt fusion machine, electric		20.85	55	165	495	199.80	
	1695	Electro fusion machine		15.35	40	120	360	146.80	
	1700	Carts, brick, hand powered, 1000 lb. capacity		.38	63	189	565	40.85	
	1800	Gas engine, 1500 lb., 7-1/2' lift		4.73	119	357	1,075	109.25	
	1822	Dehumidifier, medium, 6 lb/hr, 150 CFM		.85	51.50	155	465	37.80	
	1824	Large, 18 lb/hr, 600 CFM		1.64	100	300	900	73.10	
	1830	Distributor, asphalt, trailer mtd, 2000 gal., 38 H.P. diesel		9.15	340	1,020	3,050	277.20	
	1840	3000 gal., 38 H.P. diesel		10.40	370	1,105	3,325	304.20	
	1850	Drill, rotary hammer, electric, 1-1/2" diameter		.70	22	66	198	18.80	
	1860	Carbide bit for above		.05	9	27	81	5.80	
	1865	Rotary, crawler, 250 H.P.		122.45	2,075	6,240	18,700	2,228	
	1870	Emulsion sprayer, 65 gal., 5 H.P. gas engine		3.20	99	297	890	85	
	1880	200 gal., 5 H.P. engine		6.85	165	495	1,475	153.80	
	1930	Floodlight, mercury vapor, or quartz, on tripod, 1000 watt		.34	13	39	117	10.50	
	1940	2000 watt		.59	24.50	73	219	19.30	
	1950	Floodlights, trailer mounted with generator, 1 - 300 watt light		3.00	70	210	630	66	
	1960	2 - 1000 watt lights		4.15	107	320	960	97.20	
	2000	4 - 300 watt lights		3.65	86.50	260	780	81.20	
	2005	Foam spray rig, incl. box trailer, compressor, generator, proportioner		24.90	480	1,445	4,325	488.20	
	2020	Forklift, straight mast, 12' lift, 5000 lb., 2 wheel drive, gas		22.00	198	595	1,775	295	
	2040	21' lift, 5000 lb., 4 wheel drive, diesel		16.65	247	740	2,225	281.20	
	2050	For rough terrain, 42' lift, 35' reach, 9000 lb., 110 HP		22.50	445	1,330	4,000	446	
	2060	For plant, 4 T. capacity, 80 H.P., 2 wheel drive, gas		13.55	105	315	945	171.40	
	2080	10 T. capacity, 120 H.P., 2 wheel drive, diesel		18.65	183	550	1,650	259.20	
	2100	Generator, electric, gas engine, 1.5 KW to 3 KW		3.10	10.35	31	93	31	
	2200	5 KW		3.95	14.35	43	129	40.20	
	2300	10 KW		7.45	33.50	100	300	79.60	
	2400	25 KW		9.10	65	195	585	111.80	
	2500	Diesel engine, 20 KW		9.25	73.50	220	660	118	
	2600	50 KW		16.60	98.50	295	885	191.80	
	2700	100 KW		31.90	120	360	1,075	327.20	
	2800	250 KW		63.05	232	695	2,075	643.40	

01 54 33 | Equipment Rental

		UNIT	HOURLY OPER. COST	RENT PER DAY	RENT PER WEEK	RENT PER MONTH	EQUIPMENT COST/DAY	
40	2850	Hammer, hydraulic, for mounting on boom, to 500 ft.-lb.	Ea.	2.15	68.50	205	615	58.20	40
	2860	1000 ft.-lb.		3.65	112	335	1,000	96.20	
	2900	Heaters, space, oil or electric, 50 MBH		2.50	6.65	20	60	24	
	3000	100 MBH		4.52	9	27	81	41.55	
	3100	300 MBH		14.52	36.50	110	330	138.15	
	3150	500 MBH		29.33	45	135	405	261.65	
	3200	Hose, water, suction with coupling, 20' long, 2" diameter		.02	3	9	27	1.95	
	3210	3" diameter		.03	4.67	14	42	3.05	
	3220	4" diameter		.03	5.35	16	48	3.45	
	3230	6" diameter		.11	18.35	55	165	11.90	
	3240	8" diameter		.20	33.50	100	300	21.60	
	3250	Discharge hose with coupling, 50' long, 2" diameter		.01	1.67	5	15	1.10	
	3260	3" diameter		.02	2.67	8	24	1.75	
	3270	4" diameter		.02	3.67	11	33	2.35	
	3280	6" diameter		.06	9.65	29	87	6.30	
	3290	8" diameter		.20	33.50	100	300	21.60	
	3295	Insulation blower		.22	6	18	54	5.35	
	3300	Ladders, extension type, 16' to 36' long		.14	23	69	207	14.90	
	3400	40' to 60' long		.20	33.50	100	300	21.60	
	3405	Lance for cutting concrete		2.97	104	313	940	86.35	
	3407	Lawn mower, rotary, 22", 5HP		2.48	65	195	585	58.85	
	3408	48" self propelled		4.48	128	385	1,150	112.85	
	3410	Level, laser type, for pipe and sewer leveling		1.46	97.50	292	875	70.10	
	3430	Electronic		.75	50	150	450	36	
	3440	Laser type, rotating beam for grade control		1.17	77.50	233	700	55.95	
	3460	Builders level with tripod and rod		.08	14	42	126	9.05	
	3500	Light towers, towable, with diesel generator, 2000 watt		3.65	86.50	260	780	81.20	
	3600	4000 watt		4.15	107	320	960	97.20	
	3700	Mixer, powered, plaster and mortar, 6 C.F., 7 H.P.		2.10	19.65	59	177	28.60	
	3800	10 C.F., 9 H.P.		2.25	31.50	95	285	37	
	3850	Nailer, pneumatic		.45	30	90	270	21.60	
	3900	Paint sprayers complete, 8 CFM		.67	44.50	134	400	32.15	
	4000	17 CFM		1.17	77.50	233	700	55.95	
	4020	Pavers, bituminous, rubber tires, 8' wide, 50 H.P., diesel		35.70	1,050	3,155	9,475	916.60	
	4030	10' wide, 150 H.P.		75.65	1,700	5,095	15,300	1,624	
	4050	Crawler, 8' wide, 100 H.P., diesel		75.65	1,775	5,355	16,100	1,676	
	4060	10' wide, 150 H.P.		84.80	2,175	6,505	19,500	1,979	
	4070	Concrete paver, 12' to 24' wide, 250 H.P.		81.00	1,575	4,755	14,300	1,599	
	4080	Placer-spreader-trimmer, 24' wide, 300 H.P.		116.90	2,600	7,825	23,500	2,500	
	4100	Pump, centrifugal gas pump, 1-1/2" diam., 65 GPM		3.35	45	135	405	53.80	
	4200	2" diameter, 130 GPM		4.55	50	150	450	66.40	
	4300	3" diameter, 250 GPM		4.80	51.50	155	465	69.40	
	4400	6" diameter, 1500 GPM		25.55	175	525	1,575	309.40	
	4500	Submersible electric pump, 1-1/4" diameter, 55 GPM		.39	15.65	47	141	12.50	
	4600	1-1/2" diameter, 83 GPM		.42	18	54	162	14.15	
	4700	2" diameter, 120 GPM		1.20	22.50	67	201	23	
	4800	3" diameter, 300 GPM		2.05	41.50	125	375	41.40	
	4900	4" diameter, 560 GPM		8.90	163	490	1,475	169.20	
	5000	6" diameter, 1590 GPM		13.05	222	665	2,000	237.40	
	5100	Diaphragm pump, gas, single, 1-1/2" diameter		1.24	45.50	136	410	37.10	
	5200	2" diameter		3.65	56.50	170	510	63.20	
	5300	3" diameter		3.65	56.50	170	510	63.20	
	5400	Double, 4" diameter		4.90	78.50	235	705	86.20	
	5450	Pressure Washer 5 Ga.Pm,3000 PSI		4.00	35	105	315	53	
	5500	Trash pump, self-priming, gas, 2" diameter		3.85	20	60	180	42.80	
	5600	Diesel, 4" diameter		10.25	58.50	175	525	117	
	5650	Diesel, 6" diameter		34.00	130	390	1,175	350	
	5655	Grout Pump		13.80	93.50	280	840	166.40	

01 54 33 | Equipment Rental

		UNIT	HOURLY OPER. COST	RENT PER DAY	RENT PER WEEK	RENT PER MONTH	EQUIPMENT COST/DAY	
40	5700	Salamanders, L.P. gas fired, 100,000 BTU	Ea.	4.56	15.35	46	138	45.70	40
	5705	50,000 BTU		3.40	8.65	26	78	32.40	
	5720	Sandblaster, portable, open top, 3 C.F. capacity		.55	27	81	243	20.60	
	5730	6 C.F. capacity		.85	40	120	360	30.80	
	5740	Accessories for above		.12	19.65	59	177	12.75	
	5750	Sander, floor		.81	18.65	56	168	17.70	
	5760	Edger		.75	25.50	77	231	21.40	
	5800	Saw, chain, gas engine, 18" long		1.90	17.65	53	159	25.80	
	5900	Hydraulic powered, 36" long		.60	55	165	495	37.80	
	5950	60" long		.65	56.50	170	510	39.20	
	6000	Masonry, table mounted, 14" diameter, 5 H.P.		1.25	53	159	475	41.80	
	6050	Portable cut-off, 8 H.P.		2.05	28	84	252	33.20	
	6100	Circular, hand held, electric, 7-1/4" diameter		.19	4	12	36	3.90	
	6200	12" diameter		.26	7.35	22	66	6.50	
	6250	Wall saw, w/hydraulic power, 10 H.P		7.50	60	180	540	96	
	6275	Shot blaster, walk behind, 20" wide		4.60	300	905	2,725	217.80	
	6280	Sidewalk broom, walk-behind		2.31	57	171	515	52.70	
	6300	Steam cleaner, 100 gallons per hour		2.90	46.50	140	420	51.20	
	6310	200 gallons per hour		4.15	55	165	495	66.20	
	6340	Tar Kettle/Pot, 400 gallon		6.15	80	240	720	97.20	
	6350	Torch, cutting, acetylene-oxygen, 150' hose		.50	21.50	65	195	17	
	6360	Hourly operating cost includes tips and gas		9.45				75.60	
	6410	Toilet, portable chemical		.11	19	57	171	12.30	
	6420	Recycle flush type		.14	23	69	207	14.90	
	6430	Toilet, fresh water flush, garden hose,		.16	26.50	79	237	17.10	
	6440	Hoisted, non-flush, for high rise		.14	22.50	68	204	14.70	
	6450	Toilet, trailers, minimum		.24	39.50	118	355	25.50	
	6460	Maximum		.72	119	358	1,075	77.35	
	6465	Tractor, farm with attachment		14.40	262	785	2,350	272.20	
	6500	Trailers, platform, flush deck, 2 axle, 25 ton capacity		4.75	107	320	960	102	
	6600	40 ton capacity		6.20	150	450	1,350	139.60	
	6700	3 axle, 50 ton capacity		6.70	165	495	1,475	152.60	
	6800	75 ton capacity		8.35	218	655	1,975	197.80	
	6810	Trailer mounted cable reel for H.V. line work		4.84	231	692	2,075	177.10	
	6820	Trailer mounted cable tensioning rig		9.59	455	1,370	4,100	350.70	
	6830	Cable pulling rig		69.62	2,575	7,710	23,100	2,099	
	6900	Water tank, engine driven discharge, 5000 gallons		6.25	143	430	1,300	136	
	6925	10,000 gallons		8.50	202	605	1,825	189	
	6950	Water truck, off highway, 6000 gallons		66.95	775	2,320	6,950	999.60	
	7010	Tram car for H.V. line work, powered, 2 conductor		6.48	125	375	1,125	126.85	
	7020	Transit (builder's level) with tripod		.08	14	42	126	9.05	
	7030	Trench box, 3000 lbs. 6'x8'		.56	93	279	835	60.30	
	7040	7200 lbs. 6'x20'		1.05	175	525	1,575	113.40	
	7050	8000 lbs., 8' x 16'		.95	158	475	1,425	102.60	
	7060	9500 lbs., 8'x20'		1.16	194	581	1,750	125.50	
	7065	11,000 lbs., 8'x24'		1.27	212	637	1,900	137.55	
	7070	12,000 lbs., 10' x 20'		1.71	285	855	2,575	184.70	
	7100	Truck, pickup, 3/4 ton, 2 wheel drive		10.35	56.50	170	510	116.80	
	7200	4 wheel drive		10.65	71.50	215	645	128.20	
	7250	Crew carrier, 9 passenger		14.70	86.50	260	780	169.60	
	7290	Flat bed truck, 20,000 G.V.W.		15.55	122	365	1,100	197.40	
	7300	Tractor, 4 x 2, 220 H.P.		21.80	190	570	1,700	288.40	
	7410	330 H.P.		32.15	262	785	2,350	414.20	
	7500	6 x 4, 380 H.P.		36.85	305	920	2,750	478.80	
	7600	450 H.P.		44.90	370	1,110	3,325	581.20	
	7620	Vacuum truck, hazardous material, 2500 gallon		10.80	310	925	2,775	271.40	
	7625	5,000 gallon		16.90	435	1,300	3,900	395.20	
	7640	Tractor, with A frame, boom and winch, 225 H.P.		24.35	267	800	2,400	354.80	

01 54 33 | Equipment Rental

			UNIT	HOURLY OPER. COST	RENT PER DAY	RENT PER WEEK	RENT PER MONTH	EQUIPMENT COST/DAY	
40	7650	Vacuum, H.E.P.A., 16 gal., wet/dry	Ea.	.82	18	54	162	17.35	40
	7655	55 gal, wet/dry		.83	27	81	243	22.85	
	7660	Water tank, portable		.17	28.50	85.50	257	18.45	
	7690	Sewer/catch basin vacuum, 14 CY, 1500 Gallon		21.45	650	1,950	5,850	561.60	
	7700	Welder, electric, 200 amp		3.63	17.65	53	159	39.65	
	7800	300 amp		5.36	21.50	64	192	55.70	
	7900	Gas engine, 200 amp		12.85	25.50	77	231	118.20	
	8000	300 amp		14.70	27.50	83	249	134.20	
	8100	Wheelbarrow, any size		.07	11.65	35	105	7.55	
	8200	Wrecking ball, 4000 lb.		2.10	73.50	220	660	60.80	
50	0010	**HIGHWAY EQUIPMENT RENTAL** without operators							50
	0050	Asphalt batch plant, portable drum mixer, 100 ton/hr.	Ea.	63.70	1,450	4,330	13,000	1,376	
	0060	200 ton/hr.		71.10	1,525	4,565	13,700	1,482	
	0070	300 ton/hr.		82.55	1,800	5,395	16,200	1,739	
	0100	Backhoe attachment, long stick, up to 185 HP, 10.5' long		.32	21.50	64	192	15.35	
	0140	Up to 250 HP, 12' long		.35	23	69	207	16.60	
	0180	Over 250 HP, 15' long		.47	31	93	279	22.35	
	0200	Special dipper arm, up to 100 HP, 32' long		.96	63.50	191	575	45.90	
	0240	Over 100 HP, 33' long		1.19	79.50	238	715	57.10	
	0280	Catch basin/sewer cleaning truck, 3 ton, 9 CY, 1000 Gal		34.00	405	1,210	3,625	514	
	0300	Concrete batch plant, portable, electric, 200 CY/Hr		17.20	510	1,530	4,600	443.60	
	0520	Grader/dozer attachment, ripper/scarifier, rear mounted, up to 135 HP		3.10	65	195	585	63.80	
	0540	Up to 180 HP		3.70	83.50	250	750	79.60	
	0580	Up to 250 HP		4.05	95	285	855	89.40	
	0700	Pvmt. removal bucket, for hyd. excavator, up to 90 HP		1.65	48.50	145	435	42.20	
	0740	Up to 200 HP		1.85	70	210	630	56.80	
	0780	Over 200 HP		1.95	83.50	250	750	65.60	
	0900	Aggregate spreader, self-propelled, 187 HP		54.90	695	2,080	6,250	855.20	
	1000	Chemical spreader, 3 C.Y.		2.75	43.50	130	390	48	
	1900	Hammermill, traveling, 250 HP		92.70	1,850	5,540	16,600	1,850	
	2000	Horizontal borer, 3" diam, 13 HP gas driven		5.45	56.50	170	510	77.60	
	2150	Horizontal directional drill, 20,000 lb. thrust, 78 H.P. diesel		26.15	700	2,095	6,275	628.20	
	2160	30,000 lb. thrust, 115 H.P. diesel		32.45	1,050	3,185	9,550	896.60	
	2170	50,000 lb. thrust, 170 H.P. diesel		45.85	1,350	4,080	12,200	1,183	
	2190	Mud trailer for HDD, 1500 gallon, 175 H.P., gas		24.40	147	440	1,325	283.20	
	2200	Hydromulchers, gas power, 3000 gal., for truck mounting		17.00	258	775	2,325	291	
	2400	Joint & crack cleaner, walk behind, 25 HP		3.10	50	150	450	54.80	
	2500	Filler, trailer mounted, 400 gal., 20 HP		7.90	218	655	1,975	194.20	
	3000	Paint striper, self propelled, double line, 30 HP		6.25	162	485	1,450	147	
	3200	Post drivers, 6" I-Beam frame, for truck mounting		13.40	435	1,305	3,925	368.20	
	3400	Road sweeper, self propelled, 8' wide, 90 HP		31.95	575	1,730	5,200	601.60	
	3450	Road sweeper, vacuum assisted, 4 CY, 220 Gal		56.60	630	1,895	5,675	831.80	
	4000	Road mixer, self-propelled, 130 HP		39.70	770	2,315	6,950	780.60	
	4100	310 HP		71.00	2,225	6,640	19,900	1,896	
	4220	Cold mix paver, incl pug mill and bitumen tank, 165 HP		86.60	2,175	6,515	19,500	1,996	
	4250	Paver, asphalt, wheel or crawler, 130 H.P., diesel		83.50	2,125	6,340	19,000	1,936	
	4300	Paver, road widener, gas 1' to 6', 67 HP		40.10	825	2,480	7,450	816.80	
	4400	Diesel, 2' to 14', 88 HP		52.30	1,050	3,175	9,525	1,053	
	4600	Slipform pavers, curb and gutter, 2 track, 75 HP		36.80	765	2,300	6,900	754.40	
	4700	4 track, 165 HP		45.85	815	2,445	7,325	855.80	
	4800	Median barrier, 215 HP		46.50	845	2,535	7,600	879	
	4901	Trailer, low bed, 75 ton capacity		9.00	218	655	1,975	203	
	5000	Road planer, walk behind, 10" cutting width, 10 HP		2.90	30.50	91	273	41.40	
	5100	Self propelled, 12" cutting width, 64 HP		8.25	127	380	1,150	142	
	5120	Traffic line remover, metal ball blaster, truck mounted, 115 HP		46.45	760	2,285	6,850	828.60	
	5140	Grinder, truck mounted, 115 HP		49.55	805	2,420	7,250	880.40	
	5160	Walk-behind, 11 HP		3.25	46.50	140	420	54	
	5200	Pavement profiler, 4' to 6' wide, 450 HP		210.55	3,400	10,220	30,700	3,728	

Note at row 0050: R015433 -10

01 54 33 | Equipment Rental

		Description	UNIT	HOURLY OPER. COST	RENT PER DAY	RENT PER WEEK	RENT PER MONTH	EQUIPMENT COST/DAY	
50	5300	8' to 10' wide, 750 HP	Ea.	333.90	4,725	14,160	42,500	5,503	50
	5400	Roadway plate, steel, 1"x8'x20'		.07	11.35	34	102	7.35	
	5600	Stabilizer, self-propelled, 150 HP		39.45	610	1,830	5,500	681.60	
	5700	310 HP		66.45	1,350	4,060	12,200	1,344	
	5800	Striper, thermal, truck mounted 120 gal. paint, 150 H.P.		49.90	505	1,520	4,550	703.20	
	6000	Tar kettle, 330 gal., trailer mounted		5.77	61.50	185	555	83.15	
	7000	Tunnel locomotive, diesel, 8 to 12 ton		26.70	585	1,750	5,250	563.60	
	7005	Electric, 10 ton		23.25	665	2,000	6,000	586	
	7010	Muck cars, 1/2 C.Y. capacity		1.75	23	69	207	27.80	
	7020	1 C.Y. capacity		1.95	31.50	94	282	34.40	
	7030	2 C.Y. capacity		2.10	36.50	110	330	38.80	
	7040	Side dump, 2 C.Y. capacity		2.30	45	135	405	45.40	
	7050	3 C.Y. capacity		3.10	51.50	155	465	55.80	
	7060	5 C.Y. capacity		4.40	65	195	585	74.20	
	7100	Ventilating blower for tunnel, 7-1/2 H.P.		1.35	51.50	155	465	41.80	
	7110	10 H.P.		1.57	53.50	160	480	44.55	
	7120	20 H.P.		2.58	69.50	208	625	62.25	
	7140	40 H.P.		4.56	98.50	295	885	95.50	
	7160	60 H.P.		6.90	152	455	1,375	146.20	
	7175	75 H.P.		8.81	202	607	1,825	191.90	
	7180	200 H.P.		19.95	305	910	2,725	341.60	
	7800	Windrow loader, elevating		47.65	1,350	4,080	12,200	1,197	
60	0010	**LIFTING AND HOISTING EQUIPMENT RENTAL** without operators R015433 -10							60
	0120	Aerial lift truck, 2 person, to 80'	Ea.	25.55	725	2,180	6,550	640.40	
	0140	Boom work platform, 40' snorkel R015433 -15		15.40	260	780	2,350	279.20	
	0150	Crane, flatbed mntd, 3 ton cap.		13.50	195	585	1,750	225	
	0200	Crane, climbing, 106' jib, 6000 lb. capacity, 410 FPM R312316 -45		37.40	1,475	4,400	13,200	1,179	
	0300	101' jib, 10,250 lb. capacity, 270 FPM		43.30	1,850	5,580	16,700	1,462	
	0500	Tower, static, 130' high, 106' jib, 6200 lb. capacity at 400 FPM		40.85	1,700	5,090	15,300	1,345	
	0600	Crawler mounted, lattice boom, 1/2 C.Y., 15 tons at 12' radius		30.66	615	1,850	5,550	615.30	
	0700	3/4 C.Y., 20 tons at 12' radius		40.88	770	2,310	6,925	789.05	
	0800	1 C.Y., 25 tons at 12' radius		54.50	1,025	3,080	9,250	1,052	
	0900	1-1/2 C.Y., 40 tons at 12' radius		54.50	1,025	3,095	9,275	1,055	
	1000	2 C.Y., 50 tons at 12' radius		54.20	1,125	3,380	10,100	1,110	
	1100	3 C.Y., 75 tons at 12' radius		72.95	1,525	4,560	13,700	1,496	
	1200	100 ton capacity, 60' boom		71.55	1,725	5,195	15,600	1,611	
	1300	165 ton capacity, 60' boom		89.60	1,925	5,750	17,300	1,867	
	1400	200 ton capacity, 70' boom		111.30	2,350	7,085	21,300	2,307	
	1500	350 ton capacity, 80' boom		158.35	3,775	11,335	34,000	3,534	
	1600	Truck mounted, lattice boom, 6 x 4, 20 tons at 10' radius		44.77	1,200	3,610	10,800	1,080	
	1700	25 tons at 10' radius		48.31	1,300	3,930	11,800	1,172	
	1800	30 tons at 10' radius		52.66	1,400	4,180	12,500	1,257	
	1900	40 tons at 12' radius		56.59	1,450	4,370	13,100	1,327	
	2000	60 tons at 15' radius		64.84	1,550	4,620	13,900	1,443	
	2050	82 tons at 15' radius		73.58	1,650	4,940	14,800	1,577	
	2100	90 tons at 15' radius		83.16	1,800	5,380	16,100	1,741	
	2200	115 tons at 15' radius		94.14	2,000	6,020	18,100	1,957	
	2300	150 tons at 18' radius		77.90	2,100	6,335	19,000	1,890	
	2350	165 tons at 18' radius		112.04	2,250	6,720	20,200	2,240	
	2400	Truck mounted, hydraulic, 12 ton capacity		49.10	625	1,880	5,650	768.80	
	2500	25 ton capacity		49.30	660	1,975	5,925	789.40	
	2550	33 ton capacity		50.35	695	2,080	6,250	818.80	
	2560	40 ton capacity		48.50	670	2,015	6,050	791	
	2600	55 ton capacity		72.00	915	2,745	8,225	1,125	
	2700	80 ton capacity		85.50	1,025	3,060	9,175	1,296	
	2720	100 ton capacity		104.90	1,550	4,685	14,100	1,776	
	2740	120 ton capacity		83.95	2,000	5,975	17,900	1,867	
	2760	150 ton capacity		89.85	2,175	6,490	19,500	2,017	

01 54 33 | Equipment Rental

		UNIT	HOURLY OPER. COST	RENT PER DAY	RENT PER WEEK	RENT PER MONTH	EQUIPMENT COST/DAY	
60	2800	Self-propelled, 4 x 4, with telescoping boom, 5 ton	Ea.	14.65	235	705	2,125	258.20	**60**
	2900	12-1/2 ton capacity		34.60	540	1,625	4,875	601.80	
	3000	15 ton capacity		32.30	575	1,725	5,175	603.40	
	3050	20 ton capacity		37.95	660	1,985	5,950	700.60	
	3100	25 ton capacity		38.85	675	2,030	6,100	716.80	
	3150	40 ton capacity		50.15	895	2,680	8,050	937.20	
	3200	Derricks, guy, 20 ton capacity, 60' boom, 75' mast		33.55	360	1,078	3,225	484	
	3300	100' boom, 115' mast		51.95	615	1,850	5,550	785.60	
	3400	Stiffleg, 20 ton capacity, 70' boom, 37' mast		35.80	465	1,400	4,200	566.40	
	3500	100' boom, 47' mast		54.68	745	2,240	6,725	885.45	
	3550	Helicopter, small, lift to 1250 lbs. maximum, w/pilot		105.67	2,900	8,710	26,100	2,587	
	3600	Hoists, chain type, overhead, manual, 3/4 ton		.10	1	3	9	1.40	
	3900	10 ton		.70	9.65	29	87	11.40	
	4000	Hoist and tower, 5000 lb. cap., portable electric, 40' high		4.65	207	621	1,875	161.40	
	4100	For each added 10' section, add		.10	16.35	49	147	10.60	
	4200	Hoist and single tubular tower, 5000 lb. electric, 100' high		6.26	289	867	2,600	223.50	
	4300	For each added 6'-6" section, add		.17	27.50	83	249	17.95	
	4400	Hoist and double tubular tower, 5000 lb., 100' high		6.70	320	955	2,875	244.60	
	4500	For each added 6'-6" section, add		.19	31	93	279	20.10	
	4550	Hoist and tower, mast type, 6000 lb., 100' high		7.26	330	990	2,975	256.10	
	4570	For each added 10' section, add		.12	19.65	59	177	12.75	
	4600	Hoist and tower, personnel, electric, 2000 lb., 100' @ 125 FPM		14.74	880	2,640	7,925	645.90	
	4700	3000 lb., 100' @ 200 FPM		16.88	995	2,990	8,975	733.05	
	4800	3000 lb., 150' @ 300 FPM		18.63	1,125	3,340	10,000	817.05	
	4900	4000 lb., 100' @ 300 FPM		19.36	1,125	3,410	10,200	836.90	
	5000	6000 lb., 100' @ 275 FPM	▼	20.98	1,200	3,580	10,700	883.85	
	5100	For added heights up to 500', add	L.F.	.01	1.67	5	15	1.10	
	5200	Jacks, hydraulic, 20 ton	Ea.	.05	1.67	5	15	1.40	
	5500	100 ton		.35	10.65	32	96	9.20	
	6100	Jacks, hydraulic, climbing w/ 50' jackrods, control console, 30 ton cap		1.79	119	357	1,075	85.70	
	6150	For each added 10' jackrod section, add		.05	3.33	10	30	2.40	
	6300	50 ton capacity		2.87	191	574	1,725	137.75	
	6350	For each added 10' jackrod section, add		.06	4	12	36	2.90	
	6500	125 ton capacity		7.55	505	1,510	4,525	362.40	
	6550	For each added 10' jackrod section, add		.52	34.50	103	310	24.75	
	6600	Cable jack, 10 ton capacity with 200' cable		1.50	99.50	299	895	71.80	
	6650	For each added 50' of cable, add	▼	.17	11.35	34	102	8.15	
70	0010	**WELLPOINT EQUIPMENT RENTAL** without operators							**70**
	0020	Based on 2 months rental							
	0100	Combination jetting & wellpoint pump, 60 H.P. diesel	Ea.	21.79	295	884	2,650	351.10	
	0200	High pressure gas jet pump, 200 H.P., 300 psi	"	57.29	252	756	2,275	609.50	
	0300	Discharge pipe, 8" diameter	L.F.	.01	.48	1.44	4.32	.35	
	0350	12" diameter		.01	.71	2.12	6.35	.50	
	0400	Header pipe, flows up to 150 G.P.M., 4" diameter		.01	.43	1.29	3.87	.35	
	0500	400 G.P.M., 6" diameter		.01	.51	1.54	4.62	.40	
	0600	800 G.P.M., 8" diameter		.01	.71	2.12	6.35	.50	
	0700	1500 G.P.M., 10" diameter		.01	.74	2.22	6.65	.50	
	0800	2500 G.P.M., 12" diameter		.02	1.40	4.21	12.65	1	
	0900	4500 G.P.M., 16" diameter		.03	1.80	5.39	16.15	1.30	
	0950	For quick coupling aluminum and plastic pipe, add	▼	.03	1.86	5.58	16.75	1.35	
	1100	Wellpoint, 25' long, with fittings & riser pipe, 1-1/2" or 2" diameter	Ea.	.06	3.71	11.13	33.50	2.70	
	1200	Wellpoint pump, diesel powered, 4" diameter, 20 H.P.		8.77	170	510	1,525	172.15	
	1300	6" diameter, 30 H.P.		12.22	211	632	1,900	224.15	
	1400	8" suction, 40 H.P.		16.47	289	867	2,600	305.15	
	1500	10" suction, 75 H.P.		26.59	340	1,013	3,050	415.30	
	1600	12" suction, 100 H.P.		37.34	540	1,620	4,850	622.70	
	1700	12" suction, 175 H.P.	▼	57.89	590	1,770	5,300	817.10	

R015433 -10

01 54 33 | Equipment Rental

		UNIT	HOURLY OPER. COST	RENT PER DAY	RENT PER WEEK	RENT PER MONTH	EQUIPMENT COST/DAY	
80	**0010** MARINE EQUIPMENT RENTAL without operators	R015433 -10						**80**
	0200 Barge, 400 Ton, 30' wide x 90' long	Ea.	14.80	1,050	3,140	9,425	746.40	
	0240 800 Ton, 45' wide x 90' long		17.95	1,275	3,800	11,400	903.60	
	2000 Tugboat, diesel, 100 HP		27.75	195	585	1,750	339	
	2040 250 HP		56.05	360	1,075	3,225	663.40	
	2080 380 HP		115.60	1,075	3,195	9,575	1,564	

Crew A-1

Crew A-1	Bare Costs Hr.	Daily	Incl. Subs O & P Hr.	Daily	Bare Costs	Incl. O&P
1 Building Laborer	$20.55	$164.40	$34.75	$278.00	$20.55	$34.75
1 Concrete saw, gas manual		63.20		69.52	7.90	8.69
8 L.H., Daily Totals		$227.60		$347.52	$28.45	$43.44

Crew A-1A

Crew A-1A	Bare Costs Hr.	Daily	Incl. Subs O & P Hr.	Daily	Bare Costs	Incl. O&P
1 Skilled Worker	$28.05	$224.40	$47.35	$378.80	$28.05	$47.35
1 Shot Blaster, 20"		217.80		239.58	27.23	29.95
8 L.H., Daily Totals		$442.20		$618.38	$55.27	$77.30

Crew A-1B

Crew A-1B	Bare Costs Hr.	Daily	Incl. Subs O & P Hr.	Daily	Bare Costs	Incl. O&P
1 Building Laborer	$20.55	$164.40	$34.75	$278.00	$20.55	$34.75
1 Concrete Saw		141.00		155.10	17.63	19.39
8 L.H., Daily Totals		$305.40		$433.10	$38.17	$54.14

Crew A-1C

Crew A-1C	Bare Costs Hr.	Daily	Incl. Subs O & P Hr.	Daily	Bare Costs	Incl. O&P
1 Building Laborer	$20.55	$164.40	$34.75	$278.00	$20.55	$34.75
1 Chain saw, gas, 18"		25.80		28.38	3.23	3.55
8 L.H., Daily Totals		$190.20		$306.38	$23.77	$38.30

Crew A-1D

Crew A-1D	Bare Costs Hr.	Daily	Incl. Subs O & P Hr.	Daily	Bare Costs	Incl. O&P
1 Building Laborer	$20.55	$164.40	$34.75	$278.00	$20.55	$34.75
1 Vibrating plate, gas, 18"		31.80		34.98	3.98	4.37
8 L.H., Daily Totals		$196.20		$312.98	$24.52	$39.12

Crew A-1E

Crew A-1E	Bare Costs Hr.	Daily	Incl. Subs O & P Hr.	Daily	Bare Costs	Incl. O&P
1 Building Laborer	$20.55	$164.40	$34.75	$278.00	$20.55	$34.75
1 Vibratory Plate, Gas, 21"		39.20		43.12	4.90	5.39
8 L.H., Daily Totals		$203.60		$321.12	$25.45	$40.14

Crew A-1F

Crew A-1F	Bare Costs Hr.	Daily	Incl. Subs O & P Hr.	Daily	Bare Costs	Incl. O&P
1 Building Laborer	$20.55	$164.40	$34.75	$278.00	$20.55	$34.75
1 Rammer/tamper, gas, 8"		43.40		47.74	5.42	5.97
8 L.H., Daily Totals		$207.80		$325.74	$25.98	$40.72

Crew A-1G

Crew A-1G	Bare Costs Hr.	Daily	Incl. Subs O & P Hr.	Daily	Bare Costs	Incl. O&P
1 Building Laborer	$20.55	$164.40	$34.75	$278.00	$20.55	$34.75
1 Rammer/tamper, gas, 15"		49.40		54.34	6.17	6.79
8 L.H., Daily Totals		$213.80		$332.34	$26.73	$41.54

Crew A-1H

Crew A-1H	Bare Costs Hr.	Daily	Incl. Subs O & P Hr.	Daily	Bare Costs	Incl. O&P
1 Building Laborer	$20.55	$164.40	$34.75	$278.00	$20.55	$34.75
1 Exterior Steam Cleaner		51.20		56.32	6.40	7.04
8 L.H., Daily Totals		$215.60		$334.32	$26.95	$41.79

Crew A-1J

Crew A-1J	Bare Costs Hr.	Daily	Incl. Subs O & P Hr.	Daily	Bare Costs	Incl. O&P
1 Building Laborer	$20.55	$164.40	$34.75	$278.00	$20.55	$34.75
1 Cultivator, Walk-Behind, 5 H.P.		57.05		62.76	7.13	7.84
8 L.H., Daily Totals		$221.45		$340.76	$27.68	$42.59

Crew A-1K

Crew A-1K	Bare Costs Hr.	Daily	Incl. Subs O & P Hr.	Daily	Bare Costs	Incl. O&P
1 Building Laborer	$20.55	$164.40	$34.75	$278.00	$20.55	$34.75
1 Cultivator, Walk-Behind, 8 H.P.		91.10		100.21	11.39	12.53
8 L.H., Daily Totals		$255.50		$378.21	$31.94	$47.28

Crew A-1M

Crew A-1M	Bare Costs Hr.	Daily	Incl. Subs O & P Hr.	Daily	Bare Costs	Incl. O&P
1 Building Laborer	$20.55	$164.40	$34.75	$278.00	$20.55	$34.75
1 Snow Blower, Walk-Behind		52.70		57.97	6.59	7.25
8 L.H., Daily Totals		$217.10		$335.97	$27.14	$42.00

Crew A-2

Crew A-2	Bare Costs Hr.	Daily	Incl. Subs O & P Hr.	Daily	Bare Costs	Incl. O&P
2 Laborers	$20.55	$328.80	$34.75	$556.00	$21.02	$35.45
1 Truck Driver (light)	21.95	175.60	36.85	294.80		
1 Flatbed Truck, Gas, 1.5 Ton		194.40		213.84	8.10	8.91
24 L.H., Daily Totals		$698.80		$1064.64	$29.12	$44.36

Crew A-2A

Crew A-2A	Bare Costs Hr.	Daily	Incl. Subs O & P Hr.	Daily	Bare Costs	Incl. O&P
2 Laborers	$20.55	$328.80	$34.75	$556.00	$21.02	$35.45
1 Truck Driver (light)	21.95	175.60	36.85	294.80		
1 Flatbed Truck, Gas, 1.5 Ton		194.40		213.84		
1 Concrete Saw		141.00		155.10	13.98	15.37
24 L.H., Daily Totals		$839.80		$1219.74	$34.99	$50.82

Crew A-2B

Crew A-2B	Bare Costs Hr.	Daily	Incl. Subs O & P Hr.	Daily	Bare Costs	Incl. O&P
1 Truck Driver (light)	$21.95	$175.60	$36.85	$294.80	$21.95	$36.85
1 Flatbed Truck, Gas, 1.5 Ton		194.40		213.84	24.30	26.73
8 L.H., Daily Totals		$370.00		$508.64	$46.25	$63.58

Crew A-3A

Crew A-3A	Bare Costs Hr.	Daily	Incl. Subs O & P Hr.	Daily	Bare Costs	Incl. O&P
1 Truck Driver (light)	$21.95	$175.60	$36.85	$294.80	$21.95	$36.85
1 Pickup truck, 4 x 4, 3/4 ton		128.20		141.02	16.02	17.63
8 L.H., Daily Totals		$303.80		$435.82	$37.98	$54.48

Crew A-3B

Crew A-3B	Bare Costs Hr.	Daily	Incl. Subs O & P Hr.	Daily	Bare Costs	Incl. O&P
1 Equip. Oper. (medium)	$28.55	$228.40	$46.90	$375.20	$25.63	$42.50
1 Truck Driver (heavy)	22.70	181.60	38.10	304.80		
1 Dump Truck, 12 C.Y., 400 H.P.		533.00		586.30		
1 F.E. Loader, W.M., 2.5 C.Y.		395.80		435.38	58.05	63.85
16 L.H., Daily Totals		$1338.80		$1701.68	$83.67	$106.36

Crew A-3C

Crew A-3C	Bare Costs Hr.	Daily	Incl. Subs O & P Hr.	Daily	Bare Costs	Incl. O&P
1 Equip. Oper. (light)	$26.95	$215.60	$44.30	$354.40	$26.95	$44.30
1 Loader, Skid Steer, 78 H.P.		259.60		285.56	32.45	35.70
8 L.H., Daily Totals		$475.20		$639.96	$59.40	$80.00

Crew A-3D

Crew A-3D	Bare Costs Hr.	Daily	Incl. Subs O & P Hr.	Daily	Bare Costs	Incl. O&P
1 Truck Driver, Light	$21.95	$175.60	$36.85	$294.80	$21.95	$36.85
1 Pickup truck, 4 x 4, 3/4 ton		128.20		141.02		
1 Flatbed Trailer, 25 Ton		102.00		112.20	28.77	31.65
8 L.H., Daily Totals		$405.80		$548.02	$50.73	$68.50

Crew A-3E

Crew A-3E	Bare Costs Hr.	Daily	Incl. Subs O & P Hr.	Daily	Bare Costs	Incl. O&P
1 Equip. Oper. (crane)	$29.35	$234.80	$48.20	$385.60	$26.02	$43.15
1 Truck Driver (heavy)	22.70	181.60	38.10	304.80		
1 Pickup truck, 4 x 4, 3/4 ton		128.20		141.02	8.01	8.81
16 L.H., Daily Totals		$544.60		$831.42	$34.04	$51.96

Crew A-3F

Crew A-3F	Bare Costs Hr.	Daily	Incl. Subs O & P Hr.	Daily	Bare Costs	Incl. O&P
1 Equip. Oper. (crane)	$29.35	$234.80	$48.20	$385.60	$26.02	$43.15
1 Truck Driver (heavy)	22.70	181.60	38.10	304.80		
1 Pickup truck, 4 x 4, 3/4 ton		128.20		141.02		
1 Truck Tractor, 6x4, 380 H.P.		478.80		526.68		
1 Lowbed Trailer, 75 Ton		203.00		223.30	50.63	55.69
16 L.H., Daily Totals		$1226.40		$1581.40	$76.65	$98.84

Crew A-3G

Crew A-3G	Hr.	Daily	Hr.	Daily	Bare Costs	Incl. O&P
1 Equip. Oper. (crane)	$29.35	$234.80	$48.20	$385.60	$26.02	$43.15
1 Truck Driver (heavy)	22.70	181.60	38.10	304.80		
1 Pickup truck, 4 x 4, 3/4 ton		128.20		141.02		
1 Truck Tractor, 6x4, 450 H.P.		581.20		639.32		
1 Lowbed Trailer, 75 Ton		203.00		223.30	57.02	62.73
16 L.H., Daily Totals		$1328.80		$1694.04	$83.05	$105.88

Crew A-3H

Crew A-3H	Hr.	Daily	Hr.	Daily	Bare Costs	Incl. O&P
1 Equip. Oper. (crane)	$29.35	$234.80	$48.20	$385.60	$29.35	$48.20
1 Hyd. crane, 12 Ton (daily)		1018.00		1119.80	127.25	139.97
8 L.H., Daily Totals		$1252.80		$1505.40	$156.60	$188.18

Crew A-3I

Crew A-3I	Hr.	Daily	Hr.	Daily	Bare Costs	Incl. O&P
1 Equip. Oper. (crane)	$29.35	$234.80	$48.20	$385.60	$29.35	$48.20
1 Hyd. crane, 25 Ton (daily)		1054.00		1159.40	131.75	144.93
8 L.H., Daily Totals		$1288.80		$1545.00	$161.10	$193.13

Crew A-3J

Crew A-3J	Hr.	Daily	Hr.	Daily	Bare Costs	Incl. O&P
1 Equip. Oper. (crane)	$29.35	$234.80	$48.20	$385.60	$29.35	$48.20
1 Hyd. crane, 40 Ton (daily)		1058.00		1163.80	132.25	145.47
8 L.H., Daily Totals		$1292.80		$1549.40	$161.60	$193.68

Crew A-3K

Crew A-3K	Hr.	Daily	Hr.	Daily	Bare Costs	Incl. O&P
1 Equip. Oper. (crane)	$29.35	$234.80	$48.20	$385.60	$27.38	$44.98
1 Equip. Oper. Oiler	25.40	203.20	41.75	334.00		
1 Hyd. crane, 55 Ton (daily)		1491.00		1640.10		
1 P/U Truck, 3/4 Ton (daily)		137.80		151.58	101.80	111.98
16 L.H., Daily Totals		$2066.80		$2511.28	$129.18	$156.96

Crew A-3L

Crew A-3L	Hr.	Daily	Hr.	Daily	Bare Costs	Incl. O&P
1 Equip. Oper. (crane)	$29.35	$234.80	$48.20	$385.60	$27.38	$44.98
1 Equip. Oper. Oiler	25.40	203.20	41.75	334.00		
1 Hyd. crane, 80 Ton (daily)		1704.00		1874.40		
1 P/U Truck, 3/4 Ton (daily)		137.80		151.58	115.11	126.62
16 L.H., Daily Totals		$2279.80		$2745.58	$142.49	$171.60

Crew A-3M

Crew A-3M	Hr.	Daily	Hr.	Daily	Bare Costs	Incl. O&P
1 Equip. Oper. (crane)	$29.35	$234.80	$48.20	$385.60	$27.38	$44.98
1 Equip. Oper. Oiler	25.40	203.20	41.75	334.00		
1 Hyd. crane, 100 Ton (daily)		2399.00		2638.90		
1 P/U Truck, 3/4 Ton (daily)		137.80		151.58	158.55	174.41
16 L.H., Daily Totals		$2974.80		$3510.08	$185.93	$219.38

Crew A-3N

Crew A-3N	Hr.	Daily	Hr.	Daily	Bare Costs	Incl. O&P
1 Equip. Oper. (crane)	$29.35	$234.80	$48.20	$385.60	$29.35	$48.20
1 Tower crane (monthly)		1022.00		1124.20	127.75	140.53
8 L.H., Daily Totals		$1256.80		$1509.80	$157.10	$188.72

Crew A-3P

Crew A-3P	Hr.	Daily	Hr.	Daily	Bare Costs	Incl. O&P
1 Equip. Oper., Light	$26.95	$215.60	$44.30	$354.40	$26.95	$44.30
1 A.T. Forklift, 42' lift		446.00		490.60	55.75	61.33
8 L.H., Daily Totals		$661.60		$845.00	$82.70	$105.63

Crew A-4

Crew A-4	Hr.	Daily	Hr.	Daily	Bare Costs	Incl. O&P
2 Carpenters	$27.95	$447.20	$47.25	$756.00	$26.97	$45.15
1 Painter, Ordinary	25.00	200.00	40.95	327.60		
24 L.H., Daily Totals		$647.20		$1083.60	$26.97	$45.15

Crew A-5

Crew A-5	Hr.	Daily	Hr.	Daily	Bare Costs	Incl. O&P
2 Laborers	$20.55	$328.80	$34.75	$556.00	$20.71	$34.98
.25 Truck Driver (light)	21.95	43.90	36.85	73.70		
.25 Flatbed Truck, Gas, 1.5 Ton		48.60		53.46	2.70	2.97
18 L.H., Daily Totals		$421.30		$683.16	$23.41	$37.95

Crew A-6

Crew A-6	Hr.	Daily	Hr.	Daily	Bare Costs	Incl. O&P
1 Instrument Man	$28.05	$224.40	$47.35	$378.80	$27.73	$46.30
1 Rodman/Chainman	27.40	219.20	45.25	362.00		
1 Laser Transit/Level		70.10		77.11	4.38	4.82
16 L.H., Daily Totals		$513.70		$817.91	$32.11	$51.12

Crew A-7

Crew A-7	Hr.	Daily	Hr.	Daily	Bare Costs	Incl. O&P
1 Chief Of Party	$33.15	$265.20	$55.60	$444.80	$29.53	$49.40
1 Instrument Man	28.05	224.40	47.35	378.80		
1 Rodman/Chainman	27.40	219.20	45.25	362.00		
1 Laser Transit/Level		70.10		77.11	2.92	3.21
24 L.H., Daily Totals		$778.90		$1262.71	$32.45	$52.61

Crew A-8

Crew A-8	Hr.	Daily	Hr.	Daily	Bare Costs	Incl. O&P
1 Chief of Party	$33.15	$265.20	$55.60	$444.80	$29.00	$48.36
1 Instrument Man	28.05	224.40	47.35	378.80		
2 Rodmen/Chainmen	27.40	438.40	45.25	724.00		
1 Laser Transit/Level		70.10		77.11	2.19	2.41
32 L.H., Daily Totals		$998.10		$1624.71	$31.19	$50.77

Crew A-9

Crew A-9	Hr.	Daily	Hr.	Daily	Bare Costs	Incl. O&P
1 Asbestos Foreman	$29.60	$236.80	$50.55	$404.40	$29.16	$49.76
7 Asbestos Workers	29.10	1629.60	49.65	2780.40		
64 L.H., Daily Totals		$1866.40		$3184.80	$29.16	$49.76

Crew A-10A

Crew A-10A	Hr.	Daily	Hr.	Daily	Bare Costs	Incl. O&P
1 Asbestos Foreman	$29.60	$236.80	$50.55	$404.40	$29.27	$49.95
2 Asbestos Workers	29.10	465.60	49.65	794.40		
24 L.H., Daily Totals		$702.40		$1198.80	$29.27	$49.95

Crew A-10B

Crew A-10B	Hr.	Daily	Hr.	Daily	Bare Costs	Incl. O&P
1 Asbestos Foreman	$29.60	$236.80	$50.55	$404.40	$29.23	$49.88
3 Asbestos Workers	29.10	698.40	49.65	1191.60		
32 L.H., Daily Totals		$935.20		$1596.00	$29.23	$49.88

Crew A-10C

Crew A-10C	Hr.	Daily	Hr.	Daily	Bare Costs	Incl. O&P
3 Asbestos Workers	$29.10	$698.40	$49.65	$1191.60	$29.10	$49.65
1 Flatbed Truck, Gas, 1.5 Ton		194.40		213.84	8.10	8.91
24 L.H., Daily Totals		$892.80		$1405.44	$37.20	$58.56

Crew A-10D

Crew A-10D	Hr.	Daily	Hr.	Daily	Bare Costs	Incl. O&P
2 Asbestos Workers	$29.10	$465.60	$49.65	$794.40	$28.24	$47.31
1 Equip. Oper. (crane)	29.35	234.80	48.20	385.60		
1 Equip. Oper. Oiler	25.40	203.20	41.75	334.00		
1 Hydraulic Crane, 33 Ton		818.80		900.68	25.59	28.15
32 L.H., Daily Totals		$1722.40		$2414.68	$53.83	$75.46

Crew A-11

Crew A-11	Hr.	Daily	Hr.	Daily	Bare Costs	Incl. O&P
1 Asbestos Foreman	$29.60	$236.80	$50.55	$404.40	$29.16	$49.76
7 Asbestos Workers	29.10	1629.60	49.65	2780.40		
2 Chipping Hammers, 12 Lb., Elec.		34.00		37.40	0.53	0.58
64 L.H., Daily Totals		$1900.40		$3222.20	$29.69	$50.35

Crew No.	Bare Costs Hr.	Daily	Incl. Subs O & P Hr.	Daily	Cost Per Labor-Hour Bare Costs	Incl. O&P
Crew A-12	Hr.	Daily	Hr.	Daily	Bare Costs	Incl. O&P
1 Asbestos Foreman	$29.60	$236.80	$50.55	$404.40	$29.16	$49.76
7 Asbestos Workers	29.10	1629.60	49.65	2780.40		
1 Trk-mtd vac, 14 CY, 1500 Gal.		561.60		617.76		
1 Flatbed Truck, 20,000 GVW		197.40		217.14	11.86	13.05
64 L.H., Daily Totals		$2625.40		$4019.70	$41.02	$62.81

Crew A-13	Hr.	Daily	Hr.	Daily	Bare Costs	Incl. O&P
1 Equip. Oper. (light)	$26.95	$215.60	$44.30	$354.40	$26.95	$44.30
1 Trk-mtd vac, 14 CY, 1500 Gal.		561.60		617.76		
1 Flatbed Truck, 20,000 GVW		197.40		217.14	94.88	104.36
8 L.H., Daily Totals		$974.60		$1189.30	$121.83	$148.66

Crew B-1	Hr.	Daily	Hr.	Daily	Bare Costs	Incl. O&P
1 Labor Foreman (outside)	$22.55	$180.40	$38.15	$305.20	$21.22	$35.88
2 Laborers	20.55	328.80	34.75	556.00		
24 L.H., Daily Totals		$509.20		$861.20	$21.22	$35.88

Crew B-1A	Hr.	Daily	Hr.	Daily	Bare Costs	Incl. O&P
1 Laborer Foreman	$22.55	$180.40	$38.15	$305.20	$21.22	$35.88
2 Laborers	20.55	328.80	34.75	556.00		
2 Cutting Torches		34.00		37.40		
2 Gases		151.20		166.32	7.72	8.49
24 L.H., Daily Totals		$694.40		$1064.92	$28.93	$44.37

Crew B-1B	Hr.	Daily	Hr.	Daily	Bare Costs	Incl. O&P
1 Laborer Foreman	$22.55	$180.40	$38.15	$305.20	$23.25`	$38.96
2 Laborers	20.55	328.80	34.75	556.00		
1 Equip. Oper. (crane)	29.35	234.80	48.20	385.60		
2 Cutting Torches		34.00		37.40		
2 Gases		151.20		166.32		
1 Hyd. Crane, 12 Ton		768.80		845.68	29.81	32.79
32 L.H., Daily Totals		$1698.00		$2296.20	$53.06	$71.76

Crew B-2	Hr.	Daily	Hr.	Daily	Bare Costs	Incl. O&P
1 Labor Foreman (outside)	$22.55	$180.40	$38.15	$305.20	$20.95	$35.43
4 Laborers	20.55	657.60	34.75	1112.00		
40 L.H., Daily Totals		$838.00		$1417.20	$20.95	$35.43

Crew B-3	Hr.	Daily	Hr.	Daily	Bare Costs	Incl. O&P
1 Labor Foreman (outside)	$22.55	$180.40	$38.15	$305.20	$22.93	$38.46
2 Laborers	20.55	328.80	34.75	556.00		
1 Equip. Oper. (med.)	28.55	228.40	46.90	375.20		
2 Truck Drivers (heavy)	22.70	363.20	38.10	609.60		
1 Crawler Loader, 3 C.Y.		990.00		1089.00		
2 Dump Trucks 12 C.Y., 400 H.P.		1066.00		1172.60	42.83	47.12
48 L.H., Daily Totals		$3156.80		$4107.60	$65.77	$85.58

Crew B-3A	Hr.	Daily	Hr.	Daily	Bare Costs	Incl. O&P
4 Laborers	$20.55	$657.60	$34.75	$1112.00	$22.15	$37.18
1 Equip. Oper. (med.)	28.55	228.40	46.90	375.20		
1 Hyd. Excavator, 1.5 C.Y.		865.80		952.38	21.65	23.81
40 L.H., Daily Totals		$1751.80		$2439.58	$43.80	$60.99

Crew B-3B	Hr.	Daily	Hr.	Daily	Bare Costs	Incl. O&P
2 Laborers	$20.55	$328.80	$34.75	$556.00	$23.09	$38.63
1 Equip. Oper. (med.)	28.55	228.40	46.90	375.20		
1 Truck Driver (heavy)	22.70	181.60	38.10	304.80		
1 Backhoe Loader, 80 H.P.		345.00		379.50		
1 Dump Truck, 12 C.Y., 400 H.P.		533.00		586.30	27.44	30.18
32 L.H., Daily Totals		$1616.80		$2201.80	$50.52	$68.81

Crew B-3C	Hr.	Daily	Hr.	Daily	Bare Costs	Incl. O&P
3 Laborers	$20.55	$493.20	$34.75	$834.00	$22.55	$37.79
1 Equip. Oper. (med.)	28.55	228.40	46.90	375.20		
1 Crawler Loader, 4 C.Y.		1450.00		1595.00	45.31	49.84
32 L.H., Daily Totals		$2171.60		$2804.20	$67.86	$87.63

Crew B-4	Hr.	Daily	Hr.	Daily	Bare Costs	Incl. O&P
1 Labor Foreman (outside)	$22.55	$180.40	$38.15	$305.20	$21.24	$35.88
4 Laborers	20.55	657.60	34.75	1112.00		
1 Truck Driver (heavy)	22.70	181.60	38.10	304.80		
1 Truck Tractor, 220 H.P.		288.40		317.24		
1 Flatbed Trailer, 40 Ton		139.60		153.56	8.92	9.81
48 L.H., Daily Totals		$1447.60		$2192.80	$30.16	$45.68

Crew B-5	Hr.	Daily	Hr.	Daily	Bare Costs	Incl. O&P
1 Labor Foreman (outside)	$22.55	$180.40	$38.15	$305.20	$22.55	$37.86
3 Laborers	20.55	493.20	34.75	834.00		
1 Equip. Oper. (med.)	28.55	228.40	46.90	375.20		
1 Air Compressor, 250 cfm		162.40		178.64		
2 Breakers, Pavement, 60 lb.		16.40		18.04		
2 -50' Air Hoses, 1.5"		12.60		13.86		
1 Crawler Loader, 3 C.Y.		990.00		1089.00	29.54	32.49
40 L.H., Daily Totals		$2083.40		$2813.94	$52.09	$70.35

Crew B-5A	Hr.	Daily	Hr.	Daily	Bare Costs	Incl. O&P
1 Foreman	$22.55	$180.40	$38.15	$305.20	$22.94	$38.41
6 Laborers	20.55	986.40	34.75	1668.00		
2 Equip. Oper. (med.)	28.55	456.80	46.90	750.40		
1 Equip. Oper. (light)	26.95	215.60	44.30	354.40		
2 Truck Drivers (heavy)	22.70	363.20	38.10	609.60		
1 Air Compressor, 365 cfm		211.20		232.32		
2 Breakers, Pavement, 60 lb.		16.40		18.04		
8 -50' Air Hoses, 1"		32.80		36.08		
2 Dump Trucks, 8 C.Y., 220 H.P.		652.80		718.08	9.51	10.46
96 L.H., Daily Totals		$3115.60		$4692.12	$32.45	$48.88

Crew B-5B	Hr.	Daily	Hr.	Daily	Bare Costs	Incl. O&P
1 Powderman	$28.05	$224.40	$47.35	$378.80	$25.54	$42.58
2 Equip. Oper. (med.)	28.55	456.80	46.90	750.40		
3 Truck Drivers (heavy)	22.70	544.80	38.10	914.40		
1 F.E. Loader, W.M., 2.5 C.Y.		395.80		435.38		
3 Dump Trucks, 12 C.Y., 400 H.P.		1599.00		1758.90		
1 Air Compressor, 365 cfm		211.20		232.32	45.96	50.55
48 L.H., Daily Totals		$3432.00		$4470.20	$71.50	$93.13

Crew B-5C	Hr.	Daily	Hr.	Daily	Bare Costs	Incl. O&P
3 Laborers	$20.55	$493.20	$34.75	$834.00	$23.79	$39.66
1 Equip. Oper. (medium)	28.55	228.40	46.90	375.20		
2 Truck Drivers (heavy)	22.70	363.20	38.10	609.60		
1 Equip. Oper. (crane)	29.35	234.80	48.20	385.60		
1 Equip. Oper. Oiler	25.40	203.20	41.75	334.00		
2 Dump Trucks, 12 C.Y., 400 H.P.		1066.00		1172.60		
1 Crawler Loader, 4 C.Y.		1450.00		1595.00		
1 S.P. Crane, 4x4, 25 Ton		716.80		788.48	50.51	55.56
64 L.H., Daily Totals		$4755.60		$6094.48	$74.31	$95.23

Crew B-6	Hr.	Daily	Hr.	Daily	Bare Costs	Incl. O&P
2 Laborers	$20.55	$328.80	$34.75	$556.00	$22.68	$37.93
1 Equip. Oper. (light)	26.95	215.60	44.30	354.40		
1 Backhoe Loader, 48 H.P.		293.80		323.18	12.24	13.47
24 L.H., Daily Totals		$838.20		$1233.58	$34.92	$51.40

Crew No.	Bare Costs		Incl. Subs O & P		Cost Per Labor-Hour	

Crew B-6B

	Hr.	Daily	Hr.	Daily	Bare Costs	Incl. O&P
2 Labor Foremen (out)	$22.55	$360.80	$38.15	$610.40	$21.22	$35.88
4 Laborers	20.55	657.60	34.75	1112.00		
1 S.P. Crane, 4x4, 5 Ton		258.20		284.02		
1 Flatbed Truck, Gas, 1.5 Ton		194.40		213.84		
1 Butt Fusion Machine		199.80		219.78	13.59	14.95
48 L.H., Daily Totals		$1670.80		$2440.04	$34.81	$50.83

Crew B-7

	Hr.	Daily	Hr.	Daily	Bare Costs	Incl. O&P
1 Labor Foreman (outside)	$22.55	$180.40	$38.15	$305.20	$22.22	$37.34
4 Laborers	20.55	657.60	34.75	1112.00		
1 Equip. Oper. (med.)	28.55	228.40	46.90	375.20		
1 Brush Chipper, 12", 130 H.P.		225.60		248.16		
1 Crawler Loader, 3 C.Y.		990.00		1089.00		
2 Chainsaws, Gas, 36" Long		75.60		83.16	26.90	29.59
48 L.H., Daily Totals		$2357.60		$3212.72	$49.12	$66.93

Crew B-7A

	Hr.	Daily	Hr.	Daily	Bare Costs	Incl. O&P
2 Laborers	$20.55	$328.80	$34.75	$556.00	$22.68	$37.93
1 Equip. Oper. (light)	26.95	215.60	44.30	354.40		
1 Rake w/Tractor		257.70		283.47		
2 Chain Saws, gas, 18"		51.60		56.76	12.89	14.18
24 L.H., Daily Totals		$853.70		$1250.63	$35.57	$52.11

Crew B-8

	Hr.	Daily	Hr.	Daily	Bare Costs	Incl. O&P
1 Labor Foreman (outside)	$22.55	$180.40	$38.15	$305.20	$23.74	$39.66
2 Laborers	20.55	328.80	34.75	556.00		
2 Equip. Oper. (med.)	28.55	456.80	46.90	750.40		
2 Truck Drivers (heavy)	22.70	363.20	38.10	609.60		
1 Hyd. Crane, 25 Ton		789.40		868.34		
1 Crawler Loader, 3 C.Y.		990.00		1089.00		
2 Dump Trucks, 12 C.Y., 400 H.P.		1066.00		1172.60	50.81	55.89
56 L.H., Daily Totals		$4174.60		$5351.14	$74.55	$95.56

Crew B-9

	Hr.	Daily	Hr.	Daily	Bare Costs	Incl. O&P
1 Labor Foreman (outside)	$22.55	$180.40	$38.15	$305.20	$20.95	$35.43
4 Laborers	20.55	657.60	34.75	1112.00		
1 Air Compressor, 250 cfm		162.40		178.64		
2 Breakers, Pavement, 60 lb.		16.40		18.04		
2 -50' Air Hoses, 1.5"		12.60		13.86	4.79	5.26
40 L.H., Daily Totals		$1029.40		$1627.74	$25.73	$40.69

Crew B-9A

	Hr.	Daily	Hr.	Daily	Bare Costs	Incl. O&P
2 Laborers	$20.55	$328.80	$34.75	$556.00	$21.27	$35.87
1 Truck Driver (heavy)	22.70	181.60	38.10	304.80		
1 Water Tanker, 5000 Gal.		136.00		149.60		
1 Truck Tractor, 220 H.P.		288.40		317.24		
2 -50' Discharge Hoses, 3"		3.50		3.85	17.83	19.61
24 L.H., Daily Totals		$938.30		$1331.49	$39.10	$55.48

Crew B-9B

	Hr.	Daily	Hr.	Daily	Bare Costs	Incl. O&P
2 Laborers	$20.55	$328.80	$34.75	$556.00	$21.27	$35.87
1 Truck Driver (heavy)	22.70	181.60	38.10	304.80		
2 -50' Discharge Hoses, 3"		3.50		3.85		
1 Water Tanker, 5000 Gal.		136.00		149.60		
1 Truck Tractor, 220 H.P.		288.40		317.24		
1 Pressure Washer		53.00		58.30	20.04	22.04
24 L.H., Daily Totals		$991.30		$1389.79	$41.30	$57.91

Crew B-9D

	Hr.	Daily	Hr.	Daily	Bare Costs	Incl. O&P
1 Labor Foreman (Outside)	$22.55	$180.40	$38.15	$305.20	$20.95	$35.43
4 Common Laborers	20.55	657.60	34.75	1112.00		
1 Air Compressor, 250 cfm		162.40		178.64		
2 -50' Air Hoses, 1.5"		12.60		13.86		
2 Air Powered Tampers		48.00		52.80	5.58	6.13
40 L.H., Daily Totals		$1061.00		$1662.50	$26.52	$41.56

Crew B-10

	Hr.	Daily	Hr.	Daily	Bare Costs	Incl. O&P
1 Equip. Oper. (med.)	$28.55	$228.40	$46.90	$375.20	$28.55	$46.90
8 L.H., Daily Totals		$228.40		$375.20	$28.55	$46.90

Crew B-10A

	Hr.	Daily	Hr.	Daily	Bare Costs	Incl. O&P
1 Equip. Oper. (med.)	$28.55	$228.40	$46.90	$375.20	$28.55	$46.90
1 Roller, 2-Drum, W.B., 7.5 H.P.		145.20		159.72	18.15	19.97
8 L.H., Daily Totals		$373.60		$534.92	$46.70	$66.86

Crew B-10B

	Hr.	Daily	Hr.	Daily	Bare Costs	Incl. O&P
1 Equip. Oper. (med.)	$28.55	$228.40	$46.90	$375.20	$28.55	$46.90
1 Dozer, 200 H.P.		1082.00		1190.20	135.25	148.78
8 L.H., Daily Totals		$1310.40		$1565.40	$163.80	$195.68

Crew B-10C

	Hr.	Daily	Hr.	Daily	Bare Costs	Incl. O&P
1 Equip. Oper. (med.)	$28.55	$228.40	$46.90	$375.20	$28.55	$46.90
1 Dozer, 200 H.P.		1082.00		1190.20		
1 Vibratory Roller, Towed, 23 Ton		411.00		452.10	186.63	205.29
8 L.H., Daily Totals		$1721.40		$2017.50	$215.18	$252.19

Crew B-10D

	Hr.	Daily	Hr.	Daily	Bare Costs	Incl. O&P
1 Equip. Oper. (med.)	$28.55	$228.40	$46.90	$375.20	$28.55	$46.90
1 Dozer, 200 H.P.		1082.00		1190.20		
1 Sheepsft. Roller, Towed		465.40		511.94	193.43	212.77
8 L.H., Daily Totals		$1775.80		$2077.34	$221.97	$259.67

Crew B-10E

	Hr.	Daily	Hr.	Daily	Bare Costs	Incl. O&P
1 Equip. Oper. (med.)	$28.55	$228.40	$46.90	$375.20	$28.55	$46.90
1 Tandem Roller, 5 Ton		137.80		151.58	17.23	18.95
8 L.H., Daily Totals		$366.20		$526.78	$45.77	$65.85

Crew B-10F

	Hr.	Daily	Hr.	Daily	Bare Costs	Incl. O&P
1 Equip. Oper. (med.)	$28.55	$228.40	$46.90	$375.20	$28.55	$46.90
1 Tandem Roller, 10 Ton		233.00		256.30	29.13	32.04
8 L.H., Daily Totals		$461.40		$631.50	$57.67	$78.94

Crew B-10G

	Hr.	Daily	Hr.	Daily	Bare Costs	Incl. O&P
1 Equip. Oper. (med.)	$28.55	$228.40	$46.90	$375.20	$28.55	$46.90
1 Sheepsft. Roll., 240 H.P.		1142.00		1256.20	142.75	157.03
8 L.H., Daily Totals		$1370.40		$1631.40	$171.30	$203.93

Crew B-10H

	Hr.	Daily	Hr.	Daily	Bare Costs	Incl. O&P
1 Equip. Oper. (med.)	$28.55	$228.40	$46.90	$375.20	$28.55	$46.90
1 Diaphragm Water Pump, 2"		63.20		69.52		
1 -20' Suction Hose, 2"		1.95		2.15		
2 -50' Discharge Hoses, 2"		2.20		2.42	8.42	9.26
8 L.H., Daily Totals		$295.75		$449.29	$36.97	$56.16

Crew B-10I

Crew No.	Bare Costs Hr.	Daily	Incl. Subs O & P Hr.	Daily	Cost Per Labor-Hour Bare Costs	Incl. O&P
1 Equip. Oper. (med.)	$28.55	$228.40	$46.90	$375.20	$28.55	$46.90
1 Diaphragm Water Pump, 4"		86.20		94.82		
1 -20' Suction Hose, 4"		3.45		3.79		
2 -50' Discharge Hoses, 4"		4.70		5.17	11.79	12.97
8 L.H., Daily Totals		$322.75		$478.99	$40.34	$59.87

Crew B-10J

Crew No.	Bare Costs Hr.	Daily	Incl. Subs O & P Hr.	Daily	Bare Costs	Incl. O&P
1 Equip. Oper. (med.)	$28.55	$228.40	$46.90	$375.20	$28.55	$46.90
1 Centrifugal Water Pump, 3"		69.40		76.34		
1 -20' Suction Hose, 3"		3.05		3.36		
2 -50' Discharge Hoses, 3"		3.50		3.85	9.49	10.44
8 L.H., Daily Totals		$304.35		$458.75	$38.04	$57.34

Crew B-10K

Crew No.	Bare Costs Hr.	Daily	Incl. Subs O & P Hr.	Daily	Bare Costs	Incl. O&P
1 Equip. Oper. (med.)	$28.55	$228.40	$46.90	$375.20	$28.55	$46.90
1 Centr. Water Pump, 6"		309.40		340.34		
1 -20' Suction Hose, 6"		11.90		13.09		
2 -50' Discharge Hoses, 6"		12.60		13.86	41.74	45.91
8 L.H., Daily Totals		$562.30		$742.49	$70.29	$92.81

Crew B-10L

Crew No.	Bare Costs Hr.	Daily	Incl. Subs O & P Hr.	Daily	Bare Costs	Incl. O&P
1 Equip. Oper. (med.)	$28.55	$228.40	$46.90	$375.20	$28.55	$46.90
1 Dozer, 80 H.P.		399.20		439.12	49.90	54.89
8 L.H., Daily Totals		$627.60		$814.32	$78.45	$101.79

Crew B-10M

Crew No.	Bare Costs Hr.	Daily	Incl. Subs O & P Hr.	Daily	Bare Costs	Incl. O&P
1 Equip. Oper. (med.)	$28.55	$228.40	$46.90	$375.20	$28.55	$46.90
1 Dozer, 300 H.P.		1423.00		1565.30	177.88	195.66
8 L.H., Daily Totals		$1651.40		$1940.50	$206.43	$242.56

Crew B-10N

Crew No.	Bare Costs Hr.	Daily	Incl. Subs O & P Hr.	Daily	Bare Costs	Incl. O&P
1 Equip. Oper. (med.)	$28.55	$228.40	$46.90	$375.20	$28.55	$46.90
1 F.E. Loader, T.M., 1.5 C.Y.		367.00		403.70	45.88	50.46
8 L.H., Daily Totals		$595.40		$778.90	$74.42	$97.36

Crew B-10O

Crew No.	Bare Costs Hr.	Daily	Incl. Subs O & P Hr.	Daily	Bare Costs	Incl. O&P
1 Equip. Oper. (med.)	$28.55	$228.40	$46.90	$375.20	$28.55	$46.90
1 F.E. Loader, T.M., 2.25 C.Y.		767.40		844.14	95.92	105.52
8 L.H., Daily Totals		$995.80		$1219.34	$124.47	$152.42

Crew B-10P

Crew No.	Bare Costs Hr.	Daily	Incl. Subs O & P Hr.	Daily	Bare Costs	Incl. O&P
1 Equip. Oper. (med.)	$28.55	$228.40	$46.90	$375.20	$28.55	$46.90
1 Crawler Loader, 3 C.Y.		990.00		1089.00	123.75	136.13
8 L.H., Daily Totals		$1218.40		$1464.20	$152.30	$183.03

Crew B-10Q

Crew No.	Bare Costs Hr.	Daily	Incl. Subs O & P Hr.	Daily	Bare Costs	Incl. O&P
1 Equip. Oper. (med.)	$28.55	$228.40	$46.90	$375.20	$28.55	$46.90
1 Crawler Loader, 4 C.Y.		1450.00		1595.00	181.25	199.38
8 L.H., Daily Totals		$1678.40		$1970.20	$209.80	$246.28

Crew B-10R

Crew No.	Bare Costs Hr.	Daily	Incl. Subs O & P Hr.	Daily	Bare Costs	Incl. O&P
1 Equip. Oper. (med.)	$28.55	$228.40	$46.90	$375.20	$28.55	$46.90
1 F.E. Loader, W.M., 1 C.Y.		239.00		262.90	29.88	32.86
8 L.H., Daily Totals		$467.40		$638.10	$58.42	$79.76

Crew B-10S

Crew No.	Bare Costs Hr.	Daily	Incl. Subs O & P Hr.	Daily	Bare Costs	Incl. O&P
1 Equip. Oper. (med.)	$28.55	$228.40	$46.90	$375.20	$28.55	$46.90
1 F.E. Loader, W.M., 1.5 C.Y.		328.20		361.02	41.02	45.13
8 L.H., Daily Totals		$556.60		$736.22	$69.58	$92.03

Crew B-10T

Crew No.	Bare Costs Hr.	Daily	Incl. Subs O & P Hr.	Daily	Bare Costs	Incl. O&P
1 Equip. Oper. (med.)	$28.55	$228.40	$46.90	$375.20	$28.55	$46.90
1 F.E. Loader, W.M., 2.5 C.Y.		395.80		435.38	49.48	54.42
8 L.H., Daily Totals		$624.20		$810.58	$78.03	$101.32

Crew B-10U

Crew No.	Bare Costs Hr.	Daily	Incl. Subs O & P Hr.	Daily	Bare Costs	Incl. O&P
1 Equip. Oper. (med.)	$28.55	$228.40	$46.90	$375.20	$28.55	$46.90
1 F.E. Loader, W.M., 5.5 C.Y.		854.40		939.84	106.80	117.48
8 L.H., Daily Totals		$1082.80		$1315.04	$135.35	$164.38

Crew B-10V

Crew No.	Bare Costs Hr.	Daily	Incl. Subs O & P Hr.	Daily	Bare Costs	Incl. O&P
1 Equip. Oper. (med.)	$28.55	$228.40	$46.90	$375.20	$28.55	$46.90
1 Dozer, 700 H.P.		4232.00		4655.20	529.00	581.90
8 L.H., Daily Totals		$4460.40		$5030.40	$557.55	$628.80

Crew B-10W

Crew No.	Bare Costs Hr.	Daily	Incl. Subs O & P Hr.	Daily	Bare Costs	Incl. O&P
1 Equip. Oper. (med.)	$28.55	$228.40	$46.90	$375.20	$28.55	$46.90
1 Dozer, 105 H.P.		600.60		660.66	75.08	82.58
8 L.H., Daily Totals		$829.00		$1035.86	$103.63	$129.48

Crew B-10X

Crew No.	Bare Costs Hr.	Daily	Incl. Subs O & P Hr.	Daily	Bare Costs	Incl. O&P
1 Equip. Oper. (med.)	$28.55	$228.40	$46.90	$375.20	$28.55	$46.90
1 Dozer, 410 H.P.		1849.00		2033.90	231.13	254.24
8 L.H., Daily Totals		$2077.40		$2409.10	$259.68	$301.14

Crew B-10Y

Crew No.	Bare Costs Hr.	Daily	Incl. Subs O & P Hr.	Daily	Bare Costs	Incl. O&P
1 Equip. Oper. (med.)	$28.55	$228.40	$46.90	$375.20	$28.55	$46.90
1 Vibr. Roller, Towed, 12 Ton		469.60		516.56	58.70	64.57
8 L.H., Daily Totals		$698.00		$891.76	$87.25	$111.47

Crew B-11A

Crew No.	Bare Costs Hr.	Daily	Incl. Subs O & P Hr.	Daily	Bare Costs	Incl. O&P
1 Equipment Oper. (med.)	$28.55	$228.40	$46.90	$375.20	$24.55	$40.83
1 Laborer	20.55	164.40	34.75	278.00		
1 Dozer, 200 H.P.		1082.00		1190.20	67.63	74.39
16 L.H., Daily Totals		$1474.80		$1843.40	$92.17	$115.21

Crew B-11B

Crew No.	Bare Costs Hr.	Daily	Incl. Subs O & P Hr.	Daily	Bare Costs	Incl. O&P
1 Equipment Oper. (light)	$26.95	$215.60	$44.30	$354.40	$23.75	$39.52
1 Laborer	20.55	164.40	34.75	278.00		
1 Air Powered Tamper		24.00		26.40		
1 Air Compressor, 365 cfm		211.20		232.32		
2 -50' Air Hoses, 1.5"		12.60		13.86	15.49	17.04
16 L.H., Daily Totals		$627.80		$904.98	$39.24	$56.56

Crew B-11C

Crew No.	Bare Costs Hr.	Daily	Incl. Subs O & P Hr.	Daily	Bare Costs	Incl. O&P
1 Equipment Oper. (med.)	$28.55	$228.40	$46.90	$375.20	$24.55	$40.83
1 Laborer	20.55	164.40	34.75	278.00		
1 Backhoe Loader, 48 H.P.		293.80		323.18	18.36	20.20
16 L.H., Daily Totals		$686.60		$976.38	$42.91	$61.02

Crew B-11K

Crew No.	Bare Costs Hr.	Daily	Incl. Subs O & P Hr.	Daily	Bare Costs	Incl. O&P
1 Equipment Oper. (med.)	$28.55	$228.40	$46.90	$375.20	$24.55	$40.83
1 Laborer	20.55	164.40	34.75	278.00		
1 Trencher, Chain Type, 8' D		1777.00		1954.70	111.06	122.17
16 L.H., Daily Totals		$2169.80		$2607.90	$135.61	$162.99

Crew B-11L

Crew B-11L	Hr.	Daily	Hr.	Daily	Bare Costs	Incl. O&P
1 Equipment Oper. (med.)	$28.55	$228.40	$46.90	$375.20	$24.55	$40.83
1 Laborer	20.55	164.40	34.75	278.00		
1 Grader, 30,000 Lbs.		550.20		605.22	34.39	37.83
16 L.H., Daily Totals		$943.00		$1258.42	$58.94	$78.65

Crew B-11M

Crew B-11M	Hr.	Daily	Hr.	Daily	Bare Costs	Incl. O&P
1 Equipment Oper. (med.)	$28.55	$228.40	$46.90	$375.20	$24.55	$40.83
1 Laborer	20.55	164.40	34.75	278.00		
1 Backhoe Loader, 80 H.P.		345.00		379.50	21.56	23.72
16 L.H., Daily Totals		$737.80		$1032.70	$46.11	$64.54

Crew B-11W

Crew B-11W	Hr.	Daily	Hr.	Daily	Bare Costs	Incl. O&P
1 Equipment Operator (med.)	$28.55	$228.40	$46.90	$375.20	$23.01	$38.55
1 Common Laborer	20.55	164.40	34.75	278.00		
10 Truck Drivers (hvy.)	22.70	1816.00	38.10	3048.00		
1 Dozer, 200 H.P.		1082.00		1190.20		
1 Vibratory Roller, Towed, 23 Ton		411.00		452.10		
10 Dump Trucks, 8 C.Y., 220 H.P.		3264.00		3590.40	49.55	54.51
96 L.H., Daily Totals		$6965.80		$8933.90	$72.56	$93.06

Crew B-11Y

Crew B-11Y	Hr.	Daily	Hr.	Daily	Bare Costs	Incl. O&P
1 Labor Foreman (Outside)	$22.55	$180.40	$38.15	$305.20	$23.44	$39.18
5 Common Laborers	20.55	822.00	34.75	1390.00		
3 Equipment Operators (med.)	28.55	685.20	46.90	1125.60		
1 Dozer, 80 H.P.		399.20		439.12		
2 Roller, 2-Drum, W.B., 7.5 H.P.		290.40		319.44		
4 Vibratory Plates, Gas, 21"		156.80		172.48	11.76	12.93
72 L.H., Daily Totals		$2534.00		$3751.84	$35.19	$52.11

Crew B-12A

Crew B-12A	Hr.	Daily	Hr.	Daily	Bare Costs	Incl. O&P
1 Equip. Oper. (crane)	$29.35	$234.80	$48.20	$385.60	$24.95	$41.48
1 Laborer	20.55	164.40	34.75	278.00		
1 Hyd. Excavator, 1 C.Y.		669.20		736.12	41.83	46.01
16 L.H., Daily Totals		$1068.40		$1399.72	$66.78	$87.48

Crew B-12B

Crew B-12B	Hr.	Daily	Hr.	Daily	Bare Costs	Incl. O&P
1 Equip. Oper. (crane)	$29.35	$234.80	$48.20	$385.60	$24.95	$41.48
1 Laborer	20.55	164.40	34.75	278.00		
1 Hyd. Excavator, 1.5 C.Y.		865.80		952.38	54.11	59.52
16 L.H., Daily Totals		$1265.00		$1615.98	$79.06	$101.00

Crew B-12C

Crew B-12C	Hr.	Daily	Hr.	Daily	Bare Costs	Incl. O&P
1 Equip. Oper. (crane)	$29.35	$234.80	$48.20	$385.60	$24.95	$41.48
1 Laborer	20.55	164.40	34.75	278.00		
1 Hyd. Excavator, 2 C.Y.		1182.00		1300.20	73.88	81.26
16 L.H., Daily Totals		$1581.20		$1963.80	$98.83	$122.74

Crew B-12D

Crew B-12D	Hr.	Daily	Hr.	Daily	Bare Costs	Incl. O&P
1 Equip. Oper. (crane)	$29.35	$234.80	$48.20	$385.60	$24.95	$41.48
1 Laborer	20.55	164.40	34.75	278.00		
1 Hyd. Excavator, 3.5 C.Y.		2162.00		2378.20	135.13	148.64
16 L.H., Daily Totals		$2561.20		$3041.80	$160.07	$190.11

Crew B-12E

Crew B-12E	Hr.	Daily	Hr.	Daily	Bare Costs	Incl. O&P
1 Equip. Oper. (crane)	$29.35	$234.80	$48.20	$385.60	$24.95	$41.48
1 Laborer	20.55	164.40	34.75	278.00		
1 Hyd. Excavator, .5 C.Y.		394.60		434.06	24.66	27.13
16 L.H., Daily Totals		$793.80		$1097.66	$49.61	$68.60

Crew B-12F

Crew B-12F	Hr.	Daily	Hr.	Daily	Bare Costs	Incl. O&P
1 Equip. Oper. (crane)	$29.35	$234.80	$48.20	$385.60	$24.95	$41.48
1 Laborer	20.55	164.40	34.75	278.00		
1 Hyd. Excavator, .75 C.Y.		598.60		658.46	37.41	41.15
16 L.H., Daily Totals		$997.80		$1322.06	$62.36	$82.63

Crew B-12G

Crew B-12G	Hr.	Daily	Hr.	Daily	Bare Costs	Incl. O&P
1 Equip. Oper. (crane)	$29.35	$234.80	$48.20	$385.60	$24.95	$41.48
1 Laborer	20.55	164.40	34.75	278.00		
1 Crawler Crane, 15 Ton		615.30		676.83		
1 Clamshell Bucket, .5 C.Y.		36.60		40.26	40.74	44.82
16 L.H., Daily Totals		$1051.10		$1380.69	$65.69	$86.29

Crew B-12H

Crew B-12H	Hr.	Daily	Hr.	Daily	Bare Costs	Incl. O&P
1 Equip. Oper. (crane)	$29.35	$234.80	$48.20	$385.60	$24.95	$41.48
1 Laborer	20.55	164.40	34.75	278.00		
1 Crawler Crane, 25 Ton		1052.00		1157.20		
1 Clamshell Bucket, 1 C.Y.		46.20		50.82	68.64	75.50
16 L.H., Daily Totals		$1497.40		$1871.62	$93.59	$116.98

Crew B-12I

Crew B-12I	Hr.	Daily	Hr.	Daily	Bare Costs	Incl. O&P
1 Equip. Oper. (crane)	$29.35	$234.80	$48.20	$385.60	$24.95	$41.48
1 Laborer	20.55	164.40	34.75	278.00		
1 Crawler Crane, 20 Ton		789.05		867.96		
1 Dragline Bucket, .75 C.Y.		20.00		22.00	50.57	55.62
16 L.H., Daily Totals		$1208.25		$1553.56	$75.52	$97.10

Crew B-12J

Crew B-12J	Hr.	Daily	Hr.	Daily	Bare Costs	Incl. O&P
1 Equip. Oper. (crane)	$29.35	$234.80	$48.20	$385.60	$24.95	$41.48
1 Laborer	20.55	164.40	34.75	278.00		
1 Gradall, 5/8 C.Y.		958.80		1054.68	59.92	65.92
16 L.H., Daily Totals		$1358.00		$1718.28	$84.88	$107.39

Crew B-12K

Crew B-12K	Hr.	Daily	Hr.	Daily	Bare Costs	Incl. O&P
1 Equip. Oper. (crane)	$29.35	$234.80	$48.20	$385.60	$24.95	$41.48
1 Laborer	20.55	164.40	34.75	278.00		
1 Gradall, 3 Ton, 1 C.Y.		1130.00		1243.00	70.63	77.69
16 L.H., Daily Totals		$1529.20		$1906.60	$95.58	$119.16

Crew B-12L

Crew B-12L	Hr.	Daily	Hr.	Daily	Bare Costs	Incl. O&P
1 Equip. Oper. (crane)	$29.35	$234.80	$48.20	$385.60	$24.95	$41.48
1 Laborer	20.55	164.40	34.75	278.00		
1 Crawler Crane, 15 Ton		615.30		676.83		
1 F.E. Attachment, .5 C.Y.		52.80		58.08	41.76	45.93
16 L.H., Daily Totals		$1067.30		$1398.51	$66.71	$87.41

Crew B-12M

Crew B-12M	Hr.	Daily	Hr.	Daily	Bare Costs	Incl. O&P
1 Equip. Oper. (crane)	$29.35	$234.80	$48.20	$385.60	$24.95	$41.48
1 Laborer	20.55	164.40	34.75	278.00		
1 Crawler Crane, 20 Ton		789.05		867.96		
1 F.E. Attachment, .75 C.Y.		58.20		64.02	52.95	58.25
16 L.H., Daily Totals		$1246.45		$1595.58	$77.90	$99.72

Crew B-12N

Crew B-12N	Hr.	Daily	Hr.	Daily	Bare Costs	Incl. O&P
1 Equip. Oper. (crane)	$29.35	$234.80	$48.20	$385.60	$24.95	$41.48
1 Laborer	20.55	164.40	34.75	278.00		
1 Crawler Crane, 25 Ton		1052.00		1157.20		
1 F.E. Attachment, 1 C.Y.		64.00		70.40	69.75	76.72
16 L.H., Daily Totals		$1515.20		$1891.20	$94.70	$118.20

Crew B-120

Crew No.	Bare Costs Hr.	Bare Costs Daily	Incl. Subs O&P Hr.	Incl. Subs O&P Daily	Cost Per Labor-Hour Bare Costs	Cost Per Labor-Hour Incl. O&P
1 Equip. Oper. (crane)	$29.35	$234.80	$48.20	$385.60	$24.95	$41.48
1 Laborer	20.55	164.40	34.75	278.00		
1 Crawler Crane, 40 Ton		1055.00		1160.50		
1 F.E. Attachment, 1.5 C.Y.		74.20		81.62	70.58	77.63
16 L.H., Daily Totals		$1528.40		$1905.72	$95.53	$119.11

Crew B-12P

Crew No.	Bare Costs Hr.	Bare Costs Daily	Incl. Subs O&P Hr.	Incl. Subs O&P Daily	Cost Per Labor-Hour Bare Costs	Cost Per Labor-Hour Incl. O&P
1 Equip. Oper. (crane)	$29.35	$234.80	$48.20	$385.60	$24.95	$41.48
1 Laborer	20.55	164.40	34.75	278.00		
1 Crawler Crane, 40 Ton		1055.00		1160.50		
1 Dragline Bucket, 1.5 C.Y.		32.80		36.08	67.99	74.79
16 L.H., Daily Totals		$1487.00		$1860.18	$92.94	$116.26

Crew B-12Q

Crew No.	Bare Costs Hr.	Bare Costs Daily	Incl. Subs O&P Hr.	Incl. Subs O&P Daily	Cost Per Labor-Hour Bare Costs	Cost Per Labor-Hour Incl. O&P
1 Equip. Oper. (crane)	$29.35	$234.80	$48.20	$385.60	$24.95	$41.48
1 Laborer	20.55	164.40	34.75	278.00		
1 Hyd. Excavator, 5/8 C.Y.		517.20		568.92	32.33	35.56
16 L.H., Daily Totals		$916.40		$1232.52	$57.27	$77.03

Crew B-12S

Crew No.	Bare Costs Hr.	Bare Costs Daily	Incl. Subs O&P Hr.	Incl. Subs O&P Daily	Cost Per Labor-Hour Bare Costs	Cost Per Labor-Hour Incl. O&P
1 Equip. Oper. (crane)	$29.35	$234.80	$48.20	$385.60	$24.95	$41.48
1 Laborer	20.55	164.40	34.75	278.00		
1 Hyd. Excavator, 2.5 C.Y.		1619.00		1780.90	101.19	111.31
16 L.H., Daily Totals		$2018.20		$2444.50	$126.14	$152.78

Crew B-12T

Crew No.	Bare Costs Hr.	Bare Costs Daily	Incl. Subs O&P Hr.	Incl. Subs O&P Daily	Cost Per Labor-Hour Bare Costs	Cost Per Labor-Hour Incl. O&P
1 Equip. Oper. (crane)	$29.35	$234.80	$48.20	$385.60	$24.95	$41.48
1 Laborer	20.55	164.40	34.75	278.00		
1 Crawler Crane, 75 Ton		1496.00		1645.60		
1 F.E. Attachment, 3 C.Y.		97.60		107.36	99.60	109.56
16 L.H., Daily Totals		$1992.80		$2416.56	$124.55	$151.04

Crew B-12V

Crew No.	Bare Costs Hr.	Bare Costs Daily	Incl. Subs O&P Hr.	Incl. Subs O&P Daily	Cost Per Labor-Hour Bare Costs	Cost Per Labor-Hour Incl. O&P
1 Equip. Oper. (crane)	$29.35	$234.80	$48.20	$385.60	$24.95	$41.48
1 Laborer	20.55	164.40	34.75	278.00		
1 Crawler Crane, 75 Ton		1496.00		1645.60		
1 Dragline Bucket, 3 C.Y.		46.20		50.82	96.39	106.03
16 L.H., Daily Totals		$1941.40		$2360.02	$121.34	$147.50

Crew B-13

Crew No.	Bare Costs Hr.	Bare Costs Daily	Incl. Subs O&P Hr.	Incl. Subs O&P Daily	Cost Per Labor-Hour Bare Costs	Cost Per Labor-Hour Incl. O&P
1 Labor Foreman (outside)	$22.55	$180.40	$38.15	$305.20	$22.35	$37.56
4 Laborers	20.55	657.60	34.75	1112.00		
1 Equip. Oper. (crane)	29.35	234.80	48.20	385.60		
1 Hyd. Crane, 25 Ton		789.40		868.34	16.45	18.09
48 L.H., Daily Totals		$1862.20		$2671.14	$38.80	$55.65

Crew B-13A

Crew No.	Bare Costs Hr.	Bare Costs Daily	Incl. Subs O&P Hr.	Incl. Subs O&P Daily	Cost Per Labor-Hour Bare Costs	Cost Per Labor-Hour Incl. O&P
1 Foreman	$22.55	$180.40	$38.15	$305.20	$23.74	$39.66
2 Laborers	20.55	328.80	34.75	556.00		
2 Equipment Operators	28.55	456.80	46.90	750.40		
2 Truck Drivers (heavy)	22.70	363.20	38.10	609.60		
1 Crawler Crane, 75 Ton		1496.00		1645.60		
1 Crawler Loader, 4 C.Y.		1450.00		1595.00		
2 Dump Trucks, 8 C.Y., 220 H.P.		652.80		718.08	64.26	70.69
56 L.H., Daily Totals		$4928.00		$6179.88	$88.00	$110.36

Crew B-13B

Crew No.	Bare Costs Hr.	Bare Costs Daily	Incl. Subs O&P Hr.	Incl. Subs O&P Daily	Cost Per Labor-Hour Bare Costs	Cost Per Labor-Hour Incl. O&P
1 Labor Foreman (outside)	$22.55	$180.40	$38.15	$305.20	$22.79	$38.16
4 Laborers	20.55	657.60	34.75	1112.00		
1 Equip. Oper. (crane)	29.35	234.80	48.20	385.60		
1 Equip. Oper. Oiler	25.40	203.20	41.75	334.00		
1 Hyd. Crane, 55 Ton		1125.00		1237.50	20.09	22.10
56 L.H., Daily Totals		$2401.00		$3374.30	$42.88	$60.26

Crew B-13C

Crew No.	Bare Costs Hr.	Bare Costs Daily	Incl. Subs O&P Hr.	Incl. Subs O&P Daily	Cost Per Labor-Hour Bare Costs	Cost Per Labor-Hour Incl. O&P
1 Labor Foreman (outside)	$22.55	$180.40	$38.15	$305.20	$22.79	$38.16
4 Laborers	20.55	657.60	34.75	1112.00		
1 Equip. Oper. (crane)	29.35	234.80	48.20	385.60		
1 Equip. Oper. Oiler	25.40	203.20	41.75	334.00		
1 Crawler Crane, 100 Ton		1611.00		1772.10	28.77	31.64
56 L.H., Daily Totals		$2887.00		$3908.90	$51.55	$69.80

Crew B-13D

Crew No.	Bare Costs Hr.	Bare Costs Daily	Incl. Subs O&P Hr.	Incl. Subs O&P Daily	Cost Per Labor-Hour Bare Costs	Cost Per Labor-Hour Incl. O&P
1 Laborer	$20.55	$164.40	$34.75	$278.00	$24.95	$41.48
1 Equip. Oper. (crane)	29.35	234.80	48.20	385.60		
1 Hyd. Excavator, 1 C.Y.		669.20		736.12		
1 Trench Box		113.40		124.74	48.91	53.80
16 L.H., Daily Totals		$1181.80		$1524.46	$73.86	$95.28

Crew B-13E

Crew No.	Bare Costs Hr.	Bare Costs Daily	Incl. Subs O&P Hr.	Incl. Subs O&P Daily	Cost Per Labor-Hour Bare Costs	Cost Per Labor-Hour Incl. O&P
1 Laborer	$20.55	$164.40	$34.75	$278.00	$24.95	$41.48
1 Equip. Oper. (crane)	29.35	234.80	48.20	385.60		
1 Hyd. Excavator, 1.5 C.Y.		865.80		952.38		
1 Trench Box		113.40		124.74	61.20	67.32
16 L.H., Daily Totals		$1378.40		$1740.72	$86.15	$108.80

Crew B-13F

Crew No.	Bare Costs Hr.	Bare Costs Daily	Incl. Subs O&P Hr.	Incl. Subs O&P Daily	Cost Per Labor-Hour Bare Costs	Cost Per Labor-Hour Incl. O&P
1 Laborer	$20.55	$164.40	$34.75	$278.00	$24.95	$41.48
1 Equip. Oper. (crane)	29.35	234.80	48.20	385.60		
1 Hyd. Excavator, 3.5 C.Y.		2162.00		2378.20		
1 Trench Box		113.40		124.74	142.21	156.43
16 L.H., Daily Totals		$2674.60		$3166.54	$167.16	$197.91

Crew B-13G

Crew No.	Bare Costs Hr.	Bare Costs Daily	Incl. Subs O&P Hr.	Incl. Subs O&P Daily	Cost Per Labor-Hour Bare Costs	Cost Per Labor-Hour Incl. O&P
1 Laborer	$20.55	$164.40	$34.75	$278.00	$24.95	$41.48
1 Equip. Oper. (crane)	29.35	234.80	48.20	385.60		
1 Hyd. Excavator, .75 C.Y.		598.60		658.46		
1 Trench Box		113.40		124.74	44.50	48.95
16 L.H., Daily Totals		$1111.20		$1446.80	$69.45	$90.42

Crew B-13H

Crew No.	Bare Costs Hr.	Bare Costs Daily	Incl. Subs O&P Hr.	Incl. Subs O&P Daily	Cost Per Labor-Hour Bare Costs	Cost Per Labor-Hour Incl. O&P
1 Laborer	$20.55	$164.40	$34.75	$278.00	$24.95	$41.48
1 Equip. Oper. (crane)	29.35	234.80	48.20	385.60		
1 Gradall, 5/8 C.Y.		958.80		1054.68		
1 Trench Box		113.40		124.74	67.01	73.71
16 L.H., Daily Totals		$1471.40		$1843.02	$91.96	$115.19

Crew B-13I

Crew No.	Bare Costs Hr.	Bare Costs Daily	Incl. Subs O&P Hr.	Incl. Subs O&P Daily	Cost Per Labor-Hour Bare Costs	Cost Per Labor-Hour Incl. O&P
1 Laborer	$20.55	$164.40	$34.75	$278.00	$24.95	$41.48
1 Equip. Oper. (crane)	29.35	234.80	48.20	385.60		
1 Gradall, 3 Ton, 1 C.Y.		1130.00		1243.00		
1 Trench Box		113.40		124.74	77.71	85.48
16 L.H., Daily Totals		$1642.60		$2031.34	$102.66	$126.96

Crew B-13J

Crew No.	Bare Costs Hr.	Daily	Incl. Subs O & P Hr.	Daily	Cost Per Labor-Hour Bare Costs	Incl. O&P
1 Laborer	$20.55	$164.40	$34.75	$278.00	$24.95	$41.48
1 Equip. Oper. (crane)	29.35	234.80	48.20	385.60		
1 Hyd. Excavator, 2.5 C.Y.		1619.00		1780.90		
1 Trench Box		113.40		124.74	108.28	119.10
16 L.H., Daily Totals		$2131.60		$2569.24	$133.22	$160.58

Crew B-14

Crew No.	Bare Costs Hr.	Daily	Incl. Subs O & P Hr.	Daily	Cost Per Labor-Hour Bare Costs	Incl. O&P
1 Labor Foreman (outside)	$22.55	$180.40	$38.15	$305.20	$21.95	$36.91
4 Laborers	20.55	657.60	34.75	1112.00		
1 Equip. Oper. (light)	26.95	215.60	44.30	354.40		
1 Backhoe Loader, 48 H.P.		293.80		323.18	6.12	6.73
48 L.H., Daily Totals		$1347.40		$2094.78	$28.07	$43.64

Crew B-14A

Crew No.	Bare Costs Hr.	Daily	Incl. Subs O & P Hr.	Daily	Cost Per Labor-Hour Bare Costs	Incl. O&P
1 Equip. Oper. (crane)	$29.35	$234.80	$48.20	$385.60	$26.42	$43.72
.5 Laborer	20.55	82.20	34.75	139.00		
1 Hyd. Excavator, 4.5 C.Y.		2635.00		2898.50	219.58	241.54
12 L.H., Daily Totals		$2952.00		$3423.10	$246.00	$285.26

Crew B-14B

Crew No.	Bare Costs Hr.	Daily	Incl. Subs O & P Hr.	Daily	Cost Per Labor-Hour Bare Costs	Incl. O&P
1 Equip. Oper. (crane)	$29.35	$234.80	$48.20	$385.60	$26.42	$43.72
.5 Laborer	20.55	82.20	34.75	139.00		
1 Hyd. Excavator, 6 C.Y.		3080.00		3388.00	256.67	282.33
12 L.H., Daily Totals		$3397.00		$3912.60	$283.08	$326.05

Crew B-14C

Crew No.	Bare Costs Hr.	Daily	Incl. Subs O & P Hr.	Daily	Cost Per Labor-Hour Bare Costs	Incl. O&P
1 Equip. Oper. (crane)	$29.35	$234.80	$48.20	$385.60	$26.42	$43.72
.5 Laborer	20.55	82.20	34.75	139.00		
1 Hyd. Excavator, 7 C.Y.		3256.00		3581.60	271.33	298.47
12 L.H., Daily Totals		$3573.00		$4106.20	$297.75	$342.18

Crew B-14F

Crew No.	Bare Costs Hr.	Daily	Incl. Subs O & P Hr.	Daily	Cost Per Labor-Hour Bare Costs	Incl. O&P
1 Equip. Oper. (crane)	$29.35	$234.80	$48.20	$385.60	$26.42	$43.72
.5 Laborer	20.55	82.20	34.75	139.00		
1 Hyd. Shovel, 7 C.Y.		3090.00		3399.00	257.50	283.25
12 L.H., Daily Totals		$3407.00		$3923.60	$283.92	$326.97

Crew B-14G

Crew No.	Bare Costs Hr.	Daily	Incl. Subs O & P Hr.	Daily	Cost Per Labor-Hour Bare Costs	Incl. O&P
1 Equip. Oper. (crane)	$29.35	$234.80	$48.20	$385.60	$26.42	$43.72
.5 Laborer	20.55	82.20	34.75	139.00		
1 Hyd. Shovel, 12 C.Y.		4148.00		4562.80	345.67	380.23
12 L.H., Daily Totals		$4465.00		$5087.40	$372.08	$423.95

Crew B-14J

Crew No.	Bare Costs Hr.	Daily	Incl. Subs O & P Hr.	Daily	Cost Per Labor-Hour Bare Costs	Incl. O&P
1 Equip. Oper. (med.)	$28.55	$228.40	$46.90	$375.20	$25.88	$42.85
.5 Laborer	20.55	82.20	34.75	139.00		
1 F.E. Loader, 8 C.Y.		1497.00		1646.70	124.75	137.22
12 L.H., Daily Totals		$1807.60		$2160.90	$150.63	$180.07

Crew B-14K

Crew No.	Bare Costs Hr.	Daily	Incl. Subs O & P Hr.	Daily	Cost Per Labor-Hour Bare Costs	Incl. O&P
1 Equip. Oper. (med.)	$28.55	$228.40	$46.90	$375.20	$25.88	$42.85
.5 Laborer	20.55	82.20	34.75	139.00		
1 F.E. Loader, 10 C.Y.		2225.00		2447.50	185.42	203.96
12 L.H., Daily Totals		$2535.60		$2961.70	$211.30	$246.81

Crew B-15

Crew No.	Bare Costs Hr.	Daily	Incl. Subs O & P Hr.	Daily	Cost Per Labor-Hour Bare Costs	Incl. O&P
1 Equipment Oper. (med)	$28.55	$228.40	$46.90	$375.20	$24.06	$40.14
.5 Laborer	20.55	82.20	34.75	139.00		
2 Truck Drivers (heavy)	22.70	363.20	38.10	609.60		
2 Dump Trucks, 12 C.Y., 400 H.P.		1066.00		1172.60		
1 Dozer, 200 H.P.		1082.00		1190.20	76.71	84.39
28 L.H., Daily Totals		$2821.80		$3486.60	$100.78	$124.52

Crew B-16

Crew No.	Bare Costs Hr.	Daily	Incl. Subs O & P Hr.	Daily	Cost Per Labor-Hour Bare Costs	Incl. O&P
1 Labor Foreman (outside)	$22.55	$180.40	$38.15	$305.20	$21.59	$36.44
2 Laborers	20.55	328.80	34.75	556.00		
1 Truck Driver (heavy)	22.70	181.60	38.10	304.80		
1 Dump Truck, 12 C.Y., 400 H.P.		533.00		586.30	16.66	18.32
32 L.H., Daily Totals		$1223.80		$1752.30	$38.24	$54.76

Crew B-17

Crew No.	Bare Costs Hr.	Daily	Incl. Subs O & P Hr.	Daily	Cost Per Labor-Hour Bare Costs	Incl. O&P
2 Laborers	$20.55	$328.80	$34.75	$556.00	$22.69	$37.98
1 Equip. Oper. (light)	26.95	215.60	44.30	354.40		
1 Truck Driver (heavy)	22.70	181.60	38.10	304.80		
1 Backhoe Loader, 48 H.P.		293.80		323.18		
1 Dump Truck, 8 C.Y., 220 H.P.		326.40		359.04	19.38	21.32
32 L.H., Daily Totals		$1346.20		$1897.42	$42.07	$59.29

Crew B-17A

Crew No.	Bare Costs Hr.	Daily	Incl. Subs O & P Hr.	Daily	Cost Per Labor-Hour Bare Costs	Incl. O&P
2 Laborer Foremen	$22.55	$360.80	$38.15	$610.40	$22.65	$38.28
6 Laborers	20.55	986.40	34.75	1668.00		
1 Skilled Worker Foreman	30.05	240.40	50.70	405.60		
1 Skilled Worker	28.05	224.40	47.35	378.80		
80 L.H., Daily Totals		$1812.00		$3062.80	$22.65	$38.28

Crew B-18

Crew No.	Bare Costs Hr.	Daily	Incl. Subs O & P Hr.	Daily	Cost Per Labor-Hour Bare Costs	Incl. O&P
1 Labor Foreman (outside)	$22.55	$180.40	$38.15	$305.20	$21.22	$35.88
2 Laborers	20.55	328.80	34.75	556.00		
1 Vibratory Plate, Gas, 21"		39.20		43.12	1.63	1.80
24 L.H., Daily Totals		$548.40		$904.32	$22.85	$37.68

Crew B-19

Crew No.	Bare Costs Hr.	Daily	Incl. Subs O & P Hr.	Daily	Cost Per Labor-Hour Bare Costs	Incl. O&P
1 Pile Driver Foreman	$28.95	$231.60	$51.05	$408.40	$26.66	$46.31
4 Pile Drivers	26.95	862.40	47.55	1521.60		
1 Equip. Oper. (crane)	29.35	234.80	48.20	385.60		
1 Building Laborer	20.55	164.40	34.75	278.00		
1 Crawler Crane, 40 Ton		1055.00		1160.50		
1 Lead, 90' high		118.80		130.68		
1 Hammer, Diesel, 22k Ft-Lb		590.00		649.00	31.50	34.65
56 L.H., Daily Totals		$3257.00		$4533.78	$58.16	$80.96

Crew B-19A

Crew No.	Bare Costs Hr.	Daily	Incl. Subs O & P Hr.	Daily	Cost Per Labor-Hour Bare Costs	Incl. O&P
1 Pile Driver Foreman	$28.95	$231.60	$51.05	$408.40	$27.61	$47.42
4 Pile Drivers	26.95	862.40	47.55	1521.60		
2 Equip. Oper. (crane)	29.35	469.60	48.20	771.20		
1 Equip. Oper. Oiler	25.40	203.20	41.75	334.00		
1 Crawler Crane, 75 Ton		1496.00		1645.60		
1 Lead, 90' high		118.80		130.68		
1 Hammer, Diesel, 41k Ft-Lb		677.60		745.36	35.82	39.40
64 L.H., Daily Totals		$4059.20		$5556.84	$63.43	$86.83

Crew B-20

Crew No.	Bare Costs Hr.	Daily	Incl. Subs O & P Hr.	Daily	Cost Per Labor-Hour Bare Costs	Incl. O&P
1 Labor Foreman (out)	$22.55	$180.40	$38.15	$305.20	$21.22	$35.88
2 Laborers	20.55	328.80	34.75	556.00		
24 L.H., Daily Totals		$509.20		$861.20	$21.22	$35.88

Crew No.	Bare Costs		Incl. Subs O & P		Cost Per Labor-Hour	

Crew B-20A

Crew B-20A	Hr.	Daily	Hr.	Daily	Bare Costs	Incl. O&P
1 Labor Foreman	$22.55	$180.40	$38.15	$305.20	$25.26	$42.00
1 Laborer	20.55	164.40	34.75	278.00		
1 Plumber	32.20	257.60	52.85	422.80		
1 Plumber Apprentice	25.75	206.00	42.25	338.00		
32 L.H., Daily Totals		$808.40		$1344.00	$25.26	$42.00

Crew B-21

Crew B-21	Hr.	Daily	Hr.	Daily	Bare Costs	Incl. O&P
1 Labor Foreman (out)	$22.55	$180.40	$38.15	$305.20	$22.38	$37.64
2 Laborers	20.55	328.80	34.75	556.00		
.5 Equip. Oper. (crane)	29.35	117.40	48.20	192.80		
.5 S.P. Crane, 4x4, 5 Ton		129.10		142.01	4.61	5.07
28 L.H., Daily Totals		$755.70		$1196.01	$26.99	$42.71

Crew B-21A

Crew B-21A	Hr.	Daily	Hr.	Daily	Bare Costs	Incl. O&P
1 Labor Foreman	$22.55	$180.40	$38.15	$305.20	$26.08	$43.24
1 Laborer	20.55	164.40	34.75	278.00		
1 Plumber	32.20	257.60	52.85	422.80		
1 Plumber Apprentice	25.75	206.00	42.25	338.00		
1 Equip. Oper. (crane)	29.35	234.80	48.20	385.60		
1 S.P. Crane, 4x4, 12 Ton		601.80		661.98	15.05	16.55
40 L.H., Daily Totals		$1645.00		$2391.58	$41.13	$59.79

Crew B-21B

Crew B-21B	Hr.	Daily	Hr.	Daily	Bare Costs	Incl. O&P
1 Laborer Foreman	$22.55	$180.40	$38.15	$305.20	$22.71	$38.12
3 Laborers	20.55	493.20	34.75	834.00		
1 Equip. Oper. (crane)	29.35	234.80	48.20	385.60		
1 Hyd. Crane, 12 Ton		768.80		845.68	19.22	21.14
40 L.H., Daily Totals		$1677.20		$2370.48	$41.93	$59.26

Crew B-21C

Crew B-21C	Hr.	Daily	Hr.	Daily	Bare Costs	Incl. O&P
1 Laborer Foreman	$22.55	$180.40	$38.15	$305.20	$22.79	$38.16
4 Laborers	20.55	657.60	34.75	1112.00		
1 Equip. Oper. (crane)	29.35	234.80	48.20	385.60		
1 Equip. Oper. Oiler	25.40	203.20	41.75	334.00		
2 Cutting Torches		34.00		37.40		
2 Gases		151.20		166.32		
1 Lattice Boom Crane, 90 Ton		1741.00		1915.10	34.40	37.84
56 L.H., Daily Totals		$3202.20		$4255.62	$57.18	$75.99

Crew B-22

Crew B-22	Hr.	Daily	Hr.	Daily	Bare Costs	Incl. O&P
1 Labor Foreman (out)	$22.55	$180.40	$38.15	$305.20	$22.84	$38.35
2 Laborers	20.55	328.80	34.75	556.00		
.75 Equip. Oper. (crane)	29.35	176.10	48.20	289.20		
.75 S.P. Crane, 4x4, 5 Ton		193.65		213.01	6.46	7.10
30 L.H., Daily Totals		$878.95		$1363.42	$29.30	$45.45

Crew B-22A

Crew B-22A	Hr.	Daily	Hr.	Daily	Bare Costs	Incl. O&P
1 Labor Foreman (out)	$22.55	$180.40	$38.15	$305.20	$23.94	$40.24
1 Skilled Worker	28.05	224.40	47.35	378.80		
2 Laborers	20.55	328.80	34.75	556.00		
.75 Equipment Oper. (crane)	29.35	176.10	48.20	289.20		
.75 S.P. Crane, 4x4, 5 Ton		193.65		213.01		
1 Generator, 5 kW		40.20		44.22		
1 Butt Fusion Machine		199.80		219.78	11.41	12.55
38 L.H., Daily Totals		$1343.35		$2006.21	$35.35	$52.80

Crew B-22B

Crew B-22B	Hr.	Daily	Hr.	Daily	Bare Costs	Incl. O&P
1 Skilled Worker	$28.05	$224.40	$47.35	$378.80	$24.30	$41.05
1 Laborer	20.55	164.40	34.75	278.00		
1 Electro Fusion Machine		146.80		161.48	9.18	10.09
16 L.H., Daily Totals		$535.60		$818.28	$33.48	$51.14

Crew B-23

Crew B-23	Hr.	Daily	Hr.	Daily	Bare Costs	Incl. O&P
1 Labor Foreman (outside)	$22.55	$180.40	$38.15	$305.20	$20.95	$35.43
4 Laborers	20.55	657.60	34.75	1112.00		
1 Drill Rig, Truck-Mounted		2533.00		2786.30		
1 Flatbed Truck, Gas, 3 Ton		242.60		266.86	69.39	76.33
40 L.H., Daily Totals		$3613.60		$4470.36	$90.34	$111.76

Crew B-23A

Crew B-23A	Hr.	Daily	Hr.	Daily	Bare Costs	Incl. O&P
1 Labor Foreman (outside)	$22.55	$180.40	$38.15	$305.20	$23.88	$39.93
1 Laborer	20.55	164.40	34.75	278.00		
1 Equip. Operator (medium)	28.55	228.40	46.90	375.20		
1 Drill Rig, Truck-Mounted		2533.00		2786.30		
1 Pickup Truck, 3/4 Ton		116.80		128.48	110.41	121.45
24 L.H., Daily Totals		$3223.00		$3873.18	$134.29	$161.38

Crew B-23B

Crew B-23B	Hr.	Daily	Hr.	Daily	Bare Costs	Incl. O&P
1 Labor Foreman (outside)	$22.55	$180.40	$38.15	$305.20	$23.88	$39.93
1 Laborer	20.55	164.40	34.75	278.00		
1 Equip. Operator (medium)	28.55	228.40	46.90	375.20		
1 Drill Rig, Truck-Mounted		2533.00		2786.30		
1 Pickup Truck, 3/4 Ton		116.80		128.48		
1 Centr. Water Pump, 6"		309.40		340.34	123.30	135.63
24 L.H., Daily Totals		$3532.40		$4213.52	$147.18	$175.56

Crew B-24

Crew B-24	Hr.	Daily	Hr.	Daily	Bare Costs	Incl. O&P
1 Cement Finisher	$26.80	$214.40	$43.00	$344.00	$25.10	$41.67
1 Laborer	20.55	164.40	34.75	278.00		
1 Carpenter	27.95	223.60	47.25	378.00		
24 L.H., Daily Totals		$602.40		$1000.00	$25.10	$41.67

Crew B-25

Crew B-25	Hr.	Daily	Hr.	Daily	Bare Costs	Incl. O&P
1 Labor Foreman	$22.55	$180.40	$38.15	$305.20	$22.91	$38.37
7 Laborers	20.55	1150.80	34.75	1946.00		
3 Equip. Oper. (med.)	28.55	685.20	46.90	1125.60		
1 Asphalt Paver, 130 H.P.		1936.00		2129.60		
1 Tandem Roller, 10 Ton		233.00		256.30		
1 Roller, Pneum. Whl, 12 Ton		307.80		338.58	28.15	30.96
88 L.H., Daily Totals		$4493.20		$6101.28	$51.06	$69.33

Crew B-25B

Crew B-25B	Hr.	Daily	Hr.	Daily	Bare Costs	Incl. O&P
1 Labor Foreman	$22.55	$180.40	$38.15	$305.20	$23.38	$39.08
7 Laborers	20.55	1150.80	34.75	1946.00		
4 Equip. Oper. (medium)	28.55	913.60	46.90	1500.80		
1 Asphalt Paver, 130 H.P.		1936.00		2129.60		
2 Tandem Rollers, 10 Ton		466.00		512.60		
1 Roller, Pneum. Whl, 12 Ton		307.80		338.58	28.23	31.05
96 L.H., Daily Totals		$4954.60		$6732.78	$51.61	$70.13

Crew B-25C

Crew B-25C	Hr.	Daily	Hr.	Daily	Bare Costs	Incl. O&P
1 Labor Foreman	$22.55	$180.40	$38.15	$305.20	$23.55	$39.37
3 Laborers	20.55	493.20	34.75	834.00		
2 Equip. Oper. (medium)	28.55	456.80	46.90	750.40		
1 Asphalt Paver, 130 H.P.		1936.00		2129.60		
1 Tandem Roller, 10 Ton		233.00		256.30	45.19	49.71
48 L.H., Daily Totals		$3299.40		$4275.50	$68.74	$89.07

Crews

Crew B-26

Crew No.	Bare Costs Hr.	Bare Costs Daily	Incl. Subs O & P Hr.	Incl. Subs O & P Daily	Cost Per Labor-Hour Bare Costs	Cost Per Labor-Hour Incl. O&P
1 Labor Foreman (outside)	$22.55	$180.40	$38.15	$305.20	$23.60	$39.68
6 Laborers	20.55	986.40	34.75	1668.00		
2 Equip. Oper. (med.)	28.55	456.80	46.90	750.40		
1 Rodman (reinf.)	29.85	238.80	53.00	424.00		
1 Cement Finisher	26.80	214.40	43.00	344.00		
1 Grader, 30,000 Lbs.		550.20		605.22		
1 Paving Mach. & Equip.		2500.00		2750.00	34.66	38.13
88 L.H., Daily Totals		$5127.00		$6846.82	$58.26	$77.80

Crew B-26A

Crew No.	Bare Costs Hr.	Bare Costs Daily	Incl. Subs O & P Hr.	Incl. Subs O & P Daily	Cost Per Labor-Hour Bare Costs	Cost Per Labor-Hour Incl. O&P
1 Labor Foreman (outside)	$22.55	$180.40	$38.15	$305.20	$23.60	$39.68
6 Laborers	20.55	986.40	34.75	1668.00		
2 Equip. Oper. (med.)	28.55	456.80	46.90	750.40		
1 Rodman (reinf.)	29.85	238.80	53.00	424.00		
1 Cement Finisher	26.80	214.40	43.00	344.00		
1 Grader, 30,000 Lbs.		550.20		605.22		
1 Paving Mach. & Equip.		2500.00		2750.00		
1 Concrete Saw		141.00		155.10	36.26	39.89
88 L.H., Daily Totals		$5268.00		$7001.92	$59.86	$79.57

Crew B-26B

Crew No.	Bare Costs Hr.	Bare Costs Daily	Incl. Subs O & P Hr.	Incl. Subs O & P Daily	Cost Per Labor-Hour Bare Costs	Cost Per Labor-Hour Incl. O&P
1 Labor Foreman (outside)	$22.55	$180.40	$38.15	$305.20	$24.01	$40.28
6 Laborers	20.55	986.40	34.75	1668.00		
3 Equip. Oper. (med.)	28.55	685.20	46.90	1125.60		
1 Rodman (reinf.)	29.85	238.80	53.00	424.00		
1 Cement Finisher	26.80	214.40	43.00	344.00		
1 Grader, 30,000 Lbs.		550.20		605.22		
1 Paving Mach. & Equip.		2500.00		2750.00		
1 Concrete Pump, 110' Boom		977.00		1074.70	41.95	46.15
96 L.H., Daily Totals		$6332.40		$8296.72	$65.96	$86.42

Crew B-27

Crew No.	Bare Costs Hr.	Bare Costs Daily	Incl. Subs O & P Hr.	Incl. Subs O & P Daily	Cost Per Labor-Hour Bare Costs	Cost Per Labor-Hour Incl. O&P
1 Labor Foreman (outside)	$22.55	$180.40	$38.15	$305.20	$21.05	$35.60
3 Laborers	20.55	493.20	34.75	834.00		
1 Berm Machine		271.40		298.54	8.48	9.33
32 L.H., Daily Totals		$945.00		$1437.74	$29.53	$44.93

Crew B-28

Crew No.	Bare Costs Hr.	Bare Costs Daily	Incl. Subs O & P Hr.	Incl. Subs O & P Daily	Cost Per Labor-Hour Bare Costs	Cost Per Labor-Hour Incl. O&P
2 Carpenters	$27.95	$447.20	$47.25	$756.00	$25.48	$43.08
1 Laborer	20.55	164.40	34.75	278.00		
24 L.H., Daily Totals		$611.60		$1034.00	$25.48	$43.08

Crew B-29

Crew No.	Bare Costs Hr.	Bare Costs Daily	Incl. Subs O & P Hr.	Incl. Subs O & P Daily	Cost Per Labor-Hour Bare Costs	Cost Per Labor-Hour Incl. O&P
1 Labor Foreman (outside)	$22.55	$180.40	$38.15	$305.20	$22.35	$37.56
4 Laborers	20.55	657.60	34.75	1112.00		
1 Equip. Oper. (crane)	29.35	234.80	48.20	385.60		
1 Gradall, 5/8 C.Y.		958.80		1054.68	19.98	21.97
48 L.H., Daily Totals		$2031.60		$2857.48	$42.33	$59.53

Crew B-30

Crew No.	Bare Costs Hr.	Bare Costs Daily	Incl. Subs O & P Hr.	Incl. Subs O & P Daily	Cost Per Labor-Hour Bare Costs	Cost Per Labor-Hour Incl. O&P
1 Equip. Oper. (med.)	$28.55	$228.40	$46.90	$375.20	$24.65	$41.03
2 Truck Drivers (heavy)	22.70	363.20	38.10	609.60		
1 Hyd. Excavator, 1.5 C.Y.		865.80		952.38		
2 Dump Trucks, 12 C.Y., 400 H.P.		1066.00		1172.60	80.49	88.54
24 L.H., Daily Totals		$2523.40		$3109.78	$105.14	$129.57

Crew B-31

Crew No.	Bare Costs Hr.	Bare Costs Daily	Incl. Subs O & P Hr.	Incl. Subs O & P Daily	Cost Per Labor-Hour Bare Costs	Cost Per Labor-Hour Incl. O&P
1 Labor Foreman (outside)	$22.55	$180.40	$38.15	$305.20	$20.95	$35.43
4 Laborers	20.55	657.60	34.75	1112.00		
1 Air Compressor, 250 cfm		162.40		178.64		
1 Sheeting Driver		5.75		6.33		
2 -50' Air Hoses, 1.5"		12.60		13.86	4.52	4.97
40 L.H., Daily Totals		$1018.75		$1616.03	$25.47	$40.40

Crew B-32

Crew No.	Bare Costs Hr.	Bare Costs Daily	Incl. Subs O & P Hr.	Incl. Subs O & P Daily	Cost Per Labor-Hour Bare Costs	Cost Per Labor-Hour Incl. O&P
1 Laborer	$20.55	$164.40	$34.75	$278.00	$26.55	$43.86
3 Equip. Oper. (med.)	28.55	685.20	46.90	1125.60		
1 Grader, 30,000 Lbs.		550.20		605.22		
1 Tandem Roller, 10 Ton		233.00		256.30		
1 Dozer, 200 H.P.		1082.00		1190.20	58.29	64.12
32 L.H., Daily Totals		$2714.80		$3455.32	$84.84	$107.98

Crew B-32A

Crew No.	Bare Costs Hr.	Bare Costs Daily	Incl. Subs O & P Hr.	Incl. Subs O & P Daily	Cost Per Labor-Hour Bare Costs	Cost Per Labor-Hour Incl. O&P
1 Laborer	$20.55	$164.40	$34.75	$278.00	$25.88	$42.85
2 Equip. Oper. (medium)	28.55	456.80	46.90	750.40		
1 Grader, 30,000 Lbs.		550.20		605.22		
1 Roller, Vibratory, 25 Ton		594.60		654.06	47.70	52.47
24 L.H., Daily Totals		$1766.00		$2287.68	$73.58	$95.32

Crew B-32B

Crew No.	Bare Costs Hr.	Bare Costs Daily	Incl. Subs O & P Hr.	Incl. Subs O & P Daily	Cost Per Labor-Hour Bare Costs	Cost Per Labor-Hour Incl. O&P
1 Laborer	$20.55	$164.40	$34.75	$278.00	$25.88	$42.85
2 Equip. Oper. (medium)	28.55	456.80	46.90	750.40		
1 Dozer, 200 H.P.		1082.00		1190.20		
1 Roller, Vibratory, 25 Ton		594.60		654.06	69.86	76.84
24 L.H., Daily Totals		$2297.80		$2872.66	$95.74	$119.69

Crew B-32C

Crew No.	Bare Costs Hr.	Bare Costs Daily	Incl. Subs O & P Hr.	Incl. Subs O & P Daily	Cost Per Labor-Hour Bare Costs	Cost Per Labor-Hour Incl. O&P
1 Labor Foreman	$22.55	$180.40	$38.15	$305.20	$24.88	$41.39
2 Laborers	20.55	328.80	34.75	556.00		
3 Equip. Oper. (medium)	28.55	685.20	46.90	1125.60		
1 Grader, 30,000 Lbs.		550.20		605.22		
1 Tandem Roller, 10 Ton		233.00		256.30		
1 Dozer, 200 H.P.		1082.00		1190.20	38.86	42.74
48 L.H., Daily Totals		$3059.60		$4038.52	$63.74	$84.14

Crew B-33A

Crew No.	Bare Costs Hr.	Bare Costs Daily	Incl. Subs O & P Hr.	Incl. Subs O & P Daily	Cost Per Labor-Hour Bare Costs	Cost Per Labor-Hour Incl. O&P
1 Equip. Oper. (med.)	$28.55	$228.40	$46.90	$375.20	$28.55	$46.90
.25 Equip. Oper. (med.)	28.55	57.10	46.90	93.80		
1 Scraper, Towed, 7 C.Y.		102.20		112.42		
1.250 Dozers, 300 H.P.		1778.75		1956.63	188.10	206.90
10 L.H., Daily Totals		$2166.45		$2538.05	$216.65	$253.80

Crew B-33B

Crew No.	Bare Costs Hr.	Bare Costs Daily	Incl. Subs O & P Hr.	Incl. Subs O & P Daily	Cost Per Labor-Hour Bare Costs	Cost Per Labor-Hour Incl. O&P
1 Equip. Oper. (med.)	$28.55	$228.40	$46.90	$375.20	$28.55	$46.90
.25 Equip. Oper. (med.)	28.55	57.10	46.90	93.80		
1 Scraper, Towed, 10 C.Y.		153.60		168.96		
1.250 Dozers, 300 H.P.		1778.75		1956.63	193.24	212.56
10 L.H., Daily Totals		$2217.85		$2594.59	$221.79	$259.46

Crew B-33C

Crew No.	Bare Costs Hr.	Bare Costs Daily	Incl. Subs O & P Hr.	Incl. Subs O & P Daily	Cost Per Labor-Hour Bare Costs	Cost Per Labor-Hour Incl. O&P
1 Equip. Oper. (med.)	$28.55	$228.40	$46.90	$375.20	$28.55	$46.90
.25 Equip. Oper. (med.)	28.55	57.10	46.90	93.80		
1 Scraper, Towed, 15 C.Y.		168.80		185.68		
1.250 Dozers, 300 H.P.		1778.75		1956.63	194.76	214.23
10 L.H., Daily Totals		$2233.05		$2611.30	$223.31	$261.13

Crew No.	Bare Costs		Incl. Subs O & P		Cost Per Labor-Hour	
Crew B-33D	**Hr.**	**Daily**	**Hr.**	**Daily**	**Bare Costs**	**Incl. O&P**
1 Equip. Oper. (med.)	$28.55	$228.40	$46.90	$375.20	$28.55	$46.90
.25 Equip. Oper. (med.)	28.55	57.10	46.90	93.80		
1 S.P. Scraper, 14 C.Y.		1509.00		1659.90		
.25 Dozer, 300 H.P.		355.75		391.32	186.47	205.12
10 L.H., Daily Totals		$2150.25		$2520.22	$215.03	$252.02
Crew B-33E	**Hr.**	**Daily**	**Hr.**	**Daily**	**Bare Costs**	**Incl. O&P**
1 Equip. Oper. (med.)	$28.55	$228.40	$46.90	$375.20	$28.55	$46.90
.25 Equip. Oper. (med.)	28.55	57.10	46.90	93.80		
1 S.P. Scraper, 21 C.Y.		1924.00		2116.40		
.25 Dozer, 300 H.P.		355.75		391.32	227.97	250.77
10 L.H., Daily Totals		$2565.25		$2976.72	$256.52	$297.67
Crew B-33F	**Hr.**	**Daily**	**Hr.**	**Daily**	**Bare Costs**	**Incl. O&P**
1 Equip. Oper. (med.)	$28.55	$228.40	$46.90	$375.20	$28.55	$46.90
.25 Equip. Oper. (med.)	28.55	57.10	46.90	93.80		
1 Elev. Scraper, 11 C.Y.		1095.00		1204.50		
.25 Dozer, 300 H.P.		355.75		391.32	145.07	159.58
10 L.H., Daily Totals		$1736.25		$2064.82	$173.63	$206.48
Crew B-33G	**Hr.**	**Daily**	**Hr.**	**Daily**	**Bare Costs**	**Incl. O&P**
1 Equip. Oper. (med.)	$28.55	$228.40	$46.90	$375.20	$28.55	$46.90
.25 Equip. Oper. (med.)	28.55	57.10	46.90	93.80		
1 Elev. Scraper, 22 C.Y.		2127.00		2339.70		
.25 Dozer, 300 H.P.		355.75		391.32	248.28	273.10
10 L.H., Daily Totals		$2768.25		$3200.03	$276.82	$320.00
Crew B-33K	**Hr.**	**Daily**	**Hr.**	**Daily**	**Bare Costs**	**Incl. O&P**
1 Equipment Operator (med.)	$28.55	$228.40	$46.90	$375.20	$26.26	$43.43
.25 Equipment Operator (med.)	28.55	57.10	46.90	93.80		
.5 Laborer	20.55	82.20	34.75	139.00		
1 S.P. Scraper, 31 C.Y.		2878.00		3165.80		
.25 Dozer, 410 H.P.		462.25		508.48	238.59	262.45
14 L.H., Daily Totals		$3707.95		$4282.27	$264.85	$305.88
Crew B-34A	**Hr.**	**Daily**	**Hr.**	**Daily**	**Bare Costs**	**Incl. O&P**
1 Truck Driver (heavy)	$22.70	$181.60	$38.10	$304.80	$22.70	$38.10
1 Dump Truck, 8 C.Y., 220 H.P.		326.40		359.04	40.80	44.88
8 L.H., Daily Totals		$508.00		$663.84	$63.50	$82.98
Crew B-34B	**Hr.**	**Daily**	**Hr.**	**Daily**	**Bare Costs**	**Incl. O&P**
1 Truck Driver (heavy)	$22.70	$181.60	$38.10	$304.80	$22.70	$38.10
1 Dump Truck, 12 C.Y., 400 H.P.		533.00		586.30	66.63	73.29
8 L.H., Daily Totals		$714.60		$891.10	$89.33	$111.39
Crew B-34C	**Hr.**	**Daily**	**Hr.**	**Daily**	**Bare Costs**	**Incl. O&P**
1 Truck Driver (heavy)	$22.70	$181.60	$38.10	$304.80	$22.70	$38.10
1 Truck Tractor, 6x4, 380 H.P.		478.80		526.68		
1 Dump Trailer, 16.5 C.Y.		115.60		127.16	74.30	81.73
8 L.H., Daily Totals		$776.00		$958.64	$97.00	$119.83
Crew B-34D	**Hr.**	**Daily**	**Hr.**	**Daily**	**Bare Costs**	**Incl. O&P**
1 Truck Driver (heavy)	$22.70	$181.60	$38.10	$304.80	$22.70	$38.10
1 Truck Tractor, 6x4, 380 H.P.		478.80		526.68		
1 Dump Trailer, 20 C.Y.		130.20		143.22	76.13	83.74
8 L.H., Daily Totals		$790.60		$974.70	$98.83	$121.84

Crew No.	Bare Costs		Incl. Subs O & P		Cost Per Labor-Hour	
Crew B-34E	**Hr.**	**Daily**	**Hr.**	**Daily**	**Bare Costs**	**Incl. O&P**
1 Truck Driver (heavy)	$22.70	$181.60	$38.10	$304.80	$22.70	$38.10
1 Dump Truck, Off Hwy., 25 Ton		1280.00		1408.00	160.00	176.00
8 L.H., Daily Totals		$1461.60		$1712.80	$182.70	$214.10
Crew B-34F	**Hr.**	**Daily**	**Hr.**	**Daily**	**Bare Costs**	**Incl. O&P**
1 Truck Driver (heavy)	$22.70	$181.60	$38.10	$304.80	$22.70	$38.10
1 Dump Truck, Off Hwy., 35 Ton		1144.00		1258.40	143.00	157.30
8 L.H., Daily Totals		$1325.60		$1563.20	$165.70	$195.40
Crew B-34G	**Hr.**	**Daily**	**Hr.**	**Daily**	**Bare Costs**	**Incl. O&P**
1 Truck Driver (heavy)	$22.70	$181.60	$38.10	$304.80	$22.70	$38.10
1 Dump Truck, Off Hwy., 50 Ton		1582.00		1740.20	197.75	217.53
8 L.H., Daily Totals		$1763.60		$2045.00	$220.45	$255.63
Crew B-34H	**Hr.**	**Daily**	**Hr.**	**Daily**	**Bare Costs**	**Incl. O&P**
1 Truck Driver (heavy)	$22.70	$181.60	$38.10	$304.80	$22.70	$38.10
1 Dump Truck, Off Hwy., 65 Ton		1594.00		1753.40	199.25	219.18
8 L.H., Daily Totals		$1775.60		$2058.20	$221.95	$257.27
Crew B-34J	**Hr.**	**Daily**	**Hr.**	**Daily**	**Bare Costs**	**Incl. O&P**
1 Truck Driver (heavy)	$22.70	$181.60	$38.10	$304.80	$22.70	$38.10
1 Dump Truck, Off Hwy., 100 Ton		2047.00		2251.70	255.88	281.46
8 L.H., Daily Totals		$2228.60		$2556.50	$278.57	$319.56
Crew B-34K	**Hr.**	**Daily**	**Hr.**	**Daily**	**Bare Costs**	**Incl. O&P**
1 Truck Driver (heavy)	$22.70	$181.60	$38.10	$304.80	$22.70	$38.10
1 Truck Tractor, 6x4, 450 H.P.		581.20		639.32		
1 Lowbed Trailer, 75 Ton		203.00		223.30	98.03	107.83
8 L.H., Daily Totals		$965.80		$1167.42	$120.72	$145.93
Crew B-34L	**Hr.**	**Daily**	**Hr.**	**Daily**	**Bare Costs**	**Incl. O&P**
1 Equip. Oper. (light)	$26.95	$215.60	$44.30	$354.40	$26.95	$44.30
1 Flatbed Truck, Gas, 1.5 Ton		194.40		213.84	24.30	26.73
8 L.H., Daily Totals		$410.00		$568.24	$51.25	$71.03
Crew B-34N	**Hr.**	**Daily**	**Hr.**	**Daily**	**Bare Costs**	**Incl. O&P**
1 Truck Driver (heavy)	$22.70	$181.60	$38.10	$304.80	$22.70	$38.10
1 Dump Truck, 8 C.Y., 220 H.P.		326.40		359.04		
1 Flatbed Trailer, 40 Ton		139.60		153.56	58.25	64.08
8 L.H., Daily Totals		$647.60		$817.40	$80.95	$102.18
Crew B-34P	**Hr.**	**Daily**	**Hr.**	**Daily**	**Bare Costs**	**Incl. O&P**
1 Pipe Fitter	$32.55	$260.40	$53.40	$427.20	$27.68	$45.72
1 Truck Driver (light)	21.95	175.60	36.85	294.80		
1 Equip. Oper. (medium)	28.55	228.40	46.90	375.20		
1 Flatbed Truck, Gas, 3 Ton		242.60		266.86		
1 Backhoe Loader, 48 H.P.		293.80		323.18	22.35	24.59
24 L.H., Daily Totals		$1200.80		$1687.24	$50.03	$70.30
Crew B-34Q	**Hr.**	**Daily**	**Hr.**	**Daily**	**Bare Costs**	**Incl. O&P**
1 Pipe Fitter	$32.55	$260.40	$53.40	$427.20	$27.95	$46.15
1 Truck Driver (light)	21.95	175.60	36.85	294.80		
1 Eqip. Oper. (crane)	29.35	234.80	48.20	385.60		
1 Flatbed Trailer, 25 Ton		102.00		112.20		
1 Dump Truck, 8 C.Y., 220 H.P.		326.40		359.04		
1 Hyd. Crane, 25 Ton		789.40		868.34	50.74	55.82
24 L.H., Daily Totals		$1888.60		$2447.18	$78.69	$101.97

Crews

Crew B-34R

Crew No.	Bare Costs Hr.	Bare Costs Daily	Incl. Subs O&P Hr.	Incl. Subs O&P Daily	Cost Per Labor-Hour Bare Costs	Cost Per Labor-Hour Incl. O&P
1 Pipe Fitter	$32.55	$260.40	$53.40	$427.20	$27.95	$46.15
1 Truck Driver (light)	21.95	175.60	36.85	294.80		
1 Eqip. Oper. (crane)	29.35	234.80	48.20	385.60		
1 Flatbed Trailer, 25 Ton		102.00		112.20		
1 Dump Truck, 8 C.Y., 220 H.P.		326.40		359.04		
1 Hyd. Crane, 25 Ton		789.40		868.34		
1 Hyd. Excavator, 1 C.Y.		669.20		736.12	78.63	86.49
24 L.H., Daily Totals		$2557.80		$3183.30	$106.58	$132.64

Crew B-34S

Crew No.	Bare Costs Hr.	Bare Costs Daily	Incl. Subs O&P Hr.	Incl. Subs O&P Daily	Cost Per Labor-Hour Bare Costs	Cost Per Labor-Hour Incl. O&P
2 Pipe Fitters	$32.55	$520.80	$53.40	$854.40	$29.29	$48.27
1 Truck Driver (heavy)	22.70	181.60	38.10	304.80		
1 Eqip. Oper. (crane)	29.35	234.80	48.20	385.60		
1 Flatbed Trailer, 40 Ton		139.60		153.56		
1 Truck Tractor, 6x4, 380 H.P.		478.80		526.68		
1 Hyd. Crane, 80 Ton		1296.00		1425.60		
1 Hyd. Excavator, 2 C.Y.		1182.00		1300.20	96.76	106.44
32 L.H., Daily Totals		$4033.60		$4950.84	$126.05	$154.71

Crew B-34T

Crew No.	Bare Costs Hr.	Bare Costs Daily	Incl. Subs O&P Hr.	Incl. Subs O&P Daily	Cost Per Labor-Hour Bare Costs	Cost Per Labor-Hour Incl. O&P
2 Pipe Fitters	$32.55	$520.80	$53.40	$854.40	$29.29	$48.27
1 Truck Driver (heavy)	22.70	181.60	38.10	304.80		
1 Eqip. Oper. (crane)	29.35	234.80	48.20	385.60		
1 Flatbed Trailer, 40 Ton		139.60		153.56		
1 Truck Tractor, 6x4, 380 H.P.		478.80		526.68		
1 Hyd. Crane, 80 Ton		1296.00		1425.60	59.83	65.81
32 L.H., Daily Totals		$2851.60		$3650.64	$89.11	$114.08

Crew B-35

Crew No.	Bare Costs Hr.	Bare Costs Daily	Incl. Subs O&P Hr.	Incl. Subs O&P Daily	Cost Per Labor-Hour Bare Costs	Cost Per Labor-Hour Incl. O&P
1 Laborer Foreman (out)	$22.55	$180.40	$38.15	$305.20	$26.54	$44.26
1 Skilled Worker	28.05	224.40	47.35	378.80		
1 Welder (plumber)	32.20	257.60	52.85	422.80		
1 Laborer	20.55	164.40	34.75	278.00		
1 Equip. Oper. (crane)	29.35	234.80	48.20	385.60		
1 Welder, electric, 300 amp		55.70		61.27		
1 Hyd. Excavator, .75 C.Y.		598.60		658.46	16.36	17.99
40 L.H., Daily Totals		$1715.90		$2490.13	$42.90	$62.25

Crew B-35A

Crew No.	Bare Costs Hr.	Bare Costs Daily	Incl. Subs O&P Hr.	Incl. Subs O&P Daily	Cost Per Labor-Hour Bare Costs	Cost Per Labor-Hour Incl. O&P
1 Laborer Foreman (out)	$22.55	$180.40	$38.15	$305.20	$25.52	$42.54
2 Laborers	20.55	328.80	34.75	556.00		
1 Skilled Worker	28.05	224.40	47.35	378.80		
1 Welder (plumber)	32.20	257.60	52.85	422.80		
1 Equip. Oper. (crane)	29.35	234.80	48.20	385.60		
1 Equip. Oper. Oiler	25.40	203.20	41.75	334.00		
1 Welder, gas engine, 300 amp		134.20		147.62		
1 Crawler Crane, 75 Ton		1496.00		1645.60	29.11	32.02
56 L.H., Daily Totals		$3059.40		$4175.62	$54.63	$74.56

Crew B-36

Crew No.	Bare Costs Hr.	Bare Costs Daily	Incl. Subs O&P Hr.	Incl. Subs O&P Daily	Cost Per Labor-Hour Bare Costs	Cost Per Labor-Hour Incl. O&P
1 Labor Foreman (outside)	$22.55	$180.40	$38.15	$305.20	$24.15	$40.29
2 Laborers	20.55	328.80	34.75	556.00		
2 Equip. Oper. (med.)	28.55	456.80	46.90	750.40		
1 Dozer, 200 H.P.		1082.00		1190.20		
1 Aggregate Spreader		35.00		38.50		
1 Tandem Roller, 10 Ton		233.00		256.30	33.75	37.13
40 L.H., Daily Totals		$2316.00		$3096.60	$57.90	$77.42

Crew B-36A

Crew No.	Bare Costs Hr.	Bare Costs Daily	Incl. Subs O&P Hr.	Incl. Subs O&P Daily	Cost Per Labor-Hour Bare Costs	Cost Per Labor-Hour Incl. O&P
1 Labor Foreman (outside)	$22.55	$180.40	$38.15	$305.20	$25.41	$42.18
2 Laborers	20.55	328.80	34.75	556.00		
4 Equip. Oper. (med.)	28.55	913.60	46.90	1500.80		
1 Dozer, 200 H.P.		1082.00		1190.20		
1 Aggregate Spreader		35.00		38.50		
1 Tandem Roller, 10 Ton		233.00		256.30		
1 Roller, Pneum. Whl, 12 Ton		307.80		338.58	29.60	32.56
56 L.H., Daily Totals		$3080.60		$4185.58	$55.01	$74.74

Crew B-36B

Crew No.	Bare Costs Hr.	Bare Costs Daily	Incl. Subs O&P Hr.	Incl. Subs O&P Daily	Cost Per Labor-Hour Bare Costs	Cost Per Labor-Hour Incl. O&P
1 Labor Foreman (outside)	$22.55	$180.40	$38.15	$305.20	$25.07	$41.67
2 Laborers	20.55	328.80	34.75	556.00		
4 Equip. Oper. (medium)	28.55	913.60	46.90	1500.80		
1 Truck Driver, Heavy	22.70	181.60	38.10	304.80		
1 Grader, 30,000 Lbs.		550.20		605.22		
1 F.E. Loader, crl, 1.5 C.Y.		463.00		509.30		
1 Dozer, 300 H.P.		1423.00		1565.30		
1 Roller, Vibratory, 25 Ton		594.60		654.06		
1 Truck Tractor, 6x4, 450 H.P.		581.20		639.32		
1 Water Tanker, 5000 Gal.		136.00		149.60	58.56	64.42
64 L.H., Daily Totals		$5352.40		$6789.60	$83.63	$106.09

Crew B-36C

Crew No.	Bare Costs Hr.	Bare Costs Daily	Incl. Subs O&P Hr.	Incl. Subs O&P Daily	Cost Per Labor-Hour Bare Costs	Cost Per Labor-Hour Incl. O&P
1 Labor Foreman (outside)	$22.55	$180.40	$38.15	$305.20	$26.18	$43.39
3 Equip. Oper. (medium)	28.55	685.20	46.90	1125.60		
1 Truck Driver, Heavy	22.70	181.60	38.10	304.80		
1 Grader, 30,000 Lbs.		550.20		605.22		
1 Dozer, 300 H.P.		1423.00		1565.30		
1 Roller, Vibratory, 25 Ton		594.60		654.06		
1 Truck Tractor, 6x4, 450 H.P.		581.20		639.32		
1 Water Tanker, 5000 Gal.		136.00		149.60	82.13	90.34
40 L.H., Daily Totals		$4332.20		$5349.10	$108.31	$133.73

Crew B-36E

Crew No.	Bare Costs Hr.	Bare Costs Daily	Incl. Subs O&P Hr.	Incl. Subs O&P Daily	Cost Per Labor-Hour Bare Costs	Cost Per Labor-Hour Incl. O&P
1 Labor Foreman (outside)	$22.55	$180.40	$38.15	$305.20	$26.57	$43.98
4 Equip. Oper. (medium)	28.55	913.60	46.90	1500.80		
1 Truck Driver, Heavy	22.70	181.60	38.10	304.80		
1 Grader, 30,000 Lbs.		550.20		605.22		
1 Dozer, 300 H.P.		1423.00		1565.30		
1 Roller, Vibratory, 25 Ton		594.60		654.06		
1 Truck Tractor, 6x4, 380 H.P.		478.80		526.68		
1 Dist. Tanker, 3000 Gallon		304.20		334.62	69.81	76.79
48 L.H., Daily Totals		$4626.40		$5796.68	$96.38	$120.76

Crew B-37

Crew No.	Bare Costs Hr.	Bare Costs Daily	Incl. Subs O&P Hr.	Incl. Subs O&P Daily	Cost Per Labor-Hour Bare Costs	Cost Per Labor-Hour Incl. O&P
1 Labor Foreman (outside)	$22.55	$180.40	$38.15	$305.20	$21.95	$36.91
4 Laborers	20.55	657.60	34.75	1112.00		
1 Equip. Oper. (light)	26.95	215.60	44.30	354.40		
1 Tandem Roller, 5 Ton		137.80		151.58	2.87	3.16
48 L.H., Daily Totals		$1191.40		$1923.18	$24.82	$40.07

Crew B-38

Crew No.	Bare Costs Hr.	Bare Costs Daily	Incl. Subs O&P Hr.	Incl. Subs O&P Daily	Cost Per Labor-Hour Bare Costs	Cost Per Labor-Hour Incl. O&P
2 Laborers	$20.55	$328.80	$34.75	$556.00	$22.68	$37.93
1 Equip. Oper. (light)	26.95	215.60	44.30	354.40		
1 Backhoe Loader, 48 H.P.		293.80		323.18		
1 Hyd.Hammer, (1200 lb.)		153.00		168.30	18.62	20.48
24 L.H., Daily Totals		$991.20		$1401.88	$41.30	$58.41

Crews

Crew B-39

Crew B-39	Hr.	Daily	Hr.	Daily	Bare Costs	Incl. O&P
1 Labor Foreman (outside)	$22.55	$180.40	$38.15	$305.20	$20.88	$35.32
5 Laborers	20.55	822.00	34.75	1390.00		
1 Air Compressor, 250 cfm		162.40		178.64		
2 Breakers, Pavement, 60 lb.		16.40		18.04		
2 -50' Air Hoses, 1.5"		12.60		13.86	3.99	4.39
48 L.H., Daily Totals		$1193.80		$1905.74	$24.87	$39.70

Crew B-40

Crew B-40	Hr.	Daily	Hr.	Daily	Bare Costs	Incl. O&P
1 Pile Driver Foreman (out)	$28.95	$231.60	$51.05	$408.40	$26.66	$46.31
4 Pile Drivers	26.95	862.40	47.55	1521.60		
1 Building Laborer	20.55	164.40	34.75	278.00		
1 Equip. Oper. (crane)	29.35	234.80	48.20	385.60		
1 Crawler Crane, 40 Ton		1055.00		1160.50		
1 Vibratory Hammer & Gen.		2257.00		2482.70	59.14	65.06
56 L.H., Daily Totals		$4805.20		$6236.80	$85.81	$111.37

Crew B-40B

Crew B-40B	Hr.	Daily	Hr.	Daily	Bare Costs	Incl. O&P
1 Laborer Foreman	$22.55	$180.40	$38.15	$305.20	$23.16	$38.73
3 Laborers	20.55	493.20	34.75	834.00		
1 Equip. Oper. (crane)	29.35	234.80	48.20	385.60		
1 Equip. Oper. Oiler	25.40	203.20	41.75	334.00		
1 Lattice Boom Crane, 40 Ton		1327.00		1459.70	27.65	30.41
48 L.H., Daily Totals		$2438.60		$3318.50	$50.80	$69.14

Crew B-41

Crew B-41	Hr.	Daily	Hr.	Daily	Bare Costs	Incl. O&P
1 Labor Foreman (outside)	$22.55	$180.40	$38.15	$305.20	$21.53	$36.30
4 Laborers	20.55	657.60	34.75	1112.00		
.25 Equip. Oper. (crane)	29.35	58.70	48.20	96.40		
.25 Equip. Oper. Oiler	25.40	50.80	41.75	83.50		
.25 Crawler Crane, 40 Ton		263.75		290.13	5.99	6.59
44 L.H., Daily Totals		$1211.25		$1887.22	$27.53	$42.89

Crew B-42

Crew B-42	Hr.	Daily	Hr.	Daily	Bare Costs	Incl. O&P
1 Labor Foreman (outside)	$22.55	$180.40	$38.15	$305.20	$23.76	$39.74
4 Laborers	20.55	657.60	34.75	1112.00		
1 Equip. Oper. (crane)	29.35	234.80	48.20	385.60		
1 Welder	32.20	257.60	52.85	422.80		
1 Hyd. Crane, 25 Ton		789.40		868.34		
1 Welder, gas engine, 300 amp		134.20		147.62		
1 Horz. Boring Csg. Mch.		444.20		488.62	24.43	26.87
56 L.H., Daily Totals		$2698.20		$3730.18	$48.18	$66.61

Crew B-43

Crew B-43	Hr.	Daily	Hr.	Daily	Bare Costs	Incl. O&P
1 Labor Foreman (outside)	$22.55	$180.40	$38.15	$305.20	$20.95	$35.43
4 Laborers	20.55	657.60	34.75	1112.00		
1 Drill Rig, Truck-Mounted		2533.00		2786.30	63.33	69.66
40 L.H., Daily Totals		$3371.00		$4203.50	$84.28	$105.09

Crew B-44

Crew B-44	Hr.	Daily	Hr.	Daily	Bare Costs	Incl. O&P
1 Pile Driver Foreman	$28.95	$231.60	$51.05	$408.40	$25.90	$44.87
4 Pile Drivers	26.95	862.40	47.55	1521.60		
1 Equip. Oper. (crane)	29.35	234.80	48.20	385.60		
2 Laborers	20.55	328.80	34.75	556.00		
1 Crawler Crane, 40 Ton		1055.00		1160.50		
1 Lead, 60' high		69.60		76.56		
1 Hammer, diesel, 15K Ft.-Lbs.		578.60		636.46	26.61	29.27
64 L.H., Daily Totals		$3360.80		$4745.12	$52.51	$74.14

Crew B-45

Crew B-45	Hr.	Daily	Hr.	Daily	Bare Costs	Incl. O&P
1 Building Laborer	$20.55	$164.40	$34.75	$278.00	$21.63	$36.42
1 Truck Driver (heavy)	22.70	181.60	38.10	304.80		
1 Dist. Tanker, 3000 Gallon		304.20		334.62	19.01	20.91
16 L.H., Daily Totals		$650.20		$917.42	$40.64	$57.34

Crew B-46

Crew B-46	Hr.	Daily	Hr.	Daily	Bare Costs	Incl. O&P
1 Pile Driver Foreman	$28.95	$231.60	$51.05	$408.40	$24.08	$41.73
2 Pile Drivers	26.95	431.20	47.55	760.80		
3 Laborers	20.55	493.20	34.75	834.00		
1 Chainsaw, Gas, 36" Long		37.80		41.58	0.79	0.87
48 L.H., Daily Totals		$1193.80		$2044.78	$24.87	$42.60

Crew B-47

Crew B-47	Hr.	Daily	Hr.	Daily	Bare Costs	Incl. O&P
1 Blast Foreman	$22.55	$180.40	$38.15	$305.20	$21.55	$36.45
1 Driller	20.55	164.40	34.75	278.00		
1 Air Track Drill, 4"		850.60		935.66		
1 Air Compressor, 600 cfm		434.40		477.84		
2 -50' Air Hoses, 3"		32.40		35.64	82.34	90.57
16 L.H., Daily Totals		$1662.20		$2032.34	$103.89	$127.02

Crew B-47A

Crew B-47A	Hr.	Daily	Hr.	Daily	Bare Costs	Incl. O&P
1 Drilling Foreman	$22.55	$180.40	$38.15	$305.20	$25.77	$42.70
1 Equip. Oper. (heavy)	29.35	234.80	48.20	385.60		
1 Oiler	25.40	203.20	41.75	334.00		
1 Air Track Drill, 5"		974.00		1071.40	40.58	44.64
24 L.H., Daily Totals		$1592.40		$2096.20	$66.35	$87.34

Crew B-47C

Crew B-47C	Hr.	Daily	Hr.	Daily	Bare Costs	Incl. O&P
1 Laborer	$20.55	$164.40	$34.75	$278.00	$23.75	$39.52
1 Equip. Oper. (light)	26.95	215.60	44.30	354.40		
1 Air Compressor, 750 cfm		439.60		483.56		
2 -50' Air Hoses, 3"		32.40		35.64		
1 Air Track Drill, 4"		850.60		935.66	82.66	90.93
16 L.H., Daily Totals		$1702.60		$2087.26	$106.41	$130.45

Crew B-47E

Crew B-47E	Hr.	Daily	Hr.	Daily	Bare Costs	Incl. O&P
1 Laborer Foreman	$22.55	$180.40	$38.15	$305.20	$21.05	$35.60
3 Laborers	20.55	493.20	34.75	834.00		
1 Flatbed Truck, Gas, 3 Ton		242.60		266.86	7.58	8.34
32 L.H., Daily Totals		$916.20		$1406.06	$28.63	$43.94

Crew B-47G

Crew B-47G	Hr.	Daily	Hr.	Daily	Bare Costs	Incl. O&P
1 Laborer Foreman	$22.55	$180.40	$38.15	$305.20	$21.22	$35.88
2 Laborers	20.55	328.80	34.75	556.00		
1 Air Track Drill, 4"		850.60		935.66		
1 Air Compressor, 600 cfm		434.40		477.84		
2 -50' Air Hoses, 3"		32.40		35.64		
1 Grout Pump		166.40		183.04	61.83	68.01
24 L.H., Daily Totals		$1993.00		$2493.38	$83.04	$103.89

Crew B-48

Crew B-48	Hr.	Daily	Hr.	Daily	Bare Costs	Incl. O&P
1 Labor Foreman (outside)	$22.55	$180.40	$38.15	$305.20	$22.35	$37.56
4 Laborers	20.55	657.60	34.75	1112.00		
1 Equip. Oper. (crane)	29.35	234.80	48.20	385.60		
1 Centr. Water Pump, 6"		309.40		340.34		
1 -20' Suction Hose, 6"		11.90		13.09		
1 -50' Discharge Hose, 6"		6.30		6.93		
1 Drill Rig, Truck-Mounted		2533.00		2786.30	59.60	65.56
48 L.H., Daily Totals		$3933.40		$4949.46	$81.95	$103.11

359

Crew No.	Bare Costs		Incl. Subs O & P		Cost Per Labor-Hour	

Crew B-49

	Hr.	Daily	Hr.	Daily	Bare Costs	Incl. O&P
1 Labor Foreman (outside)	$22.55	$180.40	$38.15	$305.20	$23.17	$39.47
5 Laborers	20.55	822.00	34.75	1390.00		
1 Equip. Oper. (crane)	29.35	234.80	48.20	385.60		
2 Pile Drivers	26.95	431.20	47.55	760.80		
1 Hyd. Crane, 25 Ton		789.40		868.34		
1 Centr. Water Pump, 6"		309.40		340.34		
1 -20' Suction Hose, 6"		11.90		13.09		
1 -50' Discharge Hose, 6"		6.30		6.93		
1 Drill Rig, Truck-Mounted		2533.00		2786.30	50.69	55.76
72 L.H., Daily Totals		$5318.40		$6856.60	$73.87	$95.23

Crew B-50

	Hr.	Daily	Hr.	Daily	Bare Costs	Incl. O&P
1 Pile Driver Foreman	$28.95	$231.60	$51.05	$408.40	$24.83	$42.95
6 Pile Drivers	26.95	1293.60	47.55	2282.40		
1 Equip. Oper. (crane)	29.35	234.80	48.20	385.60		
5 Laborers	20.55	822.00	34.75	1390.00		
1 Crawler Crane, 40 Ton		1055.00		1160.50		
1 Lead, 60' high		69.60		76.56		
1 Hammer, diesel, 15K Ft.-Lbs.		578.60		636.46		
1 Air Compressor, 600 cfm		434.40		477.84		
2 -50' Air Hoses, 3"		32.40		35.64		
1 Chainsaw, Gas, 36" Long		37.80		41.58	21.23	23.35
104 L.H., Daily Totals		$4789.80		$6894.98	$46.06	$66.30

Crew B-51

	Hr.	Daily	Hr.	Daily	Bare Costs	Incl. O&P
1 Labor Foreman (outside)	$22.55	$180.40	$38.15	$305.20	$21.12	$35.67
4 Laborers	20.55	657.60	34.75	1112.00		
1 Truck Driver (light)	21.95	175.60	36.85	294.80		
1 Flatbed Truck, Gas, 1.5 Ton		194.40		213.84	4.05	4.46
48 L.H., Daily Totals		$1208.00		$1925.84	$25.17	$40.12

Crew B-52

	Hr.	Daily	Hr.	Daily	Bare Costs	Incl. O&P
1 Labor Foreman	$22.55	$180.40	$38.15	$305.20	$23.13	$39.19
1 Carpenter	27.95	223.60	47.25	378.00		
4 Laborers	20.55	657.60	34.75	1112.00		
.5 Rodman (reinf.)	29.85	119.40	53.00	212.00		
.5 Equip. Oper. (med.)	28.55	114.20	46.90	187.60		
.5 Crawler Loader, 3 C.Y.		495.00		544.50	8.84	9.72
56 L.H., Daily Totals		$1790.20		$2739.30	$31.97	$48.92

Crew B-53

	Hr.	Daily	Hr.	Daily	Bare Costs	Incl. O&P
1 Building Laborer	$20.55	$164.40	$34.75	$278.00	$20.55	$34.75
1 Trencher, Chain, 12 H.P.		59.60		65.56	7.45	8.20
8 L.H., Daily Totals		$224.00		$343.56	$28.00	$42.95

Crew B-54

	Hr.	Daily	Hr.	Daily	Bare Costs	Incl. O&P
1 Equip. Oper. (light)	$26.95	$215.60	$44.30	$354.40	$26.95	$44.30
1 Trencher, Chain, 40 H.P.		301.00		331.10	37.63	41.39
8 L.H., Daily Totals		$516.60		$685.50	$64.58	$85.69

Crew B-54A

	Hr.	Daily	Hr.	Daily	Bare Costs	Incl. O&P
.17 Labor Foreman (outside)	$22.55	$30.67	$38.15	$51.88	$27.68	$45.63
1 Equipment Operator (med.)	28.55	228.40	46.90	375.20		
1 Wheel Trencher, 67 H.P.		1009.00		1109.90	107.80	118.58
9.36 L.H., Daily Totals		$1268.07		$1536.98	$135.48	$164.21

Crew B-54B

	Hr.	Daily	Hr.	Daily	Bare Costs	Incl. O&P
.25 Labor Foreman (outside)	$22.55	$45.10	$38.15	$76.30	$27.35	$45.15
1 Equipment Operator (med.)	28.55	228.40	46.90	375.20		
1 Wheel Trencher, 150 H.P.		1753.00		1928.30	175.30	192.83
10 L.H., Daily Totals		$2026.50		$2379.80	$202.65	$237.98

Crew B-55

	Hr.	Daily	Hr.	Daily	Bare Costs	Incl. O&P
1 Laborer	$20.55	$164.40	$34.75	$278.00	$21.25	$35.80
1 Truck Driver (light)	21.95	175.60	36.85	294.80		
1 Truck-mounted earth auger		706.60		777.26		
1 Flatbed Truck, Gas, 3 Ton		242.60		266.86	59.33	65.26
16 L.H., Daily Totals		$1289.20		$1616.92	$80.58	$101.06

Crew B-56

	Hr.	Daily	Hr.	Daily	Bare Costs	Incl. O&P
2 Laborers	$20.55	$328.80	$34.75	$556.00	$20.55	$34.75
1 Air Track Drill, 4"		850.60		935.66		
1 Air Compressor, 600 cfm		434.40		477.84		
1 -50' Air Hose, 3"		16.20		17.82	81.33	89.46
16 L.H., Daily Totals		$1630.00		$1987.32	$101.88	$124.21

Crew B-57

	Hr.	Daily	Hr.	Daily	Bare Costs	Incl. O&P
1 Labor Foreman (outside)	$22.55	$180.40	$38.15	$305.20	$22.71	$38.12
3 Laborers	20.55	493.20	34.75	834.00		
1 Equip. Oper. (crane)	29.35	234.80	48.20	385.60		
1 Barge, 400 Ton		746.40		821.04		
1 Crawler Crane, 25 Ton		1052.00		1157.20		
1 Clamshell Bucket, 1 C.Y.		46.20		50.82		
1 Centr. Water Pump, 6"		309.40		340.34		
1 -20' Suction Hose, 6"		11.90		13.09		
20 -50' Discharge Hoses, 6"		126.00		138.60	57.30	63.03
40 L.H., Daily Totals		$3200.30		$4045.89	$80.01	$101.15

Crew B-58

	Hr.	Daily	Hr.	Daily	Bare Costs	Incl. O&P
2 Laborers	$20.55	$328.80	$34.75	$556.00	$22.68	$37.93
1 Equip. Oper. (light)	26.95	215.60	44.30	354.40		
1 Backhoe Loader, 48 H.P.		293.80		323.18		
1 Small Helicopter, w/pilot		2587.00		2845.70	120.03	132.04
24 L.H., Daily Totals		$3425.20		$4079.28	$142.72	$169.97

Crew B-59

	Hr.	Daily	Hr.	Daily	Bare Costs	Incl. O&P
1 Truck Driver (heavy)	$22.70	$181.60	$38.10	$304.80	$22.70	$38.10
1 Truck Tractor, 220 H.P.		288.40		317.24		
1 Water Tanker, 5000 Gal.		136.00		149.60	53.05	58.35
8 L.H., Daily Totals		$606.00		$771.64	$75.75	$96.45

Crew B-60

	Hr.	Daily	Hr.	Daily	Bare Costs	Incl. O&P
1 Labor Foreman (outside)	$22.55	$180.40	$38.15	$305.20	$23.42	$39.15
3 Laborers	20.55	493.20	34.75	834.00		
1 Equip. Oper. (crane)	29.35	234.80	48.20	385.60		
1 Equip. Oper. (light)	26.95	215.60	44.30	354.40		
1 Crawler Crane, 40 Ton		1055.00		1160.50		
1 Lead, 60' high		69.60		76.56		
1 Hammer, diesel, 15K Ft.-Lbs.		578.60		636.46		
1 Backhoe Loader, 48 H.P.		293.80		323.18	41.60	45.76
48 L.H., Daily Totals		$3121.00		$4075.90	$65.02	$84.91

Crew B-61

	Hr.	Daily	Hr.	Daily	Bare Costs	Incl. O&P
1 Labor Foreman (outside)	$22.55	$180.40	$38.15	$305.20	$20.95	$35.43
4 Laborers	20.55	657.60	34.75	1112.00		
1 Cement Mixer, 2 C.Y.		187.80		206.58		
1 Air Compressor, 160 cfm		138.80		152.68	8.16	8.98
40 L.H., Daily Totals		$1164.60		$1776.46	$29.11	$44.41

Crew B-62

	Hr.	Daily	Hr.	Daily	Bare Costs	Incl. O&P
2 Laborers	$20.55	$328.80	$34.75	$556.00	$22.68	$37.93
1 Equip. Oper. (light)	26.95	215.60	44.30	354.40		
1 Loader, Skid Steer, 30 H.P., gas		149.20		164.12	6.22	6.84
24 L.H., Daily Totals		$693.60		$1074.52	$28.90	$44.77

Crew No.	Bare Costs		Incl. Subs O & P		Cost Per Labor-Hour	

Crew B-63

	Hr.	Daily	Hr.	Daily	Bare Costs	Incl. O&P
5 Laborers	$20.55	$822.00	$34.75	$1390.00	$20.55	$34.75
1 Loader, Skid Steer, 30 H.P., gas		149.20		164.12	3.73	4.10
40 L.H., Daily Totals		$971.20		$1554.12	$24.28	$38.85

Crew B-64

	Hr.	Daily	Hr.	Daily	Bare Costs	Incl. O&P
1 Laborer	$20.55	$164.40	$34.75	$278.00	$21.25	$35.80
1 Truck Driver (light)	21.95	175.60	36.85	294.80		
1 Power Mulcher (small)		136.60		150.26		
1 Flatbed Truck, Gas, 1.5 Ton		194.40		213.84	20.69	22.76
16 L.H., Daily Totals		$671.00		$936.90	$41.94	$58.56

Crew B-65

	Hr.	Daily	Hr.	Daily	Bare Costs	Incl. O&P
1 Laborer	$20.55	$164.40	$34.75	$278.00	$21.25	$35.80
1 Truck Driver (light)	21.95	175.60	36.85	294.80		
1 Power Mulcher (large)		271.80		298.98		
1 Flatbed Truck, Gas, 1.5 Ton		194.40		213.84	29.14	32.05
16 L.H., Daily Totals		$806.20		$1085.62	$50.39	$67.85

Crew B-66

	Hr.	Daily	Hr.	Daily	Bare Costs	Incl. O&P
1 Equip. Oper. (light)	$26.95	$215.60	$44.30	$354.40	$26.95	$44.30
1 Loader-Backhoe		231.20		254.32	28.90	31.79
8 L.H., Daily Totals		$446.80		$608.72	$55.85	$76.09

Crew B-67

	Hr.	Daily	Hr.	Daily	Bare Costs	Incl. O&P
1 Millwright	$29.10	$232.80	$46.80	$374.40	$28.02	$45.55
1 Equip. Oper. (light)	26.95	215.60	44.30	354.40		
1 Forklift, R/T, 4,000 Lb.		281.20		309.32	17.57	19.33
16 L.H., Daily Totals		$729.60		$1038.12	$45.60	$64.88

Crew B-68

	Hr.	Daily	Hr.	Daily	Bare Costs	Incl. O&P
2 Millwrights	$29.10	$465.60	$46.80	$748.80	$28.38	$45.97
1 Equip. Oper. (light)	26.95	215.60	44.30	354.40		
1 Forklift, R/T, 4,000 Lb.		281.20		309.32	11.72	12.89
24 L.H., Daily Totals		$962.40		$1412.52	$40.10	$58.85

Crew B-69

	Hr.	Daily	Hr.	Daily	Bare Costs	Incl. O&P
1 Labor Foreman (outside)	$22.55	$180.40	$38.15	$305.20	$23.16	$38.73
3 Laborers	20.55	493.20	34.75	834.00		
1 Equip Oper. (crane)	29.35	234.80	48.20	385.60		
1 Equip Oper. Oiler	25.40	203.20	41.75	334.00		
1 Hyd. Crane, 80 Ton		1296.00		1425.60	27.00	29.70
48 L.H., Daily Totals		$2407.60		$3284.40	$50.16	$68.42

Crew B-69A

	Hr.	Daily	Hr.	Daily	Bare Costs	Incl. O&P
1 Labor Foreman	$22.55	$180.40	$38.15	$305.20	$23.26	$38.72
3 Laborers	20.55	493.20	34.75	834.00		
1 Equip. Oper. (medium)	28.55	228.40	46.90	375.20		
1 Concrete Finisher	26.80	214.40	43.00	344.00		
1 Curb/Gutter Paver, 2-Track		754.40		829.84	15.72	17.29
48 L.H., Daily Totals		$1870.80		$2688.24	$38.98	$56.01

Crew B-69B

	Hr.	Daily	Hr.	Daily	Bare Costs	Incl. O&P
1 Labor Foreman	$22.55	$180.40	$38.15	$305.20	$23.26	$38.72
3 Laborers	20.55	493.20	34.75	834.00		
1 Equip. Oper. (medium)	28.55	228.40	46.90	375.20		
1 Cement Finisher	26.80	214.40	43.00	344.00		
1 Curb/Gutter Paver, 4-Track		855.80		941.38	17.83	19.61
48 L.H., Daily Totals		$1972.20		$2799.78	$41.09	$58.33

Crew B-70

	Hr.	Daily	Hr.	Daily	Bare Costs	Incl. O&P
1 Labor Foreman (outside)	$22.55	$180.40	$38.15	$305.20	$24.26	$40.44
3 Laborers	20.55	493.20	34.75	834.00		
3 Equip. Oper. (med.)	28.55	685.20	46.90	1125.60		
1 Grader, 30,000 Lbs.		550.20		605.22		
1 Ripper, beam & 1 shank		79.60		87.56		
1 Road Sweeper, S.P., 8' wide		601.60		661.76		
1 F.E. Loader, W.M., 1.5 C.Y.		328.20		361.02	27.85	30.64
56 L.H., Daily Totals		$2918.40		$3980.36	$52.11	$71.08

Crew B-71

	Hr.	Daily	Hr.	Daily	Bare Costs	Incl. O&P
1 Labor Foreman (outside)	$22.55	$180.40	$38.15	$305.20	$24.26	$40.44
3 Laborers	20.55	493.20	34.75	834.00		
3 Equip. Oper. (med.)	28.55	685.20	46.90	1125.60		
1 Pvmt. Profiler, 750 H.P.		5503.00		6053.30		
1 Road Sweeper, S.P., 8' wide		601.60		661.76		
1 F.E. Loader, W.M., 1.5 C.Y.		328.20		361.02	114.87	126.36
56 L.H., Daily Totals		$7791.60		$9340.88	$139.14	$166.80

Crew B-72

	Hr.	Daily	Hr.	Daily	Bare Costs	Incl. O&P
1 Labor Foreman (outside)	$22.55	$180.40	$38.15	$305.20	$24.80	$41.25
3 Laborers	20.55	493.20	34.75	834.00		
4 Equip. Oper. (med.)	28.55	913.60	46.90	1500.80		
1 Pvmt. Profiler, 750 H.P.		5503.00		6053.30		
1 Hammermill, 250 H.P.		1850.00		2035.00		
1 Windrow Loader		1197.00		1316.70		
1 Mix Paver 165 H.P.		1996.00		2195.60		
1 Roller, Pneum. Whl, 12 Ton		307.80		338.58	169.59	186.55
64 L.H., Daily Totals		$12441.00		$14579.18	$194.39	$227.80

Crew B-73

	Hr.	Daily	Hr.	Daily	Bare Costs	Incl. O&P
1 Labor Foreman (outside)	$22.55	$180.40	$38.15	$305.20	$25.80	$42.77
2 Laborers	20.55	328.80	34.75	556.00		
5 Equip. Oper. (med.)	28.55	1142.00	46.90	1876.00		
1 Road Mixer, 310 H.P.		1896.00		2085.60		
1 Tandem Roller, 10 Ton		233.00		256.30		
1 Hammermill, 250 H.P.		1850.00		2035.00		
1 Grader, 30,000 Lbs.		550.20		605.22		
.5 F.E. Loader, W.M., 1.5 C.Y.		164.10		180.51		
.5 Truck Tractor, 220 H.P.		144.20		158.62		
.5 Water Tanker, 5000 Gal.		68.00		74.80	76.65	84.31
64 L.H., Daily Totals		$6556.70		$8133.25	$102.45	$127.08

Crew B-74

	Hr.	Daily	Hr.	Daily	Bare Costs	Incl. O&P
1 Labor Foreman (outside)	$22.55	$180.40	$38.15	$305.20	$25.34	$42.09
1 Laborer	20.55	164.40	34.75	278.00		
4 Equip. Oper. (med.)	28.55	913.60	46.90	1500.80		
2 Truck Drivers (heavy)	22.70	363.20	38.10	609.60		
1 Grader, 30,000 Lbs.		550.20		605.22		
1 Ripper, beam & 1 shank		79.60		87.56		
2 Stabilizers, 310 H.P.		2688.00		2956.80		
1 Flatbed Truck, Gas, 3 Ton		242.60		266.86		
1 Chem. Spreader, Towed		48.00		52.80		
1 Roller, Vibratory, 25 Ton		594.60		654.06		
1 Water Tanker, 5000 Gal.		136.00		149.60		
1 Truck Tractor, 220 H.P.		288.40		317.24	72.30	79.53
64 L.H., Daily Totals		$6249.00		$7783.74	$97.64	$121.62

Crew No.	Bare Costs		Incl. Subs O & P		Cost Per Labor-Hour	
Crew B-75	Hr.	Daily	Hr.	Daily	Bare Costs	Incl. O&P
1 Labor Foreman (outside)	$22.55	$180.40	$38.15	$305.20	$25.71	$42.66
1 Laborer	20.55	164.40	34.75	278.00		
4 Equip. Oper. (med.)	28.55	913.60	46.90	1500.80		
1 Truck Driver (heavy)	22.70	181.60	38.10	304.80		
1 Grader, 30,000 Lbs.		550.20		605.22		
1 Ripper, beam & 1 shank		79.60		87.56		
2 Stabilizers, 310 H.P.		2688.00		2956.80		
1 Dist. Tanker, 3000 Gallon		304.20		334.62		
1 Truck Tractor, 6x4, 380 H.P.		478.80		526.68		
1 Roller, Vibratory, 25 Ton		594.60		654.06	83.85	92.23
56 L.H., Daily Totals		$6135.40		$7553.74	$109.56	$134.89
Crew B-76	Hr.	Daily	Hr.	Daily	Bare Costs	Incl. O&P
1 Dock Builder Foreman	$28.95	$231.60	$51.05	$408.40	$27.53	$47.44
5 Dock Builders	26.95	1078.00	47.55	1902.00		
2 Equip. Oper. (crane)	29.35	469.60	48.20	771.20		
1 Equip. Oper. Oiler	25.40	203.20	41.75	334.00		
1 Crawler Crane, 50 Ton		1110.00		1221.00		
1 Barge, 400 Ton		746.40		821.04		
1 Hammer, diesel, 15K Ft.-Lbs.		578.60		636.46		
1 Lead, 60' high		69.60		76.56		
1 Air Compressor, 600 cfm		434.40		477.84		
2 -50' Air Hoses, 3"		32.40		35.64	41.27	45.40
72 L.H., Daily Totals		$4953.80		$6684.14	$68.80	$92.84
Crew B-76A	Hr.	Daily	Hr.	Daily	Bare Costs	Incl. O&P
1 Laborer Foreman	$22.55	$180.40	$38.15	$305.20	$22.51	$37.73
5 Laborers	20.55	822.00	34.75	1390.00		
1 Equip. Oper. (crane)	29.35	234.80	48.20	385.60		
1 Equip. Oper. Oiler	25.40	203.20	41.75	334.00		
1 Crawler Crane, 50 Ton		1110.00		1221.00		
1 Barge, 400 Ton		746.40		821.04	29.01	31.91
64 L.H., Daily Totals		$3296.80		$4456.84	$51.51	$69.64
Crew B-77	Hr.	Daily	Hr.	Daily	Bare Costs	Incl. O&P
1 Labor Foreman	$22.55	$180.40	$38.15	$305.20	$21.23	$35.85
3 Laborers	20.55	493.20	34.75	834.00		
1 Truck Driver (light)	21.95	175.60	36.85	294.80		
1 Crack Cleaner, 25 H.P.		54.80		60.28		
1 Crack Filler, Trailer Mtd.		194.20		213.62		
1 Flatbed Truck, Gas, 3 Ton		242.60		266.86	12.29	13.52
40 L.H., Daily Totals		$1340.80		$1974.76	$33.52	$49.37
Crew B-78	Hr.	Daily	Hr.	Daily	Bare Costs	Incl. O&P
1 Labor Foreman	$22.55	$180.40	$38.15	$305.20	$20.95	$35.43
4 Laborers	20.55	657.60	34.75	1112.00		
1 Paint Striper, S.P.		147.00		161.70		
1 Flatbed Truck, Gas, 3 Ton		242.60		266.86		
1 Pickup Truck, 3/4 Ton		116.80		128.48	12.66	13.93
40 L.H., Daily Totals		$1344.40		$1974.24	$33.61	$49.36
Crew B-78B	Hr.	Daily	Hr.	Daily	Bare Costs	Incl. O&P
2 Laborers	$20.55	$328.80	$34.75	$556.00	$21.26	$35.81
.25 Equip. Oper. (light)	26.95	53.90	44.30	88.60		
1 Pickup Truck, 3/4 Ton		116.80		128.48		
1 Line Rem., 11 H.P., walk behind		54.00		59.40		
.25 Road Sweeper, S.P., 8' wide		150.40		165.44	17.84	19.63
18 L.H., Daily Totals		$703.90		$997.92	$39.11	$55.44

Crew No.	Bare Costs		Incl. Subs O & P		Cost Per Labor-Hour	
Crew B-79	Hr.	Daily	Hr.	Daily	Bare Costs	Incl. O&P
1 Labor Foreman	$22.55	$180.40	$38.15	$305.20	$21.05	$35.60
3 Laborers	20.55	493.20	34.75	834.00		
1 Thermo. Striper, T.M.		703.20		773.52		
1 Flatbed Truck, Gas, 3 Ton		242.60		266.86		
2 Pickup Trucks, 3/4 Ton		233.60		256.96	36.86	40.54
32 L.H., Daily Totals		$1853.00		$2436.54	$57.91	$76.14
Crew B-80	Hr.	Daily	Hr.	Daily	Bare Costs	Incl. O&P
1 Labor Foreman	$22.55	$180.40	$38.15	$305.20	$21.22	$35.88
2 Laborers	20.55	328.80	34.75	556.00		
1 Flatbed Truck, Gas, 3 Ton		242.60		266.86		
1 Earth Auger, Truck-Mtd.		412.00		453.20	27.27	30.00
24 L.H., Daily Totals		$1163.80		$1581.26	$48.49	$65.89
Crew B-80A	Hr.	Daily	Hr.	Daily	Bare Costs	Incl. O&P
3 Laborers	$20.55	$493.20	$34.75	$834.00	$20.55	$34.75
1 Flatbed Truck, Gas, 3 Ton		242.60		266.86	10.11	11.12
24 L.H., Daily Totals		$735.80		$1100.86	$30.66	$45.87
Crew B-80B	Hr.	Daily	Hr.	Daily	Bare Costs	Incl. O&P
3 Laborers	$20.55	$493.20	$34.75	$834.00	$22.15	$37.14
1 Equip. Oper. (light)	26.95	215.60	44.30	354.40		
1 Crane, Flatbed Mounted, 3 Ton		225.00		247.50	7.03	7.73
32 L.H., Daily Totals		$933.80		$1435.90	$29.18	$44.87
Crew B-80C	Hr.	Daily	Hr.	Daily	Bare Costs	Incl. O&P
2 Laborers	$20.55	$328.80	$34.75	$556.00	$21.02	$35.45
1 Truck Driver (light)	21.95	175.60	36.85	294.80		
1 Flatbed Truck, Gas, 1.5 Ton		194.40		213.84		
1 Manual fence post auger, gas		7.40		8.14	8.41	9.25
24 L.H., Daily Totals		$706.20		$1072.78	$29.43	$44.70
Crew B-81	Hr.	Daily	Hr.	Daily	Bare Costs	Incl. O&P
1 Laborer	$20.55	$164.40	$34.75	$278.00	$21.63	$36.42
1 Truck Driver (heavy)	22.70	181.60	38.10	304.80		
1 Hydromulcher, T.M.		291.00		320.10		
1 Truck Tractor, 220 H.P.		288.40		317.24	36.21	39.83
16 L.H., Daily Totals		$925.40		$1220.14	$57.84	$76.26
Crew B-82	Hr.	Daily	Hr.	Daily	Bare Costs	Incl. O&P
1 Laborer	$20.55	$164.40	$34.75	$278.00	$23.75	$39.52
1 Equip. Oper. (light)	26.95	215.60	44.30	354.40		
1 Horiz. Borer, 6 H.P.		77.60		85.36	4.85	5.34
16 L.H., Daily Totals		$457.60		$717.76	$28.60	$44.86
Crew B-82A	Hr.	Daily	Hr.	Daily	Bare Costs	Incl. O&P
1 Laborer	$20.55	$164.40	$34.75	$278.00	$23.75	$39.52
1 Equip. Oper. (light)	26.95	215.60	44.30	354.40		
1 Flatbed Truck, Gas, 3 Ton		242.60		266.86		
1 Flatbed Trailer, 25 Ton		102.00		112.20		
1 Horiz. Dir. Drill, 20k lb. thrust		628.20		691.02	60.80	66.88
16 L.H., Daily Totals		$1352.80		$1702.48	$84.55	$106.41
Crew B-82B	Hr.	Daily	Hr.	Daily	Bare Costs	Incl. O&P
2 Laborers	$20.55	$328.80	$34.75	$556.00	$22.68	$37.93
1 Equip. Oper. (light)	26.95	215.60	44.30	354.40		
1 Flatbed Truck, Gas, 3 Ton		242.60		266.86		
1 Flatbed Trailer, 25 Ton		102.00		112.20		
1 Horiz. Dir. Drill, 30k lb. thrust		896.60		986.26	51.72	56.89
24 L.H., Daily Totals		$1785.60		$2275.72	$74.40	$94.82

Crew B-82C

Crew No.	Bare Costs Hr.	Bare Costs Daily	Incl. Subs O&P Hr.	Incl. Subs O&P Daily	Cost Per Labor-Hour Bare Costs	Cost Per Labor-Hour Incl. O&P
2 Laborers	$20.55	$328.80	$34.75	$556.00	$22.68	$37.93
1 Equip. Oper. (light)	26.95	215.60	44.30	354.40		
1 Flatbed Truck, Gas, 3 Ton		242.60		266.86		
1 Flatbed Trailer, 25 Ton		102.00		112.20		
1 Horiz. Dir. Drill, 50k lb. thrust		1183.00		1301.30	63.65	70.02
24 L.H., Daily Totals		$2072.00		$2590.76	$86.33	$107.95

Crew B-82D

Crew No.	Bare Costs Hr.	Bare Costs Daily	Incl. Subs O&P Hr.	Incl. Subs O&P Daily	Cost Per Labor-Hour Bare Costs	Cost Per Labor-Hour Incl. O&P
1 Equip. Oper. (light)	$26.95	$215.60	$44.30	$354.40	$26.95	$44.30
1 Mud Trailer for HDD, 1500 gallon		283.20		311.52	35.40	38.94
8 L.H., Daily Totals		$498.80		$665.92	$62.35	$83.24

Crew B-83

Crew No.	Bare Costs Hr.	Bare Costs Daily	Incl. Subs O&P Hr.	Incl. Subs O&P Daily	Cost Per Labor-Hour Bare Costs	Cost Per Labor-Hour Incl. O&P
1 Tugboat Captain	$28.55	$228.40	$46.90	$375.20	$24.55	$40.83
1 Tugboat Hand	20.55	164.40	34.75	278.00		
1 Tugboat, 250 H.P.		663.40		729.74	41.46	45.61
16 L.H., Daily Totals		$1056.20		$1382.94	$66.01	$86.43

Crew B-84

Crew No.	Bare Costs Hr.	Bare Costs Daily	Incl. Subs O&P Hr.	Incl. Subs O&P Daily	Cost Per Labor-Hour Bare Costs	Cost Per Labor-Hour Incl. O&P
1 Equip. Oper. (med.)	$28.55	$228.40	$46.90	$375.20	$28.55	$46.90
1 Rotary Mower/Tractor		273.40		300.74	34.17	37.59
8 L.H., Daily Totals		$501.80		$675.94	$62.73	$84.49

Crew B-85

Crew No.	Bare Costs Hr.	Bare Costs Daily	Incl. Subs O&P Hr.	Incl. Subs O&P Daily	Cost Per Labor-Hour Bare Costs	Cost Per Labor-Hour Incl. O&P
3 Laborers	$20.55	$493.20	$34.75	$834.00	$22.58	$37.85
1 Equip. Oper. (med.)	28.55	228.40	46.90	375.20		
1 Truck Driver (heavy)	22.70	181.60	38.10	304.80		
1 Aerial Lift Truck, 80'		640.40		704.44		
1 Brush Chipper, 12", 130 H.P.		225.60		248.16		
1 Pruning Saw, Rotary		6.50		7.15	21.81	23.99
40 L.H., Daily Totals		$1775.70		$2473.75	$44.39	$61.84

Crew B-86

Crew No.	Bare Costs Hr.	Bare Costs Daily	Incl. Subs O&P Hr.	Incl. Subs O&P Daily	Cost Per Labor-Hour Bare Costs	Cost Per Labor-Hour Incl. O&P
1 Equip. Oper. (med.)	$28.55	$228.40	$46.90	$375.20	$28.55	$46.90
1 Stump Chipper, S.P.		201.10		221.21	25.14	27.65
8 L.H., Daily Totals		$429.50		$596.41	$53.69	$74.55

Crew B-86A

Crew No.	Bare Costs Hr.	Bare Costs Daily	Incl. Subs O&P Hr.	Incl. Subs O&P Daily	Cost Per Labor-Hour Bare Costs	Cost Per Labor-Hour Incl. O&P
1 Equip. Oper. (medium)	$28.55	$228.40	$46.90	$375.20	$28.55	$46.90
1 Grader, 30,000 Lbs.		550.20		605.22	68.78	75.65
8 L.H., Daily Totals		$778.60		$980.42	$97.33	$122.55

Crew B-86B

Crew No.	Bare Costs Hr.	Bare Costs Daily	Incl. Subs O&P Hr.	Incl. Subs O&P Daily	Cost Per Labor-Hour Bare Costs	Cost Per Labor-Hour Incl. O&P
1 Equip. Oper. (medium)	$28.55	$228.40	$46.90	$375.20	$28.55	$46.90
1 Dozer, 200 H.P.		1082.00		1190.20	135.25	148.78
8 L.H., Daily Totals		$1310.40		$1565.40	$163.80	$195.68

Crew B-87

Crew No.	Bare Costs Hr.	Bare Costs Daily	Incl. Subs O&P Hr.	Incl. Subs O&P Daily	Cost Per Labor-Hour Bare Costs	Cost Per Labor-Hour Incl. O&P
1 Laborer	$20.55	$164.40	$34.75	$278.00	$26.95	$44.47
4 Equip. Oper. (med.)	28.55	913.60	46.90	1500.80		
2 Feller Bunchers, 100 H.P.		1122.80		1235.08		
1 Log Chipper, 22" Tree		588.40		647.24		
1 Dozer, 105 H.P.		600.60		660.66		
1 Chainsaw, Gas, 36" Long		37.80		41.58	58.74	64.61
40 L.H., Daily Totals		$3427.60		$4363.36	$85.69	$109.08

Crew B-88

Crew No.	Bare Costs Hr.	Bare Costs Daily	Incl. Subs O&P Hr.	Incl. Subs O&P Daily	Cost Per Labor-Hour Bare Costs	Cost Per Labor-Hour Incl. O&P
1 Laborer	$20.55	$164.40	$34.75	$278.00	$27.41	$45.16
6 Equip. Oper. (med.)	28.55	1370.40	46.90	2251.20		
2 Feller Bunchers, 100 H.P.		1122.80		1235.08		
1 Log Chipper, 22" Tree		588.40		647.24		
2 Log Skidders, 50 H.P.		1795.20		1974.72		
1 Dozer, 105 H.P.		600.60		660.66		
1 Chainsaw, Gas, 36" Long		37.80		41.58	74.01	81.42
56 L.H., Daily Totals		$5679.60		$7088.48	$101.42	$126.58

Crew B-89

Crew No.	Bare Costs Hr.	Bare Costs Daily	Incl. Subs O&P Hr.	Incl. Subs O&P Daily	Cost Per Labor-Hour Bare Costs	Cost Per Labor-Hour Incl. O&P
1 Skilled Worker	$28.05	$224.40	$47.35	$378.80	$24.30	$41.05
1 Building Laborer	20.55	164.40	34.75	278.00		
1 Flatbed Truck, Gas, 3 Ton		242.60		266.86		
1 Concrete Saw		141.00		155.10		
1 Water Tank, 65 Gal.		18.45		20.30	25.13	27.64
16 L.H., Daily Totals		$790.85		$1099.06	$49.43	$68.69

Crew B-89A

Crew No.	Bare Costs Hr.	Bare Costs Daily	Incl. Subs O&P Hr.	Incl. Subs O&P Daily	Cost Per Labor-Hour Bare Costs	Cost Per Labor-Hour Incl. O&P
1 Skilled Worker	$28.05	$224.40	$47.35	$378.80	$24.30	$41.05
1 Laborer	20.55	164.40	34.75	278.00		
1 Core Drill (large)		110.20		121.22	6.89	7.58
16 L.H., Daily Totals		$499.00		$778.02	$31.19	$48.63

Crew B-89B

Crew No.	Bare Costs Hr.	Bare Costs Daily	Incl. Subs O&P Hr.	Incl. Subs O&P Daily	Cost Per Labor-Hour Bare Costs	Cost Per Labor-Hour Incl. O&P
1 Equip. Oper. (light)	$26.95	$215.60	$44.30	$354.40	$24.45	$40.58
1 Truck Driver, Light	21.95	175.60	36.85	294.80		
1 Wall Saw, Hydraulic, 10 H.P.		96.00		105.60		
1 Generator, Diesel, 100 kW		327.20		359.92		
1 Water Tank, 65 Gal.		18.45		20.30		
1 Flatbed Truck, Gas, 3 Ton		242.60		266.86	42.77	47.04
16 L.H., Daily Totals		$1075.45		$1401.88	$67.22	$87.62

Crew B-90

Crew No.	Bare Costs Hr.	Bare Costs Daily	Incl. Subs O&P Hr.	Incl. Subs O&P Daily	Cost Per Labor-Hour Bare Costs	Cost Per Labor-Hour Incl. O&P
1 Labor Foreman (outside)	$22.55	$180.40	$38.15	$305.20	$22.94	$38.40
3 Laborers	20.55	493.20	34.75	834.00		
2 Equip. Oper. (light)	26.95	431.20	44.30	708.80		
2 Truck Drivers (heavy)	22.70	363.20	38.10	609.60		
1 Road Mixer, 310 H.P.		1896.00		2085.60		
1 Dist. Truck, 2000 Gal.		277.20		304.92	33.96	37.35
64 L.H., Daily Totals		$3641.20		$4848.12	$56.89	$75.75

Crew B-90A

Crew No.	Bare Costs Hr.	Bare Costs Daily	Incl. Subs O&P Hr.	Incl. Subs O&P Daily	Cost Per Labor-Hour Bare Costs	Cost Per Labor-Hour Incl. O&P
1 Labor Foreman	$22.55	$180.40	$38.15	$305.20	$25.41	$42.18
2 Laborers	20.55	328.80	34.75	556.00		
4 Equip. Oper. (medium)	28.55	913.60	46.90	1500.80		
2 Graders, 30,000 Lbs.		1100.40		1210.44		
1 Tandem Roller, 10 Ton		233.00		256.30		
1 Roller, Pneum. Whl, 12 Ton		307.80		338.58	29.31	32.24
56 L.H., Daily Totals		$3064.00		$4167.32	$54.71	$74.42

Crew B-90B

Crew No.	Bare Costs Hr.	Bare Costs Daily	Incl. Subs O&P Hr.	Incl. Subs O&P Daily	Cost Per Labor-Hour Bare Costs	Cost Per Labor-Hour Incl. O&P
1 Labor Foreman	$22.55	$180.40	$38.15	$305.20	$24.88	$41.39
2 Laborers	20.55	328.80	34.75	556.00		
3 Equip. Oper. (medium)	28.55	685.20	46.90	1125.60		
1 Tandem Roller, 10 Ton		233.00		256.30		
1 Roller, Pneum. Whl, 12 Ton		307.80		338.58		
1 Road Mixer, 310 H.P.		1896.00		2085.60	50.77	55.84
48 L.H., Daily Totals		$3631.20		$4667.28	$75.65	$97.23

Crews

Crew B-91

Crew B-91	Hr.	Daily	Hr.	Daily	Bare Costs	Incl. O&P
1 Labor Foreman (outside)	$22.55	$180.40	$38.15	$305.20	$25.07	$41.67
2 Laborers	20.55	328.80	34.75	556.00		
4 Equip. Oper. (med.)	28.55	913.60	46.90	1500.80		
1 Truck Driver (heavy)	22.70	181.60	38.10	304.80		
1 Dist. Tanker, 3000 Gallon		304.20		334.62		
1 Truck Tractor, 6x4, 380 H.P.		478.80		526.68		
1 Aggreg. Spreader, S.P.		855.20		940.72		
1 Roller, Pneum. Whl, 12 Ton		307.80		338.58		
1 Tandem Roller, 10 Ton		233.00		256.30	34.05	37.45
64 L.H., Daily Totals		$3783.40		$5063.70	$59.12	$79.12

Crew B-92

Crew B-92	Hr.	Daily	Hr.	Daily	Bare Costs	Incl. O&P
1 Labor Foreman (outside)	$22.55	$180.40	$38.15	$305.20	$21.05	$35.60
3 Laborers	20.55	493.20	34.75	834.00		
1 Crack Cleaner, 25 H.P.		54.80		60.28		
1 Air Compressor, 60 cfm		122.60		134.86		
1 Tar Kettle, T.M.		83.15		91.47		
1 Flatbed Truck, Gas, 3 Ton		242.60		266.86	15.72	17.30
32 L.H., Daily Totals		$1176.75		$1692.67	$36.77	$52.90

Crew B-93

Crew B-93	Hr.	Daily	Hr.	Daily	Bare Costs	Incl. O&P
1 Equip. Oper. (med.)	$28.55	$228.40	$46.90	$375.20	$28.55	$46.90
1 Feller Buncher, 100 H.P.		561.40		617.54	70.17	77.19
8 L.H., Daily Totals		$789.80		$992.74	$98.72	$124.09

Crew B-94A

Crew B-94A	Hr.	Daily	Hr.	Daily	Bare Costs	Incl. O&P
1 Laborer	$20.55	$164.40	$34.75	$278.00	$20.55	$34.75
1 Diaphragm Water Pump, 2"		63.20		69.52		
1 -20' Suction Hose, 2"		1.95		2.15		
2 -50' Discharge Hoses, 2"		2.20		2.42	8.42	9.26
8 L.H., Daily Totals		$231.75		$352.08	$28.97	$44.01

Crew B-94B

Crew B-94B	Hr.	Daily	Hr.	Daily	Bare Costs	Incl. O&P
1 Laborer	$20.55	$164.40	$34.75	$278.00	$20.55	$34.75
1 Diaphragm Water Pump, 4"		86.20		94.82		
1 -20' Suction Hose, 4"		3.45		3.79		
2 -50' Discharge Hoses, 4"		4.70		5.17	11.79	12.97
8 L.H., Daily Totals		$258.75		$381.79	$32.34	$47.72

Crew B-94C

Crew B-94C	Hr.	Daily	Hr.	Daily	Bare Costs	Incl. O&P
1 Laborer	$20.55	$164.40	$34.75	$278.00	$20.55	$34.75
1 Centrifugal Water Pump, 3"		69.40		76.34		
1 -20' Suction Hose, 3"		3.05		3.36		
2 -50' Discharge Hoses, 3"		3.50		3.85	9.49	10.44
8 L.H., Daily Totals		$240.35		$361.55	$30.04	$45.19

Crew B-94D

Crew B-94D	Hr.	Daily	Hr.	Daily	Bare Costs	Incl. O&P
1 Laborer	$20.55	$164.40	$34.75	$278.00	$20.55	$34.75
1 Centr. Water Pump, 6"		309.40		340.34		
1 -20' Suction Hose, 6"		11.90		13.09		
2 -50' Discharge Hoses, 6"		12.60		13.86	41.74	45.91
8 L.H., Daily Totals		$498.30		$645.29	$62.29	$80.66

Crew C-1

Crew C-1	Hr.	Daily	Hr.	Daily	Bare Costs	Incl. O&P
2 Carpenters	$27.95	$447.20	$47.25	$756.00	$24.32	$41.10
1 Carpenter Helper	20.85	166.80	35.15	281.20		
1 Laborer	20.55	164.40	34.75	278.00		
32 L.H., Daily Totals		$778.40		$1315.20	$24.32	$41.10

Crew C-2

Crew C-2	Hr.	Daily	Hr.	Daily	Bare Costs	Incl. O&P
1 Carpenter Foreman (out)	$29.95	$239.60	$50.65	$405.20	$24.68	$41.70
2 Carpenters	27.95	447.20	47.25	756.00		
2 Carpenter Helpers	20.85	333.60	35.15	562.40		
1 Laborer	20.55	164.40	34.75	278.00		
48 L.H., Daily Totals		$1184.80		$2001.60	$24.68	$41.70

Crew C-2A

Crew C-2A	Hr.	Daily	Hr.	Daily	Bare Costs	Incl. O&P
1 Carpenter Foreman (out)	$29.95	$239.60	$50.65	$405.20	$26.86	$45.02
3 Carpenters	27.95	670.80	47.25	1134.00		
1 Cement Finisher	26.80	214.40	43.00	344.00		
1 Laborer	20.55	164.40	34.75	278.00		
48 L.H., Daily Totals		$1289.20		$2161.20	$26.86	$45.02

Crew C-3

Crew C-3	Hr.	Daily	Hr.	Daily	Bare Costs	Incl. O&P
1 Rodman Foreman	$31.85	$254.80	$56.55	$452.40	$26.25	$45.51
3 Rodmen (reinf.)	29.85	716.40	53.00	1272.00		
1 Equip. Oper. (light)	26.95	215.60	44.30	354.40		
3 Laborers	20.55	493.20	34.75	834.00		
3 Stressing Equipment		27.60		30.36		
.5 Grouting Equipment		75.80		83.38	1.62	1.78
64 L.H., Daily Totals		$1783.40		$3026.54	$27.87	$47.29

Crew C-4

Crew C-4	Hr.	Daily	Hr.	Daily	Bare Costs	Incl. O&P
1 Rodman Foreman	$31.85	$254.80	$56.55	$452.40	$28.02	$49.33
2 Rodmen (reinf.)	29.85	477.60	53.00	848.00		
1 Building Laborer	20.55	164.40	34.75	278.00		
3 Stressing Equipment		27.60		30.36	0.86	0.95
32 L.H., Daily Totals		$924.40		$1608.76	$28.89	$50.27

Crew C-5

Crew C-5	Hr.	Daily	Hr.	Daily	Bare Costs	Incl. O&P
1 Rodman Foreman	$31.85	$254.80	$56.55	$452.40	$27.00	$46.71
2 Rodmen (reinf.)	29.85	477.60	53.00	848.00		
1 Equip. Oper. (crane)	29.35	234.80	48.20	385.60		
2 Building Laborers	20.55	328.80	34.75	556.00		
1 Hyd. Crane, 25 Ton		789.40		868.34	16.45	18.09
48 L.H., Daily Totals		$2085.40		$3110.34	$43.45	$64.80

Crew C-6

Crew C-6	Hr.	Daily	Hr.	Daily	Bare Costs	Incl. O&P
1 Labor Foreman (outside)	$22.55	$180.40	$38.15	$305.20	$21.93	$36.69
4 Laborers	20.55	657.60	34.75	1112.00		
1 Cement Finisher	26.80	214.40	43.00	344.00		
2 Gas Engine Vibrators		52.00		57.20	1.08	1.19
48 L.H., Daily Totals		$1104.40		$1818.40	$23.01	$37.88

Crew C-7

Crew C-7	Hr.	Daily	Hr.	Daily	Bare Costs	Incl. O&P
1 Labor Foreman (outside)	$22.55	$180.40	$38.15	$305.20	$22.89	$38.17
5 Laborers	20.55	822.00	34.75	1390.00		
1 Cement Finisher	26.80	214.40	43.00	344.00		
1 Equip. Oper. (med.)	28.55	228.40	46.90	375.20		
1 Equip. Oper. (oiler)	25.40	203.20	41.75	334.00		
2 Gas Engine Vibrators		52.00		57.20		
1 Concrete Bucket, 1 C.Y.		20.20		22.22		
1 Hyd. Crane, 55 Ton		1125.00		1237.50	16.63	18.29
72 L.H., Daily Totals		$2845.60		$4065.32	$39.52	$56.46

Crew No.	Bare Costs		Incl. Subs O & P		Cost Per Labor-Hour	
Crew C-8	Hr.	Daily	Hr.	Daily	Bare Costs	Incl. O&P
1 Labor Foreman (outside)	$22.55	$180.40	$38.15	$305.20	$23.76	$39.33
3 Laborers	20.55	493.20	34.75	834.00		
2 Cement Finishers	26.80	428.80	43.00	688.00		
1 Equip. Oper. (med.)	28.55	228.40	46.90	375.20		
1 Concrete Pump (small)		737.80		811.58	13.18	14.49
56 L.H., Daily Totals		$2068.60		$3013.98	$36.94	$53.82
Crew C-8A	Hr.	Daily	Hr.	Daily	Bare Costs	Incl. O&P
1 Labor Foreman (outside)	$22.55	$180.40	$38.15	$305.20	$22.97	$38.07
3 Laborers	20.55	493.20	34.75	834.00		
2 Cement Finishers	26.80	428.80	43.00	688.00		
48 L.H., Daily Totals		$1102.40		$1827.20	$22.97	$38.07
Crew C-8B	Hr.	Daily	Hr.	Daily	Bare Costs	Incl. O&P
1 Labor Foreman (outside)	$22.55	$180.40	$38.15	$305.20	$22.55	$37.86
3 Laborers	20.55	493.20	34.75	834.00		
1 Equip. Oper. (med.)	28.55	228.40	46.90	375.20		
1 Vibrating Power Screed		70.35		77.39		
1 Roller, Vibratory, 25 Ton		594.60		654.06		
1 Dozer, 200 H.P.		1082.00		1190.20	43.67	48.04
40 L.H., Daily Totals		$2648.95		$3436.05	$66.22	$85.90
Crew C-8C	Hr.	Daily	Hr.	Daily	Bare Costs	Incl. O&P
1 Labor Foreman (outside)	$22.55	$180.40	$38.15	$305.20	$23.26	$38.72
3 Laborers	20.55	493.20	34.75	834.00		
1 Cement Finisher	26.80	214.40	43.00	344.00		
1 Equip. Oper. (med.)	28.55	228.40	46.90	375.20		
1 Shotcrete Rig, 12 C.Y./hr		237.80		261.58	4.95	5.45
48 L.H., Daily Totals		$1354.20		$2119.98	$28.21	$44.17
Crew C-8D	Hr.	Daily	Hr.	Daily	Bare Costs	Incl. O&P
1 Labor Foreman (outside)	$22.55	$180.40	$38.15	$305.20	$24.21	$40.05
1 Laborer	20.55	164.40	34.75	278.00		
1 Cement Finisher	26.80	214.40	43.00	344.00		
1 Equipment Oper. (light)	26.95	215.60	44.30	354.40		
1 Air Compressor, 250 cfm		162.40		178.64		
2 -50' Air Hoses, 1"		8.20		9.02	5.33	5.86
32 L.H., Daily Totals		$945.40		$1469.26	$29.54	$45.91
Crew C-8E	Hr.	Daily	Hr.	Daily	Bare Costs	Incl. O&P
1 Labor Foreman (outside)	$22.55	$180.40	$38.15	$305.20	$24.21	$40.05
1 Laborer	20.55	164.40	34.75	278.00		
1 Cement Finisher	26.80	214.40	43.00	344.00		
1 Equipment Oper. (light)	26.95	215.60	44.30	354.40		
1 Air Compressor, 250 cfm		162.40		178.64		
2 -50' Air Hoses, 1"		8.20		9.02		
1 Concrete Pump (small)		737.80		811.58	28.39	31.23
32 L.H., Daily Totals		$1683.20		$2280.84	$52.60	$71.28
Crew C-10	Hr.	Daily	Hr.	Daily	Bare Costs	Incl. O&P
1 Laborer	$20.55	$164.40	$34.75	$278.00	$24.72	$40.25
2 Cement Finishers	26.80	428.80	43.00	688.00		
24 L.H., Daily Totals		$593.20		$966.00	$24.72	$40.25
Crew C-10B	Hr.	Daily	Hr.	Daily	Bare Costs	Incl. O&P
3 Laborers	$20.55	$493.20	$34.75	$834.00	$23.05	$38.05
2 Cement Finishers	26.80	428.80	43.00	688.00		
1 Concrete Mixer, 10 C.F.		155.20		170.72		
2 Trowels, 48" Walk-Behind		84.80		93.28	6.00	6.60
40 L.H., Daily Totals		$1162.00		$1786.00	$29.05	$44.65

Crew No.	Bare Costs		Incl. Subs O & P		Cost Per Labor-Hour	
Crew C-10C	Hr.	Daily	Hr.	Daily	Bare Costs	Incl. O&P
1 Laborer	$20.55	$164.40	$34.75	$278.00	$24.72	$40.25
2 Cement Finishers	26.80	428.80	43.00	688.00		
1 Trowel, 48" Walk-Behind		42.40		46.64	1.77	1.94
24 L.H., Daily Totals		$635.60		$1012.64	$26.48	$42.19
Crew C-10D	Hr.	Daily	Hr.	Daily	Bare Costs	Incl. O&P
1 Laborer	$20.55	$164.40	$34.75	$278.00	$24.72	$40.25
2 Cement Finishers	26.80	428.80	43.00	688.00		
1 Vibrating Power Screed		70.35		77.39		
1 Trowel, 48" Walk-Behind		42.40		46.64	4.70	5.17
24 L.H., Daily Totals		$705.95		$1090.03	$29.41	$45.42
Crew C-10E	Hr.	Daily	Hr.	Daily	Bare Costs	Incl. O&P
1 Laborer	$20.55	$164.40	$34.75	$278.00	$24.72	$40.25
2 Cement Finishers	26.80	428.80	43.00	688.00		
1 Vibrating Power Screed		70.35		77.39		
1 Cement Trowel, 96" Ride-On		166.60		183.26	9.87	10.86
24 L.H., Daily Totals		$830.15		$1226.65	$34.59	$51.11
Crew C-11	Hr.	Daily	Hr.	Daily	Bare Costs	Incl. O&P
1 Skilled Worker Foreman	$30.05	$240.40	$50.70	$405.60	$28.52	$47.95
5 Skilled Workers	28.05	1122.00	47.35	1894.00		
1 Equip. Oper. (crane)	29.35	234.80	48.20	385.60		
1 Lattice Boom Crane, 150 Ton		1890.00		2079.00	33.75	37.13
56 L.H., Daily Totals		$3487.20		$4764.20	$62.27	$85.08
Crew C-12	Hr.	Daily	Hr.	Daily	Bare Costs	Incl. O&P
1 Carpenter Foreman (out)	$29.95	$239.60	$50.65	$405.20	$27.28	$45.89
3 Carpenters	27.95	670.80	47.25	1134.00		
1 Laborer	20.55	164.40	34.75	278.00		
1 Equip. Oper. (crane)	29.35	234.80	48.20	385.60		
1 Hyd. Crane, 12 Ton		768.80		845.68	16.02	17.62
48 L.H., Daily Totals		$2078.40		$3048.48	$43.30	$63.51
Crew C-13	Hr.	Daily	Hr.	Daily	Bare Costs	Incl. O&P
2 Struc. Steel Workers	$29.95	$479.20	$57.55	$920.80	$29.28	$54.12
1 Carpenter	27.95	223.60	47.25	378.00		
1 Welder, gas engine, 300 amp		134.20		147.62	5.59	6.15
24 L.H., Daily Totals		$837.00		$1446.42	$34.88	$60.27
Crew C-14	Hr.	Daily	Hr.	Daily	Bare Costs	Incl. O&P
1 Carpenter Foreman (out)	$29.95	$239.60	$50.65	$405.20	$24.83	$41.89
3 Carpenters	27.95	670.80	47.25	1134.00		
2 Carpenter Helpers	20.85	333.60	35.15	562.40		
4 Laborers	20.55	657.60	34.75	1112.00		
2 Rodmen (reinf.)	29.85	477.60	53.00	848.00		
2 Rodman Helpers	20.85	333.60	35.15	562.40		
2 Cement Finishers	26.80	428.80	43.00	688.00		
1 Equip. Oper. (crane)	29.35	234.80	48.20	385.60		
1 Hyd. Crane, 80 Ton		1296.00		1425.60	9.53	10.48
136 L.H., Daily Totals		$4672.40		$7123.20	$34.36	$52.38

Crew No.	Bare Costs		Incl. Subs O & P		Cost Per Labor-Hour	
Crew C-14A	Hr.	Daily	Hr.	Daily	Bare Costs	Incl. O&P
1 Carpenter Foreman (out)	$29.95	$239.60	$50.65	$405.20	$27.72	$47.12
16 Carpenters	27.95	3577.60	47.25	6048.00		
4 Rodmen (reinf.)	29.85	955.20	53.00	1696.00		
2 Laborers	20.55	328.80	34.75	556.00		
1 Cement Finisher	26.80	214.40	43.00	344.00		
1 Equip. Oper. (med.)	28.55	228.40	46.90	375.20		
1 Gas Engine Vibrator		26.00		28.60		
1 Concrete Pump (small)		737.80		811.58	3.82	4.20
200 L.H., Daily Totals		$6307.80		$10264.58	$31.54	$51.32

Crew No.	Bare Costs		Incl. Subs O & P		Cost Per Labor-Hour	
Crew C-14B	Hr.	Daily	Hr.	Daily	Bare Costs	Incl. O&P
1 Carpenter Foreman (out)	$29.95	$239.60	$50.65	$405.20	$27.68	$46.96
16 Carpenters	27.95	3577.60	47.25	6048.00		
4 Rodmen (reinf.)	29.85	955.20	53.00	1696.00		
2 Laborers	20.55	328.80	34.75	556.00		
2 Cement Finishers	26.80	428.80	43.00	688.00		
1 Equip. Oper. (med.)	28.55	228.40	46.90	375.20		
1 Gas Engine Vibrator		26.00		28.60		
1 Concrete Pump (small)		737.80		811.58	3.67	4.04
208 L.H., Daily Totals		$6522.20		$10608.58	$31.36	$51.00

Crew No.	Bare Costs		Incl. Subs O & P		Cost Per Labor-Hour	
Crew C-14C	Hr.	Daily	Hr.	Daily	Bare Costs	Incl. O&P
1 Carpenter Foreman (out)	$29.95	$239.60	$50.65	$405.20	$26.17	$44.44
6 Carpenters	27.95	1341.60	47.25	2268.00		
2 Rodmen (reinf.)	29.85	477.60	53.00	848.00		
4 Laborers	20.55	657.60	34.75	1112.00		
1 Cement Finisher	26.80	214.40	43.00	344.00		
1 Gas Engine Vibrator		26.00		28.60	0.23	0.26
112 L.H., Daily Totals		$2956.80		$5005.80	$26.40	$44.69

Crew No.	Bare Costs		Incl. Subs O & P		Cost Per Labor-Hour	
Crew C-14D	Hr.	Daily	Hr.	Daily	Bare Costs	Incl. O&P
1 Carpenter Foreman (out)	$29.95	$239.60	$50.65	$405.20	$27.57	$46.66
18 Carpenters	27.95	4024.80	47.25	6804.00		
2 Rodmen (reinf.)	29.85	477.60	53.00	848.00		
2 Laborers	20.55	328.80	34.75	556.00		
1 Cement Finisher	26.80	214.40	43.00	344.00		
1 Equip. Oper. (med.)	28.55	228.40	46.90	375.20		
1 Gas Engine Vibrator		26.00		28.60		
1 Concrete Pump (small)		737.80		811.58	3.82	4.20
200 L.H., Daily Totals		$6277.40		$10172.58	$31.39	$50.86

Crew No.	Bare Costs		Incl. Subs O & P		Cost Per Labor-Hour	
Crew C-14E	Hr.	Daily	Hr.	Daily	Bare Costs	Incl. O&P
1 Carpenter Foreman (out)	$29.95	$239.60	$50.65	$405.20	$26.70	$45.85
2 Carpenters	27.95	447.20	47.25	756.00		
4 Rodmen (reinf.)	29.85	955.20	53.00	1696.00		
3 Laborers	20.55	493.20	34.75	834.00		
1 Cement Finisher	26.80	214.40	43.00	344.00		
1 Gas Engine Vibrator		26.00		28.60	0.30	0.33
88 L.H., Daily Totals		$2375.60		$4063.80	$27.00	$46.18

Crew No.	Bare Costs		Incl. Subs O & P		Cost Per Labor-Hour	
Crew C-14F	Hr.	Daily	Hr.	Daily	Bare Costs	Incl. O&P
1 Laborer Foreman (out)	$22.55	$180.40	$38.15	$305.20	$24.94	$40.63
2 Laborers	20.55	328.80	34.75	556.00		
6 Cement Finishers	26.80	1286.40	43.00	2064.00		
1 Gas Engine Vibrator		26.00		28.60	0.36	0.40
72 L.H., Daily Totals		$1821.60		$2953.80	$25.30	$41.02

Crew No.	Bare Costs		Incl. Subs O & P		Cost Per Labor-Hour	
Crew C-14G	Hr.	Daily	Hr.	Daily	Bare Costs	Incl. O&P
1 Laborer Foreman (out)	$22.55	$180.40	$38.15	$305.20	$24.41	$39.95
2 Laborers	20.55	328.80	34.75	556.00		
4 Cement Finishers	26.80	857.60	43.00	1376.00		
1 Gas Engine Vibrator		26.00		28.60	0.46	0.51
56 L.H., Daily Totals		$1392.80		$2265.80	$24.87	$40.46

Crew No.	Bare Costs		Incl. Subs O & P		Cost Per Labor-Hour	
Crew C-14H	Hr.	Daily	Hr.	Daily	Bare Costs	Incl. O&P
1 Carpenter Foreman (out)	$29.95	$239.60	$50.65	$405.20	$27.18	$45.98
2 Carpenters	27.95	447.20	47.25	756.00		
1 Rodman (reinf.)	29.85	238.80	53.00	424.00		
1 Laborer	20.55	164.40	34.75	278.00		
1 Cement Finisher	26.80	214.40	43.00	344.00		
1 Gas Engine Vibrator		26.00		28.60	0.54	0.60
48 L.H., Daily Totals		$1330.40		$2235.80	$27.72	$46.58

Crew No.	Bare Costs		Incl. Subs O & P		Cost Per Labor-Hour	
Crew C-14L	Hr.	Daily	Hr.	Daily	Bare Costs	Incl. O&P
1 Carpenter Foreman (out)	$29.95	$239.60	$50.65	$405.20	$25.55	$43.01
6 Carpenters	27.95	1341.60	47.25	2268.00		
4 Laborers	20.55	657.60	34.75	1112.00		
1 Cement Finisher	26.80	214.40	43.00	344.00		
1 Gas Engine Vibrator		26.00		28.60	0.27	0.30
96 L.H., Daily Totals		$2479.20		$4157.80	$25.82	$43.31

Crew No.	Bare Costs		Incl. Subs O & P		Cost Per Labor-Hour	
Crew C-15	Hr.	Daily	Hr.	Daily	Bare Costs	Incl. O&P
1 Carpenter Foreman (out)	$29.95	$239.60	$50.65	$405.20	$25.66	$43.16
2 Carpenters	27.95	447.20	47.25	756.00		
3 Laborers	20.55	493.20	34.75	834.00		
2 Cement Finishers	26.80	428.80	43.00	688.00		
1 Rodman (reinf.)	29.85	238.80	53.00	424.00		
72 L.H., Daily Totals		$1847.60		$3107.20	$25.66	$43.16

Crew No.	Bare Costs		Incl. Subs O & P		Cost Per Labor-Hour	
Crew C-16	Hr.	Daily	Hr.	Daily	Bare Costs	Incl. O&P
1 Labor Foreman (outside)	$22.55	$180.40	$38.15	$305.20	$25.12	$42.37
3 Laborers	20.55	493.20	34.75	834.00		
2 Cement Finishers	26.80	428.80	43.00	688.00		
1 Equip. Oper. (med.)	28.55	228.40	46.90	375.20		
2 Rodmen (reinf.)	29.85	477.60	53.00	848.00		
1 Concrete Pump (small)		737.80		811.58	10.25	11.27
72 L.H., Daily Totals		$2546.20		$3861.98	$35.36	$53.64

Crew No.	Bare Costs		Incl. Subs O & P		Cost Per Labor-Hour	
Crew C-17	Hr.	Daily	Hr.	Daily	Bare Costs	Incl. O&P
2 Skilled Worker Foremen	$30.05	$480.80	$50.70	$811.20	$28.45	$48.02
8 Skilled Workers	28.05	1795.20	47.35	3030.40		
80 L.H., Daily Totals		$2276.00		$3841.60	$28.45	$48.02

Crew No.	Bare Costs		Incl. Subs O & P		Cost Per Labor-Hour	
Crew C-17A	Hr.	Daily	Hr.	Daily	Bare Costs	Incl. O&P
2 Skilled Worker Foremen	$30.05	$480.80	$50.70	$811.20	$28.46	$48.02
8 Skilled Workers	28.05	1795.20	47.35	3030.40		
.125 Equip. Oper. (crane)	29.35	29.35	48.20	48.20		
.125 Hyd. Crane, 80 Ton		162.00		178.20	2.00	2.20
81 L.H., Daily Totals		$2467.35		$4068.00	$30.46	$50.22

Crew No.	Bare Costs		Incl. Subs O & P		Cost Per Labor-Hour	
Crew C-17B	Hr.	Daily	Hr.	Daily	Bare Costs	Incl. O&P
2 Skilled Worker Foremen	$30.05	$480.80	$50.70	$811.20	$28.47	$48.02
8 Skilled Workers	28.05	1795.20	47.35	3030.40		
.25 Equip. Oper. (crane)	29.35	58.70	48.20	96.40		
.25 Hyd. Crane, 80 Ton		324.00		356.40		
.25 Trowel, 48" Walk-Behind		10.60		11.66	4.08	4.49
82 L.H., Daily Totals		$2669.30		$4306.06	$32.55	$52.51

Header structure for each table: Crew No. | **Bare Costs** (Hr. | Daily) | **Incl. Subs O & P** (Hr. | Daily) | **Cost Per Labor-Hour** (Bare Costs | Incl. O&P)

Crew C-17C

Crew C-17C	Hr.	Daily	Hr.	Daily	Bare Costs	Incl. O&P
2 Skilled Worker Foremen	$30.05	$480.80	$50.70	$811.20	$28.48	$48.03
8 Skilled Workers	28.05	1795.20	47.35	3030.40		
.375 Equip. Oper. (crane)	29.35	88.05	48.20	144.60		
.375 Hyd. Crane, 80 Ton		486.00		534.60	5.86	6.44
83 L.H., Daily Totals		$2850.05		$4520.80	$34.34	$54.47

Crew C-17D

Crew C-17D	Hr.	Daily	Hr.	Daily	Bare Costs	Incl. O&P
2 Skilled Worker Foremen	$30.05	$480.80	$50.70	$811.20	$28.49	$48.03
8 Skilled Workers	28.05	1795.20	47.35	3030.40		
.5 Equip. Oper. (crane)	29.35	117.40	48.20	192.80		
.5 Hyd. Crane, 80 Ton		648.00		712.80	7.71	8.49
84 L.H., Daily Totals		$3041.40		$4747.20	$36.21	$56.51

Crew C-17E

Crew C-17E	Hr.	Daily	Hr.	Daily	Bare Costs	Incl. O&P
2 Skilled Worker Foremen	$30.05	$480.80	$50.70	$811.20	$28.45	$48.02
8 Skilled Workers	28.05	1795.20	47.35	3030.40		
1 Hyd. Jack with Rods		85.70		94.27	1.07	1.18
80 L.H., Daily Totals		$2361.70		$3935.87	$29.52	$49.20

Crew C-18

Crew C-18	Hr.	Daily	Hr.	Daily	Bare Costs	Incl. O&P
.125 Labor Foreman (out)	$22.55	$22.55	$38.15	$38.15	$20.77	$35.13
1 Laborer	20.55	164.40	34.75	278.00		
1 Concrete Cart, 10 C.F.		55.20		60.72	6.13	6.75
9 L.H., Daily Totals		$242.15		$376.87	$26.91	$41.87

Crew C-19

Crew C-19	Hr.	Daily	Hr.	Daily	Bare Costs	Incl. O&P
.125 Labor Foreman (out)	$22.55	$22.55	$38.15	$38.15	$20.77	$35.13
1 Laborer	20.55	164.40	34.75	278.00		
1 Concrete Cart, 18 C.F.		86.40		95.04	9.60	10.56
9 L.H., Daily Totals		$273.35		$411.19	$30.37	$45.69

Crew C-20

Crew C-20	Hr.	Daily	Hr.	Daily	Bare Costs	Incl. O&P
1 Labor Foreman (outside)	$22.55	$180.40	$38.15	$305.20	$22.58	$37.73
5 Laborers	20.55	822.00	34.75	1390.00		
1 Cement Finisher	26.80	214.40	43.00	344.00		
1 Equip. Oper. (med.)	28.55	228.40	46.90	375.20		
2 Gas Engine Vibrators		52.00		57.20		
1 Concrete Pump (small)		737.80		811.58	12.34	13.57
64 L.H., Daily Totals		$2235.00		$3283.18	$34.92	$51.30

Crew C-21

Crew C-21	Hr.	Daily	Hr.	Daily	Bare Costs	Incl. O&P
1 Labor Foreman (outside)	$22.55	$180.40	$38.15	$305.20	$22.58	$37.73
5 Laborers	20.55	822.00	34.75	1390.00		
1 Cement Finisher	26.80	214.40	43.00	344.00		
1 Equip. Oper. (med.)	28.55	228.40	46.90	375.20		
2 Gas Engine Vibrators		52.00		57.20		
1 Concrete Conveyer		179.00		196.90	3.61	3.97
64 L.H., Daily Totals		$1676.20		$2668.50	$26.19	$41.70

Crew C-22

Crew C-22	Hr.	Daily	Hr.	Daily	Bare Costs	Incl. O&P
1 Rodman Foreman	$31.85	$254.80	$56.55	$452.40	$30.11	$53.29
4 Rodmen (reinf.)	29.85	955.20	53.00	1696.00		
.125 Equip. Oper. (crane)	29.35	29.35	48.20	48.20		
.125 Equip. Oper. Oiler	25.40	25.40	41.75	41.75		
.125 Hyd. Crane, 25 Ton		98.67		108.54	2.35	2.58
42 L.H., Daily Totals		$1363.43		$2346.89	$32.46	$55.88

Crew C-23

Crew C-23	Hr.	Daily	Hr.	Daily	Bare Costs	Incl. O&P
2 Skilled Worker Foremen	$30.05	$480.80	$50.70	$811.20	$28.32	$47.55
6 Skilled Workers	28.05	1346.40	47.35	2272.80		
1 Equip. Oper. (crane)	29.35	234.80	48.20	385.60		
1 Equip. Oper. Oiler	25.40	203.20	41.75	334.00		
1 Lattice Boom Crane, 90 Ton		1741.00		1915.10	21.76	23.94
80 L.H., Daily Totals		$4006.20		$5718.70	$50.08	$71.48

Crew C-24

Crew C-24	Hr.	Daily	Hr.	Daily	Bare Costs	Incl. O&P
2 Skilled Worker Foremen	$30.05	$480.80	$50.70	$811.20	$28.32	$47.55
6 Skilled Workers	28.05	1346.40	47.35	2272.80		
1 Equip. Oper. (crane)	29.35	234.80	48.20	385.60		
1 Equip. Oper. Oiler	25.40	203.20	41.75	334.00		
1 Lattice Boom Crane, 150 Ton		1890.00		2079.00	23.63	25.99
80 L.H., Daily Totals		$4155.20		$5882.60	$51.94	$73.53

Crew C-25

Crew C-25	Hr.	Daily	Hr.	Daily	Bare Costs	Incl. O&P
2 Rodmen (reinf.)	$29.85	$477.60	$53.00	$848.00	$23.68	$42.48
2 Rodmen Helpers	17.50	280.00	31.95	511.20		
32 L.H., Daily Totals		$757.60		$1359.20	$23.68	$42.48

Crew C-27

Crew C-27	Hr.	Daily	Hr.	Daily	Bare Costs	Incl. O&P
2 Cement Finishers	$26.80	$428.80	$43.00	$688.00	$26.80	$43.00
1 Concrete Saw		141.00		155.10	8.81	9.69
16 L.H., Daily Totals		$569.80		$843.10	$35.61	$52.69

Crew C-28

Crew C-28	Hr.	Daily	Hr.	Daily	Bare Costs	Incl. O&P
1 Cement Finisher	$26.80	$214.40	$43.00	$344.00	$26.80	$43.00
1 Portable Air Compressor, Gas		19.10		21.01	2.39	2.63
8 L.H., Daily Totals		$233.50		$365.01	$29.19	$45.63

Crew D-1

Crew D-1	Hr.	Daily	Hr.	Daily	Bare Costs	Incl. O&P
1 Bricklayer	$28.35	$226.80	$47.00	$376.00	$25.43	$42.15
1 Bricklayer Helper	22.50	180.00	37.30	298.40		
16 L.H., Daily Totals		$406.80		$674.40	$25.43	$42.15

Crew D-2

Crew D-2	Hr.	Daily	Hr.	Daily	Bare Costs	Incl. O&P
3 Bricklayers	$28.35	$680.40	$47.00	$1128.00	$26.01	$43.12
2 Bricklayer Helpers	22.50	360.00	37.30	596.80		
40 L.H., Daily Totals		$1040.40		$1724.80	$26.01	$43.12

Crew D-3

Crew D-3	Hr.	Daily	Hr.	Daily	Bare Costs	Incl. O&P
3 Bricklayers	$28.35	$680.40	$47.00	$1128.00	$26.10	$43.32
2 Bricklayer Helpers	22.50	360.00	37.30	596.80		
.25 Carpenter	27.95	55.90	47.25	94.50		
42 L.H., Daily Totals		$1096.30		$1819.30	$26.10	$43.32

Crew D-4

Crew D-4	Hr.	Daily	Hr.	Daily	Bare Costs	Incl. O&P
1 Bricklayer	$28.35	$226.80	$47.00	$376.00	$23.28	$38.73
3 Bricklayer Helpers	22.50	540.00	37.30	895.20		
1 Building Laborer	20.55	164.40	34.75	278.00		
1 Grout Pump, 50 C.F./hr.		126.80		139.48	3.17	3.49
40 L.H., Daily Totals		$1058.00		$1688.68	$26.45	$42.22

Crew D-5

Crew D-5	Hr.	Daily	Hr.	Daily	Bare Costs	Incl. O&P
1 Block Mason Helper	$22.50	$180.00	$37.30	$298.40	$22.50	$37.30
8 L.H., Daily Totals		$180.00		$298.40	$22.50	$37.30

Crew D-6

Crew No.	Bare Costs Hr.	Daily	Incl. Subs O & P Hr.	Daily	Cost Per Labor-Hour Bare Costs	Incl. O&P
3 Bricklayers	$28.35	$680.40	$47.00	$1128.00	$25.43	$42.15
3 Bricklayer Helpers	22.50	540.00	37.30	895.20		
48 L.H., Daily Totals		$1220.40		$2023.20	$25.43	$42.15

Crew D-7

Crew No.	Bare Costs Hr.	Daily	Incl. Subs O & P Hr.	Daily	Cost Per Labor-Hour Bare Costs	Incl. O&P
1 Tile Layer	$26.65	$213.20	$42.75	$342.00	$23.85	$38.25
1 Tile Layer Helper	21.05	168.40	33.75	270.00		
16 L.H., Daily Totals		$381.60		$612.00	$23.85	$38.25

Crew D-8

Crew No.	Bare Costs Hr.	Daily	Incl. Subs O & P Hr.	Daily	Cost Per Labor-Hour Bare Costs	Incl. O&P
3 Bricklayers	$28.35	$680.40	$47.00	$1128.00	$26.01	$43.12
2 Bricklayer Helpers	22.50	360.00	37.30	596.80		
40 L.H., Daily Totals		$1040.40		$1724.80	$26.01	$43.12

Crew D-9

Crew No.	Bare Costs Hr.	Daily	Incl. Subs O & P Hr.	Daily	Cost Per Labor-Hour Bare Costs	Incl. O&P
3 Bricklayers	$28.35	$680.40	$47.00	$1128.00	$25.43	$42.15
3 Bricklayer Helpers	22.50	540.00	37.30	895.20		
48 L.H., Daily Totals		$1220.40		$2023.20	$25.43	$42.15

Crew D-10

Crew No.	Bare Costs Hr.	Daily	Incl. Subs O & P Hr.	Daily	Cost Per Labor-Hour Bare Costs	Incl. O&P
1 Bricklayer Foreman	$30.35	$242.80	$50.30	$402.40	$27.64	$45.70
1 Bricklayer	28.35	226.80	47.00	376.00		
1 Bricklayer Helper	22.50	180.00	37.30	298.40		
1 Equip. Oper. (crane)	29.35	234.80	48.20	385.60		
1 S.P. Crane, 4x4, 12 Ton		601.80		661.98	18.81	20.69
32 L.H., Daily Totals		$1486.20		$2124.38	$46.44	$66.39

Crew D-11

Crew No.	Bare Costs Hr.	Daily	Incl. Subs O & P Hr.	Daily	Cost Per Labor-Hour Bare Costs	Incl. O&P
2 Bricklayers	$28.35	$453.60	$47.00	$752.00	$26.40	$43.77
1 Bricklayer Helper	22.50	180.00	37.30	298.40		
24 L.H., Daily Totals		$633.60		$1050.40	$26.40	$43.77

Crew D-12

Crew No.	Bare Costs Hr.	Daily	Incl. Subs O & P Hr.	Daily	Cost Per Labor-Hour Bare Costs	Incl. O&P
2 Bricklayers	$28.35	$453.60	$47.00	$752.00	$25.43	$42.15
2 Bricklayer Helpers	22.50	360.00	37.30	596.80		
32 L.H., Daily Totals		$813.60		$1348.80	$25.43	$42.15

Crew D-13

Crew No.	Bare Costs Hr.	Daily	Incl. Subs O & P Hr.	Daily	Cost Per Labor-Hour Bare Costs	Incl. O&P
1 Bricklayer Foreman	$30.35	$242.80	$50.30	$402.40	$26.90	$44.52
2 Bricklayers	28.35	453.60	47.00	752.00		
2 Bricklayer Helpers	22.50	360.00	37.30	596.80		
1 Equip. Oper. (crane)	29.35	234.80	48.20	385.60		
1 S.P. Crane, 4x4, 12 Ton		601.80		661.98	12.54	13.79
48 L.H., Daily Totals		$1893.00		$2798.78	$39.44	$58.31

Crew E-1

Crew No.	Bare Costs Hr.	Daily	Incl. Subs O & P Hr.	Daily	Cost Per Labor-Hour Bare Costs	Incl. O&P
2 Struc. Steel Workers	$29.95	$479.20	$57.55	$920.80	$29.95	$57.55
1 Welder, gas engine, 300 amp		134.20		147.62	8.39	9.23
16 L.H., Daily Totals		$613.40		$1068.42	$38.34	$66.78

Crew E-2

Crew No.	Bare Costs Hr.	Daily	Incl. Subs O & P Hr.	Daily	Cost Per Labor-Hour Bare Costs	Incl. O&P
1 Struc. Steel Foreman	$31.95	$255.60	$61.40	$491.20	$30.18	$56.63
4 Struc. Steel Workers	29.95	958.40	57.55	1841.60		
1 Equip. Oper. (crane)	29.35	234.80	48.20	385.60		
1 Lattice Boom Crane, 90 Ton		1741.00		1915.10	36.27	39.90
48 L.H., Daily Totals		$3189.80		$4633.50	$66.45	$96.53

Crew E-3

Crew No.	Bare Costs Hr.	Daily	Incl. Subs O & P Hr.	Daily	Cost Per Labor-Hour Bare Costs	Incl. O&P
1 Struc. Steel Foreman	$31.95	$255.60	$61.40	$491.20	$30.62	$58.83
2 Struc. Steel Workers	29.95	479.20	57.55	920.80		
1 Welder, gas engine, 300 amp		134.20		147.62	5.59	6.15
24 L.H., Daily Totals		$869.00		$1559.62	$36.21	$64.98

Crew E-4

Crew No.	Bare Costs Hr.	Daily	Incl. Subs O & P Hr.	Daily	Cost Per Labor-Hour Bare Costs	Incl. O&P
1 Struc. Steel Foreman	$31.95	$255.60	$61.40	$491.20	$30.45	$58.51
3 Struc. Steel Workers	29.95	718.80	57.55	1381.20		
1 Welder, gas engine, 300 amp		134.20		147.62	4.19	4.61
32 L.H., Daily Totals		$1108.60		$2020.02	$34.64	$63.13

Crew E-5

Crew No.	Bare Costs Hr.	Daily	Incl. Subs O & P Hr.	Daily	Cost Per Labor-Hour Bare Costs	Incl. O&P
1 Struc. Steel Foreman	$31.95	$255.60	$61.40	$491.20	$30.11	$56.94
7 Struc. Steel Workers	29.95	1677.20	57.55	3222.80		
1 Equip. Oper. (crane)	29.35	234.80	48.20	385.60		
1 Lattice Boom Crane, 90 Ton		1741.00		1915.10		
1 Welder, gas engine, 300 amp		134.20		147.62	26.04	28.65
72 L.H., Daily Totals		$4042.80		$6162.32	$56.15	$85.59

Crew E-6

Crew No.	Bare Costs Hr.	Daily	Incl. Subs O & P Hr.	Daily	Cost Per Labor-Hour Bare Costs	Incl. O&P
1 Struc. Steel Foreman	$31.95	$255.60	$61.40	$491.20	$29.84	$56.30
12 Struc. Steel Workers	29.95	2875.20	57.55	5524.80		
1 Equip. Oper. (crane)	29.35	234.80	48.20	385.60		
1 Equip. Oper. (light)	26.95	215.60	44.30	354.40		
1 Lattice Boom Crane, 90 Ton		1741.00		1915.10		
1 Welder, gas engine, 300 amp		134.20		147.62		
1 Air Compressor, 160 cfm		138.80		152.68		
2 Impact Wrenches		32.40		35.64	17.05	18.76
120 L.H., Daily Totals		$5627.60		$9007.04	$46.90	$75.06

Crew E-7

Crew No.	Bare Costs Hr.	Daily	Incl. Subs O & P Hr.	Daily	Cost Per Labor-Hour Bare Costs	Incl. O&P
1 Struc. Steel Foreman	$31.95	$255.60	$61.40	$491.20	$30.11	$56.94
7 Struc. Steel Workers	29.95	1677.20	57.55	3222.80		
1 Equip. Oper. (crane)	29.35	234.80	48.20	385.60		
1 Lattice Boom Crane, 90 Ton		1741.00		1915.10		
2 Welders, gas engine, 300 amp		268.40		295.24	27.91	30.70
72 L.H., Daily Totals		$4177.00		$6309.94	$58.01	$87.64

Crew E-8

Crew No.	Bare Costs Hr.	Daily	Incl. Subs O & P Hr.	Daily	Cost Per Labor-Hour Bare Costs	Incl. O&P
1 Struc. Steel Foreman	$31.95	$255.60	$61.40	$491.20	$30.08	$57.05
9 Struc. Steel Workers	29.95	2156.40	57.55	4143.60		
1 Equip. Oper. (crane)	29.35	234.80	48.20	385.60		
1 Lattice Boom Crane, 90 Ton		1741.00		1915.10		
4 Welders, gas engine, 300 amp		536.80		590.48	25.88	28.47
88 L.H., Daily Totals		$4924.60		$7525.98	$55.96	$85.52

Crew E-9

Crew No.	Bare Costs Hr.	Daily	Incl. Subs O & P Hr.	Daily	Cost Per Labor-Hour Bare Costs	Incl. O&P
2 Struc. Steel Foremen	$31.95	$511.20	$61.40	$982.40	$29.82	$55.87
5 Struc. Steel Workers	29.95	1198.00	57.55	2302.00		
1 Welder Foreman	31.95	255.60	61.40	491.20		
5 Welders	29.95	1198.00	57.55	2302.00		
1 Equip. Oper. (crane)	29.35	234.80	48.20	385.60		
1 Equip. Oper. Oiler	25.40	203.20	41.75	334.00		
1 Equip. Oper. (light)	26.95	215.60	44.30	354.40		
1 Lattice Boom Crane, 90 Ton		1741.00		1915.10		
5 Welders, gas engine, 300 amp		671.00		738.10	18.84	20.73
128 L.H., Daily Totals		$6228.40		$9804.80	$48.66	$76.60

Crews

Crew E-10

Crew No.	Bare Costs Hr.	Bare Costs Daily	Incl. Subs O & P Hr.	Incl. Subs O & P Daily	Cost Per Labor-Hour Bare Costs	Cost Per Labor-Hour Incl. O&P
1 Struc. Steel Foreman	$31.95	$255.60	$61.40	$491.20	$30.62	$58.83
2 Struc. Steel Workers	29.95	479.20	57.55	920.80		
1 Welder, gas engine, 300 amp		134.20		147.62		
1 Flatbed Truck, Gas, 3 Ton		242.60		266.86	15.70	17.27
24 L.H., Daily Totals		$1111.60		$1826.48	$46.32	$76.10

Crew E-11

Crew No.	Bare Costs Hr.	Bare Costs Daily	Incl. Subs O & P Hr.	Incl. Subs O & P Daily	Cost Per Labor-Hour Bare Costs	Cost Per Labor-Hour Incl. O&P
2 Painters, Struc. Steel	$25.55	$408.80	$49.95	$799.20	$24.65	$44.74
1 Building Laborer	20.55	164.40	34.75	278.00		
1 Equip. Oper. (light)	26.95	215.60	44.30	354.40		
1 Air Compressor, 250 cfm		162.40		178.64		
1 Sandblaster, portable, 3 C.F.		20.60		22.66		
1 Set Sand Blasting Accessories		12.75		14.03	6.12	6.73
32 L.H., Daily Totals		$984.55		$1646.93	$30.77	$51.47

Crew E-12

Crew No.	Bare Costs Hr.	Bare Costs Daily	Incl. Subs O & P Hr.	Incl. Subs O & P Daily	Cost Per Labor-Hour Bare Costs	Cost Per Labor-Hour Incl. O&P
1 Welder Foreman	$31.95	$255.60	$61.40	$491.20	$29.45	$52.85
1 Equip. Oper. (light)	26.95	215.60	44.30	354.40		
1 Welder, gas engine, 300 amp		134.20		147.62	8.39	9.23
16 L.H., Daily Totals		$605.40		$993.22	$37.84	$62.08

Crew E-13

Crew No.	Bare Costs Hr.	Bare Costs Daily	Incl. Subs O & P Hr.	Incl. Subs O & P Daily	Cost Per Labor-Hour Bare Costs	Cost Per Labor-Hour Incl. O&P
1 Welder Foreman	$31.95	$255.60	$61.40	$491.20	$30.28	$55.70
.5 Equip. Oper. (light)	26.95	107.80	44.30	177.20		
1 Welder, gas engine, 300 amp		134.20		147.62	11.18	12.30
12 L.H., Daily Totals		$497.60		$816.02	$41.47	$68.00

Crew E-14

Crew No.	Bare Costs Hr.	Bare Costs Daily	Incl. Subs O & P Hr.	Incl. Subs O & P Daily	Cost Per Labor-Hour Bare Costs	Cost Per Labor-Hour Incl. O&P
1 Struc. Steel Worker	$29.95	$239.60	$57.55	$460.40	$29.95	$57.55
1 Welder, gas engine, 300 amp		134.20		147.62	16.77	18.45
8 L.H., Daily Totals		$373.80		$608.02	$46.73	$76.00

Crew E-16

Crew No.	Bare Costs Hr.	Bare Costs Daily	Incl. Subs O & P Hr.	Incl. Subs O & P Daily	Cost Per Labor-Hour Bare Costs	Cost Per Labor-Hour Incl. O&P
1 Welder Foreman	$31.95	$255.60	$61.40	$491.20	$30.95	$59.48
1 Welder	29.95	239.60	57.55	460.40		
1 Welder, gas engine, 300 amp		134.20		147.62	8.39	9.23
16 L.H., Daily Totals		$629.40		$1099.22	$39.34	$68.70

Crew E-17

Crew No.	Bare Costs Hr.	Bare Costs Daily	Incl. Subs O & P Hr.	Incl. Subs O & P Daily	Cost Per Labor-Hour Bare Costs	Cost Per Labor-Hour Incl. O&P
1 Structural Steel Foreman	$31.95	$255.60	$61.40	$491.20	$30.95	$59.48
1 Structural Steel Worker	29.95	239.60	57.55	460.40		
16 L.H., Daily Totals		$495.20		$951.60	$30.95	$59.48

Crew E-18

Crew No.	Bare Costs Hr.	Bare Costs Daily	Incl. Subs O & P Hr.	Incl. Subs O & P Daily	Cost Per Labor-Hour Bare Costs	Cost Per Labor-Hour Incl. O&P
1 Structural Steel Foreman	$31.95	$255.60	$61.40	$491.20	$30.07	$56.19
3 Structural Steel Workers	29.95	718.80	57.55	1381.20		
1 Equipment Operator (med.)	28.55	228.40	46.90	375.20		
1 Lattice Boom Crane, 20 Ton		1080.00		1188.00	27.00	29.70
40 L.H., Daily Totals		$2282.80		$3435.60	$57.07	$85.89

Crew E-19

Crew No.	Bare Costs Hr.	Bare Costs Daily	Incl. Subs O & P Hr.	Incl. Subs O & P Daily	Cost Per Labor-Hour Bare Costs	Cost Per Labor-Hour Incl. O&P
1 Structural Steel Worker	$29.95	$239.60	$57.55	$460.40	$29.62	$54.42
1 Structural Steel Foreman	31.95	255.60	61.40	491.20		
1 Equip. Oper. (light)	26.95	215.60	44.30	354.40		
1 Lattice Boom Crane, 20 Ton		1080.00		1188.00	45.00	49.50
24 L.H., Daily Totals		$1790.80		$2494.00	$74.62	$103.92

Crew E-20

Crew No.	Bare Costs Hr.	Bare Costs Daily	Incl. Subs O & P Hr.	Incl. Subs O & P Daily	Cost Per Labor-Hour Bare Costs	Cost Per Labor-Hour Incl. O&P
1 Structural Steel Foreman	$31.95	$255.60	$61.40	$491.20	$29.56	$54.89
5 Structural Steel Workers	29.95	1198.00	57.55	2302.00		
1 Equip. Oper. (crane)	29.35	234.80	48.20	385.60		
1 Oiler	25.40	203.20	41.75	334.00		
1 Lattice Boom Crane, 40 Ton		1327.00		1459.70	20.73	22.81
64 L.H., Daily Totals		$3218.60		$4972.50	$50.29	$77.70

Crew E-22

Crew No.	Bare Costs Hr.	Bare Costs Daily	Incl. Subs O & P Hr.	Incl. Subs O & P Daily	Cost Per Labor-Hour Bare Costs	Cost Per Labor-Hour Incl. O&P
1 Skilled Worker Foreman	$30.05	$240.40	$50.70	$405.60	$28.72	$48.47
2 Skilled Workers	28.05	448.80	47.35	757.60		
24 L.H., Daily Totals		$689.20		$1163.20	$28.72	$48.47

Crew E-24

Crew No.	Bare Costs Hr.	Bare Costs Daily	Incl. Subs O & P Hr.	Incl. Subs O & P Daily	Cost Per Labor-Hour Bare Costs	Cost Per Labor-Hour Incl. O&P
3 Structural Steel Workers	$29.95	$718.80	$57.55	$1381.20	$29.60	$54.89
1 Equipment Operator (medium)	28.55	228.40	46.90	375.20		
1 Hyd. Crane, 25 Ton		789.40		868.34	24.67	27.14
32 L.H., Daily Totals		$1736.60		$2624.74	$54.27	$82.02

Crew E-25

Crew No.	Bare Costs Hr.	Bare Costs Daily	Incl. Subs O & P Hr.	Incl. Subs O & P Daily	Cost Per Labor-Hour Bare Costs	Cost Per Labor-Hour Incl. O&P
1 Welder	$29.95	$239.60	$57.55	$460.40	$29.95	$57.55
1 Cutting Torch		17.00		18.70		
1 Gases		75.60		83.16	11.57	12.73
8 L.H., Daily Totals		$332.20		$562.26	$41.52	$70.28

Crew F-3

Crew No.	Bare Costs Hr.	Bare Costs Daily	Incl. Subs O & P Hr.	Incl. Subs O & P Daily	Cost Per Labor-Hour Bare Costs	Cost Per Labor-Hour Incl. O&P
2 Carpenters	$27.95	$447.20	$47.25	$756.00	$25.39	$42.60
2 Carpenter Helpers	20.85	333.60	35.15	562.40		
1 Equip. Oper. (crane)	29.35	234.80	48.20	385.60		
1 Hyd. Crane, 12 Ton		768.80		845.68	19.22	21.14
40 L.H., Daily Totals		$1784.40		$2549.68	$44.61	$63.74

Crew F-4

Crew No.	Bare Costs Hr.	Bare Costs Daily	Incl. Subs O & P Hr.	Incl. Subs O & P Daily	Cost Per Labor-Hour Bare Costs	Cost Per Labor-Hour Incl. O&P
2 Carpenters	$27.95	$447.20	$47.25	$756.00	$25.39	$42.60
2 Carpenter Helpers	20.85	333.60	35.15	562.40		
1 Equip. Oper. (crane)	29.35	234.80	48.20	385.60		
1 Hyd. Crane, 55 Ton		1125.00		1237.50	28.13	30.94
40 L.H., Daily Totals		$2140.60		$2941.50	$53.52	$73.54

Crew F-5

Crew No.	Bare Costs Hr.	Bare Costs Daily	Incl. Subs O & P Hr.	Incl. Subs O & P Daily	Cost Per Labor-Hour Bare Costs	Cost Per Labor-Hour Incl. O&P
2 Carpenters	$27.95	$447.20	$47.25	$756.00	$24.40	$41.20
2 Carpenter Helpers	20.85	333.60	35.15	562.40		
32 L.H., Daily Totals		$780.80		$1318.40	$24.40	$41.20

Crew F-6

Crew No.	Bare Costs Hr.	Bare Costs Daily	Incl. Subs O & P Hr.	Incl. Subs O & P Daily	Cost Per Labor-Hour Bare Costs	Cost Per Labor-Hour Incl. O&P
2 Carpenters	$27.95	$447.20	$47.25	$756.00	$25.27	$42.44
2 Building Laborers	20.55	328.80	34.75	556.00		
1 Equip. Oper. (crane)	29.35	234.80	48.20	385.60		
1 Hyd. Crane, 12 Ton		768.80		845.68	19.22	21.14
40 L.H., Daily Totals		$1779.60		$2543.28	$44.49	$63.58

Crew F-7

Crew No.	Bare Costs Hr.	Bare Costs Daily	Incl. Subs O & P Hr.	Incl. Subs O & P Daily	Cost Per Labor-Hour Bare Costs	Cost Per Labor-Hour Incl. O&P
2 Carpenters	$27.95	$447.20	$47.25	$756.00	$24.25	$41.00
2 Building Laborers	20.55	328.80	34.75	556.00		
32 L.H., Daily Totals		$776.00		$1312.00	$24.25	$41.00

Crew No.	Bare Costs Hr.	Bare Costs Daily	Incl. Subs O & P Hr.	Incl. Subs O & P Daily	Cost Per Labor-Hour Bare Costs	Cost Per Labor-Hour Incl. O&P
Crew G-1	Hr.	Daily	Hr.	Daily	Bare Costs	Incl. O&P
1 Roofer Foreman	$25.65	$205.20	$46.80	$374.40	$22.18	$40.47
4 Roofers, Composition	23.65	756.80	43.15	1380.80		
2 Roofer Helpers	17.50	280.00	31.95	511.20		
1 Application Equipment		173.20		190.52		
1 Tar Kettle/Pot		97.20		106.92		
1 Crew Truck		169.60		186.56	7.86	8.64
56 L.H., Daily Totals		$1682.00		$2750.40	$30.04	$49.11
Crew G-2	Hr.	Daily	Hr.	Daily	Bare Costs	Incl. O&P
1 Plasterer	$25.30	$202.40	$41.70	$333.60	$22.82	$37.90
1 Plasterer Helper	22.60	180.80	37.25	298.00		
1 Building Laborer	20.55	164.40	34.75	278.00		
1 Grout Pump, 50 C.F./hr.		126.80		139.48	5.28	5.81
24 L.H., Daily Totals		$674.40		$1049.08	$28.10	$43.71
Crew G-2A	Hr.	Daily	Hr.	Daily	Bare Costs	Incl. O&P
1 Roofer, composition	$23.65	$189.20	$43.15	$345.20	$20.57	$36.62
1 Roofer Helper	17.50	140.00	31.95	255.60		
1 Building Laborer	20.55	164.40	34.75	278.00		
1 Foam spray rig, trailer-mtd.		488.20		537.02		
1 Pickup Truck, 3/4 Ton		116.80		128.48	25.21	27.73
24 L.H., Daily Totals		$1098.60		$1544.30	$45.77	$64.35
Crew G-3	Hr.	Daily	Hr.	Daily	Bare Costs	Incl. O&P
2 Sheet Metal Workers	$31.15	$498.40	$52.30	$836.80	$25.85	$43.52
2 Building Laborers	20.55	328.80	34.75	556.00		
32 L.H., Daily Totals		$827.20		$1392.80	$25.85	$43.52
Crew G-4	Hr.	Daily	Hr.	Daily	Bare Costs	Incl. O&P
1 Labor Foreman (outside)	$22.55	$180.40	$38.15	$305.20	$21.22	$35.88
2 Building Laborers	20.55	328.80	34.75	556.00		
1 Flatbed Truck, Gas, 1.5 Ton		194.40		213.84		
1 Air Compressor, 160 cfm		138.80		152.68	13.88	15.27
24 L.H., Daily Totals		$842.40		$1227.72	$35.10	$51.16
Crew G-5	Hr.	Daily	Hr.	Daily	Bare Costs	Incl. O&P
1 Roofer Foreman	$25.65	$205.20	$46.80	$374.40	$21.59	$39.40
2 Roofers, Composition	23.65	378.40	43.15	690.40		
2 Roofer Helpers	17.50	280.00	31.95	511.20		
1 Application Equipment		173.20		190.52	4.33	4.76
40 L.H., Daily Totals		$1036.80		$1766.52	$25.92	$44.16
Crew G-6A	Hr.	Daily	Hr.	Daily	Bare Costs	Incl. O&P
2 Roofers Composition	$23.65	$378.40	$43.15	$690.40	$23.65	$43.15
1 Small Compressor, Electric		11.35		12.48		
2 Pneumatic Nailers		43.20		47.52	3.41	3.75
16 L.H., Daily Totals		$432.95		$750.40	$27.06	$46.90
Crew G-7	Hr.	Daily	Hr.	Daily	Bare Costs	Incl. O&P
1 Carpenter	$27.95	$223.60	$47.25	$378.00	$27.95	$47.25
1 Small Compressor, Electric		11.35		12.48		
1 Pneumatic Nailer		21.60		23.76	4.12	4.53
8 L.H., Daily Totals		$256.55		$414.25	$32.07	$51.78
Crew H-1	Hr.	Daily	Hr.	Daily	Bare Costs	Incl. O&P
2 Glaziers	$27.40	$438.40	$45.25	$724.00	$28.68	$51.40
2 Struc. Steel Workers	29.95	479.20	57.55	920.80		
32 L.H., Daily Totals		$917.60		$1644.80	$28.68	$51.40

Crew No.	Bare Costs Hr.	Bare Costs Daily	Incl. Subs O & P Hr.	Incl. Subs O & P Daily	Cost Per Labor-Hour Bare Costs	Cost Per Labor-Hour Incl. O&P
Crew H-2	Hr.	Daily	Hr.	Daily	Bare Costs	Incl. O&P
2 Glaziers	$27.40	$438.40	$45.25	$724.00	$25.12	$41.75
1 Building Laborer	20.55	164.40	34.75	278.00		
24 L.H., Daily Totals		$602.80		$1002.00	$25.12	$41.75
Crew H-3	Hr.	Daily	Hr.	Daily	Bare Costs	Incl. O&P
1 Glazier	$27.40	$219.20	$45.25	$362.00	$24.13	$40.20
1 Helper	20.85	166.80	35.15	281.20		
16 L.H., Daily Totals		$386.00		$643.20	$24.13	$40.20
Crew J-1	Hr.	Daily	Hr.	Daily	Bare Costs	Incl. O&P
3 Plasterers	$25.30	$607.20	$41.70	$1000.80	$24.22	$39.92
2 Plasterer Helpers	22.60	361.60	37.25	596.00		
1 Mixing Machine, 6 C.F.		126.60		139.26	3.17	3.48
40 L.H., Daily Totals		$1095.40		$1736.06	$27.39	$43.40
Crew J-2	Hr.	Daily	Hr.	Daily	Bare Costs	Incl. O&P
3 Plasterers	$25.30	$607.20	$41.70	$1000.80	$24.33	$39.98
2 Plasterer Helpers	22.60	361.60	37.25	596.00		
1 Lather	24.90	199.20	40.30	322.40		
1 Mixing Machine, 6 C.F.		126.60		139.26	2.64	2.90
48 L.H., Daily Totals		$1294.60		$2058.46	$26.97	$42.88
Crew J-3	Hr.	Daily	Hr.	Daily	Bare Costs	Incl. O&P
1 Terrazzo Worker	$26.40	$211.20	$42.35	$338.80	$24.00	$38.50
1 Terrazzo Helper	21.60	172.80	34.65	277.20		
1 Terrazzo Grinder, Electric		81.60		89.76		
1 Terrazzo Mixer		171.60		188.76	15.82	17.41
16 L.H., Daily Totals		$637.20		$894.52	$39.83	$55.91
Crew K-1	Hr.	Daily	Hr.	Daily	Bare Costs	Incl. O&P
1 Carpenter	$27.95	$223.60	$47.25	$378.00	$24.95	$42.05
1 Truck Driver (light)	21.95	175.60	36.85	294.80		
1 Flatbed Truck, Gas, 3 Ton		242.60		266.86	15.16	16.68
16 L.H., Daily Totals		$641.80		$939.66	$40.11	$58.73
Crew K-2	Hr.	Daily	Hr.	Daily	Bare Costs	Incl. O&P
1 Struc. Steel Foreman	$31.95	$255.60	$61.40	$491.20	$27.95	$51.93
1 Struc. Steel Worker	29.95	239.60	57.55	460.40		
1 Truck Driver (light)	21.95	175.60	36.85	294.80		
1 Flatbed Truck, Gas, 3 Ton		242.60		266.86	10.11	11.12
24 L.H., Daily Totals		$913.40		$1513.26	$38.06	$63.05
Crew L-1	Hr.	Daily	Hr.	Daily	Bare Costs	Incl. O&P
.25 Electrician	$31.50	$63.00	$51.30	$102.60	$32.06	$52.54
1 Plumber	32.20	257.60	52.85	422.80		
10 L.H., Daily Totals		$320.60		$525.40	$32.06	$52.54
Crew L-2	Hr.	Daily	Hr.	Daily	Bare Costs	Incl. O&P
1 Carpenter	$27.95	$223.60	$47.25	$378.00	$24.40	$41.20
1 Carpenter Helper	20.85	166.80	35.15	281.20		
16 L.H., Daily Totals		$390.40		$659.20	$24.40	$41.20
Crew L-3	Hr.	Daily	Hr.	Daily	Bare Costs	Incl. O&P
1 Carpenter	$27.95	$223.60	$47.25	$378.00	$28.66	$48.06
.25 Electrician	31.50	63.00	51.30	102.60		
10 L.H., Daily Totals		$286.60		$480.60	$28.66	$48.06

Crew L-3A

Crew No.	Bare Costs Hr.	Daily	Incl. Subs O & P Hr.	Daily	Cost Per Labor-Hour Bare Costs	Incl. O&P
1 Carpenter Foreman (outside)	$29.95	$239.60	$50.65	$405.20	$30.35	$51.20
.5 Sheet Metal Worker	31.15	124.60	52.30	209.20		
12 L.H., Daily Totals		$364.20		$614.40	$30.35	$51.20

Crew L-4

Crew No.	Bare Costs Hr.	Daily	Incl. Subs O & P Hr.	Daily	Cost Per Labor-Hour Bare Costs	Incl. O&P
1 Skilled Worker	$28.05	$224.40	$47.35	$378.80	$24.45	$41.25
1 Helper	20.85	166.80	35.15	281.20		
16 L.H., Daily Totals		$391.20		$660.00	$24.45	$41.25

Crew L-5

Crew No.	Bare Costs Hr.	Daily	Incl. Subs O & P Hr.	Daily	Cost Per Labor-Hour Bare Costs	Incl. O&P
1 Struc. Steel Foreman	$31.95	$255.60	$61.40	$491.20	$30.15	$56.76
5 Struc. Steel Workers	29.95	1198.00	57.55	2302.00		
1 Equip. Oper. (crane)	29.35	234.80	48.20	385.60		
1 Hyd. Crane, 25 Ton		789.40		868.34	14.10	15.51
56 L.H., Daily Totals		$2477.80		$4047.14	$44.25	$72.27

Crew L-5A

Crew No.	Bare Costs Hr.	Daily	Incl. Subs O & P Hr.	Daily	Cost Per Labor-Hour Bare Costs	Incl. O&P
1 Structural Steel Foreman	$31.95	$255.60	$61.40	$491.20	$30.30	$56.17
2 Structural Steel Workers	29.95	479.20	57.55	920.80		
1 Equip. Oper. (crane)	29.35	234.80	48.20	385.60		
1 S.P. Crane, 4x4, 25 Ton		716.80		788.48	22.40	24.64
32 L.H., Daily Totals		$1686.40		$2586.08	$52.70	$80.81

Crew L-5B

Crew No.	Bare Costs Hr.	Daily	Incl. Subs O & P Hr.	Daily	Cost Per Labor-Hour Bare Costs	Incl. O&P
1 Structural Steel Foreman	$31.95	$255.60	$61.40	$491.20	$29.98	$52.09
2 Structural Steel Workers	29.95	479.20	57.55	920.80		
2 Electricians	31.50	504.00	51.30	820.80		
2 Steamfitters/Pipefitters	32.55	520.80	53.40	854.40		
1 Equip. Oper. (crane)	29.35	234.80	48.20	385.60		
1 Common Building Laborer	20.55	164.40	34.75	278.00		
1 Hyd. Crane, 80 Ton		1296.00		1425.60	18.00	19.80
72 L.H., Daily Totals		$3454.80		$5176.40	$47.98	$71.89

Crew L-6

Crew No.	Bare Costs Hr.	Daily	Incl. Subs O & P Hr.	Daily	Cost Per Labor-Hour Bare Costs	Incl. O&P
1 Plumber	$32.20	$257.60	$52.85	$422.80	$31.97	$52.33
.5 Electrician	31.50	126.00	51.30	205.20		
12 L.H., Daily Totals		$383.60		$628.00	$31.97	$52.33

Crew L-7

Crew No.	Bare Costs Hr.	Daily	Incl. Subs O & P Hr.	Daily	Cost Per Labor-Hour Bare Costs	Incl. O&P
1 Carpenter	$27.95	$223.60	$47.25	$378.00	$23.85	$40.12
2 Carpenter Helpers	20.85	333.60	35.15	562.40		
.25 Electrician	31.50	63.00	51.30	102.60		
26 L.H., Daily Totals		$620.20		$1043.00	$23.85	$40.12

Crew L-8

Crew No.	Bare Costs Hr.	Daily	Incl. Subs O & P Hr.	Daily	Cost Per Labor-Hour Bare Costs	Incl. O&P
1 Carpenter	$27.95	$223.60	$47.25	$378.00	$25.96	$43.53
1 Carpenter Helper	20.85	166.80	35.15	281.20		
.5 Plumber	32.20	128.80	52.85	211.40		
20 L.H., Daily Totals		$519.20		$870.60	$25.96	$43.53

Crew L-9

Crew No.	Bare Costs Hr.	Daily	Incl. Subs O & P Hr.	Daily	Cost Per Labor-Hour Bare Costs	Incl. O&P
1 Skilled Worker Foreman	$30.05	$240.40	$50.70	$405.60	$25.68	$43.11
1 Skilled Worker	28.05	224.40	47.35	378.80		
2 Helpers	20.85	333.60	35.15	562.40		
.5 Electrician	31.50	126.00	51.30	205.20		
36 L.H., Daily Totals		$924.40		$1552.00	$25.68	$43.11

Crew L-10

Crew No.	Bare Costs Hr.	Daily	Incl. Subs O & P Hr.	Daily	Cost Per Labor-Hour Bare Costs	Incl. O&P
1 Structural Steel Foreman	$31.95	$255.60	$61.40	$491.20	$30.42	$55.72
1 Structural Steel Worker	29.95	239.60	57.55	460.40		
1 Equip. Oper. (crane)	29.35	234.80	48.20	385.60		
1 Hyd. Crane, 12 Ton		768.80		845.68	32.03	35.24
24 L.H., Daily Totals		$1498.80		$2182.88	$62.45	$90.95

Crew L-11

Crew No.	Bare Costs Hr.	Daily	Incl. Subs O & P Hr.	Daily	Cost Per Labor-Hour Bare Costs	Incl. O&P
2 Wreckers	$21.15	$338.40	$39.20	$627.20	$24.65	$42.73
1 Equip. Oper. (crane)	29.35	234.80	48.20	385.60		
1 Equip. Oper. (light)	26.95	215.60	44.30	354.40		
1 Hyd. Excavator, 2.5 C.Y.		1619.00		1780.90		
1 Loader, Skid Steer, 78 H.P.		259.60		285.56	58.71	64.58
32 L.H., Daily Totals		$2667.40		$3433.66	$83.36	$107.30

Crew M-1

Crew No.	Bare Costs Hr.	Daily	Incl. Subs O & P Hr.	Daily	Cost Per Labor-Hour Bare Costs	Incl. O&P
3 Elevator Constructors	$37.90	$909.60	$61.75	$1482.00	$36.00	$58.65
1 Elevator Apprentice	30.30	242.40	49.35	394.80		
5 Hand Tools		57.00		62.70	1.78	1.96
32 L.H., Daily Totals		$1209.00		$1939.50	$37.78	$60.61

Crew M-3

Crew No.	Bare Costs Hr.	Daily	Incl. Subs O & P Hr.	Daily	Cost Per Labor-Hour Bare Costs	Incl. O&P
1 Electrician Foreman (out)	$33.50	$268.00	$54.55	$436.40	$30.44	$49.91
1 Common Laborer	20.55	164.40	34.75	278.00		
.25 Equipment Operator, Medium	28.55	57.10	46.90	93.80		
1 Elevator Constructor	37.90	303.20	61.75	494.00		
1 Elevator Apprentice	30.30	242.40	49.35	394.80		
.25 S.P. Crane, 4x4, 20 Ton		175.15		192.66	5.15	5.67
34 L.H., Daily Totals		$1210.25		$1889.67	$35.60	$55.58

Crew M-4

Crew No.	Bare Costs Hr.	Daily	Incl. Subs O & P Hr.	Daily	Cost Per Labor-Hour Bare Costs	Incl. O&P
1 Electrician Foreman (out)	$33.50	$268.00	$54.55	$436.40	$30.21	$49.53
1 Common Laborer	20.55	164.40	34.75	278.00		
.25 Equipment Operator, Crane	29.35	58.70	48.20	96.40		
.25 Equipment Operator, Oiler	25.40	50.80	41.75	83.50		
1 Elevator Constructor	37.90	303.20	61.75	494.00		
1 Elevator Apprentice	30.30	242.40	49.35	394.80		
.25 S.P. Crane, 4x4, 40 Ton		234.30		257.73	6.51	7.16
36 L.H., Daily Totals		$1321.80		$2040.83	$36.72	$56.69

Crew Q-1

Crew No.	Bare Costs Hr.	Daily	Incl. Subs O & P Hr.	Daily	Cost Per Labor-Hour Bare Costs	Incl. O&P
1 Plumber	$32.20	$257.60	$52.85	$422.80	$28.98	$47.55
1 Plumber Apprentice	25.75	206.00	42.25	338.00		
16 L.H., Daily Totals		$463.60		$760.80	$28.98	$47.55

Crew Q-1A

Crew No.	Bare Costs Hr.	Daily	Incl. Subs O & P Hr.	Daily	Cost Per Labor-Hour Bare Costs	Incl. O&P
.25 Plumber Foreman (out)	$34.20	$68.40	$56.10	$112.20	$32.60	$53.50
1 Plumber	32.20	257.60	52.85	422.80		
10 L.H., Daily Totals		$326.00		$535.00	$32.60	$53.50

Crew Q-1C

Crew No.	Bare Costs Hr.	Daily	Incl. Subs O & P Hr.	Daily	Cost Per Labor-Hour Bare Costs	Incl. O&P
1 Plumber	$32.20	$257.60	$52.85	$422.80	$28.83	$47.33
1 Plumber Apprentice	25.75	206.00	42.25	338.00		
1 Equip. Oper. (medium)	28.55	228.40	46.90	375.20		
1 Trencher, Chain Type, 8' D		1777.00		1954.70	74.04	81.45
24 L.H., Daily Totals		$2469.00		$3090.70	$102.88	$128.78

Crew Q-2

Crew No.	Bare Costs Hr.	Daily	Incl. Subs O & P Hr.	Daily	Cost Per Labor-Hour Bare Costs	Incl. O&P
1 Plumber	$32.20	$257.60	$52.85	$422.80	$27.90	$45.78
2 Plumber Apprentices	25.75	412.00	42.25	676.00		
24 L.H., Daily Totals		$669.60		$1098.80	$27.90	$45.78

Crew Q-3

Crew No.	Bare Costs Hr.	Daily	Incl. Subs O & P Hr.	Daily	Cost Per Labor-Hour Bare Costs	Incl. O&P
2 Plumbers	$32.20	$515.20	$52.85	$845.60	$28.98	$47.55
2 Plumber Apprentices	25.75	412.00	42.25	676.00		
32 L.H., Daily Totals		$927.20		$1521.60	$28.98	$47.55

Crew Q-4

Crew No.	Bare Costs Hr.	Daily	Incl. Subs O & P Hr.	Daily	Cost Per Labor-Hour Bare Costs	Incl. O&P
2 Plumbers	$32.20	$515.20	$52.85	$845.60	$30.59	$50.20
1 Welder (plumber)	32.20	257.60	52.85	422.80		
1 Plumber Apprentice	25.75	206.00	42.25	338.00		
1 Welder, electric, 300 amp		55.70		61.27	1.74	1.91
32 L.H., Daily Totals		$1034.50		$1667.67	$32.33	$52.11

Crew Q-5

Crew No.	Bare Costs Hr.	Daily	Incl. Subs O & P Hr.	Daily	Cost Per Labor-Hour Bare Costs	Incl. O&P
1 Steamfitter	$32.55	$260.40	$53.40	$427.20	$29.30	$48.08
1 Steamfitter Apprentice	26.05	208.40	42.75	342.00		
16 L.H., Daily Totals		$468.80		$769.20	$29.30	$48.08

Crew Q-6

Crew No.	Bare Costs Hr.	Daily	Incl. Subs O & P Hr.	Daily	Cost Per Labor-Hour Bare Costs	Incl. O&P
1 Steamfitter	$32.55	$260.40	$53.40	$427.20	$28.22	$46.30
2 Steamfitter Apprentices	26.05	416.80	42.75	684.00		
24 L.H., Daily Totals		$677.20		$1111.20	$28.22	$46.30

Crew Q-7

Crew No.	Bare Costs Hr.	Daily	Incl. Subs O & P Hr.	Daily	Cost Per Labor-Hour Bare Costs	Incl. O&P
2 Steamfitters	$32.55	$520.80	$53.40	$854.40	$29.30	$48.08
2 Steamfitter Apprentices	26.05	416.80	42.75	684.00		
32 L.H., Daily Totals		$937.60		$1538.40	$29.30	$48.08

Crew Q-8

Crew No.	Bare Costs Hr.	Daily	Incl. Subs O & P Hr.	Daily	Cost Per Labor-Hour Bare Costs	Incl. O&P
2 Steamfitters	$32.55	$520.80	$53.40	$854.40	$30.93	$50.74
1 Welder (steamfitter)	32.55	260.40	53.40	427.20		
1 Steamfitter Apprentice	26.05	208.40	42.75	342.00		
1 Welder, electric, 300 amp		55.70		61.27	1.74	1.91
32 L.H., Daily Totals		$1045.30		$1684.87	$32.67	$52.65

Crew Q-9

Crew No.	Bare Costs Hr.	Daily	Incl. Subs O & P Hr.	Daily	Cost Per Labor-Hour Bare Costs	Incl. O&P
1 Sheet Metal Worker	$31.15	$249.20	$52.30	$418.40	$28.02	$47.05
1 Sheet Metal Apprentice	24.90	199.20	41.80	334.40		
16 L.H., Daily Totals		$448.40		$752.80	$28.02	$47.05

Crew Q-10

Crew No.	Bare Costs Hr.	Daily	Incl. Subs O & P Hr.	Daily	Cost Per Labor-Hour Bare Costs	Incl. O&P
2 Sheet Metal Workers	$31.15	$498.40	$52.30	$836.80	$29.07	$48.80
1 Sheet Metal Apprentice	24.90	199.20	41.80	334.40		
24 L.H., Daily Totals		$697.60		$1171.20	$29.07	$48.80

Crew Q-11

Crew No.	Bare Costs Hr.	Daily	Incl. Subs O & P Hr.	Daily	Cost Per Labor-Hour Bare Costs	Incl. O&P
2 Sheet Metal Workers	$31.15	$498.40	$52.30	$836.80	$28.02	$47.05
2 Sheet Metal Apprentices	24.90	398.40	41.80	668.80		
32 L.H., Daily Totals		$896.80		$1505.60	$28.02	$47.05

Crew Q-12

Crew No.	Bare Costs Hr.	Daily	Incl. Subs O & P Hr.	Daily	Cost Per Labor-Hour Bare Costs	Incl. O&P
1 Sprinkler Installer	$31.80	$254.40	$52.20	$417.60	$28.63	$46.98
1 Sprinkler Apprentice	25.45	203.60	41.75	334.00		
16 L.H., Daily Totals		$458.00		$751.60	$28.63	$46.98

Crew Q-13

Crew No.	Bare Costs Hr.	Daily	Incl. Subs O & P Hr.	Daily	Cost Per Labor-Hour Bare Costs	Incl. O&P
2 Sprinkler Installers	$31.80	$508.80	$52.20	$835.20	$28.63	$46.98
2 Sprinkler Apprentices	25.45	407.20	41.75	668.00		
32 L.H., Daily Totals		$916.00		$1503.20	$28.63	$46.98

Crew Q-14

Crew No.	Bare Costs Hr.	Daily	Incl. Subs O & P Hr.	Daily	Cost Per Labor-Hour Bare Costs	Incl. O&P
1 Asbestos Worker	$29.10	$232.80	$49.65	$397.20	$26.20	$44.70
1 Asbestos Apprentice	23.30	186.40	39.75	318.00		
16 L.H., Daily Totals		$419.20		$715.20	$26.20	$44.70

Crew Q-15

Crew No.	Bare Costs Hr.	Daily	Incl. Subs O & P Hr.	Daily	Cost Per Labor-Hour Bare Costs	Incl. O&P
1 Plumber	$32.20	$257.60	$52.85	$422.80	$28.98	$47.55
1 Plumber Apprentice	25.75	206.00	42.25	338.00		
1 Welder, electric, 300 amp		55.70		61.27	3.48	3.83
16 L.H., Daily Totals		$519.30		$822.07	$32.46	$51.38

Crew Q-16

Crew No.	Bare Costs Hr.	Daily	Incl. Subs O & P Hr.	Daily	Cost Per Labor-Hour Bare Costs	Incl. O&P
2 Plumbers	$32.20	$515.20	$52.85	$845.60	$30.05	$49.32
1 Plumber Apprentice	25.75	206.00	42.25	338.00		
1 Welder, electric, 300 amp		55.70		61.27	2.32	2.55
24 L.H., Daily Totals		$776.90		$1244.87	$32.37	$51.87

Crew Q-17

Crew No.	Bare Costs Hr.	Daily	Incl. Subs O & P Hr.	Daily	Cost Per Labor-Hour Bare Costs	Incl. O&P
1 Steamfitter	$32.55	$260.40	$53.40	$427.20	$29.30	$48.08
1 Steamfitter Apprentice	26.05	208.40	42.75	342.00		
1 Welder, electric, 300 amp		55.70		61.27	3.48	3.83
16 L.H., Daily Totals		$524.50		$830.47	$32.78	$51.90

Crew Q-17A

Crew No.	Bare Costs Hr.	Daily	Incl. Subs O & P Hr.	Daily	Cost Per Labor-Hour Bare Costs	Incl. O&P
1 Steamfitter	$32.55	$260.40	$53.40	$427.20	$29.32	$48.12
1 Steamfitter Apprentice	26.05	208.40	42.75	342.00		
1 Equip. Oper. (crane)	29.35	234.80	48.20	385.60		
1 Hyd. Crane, 12 Ton		768.80		845.68		
1 Welder, electric, 300 amp		55.70		61.27	34.35	37.79
24 L.H., Daily Totals		$1528.10		$2061.75	$63.67	$85.91

Crew Q-18

Crew No.	Bare Costs Hr.	Daily	Incl. Subs O & P Hr.	Daily	Cost Per Labor-Hour Bare Costs	Incl. O&P
2 Steamfitters	$32.55	$520.80	$53.40	$854.40	$30.38	$49.85
1 Steamfitter Apprentice	26.05	208.40	42.75	342.00		
1 Welder, electric, 300 amp		55.70		61.27	2.32	2.55
24 L.H., Daily Totals		$784.90		$1257.67	$32.70	$52.40

Crew Q-19

Crew No.	Bare Costs Hr.	Daily	Incl. Subs O & P Hr.	Daily	Cost Per Labor-Hour Bare Costs	Incl. O&P
1 Steamfitter	$32.55	$260.40	$53.40	$427.20	$30.03	$49.15
1 Steamfitter Apprentice	26.05	208.40	42.75	342.00		
1 Electrician	31.50	252.00	51.30	410.40		
24 L.H., Daily Totals		$720.80		$1179.60	$30.03	$49.15

Crew Q-20

Crew No.	Bare Costs Hr.	Daily	Incl. Subs O & P Hr.	Daily	Cost Per Labor-Hour Bare Costs	Incl. O&P
1 Sheet Metal Worker	$31.15	$249.20	$52.30	$418.40	$28.72	$47.90
1 Sheet Metal Apprentice	24.90	199.20	41.80	334.40		
.5 Electrician	31.50	126.00	51.30	205.20		
20 L.H., Daily Totals		$574.40		$958.00	$28.72	$47.90

Crew Q-21

Crew No.	Bare Costs Hr.	Daily	Incl. Subs O & P Hr.	Daily	Cost Per Labor-Hour Bare Costs	Incl. O&P
2 Steamfitters	$32.55	$520.80	$53.40	$854.40	$30.66	$50.21
1 Steamfitter Apprentice	26.05	208.40	42.75	342.00		
1 Electrician	31.50	252.00	51.30	410.40		
32 L.H., Daily Totals		$981.20		$1606.80	$30.66	$50.21

Crew Q-22

Crew No.	Bare Costs Hr.	Daily	Incl. Subs O & P Hr.	Daily	Cost Per Labor-Hour Bare Costs	Incl. O&P
1 Plumber	$32.20	$257.60	$52.85	$422.80	$28.98	$47.55
1 Plumber Apprentice	25.75	206.00	42.25	338.00		
1 Hyd. Crane, 12 Ton		768.80		845.68	48.05	52.85
16 L.H., Daily Totals		$1232.40		$1606.48	$77.03	$100.41

Crew Q-22A	Bare Costs Hr.	Bare Costs Daily	Incl. Subs O & P Hr.	Incl. Subs O & P Daily	Cost Per Labor-Hour Bare Costs	Cost Per Labor-Hour Incl. O&P
1 Plumber	$32.20	$257.60	$52.85	$422.80	$26.96	$44.51
1 Plumber Apprentice	25.75	206.00	42.25	338.00		
1 Laborer	20.55	164.40	34.75	278.00		
1 Equip. Oper. (crane)	29.35	234.80	48.20	385.60		
1 Hyd. Crane, 12 Ton		768.80		845.68	24.02	26.43
32 L.H., Daily Totals		$1631.60		$2270.08	$50.99	$70.94

Crew Q-23	Hr.	Daily	Hr.	Daily	Bare Costs	Incl. O&P
1 Plumber Foreman	$34.20	$273.60	$56.10	$448.80	$31.65	$51.95
1 Plumber	32.20	257.60	52.85	422.80		
1 Equip. Oper. (medium)	28.55	228.40	46.90	375.20		
1 Lattice Boom Crane, 20 Ton		1080.00		1188.00	45.00	49.50
24 L.H., Daily Totals		$1839.60		$2434.80	$76.65	$101.45

Crew R-1	Hr.	Daily	Hr.	Daily	Bare Costs	Incl. O&P
1 Electrician Foreman	$32.00	$256.00	$52.10	$416.80	$28.03	$46.05
3 Electricians	31.50	756.00	51.30	1231.20		
2 Helpers	20.85	333.60	35.15	562.40		
48 L.H., Daily Totals		$1345.60		$2210.40	$28.03	$46.05

Crew R-1A	Hr.	Daily	Hr.	Daily	Bare Costs	Incl. O&P
1 Electrician	$31.50	$252.00	$51.30	$410.40	$26.18	$43.23
1 Helper	20.85	166.80	35.15	281.20		
16 L.H., Daily Totals		$418.80		$691.60	$26.18	$43.23

Crew R-2	Hr.	Daily	Hr.	Daily	Bare Costs	Incl. O&P
1 Electrician Foreman	$32.00	$256.00	$52.10	$416.80	$28.22	$46.36
3 Electricians	31.50	756.00	51.30	1231.20		
2 Helpers	20.85	333.60	35.15	562.40		
1 Equip. Oper. (crane)	29.35	234.80	48.20	385.60		
1 S.P. Crane, 4x4, 5 Ton		258.20		284.02	4.61	5.07
56 L.H., Daily Totals		$1838.60		$2880.02	$32.83	$51.43

Crew R-3	Hr.	Daily	Hr.	Daily	Bare Costs	Incl. O&P
1 Electrician Foreman	$32.00	$256.00	$52.10	$416.80	$31.27	$51.00
1 Electrician	31.50	252.00	51.30	410.40		
.5 Equip. Oper. (crane)	29.35	117.40	48.20	192.80		
.5 S.P. Crane, 4x4, 5 Ton		129.10		142.01	6.46	7.10
20 L.H., Daily Totals		$754.50		$1162.01	$37.73	$58.10

Crew R-4	Hr.	Daily	Hr.	Daily	Bare Costs	Incl. O&P
1 Struc. Steel Foreman	$31.95	$255.60	$61.40	$491.20	$30.66	$57.07
3 Struc. Steel Workers	29.95	718.80	57.55	1381.20		
1 Electrician	31.50	252.00	51.30	410.40		
1 Welder, gas engine, 300 amp		134.20		147.62	3.36	3.69
40 L.H., Daily Totals		$1360.60		$2430.42	$34.02	$60.76

Crew R-5	Hr.	Daily	Hr.	Daily	Bare Costs	Incl. O&P
1 Electrician Foreman	$32.00	$256.00	$52.10	$416.80	$27.67	$45.50
4 Electrician Linemen	31.50	1008.00	51.30	1641.60		
2 Electrician Operators	31.50	504.00	51.30	820.80		
4 Electrician Groundmen	20.85	667.20	35.15	1124.80		
1 Crew Truck		169.60		186.56		
1 Flatbed Truck, 20,000 GVW		197.40		217.14		
1 Pickup Truck, 3/4 Ton		116.80		128.48		
.2 Hyd. Crane, 55 Ton		225.00		247.50		
.2 Hyd. Crane, 12 Ton		153.76		169.14		
.2 Earth Auger, Truck-Mtd.		82.40		90.64		
1 Tractor w/Winch		354.80		390.28	14.77	16.25
88 L.H., Daily Totals		$3734.96		$5433.74	$42.44	$61.75

Crew R-6	Hr.	Daily	Hr.	Daily	Bare Costs	Incl. O&P
1 Electrician Foreman	$32.00	$256.00	$52.10	$416.80	$27.67	$45.50
4 Electrician Linemen	31.50	1008.00	51.30	1641.60		
2 Electrician Operators	31.50	504.00	51.30	820.80		
4 Electrician Groundmen	20.85	667.20	35.15	1124.80		
1 Crew Truck		169.60		186.56		
1 Flatbed Truck, 20,000 GVW		197.40		217.14		
1 Pickup Truck, 3/4 Ton		116.80		128.48		
.2 Hyd. Crane, 55 Ton		225.00		247.50		
.2 Hyd. Crane, 12 Ton		153.76		169.14		
.2 Earth Auger, Truck-Mtd.		82.40		90.64		
1 Tractor w/Winch		354.80		390.28		
3 Cable Trailers		531.30		584.43		
.5 Tensioning Rig		175.35		192.88		
.5 Cable Pulling Rig		1049.50		1154.45	34.73	38.20
88 L.H., Daily Totals		$5491.11		$7365.50	$62.40	$83.70

Crew R-7	Hr.	Daily	Hr.	Daily	Bare Costs	Incl. O&P
1 Electrician Foreman	$32.00	$256.00	$52.10	$416.80	$22.71	$37.98
5 Electrician Groundmen	20.85	834.00	35.15	1406.00		
1 Crew Truck		169.60		186.56	3.53	3.89
48 L.H., Daily Totals		$1259.60		$2009.36	$26.24	$41.86

Crew R-8	Hr.	Daily	Hr.	Daily	Bare Costs	Incl. O&P
1 Electrician Foreman	$32.00	$256.00	$52.10	$416.80	$28.03	$46.05
3 Electrician Linemen	31.50	756.00	51.30	1231.20		
2 Electrician Groundmen	20.85	333.60	35.15	562.40		
1 Pickup Truck, 3/4 Ton		116.80		128.48		
1 Crew Truck		169.60		186.56	5.97	6.56
48 L.H., Daily Totals		$1632.00		$2525.44	$34.00	$52.61

Crew R-9	Hr.	Daily	Hr.	Daily	Bare Costs	Incl. O&P
1 Electrician Foreman	$32.00	$256.00	$52.10	$416.80	$26.24	$43.33
1 Electrician Lineman	31.50	252.00	51.30	410.40		
2 Electrician Operators	31.50	504.00	51.30	820.80		
4 Electrician Groundmen	20.85	667.20	35.15	1124.80		
1 Pickup Truck, 3/4 Ton		116.80		128.48		
1 Crew Truck		169.60		186.56	4.47	4.92
64 L.H., Daily Totals		$1965.60		$3087.84	$30.71	$48.25

Crew R-10	Hr.	Daily	Hr.	Daily	Bare Costs	Incl. O&P
1 Electrician Foreman	$32.00	$256.00	$52.10	$416.80	$29.81	$48.74
4 Electrician Linemen	31.50	1008.00	51.30	1641.60		
1 Electrician Groundman	20.85	166.80	35.15	281.20		
1 Crew Truck		169.60		186.56		
3 Tram Cars		380.55		418.61	11.46	12.61
48 L.H., Daily Totals		$1980.95		$2944.76	$41.27	$61.35

Crew R-11	Hr.	Daily	Hr.	Daily	Bare Costs	Incl. O&P
1 Electrician Foreman	$32.00	$256.00	$52.10	$416.80	$29.70	$48.61
4 Electricians	31.50	1008.00	51.30	1641.60		
1 Equip. Oper. (crane)	29.35	234.80	48.20	385.60		
1 Common Laborer	20.55	164.40	34.75	278.00		
1 Crew Truck		169.60		186.56		
1 Hyd. Crane, 12 Ton		768.80		845.68	16.76	18.43
56 L.H., Daily Totals		$2601.60		$3754.24	$46.46	$67.04

Crew No.	Bare Costs		Incl. Subs O & P		Cost Per Labor-Hour	

Crew R-12	Hr.	Daily	Hr.	Daily	Bare Costs	Incl. O&P
1 Carpenter Foreman	$28.45	$227.60	$48.10	$384.80	$25.54	$43.69
4 Carpenters	27.95	894.40	47.25	1512.00		
4 Common Laborers	20.55	657.60	34.75	1112.00		
1 Equip. Oper. (med.)	28.55	228.40	46.90	375.20		
1 Steel Worker	29.95	239.60	57.55	460.40		
1 Dozer, 200 H.P.		1082.00		1190.20		
1 Pickup Truck, 3/4 Ton		116.80		128.48	13.62	14.98
88 L.H., Daily Totals		$3446.40		$5163.08	$39.16	$58.67

Crew R-15	Hr.	Daily	Hr.	Daily	Bare Costs	Incl. O&P
1 Electrician Foreman	$32.00	$256.00	$52.10	$416.80	$30.82	$50.27
4 Electricians	31.50	1008.00	51.30	1641.60		
1 Equipment Oper. (Light)	26.95	215.60	44.30	354.40		
1 Aerial Lift Truck		328.80		361.68	6.85	7.54
48 L.H., Daily Totals		$1808.40		$2774.48	$37.67	$57.80

Crew R-15A	Hr.	Daily	Hr.	Daily	Bare Costs	Incl. O&P
1 Electrician Foreman	$32.00	$256.00	$52.10	$416.80	$27.18	$44.75
2 Electricians	31.50	504.00	51.30	820.80		
2 Common Laborers	20.55	328.80	34.75	556.00		
1 Equipment Operator	26.95	215.60	44.30	354.40		
1 Aerial Lift Truck		328.80		361.68	6.85	7.54
48 L.H., Daily Totals		$1633.20		$2509.68	$34.02	$52.28

Crew R-18	Hr.	Daily	Hr.	Daily	Bare Costs	Incl. O&P
.25 Electrician Foreman	$32.00	$64.00	$52.10	$104.20	$24.98	$41.42
1 Electrician	31.50	252.00	51.30	410.40		
2 Helpers	20.85	333.60	35.15	562.40		
26 L.H., Daily Totals		$649.60		$1077.00	$24.98	$41.42

Crew R-19	Hr.	Daily	Hr.	Daily	Bare Costs	Incl. O&P
.5 Electrician Foreman	$32.00	$128.00	$52.10	$208.40	$31.60	$51.46
2 Electricians	31.50	504.00	51.30	820.80		
20 L.H., Daily Totals		$632.00		$1029.20	$31.60	$51.46

Crew R-21	Hr.	Daily	Hr.	Daily	Bare Costs	Incl. O&P
1 Electrician Foreman	$32.00	$256.00	$52.10	$416.80	$31.55	$51.39
3 Electricians	31.50	756.00	51.30	1231.20		
.1 Equip. Oper. (med.)	28.55	22.84	46.90	37.52		
.1 S.P. Crane, 4x4, 25 Ton		71.68		78.85	2.19	2.40
32.8 L.H., Daily Totals		$1106.52		$1764.37	$33.74	$53.79

Crew R-22	Hr.	Daily	Hr.	Daily	Bare Costs	Incl. O&P
.66 Electrician Foreman	$32.00	$168.96	$52.10	$275.09	$27.00	$44.48
2 Helpers	20.85	333.60	35.15	562.40		
2 Electricians	31.50	504.00	51.30	820.80		
37.28 L.H., Daily Totals		$1006.56		$1658.29	$27.00	$44.48

Crew R-30	Hr.	Daily	Hr.	Daily	Bare Costs	Incl. O&P
.25 Electrician Foreman (out)	$33.50	$67.00	$54.55	$109.10	$24.92	$41.37
1 Electrician	31.50	252.00	51.30	410.40		
2 Laborers, (Semi-Skilled)	20.55	328.80	34.75	556.00		
26 L.H., Daily Totals		$647.80		$1075.50	$24.92	$41.37

Location Factors

Costs shown in *RSMeans Residential Cost Data* are based on national averages for materials and installation. To adjust these costs to a specific location, simply multiply the base cost by the factor for that city. The data is arranged alphabetically by state and postal zip code numbers. For a city not listed, use the factor for a nearby city with similar economic characteristics.

STATE	CITY	Residential
ALABAMA		
350-352	Birmingham	.87
354	Tuscaloosa	.78
355	Jasper	.72
356	Decatur	.78
357-358	Huntsville	.84
359	Gadsden	.75
360-361	Montgomery	.77
362	Anniston	.73
363	Dothan	.76
364	Evergreen	.74
365-366	Mobile	.82
367	Selma	.74
368	Phenix City	.75
369	Butler	.75
ALASKA		
995-996	Anchorage	1.25
997	Fairbanks	1.28
998	Juneau	1.24
999	Ketchikan	1.28
ARIZONA		
850,853	Phoenix	.85
852	Mesa/Tempe	.82
855	Globe	.78
856-857	Tucson	.83
859	Show Low	.80
860	Flagstaff	.85
863	Prescott	.79
864	Kingman	.83
865	Chambers	.79
ARKANSAS		
716	Pine Bluff	.80
717	Camden	.68
718	Texarkana	.73
719	Hot Springs	.69
720-722	Little Rock	.84
723	West Memphis	.79
724	Jonesboro	.77
725	Batesville	.74
726	Harrison	.76
727	Fayetteville	.71
728	Russellville	.76
729	Fort Smith	.77
CALIFORNIA		
900-902	Los Angeles	1.08
903-905	Inglewood	1.03
906-908	Long Beach	1.02
910-912	Pasadena	1.02
913-916	Van Nuys	1.05
917-918	Alhambra	1.06
919-921	San Diego	1.04
922	Palm Springs	1.02
923-924	San Bernardino	1.03
925	Riverside	1.07
926-927	Santa Ana	1.04
928	Anaheim	1.07
930	Oxnard	1.08
931	Santa Barbara	1.07
932-933	Bakersfield	1.06
934	San Luis Obispo	1.05
935	Mojave	1.03
936-938	Fresno	1.09
939	Salinas	1.11
940-941	San Francisco	1.26
942,956-958	Sacramento	1.12
943	Palo Alto	1.16
944	San Mateo	1.23
945	Vallejo	1.15
946	Oakland	1.22
947	Berkeley	1.22
948	Richmond	1.24
949	San Rafael	1.22
950	Santa Cruz	1.14
951	San Jose	1.21
952	Stockton	1.08
953	Modesto	1.08

STATE	CITY	Residential
CALIFORNIA (CONT'D)		
954	Santa Rosa	1.16
955	Eureka	1.11
959	Marysville	1.09
960	Redding	1.09
961	Susanville	1.09
COLORADO		
800-802	Denver	.93
803	Boulder	.92
804	Golden	.90
805	Fort Collins	.88
806	Greeley	.78
807	Fort Morgan	.91
808-809	Colorado Springs	.89
810	Pueblo	.90
811	Alamosa	.86
812	Salida	.89
813	Durango	.89
814	Montrose	.86
815	Grand Junction	.90
816	Glenwood Springs	.88
CONNECTICUT		
060	New Britain	1.09
061	Hartford	1.09
062	Willimantic	1.10
063	New London	1.09
064	Meriden	1.09
065	New Haven	1.10
066	Bridgeport	1.10
067	Waterbury	1.10
068	Norwalk	1.10
069	Stamford	1.11
D.C.		
200-205	Washington	.96
DELAWARE		
197	Newark	1.02
198	Wilmington	1.03
199	Dover	1.02
FLORIDA		
320,322	Jacksonville	.80
321	Daytona Beach	.89
323	Tallahassee	.77
324	Panama City	.74
325	Pensacola	.81
326,344	Gainesville	.80
327-328,347	Orlando	.89
329	Melbourne	.90
330-332,340	Miami	.86
333	Fort Lauderdale	.84
334,349	West Palm Beach	.84
335-336,346	Tampa	.91
337	St. Petersburg	.78
338	Lakeland	.88
339,341	Fort Myers	.86
342	Sarasota	.89
GEORGIA		
300-303,399	Atlanta	.90
304	Statesboro	.71
305	Gainesville	.78
306	Athens	.78
307	Dalton	.74
308-309	Augusta	.80
310-312	Macon	.81
313-314	Savannah	.81
315	Waycross	.74
316	Valdosta	.72
317,398	Albany	.77
318-319	Columbus	.82
HAWAII		
967	Hilo	1.19
968	Honolulu	1.21

STATE	CITY	Residential
STATES & POSS.		
969	Guam	.97
IDAHO		
832	Pocatello	.86
833	Twin Falls	.72
834	Idaho Falls	.74
835	Lewiston	.96
836-837	Boise	.86
838	Coeur d'Alene	.93
ILLINOIS		
600-603	North Suburban	1.11
604	Joliet	1.14
605	South Suburban	1.11
606-608	Chicago	1.20
609	Kankakee	1.00
610-611	Rockford	1.06
612	Rock Island	.97
613	La Salle	1.05
614	Galesburg	.99
615-616	Peoria	1.03
617	Bloomington	1.01
618-619	Champaign	1.03
620-622	East St. Louis	1.01
623	Quincy	.99
624	Effingham	.98
625	Decatur	1.01
626-627	Springfield	1.01
628	Centralia	.99
629	Carbondale	.95
INDIANA		
460	Anderson	.90
461-462	Indianapolis	.93
463-464	Gary	1.01
465-466	South Bend	.90
467-468	Fort Wayne	.89
469	Kokomo	.91
470	Lawrenceburg	.85
471	New Albany	.85
472	Columbus	.90
473	Muncie	.91
474	Bloomington	.92
475	Washington	.89
476-477	Evansville	.90
478	Terre Haute	.90
479	Lafayette	.91
IOWA		
500-503,509	Des Moines	.89
504	Mason City	.76
505	Fort Dodge	.75
506-507	Waterloo	.78
508	Creston	.79
510-511	Sioux City	.84
512	Sibley	.72
513	Spencer	.73
514	Carroll	.73
515	Council Bluffs	.81
516	Shenandoah	.73
520	Dubuque	.84
521	Decorah	.74
522-524	Cedar Rapids	.92
525	Ottumwa	.82
526	Burlington	.85
527-528	Davenport	.95
KANSAS		
660-662	Kansas City	.98
664-666	Topeka	.79
667	Fort Scott	.87
668	Emporia	.74
669	Belleville	.78
670-672	Wichita	.79
673	Independence	.84
674	Salina	.77
675	Hutchinson	.78
676	Hays	.81
677	Colby	.82
678	Dodge City	.81
679	Liberal	.79
KENTUCKY		
400-402	Louisville	.91
403-405	Lexington	.88

STATE	CITY	Residential
KENTUCKY (CONT'D)		
406	Frankfort	.85
407-409	Corbin	.75
410	Covington	.97
411-412	Ashland	.91
413-414	Campton	.76
415-416	Pikeville	.83
417-418	Hazard	.72
420	Paducah	.89
421-422	Bowling Green	.89
423	Owensboro	.86
424	Henderson	.90
425-426	Somerset	.76
427	Elizabethtown	.87
LOUISIANA		
700-701	New Orleans	.86
703	Thibodaux	.82
704	Hammond	.77
705	Lafayette	.80
706	Lake Charles	.82
707-708	Baton Rouge	.84
710-711	Shreveport	.78
712	Monroe	.73
713-714	Alexandria	.74
MAINE		
039	Kittery	.86
040-041	Portland	.88
042	Lewiston	.87
043	Augusta	.88
044	Bangor	.86
045	Bath	.86
046	Machias	.87
047	Houlton	.88
048	Rockland	.87
049	Waterville	.86
MARYLAND		
206	Waldorf	.85
207-208	College Park	.88
209	Silver Spring	.86
210-212	Baltimore	.90
214	Annapolis	.84
215	Cumberland	.86
216	Easton	.67
217	Hagerstown	.86
218	Salisbury	.73
219	Elkton	.79
MASSACHUSETTS		
010-011	Springfield	1.04
012	Pittsfield	1.02
013	Greenfield	1.00
014	Fitchburg	1.11
015-016	Worcester	1.12
017	Framingham	1.13
018	Lowell	1.13
019	Lawrence	1.13
020-022, 024	Boston	1.20
023	Brockton	1.12
025	Buzzards Bay	1.10
026	Hyannis	1.10
027	New Bedford	1.12
MICHIGAN		
480,483	Royal Oak	1.00
481	Ann Arbor	1.01
482	Detroit	1.06
484-485	Flint	.97
486	Saginaw	.91
487	Bay City	.92
488-489	Lansing	.96
490	Battle Creek	.92
491	Kalamazoo	.91
492	Jackson	.92
493,495	Grand Rapids	.80
494	Muskegon	.87
496	Traverse City	.78
497	Gaylord	.81
498-499	Iron Mountain	.87
MINNESOTA		
550-551	Saint Paul	1.11
553-555	Minneapolis	1.15
556-558	Duluth	1.07

Location Factors

STATE	CITY	Residential
MINNESOTA (CONT'd)		
559	Rochester	1.03
560	Mankato	1.01
561	Windom	.82
562	Willmar	.83
563	St. Cloud	1.06
564	Brainerd	.96
565	Detroit Lakes	.95
566	Bemidji	.94
567	Thief River Falls	.94
MISSISSIPPI		
386	Clarksdale	.78
387	Greenville	.84
388	Tupelo	.79
389	Greenwood	.80
390-392	Jackson	.85
393	Meridian	.83
394	Laurel	.80
395	Biloxi	.82
396	Mccomb	.77
397	Columbus	.78
MISSOURI		
630-631	St. Louis	1.03
633	Bowling Green	.95
634	Hannibal	.86
635	Kirksville	.80
636	Flat River	.94
637	Cape Girardeau	.88
638	Sikeston	.82
639	Poplar Bluff	.83
640-641	Kansas City	1.03
644-645	St. Joseph	.93
646	Chillicothe	.87
647	Harrisonville	.96
648	Joplin	.83
650-651	Jefferson City	.87
652	Columbia	.87
653	Sedalia	.85
654-655	Rolla	.87
656-658	Springfield	.87
MONTANA		
590-591	Billings	.88
592	Wolf Point	.84
593	Miles City	.86
594	Great Falls	.89
595	Havre	.82
596	Helena	.88
597	Butte	.87
598	Missoula	.85
599	Kalispell	.83
NEBRASKA		
680-681	Omaha	.91
683-685	Lincoln	.87
686	Columbus	.87
687	Norfolk	.91
688	Grand Island	.92
689	Hastings	.93
690	Mccook	.85
691	North Platte	.92
692	Valentine	.85
693	Alliance	.85
NEVADA		
889-891	Las Vegas	1.03
893	Ely	.85
894-895	Reno	.93
897	Carson City	.94
898	Elko	.91
NEW HAMPSHIRE		
030	Nashua	.94
031	Manchester	.94
032-033	Concord	.92
034	Keene	.75
035	Littleton	.81
036	Charleston	.74
037	Claremont	.75
038	Portsmouth	.93

STATE	CITY	Residential
NEW JERSEY		
070-071	Newark	1.12
072	Elizabeth	1.14
073	Jersey City	1.10
074-075	Paterson	1.11
076	Hackensack	1.10
077	Long Branch	1.11
078	Dover	1.11
079	Summit	1.11
080,083	Vineland	1.08
081	Camden	1.09
082,084	Atlantic City	1.11
085-086	Trenton	1.10
087	Point Pleasant	1.09
088-089	New Brunswick	1.11
NEW MEXICO		
870-872	Albuquerque	.85
873	Gallup	.85
874	Farmington	.85
875	Santa Fe	.86
877	Las Vegas	.85
878	Socorro	.85
879	Truth/Consequences	.84
880	Las Cruces	.83
881	Clovis	.85
882	Roswell	.85
883	Carrizozo	.85
884	Tucumcari	.86
NEW YORK		
100-102	New York	1.37
103	Staten Island	1.31
104	Bronx	1.33
105	Mount Vernon	1.14
106	White Plains	1.17
107	Yonkers	1.18
108	New Rochelle	1.18
109	Suffern	1.13
110	Queens	1.31
111	Long Island City	1.34
112	Brooklyn	1.35
113	Flushing	1.33
114	Jamaica	1.33
115,117,118	Hicksville	1.20
116	Far Rockaway	1.32
119	Riverhead	1.21
120-122	Albany	.94
123	Schenectady	.95
124	Kingston	1.02
125-126	Poughkeepsie	1.19
127	Monticello	1.04
128	Glens Falls	.88
129	Plattsburgh	.92
130-132	Syracuse	.96
133-135	Utica	.94
136	Watertown	.93
137-139	Binghamton	.93
140-142	Buffalo	1.04
143	Niagara Falls	1.00
144-146	Rochester	.96
147	Jamestown	.87
148-149	Elmira	.85
NORTH CAROLINA		
270,272-274	Greensboro	.83
271	Winston-Salem	.83
275-276	Raleigh	.84
277	Durham	.83
278	Rocky Mount	.73
279	Elizabeth City	.75
280	Gastonia	.84
281-282	Charlotte	.85
283	Fayetteville	.82
284	Wilmington	.81
285	Kinston	.74
286	Hickory	.78
287-288	Asheville	.81
289	Murphy	.73
NORTH DAKOTA		
580-581	Fargo	.78
582	Grand Forks	.75
583	Devils Lake	.78
584	Jamestown	.73
585	Bismarck	.78

Location Factors

STATE	CITY	Residential
NORTH DAKOTA (CONT'D)		
586	Dickinson	.76
587	Minot	.81
588	Williston	.76
OHIO		
430-432	Columbus	.93
433	Marion	.89
434-436	Toledo	1.00
437-438	Zanesville	.88
439	Steubenville	.93
440	Lorain	.98
441	Cleveland	1.01
442-443	Akron	.98
444-445	Youngstown	.95
446-447	Canton	.93
448-449	Mansfield	.93
450	Hamilton	.92
451-452	Cincinnati	.92
453-454	Dayton	.91
455	Springfield	.92
456	Chillicothe	.94
457	Athens	.87
458	Lima	.90
OKLAHOMA		
730-731	Oklahoma City	.79
734	Ardmore	.78
735	Lawton	.80
736	Clinton	.76
737	Enid	.76
738	Woodward	.76
739	Guymon	.67
740-741	Tulsa	.77
743	Miami	.81
744	Muskogee	.71
745	Mcalester	.73
746	Ponca City	.77
747	Durant	.77
748	Shawnee	.75
749	Poteau	.77
OREGON		
970-972	Portland	1.00
973	Salem	.98
974	Eugene	.99
975	Medford	.98
976	Klamath Falls	.98
977	Bend	1.00
978	Pendleton	.98
979	Vale	.97
PENNSYLVANIA		
150-152	Pittsburgh	.96
153	Washington	.93
154	Uniontown	.90
155	Bedford	.87
156	Greensburg	.93
157	Indiana	.90
158	Dubois	.89
159	Johnstown	.89
160	Butler	.91
161	New Castle	.91
162	Kittanning	.93
163	Oil City	.89
164-165	Erie	.93
166	Altoona	.87
167	Bradford	.89
168	State College	.90
169	Wellsboro	.90
170-171	Harrisburg	.94
172	Chambersburg	.89
173-174	York	.91
175-176	Lancaster	.91
177	Williamsport	.85
178	Sunbury	.91
179	Pottsville	.91
180	Lehigh Valley	1.01
181	Allentown	1.03
182	Hazleton	.90
183	Stroudsburg	.91
184-185	Scranton	.95
186-187	Wilkes-Barre	.92
188	Montrose	.90
189	Doylestown	1.05

STATE	CITY	Residential
PENNSYLVANIA (CONT'D)		
190-191	Philadelphia	1.16
193	Westchester	1.10
194	Norristown	1.09
195-196	Reading	.97
PUERTO RICO		
009	San Juan	.75
RHODE ISLAND		
028	Newport	1.06
029	Providence	1.06
SOUTH CAROLINA		
290-292	Columbia	.84
293	Spartanburg	.84
294	Charleston	.87
295	Florence	.80
296	Greenville	.83
297	Rock Hill	.82
298	Aiken	.97
299	Beaufort	.82
SOUTH DAKOTA		
570-571	Sioux Falls	.79
572	Watertown	.75
573	Mitchell	.77
574	Aberdeen	.77
575	Pierre	.77
576	Mobridge	.75
577	Rapid City	.78
TENNESSEE		
370-372	Nashville	.84
373-374	Chattanooga	.75
375,380-381	Memphis	.81
376	Johnson City	.70
377-379	Knoxville	.72
382	McKenzie	.72
383	Jackson	.70
384	Columbia	.71
385	Cookeville	.71
TEXAS		
750	McKinney	.73
751	Waxahackie	.74
752-753	Dallas	.83
754	Greenville	.68
755	Texarkana	.72
756	Longview	.67
757	Tyler	.73
758	Palestine	.66
759	Lufkin	.70
760-761	Fort Worth	.81
762	Denton	.75
763	Wichita Falls	.78
764	Eastland	.71
765	Temple	.74
766-767	Waco	.76
768	Brownwood	.68
769	San Angelo	.71
770-772	Houston	.85
773	Huntsville	.68
774	Wharton	.69
775	Galveston	.83
776-777	Beaumont	.80
778	Bryan	.73
779	Victoria	.73
780	Laredo	.72
781-782	San Antonio	.80
783-784	Corpus Christi	.77
785	McAllen	.75
786-787	Austin	.79
788	Del Rio	.66
789	Giddings	.69
790-791	Amarillo	.76
792	Childress	.74
793-794	Lubbock	.74
795-796	Abilene	.74
797	Midland	.75
798-799,885	El Paso	.73
UTAH		
840-841	Salt Lake City	.81
842,844	Ogden	.78
843	Logan	.79

Location Factors

STATE	CITY	Residential
UTAH (CONT'D)		
845	Price	.70
846-847	Provo	.80
VERMONT		
050	White River Jct.	.76
051	Bellows Falls	.78
052	Bennington	.80
053	Brattleboro	.80
054	Burlington	.81
056	Montpelier	.82
057	Rutland	.81
058	St. Johnsbury	.78
059	Guildhall	.77
VIRGINIA		
220-221	Fairfax	1.02
222	Arlington	1.03
223	Alexandria	1.07
224-225	Fredericksburg	.94
226	Winchester	.91
227	Culpeper	.99
228	Harrisonburg	.89
229	Charlottesville	.90
230-232	Richmond	.98
233-235	Norfolk	1.00
236	Newport News	.99
237	Portsmouth	.92
238	Petersburg	.96
239	Farmville	.88
240-241	Roanoke	.97
242	Bristol	.85
243	Pulaski	.83
244	Staunton	.90
245	Lynchburg	.95
246	Grundy	.83
WASHINGTON		
980-981,987	Seattle	1.02
982	Everett	1.04
983-984	Tacoma	1.02
985	Olympia	1.01
986	Vancouver	.97
988	Wenatchee	.92
989	Yakima	.96
990-992	Spokane	.99
993	Richland	.97
994	Clarkston	.96
WEST VIRGINIA		
247-248	Bluefield	.88
249	Lewisburg	.90
250-253	Charleston	.95
254	Martinsburg	.86
255-257	Huntington	.96
258-259	Beckley	.90
260	Wheeling	.92
261	Parkersburg	.91
262	Buckhannon	.91
263-264	Clarksburg	.91
265	Morgantown	.92
266	Gassaway	.91
267	Romney	.89
268	Petersburg	.91
WISCONSIN		
530,532	Milwaukee	1.07
531	Kenosha	1.03
534	Racine	1.02
535	Beloit	.98
537	Madison	.98
538	Lancaster	.97
539	Portage	.96
540	New Richmond	.99
541-543	Green Bay	1.00
544	Wausau	.94
545	Rhinelander	.94
546	La Crosse	.94
547	Eau Claire	.97
548	Superior	.98
549	Oshkosh	.94
WYOMING		
820	Cheyenne	.82
821	Yellowstone Nat. Pk.	.74
822	Wheatland	.74

STATE	CITY	Residential
WYOMING (CONT'D)		
823	Rawlins	.75
824	Worland	.74
825	Riverton	.73
826	Casper	.76
827	Newcastle	.74
828	Sheridan	.79
829-831	Rock Springs	.78
CANADIAN FACTORS (reflect Canadian currency)		
ALBERTA		
	Calgary	1.14
	Edmonton	1.13
	Fort McMurray	1.14
	Lethbridge	1.11
	Lloydminster	1.06
	Medicine Hat	1.07
	Red Deer	1.07
BRITISH COLUMBIA		
	Kamloops	1.05
	Prince George	1.05
	Vancouver	1.06
	Victoria	.99
MANITOBA		
	Brandon	1.02
	Portage la Prairie	1.02
	Winnipeg	1.02
NEW BRUNSWICK		
	Bathurst	.94
	Dalhousie	.94
	Fredericton	1.01
	Moncton	.95
	Newcastle	.94
	St. John	1.01
NEWFOUNDLAND		
	Corner Brook	.96
	St. Johns	.98
NORTHWEST TERRITORIES		
	Yellowknife	1.07
NOVA SCOTIA		
	Bridgewater	.97
	Dartmouth	.98
	Halifax	1.00
	New Glasgow	.97
	Sydney	.96
	Truro	.97
	Yarmouth	.97
ONTARIO		
	Barrie	1.13
	Brantford	1.14
	Cornwall	1.14
	Hamilton	1.16
	Kingston	1.14
	Kitchener	1.09
	London	1.14
	North Bay	1.11
	Oshawa	1.13
	Ottawa	1.16
	Owen Sound	1.11
	Peterborough	1.12
	Sarnia	1.14
	Sault Ste Marie	1.07
	St. Catharines	1.10
	Sudbury	1.07
	Thunder Bay	1.12
	Timmins	1.11
	Toronto	1.17
	Windsor	1.11
PRINCE EDWARD ISLAND		
	Charlottetown	.92
	Summerside	.92
QUEBEC		
	Cap-de-la-Madeleine	1.13
	Charlesbourg	1.13
	Chicoutimi	1.16
	Gatineau	1.12
	Granby	1.12

Location Factors

STATE	CITY	Residential
QUEBEC (CONT'D)		
	Hull	1.12
	Joliette	1.13
	Laval	1.12
	Montreal	1.18
	Quebec	1.18
	Rimouski	1.16
	Rouyn-Noranda	1.12
	Saint Hyacinthe	1.12
	Sherbrooke	1.12
	Sorel	1.13
	St. Jerome	1.12
	Trois Rivieres	1.13
SASKATCHEWAN		
	Moose Jaw	.94
	Prince Albert	.93
	Regina	.96
	Saskatoon	.95
YUKON		
	Whitehorse	.93

R011105-05   Tips for Accurate Estimating

1. Use pre-printed or columnar forms for orderly sequence of dimensions and locations and for recording telephone quotations.

2. Use only the front side of each paper or form except for certain pre-printed summary forms.

3. Be consistent in listing dimensions: For example, length x width x height. This helps in rechecking to ensure that, the total length of partitions is appropriate for the building area.

4. Use printed (rather than measured) dimensions where given.

5. Add up multiple printed dimensions for a single entry where possible.

6. Measure all other dimensions carefully.

7. Use each set of dimensions to calculate multiple related quantities.

8. Convert foot and inch measurements to decimal feet when listing. Memorize decimal equivalents to .01 parts of a foot (1/8″ equals approximately .01′).

9. Do not "round off" quantities until the final summary.

10. Mark drawings with different colors as items are taken off.

11. Keep similar items together, different items separate.

12. Identify location and drawing numbers to aid in future checking for completeness.

13. Measure or list everything on the drawings or mentioned in the specifications.

14. It may be necessary to list items not called for to make the job complete.

15. Be alert for: Notes on plans such as N.T.S. (not to scale); changes in scale throughout the drawings; reduced size drawings; discrepancies between the specifications and the drawings.

16. Develop a consistent pattern of performing an estimate. For example:
    a. Start the quantity takeoff at the lower floor and move to the next higher floor.
    b. Proceed from the main section of the building to the wings.
    c. Proceed from south to north or vice versa, clockwise or counterclockwise.
    d. Take off floor plan quantities first, elevations next, then detail drawings.

17. List all gross dimensions that can be either used again for different quantities, or used as a rough check of other quantities for verification (exterior perimeter, gross floor area, individual floor areas, etc.).

18. Utilize design symmetry or repetition (repetitive floors, repetitive wings, symmetrical design around a center line, similar room layouts, etc.). Note: Extreme caution is needed here so as not to omit or duplicate an area.

19. Do not convert units until the final total is obtained. For instance, when estimating concrete work, keep all units to the nearest cubic foot, then summarize and convert to cubic yards.

20. When figuring alternatives, it is best to total all items involved in the basic system, then total all items involved in the alternates. Therefore you work with positive numbers in all cases. When adds and deducts are used, it is often confusing whether to add or subtract a portion of an item; especially on a complicated or involved alternate.

R011110-10   Architectural Fees

Tabulated below are typical percentage fees by project size, for good professional architectural service. Fees may vary from those listed depending upon degree of design difficulty and economic conditions in any particular area.

Rates can be interpolated horizontally and vertically. Various portions of the same project requiring different rates should be adjusted proportionately. For alterations, add 50% to the fee for the first $500,000 of project cost and add 25% to the fee for project cost over $500,000.

Architectural fees tabulated below include Structural, Mechanical and Electrical Engineering Fees. They do not include the fees for special consultants such as kitchen planning, security, acoustical, interior design, etc.

Civil Engineering fees are included in the Architectural fee for project sites requiring minimal design such as city sites. However, separate Civil Engineering fees must be added when utility connections require design, drainage calculations are needed, stepped foundations are required, or provisions are required to protect adjacent wetlands.

Building Types	Total Project Size in Thousands of Dollars						
	100	250	500	1,000	5,000	10,000	50,000
Factories, garages, warehouses, repetitive housing	9.0%	8.0%	7.0%	6.2%	5.3%	4.9%	4.5%
Apartments, banks, schools, libraries, offices, municipal buildings	12.2	12.3	9.2	8.0	7.0	6.6	6.2
Churches, hospitals, homes, laboratories, museums, research	15.0	13.6	12.7	11.9	9.5	8.8	8.0
Memorials, monumental work, decorative furnishings	—	16.0	14.5	13.1	10.0	9.0	8.3

R012909-80   Sales Tax by State

State sales tax on materials is tabulated below (5 states have no sales tax). Many states allow local jurisdictions, such as a county or city, to levy additional sales tax.

Some projects may be sales tax exempt, particularly those constructed with public funds.

State	Tax (%)	State	Tax (%)	State	Tax (%)	State	Tax (%)
Alabama	4	Illinois	6.25	Montana	0	Rhode Island	7
Alaska	0	Indiana	6	Nebraska	5.5	South Carolina	6
Arizona	5.6	Iowa	5	Nevada	6.5	South Dakota	4
Arkansas	6	Kansas	5.3	New Hampshire	0	Tennessee	7
California	7.25	Kentucky	6	New Jersey	7	Texas	6.25
Colorado	2.9	Louisiana	4	New Mexico	5	Utah	4.65
Connecticut	6	Maine	5	New York	4	Vermont	6
Delaware	0	Maryland	6	North Carolina	4.25	Virginia	5
District of Columbia	5.75	Massachusetts	5	North Dakota	5	Washington	6.5
Florida	6	Michigan	6	Ohio	5.5	West Virginia	6
Georgia	4	Minnesota	6.5	Oklahoma	4.5	Wisconsin	5
Hawaii	4	Mississippi	7	Oregon	0	Wyoming	4
Idaho	6	Missouri	4.225	Pennsylvania	6	Average	4.91 %

Sales Tax by Province (Canada)

GST - a value-added tax, which the government imposes on most goods and services provided in or imported into Canada. PST - a retail sales tax, which five of the provinces impose on the price of most goods and some

services. QST - a value-added tax, similar to the federal GST, which Quebec imposes. HST - Three provinces have combined their retail sales tax with the federal GST into one harmonized tax.

Province	PST (%)	QST (%)	GST(%)	HST(%)
Alberta	0	0	6	0
British Columbia	7	0	6	0
Manitoba	7	0	6	0
New Brunswick	0	0	0	13
Newfoundland	0	0	0	13
Northwest Territories	0	0	6	0
Nova Scotia	0	0	0	13
Ontario	8	0	6	0
Prince Edward Island	10	0	6	0
Quebec	0	7.5	6	0
Saskatchewan	5	0	6	0
Yukon	0	0	6	0

R012909-85   Unemployment Taxes and Social Security Taxes

State Unemployment Tax rates vary not only from state to state, but also with the experience rating of the contractor. The Federal Unemployment Tax rate is 6.2% of the first $7,000 of wages. This is reduced by a credit of up to 5.4% for timely payment to the state. The minimum Federal Unemployment Tax is 0.8% after all credits.

Social Security (FICA) for 2009 is estimated at time of publication to be 7.65% of wages up to $102,000.

R013113-40   Builder's Risk Insurance

Builder's Risk Insurance is insurance on a building during construction. Premiums are paid by the owner or the contractor. Blasting, collapse and underground insurance would raise total insurance costs above those listed. Floater policy for materials delivered to the job runs $.75 to $1.25 per $100 value. Contractor equipment insurance runs $.50 to $1.50 per $100 value. Insurance for miscellaneous tools to $1,500 value runs from $3.00 to $7.50 per $100 value.

Tabulated below are New England Builder's Risk insurance rates in dollars per $100 value for $1,000 deductible. For $25,000 deductible, rates can be reduced 13% to 34%. On contracts over $1,000,000, rates may be lower than those tabulated. Policies are written annually for the total completed value in place. For "all risk" insurance (excluding flood, earthquake and certain other perils) add $.025 to total rates below.

Coverage	Frame Construction (Class 1)			Brick Construction (Class 4)			Fire Resistive (Class 6)		
	Range		Average	Range		Average	Range		Average
Fire Insurance	$.350	to $.850	$.600	$.158	to $.189	$.174	$.052	to $.080	$.070
Extended Coverage	.115	to .200	.158	.080	to .105	.101	.081	to .105	.100
Vandalism	.012	to .016	.014	.008	to .011	.011	.008	to .011	.010
Total Annual Rate	$.477	to $1.066	$.772	$.246	to $.305	$.286	$.141	to $.196	$.180

R013113-50   General Contractor's Overhead

There are two distinct types of overhead on a construction project: Project Overhead and Main Office Overhead. Project Overhead includes those costs at a construction site not directly associated with the installation of construction materials. Examples of Project Overhead costs include the following:

1. Superintendent
2. Construction office and storage trailers
3. Temporary sanitary facilities
4. Temporary utilities
5. Security fencing
6. Photographs
7. Clean up
8. Performance and payment bonds

The above Project Overhead items are also referred to as General Requirements and therefore are estimated in Division 1. Division 1 is the first division listed in the CSI MasterFormat but it is usually the last division estimated. The sum of the costs in Divisions 1 through 49 is referred to as the sum of the direct costs.

All construction projects also include indirect costs. The primary components of indirect costs are the contractor's Main Office Overhead and profit. The amount of the Main Office Overhead expense varies depending on the the following:

1. Owner's compensation
2. Project managers and estimator's wages
3. Clerical support wages
4. Office rent and utilities
5. Corporate legal and accounting costs
6. Advertising
7. Automobile expenses
8. Association dues
9. Travel and entertainment expenses

These costs are usually calculated as a percentage of annual sales volume. This percentage can range from 35% for a small contractor doing less than $500,000 to 5% for a large contractor with sales in excess of $100 million.

R013113-60 Workers' Compensation Insurance Rates by Trade

The table below tabulates the national averages for Workers' Compensation insurance rates by trade and type of building. The average "Insurance Rate" is multiplied by the "% of Building Cost" for each trade. This produces the "Workers' Compensation Cost" by % of total labor cost, to be added for each trade by building type to determine the weighted average Workers' Compensation rate for the building types analyzed.

Trade	Insurance Rate (% Labor Cost)		% of Building Cost			Workers' Compensation		
	Range	Average	Office Bldgs.	Schools & Apts.	Mfg.	Office Bldgs.	Schools & Apts.	Mfg.
Excavation, Grading, etc.	4.2 % to 17.5%	10.0%	4.8%	4.9%	4.5%	0.48%	0.49%	0.45%
Piles & Foundations	5.9 to 40.3	20.1	7.1	5.2	8.7	1.43	1.05	1.75
Concrete	5.1 to 26.8	14.6	5.0	14.8	3.7	0.73	2.16	0.54
Masonry	4.8 to 43.3	14.4	6.9	7.5	1.9	0.99	1.08	0.27
Structural Steel	5.9 to 104.1	37.9	10.7	3.9	17.6	4.06	1.48	6.67
Miscellaneous & Ornamental Metals	3.4 to 22.8	10.7	2.8	4.0	3.6	0.30	0.43	0.39
Carpentry & Millwork	5.9 to 53.2	17.8	3.7	4.0	0.5	0.66	0.71	0.09
Metal or Composition Siding	5.9 to 38	16.6	2.3	0.3	4.3	0.38	0.05	0.71
Roofing	5.9 to 77.1	31.2	2.3	2.6	3.1	0.72	0.81	0.97
Doors & Hardware	4.9 to 32	11.6	0.9	1.4	0.4	0.10	0.16	0.05
Sash & Glazing	5.9 to 32.4	13.9	3.5	4.0	1.0	0.49	0.56	0.14
Lath & Plaster	3.3 to 43.9	13.6	3.3	6.9	0.8	0.45	0.94	0.11
Tile, Marble & Floors	3.1 to 17.9	9.1	2.6	3.0	0.5	0.24	0.27	0.05
Acoustical Ceilings	2.6 to 45.6	10.6	2.4	0.2	0.3	0.25	0.02	0.03
Painting	4.7 to 29.6	12.5	1.5	1.6	1.6	0.19	0.20	0.20
Interior Partitions	5.9 to 53.2	17.8	3.9	4.3	4.4	0.69	0.77	0.78
Miscellaneous Items	2.1 to 168.2	16.0	5.2	3.7	9.7	0.83	0.59	1.55
Elevators	2.8 to 11.8	6.6	2.1	1.1	2.2	0.14	0.07	0.15
Sprinklers	2.5 to 14.3	7.8	0.5	—	2.0	0.04	—	0.16
Plumbing	2.9 to 12.4	7.8	4.9	7.2	5.2	0.38	0.56	0.41
Heat., Vent., Air Conditioning	4.3 to 23	11.6	13.5	11.0	12.9	1.57	1.28	1.50
Electrical	2.8 to 11.5	6.5	10.1	8.4	11.1	0.66	0.55	0.72
Total	2.1 % to 168.2%	—	100.0%	100.0%	100.0%	15.78%	14.23%	17.69%
	Overall Weighted Average	15.90%						

Workers' Compensation Insurance Rates by States

The table below lists the weighted average Workers' Compensation base rate for each state with a factor comparing this with the national average of 15.5%.

State	Weighted Average	Factor	State	Weighted Average	Factor	State	Weighted Average	Factor
Alabama	24.2%	156	Kentucky	18.4%	119	North Dakota	13.7%	88
Alaska	21.7	140	Louisiana	28.3	183	Ohio	14.8	95
Arizona	9.5	61	Maine	15.4	99	Oklahoma	14.6	94
Arkansas	13.2	85	Maryland	16.7	108	Oregon	13.5	87
California	19.6	126	Massachusetts	12.5	81	Pennsylvania	13.8	89
Colorado	13.2	85	Michigan	17.3	112	Rhode Island	21.2	137
Connecticut	21.0	135	Minnesota	25.7	166	South Carolina	18.3	118
Delaware	15.1	97	Mississippi	19.1	123	South Dakota	19.1	123
District of Columbia	13.7	88	Missouri	17.2	111	Tennessee	15.1	97
Florida	13.8	89	Montana	16.6	107	Texas	12.2	79
Georgia	26.2	169	Nebraska	23.0	148	Utah	12.3	79
Hawaii	17.0	110	Nevada	11.6	75	Vermont	18.0	116
Idaho	10.8	70	New Hampshire	20.0	129	Virginia	11.7	75
Illinois	21.6	139	New Jersey	13.4	86	Washington	9.8	63
Indiana	5.9	38	New Mexico	18.5	119	West Virginia	9.2	59
Iowa	11.5	74	New York	12.6	81	Wisconsin	15.0	97
Kansas	8.9	57	North Carolina	18.9	122	Wyoming	6.5	42
			Weighted Average for U.S. is	15.9% of payroll = 100%				

Rates in the following table are the base or manual costs per $100 of payroll for Workers' Compensation in each state. Rates are usually applied to straight time wages only and not to premium time wages and bonuses.

The weighted average skilled worker rate for 35 trades is 15.5%. For bidding purposes, apply the full value of Workers' Compensation directly to total labor costs, or if labor is 38%, materials 42% and overhead and profit 20% of total cost, carry 38/80 x 15.5% = 7.4% of cost (before overhead and profit) into overhead. Rates vary not only from state to state but also with the experience rating of the contractor.

Rates are the most current available at the time of publication.

R013113-60 Workers' Compensation Insurance Rates by Trade and State (cont.)

State	Carpentry — 3 stories or less 5651	Carpentry — interior cab. work 5437	Carpentry — general 5403	Concrete Work — NOC 5213	Concrete Work — flat (flr., sdwk.) 5221	Electrical Wiring — inside 5190	Excavation — earth NOC 6217	Excavation — rock 6217	Glaziers 5462	Insulation Work 5479	Lathing 5443	Masonry 5022	Painting & Decorating 5474	Pile Driving 6003	Plastering 5480	Plumbing 5183	Roofing 5551	Sheet Metal Work (HVAC) 5538	Steel Erection — door & sash 5102	Steel Erection — inter, ornam. 5102	Steel Erection — structure 5040	Steel Erection — NOC 5057	Tile Work — (interior ceramic) 5348	Waterproofing 9014	Wrecking 5701
AL	37.98	22.82	32.46	16.20	11.27	10.68	11.16	11.16	21.26	16.28	11.93	27.00	25.81	32.65	18.85	12.27	58.14	22.96	11.01	11.01	55.68	25.37	13.31	6.56	55.68
AK	14.73	12.27	17.44	11.07	9.71	8.72	12.14	12.14	32.36	27.78	9.74	43.27	21.95	32.40	43.85	9.97	35.12	8.61	9.39	9.39	43.01	33.00	6.76	6.61	43.01
AZ	11.60	6.45	15.40	8.56	4.45	4.59	4.90	4.90	6.93	12.48	6.61	6.51	6.33	11.45	5.78	5.24	13.38	7.38	10.66	10.66	21.10	15.34	3.17	2.59	21.10
AR	14.82	7.45	16.26	12.17	6.48	5.02	7.76	7.76	9.82	16.49	5.93	9.93	11.50	16.48	16.10	5.15	23.09	13.09	6.77	6.77	32.38	25.50	6.08	3.74	32.38
CA	31.95	31.95	31.95	11.89	11.89	7.98	13.79	13.79	15.74	12.50	11.24	16.21	17.08	15.27	20.53	12.03	44.43	15.60	12.46	12.46	17.17	20.88	8.54	17.08	20.88
CO	13.57	8.92	12.33	11.57	7.87	5.28	10.34	10.34	9.75	16.84	6.02	14.01	9.02	14.31	9.58	7.43	26.09	11.61	9.31	9.31	32.62	17.58	7.06	5.03	17.58
CT	15.67	14.66	28.76	24.79	10.09	8.01	12.91	12.91	22.98	15.94	21.94	22.24	17.61	22.88	14.48	11.16	41.95	15.61	18.57	18.57	42.72	23.45	11.58	5.41	42.72
DE	15.17	15.17	11.46	11.51	9.49	5.89	9.32	9.32	13.11	11.46	13.11	12.77	26.77	18.75	13.11	7.87	26.71	9.78	12.28	12.28	26.77	12.28	9.29	12.77	26.77
DC	11.01	12.78	10.19	9.66	9.14	6.01	12.54	12.54	17.02	8.18	9.40	11.97	7.51	12.11	12.48	10.38	19.21	8.93	10.61	10.61	37.21	16.49	17.86	4.00	37.21
FL	13.05	10.47	14.39	15.31	6.97	6.64	8.30	8.30	9.91	9.24	5.98	11.32	9.73	38.24	18.70	6.75	22.19	12.68	8.85	8.85	27.89	14.56	6.55	5.17	27.89
GA	35.68	20.54	25.67	16.85	13.26	10.82	17.38	17.38	18.77	23.64	15.09	21.06	27.77	27.36	22.83	11.4	54.71	20.76	19.93	19.93	64.78	42.21	12.42	8.50	64.78
HI	17.38	11.62	28.20	13.74	12.27	7.10	7.73	7.73	20.55	21.19	10.94	17.87	11.33	20.15	15.61	6.06	33.24	8.02	11.11	11.11	31.71	21.70	9.78	11.54	31.71
ID	9.16	6.04	10.35	11.25	6.13	3.70	5.72	5.72	8.70	7.67	8.89	7.59	8.45	12.13	13.37	5.36	30.27	8.89	5.72	5.72	29.16	10.28	9.79	4.75	29.16
IL	24.93	14.46	20.78	26.59	12.10	9.62	10.06	10.06	19.18	17.48	15.59	19.27	11.62	34.85	15.88	11.07	31.06	14.96	16.05	16.05	72.39	25.69	14.02	4.84	72.39
IN	9.65	4.92	7.29	5.10	3.22	2.84	4.87	4.87	6.28	7.44	2.62	4.84	4.72	6.44	3.72	2.91	10.88	4.45	3.43	3.43	12.12	7.04	3.09	2.47	12.12
IA	10.43	10.02	10.34	12.12	7.00	4.96	6.27	6.27	9.42	7.06	5.58	9.61	6.65	9.78	7.33	6.46	22.58	7.52	6.15	6.15	30.21	35.21	7.42	3.87	35.21
KS	10.05	7.85	9.89	7.19	6.42	3.37	4.41	4.41	9.22	7.45	4.68	6.91	7.57	11.31	6.25	4.85	15.06	7.14	4.82	4.82	26.19	13.21	6.35	3.46	13.21
KY	18.72	14.07	25.30	13.39	8.00	5.43	9.15	9.15	17.37	17.59	11.24	9.12	13.18	27.72	14.41	7.06	38.25	19.58	12.32	12.32	54.03	22.56	13.67	5.74	56.16
LA	22.02	24.93	53.17	26.43	15.61	9.88	17.49	17.49	20.14	21.43	24.51	28.33	29.58	31.25	22.37	8.64	77.12	21.20	18.98	18.98	51.76	24.21	13.77	13.78	66.41
ME	13.79	10.40	28.55	20.94	8.74	6.31	9.05	9.05	18.14	11.82	7.90	14.76	12.73	16.70	11.79	10.56	22.87	11.61	9.73	9.73	31.67	23.11	7.00	7.12	31.67
MD	15.05	11.06	14.11	15.67	6.67	9.24	9.20	9.20	20.03	15.02	8.83	11.43	7.48	20.09	15.02	6.11	34.74	13.54	9.51	9.51	59.03	28.60	7.60	5.36	28.60
MA	6.80	5.60	11.46	19.51	6.57	3.20	4.19	4.19	8.96	10.06	6.35	10.81	4.79	14.68	5.08	3.98	32.80	5.15	7.61	7.61	45.71	36.69	6.21	2.09	25.88
MI	18.40	11.39	20.64	19.45	8.95	4.72	10.20	10.20	12.41	14.75	13.59	15.54	13.64	40.27	15.18	6.84	30.67	10.00	10.16	10.16	40.27	22.21	11.11	4.87	40.27
MN	19.53	19.80	41.48	14.65	15.73	7.53	14.95	14.95	15.68	11.90	22.05	19.18	16.95	24.63	22.05	10.88	70.18	14.23	10.58	10.58	104.13	33.30	14.55	6.68	36.03
MS	21.07	11.11	22.29	15.49	9.26	6.28	11.91	11.91	13.07	14.85	7.91	14.09	13.14	35.34	31.72	9.80	44.11	16.91	14.23	14.23	36.11	25.03	9.84	4.18	36.11
MO	29.99	11.16	13.98	18.91	10.39	7.64	9.17	9.17	11.13	15.17	7.93	15.01	12.53	19.12	17.02	10.05	29.98	13.19	11.31	11.31	34.56	35.84	11.18	6.51	34.56
MT	17.12	11.15	22.57	12.37	12.79	6.39	15.54	15.54	10.67	32.03	11.06	13.75	9.61	25.18	10.02	8.75	37.61	11.53	8.70	8.70	27.45	15.73	7.53	7.46	15.73
NE	24.65	17.47	21.80	25.07	15.13	10.38	16.90	16.90	23.70	31.50	12.20	22.22	20.07	22.50	21.67	12.35	35.22	19.42	16.05	16.05	45.40	35.27	11.10	6.38	47.05
NV	15.25	7.14	11.58	10.76	8.23	7.17	9.77	9.77	10.19	8.12	5.27	7.50	7.97	13.33	7.30	5.98	14.08	18.10	8.67	8.67	25.10	21.54	5.39	5.55	21.54
NH	25.92	13.36	18.90	26.79	14.17	6.94	14.52	14.52	13.07	21.36	9.63	24.34	7.77	18.99	11.71	10.13	49.23	13.59	15.27	15.27	52.24	18.83	11.37	6.09	52.24
NJ	14.34	8.94	14.34	16.01	9.99	4.76	8.35	8.35	8.55	12.65	11.64	13.09	11.92	16.03	11.64	6.17	37.77	6.49	9.46	9.46	23.72	13.59	6.99	5.72	22.31
NM	18.91	7.15	20.11	17.36	10.05	7.51	11.20	11.20	21.26	13.72	8.39	15.34	11.64	21.86	13.44	8.13	36.37	13.47	13.69	13.69	55.78	38.81	6.59	6.29	55.78
NY	12.72	6.45	12.67	16.02	11.64	6.03	8.29	8.29	10.97	7.98	12.10	16.02	9.89	12.77	9.03	6.88	27.98	10.46	8.84	8.84	23.45	12.72	6.66	6.04	8.73
NC	18.76	14.56	15.36	18.15	6.88	11.50	11.70	11.70	15.21	16.00	15.28	12.51	12.90	17.50	15.68	9.16	29.70	15.90	10.44	10.44	79.66	26.21	9.28	5.55	79.66
ND	11.17	11.17	11.17	6.42	6.42	3.63	6.64	6.64	11.17	11.17	7.51	7.71	6.61	22.78	7.51	5.22	23.53	5.22	22.78	22.78	22.78	22.78	11.17	23.53	10.98
OH	11.04	8.52	11.68	12.33	9.99	6.14	8.70	8.70	7.86	16.94	45.64	12.71	12.37	19.06	3.28	6.99	31.63	9.71	7.59	7.59	29.23	16.59	9.60	6.42	16.59
OK	13.42	9.17	11.76	11.76	6.34	5.73	10.83	10.83	16.07	19.47	8.45	10.02	8.47	20.25	12.19	7.09	21.13	9.05	13.76	13.76	39.72	23.97	6.50	5.71	39.72
OR	15.43	8.10	15.69	14.22	8.97	5.14	9.33	9.33	15.22	11.80	7.60	14.07	10.63	16.17	12.73	5.50	26.30	9.41	7.93	7.93	26.94	18.78	11.04	4.54	26.94
PA	13.18	13.18	11.38	14.21	10.46	5.95	8.27	8.27	11.47	11.38	11.47	12.22	13.31	15.78	11.47	7.33	27.42	7.84	14.30	14.30	21.31	14.30	8.02	12.22	21.37
RI	19.53	11.65	18.07	18.23	16.24	4.43	10.38	10.38	12.85	22.78	11.97	25.11	24.13	37.66	17.25	8.52	33.92	10.29	14.07	14.07	59.49	37.50	14.36	7.70	59.49
SC	22.54	15.41	20.91	14.80	7.92	10.92	13.16	13.16	16.12	14.69	11.32	12.88	16.52	19.48	16.02	11.47	45.24	12.95	12.99	12.99	32.57	26.39	9.62	6.68	47.05
SD	27.77	8.13	29.58	25.56	9.55	5.52	11.01	11.01	12.16	12.05	7.70	11.31	9.53	21.32	12.44	11.75	20.73	12.48	8.98	8.98	80.86	40.98	8.74	4.98	80.86
TN	15.68	13.11	16.06	13.89	7.97	7.35	13.68	13.68	13.09	11.80	7.30	15.60	12.63	24.89	16.02	7.66	22.25	12.77	7.17	7.17	29.18	23.47	10.15	4.65	29.18
TX	11.25	8.89	11.25	9.00	7.26	6.15	8.54	8.54	9.65	12.87	6.36	10.91	8.98	22.43	8.98	6.75	19.59	12.84	9.09	9.09	29.68	13.37	7.07	6.69	9.96
UT	15.74	8.88	12.97	9.81	8.47	4.57	11.59	11.59	14.70	12.49	6.35	11.25	8.37	12.47	7.26	6.18	24.99	12.64	7.16	7.16	25.67	14.67	6.80	4.27	18.37
VT	15.70	10.75	17.37	17.27	7.21	6.98	12.59	12.59	19.75	17.78	8.33	15.82	11.62	16.74	10.78	10.44	36.43	11.59	12.51	12.51	44.95	42.80	11.24	7.28	44.95
VA	11.05	8.02	9.16	11.05	5.31	5.52	7.57	7.57	10.25	9.90	14.34	9.62	8.99	10.75	10.22	5.83	22.52	7.65	8.37	8.37	33.52	19.66	4.74	3.04	33.52
WA	8.48	8.48	8.48	7.32	7.32	2.81	6.69	6.69	12.27	9.77	8.43	9.09	10.93	16.12	9.95	4.17	16.36	4.27	7.60	7.60	7.60	7.60	8.45	16.36	7.60
WV	10.65	7.30	10.23	8.24	4.41	4.70	6.22	6.22	7.89	7.25	6.11	7.98	7.90	11.90	7.93	4.90	18.05	5.74	5.58	5.58	24.17	13.20	5.65	2.66	17.34
WI	9.99	11.82	14.45	11.15	9.69	6.02	8.19	8.19	10.44	13.58	6.25	15.31	13.03	16.47	9.96	6.24	37.36	7.21	9.36	9.36	28.23	44.93	13.80	4.86	28.23
WY	5.87	5.87	5.87	5.87	5.87	5.87	5.87	5.87	5.87	5.87	5.87	5.87	5.87	5.87	5.87	5.87	5.87	5.87	5.87	5.87	5.87	5.87	5.87	5.87	5.87
AVG.	16.63	11.62	17.80	14.58	9.14	6.46	10.01	10.01	13.89	14.44	10.63	14.37	12.49	20.09	13.60	7.84	31.18	11.57	10.74	10.74	37.94	23.15	9.13	6.69	34.13

R013113-60 Workers' Compensation (cont.) (Canada in Canadian dollars)

Province		Alberta	British Columbia	Manitoba	Ontario	New Brunswick	Newfndld. & Labrador	Northwest Territories	Nova Scotia	Prince Edward Island	Quebec	Saskat-chewan	Yukon
Carpentry—3 stories or less	Rate	7.32	4.74	4.36	4.35	3.79	10.08	3.84	7.82	5.96	15.05	6.23	8.89
	Code	42143	721028	40102	723	4226	4226	4-41	4226	401	80110	B1317	202
Carpentry—interior cab. work	Rate	2.18	4.90	4.36	4.35	4.40	4.69	3.84	4.52	3.71	15.05	3.51	8.89
	Code	42133	721021	40102	723	4279	4270	4-41	4274	402	80110	B11-27	202
CARPENTRY—general	Rate	7.32	4.74	4.36	4.35	3.79	4.69	3.84	7.82	5.96	15.05	6.23	8.89
	Code	42143	721028	40102	723	4226	4299	4-41	4226	401	80110	B1317	202
CONCRETE WORK—NOC	Rate	4.41	4.81	7.46	16.02	3.79	10.08	3.84	4.52	5.96	17.99	6.24	4.67
	Code	42104	721010	40110	748	4224	4224	4-41	4224	401	80100	B13-14	203
CONCRETE WORK—flat (flr. sidewalk)	Rate	4.41	4.81	7.46	16.02	3.79	10.08	3.84	4.52	5.96	17.99	6.24	4.67
	Code	42104	721010	40110	748	4224	4224	4-41	4224	401	80100	B13-14	203
ELECTRICAL Wiring—inside	Rate	2.13	1.77	2.51	3.79	2.17	3.01	3.52	2.31	3.71	6.12	3.51	4.67
	Code	42124	721019	40203	704	4261	4261	4-46	4261	402	80170	B11-05	206
EXCAVATION—earth NOC	Rate	2.51	3.56	3.76	4.55	2.62	4.37	3.64	3.34	3.82	7.13	3.93	4.67
	Code	40604	721031	40706	711	4214	4214	4-43	4214	404	80030	R11-06	207
EXCAVATION—rock	Rate	2.51	3.56	3.76	4.55	2.62	4.37	3.64	3.34	3.82	7.13	3.93	4.67
	Code	40604	721031	40706	711	4214	4214	4-43	4214	404	80030	R11-06	207
GLAZIERS	Rate	3.16	3.62	4.36	8.90	4.55	6.81	3.84	5.21	3.71	17.08	6.23	4.67
	Code	42121	715020	40109	751	4233	4233	4-41	4233	402	80150	B13-04	212
INSULATION WORK	Rate	2.57	8.42	4.36	8.90	4.55	6.81	3.84	5.21	5.96	15.05	5.98	8.89
	Code	42184	721029	40102	751	4234	4234	4-41	4234	401	80110	B12-07	202
LATHING	Rate	5.59	6.78	4.36	4.35	4.40	4.69	3.84	4.52	3.71	15.05	6.23	8.89
	Code	42135	721033	40102	723	4273	4279	4-41	4271	402	80110	B13-16	202
MASONRY	Rate	4.41	6.78	4.36	11.15	4.55	6.81	3.84	5.21	5.96	17.99	6.23	8.89
	Code	42102	721037	40102	741	4231	4231	4-41	4231	401	80100	B13-18	202
PAINTING & DECORATING	Rate	4.35	4.94	3.18	6.75	4.40	4.69	3.84	4.52	3.71	15.05	5.98	8.89
	Code	42111	721041	40105	719	4275	4275	4-41	4275	402	80110	B12-01	202
PILE DRIVING	Rate	4.41	5.18	3.76	6.34	3.79	9.78	3.64	4.64	5.96	7.13	6.23	8.89
	Code	42159	722004	40706	732	4221	4221	4-43	4221	401	80030	B13-10	202
PLASTERING	Rate	5.59	6.78	5.08	6.75	4.40	4.69	3.84	4.52	3.71	15.05	5.98	8.89
	Code	42135	721042	40108	719	4271	4271	4-41	4271	402	80110	B12-21	202
PLUMBING	Rate	2.13	3.11	3.05	4.02	2.32	3.37	3.52	2.43	3.71	6.93	3.51	4.67
	Code	42122	721043	40204	707	4241	4241	4-46	4241	402	80160	B11-01	214
ROOFING	Rate	8.12	8.68	7.02	12.98	4.55	10.08	3.84	10.57	5.96	23.01	6.23	8.89
	Code	42118	721036	40403	728	4236	4236	4-41	4236	401	80130	B13-20	202
SHEET METAL WORK (HVAC)	Rate	2.13	3.11	7.02	4.02	2.32	3.37	3.52	2.43	3.71	6.93	3.51	4.67
	Code	42117	721043	40402	707	4244	4244	4-46	4244	402	80160	B11-07	208
STEEL ERECTION—door & sash	Rate	2.57	14.07	11.32	16.02	3.79	10.08	3.84	5.21	5.96	25.32	6.23	8.89
	Code	42106	722005	40502	748	4227	4227	4-41	4227	401	80080	B13-22	202
STEEL ERECTION—inter., ornam.	Rate	2.57	14.07	11.32	16.02	3.79	10.08	3.84	5.21	5.96	25.32	6.23	8.89
	Code	42106	722005	40502	748	4227	4227	4-41	4227	401	80080	B13-22	202
STEEL ERECTION—structure	Rate	2.57	14.07	11.32	16.02	3.79	10.08	3.84	5.21	5.96	25.32	6.23	8.89
	Code	42106	722005	40502	748	4227	4227	4-41	4227	401	80080	B13-22	202
STEEL ERECTION—NOC	Rate	2.57	14.07	11.32	16.02	3.79	10.08	3.84	5.21	5.96	25.32	6.23	8.89
	Code	42106	722005	40502	748	4227	4227	4-41	4227	401	80080	B13-22	202
TILE WORK—inter. (ceramic)	Rate	3.63	5.62	1.97	6.75	4.40	10.08	3.84	4.52	3.71	15.05	6.23	8.89
	Code	42113	721054	40103	719	4276	4276	4-41	4276	402	80110	B13-01	202
WATERPROOFING	Rate	4.35	4.93	4.36	4.35	4.55	4.69	3.84	5.21	3.71	23.01	5.98	8.89
	Code	42139	721016	40102	723	4239	4299	4-41	4239	402	80130	B12-17	202
WRECKING	Rate	2.51	4.59	6.59	16.02	2.62	4.39	3.64	3.34	5.96	15.05	6.23	8.89
	Code	40604	721005	40106	748	4211	4211	4-43	4211	401	80110	B13-09	202

R015423-10   Steel Tubular Scaffolding

On new construction, tubular scaffolding is efficient up to 60' high or five stories. Above this it is usually better to use a hung scaffolding if construction permits. Swing scaffolding operations may interfere with tenants. In this case, the tubular is more practical at all heights.

In repairing or cleaning the front of an existing building the cost of tubular scaffolding per S.F. of building front increases as the height increases above the first tier. The first tier cost is relatively high due to leveling and alignment.

The minimum efficient crew for erecting and dismantling is three workers. They can set up and remove 18 frame sections per day up to 5 stories high. For 6 to 12 stories high, a crew of four is most efficient. Use two or more on top and two on the bottom for handing up or hoisting. They can

also set up and remove 18 frame sections per day. At 7' horizontal spacing, this will run about 800 S.F. per day of erecting and dismantling. Time for placing and removing planks must be added to the above. A crew of three can place and remove 72 planks per day up to 5 stories. For over 5 stories, a crew of four can place and remove 80 planks per day.

The table below shows the number of pieces required to erect tubular steel scaffolding for 1000 S.F. of building frontage. This area is made up of a scaffolding system that is 12 frames (11 bays) long by 2 frames high.

For jobs under twenty-five frames, add 50% to rental cost. Rental rates will be lower for jobs over three months duration. Large quantities for long periods can reduce rental rates by 20%.

Description of Component	Number of Pieces for 1000 S.F. of Building Front	Unit
5' Wide Standard Frame, 6'-4" High	24	Ea.
Leveling Jack & Plate	24	
Cross Brace	44	
Side Arm Bracket, 21"	12	
Guardrail Post	12	
Guardrail, 7' section	22	
Stairway Section	2	
Stairway Starter Bar	1	
Stairway Inside Handrail	2	
Stairway Outside Handrail	2	
Walk-Thru Frame Guardrail	2	

Scaffolding is often used as falsework over 15' high during construction of cast-in-place concrete beams and slabs. Two foot wide scaffolding is generally used for heavy beam construction. The span between frames depends upon the load to be carried with a maximum span of 5'.

Heavy duty shoring frames with a capacity of 10,000#/leg can be spaced up to 10' O.C. depending upon form support design and loading.

Scaffolding used as horizontal shoring requires less than half the material required with conventional shoring.

On new construction, erection is done by carpenters.

Rolling towers supporting horizontal shores can reduce labor and speed the job. For maintenance work, catwalks with spans up to 70' can be supported by the rolling towers.

R015423-20   Pump Staging

Pump staging is generally not available for rent. The table below shows the number of pieces required to erect pump staging for 2400 S.F. of building frontage. This area is made up of a pump jack system that is 3 poles (2 bays) wide by 2 poles high.

Item	Number of Pieces for 2400 S.F. of Building Front	Unit
Aluminum pole section, 24' long	6	Ea.
Aluminum splice joint, 6' long	3	
Aluminum foldable brace	3	
Aluminum pump jack	3	
Aluminum support for workbench/back safety rail	3	
Aluminum scaffold plank/workbench, 14" wide x 24' long	4	
Safety net, 22' long	2	
Aluminum plank end safety rail	2	

The cost in place for this 2400 S.F. will depend on how many uses are realized during the life of the equipment.

R024119-10   Demolition Defined

Whole Building Demolition - Demolition of the whole building with no concern for any particular building element, component, or material type being demolished. This type of demolition is accomplished with large pieces of construction equipment that break up the structure, load it into trucks and haul it to a disposal site, but disposal or dump fees are not included. Demolition of below-grade foundation elements, such as footings, foundation walls, grade beams, slabs on grade, etc., is not included. Certain mechanical equipment containing flammable liquids or ozone-depleting refrigerants, electric lighting elements, communication equipment components, and other building elements may contain hazardous waste, and must be removed, either selectively or carefully, as hazardous waste before the building can be demolished.

Foundation Demolition - Demolition of below-grade foundation footings, foundation walls, grade beams, and slabs on grade. This type of demolition is accomplished by hand or pneumatic hand tools, and does not include saw cutting, or handling, loading, hauling, or disposal of the debris.

Gutting - Removal of building interior finishes and electrical/mechanical systems down to the load-bearing and sub-floor elements of the rough building frame, with no concern for any particular building element, component, or material type being demolished. This type of demolition is accomplished by hand or pneumatic hand tools, and includes loading into trucks, but not hauling, disposal or dump fees, scaffolding, or shoring. Certain mechanical equipment containing flammable liquids or ozone-depleting refrigerants, electric lighting elements, communication equipment components, and other building elements may contain hazardous waste, and must be removed, either selectively or carefully, as hazardous waste, before the building is gutted.

Selective Demolition - Demolition of a selected building element, component, or finish, with some concern for surrounding or adjacent elements, components, or finishes (see the first Subdivision (s) at the beginning of appropriate Divisions). This type of demolition is accomplished by hand or pneumatic hand tools, and does not include handling, loading, storing, hauling, or disposal of the debris, scaffolding, or shoring. "Gutting" methods may be used in order to save time, but damage that is caused to surrounding or adjacent elements, components, or finishes may have to be repaired at a later time.

Careful Removal - Removal of a piece of service equipment, building element or component, or material type, with great concern for both the removed item and surrounding or adjacent elements, components or finishes. The purpose of careful removal may be to protect the removed item for later re-use, preserve a higher salvage value of the removed item, or replace an item while taking care to protect surrounding or adjacent elements, components, connections, or finishes from cosmetic and/or structural damage. An approximation of the time required to perform this type of removal is 1/3 to 1/2 the time it would take to install a new item of like kind. This type of removal is accomplished by hand or pneumatic hand tools, and does not include loading, hauling, or storing the removed item, scaffolding, shoring, or lifting equipment.

Cutout Demolition - Demolition of a small quantity of floor, wall, roof, or other assembly, with concern for the appearance and structural integrity of the surrounding materials. This type of demolition is accomplished by hand or pneumatic hand tools, and does not include saw cutting, handling, loading, hauling, or disposal of debris, scaffolding, or shoring.

Rubbish Handling - Work activities that involve handling, loading or hauling of debris. Generally, the cost of rubbish handling must be added to the cost of all types of demolition, with the exception of whole building demolition.

Minor Site Demolition - Demolition of site elements outside the footprint of a building. This type of demolition is accomplished by hand or pneumatic hand tools, or with larger pieces of construction equipment, and may include loading a removed item onto a truck (check the Crew for equipment used). It does not include saw cutting, hauling or disposal of debris, and, sometimes, handling or loading.

R024119-20   Dumpsters

Dumpster rental costs on construction sites are presented in two ways.

The cost per week rental includes the delivery of the dumpster; its pulling or emptying once per week, and its final removal. The assumption is made that the dumpster contractor could choose to empty a dumpster by simply bringing in an empty unit and removing the full one. These costs also include the disposal of the materials in the dumpster.

The Alternate Pricing can be used when actual planned conditions are not approximated by the weekly numbers. For example, these lines can be used when a dumpster is needed for 4 weeks and will need to be emptied 2 or 3 times per week. Conversely the Alternate Pricing lines can be used when a dumpster will be rented for several weeks or months but needs to be emptied only a few times over this period.

R040130-10 Cleaning Face Brick

On smooth brick a person can clean 70 S.F. an hour; on rough brick 50 S.F. per hour. Use one gallon muriatic acid to 20 gallons of water for 1000 S.F. Do not use acid solution until wall is at least seven days old, but a mild soap solution may be used after two days.

Time has been allowed for clean-up in brick prices.

Masonry | **R0405 Common Work Results for Masonry**

R040513-10 Cement Mortar (material only)

Type N - 1:1:6 mix by volume. Use everywhere above grade except as noted below. - 1:3 mix using conventional masonry cement which saves handling two separate bagged materials.

Type M - 1:1/4:3 mix by volume, or 1 part cement, 1/4 (10% by wt.) lime, 3 parts sand. Use for heavy loads and where earthquakes or hurricanes may occur. Also for reinforced brick, sewers, manholes and everywhere below grade.

Mix Proportions by Volume and Compressive Strength of Mortar

Where Used	Mortar Type	Allowable Proportions by Volume				Compressive Strength @ 28 days
		Portland Cement	Masonry Cement	Hydrated Lime	Masonry Sand	
Plain Masonry	M	1	1	—	6	2500 psi
		1	—	1/4	3	
	S	1/2	1	—	4	1800 psi
		1	—	1/4 to 1/2	4	
	N	—	1	—	3	750 psi
		1	—	1/2 to 1-1/4	6	
	O	—	1	—	3	350 psi
		1	—	1-1/4 to 2-1/2	9	
	K	1	—	2-1/2 to 4	12	75 psi
Reinforced Masonry	PM	1	1	—	6	2500 psi
	PL	1	—	1/4 to 1/2	4	2500 psi

Note: The total aggregate should be between 2.25 to 3 times the sum of the cement and lime used.

The labor cost to mix the mortar is included in the productivity and labor cost of unit price lines in unit cost sections for brickwork, blockwork and stonework.

The material cost of mixed mortar is included in the material cost of those same unit price lines and includes the cost of renting and operating a 10 C.F. mixer at the rate of 200 C.F. per day.

There are two types of mortar color used. One type is the inert additive type with about 100 lbs. per M brick as the typical quantity required. These colors are also available in smaller-batch-sized bags (1 lb. to 15 lb.) which can be placed directly into the mixer without measuring. The other type is premixed and replaces the masonry cement. Dark green color has the highest cost.

R040519-50 Masonry Reinforcing

Horizontal joint reinforcing helps prevent wall cracks where wall movement may occur and in many locations is required by code. Horizontal joint reinforcing is generally not considered to be structural reinforcing and an unreinforced wall may still contain joint reinforcing.

Reinforcing strips come in 10' and 12' lengths and in truss and ladder shapes, with and without drips. Field labor runs between 2.7 to 5.3 hours per 1000 L.F. for wall thicknesses up to 12".

The wire meets ASTM A82 for cold drawn steel wire and the typical size is 9 ga. sides and ties with 3/16" diameter also available. Typical finish is mill galvanized with zinc coating at .10 oz. per S.F. Class I (.40 oz. per S.F.) and Class III (.80 oz per S.F.) are also available, as is hot dipped galvanizing at 1.50 oz. per S.F.

R042110-10   Economy in Bricklaying

Have adequate supervision. Be sure bricklayers are always supplied with materials so there is no waiting. Place best bricklayers at corners and openings.

Use only screened sand for mortar. Otherwise, labor time will be wasted picking out pebbles. Use seamless metal tubs for mortar as they do not leak or catch the trowel. Locate stack and mortar for easy wheeling.

Have brick delivered for stacking. This makes for faster handling, reduces chipping and breakage, and requires less storage space. Many dealers will deliver select common in 2' x 3' x 4' pallets or face brick packaged. This affords quick handling with a crane or forklift and easy tonging in units of ten, which reduces waste.

Use wider bricks for one wythe wall construction. Keep scaffolding away from wall to allow mortar to fall clear and not stain wall.

On large jobs develop specialized crews for each type of masonry unit.

Consider designing for prefabricated panel construction on high rise projects.

Avoid excessive corners or openings. Each opening adds about 50% to labor cost for area of opening.

Bolting stone panels and using window frames as stops reduces labor costs and speeds up erection.

R042110-20   Common and Face Brick

Common building brick manufactured according to ASTM C62 and facing brick manufactured according to ASTM C216 are the two standard bricks available for general building use.

Building brick is made in three grades; SW, where high resistance to damage caused by cyclic freezing is required; MW, where moderate resistance to cyclic freezing is needed; and NW, where little resistance to cyclic freezing is needed. Facing brick is made in only the two grades SW and MW. Additionally, facing brick is available in three types; FBS, for general use; FBX, for general use where a higher degree of precision and lower permissible variation in size than FBS is needed; and FBA, for general use to produce characteristic architectural effects resulting from non-uniformity in size and texture of the units.

In figuring the material cost of brickwork, an allowance of 25% mortar waste and 3% brick breakage was included. If bricks are delivered palletized

with 280 to 300 per pallet, or packaged, allow only 1-1/2% for breakage. Packaged or palletized delivery is practical when a job is big enough to have a crane or other equipment available to handle a package of brick. This is so on all industrial work but not always true on small commercial buildings.

The use of buff and gray face is increasing, and there is a continuing trend to the Norman, Roman, Jumbo and SCR brick.

Common red clay brick for backup is not used that often. Concrete block is the most usual backup material with occasional use of sand lime or cement brick. Building brick is commonly used in solid walls for strength and as a fire stop.

Brick panels built on the ground and then crane erected to the upper floors have proven to be economical. This allows the work to be done under cover and without scaffolding.

R042110-50   Brick, Block & Mortar Quantities

Type Brick	Nominal Size (incl. mortar) L H W	Modular Coursing	Number of Brick per S.F.	C.F. of Mortar per M Bricks, Waste Included 3/8" Joint	1/2" Joint	Bond Type	Description	Factor
Standard	8 x 2-2/3 x 4	3C=8"	6.75	10.3	12.9	Common	full header every fifth course	+20%
Economy	8 x 4 x 4	1C=4"	4.50	11.4	14.6		full header every sixth course	+16.7%
Engineer	8 x 3-1/5 x 4	5C=16"	5.63	10.6	13.6	English	full header every second course	+50%
Fire	9 x 2-1/2 x 4-1/2	2C=5"	6.40	550 # Fireclay	—	Flemish	alternate headers every course	+33.3%
Jumbo	12 x 4 x 6 or 8	1C=4"	3.00	23.8	30.8		every sixth course	+5.6%
Norman	12 x 2-2/3 x 4	3C=8"	4.50	14.0	17.9	Header = W x H exposed		+100%
Norwegian	12 x 3-1/5 x 4	5C=16"	3.75	14.6	18.6	Rowlock = H x W exposed		+100%
Roman	12 x 2 x 4	2C=4"	6.00	13.4	17.0	Rowlock stretcher = L x W exposed		+33.3%
SCR	12 x 2-2/3 x 6	3C=8"	4.50	21.8	28.0	Soldier = H x L exposed		—
Utility	12 x 4 x 4	1C=4"	3.00	15.4	19.6	Sailor = W x L exposed		-33.3%

Concrete Blocks Nominal Size		Approximate Weight per S.F. Standard	Lightweight	Blocks per 100 S.F.	Mortar per M block, waste included Partitions	Back up
2"	x 8" x 16"	20 PSF	15 PSF	113	27 C.F.	36 C.F.
4"		30	20		41	51
6"		42	30		56	66
8"		55	38		72	82
10"		70	47		87	97
12"		85	55		102	112

Masonry
R0422 Concrete Unit Masonry

R042210-20   Concrete Block

The material cost of special block such as corner, jamb and head block can be figured at the same price as ordinary block of same size. Labor on specials is about the same as equal-sized regular block.

Bond beam and 16″ high lintel blocks are more expensive than regular units of equal size. Lintel blocks are 8″ long and either 8″ or 16″ high.

Use of motorized mortar spreader box will speed construction of continuous walls.

Hollow non-load-bearing units are made according to ASTM C129 and hollow load-bearing units according to ASTM C90.

Metals
R0531 Steel Decking

R053100-10   Decking Descriptions

General - All Deck Products

Steel deck is made by cold forming structural grade sheet steel into a repeating pattern of parallel ribs. The strength and stiffness of the panels are the result of the ribs and the material properties of the steel. Deck lengths can be varied to suit job conditions, but because of shipping considerations, are usually less than 40 feet. Standard deck width varies with the product used but full sheets are usually 12″, 18″, 24″, 30″, or 36″. Deck is typically furnished in a standard width with the ends cut square. Any cutting for width, such as at openings or for angular fit, is done at the job site.

Deck is typically attached to the building frame with arc puddle welds, self-drilling screws, or powder or pneumatically driven pins. Sheet to sheet fastening is done with screws, button punching (crimping), or welds.

Composite Floor Deck

After installation and adequate fastening, floor deck serves several purposes. It (a) acts as a working platform, (b) stabilizes the frame, (c) serves as a concrete form for the slab, and (d) reinforces the slab to carry the design loads applied during the life of the building. Composite decks are distinguished by the presence of shear connector devices as part of the deck. These devices are designed to mechanically lock the concrete and deck together so that the concrete and the deck work together to carry subsequent floor loads. These shear connector devices can be rolled-in embossments, lugs, holes, or wires welded to the panels. The deck profile can also be used to interlock concrete and steel.

Composite deck finishes are either galvanized (zinc coated) or phosphatized/painted. Galvanized deck has a zinc coating on both the top and bottom surfaces. The phosphatized/painted deck has a bare (phosphatized) top surface that will come into contact with the concrete. This bare top surface can be expected to develop rust before the concrete is placed. The bottom side of the deck has a primer coat of paint.

Composite floor deck is normally installed so the panel ends do not overlap on the supporting beams. Shear lugs or panel profile shape often prevent a tight metal to metal fit if the panel ends overlap; the air gap caused by overlapping will prevent proper fusion with the structural steel supports when the panel end laps are shear stud welded.

Adequate end bearing of the deck must be obtained as shown on the drawings. If bearing is actually less in the field than shown on the drawings, further investigation is required.

Roof Deck

Roof deck is not designed to act compositely with other materials. Roof deck acts alone in transferring horizontal and vertical loads into the building frame. Roof deck rib openings are usually narrower than floor deck rib openings. This provides adequate support of rigid thermal insulation board.

Roof deck is typically installed to endlap approximately 2″ over supports. However, it can be butted (or lapped more than 2″) to solve field fit problems. Since designers frequently use the installed deck system as part of the horizontal bracing system (the deck as a diaphragm), any fastening substitution or change should be approved by the designer. Continuous perimeter support of the deck is necessary to limit edge deflection in the finished roof and may be required for diaphragm shear transfer.

Standard roof deck finishes are galvanized or primer painted. The standard factory applied paint for roof deck is a primer paint and is not intended to weather for extended periods of time. Field painting or touching up of abrasions and deterioration of the primer coat or other protective finishes is the responsibility of the contractor.

Cellular Deck

Cellular deck is made by attaching a bottom steel sheet to a roof deck or composite floor deck panel. Cellular deck can be used in the same manner as floor deck. Electrical, telephone, and data wires are easily run through the chase created between the deck panel and the bottom sheet.

When used as part of the electrical distribution system, the cellular deck must be installed so that the ribs line up and create a smooth cell transition at abutting ends. The joint that occurs at butting cell ends must be taped or otherwise sealed to prevent wet concrete from seeping into the cell. Cell interiors must be free of welding burrs, or other sharp intrusions, to prevent damage to wires.

When used as a roof deck, the bottom flat plate is usually left exposed to view. Care must be maintained during erection to keep good alignment and prevent damage.

Cellular deck is sometimes used with the flat plate on the top side to provide a flat working surface. Installation of the deck for this purpose requires special methods for attachment to the frame because the flat plate, now on the top, can prevent direct access to the deck material that is bearing on the structural steel. It may be advisable to treat the flat top surface to prevent slipping.

Cellular deck is always furnished galvanized or painted over galvanized.

Form Deck

Form deck can be any floor or roof deck product used as a concrete form. Connections to the frame are by the same methods used to anchor floor and roof deck. Welding washers are recommended when welding deck that is less than 20 gauge thickness.

Form deck is furnished galvanized, prime painted, or uncoated. Galvanized deck must be used for those roof deck systems where form deck is used to carry a lightweight insulating concrete fill.

Wood, Plastics & Comp. — R0611 Wood Framing

R061110-30 Lumber Product Material Prices

The price of forest products fluctuates widely from location to location and from season to season depending upon economic conditions. The bare material prices in the unit cost sections of the book show the National Average material prices in effect Jan. 1 of this book year. It must be noted that lumber prices in general may change significantly during the year.

Availability of certain items depends upon geographic location and must be checked prior to firm-price bidding.

Wood, Plastics & Comp. — R0616 Sheathing

R061636-20 Plywood

There are two types of plywood used in construction: interior, which is moisture-resistant but not waterproofed, and exterior, which is waterproofed.

The grade of the exterior surface of the plywood sheets is designated by the first letter: A, for smooth surface with patches allowed; B, for solid surface with patches and plugs allowed; C, which may be surface plugged or may have knot holes up to 1″ wide; and D, which is used only for interior type plywood and may have knot holes up to 2-1/2″ wide. "Structural Grade" is specifically designed for engineered applications such as box beams. All CC & DD grades have roof and floor spans marked on them.

Underlayment-grade plywood runs from 1/4″ to 1-1/4″ thick. Thicknesses 5/8″ and over have optional tongue and groove joints which eliminate the need for blocking the edges. Underlayment 19/32″ and over may be referred to as Sturd-i-Floor.

The price of plywood can fluctuate widely due to geographic and economic conditions.

Typical uses for various plywood grades are as follows:

AA-AD Interior — cupboards, shelving, paneling, furniture

BB Plyform — concrete form plywood

CDX — wall and roof sheathing

Structural — box beams, girders, stressed skin panels

AA-AC Exterior — fences, signs, siding, soffits, etc.

Underlayment — base for resilient floor coverings

Overlaid HDO — high density for concrete forms & highway signs

Overlaid MDO — medium density for painting, siding, soffits & signs

303 Siding — exterior siding, textured, striated, embossed, etc.

Thermal & Moist. Protec. — R0731 Shingles & Shakes

R073126-20 Roof Slate

16″, 18″ and 20″ are standard lengths, and slate usually comes in random widths. For standard 3/16″ thickness use 1-1/2″ copper nails. Allow for 3% breakage.

Thermal & Moist. Protec. — R0752 Modified Bituminous Membrane Roofing

R075213-30 Modified Bitumen Roofing

The cost of modified bitumen roofing is highly dependent on the type of installation that is planned. Installation is based on the type of modifier used in the bitumen. The two most popular modifiers are atactic polypropylene (APP) and styrene butadiene styrene (SBS). The modifiers are added to heated bitumen during the manufacturing process to change its characteristics. A polyethylene, polyester or fiberglass reinforcing sheet is then sandwiched between layers of this bitumen. When completed, the result is a pre-assembled, built-up roof that has increased elasticity and weatherablility. Some manufacturers include a surfacing material such as ceramic or mineral granules, metal particles or sand.

The preferred method of adhering SBS-modified bitumen roofing to the substrate is with hot-mopped asphalt (much the same as built-up roofing). This installation method requires a tar kettle/pot to heat the asphalt, as well as the labor, tools and equipment necessary to distribute and spread the hot asphalt.

The alternative method for applying APP and SBS modified bitumen is as follows. A skilled installer uses a torch to melt a small pool of bitumen off the membrane. This pool must form across the entire roll for proper adhesion. The installer must unroll the roofing at a pace slow enough to melt the bitumen, but fast enough to prevent damage to the rest of the membrane.

Modified bitumen roofing provides the advantages of both built-up and single-ply roofing. Labor costs are reduced over those of built-up roofing because only a single ply is necessary. The elasticity of single-ply roofing is attained with the reinforcing sheet and polymer modifiers. Modifieds have some self-healing characteristics and because of their multi-layer construction, they offer the reliability and safety of built-up roofing.

R081313-20 Steel Door Selection Guide

Standard steel doors are classified into four levels, as recommended by the Steel Door Institute in the chart below. Each of the four levels offers a range of construction models and designs, to meet architectural requirements for preference and appearance, including full flush, seamless, and stile & rail. Recommended minimum gauge requirements are also included.

For complete standard steel door construction specifications and available sizes, refer to the Steel Door Institute Technical Data Series, ANSI A250.8-98 (SDI-100), and ANSI A250.4-94 Test Procedure and Acceptance Criteria for Physical Endurance of Steel Door and Hardware Reinforcements.

Level		Model	Construction	For Full Flush or Seamless		
				Min. Gauge	Thickness (in)	Thickness (mm)
I	Standard Duty	1	Full Flush	20	0.032	0.8
		2	Seamless			
II	Heavy Duty	1	Full Flush	18	0.042	1.0
		2	Seamless			
III	Extra Heavy Duty	1	Full Flush	16	0.053	1.3
		2	Seamless			
		3	*Stile & Rail			
IV	Maximum Duty	1	Full Flush	14	0.067	1.6
		2	Seamless			

*Stiles & rails are 16 gauge; flush panels, when specified, are 18 gauge.

R085216-10 Window Estimates

To ensure a complete window estimate, be sure to include the material and labor costs for each window, as well as the material and labor costs for an interior wood trim set.

R085313-20 Replacement Windows

Replacement windows are typically measured per United Inch.

United Inches are calculated by rounding the width and height of the window opening up to the nearest inch, then adding the two figures.

The labor cost for replacement windows includes removal of sash, existing sash balance or weights, parting bead where necessary and installation of new window.

Debris hauling and dump fees are not included.

R087120-10   Hinges

All closer equipped doors should have ball bearing hinges. Lead lined or extremely heavy doors require special strength hinges. Usually 1-1/2 pair of hinges are used per door up to 7'-6" high openings. Table below shows typical hinge requirements.

Use Frequency	Type Hinge Required	Type of Opening	Type of Structure
High	Heavy weight ball bearing	Entrances	Banks, Office buildings, Schools, Stores & Theaters
		Toilet Rooms	Office buildings and Schools
Average	Standard weight ball bearing	Entrances	Dwellings
		Corridors	Office buildings and Schools
		Toilet Rooms	Stores
Low	Plain bearing	Interior	Dwellings

Door Thickness	Weight of Doors in Pounds per Square Foot				
	White Pine	Oak	Hollow Core	Solid Core	Hollow Metal
1-3/8"	3psf	6psf	1-1/2psf	3-1/2 — 4psf	6-1/2psf
1-3/4"	3-1/2	7	2-1/2	4-1/2 — 5-1/4	6-1/2
2-1/4"	4-1/2	9	—	5-1/2 — 6-3/4	6-1/2

R092000-50 Lath, Plaster and Gypsum Board

Gypsum board lath is available in 3/8″ thick x 16″ wide x 4′ long sheets as a base material for multi-layer plaster applications. It is also available as a base for either multi-layer or veneer plaster applications in 1/2″ and 5/8″ thick–4′ wide x 8′, 10′ or 12′ long sheets. Fasteners are screws or blued ring shank nails for wood framing and screws for metal framing.

Metal lath is available in diamond mesh pattern with flat or self-furring profiles. Paper backing is available for applications where excessive plaster waste needs to be avoided. A slotted mesh ribbed lath should be used in areas where the span between structural supports is greater than normal. Most metal lath comes in 27″ x 96″ sheets. Diamond mesh weighs 1.75, 2.5 or 3.4 pounds per square yard, slotted mesh lath weighs 2.75 or 3.4 pounds per square yard. Metal lath can be nailed, screwed or tied in place.

Many **accessories** are available. Corner beads, flat reinforcing strips, casing beads, control and expansion joints, furring brackets and channels are some examples. Note that accessories are not included in plaster or stucco line items.

Plaster is defined as a material or combination of materials that when mixed with a suitable amount of water, forms a plastic mass or paste. When applied to a surface, the paste adheres to it and subsequently hardens, preserving in a rigid state the form or texture imposed during the period of elasticity.

Gypsum plaster is made from ground calcined gypsum. It is mixed with aggregates and water for use as a base coat plaster.

Vermiculite plaster is a fire-retardant plaster covering used on steel beams, concrete slabs and other heavy construction materials. Vermiculite is a group name for certain clay minerals, hydrous silicates or aluminum, magnesium and iron that have been expanded by heat.

Perlite plaster is a plaster using perlite as an aggregate instead of sand. Perlite is a volcanic glass that has been expanded by heat.

Gauging plaster is a mix of gypsum plaster and lime putty that when applied produces a quick drying finish coat.

Veneer plaster is a one or two component gypsum plaster used as a thin finish coat over special gypsum board.

Keenes cement is a white cementitious material manufactured from gypsum that has been burned at a high temperature and ground to a fine powder. Alum is added to accelerate the set. The resulting plaster is hard and strong and accepts and maintains a high polish, hence it is used as a finishing plaster.

Stucco is a Portland cement based plaster used primarily as an exterior finish.

Plaster is used on both interior and exterior surfaces. Generally it is applied in multiple-coat systems. A three-coat system uses the terms scratch, brown and finish to identify each coat. A two-coat system uses base and finish to describe each coat. Each type of plaster and application system has attributes that are chosen by the designer to best fit the intended use.

Gypsum Plaster Quantities for 100 S.Y.	2 Coat, 5/8″ Thick		3 Coat, 3/4″ Thick		
	Base	Finish	Scratch	Brown	Finish
	1:3 Mix	2:1 Mix	1:2 Mix	1:3 Mix	2:1 Mix
Gypsum plaster	1,300 lb.		1,350 lb.	650 lb.	
Sand	1.75 C.Y.		1.85 C.Y.	1.35 C.Y.	
Finish hydrated lime		340 lb.			340 lb.
Gauging plaster		170 lb.			170 lb.

Vermiculite or Perlite Plaster Quantities for 100 S.Y.	2 Coat, 5/8″ Thick		3 Coat, 3/4″ Thick		
	Base	Finish	Scratch	Brown	Finish
Gypsum plaster	1,250 lb.		1,450 lb.	800 lb.	
Vermiculite or perlite	7.8 bags		8.0 bags	3.3 bags	
Finish hydrated lime		340 lb.			340 lb.
Gauging plaster		170 lb.			170 lb.

Stucco–Three-Coat System Quantities for 100 S.Y.	On Wood Frame	On Masonry
Portland cement	29 bags	21 bags
Sand	2.6 C.Y.	2.0 C.Y.
Hydrated lime	180 lb.	120 lb.

R092910-10   Levels of Gypsum Drywall Finish

In the past, contract documents often used phrases such as "industry standard" and "workmanlike finish" to specify the expected quality of gypsum board wall and ceiling installations. The vagueness of these descriptions led to unacceptable work and disputes.

In order to resolve this problem, four major trade associations concerned with the manufacture, erection, finish and decoration of gypsum board wall and ceiling systems have developed an industry-wide *Recommended Levels of Gypsum Board Finish*.

The finish of gypsum board walls and ceilings for specific final decoration is dependent on a number of factors. A primary consideration is the location of the surface and the degree of decorative treatment desired. Painted and unpainted surfaces in warehouses and other areas where appearance is normally not critical may simply require the taping of wallboard joints and 'spotting' of fastener heads. Blemish-free, smooth, monolithic surfaces often intended for painted and decorated walls and ceilings in habitated structures, ranging from single-family dwellings through monumental buildings, require additional finishing prior to the application of the final decoration.

Other factors to be considered in determining the level of finish of the gypsum board surface are (1) the type of angle of surface illumination (both natural and artificial lighting), and (2) the paint and method of application or the type and finish of wallcovering specified as the final decoration. Critical lighting conditions, gloss paints, and thin wallcoverings require a higher level of gypsum board finish than do heavily textured surfaces which are subsequently painted or surfaces which are to be decorated with heavy grade wallcoverings.

The following descriptions were developed jointly by the Association of the Wall and Ceiling Industries-International (AWCI), Ceiling & Interior Systems Construction Association (CISCA), Gypsum Association (GA), and Painting and Decorating Contractors of America (PDCA) as a guide.

Level 0: No taping, finishing, or accessories required. This level of finish may be useful in temporary construction or whenever the final decoration has not been determined.

Level 1: All joints and interior angles shall have tape set in joint compound. Surface shall be free of excess joint compound. Tool marks and ridges are acceptable. Frequently specified in plenum areas above ceilings, in attics, in areas where the assembly would generally be concealed or in building service corridors, and other areas not normally open to public view.

Level 2: All joints and interior angles shall have tape embedded in joint compound and wiped with a joint knife leaving a thin coating of joint compound over all joints and interior angles. Fastener heads and accessories shall be covered with a coat of joint compound. Surface shall be free of excess joint compound. Tool marks and ridges are acceptable. Joint compound applied over the body of the tape at the time of tape embedment shall be considered a separate coat of joint compound and shall satisfy the conditions of this level. Specified where water-resistant gypsum backing board is used as a substrate for tile; may be specified in garages, warehouse storage, or other similar areas where surface appearance is not of primary concern.

Level 3: All joints and interior angles shall have tape embedded in joint compound and one additional coat of joint compound applied over all joints and interior angles. Fastener heads and accessories shall be covered with two separate coats of joint compound. All joint compound shall be smooth and free of tool marks and ridges. Typically specified in appearance areas which are to receive heavy- or medium-texture (spray or hand applied) finishes before final painting, or where heavy-grade wallcoverings are to be applied as the final decoration. This level of finish is not recommended where smooth painted surfaces or light to medium wallcoverings are specified.

Level 4: All joints and interior angles shall have tape embedded in joint compound and two separate coats of joint compound applied over all flat joints and one separate coat of joint compound applied over interior angles. Fastener heads and accessories shall be covered with three separate coats of joint compound. All joint compound shall be smooth and free of tool marks and ridges. This level should be specified where flat paints, light textures, or wallcoverings are to be applied. In critical lighting areas, flat paints applied over light textures tend to reduce joint photographing. Gloss, semi-gloss, and enamel paints are not recommended over this level of finish. The weight, texture, and sheen level of wallcoverings applied over this level of finish should be carefully evaluated. Joints and fasteners must be adequately concealed if the wallcovering material is lightweight, contains limited pattern, has a gloss finish, or any combination of these finishes is present. Unbacked vinyl wallcoverings are not recommended over this level of finish.

Level 5: All joints and interior angles shall have tape embedded in joint compound and two separate coats of joint compound applied over all flat joints and one separate coat of joint compound applied over interior angles. Fastener heads and accessories shall be covered with three separate coats of joint compound. A thin skim coat of joint compound or a material manufactured especially for this purpose, shall be applied to the entire surface. The surface shall be smooth and free of tool marks and ridges. This level of finish is highly recommended where gloss, semi-gloss, enamel, or nontextured flat paints are specified or where severe lighting conditions occur. This highest quality finish is the most effective method to provide a uniform surface and minimize the possibility of joint photographing and of fasteners showing through the final decoration.

Finishes   **R0972 Wall Coverings**

R097223-10   Wall Covering

The table below lists the quantities required for 100 S.F. of wall covering.

Description	Medium-Priced Paper	Expensive Paper
Paper	1.6 dbl. rolls	1.6 dbl. rolls
Wall sizing	0.25 gallon	0.25 gallon
Vinyl wall paste	0.6 gallon	0.6 gallon
Apply sizing	0.3 hour	0.3 hour
Apply paper	1.2 hours	1.5 hours

Most wallpapers now come in double rolls only.
To remove old paper, allow 1.3 hours per 100 S.F.

R099100-10　Painting Estimating Techniques

Proper estimating methodology is needed to obtain an accurate painting estimate. There is no known reliable shortcut or square foot method. The following steps should be followed:

- List all surfaces to be painted, with an accurate quantity (area) of each. Items having similar surface condition, finish, application method and accessibility may be grouped together.
- List all the tasks required for each surface to be painted, including surface preparation, masking, and protection of adjacent surfaces. Surface preparation may include minor repairs, washing, sanding and puttying.
- Select the proper Means line for each task. Review and consider all adjustments to labor and materials for type of paint and location of work. Apply the height adjustment carefully. For instance, when applying the adjustment for work over 8' high to a wall that is 12' high, apply the adjustment only to the area between 8' and 12' high, and not to the entire wall.

When applying more than one percent (%) adjustment, apply each to the base cost of the data, rather than applying one percentage adjustment on top of the other.

When estimating the cost of painting walls and ceilings remember to add the brushwork for all cut-ins at inside corners and around windows and doors as a LF measure. One linear foot of cut-in with brush equals one square foot of painting.

All items for spray painting include the labor for roll-back.

Deduct for openings greater than 100 SF, or openings that extend from floor to ceiling and are greater than 5' wide. Do not deduct small openings.

The cost of brushes, rollers, ladders and spray equipment are considered to be part of a painting contractor's overhead, and should not be added to the estimate. The cost of rented equipment such as scaffolding and swing staging should be added to the estimate.

R099100-20　Painting

Item	Coat	One Gallon Covers			In 8 Hours a Laborer Covers			Labor-Hours per 100 S.F.		
		Brush	Roller	Spray	Brush	Roller	Spray	Brush	Roller	Spray
Paint wood siding	prime	250 S.F.	225 S.F.	290 S.F.	1150 S.F.	1300 S.F.	2275 S.F.	.695	.615	.351
	others	270	250	290	1300	1625	2600	.615	.492	.307
Paint exterior trim	prime	400	—	—	650	—	—	1.230	—	—
	1st	475	—	—	800	—	—	1.000	—	—
	2nd	520	—	—	975	—	—	.820	—	—
Paint shingle siding	prime	270	255	300	650	975	1950	1.230	.820	.410
	others	360	340	380	800	1150	2275	1.000	.695	.351
Stain shingle siding	1st	180	170	200	750	1125	2250	1.068	.711	.355
	2nd	270	250	290	900	1325	2600	.888	.603	.307
Paint brick masonry	prime	180	135	160	750	800	1800	1.066	1.000	.444
	1st	270	225	290	815	975	2275	.981	.820	.351
	2nd	340	305	360	815	1150	2925	.981	.695	.273
Paint interior plaster or drywall	prime	400	380	495	1150	2000	3250	.695	.400	.246
	others	450	425	495	1300	2300	4000	.615	.347	.200
Paint interior doors and windows	prime	400	—	—	650	—	—	1.230	—	—
	1st	425	—	—	800	—	—	1.000	—	—
	2nd	450	—	—	975	—	—	.820	—	—

Special Construction　　R1311 Swimming Pools

R131113-20　Swimming Pools

Pool prices given per square foot of surface area include pool structure, filter and chlorination equipment, pumps, related piping, ladders/steps, maintenance kit, skimmer and vacuum system. Decks and electrical service to equipment are not included.

Residential in-ground pool construction can be divided into two categories: vinyl lined and gunite. Vinyl lined pool walls are constructed of different materials including wood, concrete, plastic or metal. The bottom is often graded with sand over which the vinyl liner is installed. Vermiculite or soil cement bottoms may be substituted for an added cost.

Gunite pool construction is used both in residential and municipal installations. These structures are steel reinforced for strength and finished with a white cement limestone plaster.

Municipal pools will have a higher cost because plumbing codes require more expensive materials, chlorination equipment and higher filtration rates.

Municipal pools greater than 1,800 S.F. require gutter systems to control waves. This gutter may be formed into the concrete wall. Often a vinyl/stainless steel gutter or gutter/wall system is specified, which will raise the pool cost.

Competition pools usually require tile bottoms and sides with contrasting lane striping, which will also raise the pool cost.

Plumbing — R2211 Facility Water Distribution

R221113-50 Pipe Material Considerations

1. Malleable fittings should be used for gas service.
2. Malleable fittings are used where there are stresses/strains due to expansion and vibration.
3. Cast fittings may be broken as an aid to disassembling of heating lines frozen by long use, temperature and minerals.
4. Cast iron pipe is extensively used for underground and submerged service.
5. Type M (light wall) copper tubing is available in hard temper only and is used for nonpressure and less severe applications than K and L.

6. Type L (medium wall) copper tubing, available hard or soft for interior service.
7. Type K (heavy wall) copper tubing, available in hard or soft temper for use where conditions are severe. For underground and interior service.
8. Hard drawn tubing requires fewer hangers or supports but should not be bent. Silver brazed fittings are recommended, however soft solder is normally used.
9. Type DMV (very light wall) copper tubing designed for drainage, waste and vent plus other non-critical pressure services.

Domestic/Imported Pipe and Fittings Cost

The prices shown in this publication for steel/cast iron pipe and steel, cast iron, malleable iron fittings are based on domestic production sold at the normal trade discounts. The above listed items of foreign manufacture may be available at prices of 1/3 to 1/2 those shown. Some imported items after minor machining or finishing operations are being sold as domestic to further complicate the system.

Caution: Most pipe prices in this book also include a coupling and pipe hangers which for the larger sizes can add significantly to the per foot cost and should be taken into account when comparing "book cost" with quoted supplier's cost.

Plumbing — R2240 Plumbing Fixtures

R224000-40 Plumbing Fixture Installation Time

Item	Rough-In	Set	Total Hours	Item	Rough-In	Set	Total Hours
Bathtub	5	5	10	Shower head only	2	1	3
Bathtub and shower, cast iron	6	6	12	Shower drain	3	1	4
Fire hose reel and cabinet	4	2	6	Shower stall, slate		15	15
Floor drain to 4 inch diameter	3	1	4	Slop sink	5	3	8
Grease trap, single, cast iron	5	3	8	Test 6 fixtures			14
Kitchen gas range		4	4	Urinal, wall	6	2	8
Kitchen sink, single	4	4	8	Urinal, pedestal or floor	6	4	10
Kitchen sink, double	6	6	12	Water closet and tank	4	3	7
Laundry tubs	4	2	6	Water closet and tank, wall hung	5	3	8
Lavatory wall hung	5	3	8	Water heater, 45 gals. gas, automatic	5	2	7
Lavatory pedestal	5	3	8	Water heaters, 65 gals. gas, automatic	5	2	7
Shower and stall	6	4	10	Water heaters, electric, plumbing only	4	2	6

Fixture prices in front of book are based on the cost per fixture set in place. The rough-in cost, which must be added for each fixture, includes carrier, if required, some supply, waste and vent pipe connecting fittings and stops. The lengths of rough-in pipe are nominal runs which would connect to the larger runs and stacks. The supply runs and DWV runs and stacks must be accounted for in separate entries. In the eastern half of the United States it is common for the plumber to carry these to a point 5′ outside the building.

Exterior Improvements — R3292 Turf & Grasses

R329219-50 Seeding

The type of grass is determined by light, shade and moisture content of soil plus intended use. Fertilizer should be disked 4″ before seeding. For steep slopes disk five tons of mulch and lay two tons of hay or straw on surface per acre after seeding. Surface mulch can be staked, lightly disked or tar emulsion sprayed. Material for mulch can be wood chips, peat moss, partially rotted hay or straw, wood fibers and sprayed emulsions. Hemp seed blankets with fertilizer are also available. For spring seeding, watering is necessary. Late fall seeding may have to be reseeded in the spring. Hydraulic seeding, power mulching, and aerial seeding can be used on large areas.

R331113-80   Piping Designations

There are several systems currently in use to describe pipe and fittings. The following paragraphs will help to identify and clarify classifications of piping systems used for water distribution.

Piping may be classified by schedule. Piping schedules include 5S, 10S, 10, 20, 30, Standard, 40, 60, Extra Strong, 80, 100, 120, 140, 160 and Double Extra Strong. These schedules are dependent upon the pipe wall thickness. The wall thickness of a particular schedule may vary with pipe size.

Ductile iron pipe for water distribution is classified by Pressure Classes such as Class 150, 200, 250, 300 and 350. These classes are actually the rated water working pressure of the pipe in pounds per square inch (psi). The pipe in these pressure classes is designed to withstand the rated water working pressure plus a surge allowance of 100 psi.

The American Water Works Association (AWWA) provides standards for various types of **plastic pipe.** C-900 is the specification for polyvinyl chloride (PVC) piping used for water distribution in sizes ranging from 4″ through 12″. C-901 is the specification for polyethylene (PE) pressure pipe, tubing and fittings used for water distribution in sizes ranging from 1/2″ through 3″. C-905 is the specification for PVC piping sizes 14″ and greater.

PVC pressure-rated pipe is identified using the standard dimensional ratio (SDR) method. This method is defined by the American Society for Testing and Materials (ASTM) Standard D 2241. This pipe is available in SDR numbers 64, 41, 32.5, 26, 21, 17, and 13.5. Pipe with an SDR of 64 will have the thinnest wall while pipe with an SDR of 13.5 will have the thickest wall. When the pressure rating (PR) of a pipe is given in psi, it is based on a line supplying water at 73 degrees F.

The National Sanitation Foundation (NSF) seal of approval is applied to products that can be used with potable water. These products have been tested to ANSI/NSF Standard 14.

Valves and strainers are classified by American National Standards Institute (ANSI) Classes. These Classes are 125, 150, 200, 250, 300, 400, 600, 900, 1500 and 2500. Within each class there is an operating pressure range dependent upon temperature. Design parameters should be compared to the appropriate material dependent, pressure-temperature rating chart for accurate valve selection.

Abbreviations

A	Area Square Feet; Ampere
ABS	Acrylonitrile Butadiene Stryrene; Asbestos Bonded Steel
A.C.	Alternating Current; Air-Conditioning; Asbestos Cement; Plywood Grade A & C
A.C.I.	American Concrete Institute
AD	Plywood, Grade A & D
Addit.	Additional
Adj.	Adjustable
af	Audio-frequency
A.G.A.	American Gas Association
Agg.	Aggregate
A.H.	Ampere Hours
A hr.	Ampere-hour
A.H.U.	Air Handling Unit
A.I.A.	American Institute of Architects
AIC	Ampere Interrupting Capacity
Allow.	Allowance
alt.	Altitude
Alum.	Aluminum
a.m.	Ante Meridiem
Amp.	Ampere
Anod.	Anodized
Approx.	Approximate
Apt.	Apartment
Asb.	Asbestos
A.S.B.C.	American Standard Building Code
Asbe.	Asbestos Worker
ASCE.	American Society of Civil Engineers
A.S.H.R.A.E.	American Society of Heating, Refrig. & AC Engineers
A.S.M.E.	American Society of Mechanical Engineers
A.S.T.M.	American Society for Testing and Materials
Attchmt.	Attachment
Avg.,Ave.	Average
A.W.G.	American Wire Gauge
AWWA	American Water Works Assoc.
Bbl.	Barrel
B&B	Grade B and Better; Balled & Burlapped
B.&S.	Bell and Spigot
B.&W.	Black and White
b.c.c.	Body-centered Cubic
B.C.Y.	Bank Cubic Yards
BE	Bevel End
B.F.	Board Feet
Bg. cem.	Bag of Cement
BHP	Boiler Horsepower; Brake Horsepower
B.I.	Black Iron
Bit., Bitum.	Bituminous
Bit., Conc.	Bituminous Concrete
Bk.	Backed
Bkrs.	Breakers
Bldg.	Building
Blk.	Block
Bm.	Beam
Boil.	Boilermaker
B.P.M.	Blows per Minute
BR	Bedroom
Brg.	Bearing
Brhe.	Bricklayer Helper
Bric.	Bricklayer
Brk.	Brick
Brng.	Bearing
Brs.	Brass
Brz.	Bronze
Bsn.	Basin
Btr.	Better
Btu	British Thermal Unit
BTUH	BTU per Hour
B.U.R.	Built-up Roofing
BX	Interlocked Armored Cable
°C	degree centegrade
c	Conductivity, Copper Sweat
C	Hundred; Centigrade
C/C	Center to Center, Cedar on Cedar
C-C	Center to Center
Cab.	Cabinet
Cair.	Air Tool Laborer
Calc	Calculated
Cap.	Capacity
Carp.	Carpenter
C.B.	Circuit Breaker
C.C.A.	Chromate Copper Arsenate
C.C.F.	Hundred Cubic Feet
cd	Candela
cd/sf	Candela per Square Foot
CD	Grade of Plywood Face & Back
CDX	Plywood, Grade C & D, exterior glue
Cefi.	Cement Finisher
Cem.	Cement
CF	Hundred Feet
C.F.	Cubic Feet
CFM	Cubic Feet per Minute
c.g.	Center of Gravity
CHW	Chilled Water; Commercial Hot Water
C.I.	Cast Iron
C.I.P.	Cast in Place
Circ.	Circuit
C.L.	Carload Lot
Clab.	Common Laborer
Clam	Common maintenance laborer
C.L.F.	Hundred Linear Feet
CLF	Current Limiting Fuse
CLP	Cross Linked Polyethylene
cm	Centimeter
CMP	Corr. Metal Pipe
C.M.U.	Concrete Masonry Unit
CN	Change Notice
Col.	Column
CO_2	Carbon Dioxide
Comb.	Combination
Compr.	Compressor
Conc.	Concrete
Cont.	Continuous; Continued, Container
Corr.	Corrugated
Cos	Cosine
Cot	Cotangent
Cov.	Cover
C/P	Cedar on Paneling
CPA	Control Point Adjustment
Cplg.	Coupling
C.P.M.	Critical Path Method
CPVC	Chlorinated Polyvinyl Chloride
C.Pr.	Hundred Pair
CRC	Cold Rolled Channel
Creos.	Creosote
Crpt.	Carpet & Linoleum Layer
CRT	Cathode-ray Tube
CS	Carbon Steel, Constant Shear Bar Joist
Csc	Cosecant
C.S.F.	Hundred Square Feet
CSI	Construction Specifications Institute
C.T.	Current Transformer
CTS	Copper Tube Size
Cu	Copper, Cubic
Cu. Ft.	Cubic Foot
cw	Continuous Wave
C.W.	Cool White; Cold Water
Cwt.	100 Pounds
C.W.X.	Cool White Deluxe
C.Y.	Cubic Yard (27 cubic feet)
C.Y./Hr.	Cubic Yard per Hour
Cyl.	Cylinder
d	Penny (nail size)
D	Deep; Depth; Discharge
Dis., Disch.	Discharge
Db.	Decibel
Dbl.	Double
DC	Direct Current
DDC	Direct Digital Control
Demob.	Demobilization
d.f.u.	Drainage Fixture Units
D.H.	Double Hung
DHW	Domestic Hot Water
DI	Ductile Iron
Diag.	Diagonal
Diam., Dia	Diameter
Distrib.	Distribution
Div.	Division
Dk.	Deck
D.L.	Dead Load; Diesel
DLH	Deep Long Span Bar Joist
Do.	Ditto
Dp.	Depth
D.P.S.T.	Double Pole, Single Throw
Dr.	Drive
Drink.	Drinking
D.S.	Double Strength
D.S.A.	Double Strength A Grade
D.S.B.	Double Strength B Grade
Dty.	Duty
DWV	Drain Waste Vent
DX	Deluxe White, Direct Expansion
dyn	Dyne
e	Eccentricity
E	Equipment Only; East
Ea.	Each
E.B.	Encased Burial
Econ.	Economy
E.C.Y	Embankment Cubic Yards
EDP	Electronic Data Processing
EIFS	Exterior Insulation Finish System
E.D.R.	Equiv. Direct Radiation
Eq.	Equation
EL	elevation
Elec.	Electrician; Electrical
Elev.	Elevator; Elevating
EMT	Electrical Metallic Conduit; Thin Wall Conduit
Eng.	Engine, Engineered
EPDM	Ethylene Propylene Diene Monomer
EPS	Expanded Polystyrene
Eqhv.	Equip. Oper., Heavy
Eqlt.	Equip. Oper., Light
Eqmd.	Equip. Oper., Medium
Eqmm.	Equip. Oper., Master Mechanic
Eqol.	Equip. Oper., Oilers
Equip.	Equipment
ERW	Electric Resistance Welded
E.S.	Energy Saver
Est.	Estimated
esu	Electrostatic Units
E.W.	Each Way
EWT	Entering Water Temperature
Excav.	Excavation
Exp.	Expansion, Exposure
Ext.	Exterior
Extru.	Extrusion
f.	Fiber stress
F	Fahrenheit; Female; Fill
Fab.	Fabricated

FBGS	Fiberglass	H.P.	Horsepower; High Pressure	LE	Lead Equivalent
F.C.	Footcandles	H.P.F.	High Power Factor	LED	Light Emitting Diode
f.c.c.	Face-centered Cubic	Hr.	Hour	L.F.	Linear Foot
f'c.	Compressive Stress in Concrete; Extreme Compressive Stress	Hrs./Day	Hours per Day	L.F. Nose	Linear Foot of Stair Nosing
		HSC	High Short Circuit	L.F. Rsr	Linear Foot of Stair Riser
F.E.	Front End	Ht.	Height	Lg.	Long; Length; Large
FEP	Fluorinated Ethylene Propylene (Teflon)	Htg.	Heating	L & H	Light and Heat
		Htrs.	Heaters	LH	Long Span Bar Joist
F.G.	Flat Grain	HVAC	Heating, Ventilation & Air-Conditioning	L.H.	Labor Hours
F.H.A.	Federal Housing Administration			L.L.	Live Load
Fig.	Figure	Hvy.	Heavy	L.L.D.	Lamp Lumen Depreciation
Fin.	Finished	HW	Hot Water	lm	Lumen
Fixt.	Fixture	Hyd.;Hydr.	Hydraulic	lm/sf	Lumen per Square Foot
Fl. Oz.	Fluid Ounces	Hz.	Hertz (cycles)	lm/W	Lumen per Watt
Flr.	Floor	I.	Moment of Inertia	L.O.A.	Length Over All
F.M.	Frequency Modulation; Factory Mutual	IBC	International Building Code	log	Logarithm
		I.C.	Interrupting Capacity	L-O-L	Lateralolet
Fmg.	Framing	ID	Inside Diameter	long.	longitude
Fdn.	Foundation	I.D.	Inside Dimension; Identification	L.P.	Liquefied Petroleum; Low Pressure
Fori.	Foreman, Inside	I.F.	Inside Frosted	L.P.F.	Low Power Factor
Foro.	Foreman, Outside	I.M.C.	Intermediate Metal Conduit	LR	Long Radius
Fount.	Fountain	In.	Inch	L.S.	Lump Sum
fpm	Feet per Minute	Incan.	Incandescent	Lt.	Light
FPT	Female Pipe Thread	Incl.	Included; Including	Lt. Ga.	Light Gauge
Fr.	Frame	Int.	Interior	L.T.L.	Less than Truckload Lot
F.R.	Fire Rating	Inst.	Installation	Lt. Wt.	Lightweight
FRK	Foil Reinforced Kraft	Insul.	Insulation/Insulated	L.V.	Low Voltage
FRP	Fiberglass Reinforced Plastic	I.P.	Iron Pipe	M	Thousand; Material; Male; Light Wall Copper Tubing
FS	Forged Steel	I.P.S.	Iron Pipe Size		
FSC	Cast Body; Cast Switch Box	I.P.T.	Iron Pipe Threaded	M²CA	Meters Squared Contact Area
Ft.	Foot; Feet	I.W.	Indirect Waste	m/hr.; M.H.	Man-hour
Ftng.	Fitting	J	Joule	mA	Milliampere
Ftg.	Footing	J.I.C.	Joint Industrial Council	Mach.	Machine
Ft lb.	Foot Pound	K	Thousand; Thousand Pounds; Heavy Wall Copper Tubing, Kelvin	Mag. Str.	Magnetic Starter
Furn.	Furniture			Maint.	Maintenance
FVNR	Full Voltage Non-Reversing	K.A.H.	Thousand Amp. Hours	Marb.	Marble Setter
FXM	Female by Male	KCMIL	Thousand Circular Mils	Mat; Mat'l.	Material
Fy.	Minimum Yield Stress of Steel	KD	Knock Down	Max.	Maximum
g	Gram	K.D.A.T.	Kiln Dried After Treatment	MBF	Thousand Board Feet
G	Gauss	kg	Kilogram	MBH	Thousand BTU's per hr.
Ga.	Gauge	kG	Kilogauss	MC	Metal Clad Cable
Gal., gal.	Gallon	kgf	Kilogram Force	M.C.F.	Thousand Cubic Feet
gpm, GPM	Gallon per Minute	kHz	Kilohertz	M.C.F.M.	Thousand Cubic Feet per Minute
Galv.	Galvanized	Kip.	1000 Pounds	M.C.M.	Thousand Circular Mils
Gen.	General	KJ	Kiljoule	M.C.P.	Motor Circuit Protector
G.F.I.	Ground Fault Interrupter	K.L.	Effective Length Factor	MD	Medium Duty
Glaz.	Glazier	K.L.F.	Kips per Linear Foot	M.D.O.	Medium Density Overlaid
GPD	Gallons per Day	Km	Kilometer	Med.	Medium
GPH	Gallons per Hour	K.S.F.	Kips per Square Foot	MF	Thousand Feet
GPM	Gallons per Minute	K.S.I.	Kips per Square Inch	M.F.B.M.	Thousand Feet Board Measure
GR	Grade	kV	Kilovolt	Mfg.	Manufacturing
Gran.	Granular	kVA	Kilovolt Ampere	Mfrs.	Manufacturers
Grnd.	Ground	K.V.A.R.	Kilovar (Reactance)	mg	Milligram
H	High Henry	KW	Kilowatt	MGD	Million Gallons per Day
H.C.	High Capacity	KWh	Kilowatt-hour	MGPH	Thousand Gallons per Hour
H.D.	Heavy Duty; High Density	L	Labor Only; Length; Long; Medium Wall Copper Tubing	MH, M.H.	Manhole; Metal Halide; Man-Hour
H.D.O.	High Density Overlaid			MHz	Megahertz
H.D.P.E.	high density polyethelene	Lab.	Labor	Mi.	Mile
Hdr.	Header	lat	Latitude	MI	Malleable Iron; Mineral Insulated
Hdwe.	Hardware	Lath.	Lather	mm	Millimeter
Help.	Helper Average	Lav.	Lavatory	Mill.	Millwright
HEPA	High Efficiency Particulate Air Filter	lb.; #	Pound	Min., min.	Minimum, minute
		L.B.	Load Bearing; L Conduit Body	Misc.	Miscellaneous
Hg	Mercury	L. & E.	Labor & Equipment	ml	Milliliter, Mainline
HIC	High Interrupting Capacity	lb./hr.	Pounds per Hour	M.L.F.	Thousand Linear Feet
HM	Hollow Metal	lb./L.F.	Pounds per Linear Foot	Mo.	Month
HMWPE	high molecular weight polyethylene	lbf/sq.in.	Pound-force per Square Inch	Mobil.	Mobilization
		L.C.L.	Less than Carload Lot	Mog.	Mogul Base
H.O.	High Output	L.C.Y.	Loose Cubic Yard	MPH	Miles per Hour
Horiz.	Horizontal	Ld.	Load	MPT	Male Pipe Thread

MRT	Mile Round Trip	Pl.	Plate	S.F.C.A.	Square Foot Contact Area
ms	Millisecond	Plah.	Plasterer Helper	S.F. Flr.	Square Foot of Floor
M.S.F.	Thousand Square Feet	Plas.	Plasterer	S.F.G.	Square Foot of Ground
Mstz.	Mosaic & Terrazzo Worker	Pluh.	Plumbers Helper	S.F. Hor.	Square Foot Horizontal
M.S.Y.	Thousand Square Yards	Plum.	Plumber	S.F.R.	Square Feet of Radiation
Mtd., mtd.	Mounted	Ply.	Plywood	S.F. Shlf.	Square Foot of Shelf
Mthe.	Mosaic & Terrazzo Helper	p.m.	Post Meridiem	S4S	Surface 4 Sides
Mtng.	Mounting	Pntd.	Painted	Shee.	Sheet Metal Worker
Mult.	Multi; Multiply	Pord.	Painter, Ordinary	Sin.	Sine
M.V.A.	Million Volt Amperes	pp	Pages	Skwk.	Skilled Worker
M.V.A.R.	Million Volt Amperes Reactance	PP, PPL	Polypropylene	SL	Saran Lined
MV	Megavolt	P.P.M.	Parts per Million	S.L.	Slimline
MW	Megawatt	Pr.	Pair	Sldr.	Solder
MXM	Male by Male	P.E.S.B.	Pre-engineered Steel Building	SLH	Super Long Span Bar Joist
MYD	Thousand Yards	Prefab.	Prefabricated	S.N.	Solid Neutral
N	Natural; North	Prefin.	Prefinished	S-O-L	Socketolet
nA	Nanoampere	Prop.	Propelled	sp	Standpipe
NA	Not Available; Not Applicable	PSF, psf	Pounds per Square Foot	S.P.	Static Pressure; Single Pole; Self-Propelled
N.B.C.	National Building Code	PSI, psi	Pounds per Square Inch		
NC	Normally Closed	PSIG	Pounds per Square Inch Gauge	Spri.	Sprinkler Installer
N.E.M.A.	National Electrical Manufacturers Assoc.	PSP	Plastic Sewer Pipe	spwg	Static Pressure Water Gauge
		Pspr.	Painter, Spray	S.P.D.T.	Single Pole, Double Throw
NEHB	Bolted Circuit Breaker to 600V.	Psst.	Painter, Structural Steel	SPF	Spruce Pine Fir
N.L.B.	Non-Load-Bearing	P.T.	Potential Transformer	S.P.S.T.	Single Pole, Single Throw
NM	Non-Metallic Cable	P. & T.	Pressure & Temperature	SPT	Standard Pipe Thread
nm	Nanometer	Ptd.	Painted	Sq.	Square; 100 Square Feet
No.	Number	Ptns.	Partitions	Sq. Hd.	Square Head
NO	Normally Open	Pu	Ultimate Load	Sq. In.	Square Inch
N.O.C.	Not Otherwise Classified	PVC	Polyvinyl Chloride	S.S.	Single Strength; Stainless Steel
Nose.	Nosing	Pvmt.	Pavement	S.S.B.	Single Strength B Grade
N.P.T.	National Pipe Thread	Pwr.	Power	sst, ss	Stainless Steel
NQOD	Combination Plug-on/Bolt on Circuit Breaker to 240V.	Q	Quantity Heat Flow	Sswk.	Structural Steel Worker
		Qt.	Quart	Sswl.	Structural Steel Welder
N.R.C.	Noise Reduction Coefficient/ Nuclear Regulator Commission	Quan., Qty.	Quantity	St.;Stl.	Steel
		Q.C.	Quick Coupling	S.T.C.	Sound Transmission Coefficient
N.R.S.	Non Rising Stem	r	Radius of Gyration	Std.	Standard
ns	Nanosecond	R	Resistance	Stg.	Staging
nW	Nanowatt	R.C.P.	Reinforced Concrete Pipe	STK	Select Tight Knot
OB	Opposing Blade	Rect.	Rectangle	STP	Standard Temperature & Pressure
OC	On Center	Reg.	Regular	Stpi.	Steamfitter, Pipefitter
OD	Outside Diameter	Reinf.	Reinforced	Str.	Strength; Starter; Straight
O.D.	Outside Dimension	Req'd.	Required	Strd.	Stranded
ODS	Overhead Distribution System	Res.	Resistant	Struct.	Structural
O.G.	Ogee	Resi.	Residential	Sty.	Story
O.H.	Overhead	Rgh.	Rough	Subj.	Subject
O&P	Overhead and Profit	RGS	Rigid Galvanized Steel	Subs.	Subcontractors
Oper.	Operator	R.H.W.	Rubber, Heat & Water Resistant; Residential Hot Water	Surf.	Surface
Opng.	Opening			Sw.	Switch
Orna.	Ornamental	rms	Root Mean Square	Swbd.	Switchboard
OSB	Oriented Strand Board	Rnd.	Round	S.Y.	Square Yard
O.S.&Y.	Outside Screw and Yoke	Rodm.	Rodman	Syn.	Synthetic
Ovhd.	Overhead	Rofc.	Roofer, Composition	S.Y.P.	Southern Yellow Pine
OWG	Oil, Water or Gas	Rofp.	Roofer, Precast	Sys.	System
Oz.	Ounce	Rohe.	Roofer Helpers (Composition)	t.	Thickness
P.	Pole; Applied Load; Projection	Rots.	Roofer, Tile & Slate	T	Temperature; Ton
p.	Page	R.O.W.	Right of Way	Tan	Tangent
Pape.	Paperhanger	RPM	Revolutions per Minute	T.C.	Terra Cotta
P.A.P.R.	Powered Air Purifying Respirator	R.S.	Rapid Start	T & C	Threaded and Coupled
PAR	Parabolic Reflector	Rsr	Riser	T.D.	Temperature Difference
Pc., Pcs.	Piece, Pieces	RT	Round Trip	Tdd	Telecommunications Device for the Deaf
P.C.	Portland Cement; Power Connector	S.	Suction; Single Entrance; South		
P.C.F.	Pounds per Cubic Foot	SC	Screw Cover	T.E.M.	Transmission Electron Microscopy
P.C.M.	Phase Contrast Microscopy	SCFM	Standard Cubic Feet per Minute	TFE	Tetrafluoroethylene (Teflon)
P.E.	Professional Engineer; Porcelain Enamel; Polyethylene; Plain End	Scaf.	Scaffold	T. & G.	Tongue & Groove; Tar & Gravel
		Sch., Sched.	Schedule		
		S.C.R.	Modular Brick	Th., Thk.	Thick
Perf.	Perforated	S.D.	Sound Deadening	Thn.	Thin
PEX	Cross linked polyethylene	S.D.R.	Standard Dimension Ratio	Thrded	Threaded
Ph.	Phase	S.E.	Surfaced Edge	Tilf.	Tile Layer, Floor
P.I.	Pressure Injected	Sel.	Select	Tilh.	Tile Layer, Helper
Pile.	Pile Driver	S.E.R., S.E.U.	Service Entrance Cable	THHN	Nylon Jacketed Wire
Pkg.	Package	S.F.	Square Foot	THW.	Insulated Strand Wire

Abbreviations

THWN	Nylon Jacketed Wire	USP	United States Primed	Wrck.	Wrecker
T.L.	Truckload	UTP	Unshielded Twisted Pair	W.S.P.	Water, Steam, Petroleum
T.M.	Track Mounted	V	Volt	WT., Wt.	Weight
Tot.	Total	V.A.	Volt Amperes	WWF	Welded Wire Fabric
T-O-L	Threadolet	V.C.T.	Vinyl Composition Tile	XFER	Transfer
T.S.	Trigger Start	VAV	Variable Air Volume	XFMR	Transformer
Tr.	Trade	VC	Veneer Core	XHD	Extra Heavy Duty
Transf.	Transformer	Vent.	Ventilation	XHHW, XLPE	Cross-Linked Polyethylene Wire
Trhv.	Truck Driver, Heavy	Vert.	Vertical		Insulation
Trlr	Trailer	V.F.	Vinyl Faced	XLP	Cross-linked Polyethylene
Trlt.	Truck Driver, Light	V.G.	Vertical Grain	Y	Wye
TTY	Teletypewriter	V.H.F.	Very High Frequency	yd	Yard
TV	Television	VHO	Very High Output	yr	Year
T.W.	Thermoplastic Water Resistant	Vib.	Vibrating	Δ	Delta
	Wire	V.L.F.	Vertical Linear Foot	%	Percent
UCI	Uniform Construction Index	Vol.	Volume	~	Approximately
UF	Underground Feeder	VRP	Vinyl Reinforced Polyester	Ø	Phase; diameter
UGND	Underground Feeder	W	Wire; Watt; Wide; West	@	At
U.H.F.	Ultra High Frequency	w/	With	#	Pound; Number
U.I.	United Inch	W.C.	Water Column; Water Closet	<	Less Than
U.L.	Underwriters Laboratory	W.F.	Wide Flange	>	Greater Than
Uld.	unloading	W.G.	Water Gauge	Z	zone
Unfin.	Unfinished	Wldg.	Welding		
URD	Underground Residential	W. Mile	Wire Mile		
	Distribution	W-O-L	Weldolet		
US	United States	W.R.	Water Resistant		

Index

Index

Index

409

Index

Index

Index

413

Index

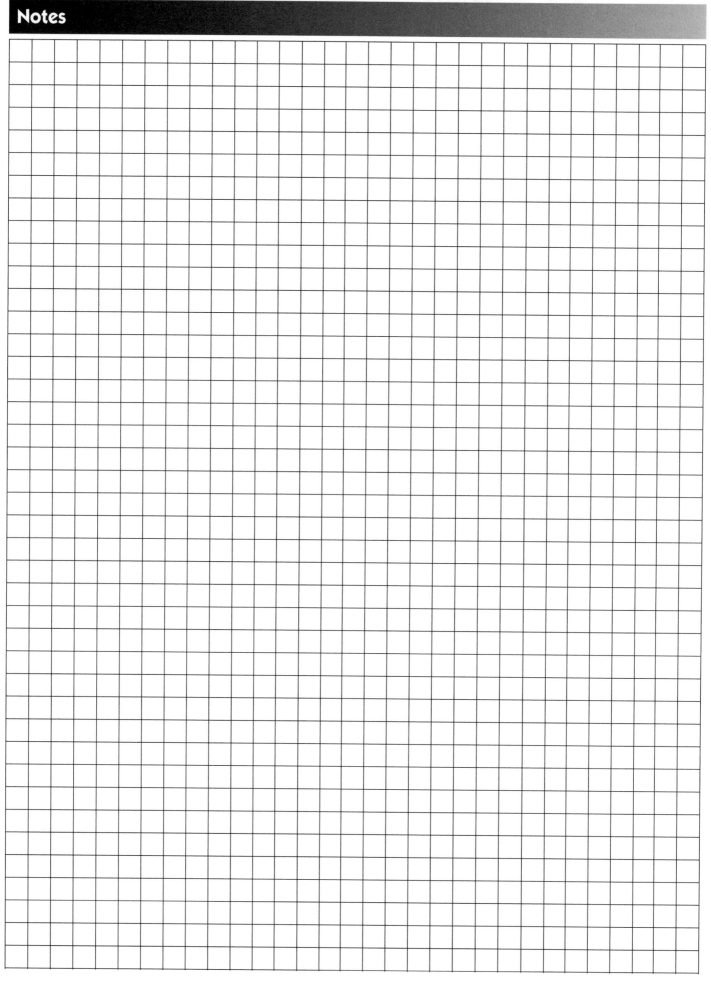

Contractor's Pricing Guide: Residential Detailed Costs 2009

Every aspect of residential construction, from overhead costs to residential lighting and wiring, is in here. All the detail you need to accurately estimate the costs of your work with or without markups—labor-hours, typical crews, and equipment are included as well. When you need a detailed estimate, this publication has all the costs to help you come up with a complete, on the money price you can rely on to win profitable work.

Unit prices organized according to MasterFormat 2004!

$39.95 per copy
Available Oct. 2008
Catalog no. 60339

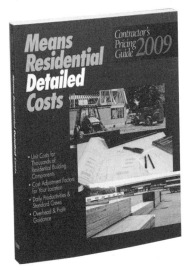

Contractor's Pricing Guide: Residential Repair & Remodeling Costs 2009

This book provides total unit price costs for every aspect of the most common repair and remodeling projects. Organized in the order of construction by component and activity, it includes demolition and installation, cleaning, painting, and more. With simplified estimating methods, clear, concise descriptions, and technical specifications for each component, the book is a valuable tool for contractors who want to speed up their estimating time, while making sure their costs are on target.

$39.95 per copy
Available Oct. 2008
Catalog no. 60349

Contractor's Pricing Guide: Residential Square Foot Costs 2009

Now available in one concise volume, all you need to know to plan and budget the cost of new homes. If you are looking for a quick reference, the model home section contains costs for over 250 different sizes and types of residences, with hundreds of easily applied modifications. If you need even more detail, the Assemblies Section lets you build your own costs or modify the model costs further. Hundreds of graphics are provided, along with forms and procedures to help you get it right.

$39.95 per copy
Available Oct. 2008
Catalog no. 60329

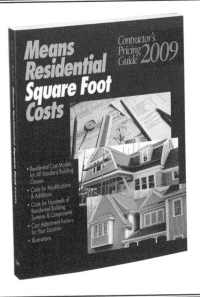

RSMeans Building Construction Cost Data 2009

Offers you unchallenged unit price reliability in an easy-to-use arrangement. Whether used for complete, finished estimates or for periodic checks, it supplies more cost facts better and faster than any comparable source. Over 20,000 unit prices for 2009. The City Cost Indexes and Location Factors cover over 930 areas, for indexing to any project location in North America. Order and get *RSMeans Quarterly Update Service* FREE. You'll have year-long access to the RSMeans Estimating **HOTLINE** FREE with your subscription. Expert assistance when using RSMeans data is just a phone call away.

$154.95 per copy
Available Oct. 2008
Catalog no. 60019

Unit prices organized according to MasterFormat 2004!

RSMeans Mechanical Cost Data 2009
- **HVAC**
- **Controls**

Total unit and systems price guidance for mechanical construction. . . materials, parts, fittings, and complete labor cost information. Includes prices for piping, heating, air conditioning, ventilation, and all related construction.

Plus new 2009 unit costs for:

- Over 2,500 installed HVAC/controls assemblies components
- "On-site" Location Factors for over 930 cities and towns in the U.S. and Canada
- Crews, labor, and equipment

$154.95 per copy
Available Oct. 2008
Catalog no. 60029

RSMeans Plumbing Cost Data 2009

Comprehensive unit prices and assemblies for plumbing, irrigation systems, commercial and residential fire protection, point-of-use water heaters, and the latest approved materials. This publication and its companion, *RSMeans Mechanical Cost Data*, provide full-range cost estimating coverage for all the mechanical trades.

Now contains more lines of no-hub CI soil pipe fittings, more flange-type escutcheons, fiberglass pipe insulation in a full range of sizes for 2-1/2" and 3" wall thicknesses, 220 lines of grease duct, and much more.

$154.95 per copy
Available Oct. 2008
Catalog no. 60219

Unit prices organized according to MasterFormat 2004!

RSMeans Facilities Construction Cost Data 2009

For the maintenance and construction of commercial, industrial, municipal, and institutional properties. Costs are shown for new and remodeling construction and are broken down into materials, labor, equipment, and overhead and profit. Special emphasis is given to sections on mechanical, electrical, furnishings, site work, building maintenance, finish work, and demolition.

More than 43,000 unit costs, plus assemblies costs and a comprehensive Reference Section are included.

$368.95 per copy
Available Nov. 2008
Catalog no. 60209

RSMeans Square Foot Costs 2009

It's Accurate and Easy To Use!

- **Updated price information** based on nationwide figures from suppliers, estimators, labor experts, and contractors

- "How-to-Use" sections, with **clear examples** of commercial, residential, industrial, and institutional structures

- Realistic graphics, offering true-to-life illustrations of building projects

- Extensive information on using square foot cost data, including sample estimates and alternate pricing methods

$168.95 per copy
Available Oct. 2008
Catalog no. 60059

RSMeans Repair & Remodeling Cost Data 2009

Commercial/Residential

Use this valuable tool to estimate commercial and residential renovation and remodeling.

Includes: New costs for hundreds of unique methods, materials, and conditions that only come up in repair and remodeling, PLUS:

- Unit costs for over 15,000 construction components
- Installed costs for over 90 assemblies
- Over 930 "on-site" localization factors for the U.S. and Canada.

Unit prices organized according to MasterFormat 2004!

$132.95 per copy
Available Nov. 2008
Catalog no. 60049

RSMeans Electrical Cost Data 2009

Pricing information for every part of electrical cost planning. More than 13,000 unit and systems costs with design tables; clear specifications and drawings; engineering guides; illustrated estimating procedures; complete labor-hour and materials costs for better scheduling and procurement; and the latest electrical products and construction methods.

- A variety of special electrical systems including cathodic protection

- Costs for maintenance, demolition, HVAC/mechanical, specialties, equipment, and more

Unit prices organized according to MasterFormat 2004!

$154.95 per copy
Available Oct. 2008
Catalog no. 60039

RSMeans Electrical Change Order Cost Data 2009

RSMeans Electrical Change Order Cost Data provides you with electrical unit prices exclusively for pricing change orders—based on the recent, direct experience of contractors and suppliers. Analyze and check your own change order estimates against the experience others have had doing the same work. It also covers productivity analysis and change order cost justifications. With useful information for calculating the effects of change orders and dealing with their administration.

$154.95 per copy
Available Dec. 2008
Catalog no. 60239

RSMeans Assemblies Cost Data 2009

RSMeans Assemblies Cost Data takes the guesswork out of preliminary or conceptual estimates. Now you don't have to try to calculate the assembled cost by working up individual component costs. We've done all the work for you.

Presents detailed illustrations, descriptions, specifications, and costs for every conceivable building assembly—240 types in all—arranged in the easy-to-use UNIFORMAT II system. Each illustrated "assembled" cost includes a complete grouping of materials and associated installation costs, including the installing contractor's overhead and profit.

$245.95 per copy
Available Oct. 2008
Catalog no. 60069

RSMeans Open Shop Building Construction Cost Data 2009

The latest costs for accurate budgeting and estimating of new commercial and residential construction. . . renovation work. . . change orders. . . cost engineering.

RSMeans Open Shop "BCCD" will assist you to:
• Develop benchmark prices for change orders
• Plug gaps in preliminary estimates and budgets
• Estimate complex projects
• Substantiate invoices on contracts
• Price ADA-related renovations

Unit prices organized according to MasterFormat 2004!

$154.95 per copy
Available Dec. 2008
Catalog no. 60159

RSMeans Residential Cost Data 2009

Contains square foot costs for 30 basic home models with the look of today, plus hundreds of custom additions and modifications you can quote right off the page. With costs for the 100 residential systems you're most likely to use in the year ahead. Complete with blank estimating forms, sample estimates, and step-by-step instructions.

Now contains line items for cultured stone and brick, PVC trim lumber, and TPO roofing.

$132.95 per copy
Available Oct. 2008
Catalog no. 60179

Unit prices organized according to MasterFormat 2004!

RSMeans Site Work & Landscape Cost Data 2009

Includes unit and assemblies costs for earthwork, sewerage, piped utilities, site improvements, drainage, paving, trees & shrubs, street openings/repairs, underground tanks, and more. Contains 57 tables of assemblies costs for accurate conceptual estimates.

Includes:
• Estimating for infrastructure improvements
• Environmentally-oriented construction
• ADA-mandated handicapped access
• Hazardous waste line items

$154.95 per copy
Available Nov. 2008
Catalog no. 60289

RSMeans Facilities Maintenance & Repair Cost Data 2009

RSMeans Facilities Maintenance & Repair Cost Data gives you a complete system to manage and plan your facility repair and maintenance costs and budget efficiently. Guidelines for auditing a facility and developing an annual maintenance plan. Budgeting is included, along with reference tables on cost and management, and information on frequency and productivity of maintenance operations.

The only nationally recognized source of maintenance and repair costs. Developed in cooperation with the Civil Engineering Research Laboratory (CERL) of the Army Corps of Engineers.

$336.95 per copy
Available Dec. 2008
Catalog no. 60309

RSMeans Light Commercial Cost Data 2009

Specifically addresses the light commercial market, which is a specialized niche in the construction industry. Aids you, the owner/designer/contractor, in preparing all types of estimates—from budgets to detailed bids. Includes new advances in methods and materials.

Assemblies Section allows you to evaluate alternatives in the early stages of design/planning.

Over 11,000 unit costs ensure that you have the prices you need. . . when you need them.

Unit prices organized according to MasterFormat 2004!

$132.95 per copy
Available Nov. 2008
Catalog no. 60189

For more information
visit the RSMeans website
at www.rsmeans.com

Annual Cost Guides

RSMeans Concrete & Masonry Cost Data 2009

Provides you with cost facts for virtually all concrete/masonry estimating needs, from complicated formwork to various sizes and face finishes of brick and block—all in great detail. The comprehensive Unit Price Section contains more than 7,500 selected entries. Also contains an Assemblies [Cost] Section, and a detailed Reference Section that supplements the cost data.

Unit prices organized according to MasterFormat 2004!

$139.95 per copy
Available Dec. 2008
Catalog no. 60119

RSMeans Labor Rates for the Construction Industry 2009

Complete information for estimating labor costs, making comparisons, and negotiating wage rates by trade for over 300 U.S. and Canadian cities. With 46 construction trades listed by local union number in each city, and historical wage rates included for comparison. Each city chart lists the county and is alphabetically arranged with handy visual flip tabs for quick reference.

$336.95 per copy
Available Dec. 2008
Catalog no. 60129

RSMeans Construction Cost Indexes 2009

What materials and labor costs will change unexpectedly this year? By how much?

• Breakdowns for 316 major cities
• National averages for 30 key cities
• Expanded five major city indexes
• Historical construction cost indexes

$335.00 per year (subscription)
Catalog no. 50149

$83.25 individual quarters
Catalog no. 60149 A,B,C,D

RSMeans Interior Cost Data 2009

Provides you with prices and guidance needed to make accurate interior work estimates. Contains costs on materials, equipment, hardware, custom installations, furnishings, and labor costs... for new and remodel commercial and industrial interior construction, including updated information on office furnishings, and reference information.

Unit prices organized according to MasterFormat 2004!

$154.95 per copy
Available Nov. 2008
Catalog no. 60099

RSMeans Heavy Construction Cost Data 2009

A comprehensive guide to heavy construction costs. Includes costs for highly specialized projects such as tunnels, dams, highways, airports, and waterways. Information on labor rates, equipment, and materials costs is included. Features unit price costs, systems costs, and numerous reference tables for costs and design.

Unit prices organized according to MasterFormat 2004!

$154.95 per copy
Available Dec. 2008
Catalog no. 60169

Value Engineering: Practical Applications

For Design, Construction, Maintenance & Operations

by Alphonse Dell'Isola, PE

A tool for immediate application—for engineers, architects, facility managers, owners, and contractors. Includes making the case for VE—the management briefing; integrating VE into planning, budgeting, and design; conducting life cycle costing; using VE methodology in design review and consultant selection; case studies; VE workbook; and a life cycle costing program on disk.

$79.95 per copy
Over 450 pages, illustrated, softcover
Catalog no. 67319A

Facilities Operations & Engineering Reference

by the Association for Facilities Engineering and RSMeans

An all-in-one technical reference for planning and managing facility projects and solving day-to-day operations problems. Selected as the official certified plant engineer reference, this handbook covers financial analysis, maintenance, HVAC and energy efficiency, and more.

$54.98 per copy
Over 700 pages, illustrated, hardcover
Catalog no. 67318

The Building Professional's Guide to Contract Documents

3rd Edition

by Waller S. Poage, AIA, CSI, CVS

A comprehensive reference for owners, design professionals, contractors, and students

- Structure your documents for maximum efficiency.
- Effectively communicate construction requirements.
- Understand the roles and responsibilities of construction professionals.
- Improve methods of project delivery.

$64.95 per copy, 400 pages
Diagrams and construction forms, hardcover
Catalog no. 67261A

Complete Book of Framing

by Scot Simpson

This straightforward, easy-to-learn method will help framers, carpenters, and handy homeowners build their skills in rough carpentry and framing. Shows how to frame all the parts of a house: floors, walls, roofs, door & window openings, and stairs—with hundreds of color photographs and drawings that show every detail.

The book gives beginners all the basics they need to go from zero framing knowledge to a journeyman level.

$29.95 per copy
352 pages, softcover
Catalog no. 67353

Cost Planning & Estimating for Facilities Maintenance

In this unique book, a team of facilities management authorities shares their expertise on:

- Evaluating and budgeting maintenance operations
- Maintaining and repairing key building components
- Applying *RSMeans Facilities Maintenance & Repair Cost Data* to your estimating

Covers special maintenance requirements of the ten major building types.

$89.95 per copy
Over 475 pages, hardcover
Catalog no. 67314

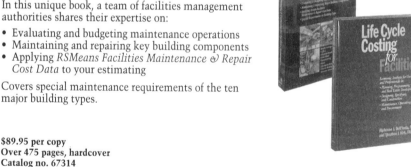

Life Cycle Costing for Facilities

by Alphonse Dell'Isola and Dr. Steven Kirk

Guidance for achieving higher quality design and construction projects at lower costs! Cost-cutting efforts often sacrifice quality to yield the cheapest product. Life cycle costing enables building designers and owners to achieve both. The authors of this book show how LCC can work for a variety of projects — from roads to HVAC upgrades to different types of buildings.

$99.95 per copy
450 pages, hardcover
Catalog no. 67341

Planning & Managing Interior Projects 2nd Edition

by Carol E. Farren, CFM

Expert guidance on managing renovation & relocation projects

This book guides you through every step in relocating to a new space or renovating an old one. From initial meeting through design and construction, to post-project administration, it helps you get the most for your company or client. Includes sample forms, spec lists, agreements, drawings, and much more!

$34.98 per copy
200 pages, softcover
Catalog no. 67245A

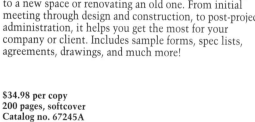

Construction Business Management

by Nick Ganaway

Only 43% of construction firms stay in business after four years. Make sure your company thrives with valuable guidance from a pro with 25 years of success as a commercial contractor. Find out what it takes to build all aspects of a business that is profitable, enjoyable, and enduring. With a bonus chapter on retail construction.

$49.95 per copy
200 pages, softcover
Catalog no. 67352

For more information
visit the RSMeans website
at www.rsmeans.com

Reference Books

Interior Home Improvement Costs 9th Edition

Updated estimates for the most popular remodeling and repair projects—from small, do-it-yourself jobs to major renovations and new construction. Includes: kitchens & baths; new living space from your attic, basement, or garage; new floors, paint, and wallpaper; tearing out or building new walls; closets, stairs, and fireplaces; new energy-saving improvements, home theaters, and more!

$24.95 per copy
250 pages, illustrated, softcover
Catalog no. 67308E

Exterior Home Improvement Costs 9th Edition

Updated estimates for the most popular remodeling and repair projects—from small, do-it-yourself jobs, to major renovations and new construction. Includes: curb appeal projects—landscaping, patios, porches, driveways, and walkways; new windows and doors; decks, greenhouses, and sunrooms; room additions and garages; roofing, siding, and painting; "green" improvements to save energy & water.

$24.95 per copy
Over 275 pages, illustrated, softcover
Catalog no. 67309E

Builder's Essentials: Plan Reading & Material Takeoff

For Residential and Light Commercial Construction

by Wayne J. DelPico

A valuable tool for understanding plans and specs, and accurately calculating material quantities. Step-by-step instructions and takeoff procedures based on a full set of working drawings.

$35.95 per copy
Over 420 pages, softcover
Catalog no. 67307

Means Unit Price Estimating Methods

New 4th Edition

This new edition includes up-to-date cost data and estimating examples, updated to reflect changes to the CSI numbering system and new features of RSMeans cost data. It describes the most productive, universally accepted ways to estimate, and uses checklists and forms to illustrate shortcuts and timesavers. A model estimate demonstrates procedures. A new chapter explores computer estimating alternatives.

$59.95 per copy
Over 350 pages, illustrated, softcover
Catalog no. 67303B

Total Productive Facilities Management

by Richard W. Sievert, Jr.

Today, facilities are viewed as strategic resources. . . elevating the facility manager to the role of asset manager supporting the organization's overall business goals. Now, Richard Sievert Jr., in this well-articulated guidebook, sets forth a new operational standard for the facility manager's emerging role. . . a comprehensive program for managing facilities as a true profit center.

$29.98 per copy
275 pages, softcover
Catalog no. 67321

Green Building: Project Planning & Cost Estimating 2nd Edition

This new edition has been completely updated with the latest in green building technologies, design concepts, standards, and costs. Now includes a 2009 Green Building *CostWorks* CD with more than 300 green building assemblies and over 5,000 unit price line items for sustainable building. The new edition is also full-color with all new case studies—plus a new chapter on deconstruction, a key aspect of green building.

$129.95 per copy
350 pages, softcover
Catalog no. 67338A

Concrete Repair and Maintenance Illustrated

by Peter Emmons

Hundreds of illustrations show users how to analyze, repair, clean, and maintain concrete structures for optimal performance and cost effectiveness. From parking garages to roads and bridges to structural concrete, this comprehensive book describes the causes, effects, and remedies for concrete wear and failure. Invaluable for planning jobs, selecting materials, and training employees, this book is a must-have for concrete specialists, general contractors, facility managers, civil and structural engineers, and architects.

$69.95 per copy
300 pages, illustrated, softcover
Catalog no. 67146

Means Illustrated Construction Dictionary Condensed, 2nd Edition

The best portable dictionary for office or field use—an essential tool for contractors, architects, insurance and real estate personnel, facility managers, homeowners, and anyone who needs quick, clear definitions for construction terms. The second edition has been further enhanced with updates and hundreds of new terms and illustrations . . . in keeping with the most recent developments in the construction industry.

Now with a quick-reference Spanish section. Includes tools and equipment, materials, tasks, and more!

$59.95 per copy
Over 500 pages, softcover
Catalog no. 67282A

Means Repair & Remodeling Estimating 4th Edition

By Edward B. Wetherill and RSMeans

This important reference focuses on the unique problems of estimating renovations of existing structures, and helps you determine the true costs of remodeling through careful evaluation of architectural details and a site visit.

New section on disaster restoration costs.

$69.95 per copy
Over 450 pages, illustrated, hardcover
Catalog no. 67265B

Facilities Planning & Relocation

by David D. Owen

A complete system for planning space needs and managing relocations. Includes step-by-step manual, over 50 forms, and extensive reference section on materials and furnishings.

New lower price and user-friendly format.

$49.98 per copy
Over 450 pages, softcover
Catalog no. 67301

Means Square Foot & UNIFORMAT Assemblies Estimating Methods 3rd Edition

Develop realistic square foot and assemblies costs for budgeting and construction funding. The new edition features updated guidance on square foot and assemblies estimating using UNIFORMAT II. An essential reference for anyone who performs conceptual estimates.

$69.95 per copy
Over 300 pages, illustrated, softcover
Catalog no. 67145B

Means Electrical Estimating Methods 3rd Edition

Expanded edition includes sample estimates and cost information in keeping with the latest version of the CSI MasterFormat and UNIFORMAT II. Complete coverage of fiber optic and uninterruptible power supply electrical systems, broken down by components, and explained in detail. Includes a new chapter on computerized estimating methods. A practical companion to *RSMeans Electrical Cost Data.*

$64.95 per copy
Over 325 pages, hardcover
Catalog no. 67230B

Means Mechanical Estimating Methods 4th Edition

Completely updated, this guide assists you in making a review of plans, specs, and bid packages, with suggestions for takeoff procedures, listings, substitutions, and pre-bid scheduling for all components of HVAC. Includes suggestions for budgeting labor and equipment usage. Compares materials and construction methods to allow you to select the best option.

$64.95 per copy
Over 350 pages, illustrated, softcover
Catalog no. 67294B

Means ADA Compliance Pricing Guide 2nd Edition

by Adaptive Environments and RSMeans

Completely updated and revised to the new 2004 *Americans with Disabilities Act Accessibility Guidelines,* this book features more than 70 of the most commonly needed modifications for ADA compliance. Projects range from installing ramps and walkways, widening doorways and entryways, and installing and refitting elevators, to relocating light switches and signage.

$79.95 per copy
Over 350 pages, illustrated, softcover
Catalog no. 67310A

Project Scheduling & Management for Construction 3rd Edition

by David R. Pierce, Jr.

A comprehensive yet easy-to-follow guide to construction project scheduling and control—from vital project management principles through the latest scheduling, tracking, and controlling techniques. The author is a leading authority on scheduling, with years of field and teaching experience at leading academic institutions. Spend a few hours with this book and come away with a solid understanding of this essential management topic.

$64.95 per copy
Over 300 pages, illustrated, hardcover
Catalog no. 67247B

For more information
visit the RSMeans website
at www.rsmeans.com

Reference Books

The Practice of Cost Segregation Analysis

by Bruce A. Desrosiers and Wayne J. DelPico

This expert guide walks you through the practice of cost segregation analysis, which enables property owners to defer taxes and benefit from "accelerated cost recovery" through depreciation deductions on assets that are properly identified and classified.

With a glossary of terms, sample cost segregation estimates for various building types, key information resources, and updates via a dedicated website, this book is a critical resource for anyone involved in cost segregation analysis.

$99.95 per copy
Over 225 pages
Catalog no. 67345

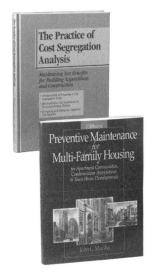

Preventive Maintenance for Multi-Family Housing

by John C. Maciha

Prepared by one of the nation's leading experts on multi-family housing.

This complete PM system for apartment and condominium communities features expert guidance, checklists for buildings and grounds maintenance tasks and their frequencies, a reusable wall chart to track maintenance, and a dedicated website featuring customizable electronic forms. A must-have for anyone involved with multi-family housing maintenance and upkeep.

$89.95 per copy
225 pages
Catalog no. 67346

How to Estimate with Means Data & CostWorks

New 3rd Edition

by RSMeans and Saleh A. Mubarak, Ph.D.

New 3rd Edition—fully updated with new chapters, plus new CD with updated *CostWorks* cost data and MasterFormat organization. Includes all major construction items—with more than 300 exercises and two sets of plans that show how to estimate for a broad range of construction items and systems—including general conditions and equipment costs.

$59.95 per copy
272 pages, softcover
Includes CostWorks CD
Catalog no. 67324B

Job Order Contracting

Expediting Construction Project Delivery

by Allen Henderson

Expert guidance to help you implement JOC—fast becoming the preferred project delivery method for repair and renovation, minor new construction, and maintenance projects in the public sector and in many states and municipalities. The author, a leading JOC expert and practitioner, shows how to:

• Establish a JOC program

• Evaluate proposals and award contracts

• Handle general requirements and estimating

• Partner for maximum benefits

$89.95 per copy
192 pages, illustrated, hardcover
Catalog no. 67348

Builder's Essentials: Estimating Building Costs

For the Residential & Light Commercial Contractor

by Wayne J. DelPico

Step-by-step estimating methods for residential and light commercial contractors. Includes a detailed look at every construction specialty—explaining all the components, takeoff units, and labor needed for well-organized, complete estimates. Covers correctly interpreting plans and specifications, and developing accurate and complete labor and material costs.

$29.95 per copy
Over 400 pages, illustrated, softcover
Catalog no. 67343

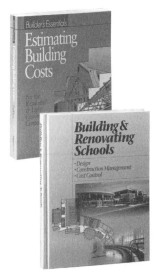

Building & Renovating Schools

This all-inclusive guide covers every step of the school construction process—from initial planning, needs assessment, and design, right through moving into the new facility. A must-have resource for anyone concerned with new school construction or renovation. With square foot cost models for elementary, middle, and high school facilities, and real-life case studies of recently completed school projects.

The contributors to this book—architects, construction project managers, contractors, and estimators who specialize in school construction—provide start-to-finish, expert guidance on the process.

$69.98 per copy
Over 425 pages, hardcover
Catalog no. 67342

Reference Books

**For more information
visit the RSMeans website
at www.rsmeans.com**

The Homeowner's Guide to Mold By Michael Pugliese

Expert guidance to protect your health and your home.

Mold, whether caused by leaks, humidity or flooding, is a real health and financial issue—for homeowners and contractors. This full-color book explains:
- Construction and maintenance practices to prevent mold
- How to inspect for and remove mold
- Mold remediation procedures and costs
- What to do after a flood
- How to deal with insurance companies

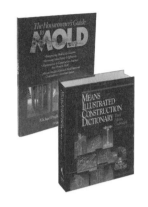

**$21.95 per copy
144 pages, softcover
Catalog no. 67344**

Means Illustrated Construction Dictionary
Unabridged 3rd Edition, with CD-ROM

Long regarded as the industry's finest, *Means Illustrated Construction Dictionary* is now even better. With the addition of over 1,000 new terms and hundreds of new illustrations, it is the clear choice for the most comprehensive and current information. The companion CD-ROM that comes with this new edition adds many extra features: larger graphics, expanded definitions, and links to both CSI MasterFormat numbers and product information.

**$99.95 per copy
Over 790 pages, illustrated, hardcover
Catalog no. 67292A**

Means Landscape Estimating Methods
New 5th Edition

Answers questions about preparing competitive landscape construction estimates, with up-to-date cost estimates and the new MasterFormat classification system. Expanded and revised to address the latest materials and methods, including new coverage on approaches to green building. Includes:
- Step-by-step explanation of the estimating process
- Sample forms and worksheets that save time and prevent errors

**$64.95 per copy
Over 350 pages, softcover
Catalog no. 67295C**

Means Plumbing Estimating Methods 3rd Edition
by Joseph Galeno and Sheldon Greene

Updated and revised! This practical guide walks you through a plumbing estimate, from basic materials and installation methods through change order analysis. *Plumbing Estimating Methods* covers residential, commercial, industrial, and medical systems, and features sample takeoff and estimate forms and detailed illustrations of systems and components.

**$29.98 per copy
330+ pages, softcover
Catalog no. 67283B**

Understanding & Negotiating Construction Contracts
by Kit Werremeyer

Take advantage of the author's 30 years' experience in small-to-large (including international) construction projects. Learn how to identify, understand, and evaluate high risk terms and conditions typically found in all construction contracts—then negotiate to lower or eliminate the risk, improve terms of payment, and reduce exposure to claims and disputes. The author avoids "legalese" and gives real-life examples from actual projects.

**$69.95 per copy
300 pages, softcover
Catalog no. 67350**

Means Estimating Handbook
2nd Edition

Updated Second Edition answers virtually any estimating technical question—all organized by CSI MasterFormat. This comprehensive reference covers the full spectrum of technical data required to estimate construction costs. The book includes information on sizing, productivity, equipment requirements, code-mandated specifications, design standards, and engineering factors.

**$99.95 per copy
Over 900 pages, hardcover
Catalog no. 67276A**

Means Spanish/English Construction Dictionary 2nd Edition
by RSMeans and the International Code Council

This expanded edition features thousands of the most common words and useful phrases in the construction industry with easy-to-follow pronunciations in both Spanish and English. Over 800 new terms, phrases, and illustrations have been added. It also features a new stand-alone "Safety & Emergencies" section, with colored pages for quick access.

Unique to this dictionary are the systems illustrations showing the relationship of components in the most common building systems for all major trades.

**$23.95 per copy
Over 400 pages
Catalog no. 67327A**

How Your House Works
by Charlie Wing

A must-have reference for every homeowner, handyman, and contractor—for repair, remodeling, and new construction. This book uncovers the mysteries behind just about every major appliance and building element in your house—from electrical, heating and AC, to plumbing, framing, foundations and appliances. Clear, "exploded" drawings show exactly how things should be put together and how they function—what to check if they don't work, and what you can do that might save you having to call in a professional.

**$21.95 per copy
160 pages, softcover
Catalog no. 67351**

Professional Development

Means CostWorks® Training

This one-day course helps users become more familiar with the functionality of *Means CostWorks* program. Each menu, icon, screen, and function found in the program is explained in depth. Time is devoted to hands-on estimating exercises.

Some of what you'll learn:
- Search the database utilizing all navigation methods
- Export RSMeans Data to your preferred spreadsheet format
- View crews, assembly components, and much more!
- Automatically regionalize the database

You are required to bring your own laptop computer for this course.

When you register for this course you will receive an outline for your laptop requirements.

Unit Price Estimating

This interactive *two-day* seminar teaches attendees how to interpret project information and process it into final, detailed estimates with the greatest accuracy level.

The most important credential an estimator can take to the job is the ability to visualize construction, and estimate accurately.

Some of what you'll learn:
- Interpreting the design in terms of cost
- The most detailed, time-tested methodology for accurate pricing
- Key cost drivers—material, labor, equipment, staging, and subcontracts
- Understanding direct and indirect costs for accurate job cost accounting and change order management

Who should attend: Corporate and government estimators and purchasers, architects, engineers... and others needing to produce accurate project estimates.

Square Foot & Assemblies Estimating

This *two-day* course teaches attendees how to quickly deliver accurate square foot estimates using limited budget and design information.

Some of what you'll learn:
- How square foot costing gets the estimate done faster
- Taking advantage of a "systems" or "assemblies" format
- The RSMeans "building assemblies/square foot cost approach"
- How to create a reliable preliminary systems estimate using bare-bones design information

Who should attend: Facilities managers, facilities engineers, estimators, planners, developers, construction finance professionals... and others needing to make quick, accurate construction cost estimates at commercial, government, educational, and medical facilities.

Repair & Remodeling Estimating

This *two-day* seminar emphasizes all the underlying considerations unique to repair/remodeling estimating and presents the correct methods for generating accurate, reliable R&R project costs using the unit price and assemblies methods.

Some of what you'll learn:
- Estimating considerations—like labor-hours, building code compliance, working within existing structures, purchasing materials in smaller quantities, unforeseen deficiencies
- Identifying problems and providing solutions to estimating building alterations
- Rules for factoring in minimum labor costs, accurate productivity estimates, and allowances for project contingencies
- R&R estimating examples calculated using unit price and assemblies data

Who should attend: Facilities managers, plant engineers, architects, contractors, estimators, builders... and others who are concerned with the proper preparation and/or evaluation of repair and remodeling estimates.

Mechanical & Electrical Estimating

This *two-day* course teaches attendees how to prepare more accurate and complete mechanical/electrical estimates, avoiding the pitfalls of omission and double-counting, while understanding the composition and rationale within the RSMeans Mechanical/Electrical database.

Some of what you'll learn:
- The unique way mechanical and electrical systems are interrelated
- M&E estimates–conceptual, planning, budgeting, and bidding stages
- Order of magnitude, square foot, assemblies, and unit price estimating
- Comparative cost analysis of equipment and design alternatives

Who should attend: Architects, engineers, facilities managers, mechanical and electrical contractors... and others needing a highly reliable method for developing, understanding, and evaluating mechanical and electrical contracts.

Green Building Planning & Construction

In this *two-day* course, learn about tailoring a building and its placement on the site to the local climate, site conditions, culture, and community. Includes information to reduce resource consumption and augment resource supply.

Some of what you'll learn:
- Green technologies, materials, systems, and standards
- Energy efficiencies with energy modeling tools
- Cost vs. value of green products over their life cycle
- Low-cost strategies and economic incentives and funding
- Health, comfort, and productivity goals and techniques

Who should attend: Contractors, project managers, building owners, building officials, healthcare and insurance professionals.

Facilities Maintenance & Repair Estimating

This *two-day* course teaches attendees how to plan, budget, and estimate the cost of ongoing and preventive maintenance and repair for existing buildings and grounds.

Some of what you'll learn:
- The most financially favorable maintenance, repair, and replacement scheduling and estimating
- Auditing and value engineering facilities
- Preventive planning and facilities upgrading
- Determining both in-house and contract-out service costs
- Annual, asset-protecting M&R plan

Who should attend: Facility managers, maintenance supervisors, buildings and grounds superintendents, plant managers, planners, estimators... and others involved in facilities planning and budgeting.

Practical Project Management for Construction Professionals

In this *two-day* course, acquire the essential knowledge and develop the skills to effectively and efficiently execute the day-to-day responsibilities of the construction project manager.

Covers:
- General conditions of the construction contract
- Contract modifications: change orders and construction change directives
- Negotiations with subcontractors and vendors
- Effective writing: notification and communications
- Dispute resolution: claims and liens

Who should attend: Architects, engineers, owner's representatives, project managers.

Facilities Repair & Remodeling Estimating

In this *two-day* course, professionals working in facilities management can get help with their daily challenges to establish budgets for all phase of a project

Some of what you'll learn:
- Determine the full scope of a project
- Identify the scope of risks & opportunities
- Creative solutions to estimating issues
- Organizing estimates for presentation & discussion
- Special techniques for repair/remodel and maintenance projects
- Negotiating project change orders

Who should attend: Facility managers, engineers, contractors, facility trades-people, planners & project managers

Professional Development

Scheduling and Project Management

This *two-day* course teaches attendees the most current and proven scheduling and management techniques needed to bring projects in on time and on budget.

Some of what you'll learn:
- Crucial phases of planning and scheduling
- How to establish project priorities and develop realistic schedules and management techniques
- Critical Path and Precedence Methods
- Special emphasis on cost control

Who should attend: Construction project managers, supervisors, engineers, estimators, contractors... and others who want to improve their project planning, scheduling, and management skills.

Understanding & Negotiating Construction Contracts

In this *two-day* course, learn how to protect the assets of your company by justifying or eliminating commercial risk through negotiation of a contract's terms and conditions.

Some of what you'll learn:
- Myths and paradigms of contracts
- Scope of work, terms of payment, and scheduling
- Dispute resolution
- Why clients love CLAIMS!
- Negotiating issues and much more

Who should attend: Contractors & subcontractors, material suppliers, project managers, risk & insurance managers, procurement managers, owners and facility managers, corporate executives.

Site Work & Heavy Construction Estimating

This *two-day* course teaches attendees how to estimate earthwork, site utilities, foundations, and site improvements, using the assemblies and the unit price methods.

Some of what you'll learn:
- Basic site work and heavy construction estimating skills
- Estimating foundations, utilities, earthwork, and site improvements
- Correct equipment usage, quality control, and site investigation for estimating purposes

Who should attend: Project managers, design engineers, estimators, and contractors doing site work and heavy construction.

Assessing Scope of Work for Facility Construction Estimating

This *two-day* course is a practical training program that addresses the vital importance of understanding the SCOPE of projects in order to produce accurate cost estimates in a facility repair and remodeling environment.

Some of what you'll learn:
- Discussions of site visits, plans/specs, record drawings of facilities, and site-specific lists
- Review of CSI divisions, including means, methods, materials, and the challenges of scoping each topic
- Exercises in SCOPE identification and SCOPE writing for accurate estimating of projects
- Hands-on exercises that require SCOPE, take-off, and pricing

Who should attend: Corporate and government estimators, planners, facility managers, and others who need to produce accurate project estimates.

2009 RSMeans Seminar Schedule

Note: call for exact dates and details.

Location	Dates
Las Vegas, NV	March
Washington, DC	April
Denver, CO	May
San Francisco, CA	June
Washington, DC	September
Dallas, TX	September
Las Vegas, NV	October
Atlantic City, NJ	October
Orlando, FL	November
San Diego, CA	December

1-800-334-3509, ext. 5115

Registration Information

Register early... Save up to $100! Register 30 days before the start date of a seminar and save $100 off your total fee. *Note: This discount can be applied only once per order. It cannot be applied to team discount registrations or any other special offer.*

How to register Register by phone today! The RSMeans toll-free number for making reservations is **1-800-334-3509, ext. 5115.**

Individual seminar registration fee - $935. *Means CostWorks®* **training registration fee - $375.** To register by mail, complete the registration form and return, with your full fee, to: RSMeans Seminars, 63 Smiths Lane, Kingston, MA 02364.

Government pricing All federal government employees save off the regular seminar price. Other promotional discounts cannot be combined with the government discount.

Team discount program for two to four seminar registrations. Call for pricing: 1-800-334-3509, ext. 5115

Multiple course discounts When signing up for two or more courses, call for pricing.

Refund policy Cancellations will be accepted up to ten business days prior to the seminar start. There are no refunds for cancellations received later than ten working days prior to the first day of the seminar. A $150 processing fee will be applied for all cancellations. Written notice of cancellation is required. Substitutions can be made at any time before the session starts. **No-shows are subject to the full seminar fee.**

AACE approved courses Many seminars described and offered here have been approved for 14 hours (1.4 recertification credits) of credit by the AACE

International Certification Board toward meeting the continuing education requirements for recertification as a Certified Cost Engineer/Certified Cost Consultant.

AIA Continuing Education We are registered with the AIA Continuing Education System (AIA/CES) and are committed to developing quality learning activities in accordance with the CES criteria. Many seminars meet the AIA/CES criteria for Quality Level 2. AIA members may receive (14) learning units (LUs) for each two-day RSMeans course.

NASBA CPE sponsor credits We are part of the National Registry of CPE sponsors. Attendees may be eligible for (16) CPE credits.

Daily course schedule The first day of each seminar session begins at 8:30 a.m. and ends at 4:30 p.m. The second day begins at 8:00 a.m. and ends at 4:00 p.m. Participants are urged to bring a hand-held calculator, since many actual problems will be worked out in each session.

Continental breakfast Your registration includes the cost of a continental breakfast, and a morning and afternoon refreshment break. These informal segments allow you to discuss topics of mutual interest with other seminar attendees. (You are free to make your own lunch and dinner arrangements.)

Hotel/transportation arrangements RSMeans arranges to hold a block of rooms at most host hotels. To take advantage of special group rates when making your reservation, be sure to mention that you are attending the RSMeans seminar. You are, of course, free to stay at the lodging place of your choice. (**Hotel reservations and transportation arrangements should be made directly by seminar attendees.**)

Important Class sizes are limited, so please register as soon as possible.

Note: Pricing subject to change.

Registration Form

ADDS-1000

Call 1-800-334-3509, ext. 5115 to register or FAX this form 1-800-632-6732. Visit our Web site: www.rsmeans.com

Please register the following people for the RSMeans construction seminars as shown here. We understand that we must make our own hotel reservations if overnight stays are necessary.

☐ Full payment of $_____enclosed.

☐ Bill me

Please print name of registrant(s)

(To appear on certificate of completion)

_____

_____

_____

_____

P.O. #: _____
GOVERNMENT AGENCIES MUST SUPPLY PURCHASE ORDER NUMBER OR TRAINING FORM.

Firm name_____

Address_____

City/State/Zip_____

Telephone no._____ Fax no._____

E-mail address_____

Charge registration(s) to: ☐ MasterCard ☐ VISA ☐ American Express

Account no._____ Exp. date_____

Cardholder's signature_____

Seminar name_____

Seminar City _____

Please mail check to: RSMeans Seminars, 63 Smiths Lane, P.O. Box 800, Kingston, MA 02364 USA

CONSTRUCTION COST INFORMATION... PRODUCTS & SERVICES